BIOFEEDBACK *and* SELF-CONTROL

EDITORS

JOE KAMIYA	*Director, Psychophysiology Lab* *The Langley Porter Institute* *University of California Medical Center* *San Francisco, California*
THEODORE X. BARBER	*Director of Psychological Research* *Medfield State Hospital* *Harding, Massachusetts*
LEO V. DICARA	*Professor of Psychiatry and Psychology* *Mental Health Research Institute* *University of Michigan* *Ann Arbor, Michigan*
NEAL E. MILLER	*Professor* *Laboratory of Physiological Psychology* *The Rockefeller University* *New York, New York*
DAVID SHAPIRO	*Associate Professor* *Department of Psychiatry* *Massachusetts Mental Health Center* *Harvard Medical School* *Boston, Massachusetts*
JOHANN STOYVA	*Associate Professor* *Department of Psychiatry* *University of Colorado Medical Center* *Denver, Colorado*

BIOFEEDBACK *and* SELF-CONTROL

an Aldine Reader on the regulation of bodily processes and consciousness

ALDINE • ATHERTON
chicago • new york

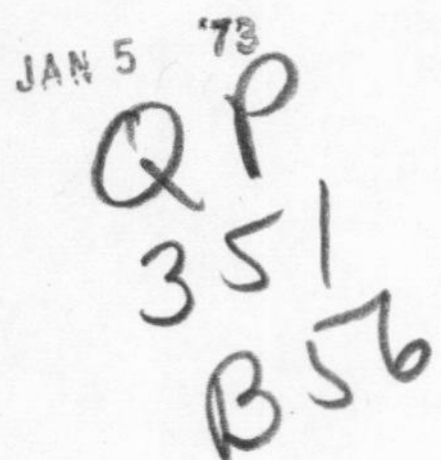

First published 1971 by
Aldine•Atherton, Inc.
529 South Wabash Avenue
Chicago, Illinois 60605

Library of Congress Catalog Card Number 71-167858
ISBN *202-25048-2*
Printed in the United States of America

Article 18
Reprinted by permission of the publisher from *Psychophysiology,* Vol. 6, pp. 283–290. Copyright 1969, The Society for Psychophysiological Research.
Article 19
Reprinted by permission of the author and publisher from *Science,* Vol. 159, pp. 1485–1486. Copyright 1968 by the American Association for the Advancement of Science.
Article 20
Copyright 1968 by the American Psychological Association, reproduced by permission. From the *Journal of Experimental Psychology,* Vol. 77, pp. 263–268.
Article 21
Reprinted by permission of Psychonomic Journals, Inc. from *Psychonomic Science,* 1964, Vol. I, pp. 147–148.
Article 22
Copyright 1968 by the American Psychological Association, reproduced by permission. From the *Journal of Comparative and Physiological Psychology,* Vol. 65, pp. 8–12.
Article 23
Copyright 1966 by the American Psychological Association, reproduced by permission. From the *Journal of Comparative and Physiological Psychology,* Vol. 62, pp. 165–166.
Article 24
Reprinted by permission of the publisher from *Psychophysiology,* Vol. 4, pp. 168–175. Copyright 1967, The Society for Psychophysiological Research.
Article 25
Copyright 1967 by the American Psychological Association, reproduced by permission. From the *Journal of Experimental Psychology,* Vol. 75, pp. 307–312.
Article 26
Reprinted by permission of the publisher from *Psychophysiology,* Vol. 4, pp. 294–299. Copyright 1968, The Society for Psychophysiological Research.
Article 27
Reprinted by permission of the publisher from *Psychophysiology,* Vol. 4, pp. 156–160. Copyright 1967, The Society for Psychophysiological Research.
Article 28
Copyright 1967 by the American Psychological Association, reproduced by permission. From the *Journal of Comparative and Physiological Psychology,* Vol. 63, pp. 1–6.
Article 29
Reprinted by permission of the American Physiological Society from the *American Journal of Physiology,* 1968, Vol. 215, pp. 677–683.
Article 30
Lapides, Sweet and Lewis, "Role of Striated Muscle in Urination," *Journal of Urology,* Vol. 77. Copyright 1957, The Williams & Wilkins Co. Reproduced by permission.
Article 31
Reprinted by permission of the American Physiological Society from the *American Journal of Physiology,* 1968, Vol. 215, pp. 684–686.
Article 32
Copyright 1968 by the American Psychological Association, reproduced by permission. From the *Journal of Comparative and Physiological Psychology,* Vol. 65, pp. 1–7.
Article 33
Reprinted by permission of the publisher from the *Archives of General Psychiatry,* Vol. II, 74. Copyright 1964, American Medical Association.
Article 34
Reprinted by permission of the publisher from the *Archives of General Psychiatry,* Vol. II, 82. Copyright 1964, American Medical Association.
Article 35
Abstract of a paper presented at the Western Psychological Association meeting, 1962. Reprinted with permission of the author.
Article 36
Reprinted by permission of the publisher from Charles Tart, ed., *Altered States of Consciousness,* pp. 489–501. Copyright © 1969 by John Wiley & Sons, Inc.
Article 37
Reprinted by permission of the publisher from *Psychophysiology,* Vol. 4, pp. 48–55. Copyright 1967, The Society for Psychophysiological Research.
Article 38
Reprinted by permission of Macmillan Journals Limited from *Nature,* 1967, Vol. 214, pp. 975–977.
Article 39
Reprinted by permission of the publisher and author from *Activitas Nervosa Superior,* 10, 4, 1968.
Article 40
Reprinted by permission of Elsevier Publishing Company from *Brain Research,* 1967, Vol. 6, pp. 143–163.
Article 41
Reprinted with permission from Wyrwicka and Sterman, "Instrumental Conditioning of Sensorymotor Cortex EEG Spindles in the Waking Cat," *Physiology and Behavior,* Vol. 3. Copyright 1968, Pergamon Press.
Article 42
Reprinted by permission of the author and publisher from *Science,* Vol. 165, pp. 821–823. Copyright 1969 by the American Association for the Advancement of Science.
Article 43
Reprinted by permission of the author and publisher from *Science,* Vol. 130, pp. 1338–1339. Copyright 1959 by the American Association for the Advancement of Science.
Article 44
Reprinted by permission of the author and publisher from *Science,* Vol. 154, pp. 1467–1468. Copyright 1966 by the American Association for the Advancement of Science.

Article 45
Reprinted by permission of the author and publisher from *Science,* Vol. 141, pp. 440–441. Copyright 1963 by the American Association for the Advancement of Science.
Article 46
Basmajian, J. V., *Muscles Alive: Their Functions Revealed by Electromyography,* (2nd ed.), "Conscious Control and Training of Motor Units and Motor Neutrons," pp. 103–114. Copyright 1967, The Williams & Wilkins Co. Reproduced by permission.
Article 47
Reprinted by permission of the author and publisher from *Science,* Vol. 155, pp. 201–203. Copyright 1967 by the American Association for the Advancement of Science.
Article 48
Reprinted by permission of the publisher from *Psychophysiology,* Vol. 6, pp. 371–377. Copyright 1969, The Society for Psychophysiological Research.
Article 49
Copyright 1967 by the American Psychological Association, reproduced by permission. From the *Psychological Bulletin,* Vol. 67, pp. 337–345.
Article 50
Copyright 1968 by the American Psychological Association, reproduced by permission. From the *Psychological Bulletin,* Vol. 70, pp. 52–68.
Article 51
Copyright 1969 by the American Psychological Association, reproduced by permission. From the *Psychological Bulletin,* Vol. 71, pp. 455–461.
Article 52
Copyright 1969 by the American Psychological Association, reproduced by permission. From the *Psychological Bulletin,* Vol. 71, pp. 462–466.
Article 53
From the *Journal of General Psychology,* 1933, Vol. 8, pp. 3–49.
Article 54
From the *Journal of Experimental Psychology,* 1941, Vol. 28, pp. 373–387.
Article 55
Reprinted by permission of Elsevier Publishing Company from *EEG and Clinical Neurophysiology,* 1969, Vol. 26, pp. 159–166.
Article 56
Copyright 1961 by the American Psychological Association, reproduced by permission. From the *Psychological Review,* Vol. 68, pp. 81–147.
Article 57
Reprinted by permission of The University of Chicago Press from *Studies from The Psychological Laboratory of the University of Chicago,* 1906, Vol. 13, pp. 258–275.
Article 58
From *Psychological Review,* 1907, Vol. 14, pp. 357–382.
Article 59
From *Psychological Review,* 1908, Vol. 15, pp. 139–149.
Article 60
Reprinted by permission of the publisher from the *Journal of the American Society of Psychosomatic Dentistry and Medicine,* 1969, Vol. 16, pp. 1–32.
Article 61
Reprinted by permission of the American Heart Association from *Circulation,* December, 1961, Vol. CCIV.
Article 62
Reprinted by permission of Elsevier Publishing Company from *EEG and Clinical Neurophysiology,* Supplement 7, 1957.
Article 63
Reprinted by permission of Elsevier Publishing Company from *EEG and Clinical Neurophysiology,* 1961, Vol. 13, pp. 452–456.
Article 64
Reprinted by permission of the publisher from *Folia Psychiatrica et Neurologica Japonica,* Vol. 20, pp. 315–336, 1966.
Article 65
Copyright 1965 by the American Psychological Association, reproduced by permission. From the *Journal of Consulting Psychology,* Vol. 29, pp. 139–145.
Article 66
Reprinted by permission of the publisher from the *American Journal of Psychotherapy,* 1963, Vol. 17, pp. 174–195.
Article 67
Reprinted by permission of the publisher from *Angiology,* Vol. 20, 1969, pp. 325–333.
Article 68
Reprinted with permission from Lader and Mathews, "A Physiological Model of Phobic Anxiety and Desensitization," *Behavior Research and Therapy,* Vol. 6, pp. 411–421. Copyright 1968, Pergamon Press.
Article 69
Reprinted by permission of Harper & Row, Publishers, from *Psychosomatic Medicine,* 1962, Vol. 24, pp. 159–169.
Article 70
Originally published in the *American Journal of Clinical Hypnosis,* 1969, 11, 155–166. Copyright 1969 by the American Society of Clinical Hypnosis.
Article 71
Reprinted by permission of the publisher from *Psychological Reports,* 1968, Vol. 22, pp. 587–608.
Article 72
Reprinted by permission of Harper & Row, Publishers, from *Psychosomatic Medicine,* 1963, Vol. 25, pp. 303–333.

PREFACE

The experiments described in this book have stirred public and scientific interest to a most unusual degree. The mass media have given extraordinary publicity to the work, with filmed interviews of leading researchers, televised scenes of laboratory sessions, feature articles in popular magazines, news coverage of scientific meetings. Some researchers have been besieged with visitors, phone calls, requests for information, and volunteers offering their services as subjects. At the scientific level there is a flurry of excitement in biofeedback in several different disciplines. The Biofeedback Research Society, created three years ago in response to a growing need among professional researchers to communicate with each other, now has more than 400 members.

Why all the excitement? The answer becomes apparent upon consideration of some examples of the work in this field. Experiments show that humans can train themselves to control their own heart rates voluntarily. The technique uses a device that is a sort of "physiological mirror." The heart rate is detected by an electronic device and fed back to the experimental subject through a signal. This keeps him continuously informed of even the slightest beat-to-beat variations in the heart rate. He tries various "internal experiments" on how to control the feedback signal and gradually achieves partial control. He may sometimes describe the "method" in terms of thoughts, feelings, and emotions; but he may also be totally unable to provide any clear verbal description of his new skill.

In other studies subjects train themselves to enter a kind of relaxed but alert mental state with the aid of a tone signal controlled by their brainwaves; or they can overcome chronic tension headaches by learning to relax their scalp muscles with the aid of a muscle tension monitor. In work with animals it has been shown that the rate of kidney output and rate of salivation can be increased by rewarding the animals whenever they spontaneously show such increases.

What these studies suggest is exciting indeed: With modern technological devices man may be able to exercise direct voluntary control over many body and brain functions that had been considered as being totally beyond such control. The range of physiological functions that can possibly be controlled is wide. Trained voluntary control of excessive sweating, high blood pressure, stomach acidity, migraine headaches and insomnia, and many other symptoms of nervous tension come to mind. The self-control of attention, moods and emotions are conceivable as possible future possibilities of biofeedback research.

Several separate disciplines are represented in this work, including physiology, psychology, psychosomatic medicine, and biomedical engineering. Since the voluntary control is achieved only by training and learning, the field of education is directly involved. Learning about the internal world and how to control it will perhaps someday be given as much emphasis in public education as learning about the external world.

These studies also raise the possibility of using technology to reach the most profoundly nontechnological, humanistic end—self exploration. By taking part in these biofeedback training sessions the subject discovers the relationship of some aspect of his *consciousness* or subjective awareness to that aspect of his *physiological* activity indicated by the feedback signal. He finds that as he learns to change his consciousness, the signal changes, or that as he learns to change the signal, his consciousness changes.

It is historically noteworthy that the perception and control of the internal world that appears to be at the core of these studies was also the concern of the early introspectionist schools which started the field of scientific psychology. To be sure, the methods are different: introspection was the chief method of observation in the earliest psychology, and analysis of the experience into the smallest number of irreducible primary dimensions was a major preoccupation. Today introspective reports are often only a secondary byproduct; the method is focused on either a conditioned discrimination of a physiological activity or the instrumental conditioning of that activity. In most laboratories today the introspective reports are of interest primarily for the hypotheses they may suggest for the physiological mechanisms of control. However the similarity of focus on the *internal* is intriguing, especially in light of the ruling behaviorism of the intervening years and its preoccupation with the externally observable. In the course of developing a methodology for dealing with the outwardly observable behaviors of organisms, behaviorism seems quite unintentionally to have evolved a most powerful analytic tool for the understanding of internal activity. Instrumental conditioning, the method chosen for control of external behavior, is also eminently well suited for modulating internal processes.

It is not possible to foresee what this means for the distant future of the science. However, all of us do seem to assume that internal processes are the substrate of consciousness. Events of consciousness such as feelings and thoughts are often relabeled "implicit behavior" or "internal behavior" by those who wish to acknowledge only behavior. Others will dare pose hypotheses relating specific physiological processes to such mental processes or internal behaviors. In this view the very processes of discriminating or controlling internal physiological states are either consciousness itself, or a close relative.

It must be made clear that the instrumental conditioning of physiological processes does not commit one to any particular view with regard to the problem of consciousness. One can ignore the subjective aspects and focus exclusively on the performance itself. In fact, most of the papers in this reader contain no reference to events of consciousness, subjective experience, or awareness. However, an increasing number of researchers consider biofeedback methods useful for examining the consciousness aspects of physiological processes, or conversely, the physiological aspects of consciousness. For example, the instrumental conditioning of elevated heart rates can be followed by both interviews and behavioral observation to determine whether this physiological parameter is related to dimensions of emotion. If the results are affirmative and tend to agree with studies of the physiological effects of experimentally induced emotions, the convergence is theoretically important. If not, the details of the divergence should help in modification of the initial concept.

Thinking of historical trends in an even broader manner, a number of perspectives related to biofeedback suggest themselves. Eastern meditation practices, the cultivation of increased personal awareness, and ecological concern are contemporary American preoccupations, especially among educated young adults. Despite superficial differences these concerns seem related to biofeedback training in an interesting way. Many of us who conduct laboratories in biofeedback, especially EEG feedback, have noted the large number of mediators who volunteer their services as subjects, apparently in the hope that they might learn some skill related to meditation, to compare themselves with other mediators, or to satisfy some specific curiosity about their meditative experiences. The similarity of aim between biofeedback and one part of many meditative practices—gaining control of bodily functions—seems to account in part for this interest. Many persons undergoing sensitivity training where other persons provide feedback have similarly shown interest in physiological or "bio"feedback training to sharpen discrimination of feelings, emotions, and tensions.

The concern about the body as a discriminable, controllable entity points to its status as an environmental ecological process. The concept of the "internal milieu" of Claude Bernard can be enlarged to include the tissues and organs. The "center" of the central nervous system most essential to consciousness can be seen as the "occupant," supported by a milieu of many layers, some more central than others, but all potentially controllable in some degree by purely internal operant behaviors.

This manner of thinking of our biological systems, first as an external ecology and then as an internal ecology surrounding a central nervous system, leads to an interesting way of comparing East and West. Western man has tended to focus on the external world, assuming the internal world to be beyond control (autonomic) except for what happens to it as a result of efforts toward goal achievement in the external world. Eastern man, however, appears to have focused more attention on achievements (knowledge and control) in his internal world, assuming the external world to be largely beyond control. The biofeedback studies can be seen as a point of convergence of these two traditions, the Western technological tradition joining forces with the more introspective traditions of the East.

The main purpose of this Reader is to bring together the scientific literature of this field. The objective laboratory studies, with the experimental hypotheses guiding them, are where the breakthroughs occurred.

The sections on the training method and on the difference between classical and operant conditioning procedures are for those readers who are unfamiliar with the research methodology of this field and may be skipped by others.

The Training Method

The training method—variously called "instrumental conditioning," "operant conditioning," "biofeedback training," or sometimes just "feedback training"—is quite simple and is based on principles that are no different from other commonly used forms of training. The only difference is that the present studies require the performance to be an internal physiological change, whereas operant conditioning as originally developed required an external act that "operated" on the external environment or at least was an externally observable movement.

The training technique has three requirements:

(1) The physiological function to be brought under control must be continuously monitored with sufficient sensitivity to detect moment-by-moment changes.

(2) Changes in the physiological measure need to be reflected immediately to the subject. In the case of animal subjects this is often done by small amounts of food or water given after appropriate deprivation, avoidance of electric shock or direct electrical stimulation of "pleasure centers" of the brain. These are delivered automatically whenever the physiological measure shows fluctuations exceeding some pre-set level. The stimulus that reflects the physiological change in human subjects is usually a visual or auditory signal. This return of physiological information via an external pathway is the feature of these studies that is denoted by the term, *feedback*.

(3) The subject must be motivated to learn. The function of the food, water, shock avoidance, or brain stimulation for animal subjects is to keep the subject interested in the task. The motivation of curiosity is often sufficient for humans, at least initially. Added to this is the motivation of a desire for mastery encouraged by instructions such as, "Try to achieve control over the feedback signal," or "Try to keep the tone as loud as you can." Rewards are sometimes given for added incentive. One effective motivator that is frequently used, especially for long-term training, is a quantitative score (given once every few minutes) that informs the subject of his performance.

The principle of biological feedback is not new; it is as old as the earliest form of biological evolution and is found at all levels of organization from intracellular processes to social communication patterns. A familiar example of feedback is seen in eye–hand coordination, where both visual and muscle information feeds back to the central nervous system at every point in the series of movements comprising the act of reaching an object, making possible precise control at every moment. The new elements added by the current studies are the external detection of physiological activity, and in many cases, augmentation of the feedback by the addition of sensory channels. As an example of such augmentation, the instantaneous heart rate of the subject may be signaled back to him via a meter display.

Classical and Instrument (Operant) Conditioning

Throughout this Reader there are references to "classical" as opposed to "instrumental" (or "operant") conditioning. A brief sketch of what these are and how they were developed is presented here. This is followed by a brief orientation to some of the theoretical issues arising from use of instrumental conditioning methods in the field of autonomic activity.

Classical conditioning got its start in Russia in the laboratory of Pavlov. In the beginning of this century, he and his colleagues observed that dogs which had been fitted with fistulas for the measurement of salivation would show salivation at the mere sight or smell of food, prior to its actual delivery. Systematic pairing of a previously neutral stimulus with delivery of meat powder to the dog's mouth led to the establishment of the conditioned salivary response. Soon thereafter classical conditioning of many other systems, both visceral and motor, was demonstrated and has continued to be the method of choice in the Soviet Union. Razran's review, "The Observable Unconscious and the Inferable Conscious in Current Soviet Psychophysiology," is included in this volume in Part IX, *Classical Conditioning,* to provide a sampling of the extensive work in that country.

In classical conditioning, an *unconditioned stimulus,* known beforehand to be a reliable elicitor of a specific response (such as meat powder for eliciting salivation in the hungry dog) is presented during or shortly after the presentation of another stimulus, the *conditioned stimulus.* The conditioned stimulus (such as a bell or metronome) does not normally elicit the response. The conditioned and unconditioned stimulus pairs are presented repeatedly, regardless of variations in the response.

If the conditioning is taking effect the conditioned stimulus begins to elicit the response that previously could only be elicited by the unconditioned stimulus. The response is then said to be classically conditioned.

If the conditioned stimulus is then repeatedly presented alone without the unconditioned stimulus, the conditioned response will eventually disappear. This procedure is called *extinction.*

After the establishment of a conditioned response with a specific conditioned stimulus, such as a bell with a given sound, the subject will show at least a partial conditioned response upon the presentation of a similar stimulus, another bell with a similar sound. This is called *stimulus generalization.* If several previously neutral stimuli are given but the unconditioned stimulus is presented repeatedly with only one of these stimuli, the subject will come to show the conditioned response only to the one, a selectivity that is called *discriminated* (or *discriminative*) responding.

The method of instrumental conditioning is usually traced to experiments done at the end of the 19th century in this country by Thorndike, who studied trial-and-error behavior of cats seeking escape from confining enclosures. Those behaviors of the cat that led to escape, such as knocking over a pole holding the cage door shut, became more probable as the animal was placed in the same situation repeatedly.

Concepts of behavioral training emerging from these and similar observations were developed and refined over the following decades into what is now known as reinforcement theory, and the procedures based on the concepts are known as instrumental or operant conditioning. In this kind of conditioning a rewarding (reinforcing) stimulus, which the subject would normally expend effort to obtain, is given immediately after the subject makes a response pre-selected by the experimenter. This results in the subject making the response more frequently and gaining rewards as he proceeds.

The phenomena of extinction, stimulus generalization, and discrimination are also seen in instrumental conditioning. The subject stops responding if he is not rewarded for making the response; he will respond in the presence of two similar stimuli as if they were equivalent unless he is rewarded for responding only to one.

The two types of procedure differ in one crucial respect— the relationship of the reinforcing stimulus to the response. In classical conditioning the reinforcing stimulus is given with the conditioned stimulus regardless of the subject's response. In instrumental conditioning the reinforcement is given contingent upon the occurrence of the pre-selected response. The reinforcer also does not by itself elicit the response to be conditioned.

As experimental procedures, classical and instrumental conditioning are easily understood and distinguished from each other. Unfortunately, however, two conflicting meanings have developed for the particular case of instrumental conditioning of physiological activity. The first definition is operational: Reward the subject for showing specific changes in physiological activity. Nothing is implied about the specific physiological or psychological processes that might account for the successful outcome of the procedure.

The second definition, however, invokes the criterion that neither skeletal muscle activity nor implicit verbal activity could possibly account for the conditioning. This developed because the autonomic physiological activity was traditionally regarded as conditionable only by the classical method. Apparently successful instances of instrumental conditioning of autonomic activity could be explained as due instead to the conditioning of a skeletal motor response or symbolic activity that then produced the change in the autonomic measure (either as an unconditioned effect or as a classically conditioned response).

This second criterion, used in discussions on whether an autonomic response was "really" or "directly" conditioned by instrumental methods, confuses the training method with the physiological or psychological mechanisms that make the method work. For the newcomer to the field, it is well to point out that all agree that instrumental conditioning as a procedure is, in fact, effective in gaining experimental control over many physiological functions. The question that is actually being raised is a different one of how, that is, by what mechanisms, the result is achieved or whether the activity of a specific system like the skeletal motor system is necessary for the conditioning.

The question of the mediating mechanism is important in its own right and deserves intensive research. At the physiological level the work done on this problem already (see the papers reporting work with curarized animals) is a significant first step. It shows that activity of the skeletal motor system is not essential for the successful instrumental conditioning of some autonomic functions. Just exactly what mechanisms do mediate the learning in intact animals and just how important the motor system is, and since the autonomic system plays an important role in emotion, whether emotionality is affected by autonomic conditioning of the instrumental sort are important questions to be pursued. DiCara and Weiss' paper (Article 14) shows such an effect can indeed occur.

The possible implicit verbal mediation of the operant conditioning of autonomic activity is, like the one of skeletal motor mediation, an additional question that goes beyond the operational definition. It raises the possibility that some verbal operant, through conditioning of a classical sort in the history of the individual, may exert its effect autonomically. (See the extract of Hudgin's early paper [Article 53] for an example of how pupillary constriction could be brought under control of a verbal command.) As Katkin and Murray (Article 51) point out in their paper, however, the question probably cannot be resolved because there is no way to prevent such mediation in humans without rendering them unconscious. Even the detection of verbal mediation or confirmation of its absence in an unequivocal fashion would be difficult if not impossible. Under the circumstances, it seems advisable to accept the reality of instrumental conditioning of voluntarily controlled autonomic activity, whether verbal processes are involved or not.

The organization of the following readings is in part according to the physiological function targeted for instrumental control, and in part according to systematic issues. Miller's widely read paper on visceral learning serves the special purpose of providing introduction with a broad review of the entire area. A series of papers on Heart Rate covers this most popular area of experimentation in chronological order. Blood Pressure and Vasomotor Responses follow, completing coverage of the cardiovascular system. The Electrodermal Responses (or Galvanic Skin Responses) come next. These responses, taken either as a change in electrical potential or resistance of the skin at the palmar or plantar areas, perhaps received most attention in earlier years with classical conditioning techniques. Conditioning of Salivation, Urine Formation and Gastric Motility are discussed next, and help extend the picture of the diverse directions that have already been taken by researchers with the basic instrumental conditioning paradigm. The papers by Stunkard and his colleagues are included to illustrate methodology in the area of discrimination of autonomic activity, which can be regarded as the opposite side of the coin to the control of autonomic activity.

The EEG defies classification as an autonomic measure. There are such autonomic-like features in the electroencephalogram as its continuous activity in the face of many external changes, thus resembling the autonomic cardiovascular and respiratory functions. But at some times the EEG is clearly reflective of sensory function, and at other times motor function. Fortunately, the feedback training of EEG activity, or the conditioned discrimination of it, as in Kamiya's paper (Article 35) and the paper by Antrobus and Antrobus (Article 37), does not depend on its classification into "autonomic" or "somatic" function.

In the section on operant conditioning of the electromyogram (EMG), we see a clear-cut case of instrumental conditioning of the skeletal motor system, where the end point is not behavior, but the electrical activity of muscle. For psychologists primarily concerned with externally observable behavior, a significant step toward joining forces with physiology occurred with this "beyond-the-skin" application of behavioral methodology, a domain long shunned by psychology.

The issues earlier discussed on the criteria of instrumental conditioning of autonomic activity are presented in several selections in the section on Methodology. The readings illustrate current academic concern for the interpretation of the results in this field. The readings in Classical Conditioning give examples of the Pavlovian method applied to pupillary constriction (Hudgins, Article 53) and the EEG alpha block (Jasper and Shagass, Article 54) and should be read together with the papers on methodology.

Voluntary control of specific physiological functions was known long before the current biofeedback experiments. Yoga, Zen meditation, and autogenic training are examples presented in Part X.

Hypnosis has also been an example of self-control, and is discussed as subject matter for psychology, with proposals for a perspective on how consciousness, physiological activity and behavior are related to each other. And the two papers presented in the last section attempt to deal conceptually with the difficult questions of the place of consciousness in the science of behavior generally, and specifically in this new field of biofeedback and self regulation. Consideration of Conscious Processes seems unavoidable in any serious study of self regulation and its voluntary control.

This Reader contains only articles published before 1970 and chosen by the editors as important to early research in Biofeedback and Self-Control. An Annual, covering research published in 1970 is already available, and in the spring of each year we plan to bring out a new Annual covering the major works published in the previous year. In order to produce the Annuals and the Reader as quickly as possible, all material was photographed and printed in its original form. We are grateful for the cooperation of the organizations, journals, publishers, and authors who granted permission to reprint their articles.

Joe Kamiya
For the Editors

CONTENTS

BIOFEEDBACK

and SELF-CONTROL

I

INTRODUCTION

1

Learning of Visceral and Glandular Responses

Neal E. Miller

There is a strong traditional belief in the inferiority of the autonomic nervous system and the visceral responses that it controls. The recent experiments disproving this belief have deep implications for theories of learning, for individual differences in autonomic responses, for the cause and the cure of abnormal psychosomatic symptoms, and possibly also for the understanding of normal homeostasis. Their success encourages investigators to try other unconventional types of training. Before describing these experiments, let me briefly sketch some elements in the history of the deeply entrenched, false belief in the gross inferiority of one major part of the nervous system.

Historical Roots and Modern Ramifications

Since ancient times, reason and the voluntary responses of the skeletal muscles have been considered to be superior, while emotions and the presumably involuntary glandular and visceral responses have been considered to be inferior. This invidious dichotomy appears in the philosophy of Plato (*1*), with his superior rational soul in the head above and inferior souls in the body below. Much later, the great French neuroanatomist Bichat (*2*) distinguished between the cerebrospinal nervous system of the great brain and spinal cord, controlling skeletal responses, and the dual chain of ganglia (which he called "little brains") running down on either side of the spinal cord in the body below and controlling emotional and visceral responses. He indicated his low opinion of the ganglionic system by calling it "vegetative"; he also believed it to be largely independent of the cerebrospinal system, an opinion which is still reflected in our modern name for it, the autonomic nervous system. Considerably later, Cannon (*3*) studied the sympathetic part of the autonomic nervous system and concluded that the different nerves in it all fire simultaneously and are incapable of the finely differentiated individual responses possible for the cerebrospinal system, a conclusion which is enshrined in modern textbooks.

Many, though not all, psychiatrists have made an invidious distinction between the hysterical and other symptoms that are mediated by the cerebrospinal nervous system and the psychosomatic symptoms that are mediated by the autonomic nervous system. Whereas the former are supposed to be subject to a higher type of control that is symbolic, the latter are presumed to be only the direct physiological consequences of the type and intensity of the patient's emotions (see, for example, *4*).

Similarly, students of learning have made a distinction between a lower form, called classical conditioning and thought to be involuntary, and a superior form variously called trial-and-error learning, operant conditioning, type II conditioning, or instrumental learning and believed to be responsible for voluntary behavior. In classical conditioning, the reinforcement must be by an unconditioned stimulus that already elicits the specific response to be learned; therefore, the possibilities are quite limited. In instrumental learning, the reinforcement, called a reward, has the property of strengthening any immediately preceding response. Therefore, the possibilities for reinforcement are much greater; a given reward may reinforce any one of a number of different responses, and a given response may be reinforced by any one of a number of different rewards.

Finally, the foregoing invidious distinctions have coalesced into the strong traditional belief that the superior type of instrumental learning involved in the superior voluntary behavior is possible only for skeletal responses mediated by the superior cerebrospinal nervous system, while, conversely, the inferior classical conditioning is the only kind possible for the inferior, presumably involuntary, visceral and emotional responses mediated by the inferior autonomic nervous system. Thus, in a recent summary generally considered authoritative, Kimble (*5*) states the almost universal belief that "for autonomically mediated behavior, the evidence points unequivocally to the conclusion that such responses can be modified by classical, but not instrumental, training methods." Upon examining the evidence, however, one finds that it consists only of failure to secure instrumental learning in two incompletely reported exploratory experiments and a vague allusion to the Russian literature (*6*). It is only against a cultural background of great prejudice that such weak evidence could lead to such a strong conviction.

The belief that instrumental learning is possible only for the cerebrospinal system and, conversely, that the autonomic nervous system can be modified only by classical conditioning has been used as one of the strongest arguments for the notion that instrumental learning and classical conditioning are two basically different phenomena rather than different manifestations of the same phenomenon under different conditions. But for many years I have been impressed with the similarity between the laws of classical conditioning and those of instrumental learning, and with the fact that, in each of these two situations, some of the specific details of learning vary with the specific conditions of learning. Failing to see any clear-cut dichotomy, I have assumed that there is only one kind of learning (*7*). This assumption has logically demanded that instrumental training procedures be able to produce the learning of any visceral responses that could be acquired through classical conditioning procedures. Yet it was only a little over a dozen years ago that I began some experimental work on this problem and a somewhat shorter time ago that I first, in published articles (*8*), made specific sharp challenges to the traditional view that the instrumental learning of visceral responses is impossible.

Some Difficulties

One of the difficulties of investigating the instrumental learning of visceral responses stems from the fact that the responses that are the easiest to measure —namely, heart rate, vasomotor responses, and the galvanic skin response —are known to be affected by skeletal responses, such as exercise, breathing, and even tensing of certain muscles, such as those in the diaphragm. Thus, it is hard to rule out the possibility that, instead of directly learning a visceral response, the subject has learned a skeletal response the performance of which causes the visceral change being recorded.

One of the controls I planned to use was the paralysis of all skeletal responses through administration of curare, a drug which selectively blocks the motor end plates of skeletal muscles without eliminating consciousness in human subjects or the neural control of visceral responses, such as the beating of the heart. The muscles involved in breathing are paralyzed, so the subject's breathing must be maintained through artificial respiration. Since it seemed unlikely that curarization and other rigorous control techniques would be easy to use with human subjects, I decided to concentrate first on experiments with animals.

Originally I thought that learning would be more difficult when the animal was paralyzed, under the influence of curare, and therefore I decided to postpone such experiments until ones on nonparalyzed animals had yielded some definitely promising results. This turned out to be a mistake because, as I found out much later, paralyzing the animal with curare not only greatly simplifies the problem of recording visceral responses without artifacts introduced by movement but also apparently makes it easier for the animal to learn, perhaps because paralysis of the skeletal muscles removes sources of variability and distraction. Also, in certain experiments I made the mistake of using rewards that induced strong unconditioned responses that interfered with instrumental learning.

One of the greatest difficulties, however, was the strength of the belief that instrumental learning of glandular and visceral responses is impossible. It was extremely difficult to get students to work on this problem, and when paid assistants were assigned to it, their attempts were so half-hearted that it soon became more economical to let them work on some other problem which they could attack with greater faith and enthusiasm. These difficulties and a few preliminary encouraging but inconclusive early results have been described elsewhere (*9*).

Success with Salivation

The first clear-cut results were secured by Alfredo Carmona and me in an experiment on the salivation of dogs. Initial attempts to use food as a reward for hungry dogs were unsuccessful, partly because of strong and persistent unconditioned salivation elicited by the food. Therefore, we decided to use water as a reward for thirsty dogs. Preliminary observations showed that the water had no appreciable effects one way or the other on the bursts of spontaneous salivation. As an additional precaution, however, we used the experimental design of rewarding dogs in one group whenever they showed a burst of spontaneous salivation, so that they would be trained to increase salivation, and rewarding dogs in another group whenever there was a long interval between spontaneous bursts, so that they would be trained to decrease salivation. If the reward had any unconditioned effect, this effect might be classically conditioned to the experimental situation and therefore produce a change in salivation that was not a true instance of instrumental learning. But in classical conditioning the reinforcement must elicit the response that is to be acquired. Therefore, conditioning of a response elicited by the reward could produce either an increase or a decrease in salivation, depending upon the direction of the unconditioned response elicited by the reward, but it could not produce a change in one direction for one group and in the opposite direction for the other group. The same type of logic applies for any unlearned cumulative aftereffects of the reward; they could not be in opposite directions for the two groups. With instrumental learning, however, the reward can reinforce any response that immediately precedes it; therefore, the same reward can be used to produce either increases or decreases.

The results are presented in Fig. 1, which summarizes the effects of 40 days of training with one 45-minute training session per day. It may be seen that in this experiment the learning proceeded slowly. However, statistical analysis showed that each of the trends in the predicted rewarded direction was highly reliable (*10*).

Since the changes in salivation for the two groups were in opposite directions, they cannot be attributed to classical conditioning. It was noted, however, that the group rewarded for increases seemed to be more aroused and active than the one rewarded for decreases. Conceivably, all we were doing was to change the level of activation of the dogs, and this change was, in turn, affecting the salivation. Although we did not observe any specific skeletal responses, such as chewing movements or panting, which might be expected to elicit salivation, it was difficult to be absolutely certain that such movements did not occur. Therefore, we decided to rule out such movements by paralyzing the dogs with curare, but we immediately found that curare had two effects which were diastrous for this experiment: it elicited such copious and continuous salivation that there were no changes in salivation to reward, and the salivation was so viscous that it almost immediately gummed up the recording apparatus.

Heart Rate

In the meantime, Jay Trowill, working with me on this problem, was displaying great ingenuity, courage, and persistence in trying to produce instrumental learning of heart rate in rats that had been paralyzed by curare to prevent them from "cheating" by muscular exertion to speed up the heart or by relaxation to slow it down. As a result of preliminary testing, he selected a dose of curare (3.6 milligrams of *d*-tubocurarine chloride per kilogram, injected intraperitoneally) which produced deep paralysis for at least 3 hours, and a rate of artificial respiration (inspiration-expiration ratio 1:1; 70 breaths per minute; peak pressure reading, 20 cm-H_2O) which maintained the heart at a constant and normal rate throughout this time.

In subsequent experiments, DiCara and I have obtained similar effects by starting with a smaller dose (1.2 milligrams per kilogram) and constantly infusing additional amounts of the drug, through intraperitoneal injection, at the rate of 1.2 milligrams per kilogram per hour, for the duration of the experiment. We have recorded, electromyographically, the response of the muscles, to determine that this dose does indeed produce a complete block of the action potentials, lasting for at least an hour after the end of infusion. We have found that if parameters of respiration and the face mask are adjusted carefully, the procedure not only maintains the heart rate of a 500-gram control animal constant but also maintains the vital signs of temperature, peripheral vasomotor responses, and the pCO_2 of the blood constant.

Since there are not very many ways

to reward an animal completely paralyzed by curare, Trowill and I decided to use direct electrical stimulation of rewarding areas of the brain. There were other technical difficulties to overcome, such as devising the automatic system for rewarding small changes in heart rate as recorded by the electrocardiogram. Nevertheless, Trowill at last succeeded in training his rats (*11*). Those rewarded for an increase in heart rate showed a statistically reliable increase, and those rewarded for a decrease in heart rate showed a statistically reliable decrease. The changes, however, were disappointingly small, averaging only 5 percent in each direction.

The next question was whether larger changes could be achieved by improving the technique of training. DiCara and I used the technique of shaping—in other words, of immediately rewarding first very small, and hence frequently occurring, changes in the correct direction and, as soon as these had been learned, requiring progressively larger changes as the criterion for reward. In this way, we were able to produce in 90 minutes of training changes averaging 20 percent in either direction (*12*).

Key Properties of Learning: Discrimination and Retention

Does the learning of visceral responses have the same properties as the learning of skeletal responses? One of the important characteristics of the instrumental learning of skeletal responses is that a discrimination can be learned, so that the responses are more likely to be made in the stimulus situations in which they are rewarded than in those in which they are not. After the training of the first few rats had convinced us that we could produce large changes in heart rate, DiCara and I gave all the rest of the rats in the experiment described above 45 minutes of additional training with the most difficult criterion. We did this in order to see whether they could learn to give a greater response during a "time-in" stimulus (the presence of a flashing light and a tone) which indicated that a response in the proper direction would be rewarded than during a "time-out" stimulus (absence of light and tone) which indicated that a correct response would not be rewarded.

Figure 2 shows the record of one of the rats given such training. Before the beginning of the special discrimination training it had slowed its heart from an initial rate of 350 beats per minute to a rate of 230 beats per minute. From the top record of Fig. 2 one can see that, at the beginning of the special discrimination training, there was no appreciable reduction in heart rate that was specifically associated with the time-in stimulus. Thus it took the rat considerable time after the onset of this stimulus to meet the criterion and get the reward. At the end of the discrimination training the heart rate during time-out remained approximately the same, but when the time-in light and tone came on, the heart slowed down

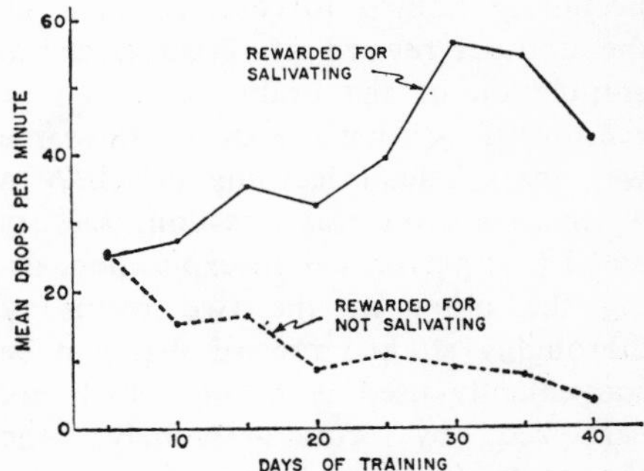

Fig. 1. Learning curves for groups of thirsty dogs rewarded with water for either increases or decreases in spontaneous salivation. [From Miller and Carmona (*10*)]

and the criterion was promptly met. Although the other rats showed less change than this, by the end of the relatively short period of discrimination training their heart rate did change reliably ($P < .001$) in the predicted direction when the time-in stimulus came on. Thus, it is clear that instrumental visceral learning has at least one of the important properties of instrumental skeletal learning—namely, the ability to be brought under the control of a discriminative stimulus.

Another of the important properties of the instrumental learning of skeletal responses is that it is remembered. DiCara and I performed a special experiment to test the retention of learned changes in heart rate (*13*). Rats that had been given a single training session were returned to their home cages for 3 months without further training. When curarized again and returned to the experimental situation for nonreinforced test trials, rats in both the "increase" and the "decrease" groups showed good retention by exhibiting reliable changes in the direction rewarded in the earlier training.

Escape and Avoidance Learning

Is visceral learning by any chance peculiarly limited to reinforcement by the unusual reward of direct electrical stimulation of the brain, or can it be reinforced by other rewards in the same way that skeletal learning can be? In order to answer this question, DiCara and I (*14*) performed an experiment using the other of the two forms of thoroughly studied reward that can be conveniently used with rats which are paralyzed by curare—namely, the chance to avoid, or escape from, mild electric shock. A shock signal was turned on; after it had been on for 10 seconds it was accompanied by brief pulses of mild electric shock delivered to the rat's tail. During the first 10 seconds the rat could turn off the shock signal and avoid the shock by making the correct response of changing its heart rate in the required direction by the required amount. If it did not make the correct response in time, the shocks continued to be delivered until the rat escaped them by making the correct response, which immediately turned off both the shock and the shock signal.

For one group of curarized rats, the correct response was an increase in heart rate; for the other group it was a decrease. After the rats had learned to make small responses in the proper direction, they were required to make larger ones. During this training the shock signals were randomly interspersed with an equal number of "safe" signals that were not followed by shock; the heart rate was also recorded during so-called blank trials—trials without any signals or shocks. For half of the rats the shock signal was a tone and the "safe" signal was a flashing light; for the other half the roles of these cues were reversed.

The results are shown in Fig. 3. Each of the 12 rats in this experiment changed its heart rate in the rewarded direction. As training progressed, the shock signal began to elicit a progressively greater change in the rewarded direction than the change recorded during the blank trials; this was a statistically reliable trend. Conversely, as training progressed, the "safe" signal came to elicit a statistically reliable change in the opposite direction, toward the initial base line. These results show learning when escape and avoidance are the rewards; this means that visceral responses in curarized rats can be reinforced by rewards other than direct electrical stimulation of the brain. These rats also discriminate between the shock

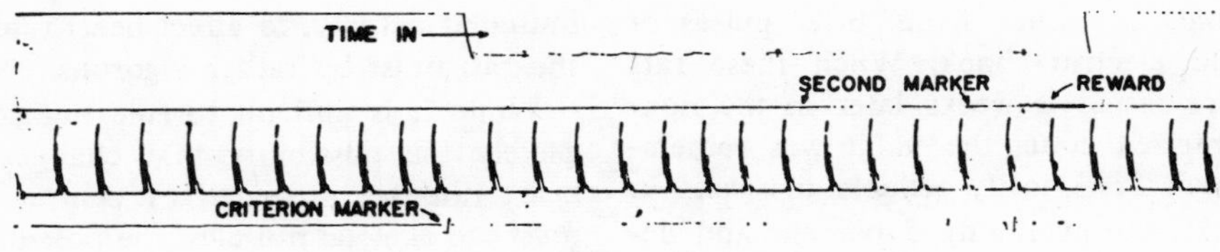

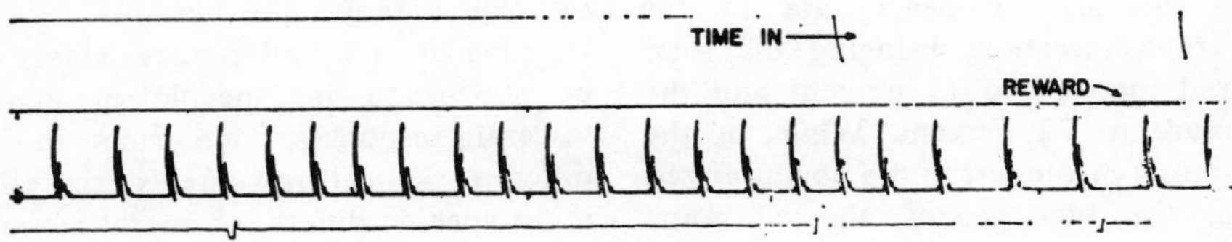

and the "safe" signals. You will remember that, with noncurarized thirsty dogs, we were able to use yet another kind of reward, water, to produce learned changes in salivation.

Transfer to Noncurarized State: More Evidence against Mediation

In the experiments discussed above, paralysis of the skeletal muscles by curare ruled out the possibility that the subjects were learning the overt performance of skeletal responses which were indirectly eliciting the changes in the heart rate. It is barely conceivable, however, that the rats were learning to send out from the motor cortex central impulses which would have activated the muscles had they not been paralyzed. And it is barely conceivable that these central impulses affected heart rate by means either of inborn connections or of classically conditioned ones that had been acquired when previous exercise had been accompanied by an increase in heart rate and relaxation had been accompanied by a decrease. But, if the changes in heart rate were produced in this indirect way, we would expect that, during a subsequent test without curare, any rat that showed learned changes in heart rate would show the movements in the muscles that were no longer paralyzed. Furthermore, the problem of whether or not visceral responses learned under curarization carry over to the noncurarized state is of interest in its own right.

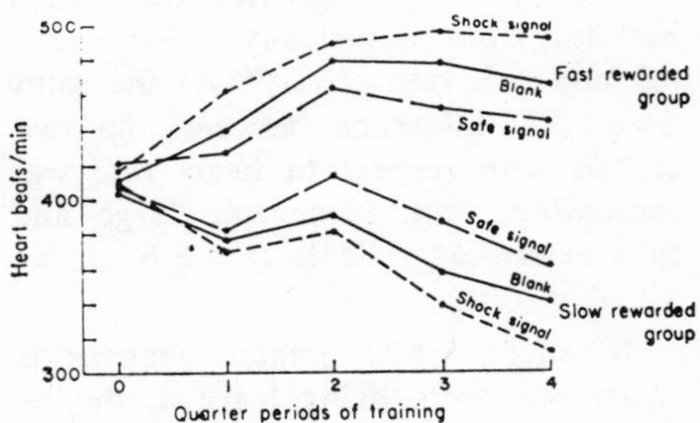

Fig. 2 (left). Electrocardiograms at the beginning and at the end of discrimination training of curarized rat rewarded for slow heart rate. Slowing of heart rate is rewarded only during a "time-in" stimulus (tone and light). [From Miller and DiCara (*12*)] Fig. 3 (above). Changes in heart rate during avoidance training. [From DiCara and Miller (*14*)]

In order to answer this question, DiCara and I (*15*) trained two groups of curarized rats to increase or decrease, respectively, their heart rate in order to avoid, or escape from, brief pulses of mild electric shock. When these rats were tested 2 weeks later in the noncurarized state, the habit was remembered. Statistically reliable increases in heart rate averaging 5 percent and decreases averaging 16 percent occurred. Immediately subsequent retraining without curare produced additional significant changes of heart rate in the rewarded direction, bringing the total overall increase to 11 percent and the decrease to 22 percent. While, at the beginning of the test in the noncurarized state, the two groups showed some differences in respiration and activity, these differences decreased until, by the end of the retraining, they were small and far from statistically reliable ($t =$ 0.3 and 1.3, respectively). At the same time, the difference between the two groups with respect to heart rate was increasing, until it became large and thus extremely reliable ($t = 8.6$, d.f. $=$ 12, $P < .001$).

In short, while greater changes in heart rate were being learned, the response was becoming more specific, involving smaller changes in respiration and muscular activity. This increase in specificity with additional training is another point of similarity with the instrumental learning of skeletal responses. Early in skeletal learning, the rewarded correct response is likely to be accompanied by many unnecessary movements. With additional training during which extraneous movements are not rewarded, they tend to drop out.

It is difficult to reconcile the foregoing results with the hypothesis that the differences in heart rate were mediated primarily by a difference in either respiration or amount of general activity. This is especially true in view of the research, summarized by Ehrlich and Malmo (*16*), which shows that muscular activity, to affect heart rate in the rat, must be rather vigorous.

While it is difficult to rule out completely the possibility that changes in heart rate are mediated by central impulses to skeletal muscles, the possibility of such mediation is much less attractive for other responses, such as intestinal contractions and the formation of urine by the kidney. Furthermore, if the learning of these different responses can be shown to be specific in enough visceral responses, one runs out of different skeletal movements each eliciting a specific different visceral response (*17*). Therefore, experiments were performed on the learning of a variety of different visceral responses and on the specificity of that learning. Each of these experiments was, of course, interesting in its own right, quite apart from any bearing on the problem of mediation.

Specificity: Intestinal versus Cardiac

The purpose of our next experiment was to determine the specificity of visceral learning. If such learning has the same properties as the instrumental learning of skeletal responses, it should be possible to learn a specific visceral response independently of other ones. Furthermore, as we have just seen, we might expect to find that, the better the rewarded response is learned, the more specific is the learning. Banuazizi and I worked on this problem (*18*). First we had to discover another visceral response that could be conveniently recorded and rewarded. We decided on intestinal contractions, and recorded them in the curarized rat with a little balloon filled with water thrust approximately 4 centimeters beyond the anal sphincter. Changes of pressure in the

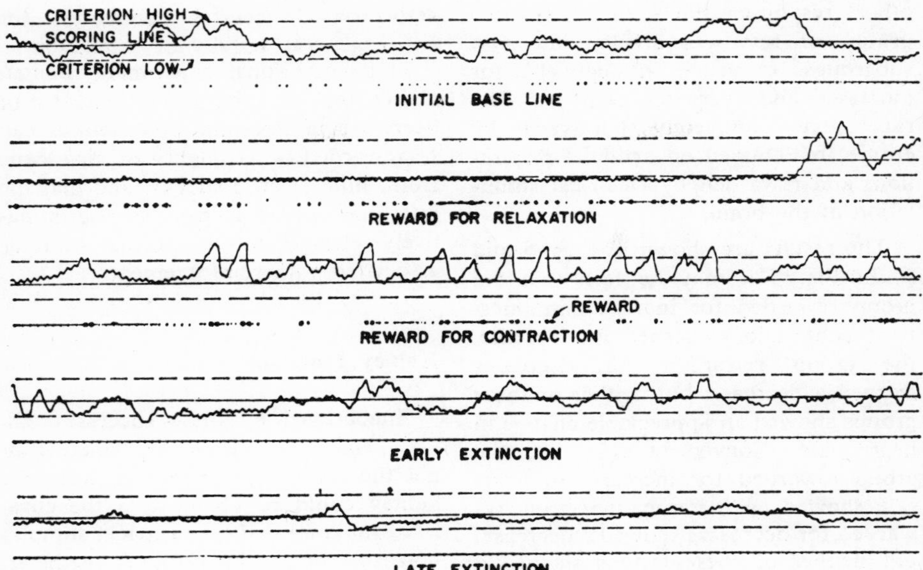

Fig. 4. Typical samples of a record of instrumental learning of an intestinal response by a curarized rat. (From top to bottom) Record of spontaneous contraction before training; record after training with reward for relaxation; record after training with reward for contractions; records during nonrewarded extinction trials. [From Miller and Banuazizi (*18*)]

balloon were transduced into electric voltages which produced a record on a polygraph and also activated an automatic mechanism for delivering the reward, which was electrical stimulation of the brain.

The results for the first rat trained, which was a typical one, are shown in Fig. 4. From the top record it may be seen that, during habituation, there were some spontaneous contractions. When the rat was rewarded by brain stimulation for keeping contractions below a certain amplitude for a certain time, the number of contractions was reduced and the base line was lowered. After the record showed a highly reliable change indicating that relaxation had been learned (Fig. 4, second record from the top), the conditions of training were reversed and the reward was delivered whenever the amplitude of contractions rose above a certain level. From the next record (Fig. 4, middle) it may be seen that this type of training increased the number of contractions and raised the base line. Finally (Fig. 4, two bottom records) the reward was discontinued and, as would be expected, the response continued for a while but gradually became extinguished, so that the activity eventually returned to approximately its original base-line level.

After studying a number of other rats in this way and convincing ourselves that the instrumental learning of intestinal responses was a possibility, we designed an experiment to test specificity. For all the rats of the experiment, both intestinal contractions and heart rate were recorded, but half the rats were rewarded for one of these responses and half were rewarded for the other response. Each of these two groups of rats was divided into two subgroups, rewarded, respectively, for increased and decreased response. The rats were completely paralyzed by curare, maintained on artificial respiration, and rewarded by electrical stimulation of the brain.

The results are shown in Figs. 5 and 6. In Fig. 5 it may be seen that the group rewarded for increases in intestinal contractions learned an increase, the group rewarded for decreases learned a decrease, but neither of these groups showed an appreciable change in heart rate. Conversely (Fig. 6), the group rewarded for increases in heart rate showed an increase, the group rewarded for decreases showed a decrease, but neither of these groups showed a change in intestinal contractions.

The fact that each type of response changed when it was rewarded rules out the interpretation that the failure to secure a change when that change was not rewarded could have been due to either a strong and stable homeostatic regulation of that response or an inability of our techniques to measure changes reliably under the particular conditions of our experiment.

Each of the 12 rats in the experiment showed statistically reliable changes in the rewarded direction; for 11 the changes were reliable beyond the $P <$ .001 level, while for the 12th the changes were reliable only beyond the .05 level. A statistically reliable negative correlation showed that the better the rewarded visceral response was learned, the less change occurred in the other, nonrewarded response. This greater specificity with better learning is what we had expected. The results showed that visceral learning can be specific to an organ system, and they clearly ruled out the possibility of mediation by any single general factor, such as level of activation or central commands for either general activity or relaxation.

In an additional experiment, Banuazizi (*19*) showed that either increases or decreases in intestinal contractions can be rewarded by avoidance of, or escape from, mild electric shocks, and that the intestinal responses can be discriminatively elicited by a specific stimulus associated with reinforcement.

Kidney Function

Encouraged by these successes, DiCara and I decided to see whether or not the rate of urine formation by the kidney could be changed in the curarized rat rewarded by electrical stimulation of the brain (*20*). A catheter, permanently inserted, was used to prevent accumulation of urine by the bladder, and the rate of urine formation was measured by an electronic device for counting minute drops. In order to secure a rate of urine formation fast enough so that small changes could be promptly detected and rewarded, the rats were kept constantly loaded with water through infusion by way of a catheter permanently inserted in the jugular vein.

All of the seven rats rewarded when the intervals between times of urine-drop formation lengthened showed decreases in the rate of urine formation, and all of the seven rats rewarded when these intervals shortened showed increases in the rate of urine formation. For both groups the changes were highly reliable ($P < .001$).

In order to determine how the change in rate of urine formation was achieved, certain additional measures were taken. As the set of bars at left in Fig. 7 shows, the rate of filtration, measured by means of ^{14}C-labeled inulin, increased when increases in the rate of

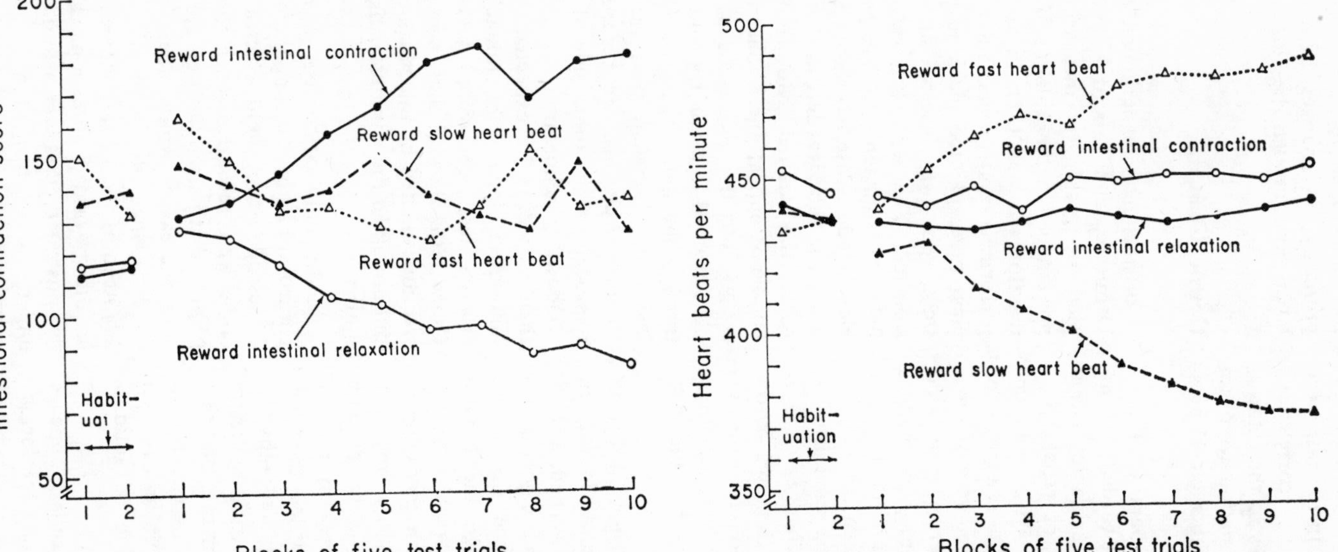

Fig. 5 (left). Graph showing that the intestinal contraction score is changed by rewarding either increases or decreases in intestinal contractions but is unaffected by rewarding changes in heart rate. [From Miller and Banuazizi (*18*)]   Fig. 6 (right). Graph showing that the heart rate is changed by rewarding either increases or decreases in heart rate but is unaffected by rewarding changes in intestinal contractions. Comparison with Fig. 5 demonstrates the specificity of visceral learning. [From Miller and Banuazizi (*18*)]

urine formation were rewarded and decreased when decreases in the rate were rewarded. Plots of the correlations showed that the changes in the rates of filtration and urine formation were not related to changes in either blood pressure or heart rate.

The middle set of bars in Fig. 7 shows that the rats rewarded for increases in the rate of urine formation had an increased rate of renal blood flow, as measured by ^{3}H-*p*-aminohippuric acid, and that those rewarded for decreases had a decreased rate of renal blood flow. Since these changes in blood flow were not accompanied by changes in general blood pressure or in heart rate, they must have been achieved by vasomotor changes of the renal arteries. That these vasomotor changes were at least somewhat specific is shown by the fact that vasomotor responses of the tail, as measured by a photoelectric plethysmograph, did not differ for the two groups of rats.

The set of bars at right in Fig. 7 shows that when decreases in rate of urine formation were rewarded, a more concentrated urine, having higher osmolarity, was formed. Since the slower passage of urine through the tubules would afford more opportunity for reabsorption of water, this higher concentration does not necessarily mean an increase in the secretion of antidiuretic hormone. When an increased rate of urine formation was rewarded, the urine did not become more diluted—that is, it showed no decrease in osmolarity; therefore, the increase in rate of urine formation observed in this experiment cannot be accounted for in terms of an inhibition of the secretion of antidiuretic hormone.

From the foregoing results it appears that the learned changes in urine formation in this experiment were produced primarily by changes in the rate of filtration, which, in turn, were produced primarily by changes in the rate of blood flow through the kidneys.

Gastric Changes

In the next experiment, Carmona, Demierre, and I used a photoelectric plethysmograph to measure changes, presumably in the amount of blood, in the stomach wall (*21*). In an operation performed under anesthesia, a small glass tube, painted black except for a small spot, was inserted into the rat's stomach. The same tube was used to hold the stomach wall against a small glass window inserted through the body wall. The tube was left in that position. After the animal had recovered, a bundle of optical fibers could be slipped snugly into the glass tube so that the light beamed through it would shine out through the unpainted spot in the tube inside the stomach, pass through the stomach wall, and be recorded by a photocell on the other side of the glass window. Preliminary tests indicated that, as would be expected, when the amount of blood in the stomach wall increased, less light would pass through. Other tests showed that stomach contractions elicited by injections of insulin did not affect the amount of light transmitted.

In the main experiment we rewarded curarized rats by enabling them to avoid or escape from mild electric shocks. Some were rewarded when the amount of light that passed through the stomach wall increased, while others were rewarded when the amount decreased. Fourteen of the 15 rats showed changes in the rewarded direction. Thus, we demonstrated that the stomach wall, under the control of the autonomic nervous system, can be modified by instrumental learning. There is strong

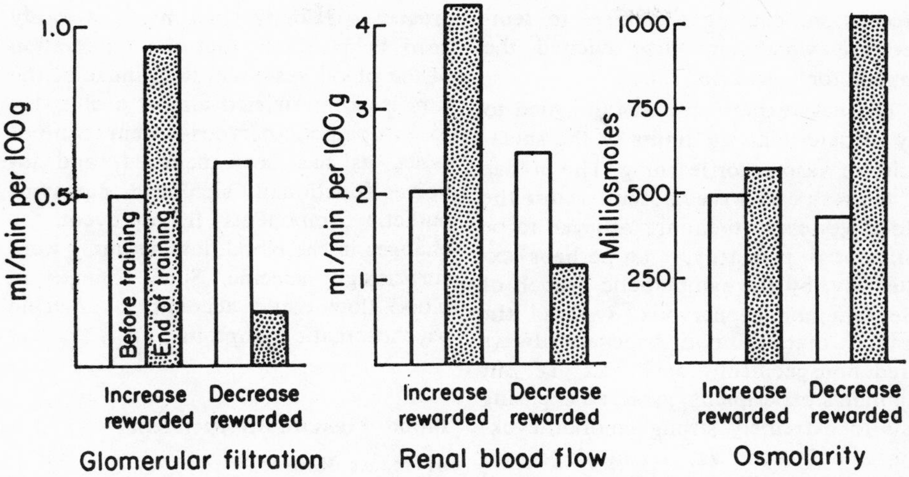

Fig. 7. Effects of rewarding increased rate of urine formation in one group and decreased rate in another on measures of glomerular filtration, renal blood flow, and osmolarity. [From data in Miller and DiCara (*20*)]

reason to believe that the learned changes were achieved by vasomotor responses affecting the amount of blood in the stomach wall or mucosa, or in both.

In another experiment, Carmona (*22*) showed that stomach contractions can be either increased or decreased by instrumental learning.

It is obvious that learned changes in the blood supply of internal organs can affect their functioning—as, for example, the rate at which urine was formed by the kidneys was affected by changes in the amount of blood that flowed through them. Thus, such changes can produce psychosomatic symptoms. And if the learned changes in blood supply can be specific to a given organ, the symptom will occur in that organ rather than in another one.

Peripheral Vasomotor Responses

Having investigated the instrumental learning of internal vasomotor responses, we next studied the learning of peripheral ones. In the first experiment, the amount of blood in the tail of a curarized rat was measured by a photoelectric plethysmograph, and changes were rewarded by electrical stimulation of the brain (*23*). All of the four rats rewarded for vasoconstriction showed that response, and, at the same time, their average core temperature, measured rectally, decreased from 98.9° to 97.9°F. All of the four rats rewarded for vasodilatation showed that response and, at the same time, their average core temperature increased from 99.9° to 101°F. The vasomotor change for each individual rat was reliable beyond the $P < .01$ level, and the difference in change in temperature between the groups was reliable beyond the .01 level. The direction of the change in temperature was opposite to that which would be expected from the heat conservation caused by peripheral vasoconstriction or the heat loss caused by peripheral vasodilatation. The changes are in the direction which would be expected if the training had altered the rate of heat production, causing a change in temperature which, in turn, elicited the vasomotor response.

The next experiment was designed to try to determine the limits of the specificity of vasomotor learning. The pinnae of the rat's ears were chosen because the blood vessels in them are believed to be innervated primarily, and perhaps exclusively, by the sympathetic branch of the autonomic nervous system, the branch that Cannon believed always fired nonspecifically as a unit (*3*). But Cannon's experiments involved exposing cats to extremely strong emotion-evoking stimuli, such as barking dogs, and such stimuli will also evoke generalized activity throughout the skeletal musculature. Perhaps his results reflected the way in which sympathetic activity was elicited, rather than demonstrating any inherent inferiority of the sympathetic nervous system.

In order to test this interpretation, DiCara and I (*24*) put photocells on both ears of the curarized rat and connected them to a bridge circuit so that only differences in the vasomotor responses of the two ears were rewarded by brain stimulation. We were somewhat surprised and greatly delighted to find that this experiment actually worked. The results are summarized in Fig. 8. Each of the six rats rewarded for relative vasodilatation of the left ear showed that response, while each of the six rats rewarded for relative vasodilatation of the right ear showed that response. Recordings from the right and left forepaws showed little if any change in vasomotor response.

It is clear that these results cannot be by-products of changes in either heart rate or blood pressure, as these would be expected to affect both ears equally. They show either that vasomotor responses mediated by the sympathetic nervous system are capable of much greater specificity than has previously been believed, or that the innervation of the blood vessels in the pinnae of the ears is not restricted almost exclusively to sympathetic-nervous-system components, as has been believed, and involves functionally significant parasympathetic components. In any event, the changes in the blood flow certainly were surprisingly specific. Such changes in blood flow could account for specific psychosomatic symptoms.

Blood Pressure Independent of Heart Rate

Although changes in blood pressure were not induced as by-products of rewarded changes in the rate of urine formation, another experiment on curarized rats showed that, when changes in systolic blood pressure are specifically reinforced, they can be learned (*25*). Blood pressure was recorded by means of a catheter permanently inserted into the aorta, and the reward was avoidance of, or escape from, mild electric shock. All seven rats rewarded for increases in blood pressure showed further increases, while all seven rewarded for decreases showed decreases, each of the changes, which were in opposite directions, being reliable beyond the $P < .01$ level. The increase was from 139 mm-Hg, which happens to be roughly comparable to the normal systolic blood pressure of an adult man, to 170 mm-Hg, which is on the borderline of abnormally high blood pressure in man.

Each experimental animal was "yoked" with a curarized partner, maintained on artificial respiration and having shock electrodes on its tail wired in series with electrodes on the tail of the experimental animal, so that it received exactly the same electric shocks and could do nothing to escape or avoid

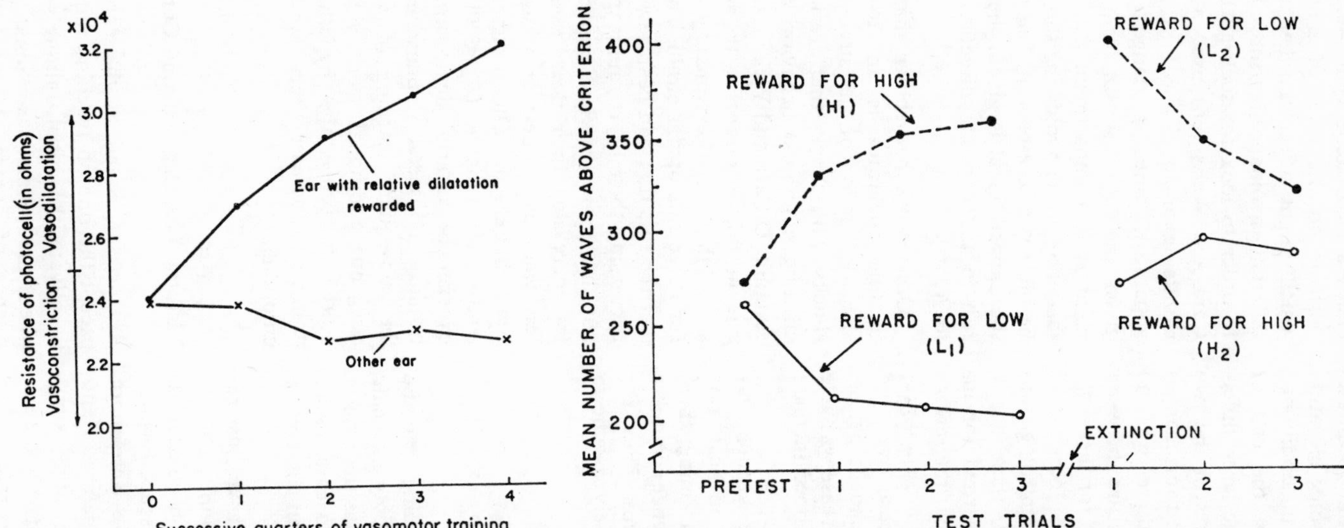

Fig. 8 (left). Learning a difference in the vasomotor responses of the two ears in the curarized rat. [From data in DiCara and Miller (*24*)] Fig. 9 (right). Instrumental learning by curarized rats rewarded for high-voltage or for low-voltage electroencephalograms recorded from the cerebral cortex. After a period of nonrewarded extinction, which produced some drowsiness, as indicated by an increase in voltage, the rats in the two groups were then rewarded for voltage changes opposite in direction to the changes for which they were rewarded earlier. [From Carmona (*29*)]

them. The yoked controls for both the increase-rewarded and the decrease-rewarded groups showed some elevation in blood pressure as an unconditioned effect of the shocks. By the end of training, in contrast to the large difference in the blood pressures of the two groups specifically rewarded for changes in opposite directions, there was no difference in blood pressure between the yoked control partners for these two groups. Furthermore, the increase in blood pressure in these control groups was reliably less ($P < .01$) than that in the group specifically rewarded for increases. Thus, it is clear that the reward for an increase in blood pressure produced an additional increase over and above the effects of the shocks per se, while the reward for a decrease was able to overcome the unconditioned increase elicited by the shocks.

For none of the four groups was there a significant change in heart rate or in temperature during training; there were no significant differences in these measures among the groups. Thus, the learned change was relatively specific to blood pressure.

Transfer from Heart Rate to Skeletal Avoidance

Although visceral learning can be quite specific, especially if only a specific response is rewarded, as was the case in the experiment on the two ears, under some circumstances it can involve a more generalized effect.

In handling the rats that had just recovered from curarization, DiCara noticed that those that had been trained, through the avoidance or escape reward, to increase their heart rate were more likely to squirm, squeal, defecate, and show other responses indicating emotionality than were those that had been trained to reduce their heart rate. Could instrumental learning of heart-rate changes have some generalized effects, perhaps on the level of emotionality, which might affect the behavior in a different avoidance-learning situation? In order to look for such an effect, DiCara and Weiss (*26*) used a modified shuttle avoidance apparatus. In this apparatus, when a danger signal is given, the rat must run from compartment A to compartment B. If he runs fast enough, he avoids the shock; if not, he must run to escape it. The next time the danger signal is given, the rat must run in the opposite direction, from B to A.

Other work had shown that learning in this apparatus is an inverted U-shaped function of the strength of the shocks, with shocks that are too strong eliciting emotional behavior instead of running. DiCara and Weiss trained their rats in this apparatus with a level of shock that is approximately optimum for naive rats of this strain. They found that the rats that had been rewarded for decreasing their heart rate learned well, but that those that had been rewarded for increasing their heart rate learned less well, as if their emotionality had been increased. The difference was statistically reliable ($P < .001$). This experiment clearly demonstrates that training a visceral response can affect the subsequent learning of a skeletal one, but additional work will be required to prove the hypothesis that training to increase heart rate increases emotionality.

Visceral Learning without Curare

Thus far, in all of the experiments except the one on teaching thirsty dogs to salivate, the initial training was given when the animal was under the influence of curare. All ot the experiments, except the one on salivation, have produced surprisingly rapid learning—definitive results within 1 or 2 hours. Will learning in the normal, noncurarized state be easier, as we originally thought it should be, or will it be harder, as the experiment on the noncurarized dogs suggests? DiCara and I have started to get additional evidence on this problem. We have obtained clear-cut evidence that rewarding (with the avoidance or escape reward) one group of freely moving rats for reducing heart rate and rewarding another group for increasing heart rate produces a difference between the two groups (*27*). That this difference was not due to the indirect effects of the overt performance of skeletal responses is shown by the fact that it persisted in subsequent tests during which the rats were paralyzed by curare. And, on subsequent retraining without curare, such differences in activity and respiration as were present earlier in training continued to decrease, while the differences in heart rate continued to increase. It seems extremely unlikely that, at the end of training, the highly reliable differences in heart rate ($t = 7.2$; $P < .0001$) can be explained by the highly unreliable differences in activity and respiration ($t =$ .07 and 0.2, respectively).

Although the rats in this experiment showed some learning when they were trained initially in the noncurarized state, this learning was much poorer than that which we have seen in our other experiments on curarized rats. This is exactly the opposite of my original expectation, but seems plausible in the light of hindsight. My hunch is that paralysis by curare improved learning by eliminating sources of distraction and variability. The stimulus situation was kept more constant, and confusing visceral fluctuations induced indirectly by skeletal movements were eliminated.

Learned Changes in Brain Waves

Encouraged by success in the experiments on the instrumental learning of visceral responses, my colleagues and I have attempted to produce other unconventional types of learning. Electrodes placed on the skull or, better yet, touching the surface of the brain record summative effects of electrical activity over a considerable area of the brain. Such electrical effects are called brain waves, and the record of them is called an electroencephalogram. When the animal is aroused, the electroencephalogram consists of fast, low-voltage activity; when the animal is drowsy or sleeping normally, the electroencephalogram consists of considerably slower, higher-voltage activity. Carmona attempted to see whether this type of brain activity, and the state of arousal accompanying it, can be modified by direct reward of changes in the brain activity (*28, 29*).

The subjects of the first experiment were freely moving cats. In order to have a reward that was under complete control and that did not require the cat to move, Carmona used direct electrical stimulation of the medial forebrain bundle, which is a rewarding area of the brain. Such stimulation produced a slight lowering in the average voltage of the electroencephalogram and an increase in behavioral arousal. In order to provide a control for these and any other unlearned effects, he rewarded one group for changes in the direction of high-voltage activity and another group for changes in the direction of low-voltage activity.

Both groups learned. The cats rewarded for high-voltage activity showed more high-voltage slow waves and tended to sit like sphinxes, staring out into space. The cats rewarded for low-voltage activity showed much more low-voltage fast activity, and appeared to be aroused, pacing restlessly about, sniffing, and looking here and there. It was clear that this type of training had modified both the character of the electrical brain waves and the general level of the behavioral activity. It was not clear, however, whether the level of arousal of the brain was directly modified and hence modified the behavior; whether the animals learned specific items of behavior which, in turn, modified the arousal of the brain as reflected in the electroencephalogram; or whether both types of learning were occurring simultaneously.

In order to rule out the direct sensory consequences of changes in muscular tension, movement, and posture, Carmona performed the next experiment on rats that had been paralyzed by means of curare. The results, given in Fig. 9, show that both rewarded groups showed changes in the rewarded direction; that a subsequent nonrewarded rest increased the number of high-voltage responses in both groups; and that, when the conditions of reward were reversed, the direction of change in voltage was reversed.

At present we are trying to use similar techniques to modify the functions of a specific part of the vagal nucleus, by recording and specifically rewarding changes in the electrical activity there. Preliminary results suggest that this is possible. The next step is to investigate the visceral consequences of such modification. This kind of work may open up possibilities for modifying the activity of specific parts of the brain and the functions that they control. In some cases, directly rewarding brain activity may be a more convenient or more powerful technique than rewarding skeletal or visceral behavior. It also may be a new way to throw light on the functions of specific parts of the brain (*30*).

Human Visceral Learning

Another question is that of whether people are capable of instrumental learning of visceral responses. I believe that in this respect they are as smart as rats. But, as a recent critical review by Katkin and Murray (*31*) points out, this has not yet been completely proved. These authors have comprehensively summarized the recent studies reporting successful use of instrumental training to modify human heart rate, vasomotor responses, and the galvanic skin response. Because of the difficulties in subjecting human subjects to the same rigorous controls, including deep paralysis by means of curare, that can be used with animal subjects, one of the most serious questions about the results of the human studies is whether the changes recorded represent the true instrumental learning of visceral responses or the unconscious learning of those skeletal responses that can produce visceral reactions. However, the able investigators who have courageously challenged the strong traditional belief in the inferiority of the autonomic nervous system with experiments at the more difficult but especially significant human level are developing ingenious controls, including demonstrations of the specificity of the visceral change, so that their cumulative results are becoming increasingly impressive.

Possible Role in Homeostasis

The functional utility of instrumental learning by the cerebrospinal nervous system under the conditions that existed during mammalian evolution is obvious. The skeletal responses mediated by the cerebrospinal nervous system operate on the external environment, so that there is survival value in the ability to learn responses that bring rewards such as food, water, or escape from pain. The fact that the responses mediated by the autonomic nervous system do not have such direct action on the external environment was one of the reasons for believing that they are not subject to instrumental learning. Is the learning ability of the autonomic nervous system something that has no normal function other than that of providing my students with subject matter for publications? Is it a mere accidental by-product of the survival value of cerebrospinal learning, or does the instrumental learning of autonomically mediated responses have some adaptive function, such as helping to maintain that constancy of the internal environment called homeostasis?

In order for instrumental learning to function homeostatically, a deviation away from the optimum level will have to function as a drive to motivate learning, and a change toward the optimum level will have to function as a reward to reinforce the learning of the particular visceral response that produced the corrective change.

When a mammal has less than the optimum amount of water in his body, this deficiency serves as a drive of thirst to motivate learning; the overt consummatory response of drinking functions as a reward to reinforce the learning of the particular skeletal responses that were successful in securing the water that restored the optimum level. But is the consummatory response essential? Can restoration of an optimum level by a glandular response function as a reward?

In order to test for the possible rewarding effects of a glandular response, DiCara, Wolf, and I (*32*) injected albino rats with antidiuretic hormone (ADH) if they chose one arm of a T-maze and with the isotonic saline vehicle if they chose the other, distinctively different, arm. The ADH permitted water to be reabsorbed in the kidney, so that a smaller volume of more concentrated urine was formed. Thus, for normal rats loaded in advance with H_2O, the ADH interfered with the excess-water excretion required for the restoration of homeostasis, while the control injection of isotonic saline allowed the excess water to be excreted. And, indeed, such rats learned to select the side of the maze that assured them an injection of saline so that their glandular response could restore homeostasis.

Conversely, for rats with diabetes insipidus, loaded in advance with hypertonic NaCl, the homeostatic effects of the same two injections were reversed; the ADH, causing the urine to be more concentrated, helped the rats to get rid of the excess NaCl, while the isotonic saline vehicle did not. And, indeed, a group of rats of this kind learned the opposite choice of selecting the ADH side of the maze. As a further control on the effects of the ADH per se, normal rats which had not been given H_2O or NaCl exhibited no learning. This experiment showed that an excess of either H_2O or NaCl functions as a drive and that the return to the normal concentration produced by the appropriate response of a gland, the kidney, functions as a reward.

When we consider the results of this experiment together with those of our experiments showing that glandular and visceral responses can be instrumentally learned, we will expect the animal to learn those glandular and visceral responses mediated by the central nervous system that promptly restore homeostasis after any considerable deviation. Whether or not this theoretically possible learning has any practical significance will depend on whether or not the innate homeostatic mechanisms control the levels closely enough to prevent any deviations large enough to function as a drive from occurring. Even if the innate control should be accurate enough to preclude learning in most cases, there remains the intriguing possibility that, when pathology interferes with innate control, visceral learning is available as a supplementary mechanism.

Implications and Speculations

We have seen how the instrumental learning of visceral responses suggests a new possible homeostatic mechanism worthy of further investigation. Such learning also shows that the autonomic nervous system is not as inferior as has been so widely and firmly believed. It removes one of the strongest arguments for the hypothesis that there are two fundamentally different mechanisms of learning, involving different parts of the nervous system.

Cause of psychosomatic symptoms. Similarly, evidence of the instrumental learning of visceral responses removes the main basis for assuming that the psychosomatic symptoms that involve the autonomic nervous system are fundamentally different from those functional symptoms, such as hysterical ones, that involve the cerebrospinal nervous system. Such evidence allows us to extend to psychosomatic symptoms the type of learning-theory analysis that Dollard and I (*7*, *33*) have applied to other symptoms.

For example, suppose a child is terror-stricken at the thought of going to school in the morning because he is completely unprepared for an important examination. The strong fear elicits a variety of fluctuating autonomic symptoms, such as a queasy stomach at one time and pallor and faintness at another; at this point his mother, who is particularly concerned about cardiovascular symptoms, says, "You are sick and must stay home." The child feels a great relief from fear, and this reward should reinforce the cardiovascular responses producing pallor and faintness. If such experiences are repeated frequently enough, the child, theoretically, should learn to respond with that kind of symptom. Similarly, another child whose mother ignored the vasomotor responses but was particularly concerned by signs of gastric distress would learn the latter type of symptom. I want to exphasize, however, that we need careful clinical research to determine how frequently, if at all, the social conditions sufficient for such theoretically possible learning of visceral symptoms actually occur. Since a given instrumental response can be reinforced by a considerable variety of rewards, and by one reward on one occasion and a different reward on another, the fact that glandular and visceral responses can be instrumentally learned opens up many new theoretical possibilities for the reinforcement of psychosomatic symptoms.

Furthermore, we do not yet know how severe a psychosomatic effect can be produced by learning. While none of the 40 rats rewarded for speeding up their heart rates have died in the course of training under curarization, 7 of the 40 rats rewarded for slowing down their heart rates have died. This statistically reliable difference (chi square $= 5.6$, $P < .02$) is highly suggestive, but it could mean that training to speed up the heart helped the rats resist the stress of curare rather than that the reward for slowing down the heart was strong enough to overcome innate regulatory mechanisms and induce sudden death. In either event the visceral learning had a vital effect. At present, DiCara and I are trying to see whether or not the learning of visceral responses can be carried far enough in the noncurarized animal to produce physical damage. We are also investigating the possibility that there may be a critical period in early infancy during which visceral learning has particularly intense and long-lasting effects.

Individual and cultural differences. It is possible that, in addition to producing psychosomatic symptoms in extreme cases, visceral learning can account for certain more benign individual and cultural differences. Lacey and Lacey (*34*) have shown that a given individual may have a tendency, which is stable over a number of years, to respond to a variety of different stresses with the same profile of autonomic responses, while other individuals may have statistically reliable tendencies to respond with different profiles. It now seems possible that differential conditions of learning may account for at least some of these individual differences in patterns of autonomic response.

Conversely, such learning may account also for certain instances in which the same individual responds to the same stress in different ways. For example, a small boy who receives a severe bump in rough-and-tumble play may learn to inhibit the secretion of tears in this situation since his peer group will punish crying by calling it "sissy." But the same small boy may burst into tears when he gets home to his mother, who will not punish weeping and may even reward tears with sympathy.

Similarly, it seems conceivable that different conditions of reward by a culture different from our own may be responsible for the fact that Homer's adult heroes so often "let the big tears fall." Indeed, a former colleague of mine, Herbert Barry III, has analyzed cross-cultural data and found that the amount of crying reported for children seems to be related to the way in which the society reacts to their tears (*35*).

I have emphasized the possible role of learning in producing the observed individual differences in visceral responses to stress, which in extreme cases may result in one type of psychosomatic symptom in one person and a different type in another. Such learning does not, of course, exclude innate individual differences in the susceptibility of different organs. In fact, given social conditions under which any form of illness will be rewarded, the symptoms of the most susceptible organ will be the most likely ones to be learned. Furthermore, some types of stress may be so strong that the innate reactions to them produce damage without any learning. My colleagues and I are currently investigating the psychological variables involved in such types of stress (*36*).

Therapeutic training. The experimental work on animals has developed a powerful technique for using instrumental learning to modify glandular and visceral responses. The improved training technique consists of moment-to-moment recording of the visceral function and immediate reward, at first, of very small changes in the desired direction and then of progressively larger ones. The success of this technique suggests that it should be able to produce therapeutic changes. If the patient who is highly motivated to get rid of a symptom understands that a signal, such as a tone, indicates a change in the desired direction, that tone could serve as a powerful reward. Instruction to try to turn the tone on as often as possible and praise for success should increase the reward. As patients find that they can secure some control of the symptom, their motivation should be strengthened. Such a procedure should be well worth trying on any symptom, functional or organic, that is under neural control, that can be continuously monitored by modern instrumentation, and for which a given direction of change is clearly indicated medically—for example, cardiac arrhythmias, spastic colitis, asthma, and those cases of high blood pressure that are not essential compensation for kidney damage (*37*). The obvious cases to begin with are those in which drugs are ineffective or contraindicated. In the light of the fact that our animals learned so much better when under the influence of curare and transferred their training so well to the normal, nondrugged state, it should be worth while to try to use hypnotic suggestion to achieve similar results by enhancing the reward effect of the signal indicating a change in the desired direction, by producing relaxation and regular breathing, and by removing interference from skeletal responses and distraction by irrelevant cues.

Engel and Melmon (*38*) have reported encouraging results in the use of instrumental training to treat cardiac arrhythmias of organic origin. Randt, Korein, Carmona, and I have had some success in using the method described above to train epileptic patients in the laboratory to suppress, in one way or another, the abnormal paroxysmal spikes in their electroencephalogram. My colleagues and I are hoping to try learning therapy for other symptoms—for example, the rewarding of high-voltage electroencephalograms as a treatment for insomnia. While it is far too early to promise any cures, it certainly will be worth while to investigate thoroughly the therapeutic possibilities of improved instrumental training techniques.

References and Notes

1. *The Dialogues of Plato*, B. Jowett, Transl.. (Univ. of Oxford Press, London, ed. 2, 1875), vol. 3, "Timaeus."
2. X. Bichat, *Recherches Physiologiques sur la Vie et le Mort* (Brosson, Gabon, Paris, 1800).
3. W. B. Cannon, *The Wisdom of the Body* (Norton, New York, 1932).
4. F. Alexander, *Psychosomatic Medicine: Its Principles and Applications* (Norton, New York, 1950), pp. 40–41.
5. G. A. Kimble, *Hilgard and Marquis' Conditioning and Learning* (Appleton-Century-Crofts, New York, ed. 2, 1961), p. 100.
6. B. F. Skinner, *The Behavior of Organisms* (Appleton-Century, New York, 1938); O. H. Mowrer, *Harvard Educ. Rev.* **17**, 102 (1947).
7. N. E. Miller and J. Dollard, *Social Learning and Imitation* (Yale Univ. Press, New Haven, 1941); J. Dollard and N. E. Miller, *Personality and Psychotherapy* (McGraw-Hill, New York, 1950); N. E. Miller, *Psychol. Rev.* **58**, 375 (1951).
8. N. E. Miller, *Ann. N.Y. Acad. Sci.* **92**, 830 (1961); ———, in *Nebraska Symposium on Motivation*, M. R. Jones, Ed. (Univ. of Nebraska Press, Lincoln, 1963); ———, in *Proc. 3rd World Congr. Psychiat., Montreal, 1961* (1963), vol. 3, p. 213.
9. ———, in "Proceedings, 18th International Congress of Psychology, Moscow, 1966," in press.
10. ——— and A. Carmona, *J. Comp. Physiol. Psychol.* **63**, 1 (1967).
11. J. A. Trowill, *ibid.*, p. 7.
12. N. E. Miller and L. V. DiCara, *ibid.*, p. 12.
13. L. V. DiCara and N. E. Miller, *Commun. Behav. Biol.* **2**, 19 (1968).
14. ———, *J. Comp. Physiol. Psychol.* **65**, 8 (1968).
15. ———, *ibid.*, in press.
16. D. J. Ehrlich and R. B. Malmo, *Neuropsychologia* **5**, 219 (1967).
17. "It even becomes difficult to postulate enough different thoughts each arousing a different emotion, each of which in turn innately elicits a specific visceral response. And if one assumes a more direct specific connection between different thoughts and different visceral responses, the notion becomes indistinguishable from the ideo-motor hypothesis of the voluntary movement of skeletal muscles." [W. James, *Principles of Psychology* (Dover, New York, new ed., 1950), vol. 2, chap. 26].
18. N. E. Miller and A. Banuazizi, *J. Comp. Physiol. Psychol.* **65**, 1 (1968).
19. A. Banuazizi, thesis, Yale University (1968).
20. N. E. Miller and L. V. DiCara, *Amer. J. Physiol.* **215**, 677 (1968).
21. A. Carmona, N. E. Miller, T. Demierre, in preparation.
22. A. Carmona, in preparation.
23. L. V. DiCara and N. E. Miller, *Commun. Behav. Biol.* **1**, 209 (1968).
24. ———, *Science* **159**, 1485 (1968).
25. ———, *Psychosom. Med.* **30**, 489 (1968).
26. L. V. DiCara and J. M. Weiss, *J. Comp. Physiol. Psychol.*, in press.
27. L. V. DiCara and N. E. Miller, *Physiol. Behav.*, in press.
28. N. E. Miller, *Science* **152**, 676 (1966).
29. A. Carmona, thesis, Yale University (1967).
30. For somewhat similar work on the single-cell level, see J. Olds and M. E. Olds, *in Brain Mechanisms and Learning*, J. Delafresnaye, A. Fessard, J. Konorski, Eds. (Blackwell, London, 1961).
31. E. S. Katkin and N. E. Murray, *Psychol. Bull.* **70**, 52 (1968); for a reply to their criticisms, see A. Crider, G. Schwartz, S. Shnidman, *ibid.*, in press.

32. N. E. Miller, L. V. DiCara, G. Wolf, *Amer. J. Physiol.* **215**, 684 (1968).

33. N. E. Miller, in *Personality Change*, D. Byrne and P. Worchel, Eds. (Wiley, New York, 1964), p. 149.

34. J. I. Lacey and B. C. Lacey, *Amer. J. Psychol.* **71**, 50 (1958); *Ann. N.Y. Acad. Sci.* **98**, 1257 (1962).

35. H. Barry III, personal communication.

36. N. E. Miller, *Proc. N.Y. Acad. Sci.*, in press.

37. Objective recording of such symptoms might be useful also in monitoring the effects of quite different types of psychotherapy.

38. B. T. Engel and K. T. Melmon, personal communication.

39. The work described is supported by U.S. Public Health Service grant MH 13189.

II

HEART RATE

Learned Stabilization of Cardiac Rate 2

Michael Hnatiow and Peter J. Lang

ABSTRACT

Human subjects learned to reduce cardiac-rate variability when a visual display provided synchronous feedback of their own heart rate. The increased stability was unaccompanied by significant changes in average heart rate, and was relatively unrelated to respiration changes.

DESCRIPTORS: Heart rate variability, Visual feedback, Respiration. (P. J. Lang)

Heart-rate variability is readily observed, in the number and depth of stylus oscillations, when the output of a cardiotachometer is written out on a strip chart. This activity is not attributed to adventitious stimuli, or to transient ideation, but is held to be a persisting biological characteristic of individuals (Lacey & Lacey, 1958; Barratt, 1962). These authors present evidence that the extent of such variability is related to individual differences in psychomotor behavior. An association between cardiac variability, perceptual responses, and the personality trait of impulsivity has also been demonstrated (Barratt, 1961; Levy & Lang, 1965).

Autonomic variability is distinguished clearly from activity level. The latter phenomenon is measured by average heart rate and in the resting organism it is relatively uncorrelated with rate stability (Lacey & Lacey, 1958). Shearn (1962) attempted to control average rate by operant conditioning methods. He demonstrated that cardiac accelerations increase when they provide the contingency for shock avoidance.

This report describes a first attempt to bring heart-rate variability under experimental control. No aversive stimuli were used in this procedure. *S* was simply instructed to keep his heart rate as steady as possible. He observed a visual display synchronized with his own cardiac rate. The only reinforcer was the immediate feedback of success or failure responses. The experiment was designed to assess the degree of heart-rate stability attained when *S*s received meaningful information about their heart rate, compared with conditions in which no auxiliary information was provided.

These data were collected by the first author in preparing a master's thesis, University of Pittsburgh, 1964. The research was directed by the second author, and supported by Public Health Service Grant M-3880, from the National Institute of Mental Health. All statistics were calculated by the IBM 7090 Computer at the University of Pittsburgh Data Processing Center.

Method

Subjects. Forty students at the University of Pittsburgh participated in the experiment, as part of an introductory psychology research requirement. *S*s had no prior information about the experiment, except that physiological responses would be monitored. All *S*s were male and in good physical health; there were no other subject requirements.

Display Apparatus. The cardiac rate display was presented with an opaque projector. *S*s observed a pointer whose lateral movements were synchronized with the output of a Fels cardiotachometer. The pointer moved on a white field, the center of which was defined by a red stripe, flanked by blue stripes on either side. The width of each stripe was equal to a pointer movement of 2 heartbeats per minute. Prior to each test period the apparatus was adjusted so that the *S*'s average heart rate fell in the center of the red stripe. The entire display occupied the center square foot of a large, reverse projection screen.

Procedure. S was seated in a reclining lounge chair in a darkened, sound- and electrically-shielded room. Electrocardiograph (EKG) electrodes of stainless steel mesh were attached to either side of the rib cage and a Phipps and Bird respiration bellows was fixed around *S*'s chest. Twenty randomly selected *S*s were instructed to try to keep their heart rate as steady as possible during the experiment. They were further instructed to observe the synchronized heart-rate display, and were informed that maintaining the pointer in the red, center stripe would help them in the task. The remaining 20 *S*s were also instructed to try to keep their bodily processes at a steady level during the experiment. They were also instructed to watch the display and visually track the pointer movements. However, pointer movements merely simulated cardiac changes for the control *S*s, and there was no actual synchronization with the *S*'s heart rate. The simulated heart rate display was produced by the experimenter, who, using an experimental *S*'s actual record as a guide, manipulated the display amplifier controls to produce pointer movements the same as those initiated "on line."[1] During the experiment all *S*s observed the projected display for three 5-min trials, and 7 min intervened between display trials. The initial presentation followed an adaptation period of 16 min. The stimulus screen was illuminated by a homogeneous green light for the nondisplay periods. Stimuli were controlled and *S*'s behavior measured from an apparatus module, adjacent to *S*'s shielded room.[2]

Somatic Measures. Increments in cardiac rate stability were measured in 2

[1] Control *S*s were told to track pointer movements as part of the experimental task to ensure that the same physical stimuli were being presented and that both groups were attending to these inputs. Those in the control group were not told that the display represented their heart rate, because this would have constituted misinformation—input in conflict with natural feedback—and voided the necessary no-information control.

[2] Half of the experimental *S*s and half of the control *S*s were given an intermittent auditory stimulus (varying unpredictably in frequency, intensity, and duration) throughout the experiment. Analysis of variance revealed no significant effects attributable to this variable, and it is therefore not considered in this report.

ways: time on target and heart-rate standard deviation. Average heart rate and respiration were also recorded.

Time on Target. Heart rate and respiration were continuously recorded on separate channels of a Grass Model 5 polygraph. In addition, an expanded representation of pointer movements on the display was written out on a third channel. *S* was considered to be in the target area when this record indicated he was either on or between the 2 blue stripes. Total time on target was measured for each 5-min display period, and the first 5 min from each following non-display period.

Heart-Rate Mean and Standard Deviation. During the test periods, the R wave of each cardiac cycle initiated a time-based scan of the 400 addresses in a CAT 400B laboratory computer. The subsequent cardiac cycle stopped the scan, recorded a count at that address, and reinitiated a new scan within 50 μsec. This process was continuous during each 5-min display period, and during the middle 5 min of the preceding and following non-display periods. This time segment was selected so that a free minute was available to the computer after each test period, which enabled it to write out the frequency distribution of interpulse intervals on a wide channel recorder. From these distributions *S*'s display and non-display means and standard deviations were calculated.

Respiration. The number of respiration cycles was totaled for the same display and non-display periods from which the heart-rate measures were calculated. As the maximum effects of learning were expected on the final display trial, this period and the preceding non-display period were subjected to a closer analysis. Two additional respiration scores were calculated: the ratio of inspiration time to the entire length of a respiratory cycle, and depth of respiration.[3]

Results

Time on Target. Table 1 presents the average time that the 20 experimental *S*s remained within the striped target area for each display period, and the average time that they remained in the target area for the subsequent 5-min, after the display was withdrawn. Analysis of variance revealed that the total time within the criterion range during display periods was significantly greater than during non-display periods ($F = 5.89$, $df = 5/90$, $p < 0.01$). Furthermore, the over-all curve was evaluated in an orthogonol polynomial analysis, which confirmed the significant quintic trend ($F = 21.23$, $df = 1/18$, $p < 0.01$) evident in successive high display and low non-display scores.

Heart Rate. Frequency distributions of interpulse intervals are presented in Figure 1, as they were drawn by the computer. This is the record of an experimental *S* for the third non-display and the subsequent display period. The more narrow distribution under display conditions was typical of the group that received the meaningful display.

Standard deviations for the 5-min periods were calculated from these records. Mean deviations for the experimental and control groups for all display and non-

[3] The latter measure was in a sense arbitrary. An external recording bellows (Phipps and Bird) was used to measure respiration, and comparability between *S*s cannot be assured. However, this represents an initial screening for the phenomenon. More complicated measurement of changes in air volume would be justified if a relationship were suggested by this experiment.

TABLE 1

Average time on target for the experimental subjects and average heartrate for the experimental and control groups[a]

Time Period	Time on Target	Mean Heart Rate	
	Experimental	Experimental	Control
	sec	*beats/min*	
ND-1		77.7	76.6
D-1	22.8	76.6	76.1
ND-2	18.2	77.4	76.7
D-2	25.2	76.3	76.3
ND-3	20.0	76.9	76.1
D-3	26.1	76.0	76.1
ND-4	19.4	76.5	75.7

[a] Non-display (ND) and display (D) periods are in temporal order. Non-display periods for the time on target measure are the 5 min immediately following withdrawal of the display. For mean heart rate, standard deviation, and the respiration measures, non-display periods are the middle 5 min from the 7-min period preceding the first display, and from each successive 7-min non-display period.

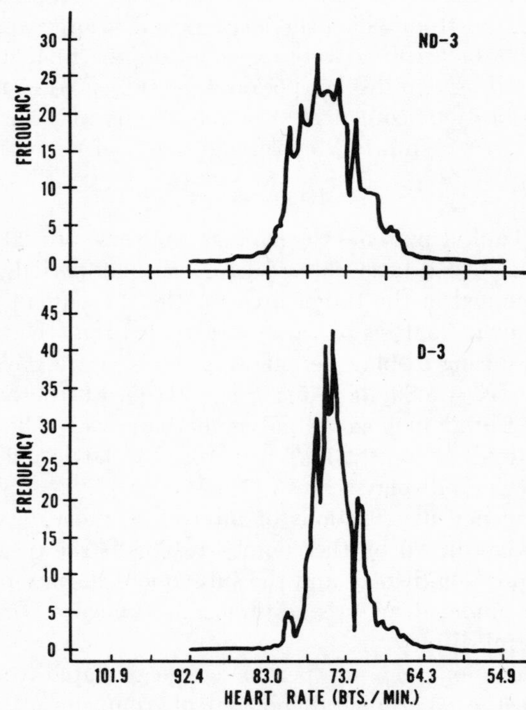

FIG. 1. Frequency histograms of interpulse intervals (reported in beats/min) taken from a single S on consecutive 5-min test periods. The originals were plotted by a 400B computer of average transients and written out on a Leeds and Northrup recorder. Curve ND-3 shows the pulse distribution when S received no information about his cardiac rate; curve D-3 was obtained during the succeeding trial in which S observed a visual display of his concurrent heart rate.

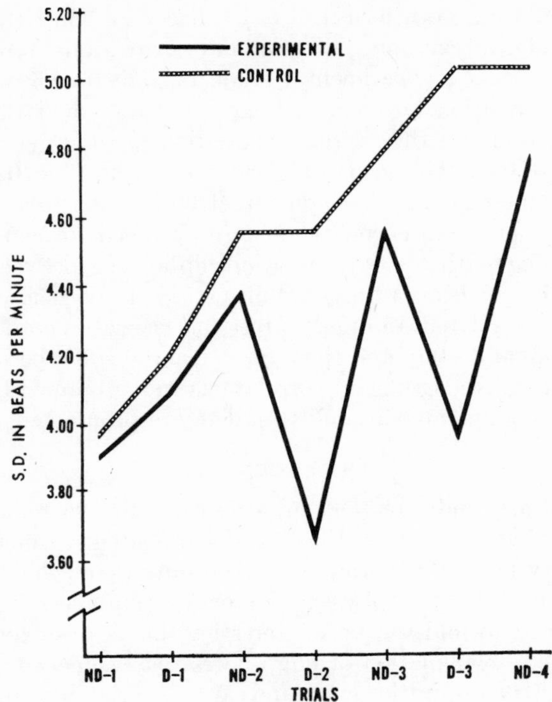

FIG. 2. Average heart-rate standard deviations of experimental and control subjects for consecutive non-display (ND) and information display (D) trials. Little change was observed during the first display period, but a significant reduction in standard deviation occurred for experimental subjects at the second and third display trials.

display periods are presented in Figure 2. It will be noted that display trials 2 and 3 yielded markedly reduced standard deviations when compared either with their own preceding non-display scores or with the coincident control-group scores. These results were confirmed in an analysis of variance by a significant over-all interaction between display information and trials ($F = 2.36$, $df = 6/216$, $p < 0.05$), and a significant sextic interaction trend obtained in a polynomial analysis of the curves ($F = 5.02$, $df = 1/36$, $p < 0.05$).

Average heart rate provided a measure of activity level. Mean rates for all test periods are presented in Table 1. Both groups yielded significantly slower cardiac rate during the display trials. However, there was no difference between the experimental (meaningful display) and control groups on this measure. The interaction F between display information and trials was less than 1.

Respiration. Respiration rate was evaluated for the same display and non-display periods as was heart rate standard deviation. Analysis of variance revealed no significant interaction between trials and display information for respiration rate ($F = 1.96$, $df = 6/216$).

For the final display and preceding non-display periods, Pearson product-moment correlation coefficients were calculated between all repiration measures (rate, the ratio of inspiration time to the entire length of a respiratory cycle, and depth of respiration), and both measures of cardiac variability (time on target, and heart-rate standard deviation). Twenty-four correlations were obtained for the 2 test periods. Among experimental subjects, only one r was significant ($p < 0.05$): the correlation between average heart-rate deviation during the final non-display period and the average inspiration to total cycle ratio for the display period was -0.62. The second highest correlation ($r = 0.43$, $p < 0.10$) was between this same ratio and time on target both taken from the final non-display period. No significant correlations were obtained for control subjects. While these data suggest that a weak relationship existed between the form of the respiration cycle and heart-rate variability, they do not show that changes in the 2 variables were related. In order to test this crucial hypothesis, difference scores between the final display and the adjacent non-display periods were then found for all measures. No significant correlations were obtained between either measure of change in heart-rate variability, and any respiration change measure.

Discussion

These findings clearly indicate that subjects can learn to reduce variability in cardiac rate, and that this stabilization effect is unrelated to coincident changes in heart-rate activity level. Furthermore, these results were achieved with naive subjects, and without resort to a physically aversive reinforcer.

All subjects seemed to find the task interesting, but a postexperiment interview yielded no useful explanation of how success was achieved. A slight relationship existed between respiration cycle and cardiac variability, but none of the 3 respiration difference score measures was significantly related to changes in heart-rate variability from non-display to display periods. Thus, the present findings lend little support to the thesis that respiration change plays a major role in learned control of heart rate.

Wenger, Bagchi, and Anand (1961) suggest an alternate control mechanism. On the basis of polygraph studies, they proposed that Indian yogis slow their heart rate via a striated muscular mechanism. One of their subjects was apparently able to stimulate and interrupt vagus nerve output to the sinoatrial node, and thereby interrupt regular cardiac cycles, briefly establishing a nodal rhythm. Further study of this system as well as of the cardiac-respiratory mechanism is needed.

Despite a lengthy initial adaptation period, heart-rate standard deviation tended to increase for all subjects throughout the experiment (Fig. 1). This general trend may account for the fact that the stability achieved by the experimental group did not persist during subsequent non-display periods. However, it must also be recalled that this stability was attained after only 15 min of training. Further trials may well produce a more lasting effect.

Lacey & Lacey (1958) have shown that resting rates of autonomic fluctuation were positively related to the frequency of occurrence of erroneous and impulsive motor responses. Barratt (1961) reported a significant positive relationship between variability of physiological measures and impulsiveness as measured by the Barratt impulsivity scale. Since *S*s in the present study had previously taken the Wilkinson impulsivity scale (1962), as part of an introductory psychology classroom assessment, the relationship between this instrument and cardiac variability was tested.

The correlation for the third non-display period between this scale and heart-rate standard deviation was 0.53 ($p < 0.02$), and with time on target it was -0.46 ($p < 0.05$). These results confirm the fact that a relationship exists between low inhibition and both measures of heart-rate variability used in the research. The present experiment establishes a method for gaining experimental control over the latter variable, and provides a basis for the exploration of possible causal relationships between cardiovascular responses, impulsivity, and other aspects of human behavior.

REFERENCES

Barratt, E. S. ANS activity related to intra-individual variability. Unpublished terminal report of project M-4534, NIMH, 1961.

Barratt, E. S. CNS correlates of intra-individual variability of ANS activity. Paper read at Amer. Psychol. Ass., St. Louis, Sept., 1962.

Lacey, J. I., & Lacey, Beatrice C. The relationship of resting autonomic activity to motor impulsivity. In Solomon, H., Cobb, S., & Penfield, W. (Eds.) *The brain and human behavior*. Baltimore: Williams & Wilkins, 1958. Pp. 144–209.

Levy, P., & Lang, P. J. Activation, control, and the spiral aftermovement. *J. pers. & soc. Psychol.*, 1965, in press.

Shearn, D. W. Operant conditioning of heart rate. *Science*, 1962, *137:* 530–531.

Wenger, M. A., Bagchi, B. K., & Anand, B. K. Experiments in India on "voluntary" control of the heart and pulse. *Circulation*, 1961, *24:* 1319–1325.

Wilkinson, H. Jean. Impulsivity-deliberativeness and time estimation. Unpublished doctoral dissertation, Univer. of Pittsburgh, 1962.

Address requests for reprints to: Dr. P. J. Lang, Psychology Dept., University of Wisconsin, Madison, Wisconsin 53706.

3 Operant Conditioning of Heart Rate Slowing

Bernard T. Engel and Stephen P. Hansen

ABSTRACT

The purpose of this study was to see if heart rate (HR) slowing could be operantly conditioned. Ten experimental *S*s and five yoked-control *S*s were studied. Experimental *S*s were positively reinforced for slowing their HR on a beat-by-beat basis, whereas yoked-control *S*s were reinforced in a pattern based on the performance of paired experimental *S*s. The data showed that: some *S*s can be taught to slow their HR by means of an operant conditioning procedure; *S*s appear to learn better when they do not infer correctly what the response is that they are controlling; the conditioned HR response is apparently not mediated by changes in breathing; and reinforcement, per se, is not adequate to lower HR.

DESCRIPTORS: Operant conditioning; Heart rate; Yoked-control *S*s.

Despite the dogma prevalent in experimental psychology that autonomic responses can not be operantly conditioned (e.g., Kimble,[1] 1961), a number of investigators have recently shown that these responses can be conditioned (Shearn, 1962; Kimmel and Kimmel, 1963; Hnatiow and Lang, 1965).

Although we do not wish to become involved in polemics about why one or another investigator fails to find evidence for learning while others succeed, we believe that one technical consideration stands out as a likely factor. Although the autonomic nervous system mediates changes in smooth muscle which can be relatively slow acting, many autonomic effects, such as heart beats, recur frequently. If an investigator attempts to condition an autonomically mediated effect, he must be prepared to develop instruments that are capable of deciding whether each of the responses *S* emits are to be rewarded. Until recently, with the advent of digital logic modules, such equipment did not exist. It would seem to us that all the previous research which reported finding no evidence of conditioning must be reexamined in the light of this consideration.

The substance of this article was reported by Dr. Engel at the annual meeting of the Society for Psychophysiological Research in Houston, Texas, on October 19, 1965.

Supported in part by USPHS grant FR-00122, for computing services at San Francisco Medical Center, the Interdisciplinary Training Program of the National Institute of Mental Health (5T1-MH-7082), and USPHS grant HE-06285.

S. P. H. is now at the Sacramento City College, Sacramento, California.

Address requests for reprints to B. T. Engel, Cardiovascular Research Institute, University of California San Francisco Medical Center, San Francisco, California.

[1] "Thus, for autonomically mediated behavior, the evidence points unequivocally to the conclusion that such responses can be modified by classical, but not instrumental, training methods." (Kimble, 1961, p. 100.)

Harwood (1962) attempted to condition heart rate (HR) slowing operantly; however, he apparently did not have equipment available to him to enable him to reinforce *S* on a beat-by-beat basis, and he reported that he was unable to obtain any evidence of conditioning.

Materials and Methods

Ss

Fifteen male college students were studied. Ten were experimental *S*s for whom response and reinforcement were contiguously associated; five were "yoked-control *S*s" for whom response and reinforcement were dissociated.

Procedure

Instructions. Each *S* was told that he was participating in a conditioning study; however, he was not told that HR slowing was the response that was being reinforced. He was told that the correct response was not related to breathing. He was also instructed to breathe normally because "abnormal breathing will foul up our measurements."

S was also told that a correct response would turn on a light and a clock, both of which he could see. The light served as *S*'s cue for correct response. The clock, which accumulated time, served as *S*'s reinforcement, and he was told that he would be paid at the rate of ½ cent per sec.

Experimental design. Each experimental session for each *S* was divided into three phases. During the first two phases, neither the clock nor the light was operating. Phase 1 was about 30 min and was an adaptation period designed to allow *S*'s HR to stabilize. Phase 2 was about 5 to 8 min and was used to establish *S*'s operant-level HR. Phase 3 was a 25-min training period during which *S* was told that the light and clock were now under his control.

Each *S* was tested during six experimental sessions. During sessions 1 and 2, an operant-level HR was found which would keep *S*'s light on 80% of the time, i.e., a level of HR below which *S*'s rate fell 80% of the time. During sessions 3 to 6 the operant-level HR was chosen so that *S*'s HR would keep the light on 50% of the time. The 80% level was chosen because it gave *S* a high rate of reinforcement and seemed analogous to the "shaping schedules" used in traditional operant-conditioning studies. The 50% levels, i.e., sessions 3 to 6, were the critical ones and learning was determined on the basis of performance during these periods.

Apparatus

A Fels cardiotachometer recorded HR.[2] The output of the cardiotachometer was fed into an Offner[3] amplifier and dynograph. The output from the amplifier was also fed into a Schmitt trigger,[4] which was used to control the experimental

[2] Yellow Springs Instrument Co., Inc., Yellow Springs, Ohio.

[3] Beckman Offner, 3900 River Road, Shiller Park, Illinois.

[4] The logic equipment used to regulate the lights and clocks was manufactured by BRS Electronics, 5451 Holland Drive, Beltsville, Maryland. In order to reduce the hysteresis in the Schmitt triggers, we modified the standard unit by grounding the emitters.

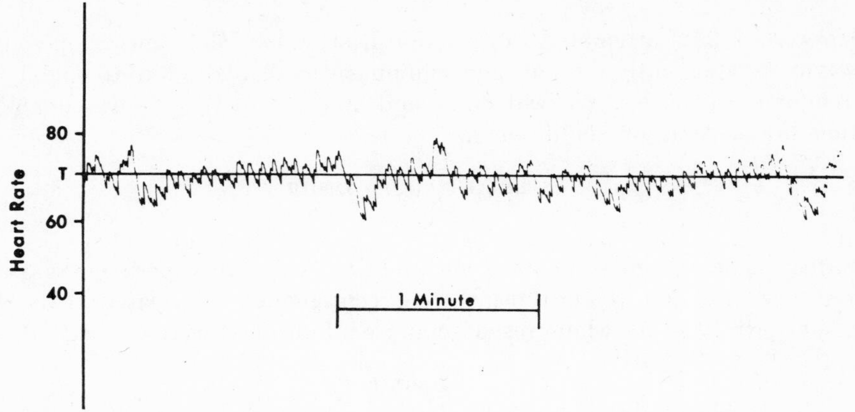

FIG. 1. Cardiotachometer record showing HR fluctuations around triggering level.

TABLE 1

Performance measures

Measure	Operant	Training
Change in percentage time HR below trigger level	10 samples each ~ 25 sec	40 samples each ~ 25 sec
Change in HR	Sample of 100 beats	Sample of 100 beats
Change in number of beats < or = operant mean HR	Sample of 100 beats	Sample of 100 beats

*S*s' reinforcements. The input to the Schmitt trigger was regulated by a zero-suppression circuit on the Offner amplifier to permit adjustment of the trigger point. Fig. 1 shows a cardiotachometer tracing during which the trigger was set so that 50% of *S*'s beats were below the trigger point. It should be clear that this system made it possible to reinforce *S* on a beat-by-beat basis.

For the yoked-control *S*s, the light and clock were controlled by a tape recording based on the performances of five of the experimental *S*s. Each of the six training sessions for five experimental *S*s was tape recorded by means of a switching circuit which passed a 60-cycle current when the light was on and no current when the light was off. The output from this tape was integrated and fed back into another Schmitt trigger to control the light and clock for a paired control *S*. Thus, the control *S*s received reinforcement which was identical in duration and pattern to the reinforcement received by the paired experimental *S*s.

Breathing patterns were measured by means of a strain gauge cemented on a belt strapped around *S*'s chest.

Analytical Procedures

Three criteria of learning were used (Table 1). The percentage time criterion was based upon successively timed samples of performance. (Each of the samples was about 25 sec; however, because the timing circuit was inconsistent, these

TABLE 2

Changes in performances for all subjects (see text for discussion)

Ss	Percentage Time	HR	Frequency Distributions	Breathing Rates
A. Experimental	*beats/min*	*no./beats*		*breaths/min*
1	26.3[a]	−2.6[a]	20.2[a]	2.9
2	15.4[a]	−1.8[a]	15.0[a]	0.6
3	−7.6	1.2[a]	−7.8[a]	0.7
4	8.6[b]	0.2	2.0	3.7
5	2.2	−1.0[a]	2.2	−0.9
6	−18.4[a]	2.6[a]	−13.5[a]	−1.9
7	−22.0[a]	1.8[a]	−9.0[a]	−1.4
8	10.2[a]	−1.2[a]	7.2[a]	−1.7
9	12.5[a]	−2.2[a]	12.2[a]	−1.9
10	10.3[a]	−2.0[a]	8.8[a]	−0.6
B. Control				
1	−2.6	−0.1	−3.0	0.4
2	−29.3[a]	6.8[a]	−27.2[a]	−2.9
3	−25.4[a]	9.3[a]	−31.5[a]	0.7
4	−10.7[a]	1.6[a]	−11.2[a]	1.6
5	−42.80[a]	10.2[a]	−38.8[a]	3.1

[a] $p < 0.01$.
[b] $p < 0.05$.

samples were timed by two electronically controlled clocks, one of which measured the duration of the sampling period, and the other of which measured the time during which *S*'s HR was below the trigger level.) The criteria for HR and the shift in frequency distribution were based on samples of 100 beats. These samples were taken at equal time intervals throughout the measurement periods.

In the case of the control *S*s, we measured not only the time *S* was being reinforced but also the time *S*'s HR was below his trigger level and the time that *S*'s HR was below his trigger level while he was being reinforced. The last measure of coincidence permitted us to determine *S*'s performance as a function of his reward.

In order to determine whether an *S* had learned, we computed separate means and standard deviations for each of the 50% operant-level sessions (i.e., sessions 3 to 6) and then combined these scores for a single significance test. The change in percentage time and the change in HR (i.e., change from operant level to training level) were tested by means of a *t* test. The significance of the shifts in frequency of beats at or below the operant level mean was tested by means of the normal approximation to the binomial distribution.

Results and Comments

Experimental Ss

Table 2A presents the difference scores for each of the criterion measures for each of the experimental *S*s for sessions 3 to 6. Five *S*s (nos. 1, 2, 8, 9, and 10) showed evidence of learning as measured by all three criteria: increase in per-

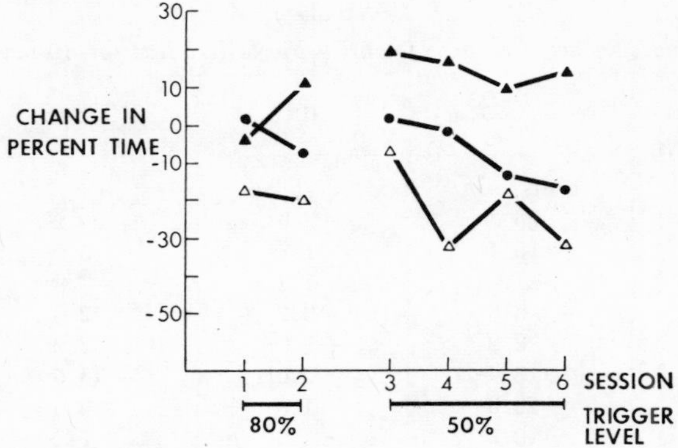

FIG. 2. Average changes in percentage time during training for learners (▲), nonlearners (●), and yoked-control *S*s (△).

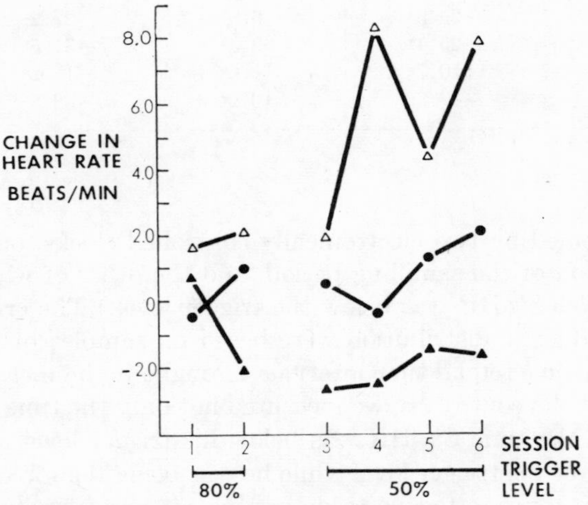

FIG. 3. Average changes in HR during training for learners (▲), nonlearners (●), and yoked-control *S*s (△).

centage times light on, decreases in average HR, and shifts in frequency distributions (i.e., number of beats at or below average operant period HR). *S*4 showed a significant increase only in percentage time light on, and *S*5 showed a significant decrease only in HR. *S*s 3, 6, and 7 showed significantly poorer performances. Thus, by our criterion of learning (significant performance changes in all three indices), five of the ten *S*s learned.

Figs. 2 to 4 show the data for the learners and nonlearners for each of the six experimental sessions for each of the three criterion indices. The poorer performance among the learners in sessions 5 and 6 is noteworthy, since this change seems inconsistent with the teaching of special skills. This apparent paradox seems to be explainable, however, by reference to the absolute HR during each of the six training sessions (Fig. 5). The learners' HR fell progressively through

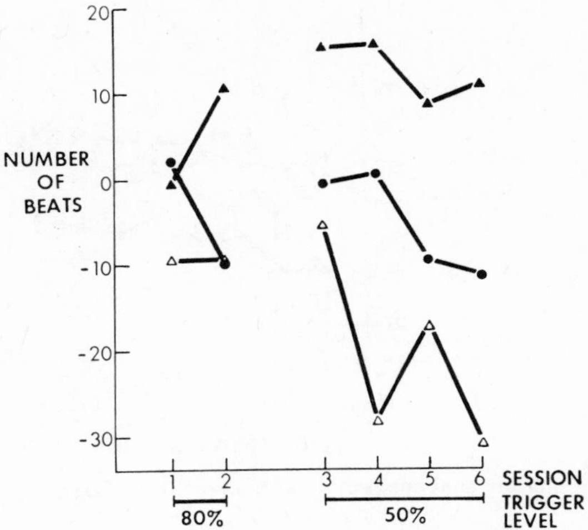

FIG. 4. Average shifts in frequency of beats at or below mean HR for learners (▲), nonlearners (●), and yoked-control *S*s (△).

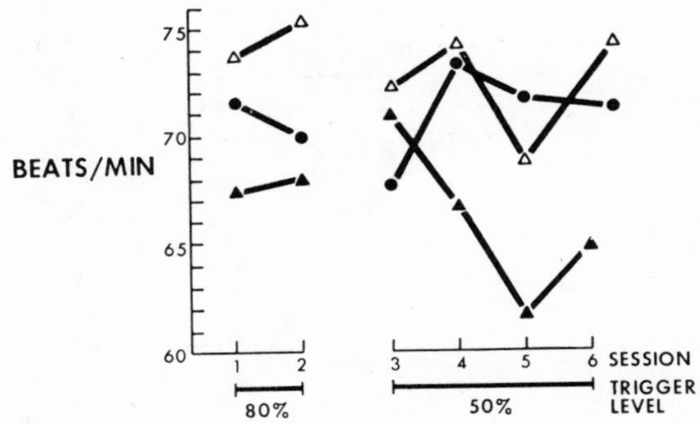

FIG. 5. Average HR during training for learners (▲), nonlearners (●), and yoked-control *S*s (△).

session 5 so that by this session the average HR for the group was 61.6 beats/min. Since this group average is one which is reported for *S*s during stages 3 and 4 sleep, it seems likely that the poor performance which they showed during the final sessions is in part an artifact of the experimental design. Since we based the trigger point on the pretraining level for each day, and since the learners probably generalized their responses to other cues in the environment, it seems likely that they were being unduly penalized. It is probable that the learners were at or near their basal HR by session 5, and since the experimental design required them to lower their rates from this already near-basal rate, it is likely that this session was a partial extinction session. It should be noted, however, that despite this difficulty, these *S*s were still able to lower their rates from their operant levels. It should also be noted from Fig. 5 that the nonlearners did not show the same trend toward decreased HR that the learners did.

We differentiated between learners and nonlearners on the basis of the *S*s'

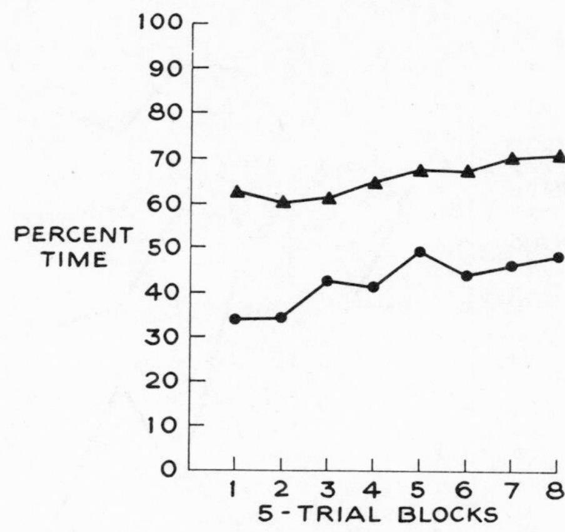

FIG. 6. Period-by-period analysis across sessions 3 to 6. Learners (▲) vs nonlearners (●).

TABLE 3

Changes in HR and standard deviations of HR (see text)

Standard Deviations of HR	HR	
	Increase	Decrease
A. Experimental		
Learners		
Increase	2	6
Decrease	6	6
Non-Learners		
Increase	6	6
Decrease	5	3
B. Control		
Increase	15	2
Decrease	3	0

performances during the training period relative to the operant period. However, if we consider the trial-by-trial changes across sessions 3 to 6, another picture emerges. Fig. 6 is a graph of percentage time that HR was below trigger level for consecutive 2½-min (approximately) blocks across sessions 3 to 6. As can be seen, the rate of change of performance is the same for both groups. The difference between them is that the learners get below the operant-level HR during the first trial block, whereas the nonlearners start above their operant level HR.

We also compared the changes in average HR from operant to training periods in sessions 3 to 6 with the changes in standard deviations of heart rates. These data are summarized in Table 3A, comparing increases and decreases in average rates and standard deviations of rates. The data in Table 3A, were obtained by comparing rates and standard deviations during sessions 3 to 6 for each of the *S*s, and the results are tabulated for the learners and nonlearners separately.

TABLE 4

Subjective reports

Response which Controlled Light	Technique for Keeping Light On
A. Experimental	
1. Current emanating from body	Complete relaxation
2. Increased HR	Concentration on problems at home, school or work
3. Nervous activity or diminished pulse rate	Relaxation
4. HR slowing	Relaxation and ignoring light and clock
5. HR slowing or blood pressure lowering	Relaxation
6. Decreased HR or reduced tension in stomach muscles	Relaxation
7. Salivation-swallowing	Tried to reduce salivation
8. No idea	Concentration on objects in room—aligning holes in acoustical tiles
9. No idea	Relaxation and "self control"
10. Something related to blood pressure	Concentrated "hard" on relaxing
B. Control	
1. Decreased HR	Relaxation
2. Brain waves	Concentration on light
3. Pulse rate or nervous impulses	Maintenance of anxious, excited state
4. Breathing or pulse rate	Relaxation
5. Cardiac output or pulse rate	Tension in chest area (sensation is similar to "mild heartburn")

It seems clear from these data that there are not systematic variations in changes in HR and changes in standard deviation of HR, since for both groups there are more instances when these changes did not agree.

*S*s were instructed to breathe normally throughout the training sessions. Table 2A, reports the average changes in breathing rates from operant to training periods for each of the experimental *S*s across all sessions. The average difference for these *S*s is 0.8 breaths/min, which is not significant ($t = 1.31$). Furthermore, these changes are in the wrong direction, since decreases in HR are usually associated with decreases in breathing rate.

Thus, to the extent that this index measures respiratory activity, it is clear that breathing is not the mediator of the learning effect. Although our technique of measuring breathing precludes an accurate estimate of any systematic variation in breathing pattern (such as the ratio of inspiration to expiration or the depth of breathing) from operant to training periods, as far as we can judge, the changes in HR we observed were not mediated by changes in respiration.

Each subject was interviewed after each training session to learn which responses he believed controlled the light, and what he did to control the light. These subjective reports are summarized in Table 4A. Especially noteworthy is the report of *S*2, who met all our criteria of learning. He said that he kept the light on by thinking of examinations in which he had performed poorly and of quarrels he had had with his wife. It is also interesting that the four *S*s who guessed that decreased HR kept the light on were all nonlearners.

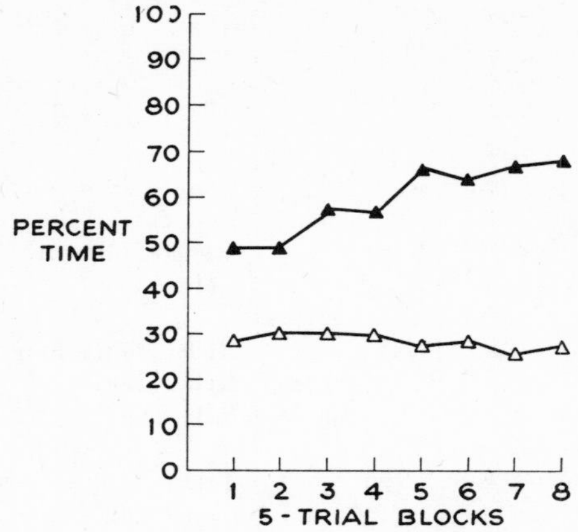

FIG. 7. Period-by-period analysis across sessions 3 to 6 for experimental (▲) vs control (△) *S*s.

Control Ss

The changes in performance in each of the three learning measures during sessions 3 to 6 are reported in Table 2B. In contrast with the experimental *S*s, four of the five control *S*s showed a significant fall in performance despite identical patterns of reward. A direct comparison of paired experimental (E) and control (C) *S*s can be made by noting that C1 was paired with E1, C2 with E2, etc.

The session-by-session changes in each of the three criterion measures are reported in Figs. 2 to 4. The yoked-control *S*s performed significantly more poorly than either the learners or nonlearners during each of the criterion sessions (sessions 3 to 6).

Fig. 5, which reports the average group HR during each of the six training sessions, includes the data for the control *S*s. It can be seen from this figure that the HR for these *S*s are not different from those of the nonlearners. Only the *S*s who learned showed declining HR over the course of this study.

Fig. 7 is a graph of percentage time HR below trigger level for consecutive 2½-min (approximately) blocks across sessions 3 to 6. It includes the data for the control *S*s and for the matched experimental *S*s. As can be seen, the experimental *S*s improve throughout the training session, whereas the controls do not change.

The comparison of HR changes and changes in standard deviations of HR among the control *S*s is given in Table 3B. In contrast to the experimental *S*s, the control *S*s do show covariance in these two measures: most increased both their rates and standard deviations.

The changes in breathing rate are reported in Table 2B. These *S*s increased their rates by 0.7 cycles/min. Just as in the case of the experimental *S*s, visual inspection of individual tracings does not indicate any systematic variations in respiratory activity.

Table 4B, reports the subjective findings. Several findings are noteworthy.

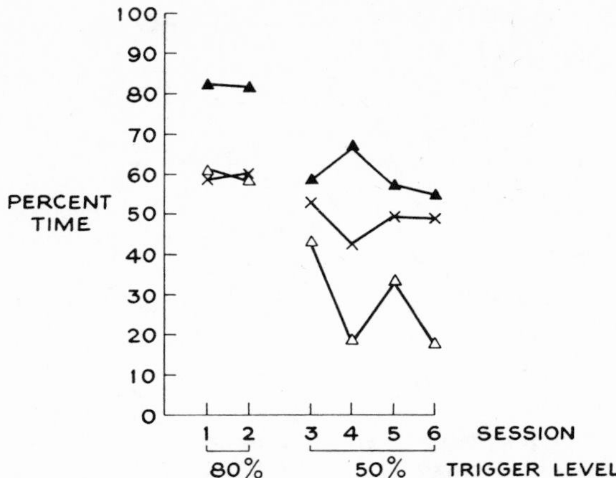

FIG. 8. Relationship between performance and reinforcement among the yoked-control *S*s. X = concordance; △ = performance; ▲ = paired experimental *S*s.

TABLE 5

Relationships between slow HR and reward/punishment among the yoked-control Ss

Session	*S* Rewarded		*S* Punished	
	HR slow	HR fast	HR slow	HR fast
	percentage time			
1	51.2	31.4	9.7	7.7
2	49.4	32.2	8.7	9.7
3	27.0	31.4	16.1	25.5
4	14.3	52.8	4.4	28.4
5	20.2	37.4	13.4	29.0
6	10.5	44.0	7.1	38.4

First, four of the five control *S*s guessed that pulse rate might be the response that controlled the light; second, all the techniques employed by these *S*s were used by the successful *S*s; and third, none of the control *S*s believed that the light was not under his control. Most of them told us not only what they did to keep the light on but also how they could turn it off.

Fig. 8 and Table 5 report the relationships between performance and reinforcement for the control *S*s during each of the six sessions. We have also included the performance data for the paired experimental *S*s in this figure. The curve labeled *concordance* is a measure of the proportion of time the light and the control *S*s' HR are in agreement (i.e., the light is on and *S*s' HR are slow, or the light is off and *S*s' HR are fast). The curve labeled *performance* shows the actual time that *S*s kept their HR below the trigger level. In sessions 1 and 2 concordance is largely determined by *S* slowing his HR (i.e., the two curves are identical); however, by session 3 the periods of agreement become more and more determined by *S*'s HR-speeding behavior. Despite this shift in determinant, concordance remains virtually constant throughout the study.

Conclusions and Discussion

The results of this study point to four conclusions.

1. Some normal *S*s can be taught to slow their HR by means of an operant conditioning procedure.
2. *S*s seem to learn this response better when they do not infer correctly what response they are controlling.
3. The effect does not seem to be mediated through changes in breathing.
4. Reinforcement, per se, is not adequate to lower HR.

Our data show clearly that five of the 10 *S*s learned to slow their HR. Why only five learned, however, is not clear. It is interesting that among the five nonlearners, four inferred correctly that slow HR was the appropriate response, and that none of the learners made this association. It is interesting, therefore, to speculate that the effect that we saw is related to the one reported by Hefferline, Keenan, and Harford (1962), that *S*s who were informed that a small muscle twitch would be reinforced were less efficient performers than *S*s not so informed.

We suspect that there is much misinformation among laymen (and perhaps among scientists as well) on the factors that mediate autonomic activity generally and HR specifically. If so, it may be that at least part of the failure of the nonlearners might be attributed to their trying to slow their HR rather than by trying to keep the light on. It is certainly clear that the subject who kept the light on by concentrating on exciting things would not have done so if he thought he was supposed to slow his heart. As a matter of fact, he believed that he was speeding his HR.

Although we have asserted throughout this paper that five *S*s learned and five did not, we must call attention to the data in Figs. 6 and 7. Fig. 6 shows that both the learners and the nonlearners improved in their performances through the course of the training sessions, and Fig. 7 shows that the control *S*s did not change. If one considered only the trend in performance, irrespective of the change from operant to training periods, then one might conclude that most, if not all, of the experimental *S*s learned. It is quite possible that our criteria of learning are too conservative.

The question of a possible somatic mediator of an autonomic effect is a vexing one. We attempted to control breathing by means of our instructions, and we were apparently successful in doing so. However, one can also ask about other somatic responses, such as muscle tension. From our observations of our *S*s it is not possible for us to infer any somatic mediators; for example, the *S* who thought about exciting things and the *S* who aligned holes in the tiles both seemed to have increased their muscle tension from operant to training periods, whereas the other three learners seemed to relax more. We prefer to reply to the question of somatic mediators by another question, "What is the evidence that a somatic response can be learned in the absence of an autonomic mediator?"

The yoked-control *S*s were included in this study to permit us to assess the effect of participation in a learning situation on HR. These *S*s showed clearly that HR does not decrease under such conditions. Furthermore, their performances indicated the critical role of contiguity of response and reward. During the early sessions their performances were fairly good (by our criteria of learning) and there was close agreement between performance and reward. As the performance deteriorated, the agreement between low HR and reward dropped and the agreement between high HR and nonreward increased (Fig. 8, Table 5). The net effect of this interplay between performance and reinforcement was that the control *S*s' HR rose from the operant to training periods (Fig. 3), and their performances did not change from the beginning to the end of the training session (Fig. 7).

These results from the control group document the technical criticisms that we raised in the introduction to this paper of earlier attempts to condition autonomic responses operantly. HR responses occur about 70 times/min. Unless the experimenter has a means of determining whether each of these responses meets his criterion, *S* will always be in a situation in which incorrect responses as well as correct responses are being rewarded. In this study, that sort of situation led to the emergence of a response tendency which was opposite to that which we were trying to achieve.

REFERENCES

Harwood, C. W. Operant heart rate conditioning. *Psychol. Rec.*, 1962, *12:* 279–284.

Hefferline, R. W., Keenan, B., and Harford, R. A. Escape and avoidance conditioning in human subjects without their observation of the response. *Science*, 1962, *130:* 1338–1339.

Hnatiow, M., and Lang, P. J. Learned stabilization of heart rate. *Psychophysiology*, 1965, *1:* 330–336.

Kimble, G. A. *Hilgard and Marquis' conditioning and learning.* (2nd ed.) New York: Appleton-Century-Crofts, 1961, Pp. 100–102.

Kimmel, E., and Kimmel, H. D. A replication of operant conditioning of the GSR. *J. exp. Psychol.*, 1963, *65:* 212–213.

Shearn, D. W. Operant conditioning of heart rate. *Science*, 1962, *137:* 530–531.

Heart Rate Control Under Conditions of Augmented Sensory Feedback 4

Jasper Brener and David Hothersall

ABSTRACT

Five human *S*s were presented with a high frequency tone on the emission of each short inter-heartbeat interval and a low frequency tone on the emission of each long inter-heartbeat interval. Under these conditions, all *S*s learned within a short period of time to produce significantly lower heart rates in the presence of one visual stimulus than in the presence of another. On the basis of this finding, it is suggested that an important determinant of where a given response falls on the voluntary/involuntary continuum is the availability of specific feedback from the response in question.

DESCRIPTORS: Response feedback, Heart rate, Voluntary control. (J. Brener)

Results showing instrumental conditioning of autonomic responses have been reported by a number of investigators (Shearn, 1960; Lisina, 1961; Kimmel & Kimmel, 1963; Brener, 1964). Since autonomic responses are usually classified as involuntary, such results might be considered contrary to the hypothesis of Konorski and Miller (1937) that only voluntary responses may be conditioned with the instrumental procedure. In defense of this hypothesis, it might be claimed that autonomic responses are subject to voluntary control. The evidence opposed to this claim is, however, overwhelming, and it seems safe to conclude that under normal conditions voluntary control of autonomic behavior is not possible. An additional hypothesis might be put forward to resolve the conflict between the experimental results and Konorski and Miller's hypothesis; it is that voluntary control of autonomic responses is only possible under the special conditions imposed by an effective reinforcement schedule. The critical feature of such a schedule is that the organism must be provided with differential feedback of its behavior, a state of affairs which does not exist for autonomic responses under normal conditions.

The importance of proprioception and feedback generally has been demonstrated by the studies of Hefferline (1958), Hefferline et al. (1959), and Basmajian (1963). Lisina (1961) has shown that instrumental conditioning of vasomotor responses is only possible when *S*s are provided with exteroceptive feedback from the relevant smooth muscles. Hnatiow and Lang (1965) further demonstrated the importance of feedback in the control of autonomic behavior. In their experiment, *S*s provided with visual feedback of heart rate performance were able to decrease heart rate variability.

In the light of this evidence, the working hypothesis was adopted that, given appropriate response feedback, autonomic behavior, and in particular the behavior of the heart, may be subject to voluntary control. A test of this hypothesis is provided in the following experiment.

Method

Subjects. Five university students (three males and two females) were drawn from a group of volunteers. The only selection criterion employed was that they should have no recent history of cardiac, respiratory, or psychological disorder. Subjects were paid at the rate of $1.50 per hr.

Apparatus. The cardiotachometer used to record heart rates (Brener, 1965) converts the EKG wave form to a standard square-wave pulse and records the intervals between successive heart beats [(inter-heart beat IBI distribution)] on a bank of 30 electromagnetic counters. A heart rate analysis circuit was used to classify each IBI in terms of a prespecified criterion. This criterion was set separately for each subject at the mode of his basal IBI distribution. Each IBI was then classified according to whether it was longer or shorter than this criterion.

When the experimental contingencies were in force, the subject was presented with a low frequency tone on the emission of each long IBI and a high frequency tone on the emission of each short IBI. These tones represented differential feedback of heart rate performance: the duration of each tone was 100 msec. Respiration was recorded on a San'ei Physiograph.

*Procedure. S*s were seated on a reclining chair in a soundproofed cubicle adjoining the control room. EKG electrodes were attached to both ankles and to the right wrist and a strain gauge was attached about *S*'s chest to record respiratory movements. All *S*s were told that *E* would leave the cubicle and that shortly thereafter a green lamp situated on *S*'s right would come on. This lamp would stay on for approximately 1 min (in fact it stayed on for the period of time it took *S* to emit 50 IBIs) and would be followed by a period of darkness which would last for about 30 sec. At the end of this period, a red lamp situated on *S*'s left would come on. When and only when one of these two lamps was on, *S* would hear a succession of high and low frequency tones through a pair of stereophonic headphones. *S* was told that these tones were associated with some aspect of his internal behavior but no indication was given that his cardiac behavior was of special importance. While the green lamp was on *S* was instructed to try to produce only high tones and to inhibit all low tones: during periods when the red lamp was on he was told to try to produce only low tones and to inhibit all high tones. *S*s were questioned in order to ensure that they understood the instructions and were specifically told not to engage in any bodily movements in their attempts to control the tones. Control was to be exercised by "mental processes" only.

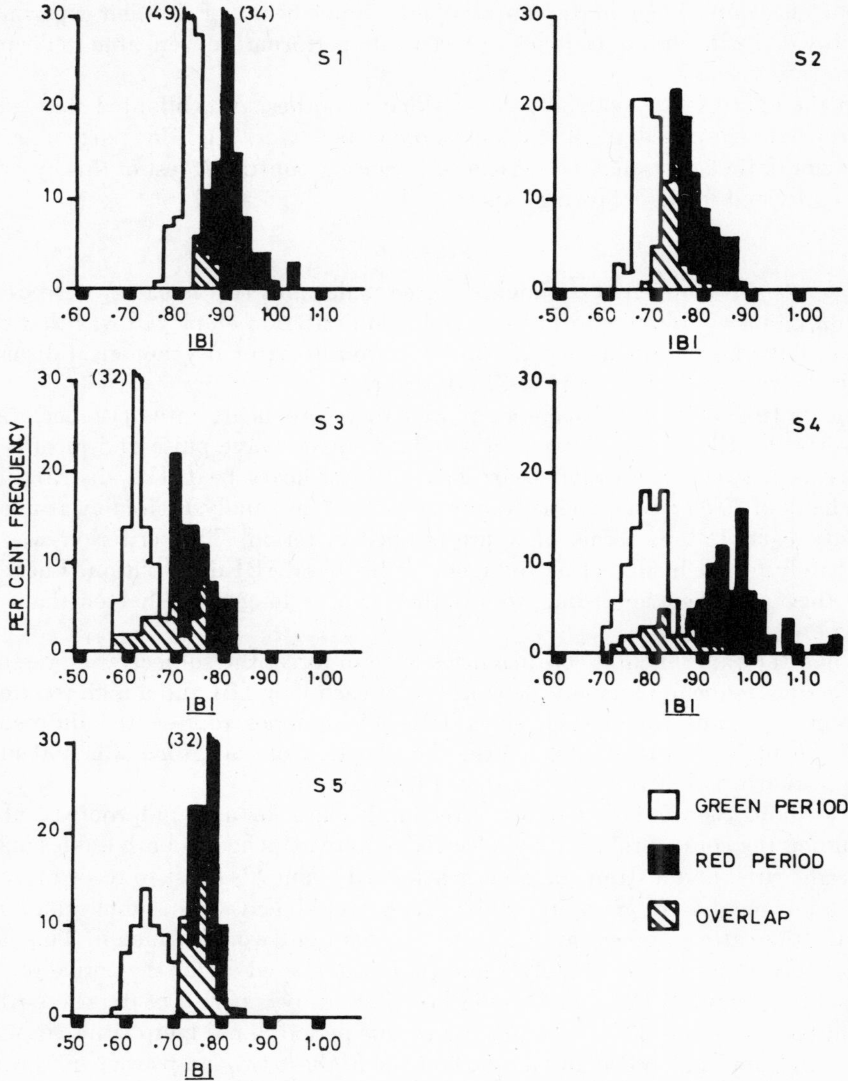

Fig. 1. Final IBI distributions

The experimental session was composed of alternating green and red periods[3] for four presentations of each, with the first period being a green period. After the final red period, *S*s were asked whether they felt that they were able to control the tones and if so, how they had done it.

[3] No systematic relationship was observed between signal light color and heart rate in this situation.

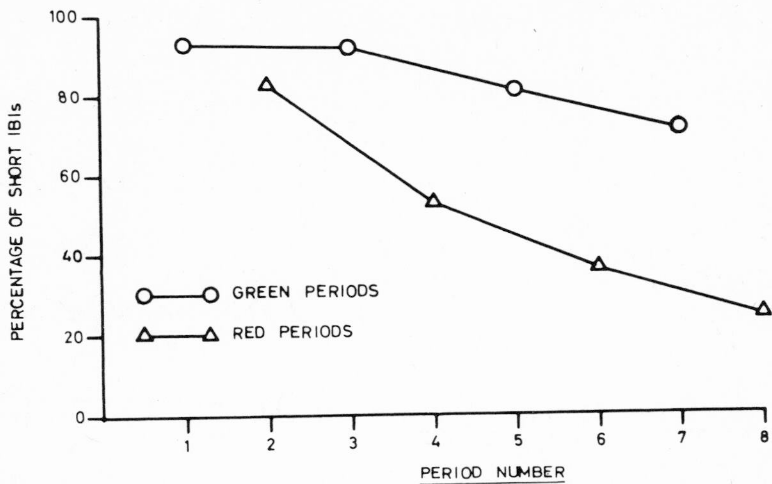

FIG. 2. Percentage of short inter-heart beat interval (IBIs) in successive periods. Mean for all 5 *S*s.

RESULTS

The IBI distributions recorded in the final green and red periods for all five *S*s are shown in Fig. 1. It is clear that in all cases the distribution of IBIs recorded in the final green period (period 7) is to the left of the distribution of IBIs recorded in the final red period (period 8). Since the IBI is the reciprocal of heart rate, this figure indicates that all *S*s displayed higher heart rates in the final green period than they did in the final red period. The degree to which each *S*'s heart rate performance differed in these two periods is shown by the amount of overlap in the two distributions. For the group as a whole, the difference between mean IBIs recorded in the final green and red periods was found to be highly significant ($t = 6.197$; $df = 4$; $p < 0.01$).

Fig. 2 presents the average percentage of short IBIs recorded in successive periods for the group as a whole. Two principal phenomena will be noticed:

1. A very high percentage of short IBIs was recorded in the first green period. Since the criterion was set at the mode of each *S*'s basal distribution (Mo approximately equal to Mdn), this change demonstrates the increase in heart rate which follows *S*'s introduction to the experimental procedure. It will be seen furthermore that the percentage of short IBIs decreases systematically over the session. In other words, heart rate tends to decrease over the session.

2. The difference between the mean percentage of short IBIs emitted in successive green and red periods tends to increase over the session. From the fourth (red) period onward, *S*s show a clearcut heart rate discrimination, with the percentage of short IBIs being far greater in green than in red light periods.

Representative examples of respiratory activity recorded in the final green and red periods for each *S* are shown in Fig. 3. It will be seen that all *S*s showed differences in respiratory behavior under these conditions. In general the respira-

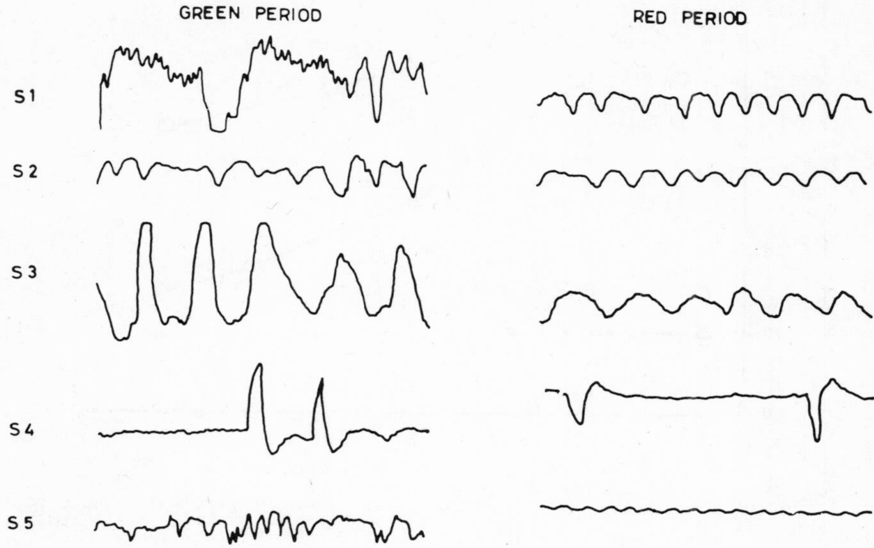

Fig. 3. Respiration records

tory patterns recorded in the green light period were more erratic than those recorded in the red light period.

Discussion

These results present clear evidence that under conditions of augmented sensory feedback, *S*s rapidly learn to control their heart rates. They do not, however, permit a firm conclusion as to the possible mechanisms mediating this control. The gross differences in respiratory patterns observed between green and red light periods suggest the definite possibility that the observed heart rate control was mediated by changes in respiration. However, it is important to note that, although this procedure produced similar changes in heart rate in all *S*s, it produced dissimilar changes in respiration. The verbal reports of *S*s on how they had achieved control of the tones was likewise extremely variable. Only one *S* suspected respiration and she (*S* 4) rejected this hypothesis well before the end of the session. One *S* reported feeling tense in the presence of the green lamp and relaxed in the presence of the red; another was happy in the green and sad in the red. Two *S*s reported that they had simply thought about high and low tones in the presence of the appropriate visual stimuli. In no case did a *S* discover that the tones were controlled by heart rate.

This experiment supports and extends the findings of Hnatiow and Lang (1965), viz. when they are provided with exteroceptive feedback of cardiac behavior, *S*s are able to control their heart rates. These findings are taken as support for a theory which postulates that the principal discriminator between so-called voluntary and involuntary behaviors is the availability of specific feedback from the muscle systems in question.

REFERENCES

Basmajian, J. V. Conscious control of single nerve cells. *New Scientist*, 1963, *369:* 661–664.

Brener, J. M. The characteristics of heart rate during various conditioning procedures. Unpublished doctoral dissertation, University of London, 1964.

Brener, J. M. The measurement of heart rate. In P. H. Venables and I. Martin (Eds.), *Manual of Psychophysiological Methods*. Amsterdam: North Holland Publishing Co., 1965.

Hefferline, R. F. The role of proprioception in the control of behavior. *Trans. N. Y. Acad. Sci.*, 1958, *20:* 739–764.

Hefferline, R. F., Keenan, B., & Harford., R. A. Escape and avoidance conditioning in human subjects without observation of their response. *Science*, 1959, *130:* 1338–1339.

Hnatiow, M., & Lang, P. J. Learned stabilization of heart rate. *Psychophysiology*, 1965, *1:* 330–336.

Kimmel, E., & Kimmel, H. D. A replication of operant conditioning of the GSR. *J. exp. Psychol.*, 1963, *65:* 212–213.

Konorski, J., & Miller, S. On two types of conditioned reflex. *J. gen. Psychol.*, 1937, *16:* 264–272.

Lisina, M. I. In Razran, G. (Ed.) The observable unconscious and the inferable conscious in current Soviet psycho-physiology: interoceptive conditioning, semantic conditioning and the orienting reflex. *Psychol. Rev.*, 1961, *68:* 81–148.

Shearn, D. W. Operant conditioning of heart rate. Unpublished doctoral dissertation, University of Indiana, 1960.

Address requests for reprints to: Jasper Brener, Department of Psychology, University of Tennessee, Knoxville, Tennessee.

[1] This research was supported by a grant made available by the Foundation's Fund for Research in Psychiatry.

[2] Both authors are now at the Department of Psychology, University of Tennessee, Knoxville, Tennessee.

5

Paced Respiration and Heart Rate Control

Jasper Brener and David Hothersall

ABSTRACT

Five human subjects were presented with a high frequency tone on each short inter-heartbeat interval (IBI) and a low frequency tone on each long IBI. They were instructed to produce high tones in the presence of one visual stimulus and low tones in the presence of another under conditions of paced and unpaced respiration. The results indicated that control of heart rate is not dependent upon respiratory mediation.

DESCRIPTORS: Heart rate, Respiration, Voluntary control, Response feedback. (J. Brener)

Brener and Hothersall (1966) have demonstrated that when subjects are provided with differential auditory feed-back of short and long inter-heartbeat intervals, they rapidly learn to control their heart rates. A similar finding has been reported by Hnatiow and Lang (1965) using visual feedback. In both experiments the observed changes in cardiac behavior were correlated with respiratory changes. In the Brener and Hothersall study high heart rates were associated with quick erratic breathing and low heart rates with normal cyclical breathing, while Hnatiow and Lang observed a "weak relationship" between respiratory cycle time and heart rate variability.

A question raised by these observations concerns the degree to which respiratory involvement is necessary in the control of heart rate. In these procedures, do individuals learn to control respiratory behavior and thereby influence cardiac activity, or is some other mediational process involved? An answer to this question is sought in the present experiment. Using the technique of paced respiration (Wood and Obrist, 1964), an attempt was made to teach *S*s to breathe at a fixed rate and amplitude while learning to produce low and high heart rates.

Method

Subjects

Five university students (four females and one male) were drawn from a group of volunteers. The only selection criterion employed was that they should have no recent history of cardiac, respiratory, or psychological disorder. *S*s were paid at the rate of $1.50/hr.

This research was supported by a grant made available by the Foundation's Fund for Research in Psychiatry and carried out at The Department of Psychology, Birkbeck College, University of London.

Address requests for reprints to: Jasper Brener, Department of Psychology, The University of Tennessee, Knoxville, Tennessee, 37916.

Apparatus

The specialized cardiotachometer employed in this experiment has been decsribed elsewhere (Brener, 1966). A heart rate analyzer classified each interheartbeat interval (IBI) according to whether it was longer or shorter than a criterion prespecified for each subject. This criterion was set at the mode of the *S*'s basal IBI distribution. When the experimental contingencies were in force, *S* was presented with a low frequency tone on the emission of each long IBI and a high frequency tone on the emission of each short IBI. These tones represented differential feedback of cardiac performance. The duration of each tone was 100 msec. Respiratory movements were picked up by a strain gauge transducer fixed about *S*'s chest and recorded on a San 'ei Physiograph.

Procedure

Each *S* was run for two sessions separated by between two and four days.

Session 1. On the first session the procedure was identical to that described in an earlier paper (Brener and Hothersall, 1966). *S*s were instructed to attempt to produce only high frequency tones in the presence of a green light and only low frequency tones in the presence of the red light. Red and green lights alternated, and each light stayed on for the period of time it took *S* to emit 50 IBIs. An inter-trial interval of approximately 30 sec, during which no experimental stimuli were presented, intervened between red and green light trial periods. *S*s were instructed not to engage in any bodily movement in their attempts to control the tones and were told that control was to be exercised by "mental processes" only. Although *S*s were told that the tones were associated with some aspect of their internal behaviors, no indication was given that their cardiac behavior was of special importance.

Following the instructions, *E* left the experimental cubicle and 1 min thereafter the green signal light came on. Green and red stimulus periods then alternated for at least 13 presentations of each. If *S* obtained less than 10 % of IBIs in the same IBI category (either short or long) on two successive trial periods, the IBI criterion was either increased or decreased by 0.02 seconds so as to make the distribution of IBIs in these two categories more equal. Adjustment of this criterion occurred in the first session only.

At the end of each experimental session, *S* was asked whether he was satisfied that he could control the tones and if so how he had done it.

Session 2. The second session was divided into four phases.

Phase A: A procedure identical to that of the first session was employed. Green and red light periods alternated for four presentations of each, *S* being told to produce high frequency tones in the presence of the green light and low frequency tones in the presence of the red light.

Following the fourth red period, *S* was told to relax and his respiration was monitored. When *E* was satisfied that *S*'s respiration had stabilized, a resting respiration rate was determined. A white signal lamp situated near *S*'s left eye was then set to flash at this rate. *S* was instructed to pace his breathing by this signal, i.e. to complete one respiratory cycle between each flash of the light. *E* monitored *S*'s breathing pattern and advised *S* on how to moderate his respiratory amplitude in order to achieve a stable rate. When *E* was satisfied that *S* had achieved stable respiratory behavior, the next phase began.

Phase PR: *S* was instructed to control the tones as before, but to continue breathing at the rate indicated by the flashing light which continued to flash throughout this phase. Green and red periods alternated for six presentations of each.

Phase B: Following phase PR, S was told that the respiratory pacing stimulus would be withdrawn and that he should no longer concern himself with respiration but should continue to control the tones as before. Red and green light periods alternated for three presentations of each.

Extinction Phase: In the final phase of the experiment, *S* was instructed to discontinue his attempts to control the tones. In this phase, green and red light periods alternated for two presentations of each.

Results and Conclusions

Heart Rate

The principal measure of heart rate control is the percentage of short IBIs recorded in each experimental period. Given the instructions, heart rate control

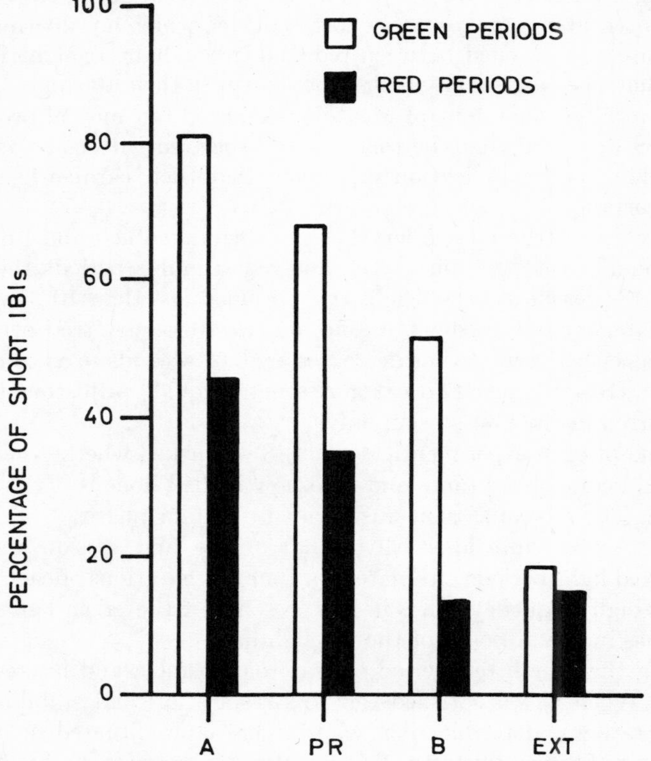

Fig. 1. Mean percentage of short IBIs recorded in each phase of Session 2. *S*s were instructed to produce feedback associated with short IBIs in green periods and feedback associated with long IBIs in red periods.

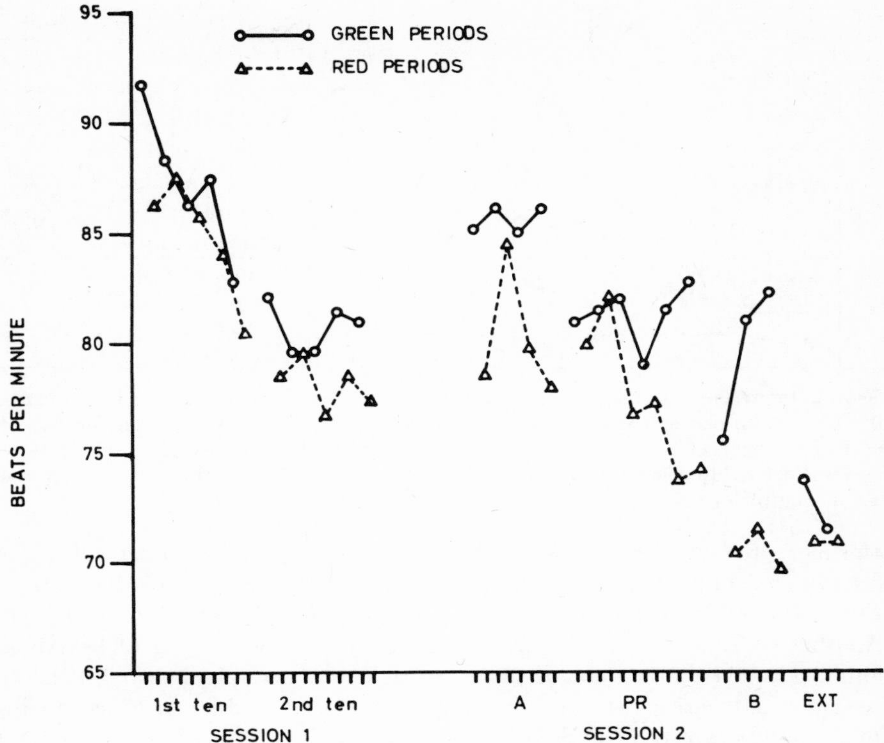

FIG. 2. Mean heart rates recorded in successive green and red trial periods over Sessions 1 and 2. Session 1 data are presented for first 10 and last 10 trials only.

should be evidenced in a greater percentage of short IBIs in green than in red light periods. Because of IBI criterion changes, the percentage of short IBIs measure is confounded for the first session and these data are not presented. During Session 2, however, the IBI criterion was not changed and remained at the final Session 1 value for each S.

Following a 2 arcsin $\sqrt{X}$ transformation, the percentage of short IBIs data for each of the four phases of Session 2 were submitted to analysis of variance. The degrees of freedom for this analysis of variance were partitioned by the procedure of Greenhouse and Geisser (Winer, 1962) so as to avoid assumptions about equal covariances in the pooled variance-covariance matrix. It was found that *S*s displayed a significantly ($p < 0.05$) greater percentage of short IBIs in green than in red periods during Phase A ($F = 12.7387$; df $= 1/4$), Phase PR ($F = 10.5358$; df $= 1/4$), and Phase B ($F = 12.4783$; df $= 1/4$), but not during the Extinction phase. No significant Trial Number effect was found in any phase except Phase B where significant ($p < 0.05$) Trial Number ($F = 12.4783$; df $= 1/4$) and Trial Number by red/green interaction ($F = 9.9335$; df $= 1/4$) effects were observed. The mean percentage of short IBIs for each phase of Session 2 are presented in Fig. 1. It will be seen that a considerable difference exists between red and green light performances in phases A, PR, and B. The overall decreasing trend

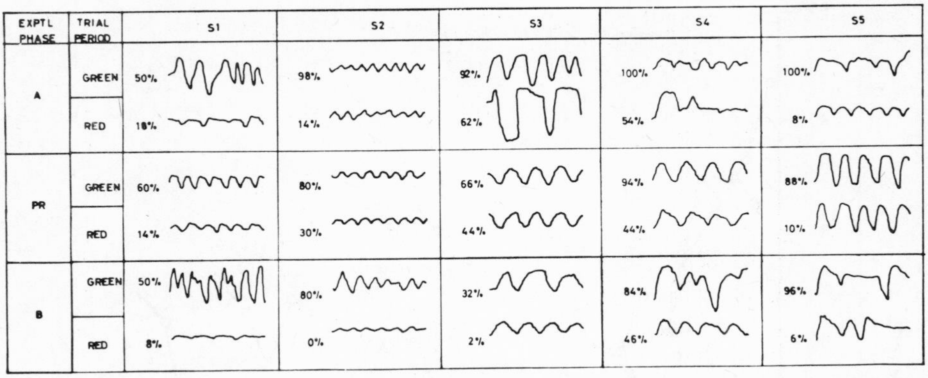

EXPTL PHASE	TRIAL PERIOD	S1	S2	S3	S4	S5
A	GREEN	50%	98%	92%	100%	100%
	RED	18%	14%	62%	54%	8%
PR	GREEN	60%	80%	66%	94%	88%
	RED	14%	30%	44%	44%	10%
B	GREEN	50%	80%	32%	84%	96%
	RED	8%	0%	2%	46%	6%

FIG. 3. Representative respiration records classified by subject, green or red trial period, and experimental phase of Session 2. The figures appearing to the left of each record indicate the percentage of short IBIs emitted during the trial period from which the respiratory record was drawn. Red/green differences in respiration, apparent in phases A and B, do not occur during phase PR.

in this measure may be accounted for by the processes of habituation to the experimental environment and the cumulative effects of immobility on cardiac activity over the session.

Another measure of cardiac performance used was the number of heartbeats counted between the 5th and 25th sec of each trial period converted to beats/min. These data are presented for Sessions 1 and 2 in Fig. 2. Since *S*s had different numbers of trials in Session 1, the averaged data take into account only the first 10 and last 10 trials for each *S*.

It will be observed that the difference between the heart rates displayed in red and green periods in Session 1 diverge as a function of training. A general decrement in heart rate may be observed in both sessions.

In the second session, the data of principle interest are those recorded during phase PR. Here it will be seen that following an initial breakdown of heart rate control, successively better performances are recorded over this phase. In other words, heart rate control is possible under the conditions imposed by the paced respiration procedure.

It should also be noted (a) that there is a good carryover effect from Sessions 1 to 2 and (b) that although the extinction instructions reduce performance markedly, they do not eliminate heart rate control immediately.

Respiration

Respiration amplitude and cycle time data were submitted to analysis of variance for each phase of Session 2. Surprisingly these analyses revealed that neither respiratory cycle time nor amplitude differed significantly between red and green light periods during any phase of this session. This may be attributed to wide between-*S* variability in respiratory behavior. In Fig. 3, representative examples of respiration recorded from each *S* during red and green periods of Phases A, PR, and B are shown. Visual inspection of these data indicates that in Phases A and B there are considerable within-*S* differences in red and green period respiration patterns but that the respiration displayed during red and green periods of Phase PR are very similar. The percentage figure written to the left of each respiratory record represents the percentage of short IBIs recorded during the trial period from which the respiratory record was drawn. It is quite evident that in Phase PR, the considerable differences between red and green period cardiac performances are not paralleled by clear differences in respiratory performance.

These findings confirm and extend the observation of Brener and Hothersall (1966) that when provided with augmented sensory feedback, humans learn to control their heart rates. It is also concluded that heart rate control may occur independently of changes in respiratory behavior. However, the possibility that the observed cardiac control was mediated by learned changes in muscle tension remains a problem worthy of empirical investigation.

REFERENCES

Brener, J. M., & Hothersall, D. Heart rate control under conditions of augmented sensory feedback. *Psychophysiology*, 1966, *3:* 23–28.

Brener, J. M. The measurement of heart rate. In P. H. Venables and I. Martin (Eds.), *Manual of spychophysiological methods*. Amsterdam: North Holland Publishing Co., 1966.

Hnatiow, M., & Lang, P. J. Learned stabilization of heart rate. *Psychophysiology*, 1965, *1:* 330–336.

Winer, B. J. *Statistical principles in experimental design*. New York: McGraw-Hill, 1962.

Wood, D. M., & Obrist, P. A. Effects of controlled and uncontrolled respiration on the conditioned heart rate response in humans. *J. Exp. Psychol.*, 1964, *68:* 221–229.

6 Operant Conditioning of Changes in Heart Rate in Curarized Rats

David Hothersall and Jasper Brener

One group of curarized rats was reinforced for heart-rate increases and another group for heart-rate decreases by electrical stimulation of the medial forebrain bundle. All *S*s were run for four sessions each of which consisted of an adaptation, conditioning, and extinction phase. During the conditioning phase of each session a feedback stimulus was presented whenever *S*'s heart rate met some criterion. Twenty-five successive-criterion heart beats produced the reinforcement. The criterion was made more stringent as *S*'s heart rate changed in the appropriate direction. The heart rates of the two groups diverged progressively over the first three conditioning phases and converged during extinction.

The traditional view that autonomically mediated responses may only be conditioned by the procedures of classical conditioning (Kimble, 1961) has been sharply challenged by numerous recent publications summarized by Kimmel (1967) and Katkin and Murray (1968). Amongst these reports have been several experiments which provide evidence of operant heart-rate conditioning. The earlier experiments in this area of research (Brener, 1966; Engel & Hansen, 1966; Shearn, 1962) used human subjects and may be criticized on the grounds that the observed heart-rate changes were an artifact of conditioned changes in striate muscle activity (Smith, 1954). To counteract such criticism, Trowill (1967) and Miller and DiCara (1967) employed tubocurarine chloride to produce complete flaccid paralysis of the striate musculature during instrumental conditioning of heart rate in rats. These investigators were able to show good evidence of conditioned heart-rate increases and decreases using electrical stimulation of the medial forebrain bundle as a reinforcing stimulus, and their main effects have been replicated by Miller and Banuazizi (1968). The present experiment aims at replicating and extending the results reported by Miller and his co-workers.

Following Brener's (1964) experiment on avoidance conditioning of heart-rate changes in human *S*s, the present authors began a series of experiments on operant conditioning of changes in heart rate in curarized rats using both positive and negative reinforcement. After the initial experiments yielded only marginal evidence of conditioning, it was decided to adopt a procedure similar to that reported by Trowill (1967) and Miller and DiCara (1967). The major differences between the experiment reported in this paper and the experiment of Miller were as follows. (*a*) Four conditioning sessions during which heart-rate changes were reinforced were given, whereas Miller used only one conditioning session. Repeated conditioning sessions were used since it was felt that extent of training might well have an important influence on the results of this type of experiment. In addition, the repeated conditioning sessions allowed an investigation of the short-term retention and transfer of the conditioned heart-rate response. (*b*) A feedback stimulus was presented during the conditioning trials. This stimulus was contingent on criterion heart rates, i.e., heart rates which if sus-

[1] This research is based on a dissertation submitted by David Hothersall to the Graduate School of the University of Tennessee in partial fulfillment of the requirements for the PhD degree. This investigation was supported by Predoctoral Fellowship 5-F1-MH-31, 428,03 to David Hothersall and by Grant MH-12105 to Jasper Brener, both from the National Institute of Mental Health.

[2] Now at the Ohio State University.

[3] Requests for reprints should be sent to Jasper Brener, Department of Psychology, University of Tennessee, Knoxville, Tennessee 37916.

tained for a fixed number of beats would produce the brain-stimulation reinforcement. This feedback stimulus was used since the results of a number of previous investigators (Brener & Hothersall, 1966, 1967; Hnatiow & Lang, 1965; Lang, Sroufe, & Hastings, 1967; Lisina, 1961) have shown the importance of feedback for operant conditioning and control of autonomic responses. (*c*) Finally, heart rates are reported for the adaptation and extinction phases of each curare conditioning session.

Method

Subjects

The *S*s were 12 naive male albino rats (Sprague-Dawley strain). In each rat a bipolar electrode was stereotaxically implanted in the medial forebrain bundle in the posterior portion of the lateral hypothalamus. The rats were randomly assigned to one of two experimental groups. One of the groups was reinforced during the curare conditioning sessions for increases in heart rate, and the other was reinforced for decreases in heart rate.

Apparatus

Tests for the reinforcing effects of brain stimulation were given in a 6 × 8 × 12 in. Plexiglas box with a lever on one wall, 3 in. above the floor. The reinforcing stimulus consisted of a ½-sec. train of 60-cps square-wave pulses delivered from a Grass SD5 stimulator through a 100-K potentiometer in series with *S*. During curare sessions the rat lay on a small hammock in a sound-shielded chamber, which in turn was inside an electrically shielded enclosure. While curarized the rat was artificially respirated with a small-animal respirator (E&M Instrument Co., Model V5KG). The rat was fitted into a specially moulded nylon mask, which was connected to the respirator via plastic tubing. The EKG was recorded on a Grass Model 7 polygraph employing Grass ear-clip electrodes attached to the surface of *S*'s skin in a Lead II configuration. The amplified EKG signal was taken from one of the polygraph's J6 output jacks and fed through a pulse-shaping circuit which converted the EKG wave form to a square-wave pulse of fixed duration. This pulse provided the input to solid-state control and analysis circuits.

Pretesting

Ten to 14 days after the implantation procedure each *S* was trained to press the lever for brain stimulation. A continuous schedule of reinforcement was used following initial shaping of the lever-press response. Once *S* was pressing the lever consistently the intensity of the current was adjusted to a value which yielded the fastest rate of responding without any gross motor side effects. This current intensity was used on all Skinner-box sessions and also to reinforce heart-rate changes in the subsequent curare-conditioning sessions. Prior to the first curare session each *S* received 500 response-contingent reinforcements in each of two sessions. Each *S* was also tested in the Skinner box at least once between each curare session.

Heart-Rate Training

At the beginning of each curare session the rat was given an ip injection of 12 units/kg d-tubocurarine chloride. This dosage was found to be sufficient to maintain the rats used in a completely immobile state for over 2 hr. Once paralysis began, usually within 2 min. after injection, *S* was fitted with the mask and was artificially respirated. Respirator settings of 1:1 inspiration/expiration ratio and a respiration rate of 70 breaths per minute were used. During the first minutes following curarization the rat's heart rate was monitored and adjustments made in respiration rate and volume so as to achieve a stable and regular heart rate. Once such a stable heart rate had been found, respiration rate and volume were fixed and no further adjustments were made during the remainder of the experimental session. During an initial adaptation period which lasted for approximately 30 min., the rat's heart rate was recorded and the median interheart beat interval (IBI) selected as the reinforcement criterion. Following the adaptation period the conditioning phase of the session was begun. The conditioning phase also lasted for approximately 30 min. During conditioning the rats in Group 1 were reinforced for increases in heart rate, and the rats in Group 2 were reinforced for decreases in heart rate. The procedure during the conditioning trials for a rat being reinforced for heart-rate decreases has been described in Figure 1. At the beginning of the conditioning trial no stimuli were present, but the rat's heart rate was recorded and analysed. If the first five IBIs emitted by the rat met the criterion which had been selected on the basis of the rat's adaptation heart rate, a light in the experimental chamber was switched on. The light was an informative

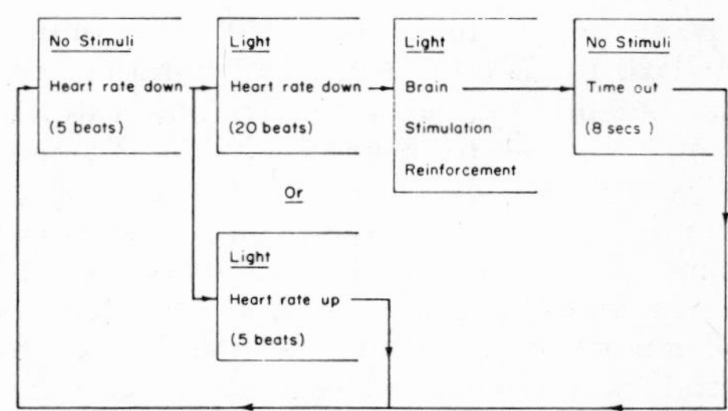

Fig. 1. The procedure used during conditioning for a rat being reinforced for heart-rate decreases

or feedback stimulus. If the rat's heart rate did not meet the criterion, no stimuli were presented and this condition continued until a sample of five IBIs was emitted which did meet the criterion. Once the light was switched on, provided that the rat's IBIs continued to meet the criterion, it remained on. When 20 successive, additional, criterion IBIs occurred, the light terminated with the delivery to the rat of brain stimulation. This brain stimulation was followed by an 8-sec. time-out period during which none of the stimuli was presented and none of the contingencies was in effect. This time-out period was used to allow the rat's heart rate to recover from the effects of the brain stimulation, since this stimulation was found to evoke a consistent decrease in heart rate. At the end of the time-out period the contingencies were reinstated and the procedure repeated until the end of the conditioning phase of the session. If the rat's heart rate increased while the light was on, it terminated and the procedure recycled. If it was found that the rat was not meeting the criterion, it was made less stringent until it was felt that a change to a more stringent criterion value could be made. If, on the other hand, it was found that the rat was easily meeting the criterion, the criterion was shifted to a more stringent value, thus making it more difficult for the rat to obtain the brain-stimulation reinforcement. Changes in the criterion in the appropriate direction were accomplished in steps of .01 sec. The procedure for conditioning an increase in heart rate would of course be opposite the procedure which has been described. An extinction phase which lasted for approximately 30 min. followed the conditioning phase of the session. During the adaptation, conditioning, and extinction phases of the experiment the number of heart beats in successive 1-min. periods was recorded. Each rat was given four curare conditioning sessions.

Results

The duration of each phase of the procedure varied slightly between *S*s with the minimum and maximum lengths of the adaptation, conditioning, and extinction phases of the curare sessions being, respectively, 20 to 31 min., 20 to 31 min., and 20 to 33 min. The mean durations of the adaptation, conditioning, and extinction phases were, respectively, 25, 27, and 26 min. Each of the three phases of each curare session were divided into four equal time periods for each *S* and the mean heart rates displayed in the four blocks of adaptation, conditioning, and extinction trials were calculated separately for each session. These individual means were then combined to give group mean heart rates for blocks of adaptation, conditioning, and extinction periods in each of the four curare sessions. These data are shown in Figure 2.

Critical data are the differences in heart rate between the last block of adaptation trials and the subsequent blocks of conditioning trials on each session for each group of *S*s. In accordance with the principle of reinforcement it will be seen that Group 1 (reinforced for heart-rate increases) displays successively greater increments in heart rate and Group 2 (reinforced for heart-rate decreases) successively greater decrements in heart rate during conditioning as a function of sessions. This effect is somewhat attentuated on Session 4. In the statistical analysis of these results differences were computed between each rat's heart rate during the last block of adaptation periods and the last block of conditioning periods for each session. These individual differences were then combined to give group mean differences. A complex analysis of variance performed on these data showed a significant treatments effect ($F = 47.86$, $df = 1/10$, $p < .001$), indicating that in terms of this measure the two groups showed significantly different changes in heart rate. The sessions effect was not found to be significant ($F = 2.89$, $df = 3/30$, $p > .05$), but the Treatments × Sessions interaction was found to be significant ($F = 5.14$, $df = 3/30$, $p < .01$). This significant interaction reflects the data presented in Figure 2, viz, Groups 1 and 2 show consecutive divergence in heart rates from Sessions 1 to 3 with a slight convergence in Session 4 due to the diminution of the conditioned heart-rate change in Group 1.

The effects of the extinction procedure may be observed in Figure 2. Data of interest here are the changes in heart rate from the last block of conditioning periods through the blocks of extinction periods on successive trials. It will be seen that the mean heart rates of the two groups tend to converge during extinction, so that the final performances on Sessions 3 and 4 closely approximate the final adaptation

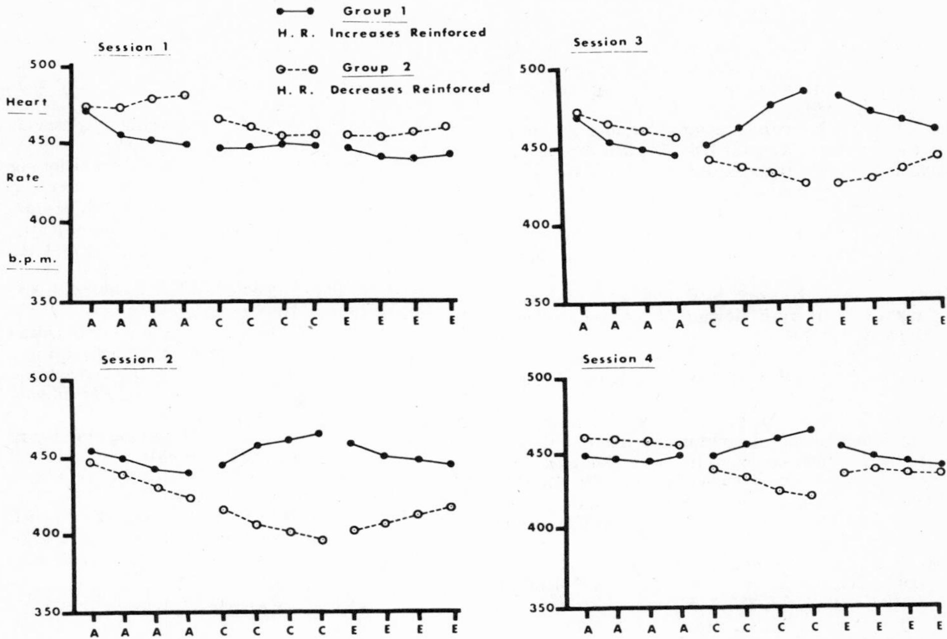

FIG. 2. Group mean heart rates for successive blocks of adaptation (A), conditioning (C), and extinction (E) periods over four sessions.

heart rates on those sessions for both groups of *S*s.

Discussion

The data presented here support those previously reported by Trowill (1967), Miller and DiCara (1967), and Miller and Banuazizi (1968), viz, heart-rate increments and decrements may be conditioned when brain-stimulation reinforcement is made contingent on the appropriate heart-rate response. Although the main conditioning effect observed in the present experiment is in agreement with the effects reported by these investigators, it should be noted that the magnitude of the conditioned response is considerably smaller than that reported in the Miller studies. This difference in the magnitude of the conditioned response is probably attributable to differences in the specification of a criterion response. In the present experiment *S*s were required to maintain criterion heart rates for a period approximately three times as long as that stipulated in the Miller procedures for the production of a reinforcing stimulus.

In the classification of heart rate as an operant it is important not only to show that the frequency of this response may be systematically influenced by reinforcing stimuli but also to demonstrate that this reponse displays other characteristics generally ascribed to operants. Miller and DiCara (1967) have demonstrated that heart rate is subject to discriminative control. The present experiment provides evidence of successive improvement of heart-rate performance as a function of training session, thereby indicating retention and transfer of the operant heart-rate response. Evidence of extinction is also provided. During extinction, depending on the direction of the conditioned response, heart rate displays either an increase or a decrease to the preconditioning level.

REFERENCES

Brener, J. M. The characteristics of heart rate during various conditioning procedures. Unpublished doctoral dissertation, University of London, 1964.

Brener, J. M. Heart rate as an avoidance response. *Psychological Record,* 1966, **16,** 329–336.

Brener, J. M., & Hothersall, D. Heart rate control under conditions of augmented sensory feedback. *Psychophysiology,* 1966, **3,** 23–28.

Brener, J. M., & Hothersall, D. Paced respiration and heart rate control. *Psychophysiology,* 1967, **4,** 1–6.

Engel, B. T., & Hansen, S. P. Operant conditioning of heart rate slowing. *Psychophysiology,* 1966. **3,** 176–187.

Hnatiow, M., & Lang, P. J. Learned stabilization of heart rate. *Psychophysiology,* 1965, **1,** 330–336.

Katkin, E. S., & Murray, E. N. Instrumental conditioning of autonomically mediated behavior: Theoretical and methodological issues. *Psychological Bulletin,* 1968, **70,** 52–68.

Kimble, G. A. *Hilgard and Marquis' conditioning and learning.* New York: Appleton-Century-Crofts, 1961.

Kimmel, H. D. Instrumental conditioning of autonomically mediated behavior. *Psychological Bulletin,* 1967, **67,** 337–345.

Lang, P. J., Sroufe, L. A., & Hastings, J. E. Effects of feedback and instructional set on the control of cardiac rate variability. *Journal of Experimental Psychology,* 1967, **75,** 425–431.

Lisina, M. I. In Razran, G. Observable unconscious and the inferable conscious in current Soviet psychophysiology: Interoceptive conditioning and sematic conditioning, and the orienting reflex. *Psychological Review,* 1961, **68,** 81–147.

Miller, N. E., & Banuazizi, A. Instrumental learning by curarized rats of a specific visceral response, intestinal or cardiac. *Journal of Comparative and Physiological Psychology,* 1968, **65,** 12–19.

Miller, N. E., & DiCara, L. Instrumental learning of heart rate changes in curarized rats: Shaping and specificity to discriminative stimulus. *Journal of Comparative and Physiological Psychology,* 1967, **63,** 12–19.

Shearn, D. W. Operant conditioning of heart rate. *Science,* 1962, **137,** 530–531.

Smith, K. Conditioning as an artifact. *Psychological Review,* 1954, **61,** 217–225.

Trowill, J. A. Instrumental conditioning of heart rate in the curarized rat. *Journal of Comparative and Physiological Psychology,* 1967, **63,** 7–11.

(Received August 15, 1968)

7

Differential Operant Conditioning of Heart Rate

Howard I. Levene, Bernard T. Engel, and John A. Pearson

Five normal female subjects (*Ss*) were taught to increase and decrease cyclically their heart rate (HR) by means of a differential operant conditioning technique. Discrimination between slowing and speeding was easily demonstrated; however, the ability significantly to increase and decrease HR relative to resting rate was less apparent. Although each *S* could do this on at least one occasion, only 2 *Ss* did so with consistency. Breathing or musculoskeletal responses did not seem to mediate the learned responses.

PREVIOUSLY, we reported that subjects (*Ss*) could be taught by means of an operant conditioning technique either to slow or to speed their heart rates (HR) significantly from resting levels.[1, 2] This paper reports the results of a differential operant conditioning study in which *Ss* were taught to increase and decrease cyclically their HR from resting rates.

From the Cardiovascular Research Institute and Langley Porter Neuropsychiatric Institute, University of California San Francisco Medical Center, San Francisco, Calif.

Supported in part by Grants FR-00122, for computing services at San Francisco Medical Center, 5TI-MH-7082, for the Interdisciplinary Training Program of the National Institute of Mental Health, and HE-06285, from the National Heart Institute, U. S. Public Health Service.

We wish to express our appreciation to Dr. E. Callaway, III, for allowing us to use his equipment for this analysis.

Received for publication May 20, 1968.

*Present address: Gerontology Branch, National Institute of Child Health and Human Development, Baltimore City Hospitals, Baltimore, Md.

Materials and Methods

Five normal college women were recruited to participate in a "psychophysiological experiment." *Ss* were never told that they were being taught to control their HR.

Instructions and Procedures

Each experimental session lasted about 2 hr. and was held on consecutive weekdays. The total number of sessions was 6–10, depending on the rapidity with which *Ss* learned the required responses.

Ss were semireclined on a bed in a soundproof, dark room adjoining the experimenter's room. They could be watched through a glass window from behind their positions. Electrodes were placed on the arms and chest to pick up cardiac potential from a unipolar lead in the right fourth intercostal space; a strain gage, cemented to a belt, was placed around the chest to record breathing. At the foot of the bed, and visible to the subject, was a vertical array of three lights and an electronically controlled clock. The upper and lower lights served as discriminative cues. The upper light always called for increased HR, the lower light for decreased HR. The middle light and the clock were immediate reinforcers for the correct response.

On the first day of the experiment, instructions were played to *Ss* by a magnetic tape. They were told that they would be participating in a conditioning study, but that we would not tell them what response was being conditioned. The instructions were to lie as quietly as possible on the bed and not change their breathing because movement and abnormal breathing would "foul up our measurements." The three lights and clock at the foot of the bed were then called to their attention; they were told that after about 30 min. either the upper or lower light in the array would come on and remain on continuously for about 25 min. During the time that the upper or lower light was lit, the middle light in the array (painted red) would be under their "control." When they made the correct response, the middle light would be on. When they made the incorrect response, the middle light would go off. The clock would accumulate the time the middle light was on, and they would be paid one-fourth cent per second for the time accumulated on the clock.

On each of the first several days of the experiment, training in HR slowing and HR speeding were conducted separately from one another. Each separate training period, which lasted 25 min., was preceded by a resting period of about one-half hour. A 10-min. break was taken between the two parts of the experiment, and S was encouraged to walk about or go to a nearby restroom.

A procedure similar to a shaping schedule in the usual operant conditioning study was used. On the first day, the equipment was adjusted toward the end of each resting period so that S's resting HR would automatically trigger the reward light and clock about 80% of the time. This 80% "trigger level" was chosen so that S could easily keep the reward light on during the following training period. On each subsequent day we adjusted the "trigger level" so that the resting HR just prior to the training-periods would keep the light on for a progressively smaller percentage of the time—at first 60–80% and then only 50% of the time. This 50% "trigger level" was maintained until S was able to perform well enough to keep the reward light on significantly more ($p < 0.05$) than 50% of the time during both the speeding and slowing training periods. S was then informed that instead of having separate training sessions with the upper and lower (cue) lights, she would have one continuous session with the cue lights alternately flashing back and forth. As before, the middle light would go on when she was making the correct response; however, now a separate clock would accumulate the time that the middle light was on for each cue light. There was one 30-min. resting period. In these cyclical training sessions, the apparatus was adjusted at the end of the rest period so that S's HR would trigger the reward light and clock 50% of the time under either cue condition. S was informed that the reward light and clocks were under her control, and the cue lights automatically flashed back and forth, at about 1-min. intervals for 1 hr. It should be clear that by the end of the experiment, in order to keep the reward light and clocks on more than 50% of the time, S, on cue, had to increase and decrease cyclically HR relative to resting rate at 1-min. intervals.

Apparatus

A Fels cardiotachometer* continuously recorded HR, and the output of the cardiotachometer was fed into an Offner amplifier and dynograph.† The output from the amplifier was fed into a Schmitt trigger,‡ which was used to control S's reinforcements. The input to the Schmitt trigger was regulated by a zero-suppression circuit on the Offner amplifier to permit adjustment of the trigger point. With this system, it was possible to reinforce S on a beat-by-beat basis.

*Yellow Springs Instrument Co., Inc., Yellow Springs, Ohio.

†Beckman Offner, Schiller Park, Ill.

‡The logic equipment used to regulate the lights and clocks was manufactured by BRS Electronics, Beltsville, Md. In order to reduce the hysteresis in the Schmitt triggers, we modified each standard unit by grounding the emitter of the input transistor.

Performance Measures and Data Analysis

The trigger level was set each day during the last 4–8 min. of each 30-min. resting-period. This level was confirmed by analyzing 10 successive 25-sec. samples with 3 electronically controlled clocks. One of the clocks measured the duration of the sampling period, while the others measured the time during which S's HR was above or below the trigger level.

The mean percentage of time that S's HR was above and below the trigger level during these resting samples was computed, and compared by means of a *t* test to the mean percentage of time that S's HR was above or below the trigger level during 30 successive 25-sec. samples of the subsequent training period. This gave us a rough estimate of S's daily performance and allowed us to decide when to change from separate speeding and slowing training sessions to the cyclical procedure.

In the days when the cue lights switched back and forth alternately in 1-min. cycles, the percentage of time that S's HR was above or below trigger level was analyzed on each cycle. This allowed us to compute an acquisition curve for the responses on a trial-by-trial basis. The analysis of percentage-time began automatically about 15 sec. after the light switched and ran for about the next 25 sec.

The most important criterion of learning, however, was a significant change (measured by means of a *t* test) in the HR from the last 4–8 min. of the resting period to the training period. This was analyzed by recording ECG's on magnetic tape (Ampex SP 300*). The pulses on these tapes were used to trigger a Computer of Average Transients (CAT)† which measured each interbeat interval and drew a frequency histogram. The output of the CAT was then recorded onto digital tape and analyzed by a computer. Thus, we were able to analyze all the interbeat intervals for each session. Although the number of interbeat intervals varied from condition to condition and session to session, the resting period usually comprised about 450 beats and each of the training conditions usually included about 2000 beats.

Results

Each S was trained in slowing and speeding separately until she was able to perform well enough to keep the reward light on significantly more ($p < 0.05$) than 50% of the time. She was then switched from the separate training sessions with each cue light to one continuous training session with alternating cue lights. *S1* was switched after 4 sessions, *S2* after 4 sessions, *S3* after 2 sessions, *S4* after 3 sessions, and *S5* after 2 sessions.

Discrimination between the alternating cue lights in the cyclical training-sessions was easily conditioned. Figure 1 shows representative resting and training cardiotachometer tracings from each subject. The HR oscillates on either side of the trigger level (*T*) in association with the alternating cue lights (fast and slow). The mean heart rate during the fast cue periods was greater than the mean heart rate during the slow cue periods in every S during every cyclical training session. The differences were significant beyond the 1% level of confidence in each case (Table 1).

The ability to increase and decrease HR cyclically and significantly relative to resting levels was a more difficult task. Although each S could do this on at least one occasion, only *S4* and *S5* performed with consistency (Table 1). Of those *S*s who were unable to change consistently with respect to resting levels, it appears that *S2* and *S3* found difficult in slowing, whereas only *S1* found difficulty in speeding.

It seems that the slowing response is more difficult and takes more time to master. This impression is enhanced by the acquisition curves for the two responses. These curves (Fig. 2) were computed by analyzing the mean per-

*Ampex Corporation, Redwood City, Calif.

†Technical Measurements Corporation, North Haven, Conn.

centage of time, in 3-trial blocks, that *Ss* made the correct response. They are based on averages of all *Ss'* cyclical training sessions. During the slowing trials, *Ss* gradually increased the percentage of time that HR was below T level, whereas during speeding trials, the percentage of time that HR was above T level began at 75–80% and showed no consistent trend thereafter. Thus, while the speeding response was well established, most *Ss* appeared to be still learning the slowing response.

There are no systematic trends in the absolute levels of HR from session to session. During rest, *S1*, *S2*, and *S4* showed no trends, while *S3* showed an increasing trend, and *S5* showed a decreasing trend. During the slowing trials, *S1* and *S4* showed no trends, whereas

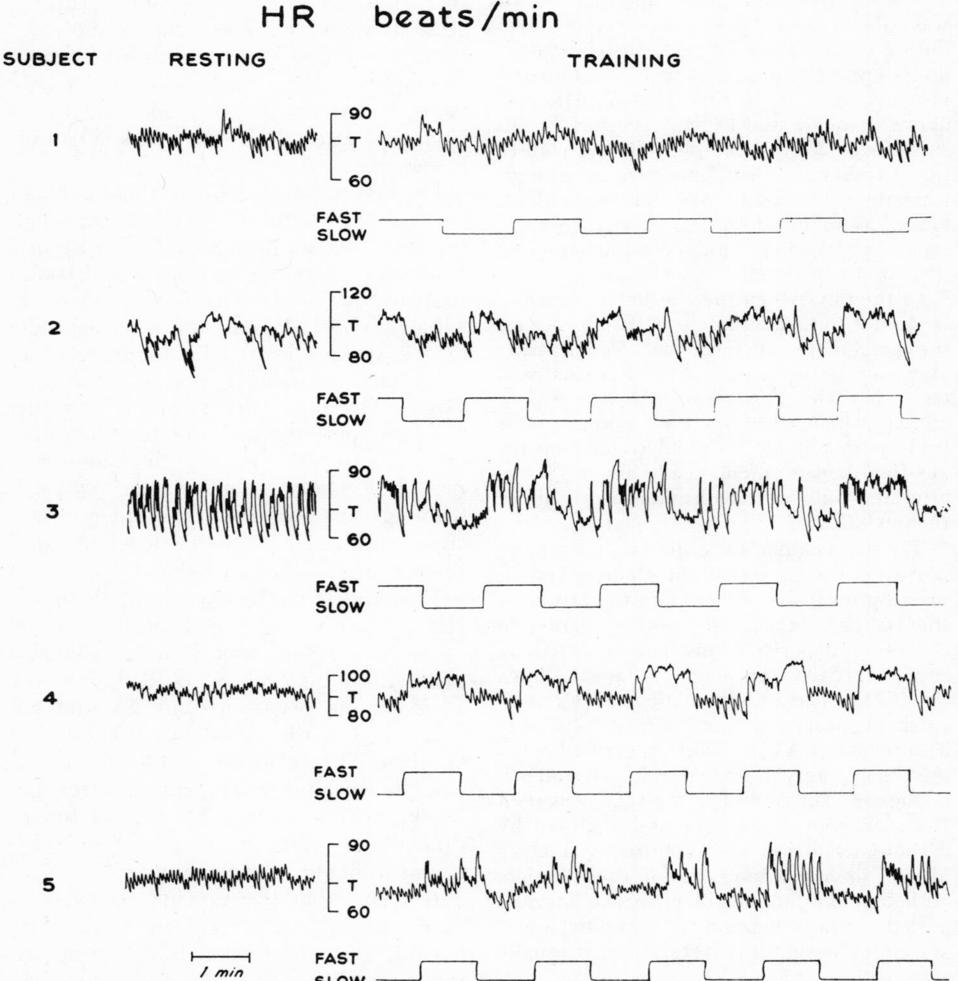

FIG. 1. Representative cardiotachometer records for each *S* at rest and during cyclical (fast and slow) training, showing HR fluctuation around triggering level (*T*).

TABLE 1. Mean HR (beats/min.) for Resting and Cyclical Fast and Slow Training

Subj.	*Period*	*Session* 1	2	3	4	5	6	$\bar{X}$
1	Rest	78.3	86.2	76.2	77.4	77.0	82.7	79.6
	Slow	73.6*	79.4*	75.5*	81.3	72.6*	76.4*	76.5
	Fast	76.2	82.7	78.0*	83.5*	76.7	82.1	79.9
2	Rest	80.8	90.8	74.1	74.0	70.9	84.5	79.2
	Slow	81.4	87.2*	79.7	78.6	74.2	79.7*	80.1
	Fast	89.1*	92.9*	80.6*	80.3*	78.2*	84.5	84.3
3	Rest	68.8	68.0	65.4	75.3	73.2	73.7	70.7
	Slow	69.3	68.7	72.7	75.2	71.9*	73.0	71.8
	Fast	72.6*	72.9*	73.5*	83.9*	83.1*	83.7*	78.3
4	Rest	91.1	84.0	88.1	—	—	—	87.7
	Slow	87.8*	80.6*	84.6*	—	—	—	84.3
	Fast	101.7*	93.0*	97.2*	—	—	—	97.3
5	Rest	78.0	82.5	82.5	74.3	73.4	—	78.1
	Slow	72.7*	79.7*	81.1*	70.4*	70.0*	—	74.8
	Fast	76.7	84.5*	84.3*	75.7*	76.1*	—	79.5

* Significant change in HR in correct direction ($p < 0.01$).

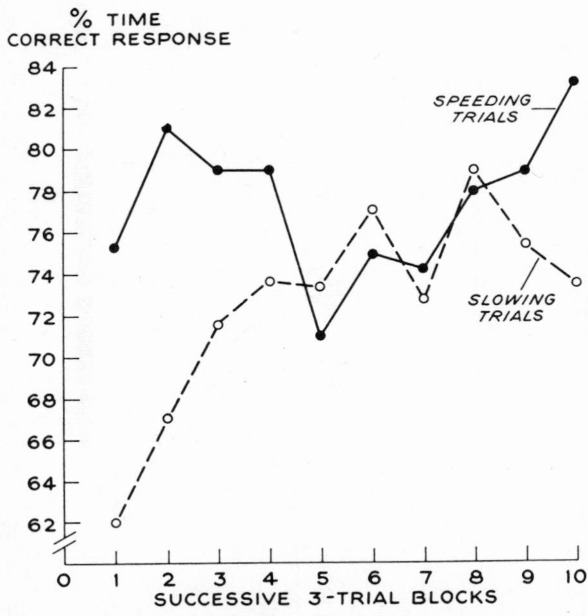

FIG. 2. Acquisition curves during cyclical training periods.

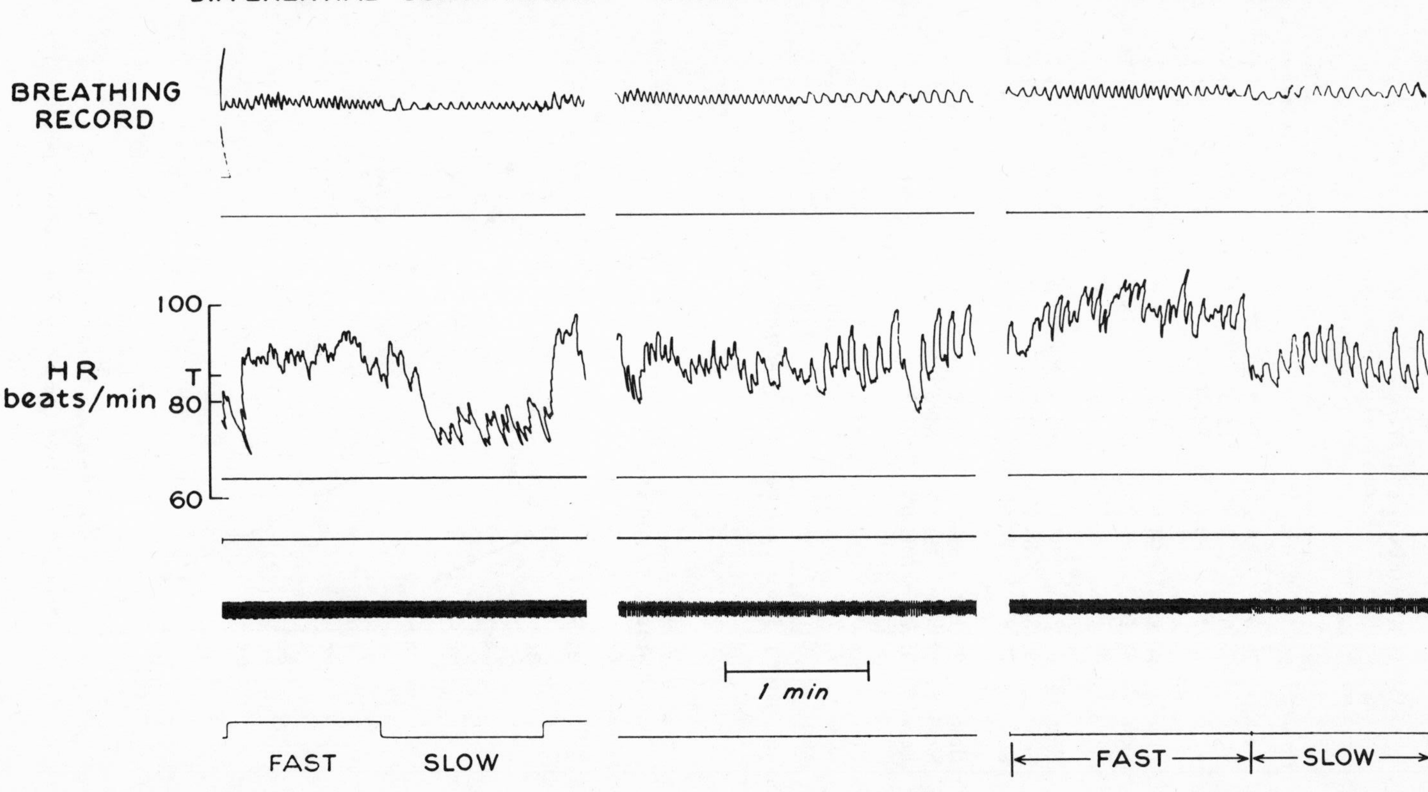

FIG. 3. Cardiotachometer and breathing records during differential conditioning, paced breathing, and recall of discriminative cues (S2).

S2 and S5 tended to decrease their HR and S3 tended to increase her HR. During the speeding trials, *S1*, *S4*, and S5 showed no trends, S2 tended to decrease, and S3 tended to increase HR.

In spite of our explicit instructions that S not change breathing during the cyclical training trials, some did so. When we commented on their altered breathing patterns and reminded them of the instructions, they stated that they were unaware that their breathing was changing. We therefore duplicated spontaneous breathing rates by pacing two of the *Ss* (S2 and S3) at the end of their training. Although some changes in HR variability were noticed, there were no significant changes in HR from slow- to fast-paced breathing. We then asked each S to imagine the top cue light and then the bottom cue light in the array. Marked changes in the HR and breathing were again noticed. Both *Ss*, when questioned after this, denied being aware of changes in breathing.

A demonstration of these effects can be seen in Fig. 3, which shows HR and breathing patterns during training, paced breathing, and imaginary recall of the state this S was in when the cue lights were available to her. The tracings of the other S were just as dramatic

Discussion

The evidence in this study points to the fact that *Ss* can gain remarkable minute-to-minute control of HR when placed on an appropriate reinforcement schedule. All *Ss* discriminated between cues rather easily and consistently. This finding is similar to the results of Brener and Hothersall[3] who demonstrated discriminative HR responses to two cues in a one-session experiment of approximately 10 min. It was not clear from their data, however, whether the increases and decreases in HR were significantly different from resting rates, as they were in our *Ss* 4 and 5.

From the acquisition curves, it seems that the speeding response is more easily acquired than the slowing response. This finding agrees with other results of separate speeding and slowing studies.[1, 2] Engel and Chism[2] point out that HR speeding from rest is an easier response to emit than HR slowing. In addition, they suggest that most *Ss* may react to the experimental conditions with an autonomic pattern that includes increased HR. *S1*'s difficulty with speeding might be accounted for by the fact that her individual response specificity includes a pattern of HR slowing.

In Engel and Hansen's study[1] of HR slowing, HR tended to fall across the days. In Engel and Chism's study[2] of HR speeding, HR tended to increase across the days. Our data show no systematic trends in either the resting, speeding, or slowing periods of the experiment. This may be explained by the fact that our *Ss* were learning two responses simultaneously, and any systematic change in HR would go unrewarded under the opposite contingency.

We wish to call attention to S3's record in Fig. 1. The marked sinus arrhythmia at rest was not due to breathing. Her breathing rate was about 12–14 cycles per minute, whereas the arrhythmia was about 4–5 cycles per minute. It can be seen from the record that she was able to change her basic rhythm as well as her rate during training. Recently we have studied several patients with cardiac arrhythmias (atrial and ventricular tachycardia, atrial fibrillation, and premature ventricular contractions), some of whom in the course of learning to control their rates acquired control over their arrhythmias as well. The potential clinical application of these observations is presently under study.

Only S3 guessed that we were measuring heart rate. S2 was not sure but thought we were measuring blood flow. It is interesting to note that both S2 and

S3 had difficulty in learning the HR slowing response. This finding agrees with the one Engel and Hansen[1] found in their study on the operant conditioning of HR slowing that *Ss* who guessed the correct response did poorly.

We are unable at this point to identify any somatic mediators for the control of HR. We did not see any gross musculoskeletal maneuvers, and all *Ss* reported that they were able to cooperate with the instructions on this point. Although several *Ss* showed phasic changes in breathing, we were unable to duplicate the HR changes by paced breathing, and *Ss* denied being aware of the breathing-changes during the experiment. It seems possible that a common mediator was responsible for the concurrent breathing and cardiac changes. This hypothesis is supported by the findings of Frazier,[4] Engel and Hansen,[1] Engel and Chism,[2] and Hnatiow and Lang,[5] who found no breathing changes that could account for the HR changes they reported. In addition, Engel and Chism,[6] who studied paced breathing and HR systematically, could not find any effect on HR, although they did see some effect on HR variability. Further evidence that breathing or skeletal muscle response do not account for mediation of these operantly learned responses comes from the animal studies of Trowill,[7] and Miller and DiCara,[8] who showed that rats who were completely paralyzed by curare and maintained on steady artificial respiration could learn both to increase and decrease HR for the reward of electrical stimulation of the medial forebrain bundle.

Summary

Five normal female subjects were taught cyclically to increase and decrease their heart rates from resting rates by means of a differential operant-conditioning procedure. Reinforcement for the correct response (alternately slowing or speeding heart rate for 1-min. intervals) was given on a beat-by-beat basis in the presence of discriminative cues. All subjects learned to discriminate between slowing and speeding; the mean heart rate for every subject during 1-min. periods where a speeding response was required for reinforcement was significantly greater ($p < 0.01$) than the mean heart rate during 1-min. periods where slowing was required for reinforcement.

The ability to increase and decrease heart rate cyclically and significantly relative to resting rate was a more difficult task to learn. Although each subject could do this on at least one occasion, only 2 subjects did so consistently.

The speeding response seems more easily acquired than the slowing response. This finding is in agreement with results of previous studies.

Breathing or musculoskeletal responses did not seem to mediate the learned heart rate responses. Although some subjects changed breathing rates cyclically with changes in heart rate, they were unaware that their breathing was changing, and we were unable to produce the cyclical changes in heart rate with paced breathing experiments.

Langley Porter Neuropsychiatric Institute
University of California
San Francisco Medical Center
San Francisco, Calif. 94122

References

1. Engel, B. T., and Hansen, S. P. Operant conditioning of heart rate slowing. *Psychophysiology 3:*176, 1966.
2. Engel, B. T., and Chism, R. A. Operant conditioning of heart speeding. *Psychophysiology* 3:418, 1967.
3. Brener, J., and Hothersall, D. Heart rate control under conditions of augmented sensory feedback. *Psychophysiology 3:*23, 1966.
4. Frazier, T. W. Avoidance conditioning of heart rate in humans. *Psychophysiology 3:*188, 1966.
5. Hnatiow, M., and Lang, P. J. Learned stabilization of cardiac rate. *Psychophysiology 1:*330, 1966.

6. Engel, B. T., and Chism, R. A. Effect of increases and decreases in breathing rate on heart rate and finger pulse volume. *Psychophysiology 4*:83, 1967.
7. Trowill, J. A. Instrumental conditioning of the heart rate in the curarized rat. *J Comp Physiol Psychol 63*:7, 1967.
8. Miller, N. E., and DiCara, L. Instrumental learning of heart rate changes in curarized rats: Shaping and specificity to discriminative stimulus. *J Comp Physiol Psychol 63*:12, 1967.

8

Instrumental Conditioning of the Heart Rate in the Curarized Rat

Jay A. Trowill

Rats, previously trained to bar press for electrical stimulation of the brain (ESB), received ESB on an FI schedule for either relatively fast or relatively slow heart rate levels while their skeletal musculature was completely paralyzed by d-tubocurarine chloride. 15 of 19 Ss rewarded for fast heart rates increased their heart rates during a training period, while 15 of 17 Ss rewarded for slow heart rates decreased their heart rates. Rats yoked to experimental Ss in either reinforced group failed to show similar changes. The results suggest that instrumental learning of the heart rate is possible even when the possibility of skeletal muscle mediation is eliminated.

In the past few years there has been a revival of interest in the problem of instrumental learning of autonomically controlled, visceral responses. Miller (1961, 1963) has challenged the idea that such responses cannot be instrumentally learned. A number of recent experimental studies have suggested that various such responses (GSR—Crider, Shapiro, & Tursky, 1966; Fowler & Kimmel, 1962; Johnson, 1963; Kimmel & Hill, 1960; Kimmel & Kimmel, 1963; Shapiro, Crider, & Tursky, 1964; heart rate—Shearn, 1962; salivation—Miller & Carmona, 1967) may be modified by instrumental training procedures. However, others (GSR—Mandler, Preven, & Kuhlman, 1962; heart rate—Harwood, 1962) have reported negative results, while a study by Fromer (1963) raises doubt about instrumental learning of vasomotor changes.

Furthermore, none of the studies yielding positive results has conclusively ruled out the possibility that *S*s learned skeletal responses which had an unlearned tendency to elicit the visceral responses recorded (Skinner, 1938). While the correlational procedures used by some of the authors cited can rule out mediation by the particular skeletal responses recorded, the possibility always remains that the visceral responses were mediated by some skeletal response that was not recorded. Therefore, a more conclusive control is to completely paralyze all skeletal responses by curare (Miller & Carmona, 1967).

The aim of this study was to investigate the possibility of the instrumental learning of a visceral response, a fast or a slow heart rate, in rats which were maintained at a constant level of artificial respiration while all skeletal muscles were paralyzed by curare. The paralyzed *S*s were given positively rewarding electrical stimulation of the brain (ESB) on an FI schedule for either slow or fast heart rates. The changes from a previous base rate under the two conditions of reward were compared with each other and with those of yoked controls.

Method

Subjects

The *S*s were 56 120-day-old, naive, male albino rats (Sprague-Dawley strain). Each *S* was stereotactically implanted with monopolar electrodes aimed at the medial forebrain bundle in the posterior portion of the lateral hypothalamus (Krieg coordinates, 1.5 mm. posterior to bregma, 8.5 mm. vertical at site of entry, and 1.7 mm. left and 1.3 mm. right, lateral to the midline) and showed a reward effect from at least one of the electrodes.

[1] Based on a dissertation submitted to the Graduate School of Yale University in partial fulfillment of the requirements for the PhD degree. The author gratefully acknowledges the guidance and encouragement of the major adviser, Neal E. Miller, and the other members of his thesis committee. An abbreviated version of this paper was presented at the 1966 Eastern Psychological Association meeting in New York City. This investigation was partially supported by Predoctoral Fellowship No. 4F 1 MH-17, 104-03 and Grant MH 00647, both from the National Institute of Mental Health.

[2] Now at University of Massachusetts.

Apparatus

Pretesting of the rewarding effect of ESB was given in a Plexiglas box with a grid floor and a Grayson-Stadler bar. The stimulation was 0.5 sec. of 60-cycle sine-wave current delivered between one of the implanted electrodes and an indifferent surface screw. The current was regulated by a micropot used as a voltage divider and was monitored by an ac microammeter in series with *S*. Fluctuations in brain resistance were minimized by an 820 K ohm resistor in series with *S*. Two similar stimulators were alternated during the pretesting.

An unpainted wooden box separated into halves by a wooden partition to provide similar but separate environmental conditions served as the experimental chamber for heart rate training. This was inserted into a standard sound-attenuated and ventilated picnic ice chest. Each of the two chambers had the leads for the EKG, ESB, the ground connection, a loudspeaker, a house light, and the tubing for artificial respiration.

The *S* was respirated by means of a small-animal respirator (E & M Equipment Company, model V5KG). A rubber mask made from the mouthpiece end of a rubber balloon was secured to the nose. The lower portion of the mask was secured by placing it behind the upper incisors, and the upper portion was clamped onto the nose by means of a hemostat after an injection of Xylocaine had been given to prevent pain. The mask was connected to the respiration tubing via a rubber stopper with a hole drilled through the middle.

The EKG electrodes were stainless steel surgical needles which were inserted subcutaneously on the back and chest after an injection of Xylocaine. A Grass instrument durable disk electrode was placed on the tongue as a ground lead.

The EKG was recorded by means of a Grass polygraph. The amplified EKG signal was taken from the driver amplifier via the J-5 output jacks. This signal was connected to a sensitive relay through a 50-μf. condenser and a 20 K ohm variable resistor. The variable resistor, as well as the preamplifier and driver amplifier controls, was adjusted so that only the QRS spike of each heartbeat activated the relay. The output of the relay was then put through a Grayson-Stadler pulse former and the formed signal was used to operate the counters and reinforcement circuitry.

The reinforcement circuit was composed of standard relay equipment programmed so that any predetermined number of heartbeats in a specified interval could be detected. A predetermining counter driven by a 10-pps pulse generator was used to time the interval. The specified number of heartbeats was set on a second predetermining counter which was recycled by the output of the first counter. Relays were programmed so that the rewarding ESB was delivered (and the time-out period was started) whenever the number of heartbeats registered on the second counter was above the specified criterion number for the fast condition or below the criterion number for the slow one. The timing interval and the number of heartbeats could be varied in order to change the reinforcement criterion. Two identical reinforcement circuits were operated alternately so that one was reset while the other counted.

Procedure for Pretesting

In order to establish independently the existence of a rewarding effect of ESB, 7–10 days after surgery each *S* was trained to bar press for ESB. The stimulation level was adjusted until *S* showed signs of sniffing and increased exploration. Then *S* was shaped to bar press. Once trained, 1 hr. of continuous reinforcement (CRF) was given to allow stabilization of the bar-pressing rate.

One day later (Day 2), *S* was replaced in the box and given 10 min. of CRF at the same current intensity used on Day 1. The following procedure was then used to ascertain that the 0.5-sec. periods of ESB would continue to be an effective reward even if separated by the time-out periods used later in the main experiment: 10 min. of bar pressing on an FI 4-sec. schedule (FI 4), 10 min. on an FI 8 schedule, and 10 min. on an FI 12 schedule. If *S* failed to sustain performance on any of the FI schedules (fewer than 10 bar presses in the last 5 min. of the FI periods), the stimulating current was increased by 4 μa. The *S* was then retrained and started again on the FI schedule procedure. This was repeated until each *S* performed on all schedules. Rats which could not be trained to make 10 presses in the last 5 min. were discarded.

Several days after Day 2, *S* was run through the same FI schedule procedure. The performance during the second 5 min. of the FI 8 period on this day was used to match those pairs of *S*s used in the yoked control procedure in order to equate the reinforcing and motivational properties of ESB; *S*s within 10 bar presses of one another for the 5-min. test were matched as a pair. Two similar brain stimulators were interchanged during the 3 days of bar-press training in order to insure similar ESB effects when either stimulator was in use.

Procedure for Heart Rate Training

Preliminary work showed that an ip injection of 8 units/kg of d-tubocurarine chloride would maintain complete paralysis in a rat for approximately 1 hr. Furthermore, there was an approximately linear relationship between the initial dose and the duration of paralysis with each additional 8 units yielding about 1 additional hr. of paralysis. Thus, the dosages used were calculated to maintain paralysis for at least 1 hr. longer than the average length of the experimental session, i.e., for at least 3 hr. or a dosage of 24 units/kg.[3]

Once paralysis began, *S* was fitted with the face mask and artificial respiration was begun. Res-

[3] A few *S*s are included which were run for a 10-min. period under the dosage of 16 units/kg.

pirator settings of a 1:1 I/E ratio and 70 breaths per minute were used for all *S*s since extensive preliminary observations indicated that these settings would maintain a rat in a healthy condition for several hours with a minimal amount of long-term changes in the baseline heart rate.

After *S* was curarized and artificial respiration was begun, the heart rate was allowed to stabilize within a range of 300–450 bpm. During this period of 30–60 min., the respiratory volume was manipulated in order to bring the heart rate within this range.

Next, the criterion for the condition of reward, either slow or fast heart rate, was determined by adjusting the automatic counting system. Specific heart rates were evaluated on the basis of the proportion of time in two successive 1-min. periods that the chosen heart rate occurred. The criterion chosen was a heart rate sufficiently above or below the average so that it occurred once every 3–5 sec. Pairs of rats which had been matched on the basis of rate of pressing for ESB were assigned randomly to the fast or slow conditions and then the members of each pair were assigned randomly to the experimental or yoked control treatment.

For some rats, a period of noncontingent ESB for use as a baseline heart rate measurement was given immediately after the criterion selection period. The ESB was delivered for a period of 2 min. on a VI schedule with an average of 12 sec. and a range of 8–16 sec. between stimulations. For *S*s run before this procedure was adopted, the first minute of training was used as the baseline.[4]

Immediately after the criterion selection period or the noncontingent ESB period, instrumental training was begun. Since preliminary observations indicated that ESB at the electrode sites used here sometimes elicited an unconditional heart rate change, an FI schedule (5 sec. for the first 9 *S*s and 8 sec. for the remaining 47) was used to minimize short-term influences of such changes during the instrumental training. The same response criterion was used during the entire training period (10 min. for the first 9 *S*s and 30 min. for the remaining).

Results and Discussion

Due to the frequency and size of unconditional heart rate reactions to ESB throughout the experiment, it was decided to use for baseline measurements heart rate scores occurring while ESB was being applied, so that differences in the size of the cumulative UR effect would not mask changes during the training period. For *S*s to which no VI ESB period was given prior to training, the heart rate during the first minute of training—a time when the UR effect could occur without being greatly masked by the possible effects of learning—was selected as the baseline measure. For all other *S*s, the heart rate during the VI ESB period was used as the baseline measure. The final measure of the training effect was the heart rate during the last 2 min. for the first *S*s run with 10 min. of training and the last 10 min. for those run with 30 min. of training. This decision, as well as the one concerning the time to be used for the initial baseline, was made before the data were analyzed. The heart rates were counted for 10-sec. periods in a random manner and were cross-checked with 30-sec. counts recorded during the training period.

Fifteen of 19 *S*s rewarded for fast heart rates increased their rates ($p < .01$), and the mean increase for all 19 *S*s was 3.03 beats per 10 sec. ($t = 2.21$, $df = 18$, $p < .05$). Fifteen of 17 *S*s rewarded for slow heart rates decreased their rates ($p < .001$) and the mean decrease for all 17 *S*s was 3.21 beats per 10 sec. ($t = 4.39$, $df = 16$, $p < .005$). The difference in the changes shown by the two groups was highly significant ($t = 3.69$, $df = 34$, $p < .001$).

The results are summarized in Figure 1, which shows that the differences in change of heart rate, as represented by the direction and length of the arrows, were indeed the product of being rewarded for fast or slow rates rather than of the other factors represented in the diagram. None of the other factors contributed systematically to the changes in rate. These were: (*a*) the different replications, which include all of the differences in procedure, (*b*) the initial baseline heart rates, and (*c*) the directions of the unconditioned effect of ESB.

A yoked control procedure was used during later replications to control for the effects of any differences between the experimental treatments in the number and patterning of the ESB presentations. The results are summarized in Table 1. A disconcerting feature of these results is that the two yoked groups showed apparent changes

[4] The minor differences in procedure which were introduced as refinements in later replications, along with the raw data on the individual *S*s involved in each of them, are presented in detail in Trowill (1966). As can be seen from Figure 1, virtually equal numbers of fast and slow *S*s were run under each condition (i.e., replication). All *S*s run were included in the analysis.

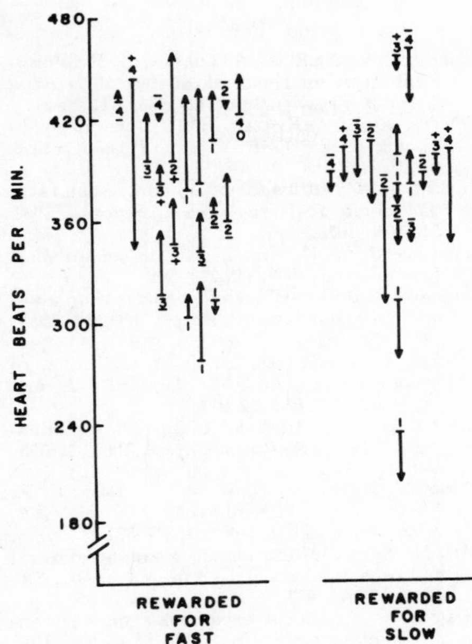

FIG. 1. Arrow diagram showing the magnitude and direction of experimental heart rate change during training for all experimental *S*s for all replications. (The base of the arrow represents the mean heart rate during the baseline period; the tip of the arrow represents the mean heart rate at the end of training. The numbers above or below the base of the arrow represent the replication number for the *S*; the symbols +, −, and 0 represent the direction of change of the heart rate from the pre-ESB to the baseline period—increase, decrease, and no change, respectively.)

during training opposite to those of their experimental partners and hence to those of each other. Because of the variability in the yoked groups, however, none of these differences was reliable. Since intensive analysis of the data failed to reveal any basis for the apparent effects in the yoked groups, they may have been produced by chance. The fact that the effects in the yoked control groups are in the opposite direction to those in the experimental ones makes it less likely that the differences in the experimental groups were due to unconditioned effects of the number and pattern of reinforcements. For the experimental rats run with yoked partners and thus represented in Table 1,

TABLE 1

MEAN HEART RATES PER 10 SEC. DURING INSTRUMENTAL TRAINING FOR YOKED PAIRS

Group	*N*	Pre-ESB	Base-line	1st 10 min.	2nd 10 min.	3rd 10 min.
			Rewarded for fast			
Experimental	11	67.1	66.8	67.3	65.7	68.1
Yoked	11	65.9	64.6	63.2	63.2	62.5
			Rewarded for slow			
Experimental	9	68.4	67.7	65.6	64.7	63.1
Yoked	9	69.2	68.7	68.9	68.8	70.3

the change in the fast group was reliably different from that in the slow one ($t = 2.43$, $df = 18$, $p < .02$).

Table 2 shows the mean numbers of responses and ESB for the two groups as a function of 10-min. training periods. The increase in the number of correct responses,

TABLE 2

MEAN NUMBER OF CORRECT RESPONSES AND OF ESB REINFORCEMENTS FOR YOKED PAIRS

Measure	5-min. interval					
	1st	2nd	3rd	4th	5th	6th
			11 pairs rewarded for fast			
Responses	122	137	121	120	140	130
Reinforcements	27.9	28.3	26.5	26.0	29.6	27.9
			9 pairs rewarded for slow			
Responses	132	165	150	175	178	184
Reinforcements	27.6	31.9	27.6	30.8	31.5	32.3

Note.—Correct responses were recorded only for experimental *S*s. The count includes time-out periods and hence is larger than the number of reinforcements which were given simultaneously to both members of the pair.

including those that occurred during the time-out portion of the FI schedule, from the first 5 min. to the last 5 min. of training was significant for the slow experimental *S*s (t = 4.86, df = 8, p = .05) and was nonsignificant for the fast experimental *S*s (t = 1.20, df = 9).

The difference in the mean number of reinforcements received by the two experimental groups was significant (t = 2.53, df = 10, $p < .05$). However, the correlation between the total number of reinforcements received and the magnitude and direction of the experimental change was only .02 for the experimental groups and only .06 for the yoked groups, showing that the difference in the number of reinforcements between the fast and slow groups did not account for the differences in heart rate during training.

The curarized rats rewarded for fast heart rates showed a reliable increase in rate, while the rats rewarded for slow ones showed a statistically reliable decrease. These differences could not be explained in terms of differences in initial base rates, differences in the unconditioned effect of introducing ESB used as a reward, or differences in the amount of ESB received.

Yoked control rats failed to show changes similar to those of their experimental partners, and in fact showed relatively large but statistically unreliable differences in the opposite direction. This unexplained but perhaps chance difference between the yoked groups, together with the smallness of the changes in the experimental groups, suggests caution. It is concluded that fairly strong evidence for, but not definitive proof of, the instrumental learning of a visceral response in a curarized animal has been secured.

REFERENCES

CRIDER, A., SHAPIRO, D., & TURSKY, B. Reinforcement of spontaneous electrodermal activity. *J. comp. physiol. Psychol.*, 1966, **61,** 20–27.

FOWLER, R. L., & KIMMEL, H. D. Operant conditioning of the GSR. *J. exp. Psychol.*, 1962, **63,** 563–567.

FROMER, R. Conditioned vasomotor responses in the rabbit. *J. comp. physiol. Psychol.*, 1963, **56,** 1050–1055.

HARWOOD, C. W. Operant heart rate conditioning. *Psychol. Rev.*, 1962, **12,** 279–284.

JOHNSON, R. J. Operant reinforcement of an automatic response. *Dissert. Abstr.*, 1963, **24,** 1255–1256. (Abstract)

KIMMEL, E., & KIMMEL, H. D. A replication of operant conditioning of the GSR. *J. exp. Psychol.*, 1963, **65,** 212–213.

KIMMEL, H. D., & HILL, F. A. Operant conditioning of the GSR. *Psychol. Rep.*, 1960, **7,** 555–562.

MANDLER, G., PREVEN, D. W., & KUHLMAN, C. K. Effects of operant reinforcement on the GSR. *J. exp. Anal. Behav.*, 1962, **5,** 317–321.

MILLER, N. E. Integration of neurophysiological and behavioral research. *Ann. N.Y. Acad. Sci.*, 1961, **92,** 830–839.

MILLER, N. E. Animal experiments on emotionally-induced ulcers. *Proc. World Congr. Psychiat., June 4–10, 1961, Montreal*, 1963, **3,** 213–219.

MILLER, N. E., & CARMONA, A. Modification of a visceral response, salivation in thirsty dogs, by instrumental training with water reward. *J. comp. physiol. Psychol.*, 1967, **63,** 1–6.

SHAPIRO, D., CRIDER, A., & TURSKY, B. Differentiation of an autonomic response through operant reinforcement. *Psychon. Sci.*, 1964, **1,** 147–148.

SHEARN, D. W. Operant conditioning of heart rate. *Science*, 1962, **137,** 530–531.

SKINNER, B. F. *The behavior of organisms: An experimental analysis.* New York: Appleton-Century, 1938.

TROWILL, J. A. *Instrumental conditioning of the heart rate.* (Doctoral dissertation, Yale University) Ann Arbor, Mich.: University Microfilms, 1966. No. 66–439.

(Early publication received June 22, 1966)

9

Instrumental Learning of Heart Rate Changes in Curarized Rats: Shaping, and Specificity to Discriminative Stimulus

Neal E. Miller and Leo V. DiCara

Artificially respirated rats with skeletal muscles completely paralyzed by curare were rewarded by electrical stimulation of the medial forebrain bundle for either increasing or decreasing their heart rates. After achieving the easy criterion of a small change, they were required to meet progressively more difficult criteria for reward. Different groups learned increases or decreases, respectively, of 20%; 21 of 23 rats showed highly reliable changes. The electrocardiogram indicated that decreased rates involved vagal inhibition. Rats learned to respond discriminatively to the stimuli signaling that cardiac changes would be rewarded.

The problem of whether or not visceral responses are subject to instrumental learning (also called operant conditioning or trial-and-error learning) has basic significance for both the theory of psychosomatic symptoms and the theory and neurophysiology of learning.

Previous papers (Miller, 1961, 1963a, 1963b; Miller & Carmona, 1967) have discussed this significance, the traditional view that visceral responses can be modified only by classical conditioning, the results of recent studies challenging this view,[3] and the fact that such studies have not been conclusive since the visceral changes observed could have been mediated via skeletal responses.

The one exception to this criticism is Trowill's (1967) study in this laboratory which used the control of paralyzing skeletal responses by curare and showed that the heart rate of a rat maintained on artificial respiration can be either increased or decreased slightly if rates slightly faster or slower than normal, respectively, are rewarded by electrical stimulation of the brain (ESB). Trowill's controls were rigorous; he used curarized rats to rule out mediation by overt skeletal responses, and rewarded increases in some rats and decreases in others to rule out possible unconditioned effects of ESB (Malmo, 1961). But the changes produced, approximately 5%, were very small and hence were not completely convincing.

The purpose of the present study was to see (*a*) whether larger changes in the heart rates of curarized rats can be achieved by "shaping" the response, i.e., progressively shifting rats to a more difficult criterion after they have learned to meet an easier one, and (*b*) whether a visceral discrimination can be learned so that the response will be more likely to occur in the stimulus situation in which it is rewarded than in the one in which it is not.

Experiment 1

Method

The apparatus and procedure were the same as that used by Trowill (1967), with the following exceptions: (*a*) the time-out period was increased from 8 to 20 sec. to allow more complete recovery from any effects of ESB; (*b*) a light and tone were used as a time-in stimulus for all *S*s; (*c*) more trials with this stimulus were given during the bar-press training used to test for the rewarding effects of ESB; (*d*) when *S*s successfully met each criterion of training, they were shifted to the next more difficult one; and (*e*) a longer period of training was used.

Subjects. The *S*s were 24 adult male albino

[1] This study was supported by National Institute of Mental Health Research Grant MH00647. Research was conducted while the second author was a National Science Foundation Postdoctoral Research Fellow. Preliminary reports were given by DiCara at the April meetings of the Eastern Psychological Association and by Miller (1966) at the National Academy of Sciences.

[2] Both authors now at Rockefeller University, New York City.

[3] Such studies published to date are: Crider, Shapiro, & Tursky, 1966; Fowler & Kimmel, 1962; Johnson, 1963; Kimmel & Hill, 1960; Kimmel & Kimmel, 1963; Shapiro, Crider, & Tursky, 1964; Shearn, 1961, 1962; Shearn & Clifford, 1964.

Sprague-Dawley rats weighing 380–400 gm.; electrodes were implanted bilaterally into *S*'s medial forebrain bundle, and reward effect from stimulation showed in at least one such electrode. Twelve *S*s were rewarded for slow heart rates and the other 12 for fast heart rates; however, since one *S* rewarded for slow heart rate died during training, its results (although showing excellent slowing) were discarded, reducing *N* to 23.

Apparatus. For preliminary testing of reward effect the apparatus was a sheet-metal box approximately 8 in. on a side with a grid floor, a Plexiglas door, and a Grason-Stadler bar in the center of one side 3 in. above floor level. Each press on the bar delivered 0.5 sec. of 60-cycle ac stimulation, regulated by a micropot used as a voltage divider and delivered from the stimulating electrode to an indifferent one on the skull. An 820 K ohm resistor in series with *S* minimized the effects of fluctuations in brain resistance. A 10-w. light near the Plexiglas door and a speaker delivering a 1,000-cycle tone at an intensity of 82 db. were on whenever the bar was connected to deliver rewarding stimulation, except during the initial training when each bar press was rewarded. The bar-pressing apparatus was enclosed in a soundproof ventilated picnic lunch box (Grason-Stadler).

During heart rate conditioning *S* lay on a folded towel in a similar picnic lunch box with similar sources of light and tone and was artificially respirated by means of a mask made from the mouthpiece end of a rubber balloon, the lower lip of which was secured by placing it behind the upper incisors, and the upper portion of which was secured by slightly pinching the skin and the balloon together with a hemostat on the top of *S*'s snout after a prior injection of Xylocaine had been given to prevent pain. The balloon end of the mask was connected via a rubber stopper with a hole drilled through its middle to a small-animal artificial respirator (ENM Instrument Company, Model VE5AG).

The EKG electrodes were stainless steel surgical needles inserted subcutaneously in the middle of the back at shoulder level and on the chest over the sternum after an injection of Xylocaine. A Grass instrument durable disk electrode was placed on the tongue as a ground lead.

A Grass EEG polygraph was used to record the EKG. The J-5 output jacks of the driver amplifier were connected via a 50-μf. condenser and a 20 K ohm variable series resistor to a 10 K ohm sensitive relay. The variable resistor was adjusted so that only the spike component of the QRS complex of each heartbeat operated the relay, which in turn operated the counters and the reinforcement circuit. One counter registered the heart rate for 30 sec., then another counter registered it for the next 30 sec., after which there was a 30-sec. period without counting during which the scores were transcribed and the counters were reset.

The reinforcement circuit involved a predetermining counter driven by a 10-pps pulse generator (Foringer) so that it could be used as an accurate timer. A second predetermining counter registered the heartbeats in a given time and was connected to programming equipment which could be set to deliver reward either when a criterion of more than a specified number of counts was achieved by *S* being rewarded for fast rates or when a criterion of less than a specified number of counts was achieved by *S*s being rewarded for slow rates. The timing counter was set at approximately 1 sec., but in order to change the reinforcement criterion in units smaller than 1 bps, it was necessary to vary the timer, as well as the specified number of beats, in a way indicated by a simple table. Two exactly similar reinforcement circuits operated alternately so that one was reset while the other counted.

Automatic programming equipment disconnected the ESB stimulator as well as the time-in stimuli of tone and light for 20 sec. after each reinforcement was achieved. Separate pens on the polygraph recorded the EKG pattern, the time-in stimulus, and the achievement of criterion during the time-out period, when achieving it activated only the recording pen, as well as during the time-in period when achieving it caused 0.5 sec. of ESB to be delivered and initiated the time-out period. One counter recorded the total number of criterion responses achieved, while another recorded the criterion responses during the time-in period, i.e., the number of rewards delivered.

Procedure for implantation and pretesting. The electrodes were implanted by the procedure described by Miller, Lewis, and Jensen (1961) and illustrated in Figure 7B of that article. A pair of enameled stainless steel surgical needles (Anchor Brand No. 1843) with straight taper points bare for 0.5 mm. at the tip were used as monopolar electrodes, and a stainless steel screw on the skull as an indifferent one. The monopolar electrodes were aimed at the medial forebrain bundle in the lateral hypothalamus at the A-P level of the ventromedial nucleus, with Krieg stereotactic coordinates of 1.5 mm. posterior to bregma, 8.5 mm. below the skull top at point of entry, and 1.7 mm. and 1.3 mm. lateral to the midline on the left and right sides, respectively. The *S*s were allowed 10–14 days to recover from electrode implantation and were kept on ad-lib food and water throughout the experiment.

A current ranging 30–80 μa. was adjusted for the individual *S*, to elicit alertness and sniffing initially and then maximum bar pressing without undue motor side effects or jumping back from the bar. The *S*s were shaped to press the bar to secure 0.5 sec. of 60-cycle ac ESB. After shaping, the current was held constant throughout the rest of the experiment. After shaping on Day 1 each *S* was further trained for two 30-min. sessions, separated by a 30-min. rest with each bar press rewarded. Rats pressing fewer than 500 times during the first of these 30-min. sessions were discarded. On Day 2, *S* received 10 min. of training with each bar press rewarded, and then 10 min. under each of four FI schedules, with the reward

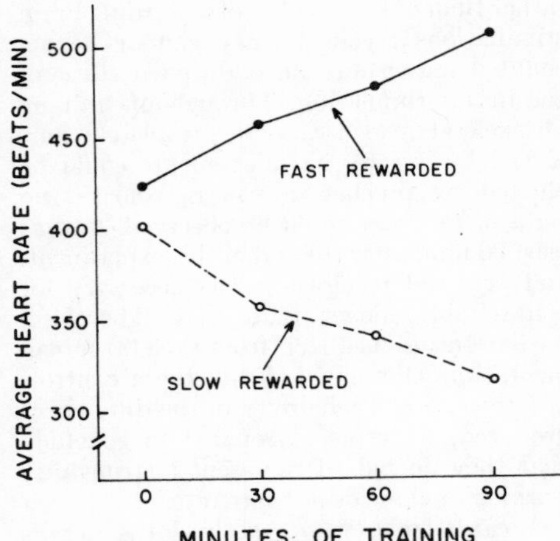

FIG. 1. Instrumental learning by heart in groups rewarded for fast or for slow rates. (Each point represents average of beats per minute during 5 min.)

turned off after each delivery for 5, 10, 20, and 30 sec., respectively, during the successive training conditions. Throughout this training a light and tone, serving as time-in stimuli, were on whenever pressing the bar would deliver the reward, and were off whenever it would not (during the time-out period). The number of bar presses made during the time-in and the time-out periods was recorded on separate counters. Day 3 of training repeated the 50-min. sequence given on Day 2.

Procedure for heart rate training. The following day *S*s were injected ip with 3.6 mg (24 units)/kg of d-tubocurarine chloride, a dose sufficient to completely paralyze skeletal muscles for approximately 3 hr. The *S*s were fitted with the face mask and respirated at 1:1 I/E ratio, 70 breaths per minute and a peak reading of 20 cm. of water pressure. The levels of respirator adjustments had been found to maintain a rat's heart rate within the normal range of approximately 300–500 bpm. Stimulating and EKG electrodes were attached to *S*, which was placed within the experimental chamber.

After a period of 30 min. during which the heart rate was allowed to stabilize and during which no respirator adjustments were made,. a coin was flipped to determine whether *S* would be rewarded for fast or slow heart rate. No adjustments were ever made in the artificial respirator during this 30 min. or thereafter, although minor adjustments in the amplifier control were made to keep the EKG pattern at approximately the same amplitude on the recording paper. The random assignment to fast or slow groups was made after *S*s had been set up for the experiment in order to eliminate any conceivable unconscious experimenter bias. However, since it resulted in two extra *S*s in one group, the last two were arbitrarily assigned to the other one.

After the heart rate had stabilized, a criterion was selected, enough above (or below) the normal heart rate so that it would be achieved on the average approximately once every 5 sec. Approximately 40 min. after the initial curare injection, the experiment proper began; reward circuit and the time-in signal of light and tone were turned on. As soon as *S* achieved criterion, the time-in cues and the reward circuit were automatically turned off for 20 sec., after which they were automatically turned on and the next "trial" began. If *S* improved enough so that it met criterion in approximately half the time (only a slight change in actual rate), it was moved on to one that was approximately 2% faster if it was being shaped for a high rate, or slower if it was being shaped for a low one. If for a period of several minutes it took *S* more than 10 sec. on the average to meet criterion after the time-in signal, *S* was shifted down to the next easier criterion, but the need for such "remedial retraining" was quite rare. The foregoing procedure was continued for a total of 90 min.

Results

The results are shown in Figure 1. It can be seen that *S*s rewarded for increasingly faster rates progressively increased their heart rates from an initial rate of 422 to a final one of 510 bpm, while those rewarded for increasingly slower ones progressively decreased their heart rates from an initial rate of 400 to a final one of 316 bpm. The initial difference between the two groups did not approach significance ($t = 0.6$, $df = 21$), but the difference at the end of training was highly significant ($t = 4.1$, $df = 21$, $p < .001$). The changes were also highly reliable for all but 2 of the 23 individual rats. Comparing for each *S* the number of heartbeats on the first five successive 30-sec. measures with those of the last five ($df = 11$), the increases in the fast group were reliable beyond the .01 level for 4 rats, and beyond the .001 level for 7, while the remaining *S* showed no appreciable change. Similarly, for the slow group the decreases were reliable at beyond the .01 level for 4 rats and beyond the .001 level for 6, while 1 showed no appreciable change.

Throughout this experiment the heart rates were remarkably regular. At the beginning of training the standard deviations of the heart rates of individual *S*s measured over 5 min. averaged 2 bpm (or only 0.5%)

for both groups. By the end of training the standard deviations of 22 out of the 23 rats had increased, but the average was only 0.7% for the slow and 1.5% for the fast groups.

That the marked differences in the overall rate were not artifacts of the number of brain stimulations received is demonstrated in Figure 2, in which the percentage and direction of change in heart rate is plotted against the number of brain stimulations received during the last 5 min. of training. It can be seen that there was a great overlap between the two groups in number of brain stimulations received, but no overlap in change of heart rate. For the fast group, the average number of rewards during the first 5 min. of training was 11.9, and during the last 5 min., 12.5. For the slow group the averages are 11.3 and 10.8, respectively. None of these differences within or between groups was statistically significant.

Discussion

Eleven out of 12 rats rewarded for fast heart rate showed significant increases, while 10 out of 11 rewarded for slow heart rate showed significant decreases. Since these changes were in opposite directions and were a function of the type of response rewarded,

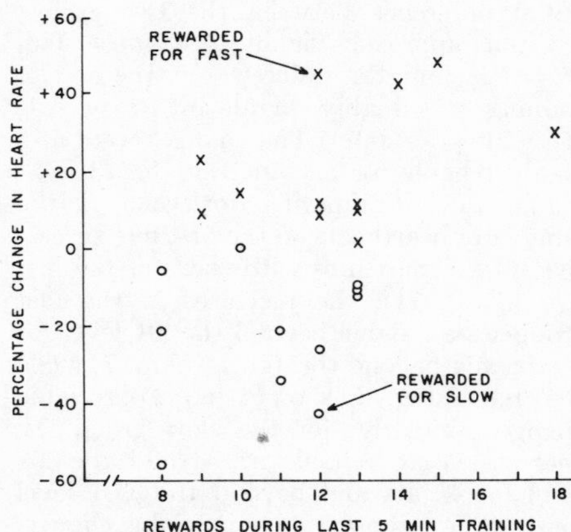

FIG. 2. Change during training as a function of type of response rewarded rather than number of rewards. (The change scores are based on the difference between the first and last 5 min. of training, the reward scores on the last 5 min. of training.)

rather than of the number of rewarding brain stimulations received, they cannot be accounted for by any unconditioned effect of the brain stimulation. Throughout training all skeletal responses were completely paralyzed by curare. No movements could be elicited by pinches or electric shocks; no signs of recovery could be observed until at least 90 min. after the end of this experiment, and artificial respiration was necessary for considerably longer than that. Therefore, mediation via feedback from skeletal movements was eliminated. From these controls and the size and reliability of the differences produced, it seems reasonable to conclude that they do indeed represent instrumental learning of changes in heart rate.

A cardiologist who examined the details of the EKGs of rats rewarded for slow heart rates reported that during the latter part of training the reduction in the amplitude of the P wave and the increased interval between it and the QRS complex were typical of the effects of vagal inhibition.[4]

EXPERIMENT 2

One of the characteristics of the instrumental learning of skeletal responses is that a discrimination can be learned so that the responses are made in the stimulus situations in which they are rewarded and not in those in which they are not. The purpose of Experiment 2 was to see whether or not the instrumental learning of a visceral response, change in heart rate, would show a similar capability for discrimination.

During the period of shaping in the first experiment, most of the rats responded by changing their overall rates in the rewarded direction during both the time-in and time-out conditions; they showed relatively little discrimination between these two conditions. After the results on the first rats which were run had made it clear that relatively large changes in heart rate could be produced by instrumental learning, we decided to give the remainder of the rats 45 min. of additional training immediately after the 90 min. involved in the first experiment. During this training they were held at the same constant

[4] We thank A. V. N. Goodyer of the Yale Medical School for this and other valuable help.

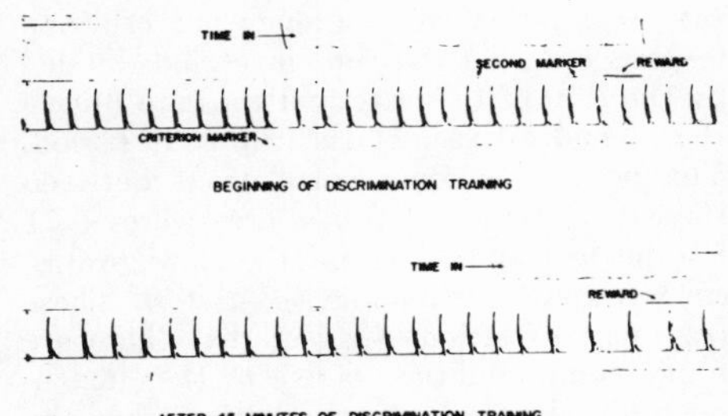

FIG. 3. Electrocardiograms at beginning and at end of discrimination training of *S* rewarded for slow heart rate.

difficult criterion in order to see whether they would learn to respond discriminatively to the time-in stimulus.

Method

The *S*s were 16 rats which had just completed the training in Experiment 1, 8 having been rewarded for fast heart rate and 8 for slow.

They were kept running in the same general apparatus with the same general procedure. Most of them were kept on exactly the same criterion, but the few who were achieving this criterion quite rapidly were moved up to a more difficult one between the end of Experiment 1 and the beginning of Experiment 2, just as they would have been if the shaping had been continued a little longer. All *S*s were run for 45 min., except 2 in the group rewarded for slow rate which died after 22 and 18 min. of discrimination training and hence were discarded from Experiment 2.[5] The criterion was held constant throughout Experiment 2.

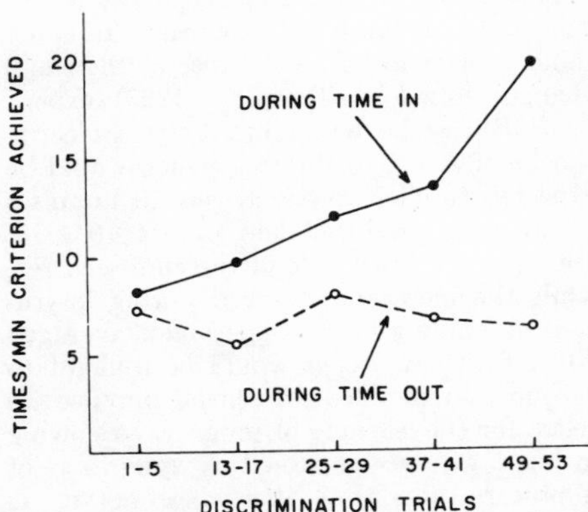

FIG. 4. Learning the discrimination of making heart rate responses to the time-in stimulus, a pattern of tone and light.

Results

The record for one of the rats rewarded for slowing is presented in Figure 3. Before discrimination training, this *S* had already slowed its heart rate from 530 to 230 bpm. From the top record one can see that at the beginning of discrimination training there was no appreciable reduction in rate that was specific to the time-in stimulus, and that it took *S* considerable time after the onset of this stimulus to meet criterion. At the end of discrimination training, the rate during time out remained approximately the same, but when the time-in stimulus came on, the heart slowed down so that criterion was promptly achieved.

Figure 4 shows a similar result for the group averages. The occurrence of criterion responses to the last 10 sec. of the time-out stimulus (first 10 sec. discarded to eliminate aftereffects of ESB) remained approximately the same throughout discrimination training. At the beginning of discrimination training the rate of occurrence of the criterion response to the time-in stimulus (calculated from the latency) was approximately the same as to the time out; but as discrimination training continued, the criterion response occurred progressively faster to the time-in stimulus. The results of the groups rewarded for fast and for slow heart rates were plotted separately, but since the trends toward discrimination were similar, their results were combined in Figure 4. The differences of the slopes of the two curves in Figure 4 and also the differences between the last points in the two curves were both highly significant ($p < .001$). Thus it is clear that the rats learned the discrimination of changing their heart rate enough to meet criterion more frequently during the time-in stimulus, when such a response would be rewarded, than during the time-out stimulus, when it would not.

It is interesting to note that the only rats

[5] If these rats had also been discarded from Experiment 1, they would not have changed materially the results since their percentage decreases during that experiment were 11.5 and 25.8, respectively, as compared to the mean decrease of 21.1% for the entire group. Their reliabilities in Experiment 1 were .01 and .001, respectively.

to die during the experiments were three in the group rewarded for slow heart rates. Since the probability of such an occurrence is between 8 and 9 times in 100, it should not be taken too seriously; but the fact that autopsies showed no tissue pathology except nonspecific acute passive congestion which is often found with terminal cardiac failure[6] raises the interesting possibility that their hearts may have been slowed up enough, perhaps with the aid of the effects of curare, to initiate cardiac failure.

Comparison between discriminations during bar pressing and during heart rate training. Walker (1942) trained hungry rats on the discrimination of running faster on runway trials with a tone signaling food reward than on ones without the tone and without reward. Walker then found that this discrimination transferred from the instrumental running response to a different instrumental response of pressing a bar for food. Trapold and Odom (1965) have secured similar transfer of a discrimination from pressing a horizontal bar to pressing a vertical bar.

We analyzed our data to see if there was any analogous relationship between performance of (*a*) the discrimination of performing the skeletal response of bar pressing to the time-in tone and light during the initial training, and (*b*) the discriminatory performance of the visceral response of changed heart rate to these same stimuli at the end of Experiment 2. In order to get a measure of the degree to which the *S*s in this experiment had learned to respond discriminatively to the time-in stimulus during the preliminary bar-pressing training, each *S* was given a score for the last 10 min. of training on Day 2, when *S* was on the most difficult schedule at the end of its bar-pressing training. This score was the number of rewarded bar presses made during the time-in period, divided by the sum of this number and the number of bar presses made during the time-out period. As a measure of its discrimination during the final five trials of heart rate conditioning, each *S* was given a similar score, namely the rate (computed from latency) of criterion responses during the time-in period divided by the sum of this rate and the rate during the second 10 sec. of the time-out period. The product-moment correlations between these two discrimination scores were +.71 for the fast and +.87 for the slow groups, each statistically reliable ($p < .05$). These positive correlations suggest that there are some communalities between the instrumentally learned discriminations of the skeletal and the visceral responses, but we cannot say whether these are abilities which favor the learning of both or a transfer of training from one to the other.

[6] We thank Stuart Wyand from the Yale Division of Animal Care for doing the pathology on these animals.

Discussion

The relatively large sizes and the high statistical reliabilities of the changes secured in 21 of the 23 rats trained show clearly that heart rate can be either increased or decreased by rewarding the desired change. The fact that these changes were secured in rats with complete paralysis of the skeletal musculature rules out any possibility that this visceral response was mediated via instrumentally learned overt skeletal responses.

It might be considered to be barely conceivable that *S*s learned to send out from the motor cortex central impulses for skeletal responses such as struggling, and that these impulses elicited innate or classically conditioned changes in heart rate. Such an effect could conceivably be the cause of the small changes found in Trowill's (1967) experiment. But we have found that strong electric shocks of 3 ma. to the tail, which would be expected to elicit maximal central impulses to struggle, produced less than a 10% increase in the heart rate of the curarized rat, while the increases observed during instrumental training in our experiment averaged 20%. Furthermore, it would be difficult to account, on the basis of central impulses to relax, for the learning of progressive slowing in heart rate accompanied by symptoms of inhibitory action of the vagal nerve, as shown by the reduced amplitude of the P wave and the increased P-R interval, since such vagal stimulation is unlikely to be produced by mere relaxation.

The fact that the same rewarding ESB

can be used to reinforce either increased or decreased heart rate, plus evidence that there is considerable overlapping between the number of such rewards delivered to the groups learning these opposite changes, rules out either any immediate or any cumulative unconditioned effects of the ESB as the reward. Nevertheless, it will be desirable to show that instrumental learning of visceral responses can be reinforced by other types of rewards, such as escape from (or avoidance of) painful electric shocks.

The fact that additional training can teach the different groups of curarized rats the discrimination of making the instrumentally rewarded increases or decreases, respectively, more specifically to the stimulus condition in which such changes are rewarded is further evidence against any cumulative effects of ESB used as a reward, and also shows that this type of visceral instrumental learning displays one of the important characteristics of the instrumental learning of skeletal responses. It is interesting to note that there is a positive correlation between learning to respond discriminatively to the time-in stimulus with the skeletal response of bar pressing during the preliminary training and the visceral response of heart rate during the last part of Experiment 2.

In a previous study demonstrating that salivary secretion could be modified by instrumental training procedures, Miller and Carmona (1967) raised the possibility that the learned increases or decreases in secretion could be part of a general pattern of activation or inhibition. Thus another problem is to determine the degree to which the changes induced in a variety of visceral responses are specific to the response rewarded, and, if a general effect is observed, the degree to which it can be counteracted by differentiation training in which an increasingly stringent criterion for specificity is progressively applied as a condition for reward.

As has been pointed out in previous articles (Miller, 1961, 1963a, 1963b), the demonstration that visceral responses can be modified by instrumental learning provides a much more flexible basis for their modification than would be the case if such responses were limited to modification by classical conditioning. This is because classical conditioning can be reinforced only by stimuli that elicit a UR similar to the specific response to be learned, while any instrumental response can be reinforced by any reward. The instrumental learning of visceral responses provides a theoretical possibility for explaining some of the individual differences in patterns of visceral response (Lacey & Lacey, 1958, 1962), as well as for the acquisition of psychosomatic symptoms. The degree to which such learning actually does practically contribute to such differences and symptoms depends, of course, on the degree to which suitable conditions for rewarding such learning exist as well as upon the possibilities for pushing such changes far enough to create organic damage.

Finally, the demonstration that visceral responses are subject to instrumental learning removes one of the main arguments in favor of assuming two functionally distinct types of learning mediated by different neuroanatomical systems, the somatic and the autonomic, as maintained by many authors (e.g., Kimble, 1961; Konorski & Miller, 1937; Mowrer, 1947, 1950; Skinner, 1938; Solomon & Wynne, 1954), although it certainly does not by itself prove the unitary nature of the learning process.

REFERENCES

CRIDER, A., SHAPIRO, D., & TURSKY, B. Reinforcement of spontaneous electrodermal activity. *J. comp. physiol. Psychol.*, 1966, **61**, 20–27.

FOWLER, R. L., & KIMMEL, H. D. Operant conditioning of the GSR. *J. exp. Psychol.*, 1962, **63**, 563–567.

JOHNSON, R. L. Operant reinforcement of an autonomic response. *Dissert. Abstr.*, 1963, **24**, 1255–1256. (Abstract)

KIMBLE, G. A. *Hilgard and Marquis' conditioning and learning.* New York: Appleton-Century-Crofts, 1961.

KIMMEL, E., & KIMMEL, H. D. A replication of operant conditioning of the GSR. *J. exp. Psychol.*, 1963, **65**, 212–213.

KIMMEL, H. D., & HILL, F. A. Operant conditioning of the GSR. *Psychol. Rep.*, 1960, **7**, 555–562.

KONORSKI, J., & MILLER, S. Further remarks on two types of conditioned reflex. *J. gen. Psychol.*, 1937, **17**, 405–407.

LACEY, J. I., & LACEY, B. C. Verification and extension of the principle of autonomic response stereotypy. *Amer. J. Psychol.*, 1958, **71**, 50–73.

LACEY, J. I., & LACEY, B. C. The law of initial value in the longitudinal study of autonomic constitution: Reproducibility of autonomic responses and response patterns over a four-year interval. *Ann. N.Y. Acad. Sci.,* 1962, **98,** 1257–1290.

MALMO, R. B. Slowing of heart rate after septal self-stimulation in rats. *Science,* 1961, **133,** 1128.

MILLER, N. E. Integration of neurophysiological and behavioral research. *Ann. N. Y. Acad. Sci.,* 1961, **92,** 830–839.

MILLER, N. E. Animal experiments on emotionally-induced ulcers. *Proc. World Congr. Psychiat., June 4–10, 1961, Montreal,* 1963, **3,** 213–219. (a)

MILLER, N. E. Some reflections on the law of effect produce a new alternative to drive reduction. In M. R. Jones (Ed.), *Nebraska symposium on motivation: 1963.* Lincoln: University of Nebraska Press, 1963. Pp. 65–112. (b)

MILLER, N. E. Extending the domain of learning. *Science,* 1966, **152,** 676. (Abstract)

MILLER, N. E., & CARMONA, A. Modification of a visceral response, salivation in thirsty dogs, by instrumental training with water reward. *J. comp. physiol. Psychol.,* 1967, **63,** 1–6.

MILLER, N. E., LEWIS, M., & JENSEN, D. D. A simple technique for use with the rat. In D. E. Sheer (Ed.), *Electrical stimulation of the brain.* Austin: University of Texas Press, 1961. Pp. 51–54.

MOWRER, O. H. On the dual nature of learning—a reinterpretation of "conditioning" and "problem solving". *Harv. educ. Rev.,* 1947, **17,** 102–148.

MOWRER, O. H. *Learning theory and personality dynamics.* New York: Ronald, 1950.

SHAPIRO, D., CRIDER, A. B., & TURSKY, B. Differentiation of an autonomic response through operant reinforcement. *Psychon. Sci.,* 1964, **1,** 147–148.

SHEARN, D. W. Does the heart learn? *Psychol. Bull.,* 1961, **58,** 452–458.

SHEARN, D. W. Operant conditioning of heart rate. *Science,* 1962, **137,** 530–531.

SHEARN, D. W., & CLIFFORD, C. D. Cardiac adaptation and contingent stimulation. *Amer. Psychologist,* 1964, **19,** 491. (Abstract)

SKINNER, B. F. *The behavior of organisms: An experimental analysis.* New York: Appleton-Century, 1938.

SOLOMON, R. L., & WYNNE, L. C. Traumatic avoidance learning: The principles of anxiety conservation and partial irreversibility. *Psychol. Rev.,* 1954, **61,** 353–385.

TRAPOLD, M. A., & ODOM, P. B. Transfer of a discrimination and a discrimination reversal between two manipulandum defined responses. *Psychol. Rep.,* 1965, **16,** 1213–1221.

TROWILL, J. A. Instrumental conditioning of the heart rate in the curarized rat. *J. comp. physiol. Psychol.,* 1967, **63,** 7–11.

WALKER, K. C. The effect of a discriminative stimulus transferred to a previously unassociated response. *J. exp. Psychol.,* 1942, **31,** 312–321.

(Early publication received June 20, 1966)

10

Changes in Heart Rate Instrumentally Learned by Curarized Rats as Avoidance Responses

Leo V. DiCara and Neal E. Miller

2 groups of curarized rats learned to increase or decrease, respectively, their heart rates in order to escape or avoid mild electric shocks. Responses in the appropriate direction were greater during the stimulus preceding shock than during control intervals between shock; they changed in the opposite direction, toward the initial pretraining level, during the different stimulus preceding nonshock. Electromyograms indicated complete paralysis of the gastrocnemius muscle throughout training and for a period of at least 1 hr. thereafter.

Previous papers in the present series presented evidence that the instrumental learning of cardiac and intestinal responses can occur in curarized rats (Miller & Banuazizi, 1968; Miller & DiCara, 1967; Trowill, 1967) when reward is provided by electrical stimulation of the brain. The main purpose of the present study was to determine whether something unique about direct electrical stimulation of the brain makes it the only kind of reward that can reinforce visceral learning; to this end, the possibility of reinforcing changes in heart rate by escape from or avoidance of mild electric shock was investigated. A subsidiary purpose was to determine whether a discrimination can be learned so that a greater change in heart rate occurs to a stimulus always preceding the shock than to one never preceding it. A second subsidiary purpose was to secure electromyographic records showing that skeletal responses were indeed completely paralyzed by curare during the experiment. Although there is every reason to believe that the rats in the previous experiments were completely paralyzed by curare, additional evidence is desirable in the light of Black's (1966) demonstration that, in dogs partially paralyzed by marginal doses of curare, the learning of skeletal responses, which show up in the EMG but not as gross overt movements, can influence heart rate.

[1] This study was supported by Research Grant MH 13189 from the National Institute of Mental Health.

Method

Subjects

Twelve naive, adult male, albino Sprague-Dawley rats weighed 390–470 gm.

Apparatus

The apparatus was similar to that described by Miller and DiCara (1967). During heart-rate conditioning, *S* lay on a folded towel in a soundproof, ventilated enclosure (Grason-Stadler) and was maintained at a constant rate of artificial respiration by means of a mask, made from the mouthpiece of a rubber balloon, placed over *S*'s snout.

The experimental chamber was equipped with a 60-w. light bulb and an Ameco oscillator capable of delivering a 1,000-cps tone at an intensity of 82 db. The shock source consisted of a voltage divider on the secondary of a 500-v. ac transformer, with the current delivered to *S* through a fixed resistance of 220 K ohms. Discontinuous electric shock consisted of a 0.1-sec. pulse of 0.3 ma. ac delivered at 2-sec. intervals through an electrode fitted snugly around the base of *S*'s tail (Weiss, 1967).

The *S*'s electrocardiogram (EKG) and electromyogram (EMG) were recorded on a Grass Model 5 polygraph; EMG recordings were taken at a sensitivity of 150 μv/cm; EKG electrodes consisting of stainless steel wire were inserted 1 wk. prior to the experiment, under ether anesthesia, subcutaneously in the middle of the back at shoulder level and on the chest over the sternum. At the same time, insulated unipolar electrodes for recording the EMG were inserted approximately 5 mm. apart on the lateral surface of the right gastrocnemius. A large flat electrode placed on the shaved medial surface of the thigh was used as a ground.

The reinforcement circuit for automatically terminating the shock signal or the electric shock when *S*'s heart rate reached a predetermined criterion was the same as that described by Miller and DiCara (1967).

Procedure

Rats were injected intraperitoneally with 1.2 mg. (8 units) per kilogram of *d*-tubocurarine chloride (Squibb) in a solution containing 3 mg/cc. As soon as *S* showed signs of difficulty in breathing, it was fitted with the face mask and respirated at 1:1 I/E ratio, 70 respiratory cycles per minute, with a peak reading of 20 cm. of water pressure. The EKG, EMG, and tail electrodes were attached to *S*, as was a 30-g. needle inserted subcutaneously on the lateral surface of the abdomen after that area had been locally anesthetized by .15 cc of a 2% solution of xylocaine hydrochloride. The needle was connected by polyethylene tubing to a Harvard infusion pump and *d*-tubocurarine chloride was infused at the rate of 1.2 mg/kg/hr for the duration of the experiment, a dosage sufficient to maintain complete paralysis of all skeletal muscles.

After a period of 30 min. during which the heart rate was allowed to stabilize, *S*s were randomly assigned to fast or slow heart-rate reward groups. Within each of these groups, *S*s were again randomly assigned to light-shock or to tone-shock groups. In the first group the onset of a flashing light (5 sec.) signaled a shock trial, and the onset of a steady tone (1,000-cps, 82-db.) signaled a nonshock trial. In the second group the functions of these cues were reversed.

On shock trials, if the criterion heart rate was achieved within 5 sec. of onset of the shock signal, the signal was turned off and no shock was delivered. If *S* failed to meet the criterion rate within the 5 sec., the shock was turned on and *S* received a 0.1-sec. pulse of 0.3 ma. once every 2 sec. until the criterion was achieved, at which time both the shock signal and the shock were turned off. Every tenth shock trial throughout training was a special test trial, during the first 5 sec. of which both the criterion and shock circuits were turned off so that heart rate could be measured for the same constant intervals throughout training without any possible contaminating effects from having the shock signal turned off or the shock turned on. At the end of this 5-sec. interval, the scoring circuit was turned off and the criterion circuit was turned on. If *S* did not achieve the criterion within the next 5 sec., the shock was turned on. In short, test trials involved an unsignaled 5-sec. delay between the onset of the shock signal and the initiation of a normal training trial, a procedure which was designed to yield less interference with learning than the use of nonreinforced test trials.

On the nonshock trials, the nonshock signal remained on for 5 sec., irrespective of *S*'s heart rate, and was never followed by shock. During the intervals between trials, *S* was in silence and darkness. To compare the effects of training on heart rate when no signals were present, all *S*s received

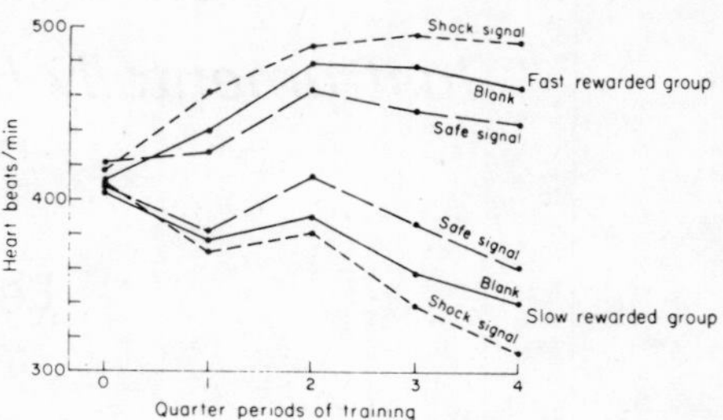

FIG. 1. Heart-rate changes during avoidance training.

so-called "blank" trials, each lasting 5 sec., during which heart rate was recorded but nothing else happened. The shock trials, the nonshock trials, and the blank trials were given in a counterbalanced sequence in which each combination of 2 trials, including 2 of the same kind, occurred an equal number of times in each cycle of 18 trials.

Initial criteria were selected enough above or below the average heart rate so that they would be achieved approximately once every 5 sec. During training each block of 10 shock trials was analyzed and the criterion made more difficult (2% faster or slower depending on group) if *S* changed its heart rate enough so that it was meeting the criterion in less than an average of 3 sec.; similarly, the criterion was made easier (by 2%) if *S* was taking more than approximately 10 sec. on the average to meet the criterion on the shock trials.

Avoidance training consisted of 300 trials divided equally among shock, safe, and blank trials presented on a VI 30-sec. schedule (range, 15–60 sec.). After 300 trials all circuits were disconnected and the rats were maintained on artificial respiration in the dark, silent experimental chamber until recovery, usually a minimum of 120 min. The EMG responses were recorded throughout this period.

RESULTS

Figure 1 shows mean heart rates for fast and slow groups computed from individual test shock, nonshock, and blank trials obtained at the beginning of and at quarterly intervals throughout the 300 training trials. It is clear that for each of the three types of trials, heart rate increased during training for the fast group and decreased for the slow one. All of the changes were highly significant ($p < .001$). Furthermore, each individual *S* changed in the predicted direction; *t* tests for the reliabilities of the changes of individual

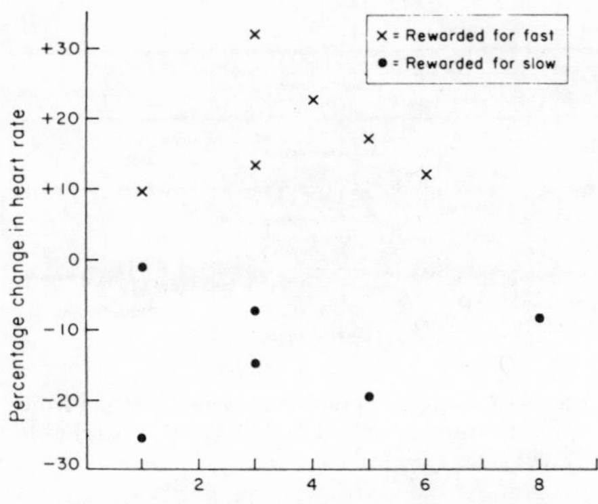

FIG. 2. Heart-rate changes as a function of type of response rewarded and number of shocks received.

rats yielded only one nonsignificant difference (for 3, $p < .01$; for 8, $p < .001$).

Trend analyses of the data (Winer, 1962, p. 298) in Figure 1 indicated that the curves for the shock, nonshock, and blank trials within each group diverged significantly from one another (fast group, $F = 8.1$, $p < .001$; slow group, $F = 6.4$, $p < .001$). When the results on the blank and signaled trials were compared, it was clear that the heart rate changed in the rewarded direction during the shock signal (shock vs. blank trials: fast group, $t = 4.9$, $p < .01$; slow group, $t = 5.1$, $p < .01$). Conversely, during the nonshock signal, the heart rate changed in the opposite direction, i.e., back toward the level before training (nonshock vs. blank trials: fast group, $t = 4.1$, $p < .01$; slow group, $t = 5.3$, $p < .01$).

The number and total duration of electric shocks received by *S*s were analyzed to determine whether the difference in the direction of the heart-rate changes could be accounted for by differences in the number or the amount of shock received by *S*s in the fast and slow groups. In Figure 2, the percentage of change in heart rate is plotted against the number of shocks received during the last 10 shock trials. There was a great overlap between the two groups in number of shocks received, but no overlap in change of heart rate achieved. For the fast group, the average number of shocks received during the first 10 shock trials of training was 4.00 and during the last 10 shock trials, 3.67. For the slow group, the averages are 4.17 and 3.83, respectively. None of these differences was significant either within or between groups.

Figure 3 presents similar data on the effects of the amount (i.e., total number of seconds) of shock received by each *S* during the last 10 shock trials. Here also there was great overlap between fast and slow groups in amount of shock received, but no overlap in change of heart rate. However, there was a significant decrease in the amount of shock received over the course of training. Over the first 10 shock trials, the fast group received a total of 216 sec. of shock, which dropped to 107 sec. over the last 10 shock trials ($t = 9.7$, $p < .001$). For the slow group, totals were 190 and 94, respectively ($t = 6.9$, $p < .001$). These data indicate that *S*s learned to escape more rapidly from shock over the course of training, despite the fact that the shaping procedure used in training involved making the criteria for avoidance progressively more difficult so that the percentage of trials on which *S*s were shocked remained relatively constant during the course of training.

Analyses of the muscle action potential (EMG) records of each *S*, at a high sensitivity of 150 μv/cm, showed that

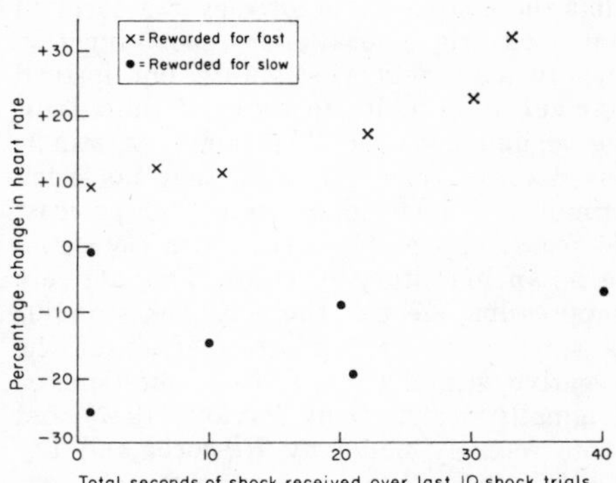

FIG. 3. Heart-rate changes as a function of type of response rewarded and amount of shock received.

within 15 min. after the initial curare injection all skeletal muscle activity ceased; the record was flat except for occasional, very small denervation potentials, the form of which could easily be distinguished from true muscle action potentials. The first minute signs of true muscle action potentials appeared at least 1 hr. after the end of training.

Figure 4 presents typical EMG records of an *S* that increased its heart rate from 376 to 440 bpm ($p < .001$) during training. Individual spikes did not appear until 90 min. after the end of training, and full spindle bursts were not noted until after 120 min.[2]

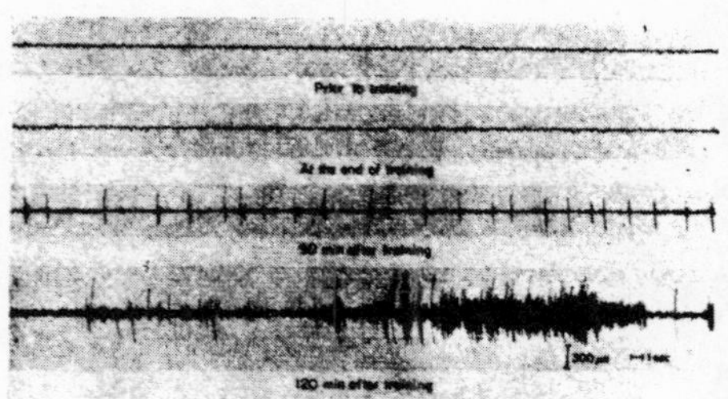

Fig. 4. Electromyographic records obtained from a representative curarized *S* that learned a reliable ($p < .001$) increase in heart rate.

Discussion

The results of the present experiment show that escape from or avoidance of mild electric shock can be used to reinforce the instrumental learning of either increases or decreases in heart rate by rats that are completely paralyzed with curare. Thus, the instrumental learning of a visceral response is not limited to any unique property of the direct electrical stimulation of the brain as a reward.

In addition, the results show that the learned visceral responses are at least partially under the control of discriminative stimuli, with the largest changes being elicited in response to the stimulus used as a signal for shock. It is interesting to note that the control stimulus, never associated with electric shock, does not function merely as a neutral stimulus, but instead has effects opposite to those of the stimulus signaling shock. The nonshock stimulus does, of course, predict that no shock stimulus will be administered for at least 15 sec. It apparently acts as a safety signal or as an inhibitory stimulus. This actively suppressing effect of the nonshock stimulus is similar to the inhibitory effect of the negative stimulus in a discrimination, as originally described by Pavlov (1927) and more recently noted by Rescorla and LoLordo (1965).

Black (1966) has shown that partially curarized dogs, unable to make gross muscular movements, can be trained to make skeletal responses that can be recorded as action potentials of the muscle (EMG), and that such training produces an increase in the heart rate. His results show the need to demonstrate that skeletal responses are completely ruled out before concluding that a visceral response has been instrumentally learned. In our experiments, however, we have used large doses of curare and the different procedure of starting out by rewarding the visceral response rather than the EMG. In the present experiment we have shown that changes in heart rate can be achieved by instrumental training techniques when the animals are so deeply curarized that muscle action potentials are completely abolished. Although EMG records were not taken in our previous experiments, the rats were curarized at a comparable depth, as indicated by the fact that it took them a similar period (approximately 2 hr.) to recover enough so that they could be removed from artificial respiration. Therefore, it seems highly probable that muscle action potentials were also absent during the training in our earlier experiments.

While our results cannot be explained on the basis of failure to paralyze skeletal muscles by curare, the possibility mentioned in a previous paper (Miller & DiCara, 1967) remains: namely, that impulses from the motor cortex have a conditioned or unconditioned effect on the heart, even though the skeletal muscles

[2] We thank J. Goodgold of the New York University Institute of Rehabilitative Medicine for advice regarding electromyography.

that they normally would activate are completely paralyzed. Although this possibility cannot be ruled out entirely until we have recorded from the nerves to such muscles, it is difficult for it to account for the following results. Miller and DiCara (1967) showed that the slowing of heart rate produced by rewarded training was accompanied by changes in the pattern of the EKG of the kind that are observed during vagal inhibition of the heart; and that the increase in heart rate produced by reward was greater than that produced by a strong electric shock which would be expected to induce maximum activity of the skeletal muscles. Miller and Banuazizi (1968) showed that instrumental training procedures can produce effects which are specific to the visceral response, either heart rate or intestinal contraction, that is rewarded. To explain this last result as a mediated response, one would have to assume that impulses sent from the motor cortex toward certain skeletal muscles affect intestinal contractions but not heart rate, while impulses to other skeletal muscles affect heart rate but not intestinal contractions.

REFERENCES

Black, A. H. The operant conditioning of heart rate in curarized dogs: Some problems of interpretation. Paper presented at the meeting of the Psychonomic Society, St. Louis, October 1966.

Miller, N. E., & Banuazizi, A. Instrumental learning by curarized rats of a specific visceral response, intestinal or cardiac. *J. comp. physiol. Psychol.*, 1968, **65,** 1–7.

Miller, N. E., & DiCara, L. Instrumental learning of heart-rate changes in curarized rats: Shaping, and specificity to discriminative stimulus. *J. comp. physiol. Psychol.*, 1967, **63,** 12–19.

Pavlov, I. P. *Conditioned reflexes.* (Trans. by G. V. Anrep) London: Oxford University Press, 1927.

Rescorla, R. A., & LoLordo, V. M. Inhibition of avoidance behavior. *J. comp. physiol. Psychol.*, 1965, **59,** 406–413.

Trowill, J. A. Instrumental conditioning of the heart rate in the curarized rat. *J. comp. physiol. Psychol.*, 1967, **63,** 7–11.

Weiss, J. A tail electrode for unrestrained rats. *J. exp. Anal. Behav.*, 1967, **10,** 85–86.

Winer, B. J. *Statistical principles in experimental design.* New York: McGraw-Hill, 1962.

(Received April 6, 1967)

11

Long Term Retention of Instrumentally Learned Heart-Rate Changes in the Curarized Rat

Neal E. Miller and Leo V. DiCara

Artificially respirated rats with skeletal muscles paralyzed by curare were rewarded by escape from and/or avoidance of mild electric shock for either increasing or decreasing their heart rates. Separate groups of 6 rats each learned significant increases and decreases, respectively, of 12 percent and 23 percent. Retention tests indicate that instrumentally learned changes in heart rate can be retained for a 3-month period without intervening practice.

Introduction

The traditional view that autonomically mediated behavior can be modified by classical, but not instrumental training methods has been recently challenged by experiments from this laboratory. In each of these experiments, increases or decreases, respectively, have been produced by rewarding them, in the following autonomically mediated responses: heart rate (Miller and DiCara, 1967, DiCara and Miller, 1968a, Trowill, 1967), blood pressure (DiCara and Miller, 1968c), urine formation (Miller and DiCara, 1968), peripheral vasomotor responses (DiCara and Miller, 1968d, e), and gastrointestinal motility (Miller and Banuazizi, 1968).

Previous experiments have shown that rats can learn a visceral dis-

*This work was supported by USPHS Research Grant MH 13189 from the National Institute of Mental Health.

crimination, so that the response will be more likely to occur in the stimulus situation, in which it is rewarded, than in the one in which it is not. These findings indicate that instrumentally learned visceral responses display one of the important characteristics of the instrumental learning of skeletal responses, namely, that of discrimination learning. The purpose of the present experiment was to determine if the instrumental learning of heart-rate changes rewarded by escape from or avoidance of electric shock displays another important characteristic of the instrumental learning of skeletal responses, namely, long-term retention. Original learning of heart rate and the test for retention were each carried out with the rat under curare. A 3-month period intervened between these tests, during which the rats were housed undisturbed in home cages.

Method

SUBJECTS

The subjects were 12 naive, adult male, albino Sprague-Dawley rats weighing from 371 to 498 gm. Six subjects were rewarded by avoidance and/or escape from electric shock for slow heart rates and the other 6 for fast heart rates.

APPARATUS

During initial heart-rate training, test for retention, and retraining, the subject lay on a folded towel in a soundproof, ventilated enclosure and was maintained at a constant rate of artificial respiration. The experimental chamber was equipped with a 60 watt light bulb and a loudspeaker capable of delivering a 1,000 cps tone at an intensity of 82 db. Discontinuous electric shock consisting of a 0.1 sec. pulse of 0.3 ma, 60 cycle AC was delivered at 2-sec. intervals through an electrode fitted around the base of the subject's tail. Details of the apparatus and the reinforcement circuit for automatically terminating the shock signal and/or the electric shock when the subject's heart rate reached a predetermined criterion can be found in an earlier report by DiCara and Miller (1968a). A Grass polygraph, Model 5, was used to record the subject's electrocardiogram.

PROCEDURE

The procedure used to establish the learned heart-rate changes was the same as that used by DiCara and Miller (1968a), with the following exceptions: (a) electromyograms were not taken, (b) rats were initially injected with 3.0 mg. per kg. d-tubocurarine chloride and were infused at the rate of 1.0 mg. per kg. per hr. throughout the experiment, and (c) instead of using different stimuli to signal shock and nonshock trials, all trials in this experiment, except blank ones, were shock trials signalled by the onset of a light and tone.

Following curarization, subjects were allowed a period of 30 min. during which the heart rate stabilized. Subjects were then randomly assigned to fast or slow heart-rate reward groups. Avoidance training consisted of 300 trials presented on a VI 30-sec. schedule (range: 15-50 sec.). Each trial was signalled by the onset of the light and tone. If the criterion heart rate was achieved within 5 sec. of onset of the shock signal, it was turned off and no shock was delivered. If the subject failed to meet the criterion rate within the 5 sec., the shock was turned on until the criterion was achieved, at which time both the shock signal and the shock were turned off. The fifth and every tenth trial thereafter throughout training was a blank trial, lasting 5 sec., during which heart rate was recorded but nothing else happened.

Initial criteria were selected enough above or below the average heart rate so that they would be achieved approximately once every 5 sec. During training each block of 10 shock trials was analyzed and the criterion made more difficult if the rat changed his heart rate enough so that it was meeting the criterion in less than an average of 3 sec.; similarly, the criterion was made easier if the subject was taking more than approximately 10 sec. on the average to meet the criterion.

Following an identical procedure for curarization and artificial respiration, each subject was allowed an undisturbed period of 30 min. in the experimental chamber before the retention test began. Retention testing consisted of the presentation of 32 trials under the same conditions as in training under curare, except that the light and tone remained on for 5.0 sec. irrespective of the rat's heart rate, and no electric shocks were given. Four blank trials were given at equal intervals during the retention test.

Following the retention test, subjects received an additional 300 trials of avoidance training. The training procedures were identical to those used in the first training session.

Results

Heart rate was computed from individual blank trials at the beginning of and at quarterly intervals throughout avoidance training and retention testing.

HEART-RATE TRAINING

The results are presented in Figure 1a. As can be seen, subjects rewarded for fast heart rate increased their heart rates from an average rate of 405 to 461 ($t = 9.80$, $df = 5$, $p = .001$) while those rewarded for slow heart rate decreased their heart rates from 418 to 347 ($t = 13.31$, $df = 5$, $p = .001$). The difference in the amount of shock received by rats rewarded for increases and those rewarded for decreases in heart rate did not approach significance ($t = .61$), indicating that the differences in the heart rates of the rats rewarded for increases and those rewarded for decreases in heart rate were not artifacts of the amount of shock received by them during heart-rate training. These results confirm those of our previous experiment (DiCara and Miller, 1968a), by showing that heart rate can be either increased or decreased, respectively, by avoidance training.

RETENTION TEST

The results are presented in Figure 1b. The 30-min. period between curarization and the retention test, during which the subject was in darkness and silence, was designed to allow the subjects' heart rate to stabilize; however, the similarities between this condition and the time-out periods of darkness

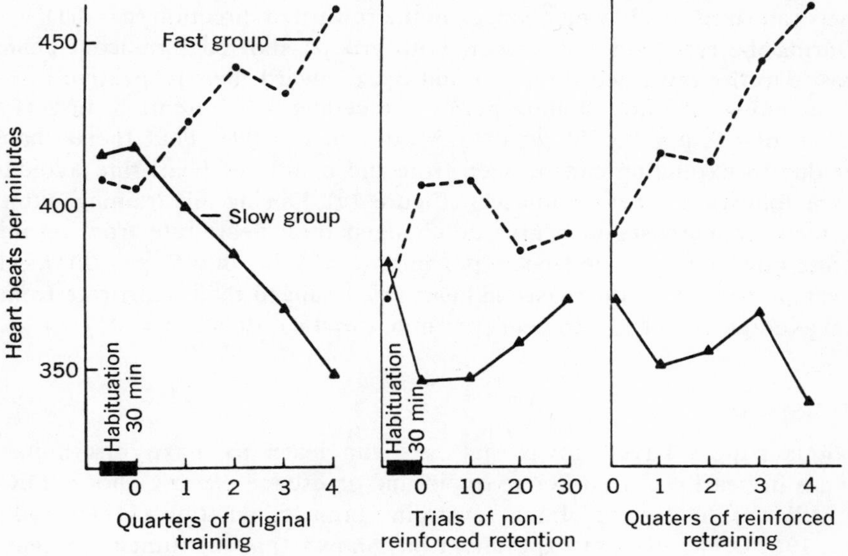

Figure 1. Changes in heart rate during initial heart-rate training under curare and test of retention and retraining 3 months later, also under curare

and silence during the preceding session of heart-rate training made it function as a retention test. While there were no significant differences between fast and slow groups at the start of the 30-min. period preceding the retention test ($t = .92$, $df = 10$, $p = .40$), there was a significant increase in heart rate over this 30-min. period in the group originally rewarded for fast heart rate, from an initial rate of 372 to 407 beats per min. ($t = 3.64$, $df = 5$, $p = .02$), and a significant decrease in heart rate in the group originally rewarded for slow heart rates, from an initial rate of 391 to 347 beats per min. ($t = 5.46$, $df = 5$, $p = .001$).

Comparing the heart-rate changes occurring during the 30-min. period preceding the retention test with the heart-rate changes which occurred in the 30-min. period preceding heart-rate training 3 months earlier, it can be seen that during the 30-min. period prior to the start of the first training session the changes in heart rate of neither group approached significance (Fast: t = .18, Slow: t = .43). Furthermore, as indicated by a Fisher exact probability test, there was no tendency whatsoever for the directional change in heart rate which occurred during the 30-min. session preceding initial heart-rate training to be associated with changes in heart rate during the 30-min. session preceding the retention test. However, there was a highly significant association obtained between the direction of change in heart rate obtained during the 30-min. period preceding the test for retention and original heart-rate learning, namely, all subjects showed changes in the rewarded direction (p = .01).

During the retention test proper, both groups showed significant changes opposite to the rewarded direction and back toward their respective rates at the beginning of the 30-min. period preceding the retention test (Fast: t = 2.84, df = 5, p = .05; Slow: t = 3.54, df = 5, p = .02). That these changes were due to extinction can be seen from the results of heart-rate avoidance training following the retention test (Figure 1c). During this training, subjects rewarded for increases in heart rate changed their heart rate from an initial average rate of 388 to 462 beats per min. (t = 5.34, df = 5, p = .001), while subjects rewarded for decreases in heart rate changed their heart rate from an initial average of 371 to 340 beats per min. (t = 4.41, df = 5, p = .01).

Discussion

Previous papers have shown that rats can learn to make discriminative changes in heart rate in order to avoid and/or escape electric shock (DiCara and Miller, 1968a) or to obtain rewarding brain stimulation (Miller and DiCara, 1967). The present experiment has shown that instrumentally learned increases and decreases in heart rate can be retained for at least a 3-month period without intervening practice. Taken as a group, these experiments indicate that the instrumental learning of heart rate displays two very important characteristics of the instrumental learning of skeletal responses, namely, capability for discriminative learning and retention. The demonstration that learned heart-rate changes can be retained over long periods of time, and the results of recent experiments showing that heart-rate changes learned in the curarized state transfer to the noncurarized state (DiCara and Miller, 1968b) as well as in the return direction (DiCara and Miller, 1969), encourages the experimental investigation of the interaction between instrumental learning of visceral and skeletal responses.

For example, in an experiment by DiCara and Weiss (1969), it was shown that heart-rate changes learned under curare have a significant effect upon subsequent noncurarized skeletal avoidance learning. Subjects which learned fast heart rate under curare showed poor avoidance learning, while subjects which learned slow heart rate showed significantly better avoidance learning.

References

DiCara, L.V. and Miller, N.E. Changes in heart rate instrumentally learned by curarized rats as avoidance responses. J. Comp. Physiol. Psychol., 65, 8-12, 1968 (a).

DiCara, L. V. and Miller, N. E. Transfer of instrumentally learned heart-rate changes from curarized to noncurarized state: Implications for a mediational hypothesis. J. Comp. Physiol Psychol., 68, 159-162, 1968(b).

DiCara, L. V. and Miller, N.E. Instrumental learning of systolic blood pressure responses by curarized rats: Dissociation of cardiac and vascular changes. Psychosom. Med., 30, 489-494, 1968 (c).

DiCara, L.V. and Miller, N.E. Instrumental learning of peripheral vasomotor responses by the curarized rat. Commun. Behav. Biology, 1, 209-212, 1968 (d).

DiCara, L. V. and Miller, N. E. Heart-rate learning in the noncurarized state, transfer to the curarized state, and subsequent retraining in the noncurarized state. Physiol. Behav., 4, 621-624, 1969.

DiCara, L.V. and Weiss, J.M. Effect of heart-rate learning under curare on subsequent noncurarized avoidance learning J. Comp. Physiol. Psychol., 69, 368-374, 1969.

DiCara, L. V. and Miller, N. E. Instrumental learning of vasomotor response by rats: learning to respond differentially in the two ears. Science, 159, 1485-1486, 1968. (e).

Miller, N.E. and Banuazizi, A. Instrumental learning by curarized rats of a specific visceral response, intestinal or cardiac. J. Comp. Physiol. Psychol., 65, 1-7, 1968.

Miller, N.E. and DiCara, L.V. Instrumental learning of heart-rate changes in curarized rats: Shaping, and specificity to discriminativ. stimulus. J. Comp. Physiol. Psychol., 63, 12-19, 1967.

Miller, N.E. and DiCara, L.V. Instrumental learning of urine formation by rats; changes in renal blood flow. Am. J. Physiol., 215, 677-683, 1968.

Trowill, J. A. Instrumental conditioning of the heart rate in the curarized rat. J. Comp. Physiol. Psychol., 63, 7-11, 1967.

12

Transfer of Instrumentally Learned Heart-Rate Changes from Curarized to Noncurarized State: Implications for a Mediational Hypothesis

Leo V. DiCara and Neal E. Miller

Two groups of curarized rats learned to increase or decrease, respectively, their heart rates to escape and/or avoid brief pulses of mild electric shock. Two weeks later, a statistically significant increase of 5% and decrease of 16% transferred to the noncurarized state. Immediately subsequent retraining without curare produced additional significant increases and decreases, bringing the total overall changes to 11% and 22%, respectively. Initial significant differences between the two groups in respiration and activity decreased until they were far from statistically reliable. It is difficult to account for the pattern of these results by the hypothesis that all of the heart-rate changes in these rats were mediated by learned changes in central commands for activity or breathing.

The primary purpose of the present experiment was to determine if increases and decreases in heart rate instrumentally learned by rats paralyzed by curare will transfer to the free-moving noncurarized state. One secondary purpose was to determine whether any training that did transfer could be augmented by additional training in the free-moving noncurarized state.

Another secondary purpose was to see what differences in respiration or activity occurred during tests for transfer and for retraining. In a previous paper, Miller and DiCara (1967) raised the possibility that the rats might conceivably learn to change their heart rates indirectly by learning to send out from the motor cortex to the skeletal muscles central commands for activity or for relaxation. Such central commands might elicit innate or classically conditioned changes in heart rate. If this should be the case, one would expect movements of the muscles to appear when *S*s trained under curare are later tested and retrained without paralysis by curare.

Method

Subjects

A total of 16 naive male albino Sprague-Dawley rats weighing 478–532 gm. were trained. Eight *S*s were rewarded by avoidance and/or escape from electric shock for increasing heart rate and eight for decreasing it.

Apparatus

Heart-rate learning under curare. During heart-rate training under curare, *S* was placed in a soundproof ventilated enclosure (Grason-Stadler) and was maintained on artificial respiration. The experimental chamber was equipped with a 60-w. light and a loudspeaker delivering a 1,000-cps tone at an intensity of 82 db. Discontinuous shock consisting of a .1-sec. pulse of .3-ma. 60-cycle ac was delivered at 2.0-sec. intervals through an electrode fitted around the base of *S*'s tail (Weiss, 1967). A detailed description of the apparatus, including the reinforcement circuits for automatically terminating the shock signal and/or the electric shock when *S*'s heart rate reached a predetermined criterion, has been given in previous reports (DiCara & Miller, 1968a; Miller & DiCara, 1967).

Tests and retraining without curare. One week after heart-rate training under curare, each *S* had the female end of an amphenol plug cemented to its skull by means of jeweler's screws and dental cement. Electric shock was delivered to the free-moving *S* through the amphenol plug which was connected to insulated wires implanted subcutaneously along *S*'s back and emerging at the base of the tail, where they were connected to the tail electrode. Heart rate was measured by insulated steel wires implanted subcutaneously in the mid-

---

[1] This study was supported by United States Public Health Service Research Grant MH-13189 from the National Institute of Mental Health.

[2] Advanced Research Fellow of the American Heart Association. Now at Department of Psychiatry, Albert Einstein College of Medicine, Bronx, New York.

[3] Requests for reprints should be sent to Neal E. Miller, Rockefeller University, New York, New York 10021.

dle of the back at shoulder level and over the heart. Respiration was measured by means of a thermistor implanted in the right nostril of *S*. Recording heart rate in the free-moving rat required the construction of a special amplifier and filter system. This system consisted of a combination of low-pass and high-pass filtering to detect the onset of the QRS complex. The EKG signal from the Grass recorder was first differentiated by a high-pass filter. The resultant signal was rectified, producing a unidirectional wave shape. Additionally, this rectifier provided suppression of low-level noise voltages and partial discrimination against the P and T waves. The rectified signal was amplified and "smoothed" in a lag network. The resultant signal passed through an adjustable level detector which produced an output pulse only when the R wave occurred. This signal triggered a delay element which prevented noise spikes from passing to the output for 80 msec. after the onset of the R wave. The detector also had a provision for being disconnected during the 100-msec. pulses of electric shock, so as to prevent artifacts from distorting detection of the QRS complex.

Subjects were tested for transfer of heart-rate learning in a metal chamber 8 × 8 × 16 in. high, equipped with an eight-channel mercury commutator (Scientific Prototype) and light and tone sources identical to those used during heart-rate training under curare. The metal chamber was mounted on a platform which measured the amount of *S*'s movement by making and breaking an electrical contact with each movement (Davis & Ellison, 1964). Both chamber and platform were placed in a soundproof ventilated enclosure. A Grass Model 5 polygraph was used for all recordings.

Procedure

Heart-rate training under curare. Rats were injected ip with 3.0 mg/kg of d-tubocurarine chloride (Squibb) in a solution containing 3.0 mg/ml. As soon as *S* showed signs of difficulty in breathing, it was fitted with a specially constructed face mask and respirated at a 1:1 inspiration-expiration ratio, 70 cycles/min with a peak reading of 20 cm. of water pressure. The EKG and tail electrodes were attached to *S* as was a 30-gauge needle inserted ip after the area had been locally anesthetized by .15 ml. of a 2% solution of Xylocaine hydrochloride. Additional d-tubocurarine was infused at the rate of 1.0 mg/kg per hour for the duration of the experiment. A previous experiment had shown that this dosage is sufficient to maintain complete paralysis of all skeletal muscles (DiCara & Miller, 1968a).

After a period of 30 min. during which heart rate was allowed to stabilize, *S*s were randomly assigned to fast or slow heart-rate reward groups. Heart-rate avoidance training consisted of trials signaled by the onset of a light and tone. If a criterion heart-rate response was achieved within 5 sec. of the onset of this CS, the CS was terminated and shock was not delivered. If *S* failed to achieve a criterion response within the 5 sec., the shock was turned on and *S* received a .1-sec. pulse of .3-ma. 60-cycle ac once every 2 sec. until the criterion was achieved, at which time both the CS and the shock were turned off. The fifth and every tenth trial thereafter throughout training was a test trial, during the first 5 sec. of which it was not possible to terminate the trial. At the end of this 5 sec., the trial could be terminated. However, if *S* did not achieve the criterion within an additional 5 sec., the shock was turned on until it did. Test trials were used so that heart rate could be measured in the presence of the CS for the same constant intervals throughout training. In order to be able to compare the effects of training on heart rate when no signals were present, the first and every tenth trial thereafter throughout training was a blank trial, each lasting 5 sec., during which heart rate was recorded but nothing else occurred. Avoidance training consisted of 300 trials presented on a VI 30-sec. schedule (range: 15–50 sec.). During training, each block of 10 trials was analyzed and the criterion made more difficult if average latency of achieving it was less than 3 sec. and easier if it was more than 10 sec. In either case, the change in the criterion was approximately 2%.

After heart-rate training all circuits were disconnected and the rats were maintained on artificial respiration in the dark silent experimental chamber until recovery. Subjects were then returned to individual cages.

Test for transfer and relearning without curare. Two weeks after heart-rate training under curare, *S*s were tested for transfer of heart-rate learning in the noncurarized state. Subjects were placed in the testing chamber and allowed a period of 5 min., during which they remained undisturbed in silence and darkness. Transfer testing consisted of the presentation of 32 trials under the same conditions as in training under curare, except that the light and tone remained on for 5.0 sec. irrespective of the rat's heart rate and no electric shocks were given. Four blank trials were given at equal intervals during the transfer test.

After the test for transfer of heart-rate learning, *S*s were trained in the noncurarized state to change their heart rates in the same direction as before. The training procedures were identical to those used earlier under curare.

Results and Discussion

Heart rate, respiration rate, and activity scores were computed from individual blank and test trials at the beginning of and at quarterly intervals throughout heart-rate training and transfer testing.

Heart-rate Training under Curare

Figure 1a shows mean heart rates for fast and slow groups. All *S*s showed heart-rate changes in the rewarded direction. The group rewarded for increases increased their heart rates from an initial rate of 415 to 441 ($t = 7.01$, $df = 7$, $p < .001$), while the group rewarded for decreases decreased their heart rates from 426 to 347 ($t = 4.66$, $df = 7$, $p < .001$). Analyses of heart rate on blank and test trials over the last quarter of training indicated that both the group that had been rewarded for increases and the group that had been rewarded for decreases showed a significant difference in the expected direction between heart-rate changes occurring during the presentation of avoidance trials signaled by the onset of light and tone stimuli and those occurring during the presentation of blank trials (fast group: $t = 3.01$, $df = 7$, $p < .02$; slow group: $t = 2.79$, $df = 7$, $p < .05$). These results indicate discrimination learning and confirm results reported in previous studies from this laboratory (DiCara & Miller, 1968a; Miller & DiCara, 1967).

The difference in the amount of shock received by *S*s rewarded for increases and *S*s rewarded for decreases in heart rate was not significant, and a low nonsignificant correlation of .20 was obtained for the overall changes in heart rate learned and the amount of shock received. These results indicate that the differences in heart rate of *S*s rewarded for increases and those rewarded for decreases were not artifacts of the amount of shock received by them during heart-rate training.

Tests for Transfer and Relearning without Curare

After the avoidance cues were introduced into the testing situation, each individual *S* showed transfer of heart-rate changes in the rewarded direction. These results occurred in spite of the fact that the offset of the avoidance cues was not contingent on heart rate during these tests and that no electric shocks were given during them.

As can be seen from the top curve in Figure 1b, the group previously rewarded for increases in heart rate under curare increased their heart rate, on blank trials, during the transfer tests from an initial

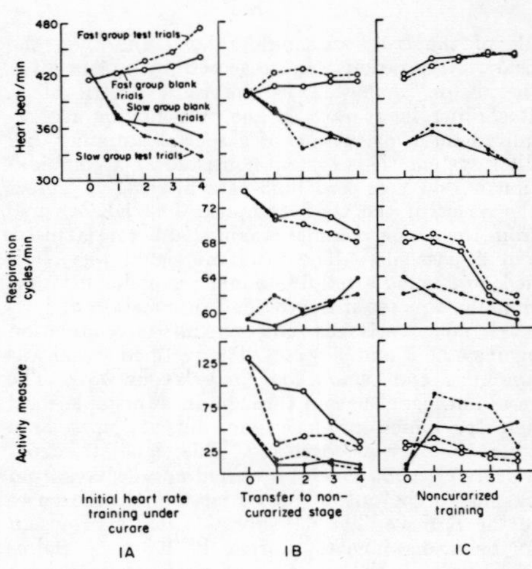

FIG. 1. Initial heart-rate learning under curare and changes in heart rate, respiration rate, and activity during test for transfer of heart-rate learning to the noncurarized state and subsequent heart-rate training in the noncurarized state. (Measures were obtained at quarterly intervals during training from individual blank and test trials.)

average of 395 to a final one of 412 beats/min ($t = 6.5$, $df = 7$, $p < .001$). Conversely, the group previously rewarded for decreases in heart rate decreased their rate, on blank trials, during transfer from an initial average of 401 to a final one of 338 beats/min ($t = 10$, $df = 7$, $p < .001$). A comparison of heart rate on test and blank trials did not indicate significant discriminative heart-rate changes in either fast or slow groups.

The results of additional heart-rate training in the noncurarized state are shown in Figure 1c. The *S*s rewarded for increases in heart rate showed an additional increase on blank trials, from 412 to 440 beats/min ($t = 3.6$, $df = 7$, $p < .01$), while those rewarded for decreases showed an additional decrease on blank trials, from an average of 338 to one of 311 beats/min ($t = 3.9$, $df = 7$, $p < .01$). Thus, it is clear that the rats were capable of additional base-line learning in the free-moving noncurarized state. However, the results did not indicate significant discrimination learning.

The average respiration rates and activity scores, as determined on blank and test trials, are shown in the lower parts of Figure 1b and 1c. Preceding the transfer test, *S*s previously trained to increase heart rate showed significantly less activity ($t = 5.0$, $df = 14$, $p < .001$) and a significantly higher rate of respiration ($t = 6.4$, $df = 14$, $p < .001$) than *S*s previously trained to decrease their heart rate. During transfer testing and retraining, the differences in heart rate between the two groups became progressively greater due to significant increases in the heart rate in the increase group and significant decreases in the decrease group. Meanwhile, the differences in both respiration and activity became significantly smaller, until by the end of retraining there were no significant differences between fast and slow groups in either respiration or activity on blank or test trials.

The two main features of the results were (*a*) a decrease in the difference between the groups in both respiration rate and activity while the difference in heart rate was increasing, and (*b*) a small and statistically far from reliable difference in respiration and in activity at the end of retraining when the difference in heart rate was large and highly reliable. It is difficult to reconcile in any simple way either of these features of the results with the hypothesis that the differences in heart rate were mediated primarily by differences in either rate of respiration or amount of general activity. Furthermore, in another study (DiCara & Miller, 1969), rats which were first trained in the noncurarized state, given additional training under curare, and then retrained in the noncurarized state showed at the end of training similar large and reliable differences in heart rate and small and highly unreliable differences in respiration rate and activity. Finally, it is difficult to account in terms of central impulses to skeletal responses for the variety and specificity of visceral learning that has been demonstrated in curarized rats: intestinal contractions independent of heart rate (Miller & Banuazizi, 1968), formation of urine by the kidney independent of heart rate and blood pressure (Miller & DiCara, 1968), blood pressure independent of heart rate (DiCara & Miller, 1968b), and differential vasomotor responses in the two ears (DiCara & Miller, 1968c).

REFERENCES

Davis, J. D., & Ellison, G. D. A general purpose activity recorder with variable sensitivity. *Journal of the Experimental Analysis of Behavior*, 1964, **7**, 117–118.

DiCara, L. V., & Miller, N. E. Changes in heart rate instrumentally learned by curarized rats as avoidance responses. *Journal of Comparative and Physiological Psychology*, 1968, **65**, 8–12. (a)

DiCara, L. V., & Miller, N. E. Instrumental learning of systolic blood pressure by curarized rats: Dissociation of cardiac and vascular changes. *Psychosomatic Medicine*, 1968, **30**, 489–494. (b)

DiCara, L. V., & Miller, N. E. Instrumental learning of vasomotor responses by rats: Learning to respond differentially in the two ears. *Science*, 1968, **159**, 1485–1486. (c)

DiCara, L. V., & Miller, N. E. Heart-rate learning in the noncurarized state, transfer to the curarized state, and subsequent retraining in the noncurarized state. *Physiology & Behavior*, 1969, in press.

Miller, N. E., & Banuazizi, A. Instrumental learning by curarized rats of a specific visceral response, intestinal or cardiac. *Journal of Comparative and Physiological Psychology*, 1968, **65**, 1–7.

Miller, N. E., & DiCara, L. V. Instrumental learning of heart-rate changes in curarized rats: Shaping, and specificity to discriminative stimulus. *Journal of Comparative and Physiological Psychology*, 1967, **63**, 12–19.

Miller, N. E., & DiCara, L. V. Instrumental learning of urine formation by rats; changes in renal blood flow. *American Journal of Physiology*, 1968, **215**, 677–683.

Weiss, J. M. A tail electrode for unrestrained rats. *Journal of the Experimental Analysis of Behavior*, 1967, **10**, 85–86.

(Received June 20, 1968)

13

Heart-Rate Learning in the Noncurarized State, Transfer to the Curarized State, and Subsequent Retraining in the Noncurarized State

Neal E. Miller and Leo V. DiCara

One group of freely-moving rats was trained to increase, and another to decrease, heart rate in order to avoid and/or escape electric shock. This training produced a statistically reliable difference between the groups. This difference persisted when the rats subsequently were tested and then given additional training while paralyzed with curare and maintained on artificial respiration. The same difference also persisted during subsequent nonreinforced tests without curare and was further increased by additional reinforced training in the normal, free-moving state. During the initial session of noncurarized training, the group rewarded for increases in heart rate showed more activity than the one rewarded for decreases, but there were no reliable differences in respiration rate. At the end of the second session of noncurarized training, the difference between the groups did not approach statistical reliability in activity ($t = 0.7$), respiration rate ($t = 0.2$), or the variability of that rate ($t = 0.6$), but was highly reliable in heart-rate ($t = 7.2$, $df = 12$, $p < 0.001$).

*Supported by U.S. Public Health Service Grants MH 13189 and MH 16569.

Previous papers have shown that curarized rats can be instrumentally trained to increase or decrease heart rate [2, 9, 11], blood pressure [4], gastrointestinal motility [8], vasomotor tone [5, 6], and rate of urine formation [10]. It has also been shown that instrumentally trained increases and decreases in heart rate can be retained for 3 months without intervening practice [7], and that these learned changes in heart rate transfer from the curarized to the noncurarized state [3].

In view of the possibility of using the learning of visceral responses in the therapy of psychosomatic disorders of the cardiovascular system, the primary purpose of the present experiment was to determine if increases and decreases in heart-rate can be learned in a noncurarized state.

A second purpose of the present experiment, if heart-rate learning is possible in the noncurarized state, was to determine if such learning transfers to the curarized state, since the possibility exists that the heart-rate changes would be mediated by changes in respiration or skeletal activity. However, if heart-rate learning in the noncurarized state is mediated by feedback from the performance of such skeletal responses, one would expect that the learned heart-rate changes would disappear when the subjects are paralyzed by curare so that they could not execute them.

A third purpose was to determine the effect on heart rate of additional training (after that with curare) in the noncurarized state.

Method

The overall design of the experiment was to train rats to increase or decrease heart-rate in order to avoid electric shock in the noncurarized state and then to test for transfer of heart-rate learning to the curarized state. While the rats were curarized, they received further heart-rate training and were subsequently tested again in the noncurarized state for transfer and/or retention of heart rate was well as for further heart-rate learning.

SUBJECTS

A total of 14 naive male Sprague-Dawley rats weighing from 375-514 g were trained, 7 to increase heart rate and 7 to decrease it.

APPARATUS

One week prior to heart-rate training, each rat had the female end of an amphenol plug cemented to its skull by means of jeweler's screws and dental cement. Discontinuous electric shock consisting of 0.1-sec pulses of 0.3 mA 60 cycle a.c. was delivered at 2-sec intervals to the free-moving rat through insulated wires connected to the plug and implanted subcutaneously along the rat's back emerging at the base of the tail where they were connected to a specially constructed electrode [12]. Heart rate was measured by insulated stainless steel wires connected to the amphenol plug and implanted subcutaneously in the middle of the back at shoulder level and over the heart. Respiration was determined by means of a thermistor implanted in the right nostril of the rat. A special amplifier and filter system had to be constructed to record heart rate in the free-moving rat [3]. A Grass polygraph, Model 5, was used for all recordings.

Heart-rate training in the noncurarized state. Animals were trained in a metal chamber 8 in. × 8 in. × 16 in. high, equipped with an 8-channel mercury swivel commutator (Scientific Prototype), a 60 W light bulb, and a loudspeaker delivering a 1,000 Hz tone at 82 dB. The metal chamber was mounted on a platform 12 in^2 which measured the amount of the rat's movement by making and breaking an electrical contact with each movement [1]. Both chamber and platform were placed in a soundproof, ventilated enclosure.

The reinforcement circuits for automatically terminating the shock signal and/or the electric shock when the rat's heart reached a predetermined criterion were the same as those described in a previous report [2].

Test for transfer under curare. The apparatus used in this phase of the experiment was identical to that used by Miller and DiCara [9]. During the test for transfer of learned heart-rate changes to the curarized state, the animal was placed on a towel in a soundproof, ventilated chamber and maintained on artificial respiration. The experimental chamber was equipped with sources of light and tone similar to those in the noncurarized training.

Procedure

HEART-RATE TRAINING IN THE NONCURARIZED STATE

Prior to heart-rate training, all rats were anesthetized with Sodium Pentobarbital (35 mg/kg) and the recording and shock wires implanted and connected to the amphenol plug which was cemented to the animal's skull.

One week later, rats were placed in the training chamber and allowed a period of 30 min, during which they remained undisturbed in silence and darkness. Animals were randomly assigned to fast or slow heart-rate groups. Heart-rate training consisted of the presentation of 300 trials on a variable-interval (VI) schedule with a mean intertrial interval of 30 sec (range: 15-50 sec). Each trial was signalled by the onset of a compound CS of a 60 W light and a 1,000 Hz tone at 82 dB.

This CS was terminated and shock was not delivered if the rat made the proper heart-rate response within 5 sec of the onset of the light and tone. However, if the animal failed to make the proper response within the 5 sec, the shock was turned on and the rat received a 0.1-sec pulse of 0.3 mA 60 cycle a.c. current every 2 sec until the criterion was achieved, at which time both the CS and the shock were turned off.

The fifth and every tenth trial thereafter throughout training was a test trial, during the first 5 sec of which it was not possible for the rat to terminate the trial regardless of its heart rate. At the end of this 5-sec period, the trial could be terminated whenever the rat achieved the heart-rate criterion. However, if the animal did not achieve the criterion within 5 sec after the test period was over, the shock was turned on until it did. Test trials were used so that heart rate could be measured in the presence of the CS for the same constant intervals throughout training without any possibly contaminating effects from having the CS turned off or the shock turned on.

In order to be able to compare the effects of training on heart rate when no signals were present, all animals received blank trials, each lasting 5 sec, during which heart rate was recorded but nothing else occurred. The tenth and every tenth trial thereafter throughout training was a blank trial.

Initial criteria were selected enough above or below the average heart rate so that they would be achieved approximately every 5 sec. During training, each block of 10 trials was analyzed and the criterion made more difficult, by approximately 2 percent, if average latency was less than 3 sec and easier by 2 per cent if more than 10 sec.

After heart-rate training, rats were returned to their individual home cages.

TEST FOR TRANSFER AND ADDITIONAL HEART-RATE TRAINING

Two weeks after heart-rate training, subjects were tested under curare for retention and transfer of heart-rate learning.

Rats were injected i. p. with 3.0 mg/kg of *d*-tubocurarine chloride (Squibb) in a solution containing 3 mg/ml. As soon as the animal showed signs of difficulty in breathing, it was fitted with a specially constructed face mask and artificially respirated at a 1:1 I/E ratio, 70 cycle/min with a peak reading of 20 cm of water pressure. Additional *d*-tubocurarine was infused through a 30 gauge needle i. p. at the rate of 1.0 mg/kg per hr for the duration of the experiment. Animals were allowed a period of 30 min during which they remained undisturbed in silence and darkness.

Transfer testing consisted of the presentation of 32 trials under the same conditions as training except that the light and tone remained on for 5.0 sec, irrespective of the rat's heart rate, and no electric shocks were given. Four blank trials were given at equal intervals during the transfer test.

After the test for transfer of heart-rate learning, animals were trained under curare to increase or decrease their heart rate. The training procedures were identical to those used earlier during noncurarized training, and all rats received 300 trials.

After this heart-rate training, the rats were maintained on artificial respiration in the dark, silent experimental chamber until recovery.

One week after the curare training, rats were tested in a noncurarized state. They were placed in the test chamber and allowed a period of 30 min during which they remained undisturbed. After this, they received 32 transfer test trials identical to those given in the curarized state. After the transfer test, animals received a final session of 300 trials for heart-rate avoidance training identical to those used during the first noncurarized session.

Results and Discussion

Heart rate, respiration rate, and activity scores obtained during the various phases of the experiment are presented in Fig. 1. Heart rate was computed from a 30-sec count prior to the first trial and from individual blank and test trials at quarterly intervals throughout heart-rate training and transfer testing. Respiration rate and activity scores were determined from 30-sec periods overlapping the blank test. Because of the relatively short duration of test trials and the complex respiratory changes which took place during the tests it was not possible to obtain an accurate measure of respiration during test trials.

HEART-RATE TRAINING IN THE NONCURARIZED STATE

As can be seen in Fig. 1A, animals rewarded for increases in heart rate, as well as those rewarded for decreases in heart rate, showed a decrease in heart rate. However, while the initial difference in heart rate between the two groups did not approach significance ($t = 0.66$), the difference at the end of the training was significant ($t = 2.28$, $df = 12$, $p < 0.05$). Furthermore, the decrease in heart rate in subjects rewarded for increases in heart rate was not significant ($t = 1.17$, $df = 6$, $p < 0.30$), while the decrease of approximately 16 per cent in the heart rate of animals rewarded for decreases in heart rate was highly significant ($t = 10.31$, $df = 6$, $p < 0.001$). It appears that heart-rate learning (at least of increases) is more difficult in the noncurarized than in the curarized state, since previous experiments on curarized rats have secured learning of heart-rate changes of approximately 15 per cent in both directions [2, 9].

A comparison of heart rate on blank and test trials did not indicate any evidence of discriminative heart-change changes in rats rewarded for either increases or decreases in heart rate. In fact, there was a slight, nonsignificant tendency for heart rate to decrease on test trials as compared to blank trials in subjects rewarded for increases in heart rate.

In previous experiments on the learning of heart rate, it was shown that, during a comparable period of training, curarized rats could learn to make discriminative changes in heart rate in order to obtain rewarding brain stimulation [9] or to avoid and/or escape electric shock to the tail [2]. The inability of rats to learn discriminative heart-rate changes in the present experiment is a further indication of poorer learning in the noncurarized state.

Analyses of activity indicated that there were no significant differences in initial activity between subjects trained to increase and those trained to decrease heart rate (t = 0.71). However, during heart-rate training, animals rewarded for increases in heart rate showed a significant increase in activity (t = 3.36, df = 6, $p < 0.02$) in spite of the fact that they did not increase their heart rates, while rats rewarded for decreases in heart rate showed a signifi-

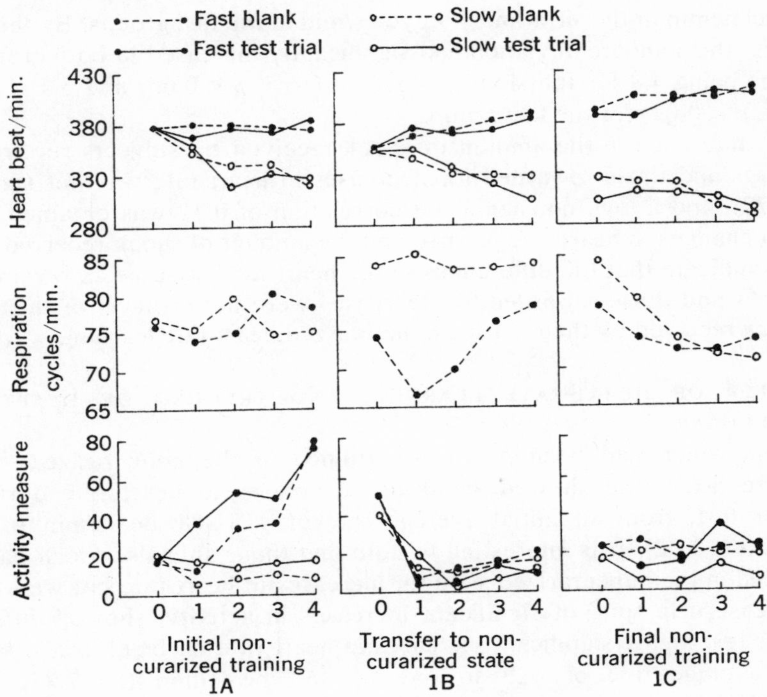

Figure 1. Average heart rate, respiration rate, and activity score during initial noncurarized training, test for transfer and/or retention from a curarized to a noncurarized state, and subsequent heart-rate training in the noncurarized state. Measures were obtained at pre-test and at quarterly intervals during training.

cant decrease in activity (t = 4.02, df = 6, $p < 0.01$). A comparison of activity scores obtained on blank and test trials did not indicate significant differences in either the fast or slow rewarded groups. However, there was a tendency for animals rewarded for increases in heart rate to be more active on test trials as compared to blank trials.

Analyses of respiration data did not indicate any significant difference in rate of respiration between animals rewarded for increases and those rewarded for decreases in heart rate, either at the beginning (t = 0.03) or at the end of training ($t = 0.21$), nor was there any change in respiration rate within groups during training (Increase: $t = 0.17$; Decrease: $t = 0.13$). However, there were complex changes in the pattern of respiration during heart-rate training, which resulted in an increase in variability of rate of respiration in both groups. At the beginning of training, the standard deviation of the rates of individual rats measured over ten successive 15-sec periods averaged 2.6 cycles/min in the slow and 1.9 cycles/min in the fast groups. By the end of training, the standard deviation had significantly increased in both groups, the average being 4.8 for the slow ($t = 3.17$, $df = 5$, $p < 0.05$) and 5.1 ($t = 4.63$, $df = 5$, $p < 0.01$) for the fast group.

The difference in the amount of shock received by subjects rewarded for increases and those rewarded for decreases in heart rate was not significant ($t = 0.77$), and a low, nonsignificant correlation of 0.12 was obtained for the overall changes in heart rate learned and the amount of shock received. These results indicate that the differences in the heart rate of subjects rewarded for increases and those rewarded for decreases were not artifacts of the amount of shock received by them during noncurarized heart-rate learning.

TRANSFER OF HEART-RATE LEARNING AND TRAINING OF HEART RATE UNDER CURARE

Subjects which had been previously trained in the noncurarized state to decrease heart rate showed significant decreases in heart rate during the transfer test, from an initial average rate of 418-381 beats/min ($t = 3.22$, $df = 6$, $p < 0.02$). It is interesting to note that those animals, which had been rewarded in the noncurarized state for increases in heart rate but were unable to increase it in spite of significant increases in activity, showed during the transfer test highly significant increases in heart rate on blank trials, from an initial average rate of 429 to one of 462 beats/min ($t = 7.87$, $df = 6$, $p < 0.001$). However, a comparison of heart rate on test trials and blanks did not indicate any evidence of discriminative changes.

The significant increase in baseline heart rate was an unexpected result since these increases in heart rate were the first ones to occur during the experiment and they persisted during the transfer test despite the fact that they were not reinforced.

During subsequent heart-rate training under curare, animals rewarded for increases did not show significant additional increases in heart rate (462-467 beats/min; $t = 0.73$), nor did rats rewarded for decreases in heart rate show significant additional decreases in heart rate (381-366 beats/min; $t = -1.52$) This result also is surprising, since other experiments [2, 3, 7, 8, 9, 11] have established unequivocal evidence for significant heart-rate learning in curarized rats. However, in each of the latter experiments naive rats were used, while in the present experiment each subject had a session of heart-rate learning in the noncurarized state prior to the curarized one.

TRANSFER OF HEART-RATE LEARNING AND FURTHER HEART-RATE TRAINING IN THE NON-CURARIZED STATE

These results are presented in Fig. 1B and 1C. Note that trials represented in Fig. 1B span a total time duration of approximately 15 min; in Fig. 1C the time duration is approximately 150 min. Animals previously trained to increase heart rate showed significant increases in heart rate when tested in the post-curare, noncurarized state ($t = 6.44$, $df = 6$, $p < 0.001$), while rats trained to decrease heart rate showed significant decreases in heart rate ($t = 4.18$, $df = 6$, $p < 0.01$). However, as before, there was no evidence of discriminative heart-rate changes. During the test for transfer, the changes in respiration in the two groups were not significantly different, and at the end of the test there was no significant difference between them ($t = 1.32$). Similarly, there was no significant difference in activity between the two groups at the end of transfer ($t = 0.55$). However, there was a significant decrease in activity in both groups (Increase: $t = 2.71$, $df = 6$, $p < 0.05$; Decrease: $t = 5.17$, $df = 6$, $p < 0.001$) during transfer testing. The heart-rate changes in the nonreinforced transfer test were not significantly correlated with those during training with curare but were with those during the initial training without curare (Fast: rho $= 0.79$; Slow: rho $= 0.78$). These facts suggest that the performance in the transfer test represented primarily retention of the original learning during training without curare.

During additional heart-rate training in the noncurarized state, animals rewarded for increases showed additional significant increases in heart rate of approximately 6 per cent ($t = 3.56$, $df = 6$, $p < 0.02$), while subjects rewarded for decreases showed significant decreases of approximately 13 per cent ($t = 3.21$, $df = 6$, $p < 0.02$). The differences in heart rate between increase and decrease groups was highly significant at the end of training ($t = 7.19$, $df = 12$, $p < 0.001$). However, there was no evidence of discriminative learning.

In contrast to the foregoing highly significant difference between the groups in heart rate, the differences at the end of training in activity ($t = 0.71$), respiration rate, ($t = 0.18$), and variability in respiration rate ($t = 0.62$) did not approach statistical reliability. These results are in line with those of a previous experiment, in which DiCara and Miller [3] found that initial differences in activity and respiration, appearing during a test for transfer from the curarized to the noncurarized state, disappeared with additional noncurarized training. At the end of this training, there was a large and highly reliable difference in heart rate between rats trained to increase and those trained to decrease it, and highly unreliable differences in activity and respiration. It does not seem reasonable in either of these two experiments to try to explain the large and highly reliable differences in heart rate as being mediated by the small and highly unreliable differences in respiration or in activity. However, it is not clear why naive rats cannot learn heart-rate changes as well in the noncurarized as in the curarized state.

The results of the present experiment and the one just cited [3] are in line with those of Miller and Banuazizi [8], which found greater specificity of instrumental learned heart-rate and intestinal responses in the rats that showed the better learning of such responses. They are also in line with the greater specificity that usually appears during the latter stages of the training of skeletal responses, and thus are what would be expected if visceral and skeletal learning obey the same general laws. They are not the results that would be predicted by the hypothesis that instrumental training can change visceral responses only via the mediation of learned skeletal responses.

Since the primary purpose of the present experiment was to determine if increases and decreases in heart rate can be learned in a noncurarized state, certain controls were omitted from the design which would be relevant to a clarification in the transfer data, such as a noncurare-noncurare group.

References

1. Davis, J. D. & G. D. Ellison: A general purpose activity recorder with variable sensitivity. *J. exp. Analysis Behav.* 7: 117-118, 1964.
2. DiCara, L. V. & N. E. Miller. Changes in heart rate instrumentally learned by curarized rats as avoidance responses. *J. comp. physiol. Psychol.* 65: 8-12, 1968.
3. DiCara, L. V. & N. E. Miller. Transfer of instrumentally learned heart-rate changes from curarized to noncurarized state: Implications for a mediational hypothesis. *J. comp. physiol. Psychol.* 68: 159-162, 1969.
4. DiCara, L. V. & N. E. Miller. Instrumental learning of systolic blood pressure responses by curarized rats: Dissociation of cardiac and vascular changes. *Psychosom. Med.* 30: 489-494, 1968.
5. DiCara, L. V. & N. E. Miller. Instrumental learning of peripheral vasomotor responses by the curarized rat. *Commun. behav. Biol.* Part A, 1: 209-212, 1968.
6. DiCara, L. V. & N. E. Miller. Instrumental learning of vasomotor responses by rats: Learning to respond differentially in the two ears. *Science* 159: 1485-1486, 1968.
7. DiCara, L. V. & N. E. Miller. Long term retention of instrumentally learned heart-rate changes in the curarized rat. *Commun. behav. Biol.* Part A, 2: 19-23, 1968.
8. Miller, N. E. & A. Banuazizi. Instrumental learning of curarized rats of a specific visceral response, intestinal or cardiac. *J. comp. physiol. Psychol.* 65: 1-7, 1968.
9. Miller, N. E. & L. V. DiCara. Instrumental learning of heart-rate changes in curarized rats: Shaping, and specificity of discriminative stimulus. *J. comp. physiol. Psychol.* 63: 12-19, 1967.
10. Miller, N. E. & L. V. DiCara. Instrumental learning of urine formation by rats; changes in renal blood flow. *Am. J. Physiol.*, 215, 677-683, 1968.
11. Trowill, J. A. Instrumental conditioning of heart rate in the curarized rat. *J. comp. physiol. Psychol.* 63: 7-11, 1967.
12. Weiss, J. M. A tail electrode for unrestrained rats. *J. exp. Analysis Behav.* 10: 85-86, 1967.

14

Effect of Heart-Rate Learning Under Curare on Subsequent Noncurarized Avoidance Learning

Leo V. DiCara and Jay M. Weiss

Rats which learned to decrease heart rate under curare in order to avoid electric shock showed good subsequent learning of a free-moving skeletal-avoidance response in a modified shuttle device, while subjects which learned to increase heart rate under curare showed poor subsequent avoidance and escape learning. Similar poor avoidance was shown by naive subjects trained with strong electric shock. These results suggest that instrumental learning of heart rate may alter emotional responses. An alternative explanation for poor avoidance performance based on learned "helplessness" is ruled out.

Recent studies have firmly established that autonomic responses can be modified by instrumental learning (DiCara & Miller, 1968a, 1968b, 1968c, 1968d; Miller & Banuazizi, 1968. Miller & DiCara, 1967, 1968; Trowill, 1967). In each of these experiments, the mediation of the autonomic responses by respiratory changes or overt skeletal responses was ruled out by training rats paralyzed by curare and maintained on artificial respiration. Further studies have shown that heart-rate changes learned under curare can be retained over a 3-mo. period (DiCara & Miller, 1968e), and transfer to the noncurarized state (DiCara & Miller, 1969).

One of the most prominent features of adaptive reactions is that they involve autonomic changes. The following question then arises: If an animal learns to regulate autonomic responses, will this affect subsequent adaptive reactions? In the present experiment, we investigated how instrumentally learning to increase or decrease heart rate to avoid and/or escape electric shock would affect subsequent acquisition of one type of adaptive reaction, a skeletal-avoidance response. Rats were initially trained to increase or decrease their heart rate under curare, then tested in the free-moving state for transfer of heart-rate learning, and finally trained in a modified one-way avoidance situation.

[1] This study was supported by United States Public Health Service Grant MH-13189 from the National Institute of Mental Health.

[2] Requests for reprints should be sent to Leo V. DiCara, who is now at the Department of Psychiatry, Albert Einstein College of Medicine, Bronx, New York 10467.

Experiment 1

Method

Subjects

The *S*s were 14 male Sprague-Dawley rats weighing 410–508 gm.

Apparatus

A detailed description of the apparatus used during heart-rate conditioning is given elsewhere (DiCara & Miller, 1968a).

During heart-rate conditioning, *S* lay on a folded towel in a soundproof ventilated enclosure (Grason-Stadler) and was maintained at a constant rate of artificial respiration by means of a mask made from the mouthpiece of a rubber balloon, placed over *S*'s snout.

The experimental chamber was equipped with an Ameco oscillator capable of delivering a 1,000-cps tone at an intensity of 82 db. The shock source consisted of a voltage divider on the secondary of a 500-v. ac transformer, with the current delivered to *S* through a fixed resistance of 200 K ohms. Discontinuous electric shock consisting of a .1-sec. pulse of .3 ma. ac was delivered at 2-sec. intervals through an electrode fitted snugly around the base of *S*'s tail (Weiss, 1967). The reinforcement circuit for automatically terminating the shock signal and/or the electric shock when the *S*'s heart rate reached a predetermined criterion was the same as that described by Miller and DiCara (1967).

One week after heart-rate conditioning, each *S* had an amphenol plug cemented to its skull, through which heart rate and respiration could be measured. Heart rate was measured by stainless-steel wires, implanted subcutaneously in the middle of the back at shoulder level and over the heart. Respiration was measured by means of a thermistor implanted in the right nostril of *S*.

Subjects were tested for transfer of heart-rate learning in a shielded metal chamber 16 in. high × 8 in. wide × 8 in. deep, equipped with a mercury swivel commutator and a loudspeaker capable of delivering a 1,000-cps tone at 82 db. The chamber was mounted on an electromechanical platform which measured the amount of *S*'s movement (Davis & Ellison, 1964). Both chamber and platform were placed in a soundproof, ventilated enclosure (Grason-Stadler). A Grass polygraph (Model 5) was used for all recordings.

The avoidance apparatus was a modified shuttle-avoidance box consisting of two compartments 8 × 8 × 12 in. placed end to end. The front and rear walls of each compartment were mounted on runners and could be moved to open or close the compartment, and the walls at the center of the apparatus could be withdrawn so that the rat could move from one compartment into the other. The safe compartment had a 15-w. light bulb placed below its floor so that it would be distinctive to the rat. An Ameco oscillator capable of delivering a tone of 1,000 cps at 82 db. was used as the CS. It was not possible to record heart rate or respiration since the wires were no longer patent.

Procedure

Heart-rate conditioning. Prior to heart-rate conditioning, each *S* was injected ip with 3.0 mg/kg of d-tubocurarine chloride in a solution containing 3 mg/ml. As soon as *S* showed signs of difficulty in breathing, it was fitted with the face mask and respirated at a 1:1 I/E ratio, 70 cycles/min, with a peak reading of 15 cm. of H_2O pressure.

Electrocardiogram and tail electrodes were attached to *S* as was a 30-gauge needle inserted ip after the area had been locally anesthetized by .15 ml. of a 2% solution of Xylocaine hydrochloride. Additional d-tubocurarine was infused at the rate of 1.0 mg/kg/hr for the duration of the experiment. A previous experiment (DiCara & Miller, 1968a) has shown that this dosage is sufficient to maintain complete paralysis of all skeletal muscles.

After a period of 30 min., during which the heart rate was allowed to stabilize, *S*s were randomly assigned to fast or slow heart-rate reward groups. A tone of 1,000 cps and 82-db. intensity was used as a warning CS for both groups. Heart-rate training consisted of trials signaled by the onset of the tone. If the criterion heart rate was achieved within 5 sec. of onset of the CS, it was turned off and no shock was delivered. If *S* failed to meet the criterion rate within the 5 sec., the shock was turned on and *S* received a .1-sec. pulse of .3 ma. once every 2 sec. until the criterion was achieved, at which time both the CS and the shock were turned off. The fifth and every tenth trial thereafter throughout training was a special test trial, during the first 5 sec. of which both the criterion and shock circuits were turned off so that heart rate could be measured for the same constant intervals throughout training without any possible contaminating effects from having the CS turned off or the shock turned on. At the end of this 5-sec. interval the scoring circuit was turned off and the criterion was turned on. If *S* did not achieve the criterion within the next 5 sec., the shock was turned on.

In order to be able to compare the effects of training on heart rate when no signals were present, the first and every tenth trial thereafter throughout training was a blank trial, each lasting 5 sec., during which heart rate was recorded but nothing else happened. During the intervals between trials, *S* was in silence and darkness.

Initial criteria were selected enough above or below the average heart rate so that they would be achieved approximately once every 5 sec. During training each block of 10 shock trials was analyzed and the criterion made more difficult if the rat changed his heart rate enough so that it was meeting the criterion in less than an average of 3 sec.; similarly, the criterion was made easier if *S* was taking more than approximately 10 sec. on the average to meet the criterion.

Avoidance training consisted of 300 trials presented on a VI 30-sec. schedule (range: 15–60 sec.). After 300 trials all circuits were disconnected and the rats maintained on artificial respiration in the dark silent experimental chamber until recovery, usually a minimum of 120 min.

Test for transfer. Two weeks after heart-rate conditioning, *S*s were tested in free-moving conditions for transfer of heart-rate learning. Heart rate and respiration were measured by the previously implanted equipment. Subjects were placed in the testing chamber and allowed a period of 10 min., during which they remained in silence and darkness. Ten tone presentations were then given on a VI schedule with a mean of 30 sec. (range: 10–50 sec.). Each tone presentation remained on for 5.0 sec. The *S* was never shocked.

Skeletal-avoidance learning. One month later, *S*s were tested for skeletal-avoidance learning in the modified shuttle-avoidance situation. At the start of each trial, the tone was turned on and the walls at the center of the apparatus were withdrawn so that the rat could move from one compartment into the other. The safe compartment had a 15-w. light bulb placed below its floor so that it would be distinctive to the rat. If the subject moved to the opposite, safe, compartment within 12 sec., the tone was terminated and shock was not presented; if the subject did not cross within 12 sec., shock remained on until *S* made the correct response or until 60 sec. had elapsed, at which time the trial was terminated and the walls returned to the closed position. The electric shock, delivered through the tail electrode as before, was the same intensity, etc., as that used during heart-rate conditioning.

Handling of the animals between trials was eliminated by interchanging the two compartments; when the rat ran from Compartment A to B, B (which now contained the rat) was moved slowly to the original position of A, and A was placed in the B position, thus setting up the next trial for the rat again to run from Position A to B. In the intertrial interval after a trial in which the rat did not escape from A to B, the boxes were mock-moved as if it had. The intertrial interval was 30 sec. Two sessions of 20 trials each, separated by 48 hr., were given.

Results

Heart-Rate Conditioning and Transfer

The results are shown in Figure 1. During heart-rate conditioning, all 14 *Ss* showed heart-rate changes in the rewarded direction, as indicated by blank trials at the beginning and at quarterly intervals throughout training. The *Ss* rewarded for increases increased their heart rate from an initial average rate of 395 to 430 ($t = 7.6$, $df = 6$, $p < .001$), while those rewarded for decreases decreased their average rate from 400 to 369 ($t = 9.2$, $df = 6$, $p < .001$).

In order to shape the response, to produce larger changes in heart rate, and also to control for possible effects of different frequencies of electric shock by holding these relatively constant, the avoidance criterion was progressively increased throughout training. Therefore, changes in the number of successful avoidances during training cannot be used as an indication of avoidance learning. However, a comparison of heart rate on blank and test trials over the last quarter of training indicated that *Ss* in the fast and slow groups had learned significant discriminative heart-rate changes in the rewarded direction (fast: $t = 2.9$, $df = 6$, $p < .05$; slow: $t = 2.5$, $df = 6$, $p < .05$) during the signal for shock. These results are in agreement with previous studies reported from this laboratory (DiCara & Miller, 1968a). As a result of the constant adjustment of criterion there was no significant difference in the amount or number of shocks received by *Ss* in the fast and slow groups at the beginning, middle, or end of training.

In the free-moving test, the rats showed transfer of heart-rate learning from the curarized state. Heart rate, respiration rate, and activity scores were computed from individual trials throughout transfer testing. The group previously rewarded for increases in heart rate under curare increased their heart rate from an initial 350

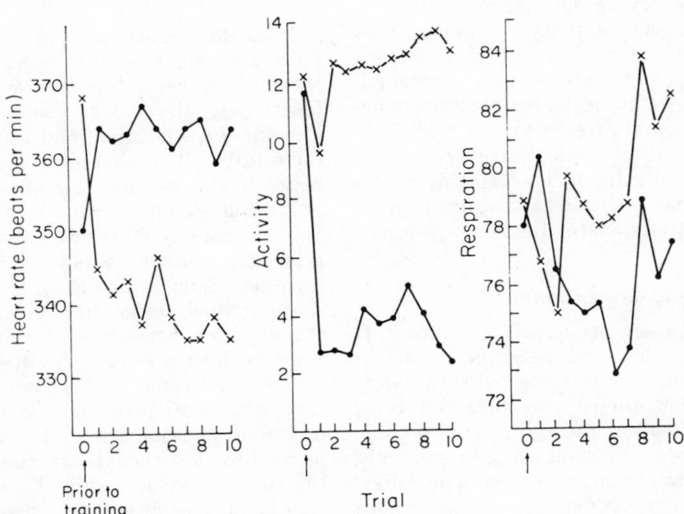

FIG. 1. Heart rate during training under curare and test for transfer. (Respiration and activity measures obtained during the test for transfer are presented.)

prior to the transfer test to 363 after 10 tone presentations ($t = 4.8$, $df = 6$, $p < .01$), and the group previously rewarded for decreases in heart rate under curare decreased their heart rate from 368 to 336 during the transfer test ($t = 4.1$, $df = 6$, $p < .01$).

Neither before, during, nor at the end of the test for transfer of heart rate were there any significant differences in respiration between the rats that had learned to speed up their heart rate and those that had learned to slow it down. The average rate before the test was 78 cycles/min for the heart-rate increase group and 79 cycles/min for the decrease group. The rates at the end of the 10 trials were, respectively, 77 and 82 cycles/min. These results are similar to those of a previous experiment (DiCara & Miller, 1969). Although there was no difference in activity between the two heart-rate condition groups prior to the first tone presentation ($t = .04$), *S*s trained to increase heart rate showed a significant decrease in activity as measured during tone trials ($t = 2.5$, $df = 6$, $p < .05$), while *S*s trained to decrease heart rate did not. These differences in activity are exactly the opposite of what would be expected on the hypothesis that the *S*s were trying to change their heart rate by skeletal activity; however, they are what one would expect if heart-rate conditioning had increased the emotionality of the *S*s in the increase group, since albino rats tend to freeze when afraid.

The skeletal-avoidance learning test was given 1 month later. During this training, it was not possible to measure respiration or activity since the wires were no longer intact.

Skeletal Avoidance Learning

The results are presented in Figure 2. It can be seen that large, significant differences in avoidance learning between heart-rate condition groups were obtained. Slow heart-rate learners acquired the skeletal avoidance rapidly, while fast *S*s generally failed to learn even to escape. The difference in latency between heart-rate groups is highly significant ($p < .001$) and latency was significantly correlated with the amount of heart-rate transfer shown (correlation based on *r*'s from fast and slow groups combined by using z score $= .50$, $p < .05$).

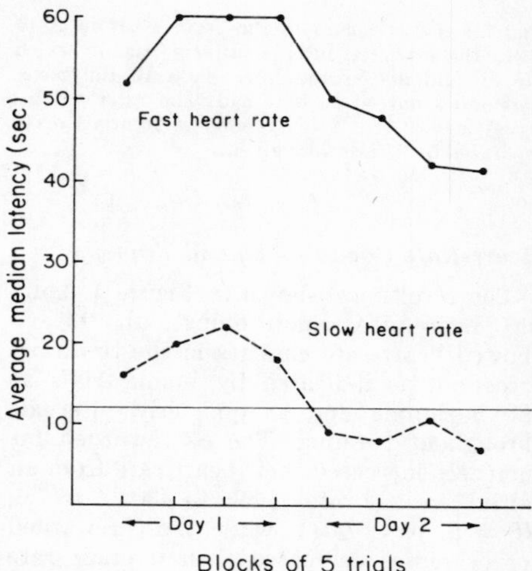

Fig. 2. Latency to cross during 2 days of avoidance training in a modified shuttle in rats that had learned to either increase or decrease their heart rates. (Latencies falling below 12 sec. are avoidance responses.)

Discussion

A striking aspect of the skeletal-avoidance learning was the rats' reaction to shock. Fast heart-rate *S*s were extremely reactive to the shock, generally jumping into the air, squealing, and turning toward their tail with each pulse of shock, while between shocks they remained immobile, or frozen. This sequence of responses greatly reduced forward locomotion and prevented either escape or avoidance learning. Slow heart-rate *S*s, in contrast, showed much more inhibited reactions to shock pulses, consisting of mild jerks forward, with slow walking between shocks.

These patterns of behavior suggested that fast heart-rate animals were hyperexcitable in comparison to *S*s that had learned to slow their heart rate. In view of the inverse relationship between strong emotional responses and many types of avoidance learning, it seems likely that the fast heart-rate *S*s were too fearful to learn efficiently, whereas slow heart-rate *S*s were less fearful and learned well. The inability of fast heart-rate *S*s to perform the correct response is consistent with the notion that poor avoidance learning under conditions of strong fear is often produced by interfering responses (Weiss, Krieckhaus, & Conte, 1968). One test of this hypothesis is the following: If fast heart-rate *S*s had learned the skeletal-avoidance response poorly because they were hyperreactive or hyperemotional, then learning of the present avoidance task, with the shock parameters used, should be optimal at a moderate fear level and become progressively poorer at higher fear levels. In order to test this hypothesis, naive rats were given avoidance training in the same apparatus at different shock intensities.

EXPERIMENT 2

Method

Subjects

The *S*s were 18 albino rats as in Experiment 1, weighing approximately 450 gm., which was the average weight of *S*s in Experiment 1 at the time of avoidance training.

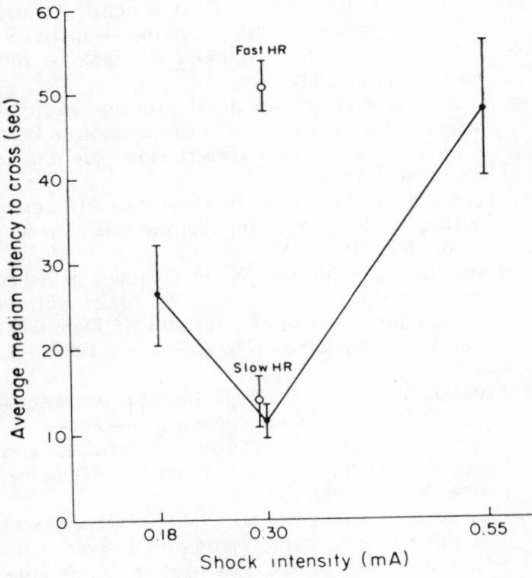

FIG. 3. Latency to cross during 2 days of avoidance training in naive rats trained at different shock intensities (closed circles). (Open circles represent rats with previous training to modify heart rate [run at .30-ma. shock intensity]. Latencies falling below 12 sec. are avoidance responses.)

Procedure

The *S*s were given two 20-trial sessions of avoidance training, as described above. The *S*s were divided into three groups of six rats each. One group was trained with shock of .18 ma. intensity, a second with .30 ma. intensity, and the third group with .55 ma. intensity. Thus, one group received the same shock intensity (.30 ma.) as was used in Experiment 1, one a lower, and one a higher intensity.

Results

The average median latency to cross for the three groups, shown in Figure 3, clearly indicates that learning for this avoidance task was best at the moderate (.30-ma.) intensity, poor at the low intensity, and very poor at the high intensity. An analysis of variance showed that the three shock groups differed significantly from one another ($F = 10.1$, $df = 2/15$, $p < .001$), and, most importantly, that the group means fit a highly significant quadratic function ($F = 39.45$, $df = 1$, $p < .001$). The *S*s trained at .55 ma. showed hyperreactivity to the shock—jumping and turning to the US, and freezing between shock pulses. This response pattern was similar to that shown by fast heart-rate *S*s in Experiment 1. The average median latency for all groups is shown in Figure 3, and shows clearly that the fast heart-rate *S*s, while trained at .30-ma. shock intensity, performed as did naive *S*s trained at .55 ma.

Discussion

The results of Experiment 2 showed that learning of the present avoidance task was quite poor at the high (.55-ma.) shock level, and that the performance of fast heart-rate *S*s at .30-ma. intensity resembled that of naive *S*s trained at the higher level. These results strongly suggest that learning to speed up heart rate in a shock-avoidance situation had the effect of increasing fear or excitability.

But how can the learning of fast heart rate result in strong fear when a number of studies on the classical conditioning of heart rate (deToledo & Black, 1966;

Holdstock & Schwartzbaum, 1965) have shown that fear in the albino rat is correlated with reduced heart rate? Recent studies suggest an answer. DeToledo (1968) has shown that the heart rate of a rat is indeed reduced by the *S*'s fear of a moderately intense electric shock, but that as shock intensity is increased this relationship progressively deteriorates, until at high shock intensity there is almost no suppression of heart rate. These results suggest that the control of heart rate in a fear situation reflects a balance between inhibitory and facilitatory influences and, most important for the present results, factors which tend to increase heart rate exert progressively more influence as fear is increased. The learning of fast heart rate therefore represents acquisition of one component of autonomic responding associated with strong fear reactions.

It is not possible, however, to determine from the present data exactly how fast heart-rate learning resulted in strong fear or high excitability. It may be that under curare, *S*s learned to raise or lower their general level of excitability in order to accomplish the heart-rate change. In this case, the hyperexcitability or hyperreactivity observed during skeletal-avoidance learning would have been specifically learned during heart-rate training under curare. However, Miller (1969) has shown that since rats can learn to regulate specific autonomic responses differentially, they do not learn to change autonomic patterns by such gross methods as altering general excitability. Another possibility is that *S*s did not mediate heart-rate changes by changing excitability but learned the heart-rate directly. This would mean that because *S*s had learned to speed up their heart rate, they became hyperreactive or hyperfearful in the fear-avoidance situation.

The present avoidance results appear similar to those reported by Weiss et al. (1968), who showed that prior fear conditioning resulted in poor subsequent shuttle avoidance learning. In the present experiment, fast heart-rate learning similarly resulted in poor subsequent avoidance learning. Weiss et al. suggested that prior fear conditioning produced strong fear which elicited unconditioned competing responses, resulting in poor avoidance learning. The present results suggest that fast heart-rate learning likewise produced strong fear which led to unconditioned competing responses. Thus, the poor avoidance learning reported by Weiss et al. and in the present experiment can be explained effectively in the same way; only the means by which strong fear was induced differs—via classical conditioning in the experiments by Weiss et al. and by instrumental conditioning of an autonomic response in the present study. The present experiment shows clearly that the concept of "helplessness" (Seligman & Maier, 1967) need not be invoked to explain such results. A helplessness explanation was specifically ruled out in the present experiment by giving both fast and slow heart-rate groups an equally effective coping response to avoid shock, and marked differences in avoidance learning still emerged, based apparently on a difference in fear level.

REFERENCES

Davis, J. D., & Ellison, G. D. A general purpose activity recorder with variable sensitivity. *Journal of the Experimental Analysis of Behavior,* 1964, **7,** 117–118.

de Toledo, L. Changes in heart rate and skeletal activity during conditioned suppression in rats. Unpublished doctoral dissertation, McMaster University, 1968.

de Toledo, L., & Black, A. H. Heart rate: Changes during conditioned suppression in rats. *Science,* 1966, **152,** 1404–1406.

DiCara, L. V., & Miller, N. E. Changes in heart rate instrumentally learned by curarized rats as avoidance responses. *Journal of Comparative and Physiological Psychology,* 1968, **65,** 8–12. (a)

DiCara, L. V., & Miller, N. E. Instrumental learning of systolic blood pressure responses by curarized rats: Dissociation of cardiac and vascular changes. *Psychosomatic Medicine,* 1968, **30,** 489–494. (b)

DiCara, L. V., & Miller, N. E. Instrumental learning of peripheral vasomotor responses by the curarized rat. *Communications in Behavioral Biology,* Part A, 1968, **1,** 209–212. (c)

DiCara, L. V., & Miller, N. E. Instrumental learning of vasomotor responses by rats: Learning to respond differentially in the two ears. *Science,* 1968, **159,** 1485–1486. (d)

DiCara, L. V., & Miller, N. E. Long term retention of instrumentally learned heart-rate changes in the curarized rat. *Communications in Behavioral Biology*, Part A, 1968, **2,** 19–23. (e)

DiCara, L. V., & Miller, N. E. Transfer of instrumentally learned heart-rate changes from curarized to noncurarized state: Implications for a mediational hypothesis. *Journal of Comparative and Physiological Psychology*, 1969, **68,** 159–162.

Holdstock, T. L., & Schwartzbaum, J. S. Classical conditioning of heart rate and galvanic skin response in the rat. *Psychophysiology*, 1965, **2,** 25–38.

Miller, N. E. Psychosomatic effects of specific types of training. *Proceedings of the New York Academy of Science*, 1969, in press.

Miller, N. E., & Banuazizi, A. Instrumental learning by curarized rats of a specific visceral response, intestinal or cardiac. *Journal of Comparative and Physiological Psychology*, 1968, **65,** 1–7.

Miller, N. E., & DiCara, L. V. Instrumental learning of heart-rate changes in curarized rats: Shaping, and specificity to discriminative stimulus. *Journal of Comparative and Physiological Psychology*, 1967, **63,** 12–19.

Miller, N. E., & DiCara, L. V. Instrumental learning of urine formation by rats; changes in renal blood flow. *American Journal of Physiology*, 1968, **215,** 677–683.

Seligman, M. E. P., & Maier, S. F. Failure to escape traumatic shock. *Journal of Experimental Psychology*, 1967, **74,** 1–9.

Trowill, J. A. Instrumental conditioning of the heart rate in the curarized rat. *Journal of Comparative and Physiological Psychology*, 1967, **63,** 7–11.

Weiss, J. A tail electrode for unrestrained rats. *Journal of the Experimental Analysis of Behavior*, 1967, **10,** 85–86.

Weiss, J. M., Krieckhaus, E. E., & Conte, R. Effects of fear conditioning on subsequent avoidance behavior. *Journal of Comparative and Physiological Psychology*, 1968, **65,** 413–421.

(Received January 21, 1969)

III

BLOOD PRESSURE AND VASOMOTOR RESPONSES

15

Instrumental Learning of Systolic Blood Pressure Responses by Curarized Rats: Dissociation of Cardiac and Vascular Changes

Leo V. DiCara and Neal E. Miller

Rats with skeletal muscles paralyzed by curare to rule out the effect of muscular activity were given artificial respiration and were rewarded by escape and/or avoidance of mild electric shock for either increasing or decreasing their systolic blood pressure. A chronic catheter was implanted within the abdominal aorta for measurement of same. Each trial was signaled by the onset of a light and tone; each experimental rat was yoked to a control rat. The yoked control received exactly the same treatment as the experimental S, except that it could do nothing to avoid being shocked, and was shocked whenever the experimental S was shocked. Over-all group increases and decreases of 22.3% and 19.2%, respectively, were obtained for the experimental groups, and were significantly different from the changes in the control groups. All experimental *Ss* without exception changed their blood pressure in the rewarded direction. Analyses of heart rate and temperature did not reveal any significant changes either between experimental groups at the beginning, at the end of training, or within experimental groups during training. Implications for learning theory, psychosomatic pathology, and treatment are discussed.

IN CLASSICAL CONDITIONING, the reinforcement must be a stimulus which has the unconditioned property of eliciting the specific response to be conditioned. The prevailing view has been that responses mediated by the autonomic nervous system can be modified only by classical conditioning, and not by instrumental learning. In instrumental learning, it is not necessary that the reinforcement have the property of eliciting the response to be learned; instead, it strengthens any immediately preceding response. This results in greater flexibility of learning, since a given response can be reinforced by a variety of rewards and a given reward can reinforce a variety of responses.

The question of whether the autonomic nervous system can be instrumentally conditioned is of fundamental importance for the functioning of this system in maintaining homeostasis, its malfunction in producing psychosomatic symptoms and for the possibility of using conditioning techniques in the therapy of such symptoms. A number of recent experiments have summarized the literature and presented evidence that heart rate, peripheral vasomotor responses, gastrointestinal motility, and urine formation can be modified by instrumental learning—using either onset of rewarding brain stimulation or the offset of painful electric shock as reinforcement.[2–4, 8–12] In each of these experiments, the media-

From The Rockefeller University, New York, N. Y. 10021.

Supported by Grant MH 13189 from the National Institute of Mental Health, U. S. Public Health Service.

Presented at Annual Meeting, American Psychosomatic Society, Boston, Mass., Mar. 29, 1968.

Received for publication Feb. 7, 1968.

*Advanced Research Fellow of the American Heart Association.

tion of the visceral response by respiratory changes or of overt skeletal responses was ruled out by training rats paralyzed by *d*-tubocurarine and maintained on artificial respiration.

The purpose of the present experiment was to determine whether conditioned increases and decreases of systolic blood pressure could be obtained in curarized rats, using as a reward escape from, and/or avoidance of, an electric shock to the tail.

Material and Method

The *Ss* were 28 naïve, male Sprague-Dawley rats weighing from 408 to 615 gm., divided into experimental and yoked control groups. Three days prior to training, a catheter was implanted into the lower third of the abdominal aorta of each S; the distal end was threaded under the skin and emerged at the back of the neck, using a modification of a procedure described by Herd and Barger.[5] Heparinized Ringer's solution (500 U./ml.) was infused through a fluid swivel at the rate of 0.5 cc./hr. for the next 72 hr. On the day of conditioning, *Ss* were given I.P. injections of 3.0 mg./kg. of *d*-tubocurarine chloride (Squibb) in a solution containing 3 mg./ml. As soon as the rat showed signs of difficulty in breathing, it was fitted with a face mask made from the mouthpiece of a rubber balloon, the lower portion of which was secured by placing it behind the upper incisors and the upper portion fitted snugly over the top of the snout. The balloon end of the mask was connected, via a one-holed rubber stopper to the tubes from a small-animal respirator. Respiration was administered at a 1:1 I/E ratio, 70 cycles per minute, with a peak reading of 20 cm. water pressure. In previous experiments, such respiration had been found to maintain the rats' heart rate and temperature at stable normal levels. After the lateral abdomen had been locally anesthetized by 0.15 ml. of 2% Xylocaine hydrochloride, a 30-gauge needle was inserted I.P., and additional *d*-tubocurarine was infused at the rate of 1 mg./kg./hr. throughout the rest of the experiment. Blood pressure was measured by means of a Statham pressure-transducer (P23 Dd) connected to the catheter at the level of the S's heart. The distance from the exit of the catheter from the S to the pressure transducer was 5 cm. Recordings were made at a sensitivity of 5.0 mv/cm. using a Grass Model 5PI low-level DC preamplifier.

In order to reward specific levels of systolic blood pressure, the amplified signal from the pressure transducer was fed parallel into two pens of a Grass polygraph. The first pen had a brass ball-point tip, which made contact with the surface of a plate constructed of 25 strips of brass, each 1.7 mm. wide, separated from each other by 0.3 cm., and inlaid into a flat piece of Plexiglas. By connecting these strips appropriately to programming equipment, reward could be delivered whenever the pen was on, above, or below a specified strip. The second pen traced an ink record of the exact movement of the first pen.

Heart rate was measured by stainless steel wires which had been previously implanted subcutaneously in the middle of the back at shoulder level and over the heart. The S's temperature was measured by a thermistor probe inserted 4 cm. into the rectum.

The shock source consisted of a voltage divider on the secondary of a 500-v AC transformer, with the current delivered to S through a fixed resistance of 220 K ohm. Discontinuous electric shock consisting of 0.1 sec. pulse of 0.3 mamp. AC was delivered at 2-sec. intervals through an electrode fitted snugly around the base of the tail.

During training, the S lay on a folded towel on its right side in a dark, soundproof, ventilated chamber (Grason Stadler) equipped with a speaker delivering a tone of 1000 cps at 82 db. After a period of 60 min. following curarization, during which heart rate and blood pressure were allowed to stabilize, blood pressure conditioning began.

Seven of the 14 experimental rats were rewarded for blood pressure increases, and the other 7 were rewarded for blood pressure decreases. Assignment to group was random, except for the last S, which had to be assigned to the group rewarded for blood pressure increases. Each of the experimental *Ss* was paired with a yoked partner. The yoked S received the same treatment as the experimental S, except that it was shocked whenever the experimental S was shocked and could do nothing to escape or avoid the shock.

Training consisted of the presentation of trials signaled by the onset of the 1000 cps tone, which was followed after 7.5 sec. by brief pulses of mild electric shock delivered to the tail. The reward was avoidance of, or escape from, the shock. If S made the desired-criterion blood-pressure response within 7.5 sec. of the onset of the warning tone, the tone was turned off and no shock was delivered. If S failed to make the desired-criterion response within the 7.5 sec., the shock was turned on until the criterion was achieved, at which time both the shock signal and shock were terminated. However, duration of shock on escape trials rarely, if ever, exceeded 60 sec., owing to the fact that during training, every successive block of 20 trials was analyzed for mean latency, and the level of blood pressure which S had to achieve in order to escape or avoid shock was increased if average trial latencies fell below 5 sec. and decreased if average latencies went above 10 sec. Training consisted of 200 time-in periods separated by time-out periods averaging 30 sec. (range: 10 to 50 sec.) during which S remained in silence. Training for the increase group required an average of 144 min. as compared to 152 min. for the decrease group ($t = .83$). The fifth and every tenth trial, thereafter, throughout training was a special test trial. During the first 7.5 sec. of the special trial both the criterion and the shock circuits were turned off, so that blood pressure could be measured for the same constant intervals throughout training without any possible contaminating effects from having the shock signal turned off, or the shock turned on. The scoring circuit was turned off and the criterion circuit was turned on at the end of this 7.5-sec. interval. If S did not achieve the criterion within the next 7.5 sec., the shock was turned on. In order to compare the effects of training on blood pressure, when no signals were present, all *Ss* received blank trials on the tenth trial, and every tenth one thereafter throughout training, each lasting 7.5 sec., during which blood pressure was measured but nothing else occurred.

Results

The results are presented in Fig. 1 for changes in blood pressure in experimental and yoked groups over training. It can be seen that *Ss* rewarded for increasingly higher levels of blood pressure increased their blood pressure from an initial 139 mm. Hg ($\pm$ 9.9) to a final one of 170 mm. Hg ($\pm$ 10.2), while those rewarded for increasingly lower levels progressively decreased their blood pressure from an initial level of 140 mm. Hg ($\pm$ 10.8) to a final one of 113 mm. Hg ($\pm$ 9.5). Since these learned changes are in opposite directions and each highly significant ($p < 0.01$), the yoked control results are not as crucial as they might have been if the differences in blood pressure between the two experimental groups did not change in opposite directions over training. Subjects yoked to the experimental *Ss* rewarded for increases in blood pressure showed a significant increase in their blood pressure ($t = 4.21$, df $= 6$, $p < 0.01$). However, the increase in the experimental group was significantly greater than the increase in the yoked group ($t = 3.33$, df $= 12$, $p < 0.01$). The *Ss* yoked to the experimental *Ss* rewarded for decreases in blood pressure also showed a significant increase in blood pressure ($t = 4.03$, df $= 6$, $p < 0.01$). Thus, there was a highly significant difference between experimental *Ss* rewarded for decreases and their yoked partners ($t = 11.48$, df $= 12$, $p < 0.001$).

Blood pressure on blank and test trials was analyzed to determine if the rats learned the discrimination of changing their blood pressure during the time-in stimulus, when such a response would be rewarded, rather than during the time-out intervals between trials, when it would not. The results did not indicate any evidence of discrimination learning, in either the group rewarded for increases or the group rewarded for decreases in blood pressure. This finding is in contrast to previous experiments on heart rate, in which rats learned discriminating responses to the time-in stimuli.[2, 10] Since a very low correlation was obtained between over-all changes in blood pressure and heart rate (0.08), in this experiment, rapid changes in heart rate—which can be learned as a discrimination—probably do not produce fast enough changes in blood pressure to demonstrate discrimination learning.

Analyses of heart rate and temperature did not show any significant over-all

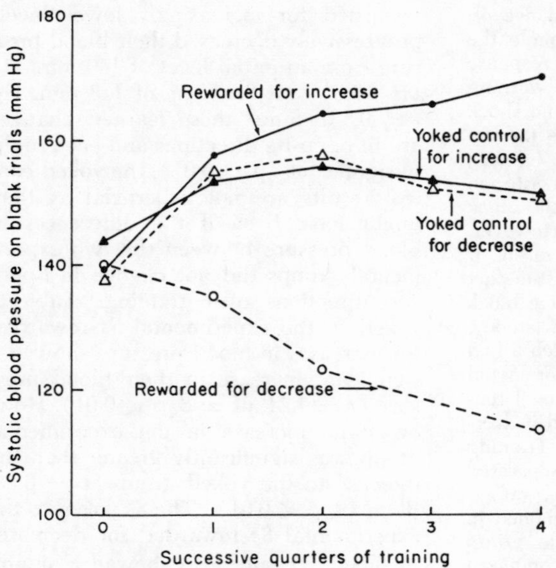

FIG. 1. Changes in blood pressure in experimental and yoked control groups during training.

changes in heart rate or temperature between the experimental groups at the beginning or end of training, or within the groups during training. Likewise, there were no significant differences in heart rate or temperature between the two yoked control groups at the beginning or end of training, or within the groups during training. In all individual *Ss*, heart rate and temperature were within normal ranges.

A nonsignificant correlation of 0.39 was obtained between changes in heart rate and blood pressure over training in the yoked control group, as compared to 0.08 for the experimental groups. A nonsignificant correlation of 0.14 was obtained for the over-all change in blood pressure learned and the amount of shock received, indicating that the marked differences in the blood pressure obtained in the experimental groups were not artifacts of the amount of shock received by *Ss* during training.

Discussion

The results of the present experiment indicate that it is possible to instrumentally train increases and decreases in systolic blood pressure, independently of changes in heart rate. The specificity of learning and dissociation of heart rate and blood pressure, exhibited in the present experiment, is in agreement with results from previous experiments from this laboratory. In one of these experiments, DiCara and Miller[4] showed that vasomotor changes in the cutaneous circulation can be learned and made specific to a given structure, as for example, one ear as opposed to the other. In another experiment, vasomotor changes affecting the blood flow to the kidney have been learned independently of changes in heart rate, blood pressure, or peripheral vasomotor activity.[11] In addition, changes in either heart rate or intestinal contraction have been learned without correlated changes in the nonrewarded response, indicating that the instrumental learning of a visceral response can be made specific to the type of response that is rewarded and is not limited to changes in the general pattern of activation, arousal, or parasympathetic-sympathetic balance.[9]

The fact that visceral responses are subject to instrumental learning means that the reinforcement of changes in them is not limited to unconditioned stimuli eliciting as an unconditioned response the specific change to be learned. Instead, it is highly probable that they can be modified by any one of the great variety of rewards and punishments known to be able to produce the learning of skeletal responses. Psychiatrists commonly refer to such rewards as secondary gains. It is clearly possible for the visceral responses involved in psychosomatic symptoms to be learned, as well as to be innately elicited, in the innate hierarchy of responses to a strong emotion, such as fear. The degree to which psychosomatic symptoms, and the individual differences in profiles studied by Lacey and Lacey,[6, 7] actually are learned will depend on the degree to which suitable conditions for reinforcement are present in the life of the individual.

The instrumental learning of visceral responses also has interesting therapeutic possibilities for clinical medicine. By recording changes in blood pressure, it should be possible to teach a well-motivated patient to change undesirable responses to more desirable ones. Engel[13] has recently tried a similar approach, and has reported success in the treatment of cardiac arrhythmias. Since other experiments from this laboratory have shown that the voltage of brain waves of curarized rats can be increased by rewarding increases, or decreased by rewarding decreases, it is conceivable that in some cases the responses reflected in abnormal EEG activity can be modified by learning.[1]

Summary

The traditional view is that visceral responses can be modified only by classical conditioning, and not by trial-and-error or instrumental learning. This problem is of critical importance, both for theories of learning and for understanding psychosomatic pathology. The present experiment has demonstrated that it is possible to obtain conditioned increases, and decreases of systolic blood pressure in rats. Each S was paralyzed by curare and maintained on artificial respiration to rule out the effects of skeletal muscle, and respiratory changes on blood pressure. Separate groups of *Ss* were rewarded by escape from, and/or avoidance of, mild electric shock for either increasing or decreasing their levels of blood pressure, on trials signaled by the onset of a light and tone. Blood pressure was measured by means of a chronic catheter in the abdominal aorta. Each experimental rat was yoked to a control rat. The yoked control received exactly the same treatment as the experimental S, except that it could do nothing to avoid being shocked, and was shocked whenever the experimental S was shocked. Over-all average increases and decreases in blood pressure of 22.3% and 19.2%, respectively, were obtained in the experimental groups, and each were significantly different from the changes in the yoked control groups. All experimental *Ss*, without exception, changed their blood pressure in the rewarded direction. Analyses of heart rate and temperature did not reveal any significant changes either between experimental groups at the beginning or at the end of training, or within experimental groups during the period of training. Likewise, there were no significant differences in heart rate or temperature between the two yoked control groups, either at the beginning or end of training, or within the groups during training. A correlation of 0.39 was obtained between changes in heart rate and blood pressure in the yoked control groups as compared to 0.08 for the experimental groups.

The demonstration that it is possible to instrumentally condition blood pressure levels raises the possibility that abnormal cardiovascular responses may be learned as psychosomatic symptoms and contribute to the development of atypical profiles of autonomic reactivity to stress and cardiovascular-renal pathology. The instrumental learning of blood pressure responses has interesting therapeutic possibilities for clinical medicine, since it may be possible to teach hypertensive patients to permanently lower their blood pressure levels.

The Rockefeller University
New York, N. Y. 10021

References

1. Carmona, A. Trial-and-error learning of the voltage of the cortical EEG activity. Ph.D. thesis, Yale University, 1967.
2. DiCara, L. V., and Miller, N. E. Changes in heart rate instrumentally learned by curarized rats as avoidance responses. *J Comp Physiol Psychol 65:*8, 1968.
3. DiCara, L. V., and Miller, N. E. Instrumental learning of peripheral vasomotor responses by the curarized rat. *Commun Behav Biol 1:*209, 1968.
4. DiCara, L. V., and Miller, N. E. Instrumental learning of vasomotor responses by rats: learning to respond differentially in the two ears. *Science 159:*1485, 1968.
5. Herd, H. A., and Barger, A. C. Simplified technique for chronic catheterization of blood vessels. *J Appl Physiol 19:*791, 1964.
6. Lacey, J. I., and Lacey, B. C. Verification and extension of the principle of autonomic response stereotypy. *Amer J Psychol 71:*50, 1958.
7. Lacey, J. I., and Lacey, B. C. The law of initial value in the longitudinal study of autonomic constitution: reproducibility of autonomic responses and response patterns over a four-year interval. *Ann NY Acad Sci 98:*1257, 1962.
8. Miller, N. E. Psychosomatic effects of specific types of training. *Proc NY Acad Sci* (In press).
9. Miller, N. E., and Banuazizi, A. Instrumental learning by curarized rats of a specific visceral response, intestinal or cardiac. *J Comp Physiol Psychol 65:*1, 1968.
10. Miller, N. E., and DiCara, L. V. Instrumental learning of heart-rate changes in curarized rats: shaping, and specificity to discriminative stimulus. *J Comp Physiol Psychol 63:*12, 1967.
11. Miller, N. E., and DiCara, L. V. Instrumental learning of urine formation by rats: changes in filtration, renal blood flow and osmolarity. *Amer J Physiol* (In press).
12. Trowill, J. A. Instrumental conditioning of the heart rate in the curarized rat. *J Comp Physiol Psychol 63:*7, 1967.
13. Engel, B. T. Personal communication.

16

Effects of Feedback and Reinforcement on the Control of Human Systolic Blood Pressure

David Shapiro, Bernard Tursky, Elliot Gershon and Melvin Stern

Abstract. *An automatic procedure providing information about human systolic blood pressure at each successive heartbeat under routine laboratory conditions is described. Twenty normal male subjects were given feedback of their own systolic pressure, half operantly reinforced for increasing and half reinforced for decreasing their pressure. Significant differences in pressure were obtained in a single session. The apparatus and results suggest a possible approach to the treatment of essential hypertension.*

Learning theorists have assumed that autonomic responses cannot act on the external environment and are therefore incapable of being modified by their consequences, such as reward or punishment (*1*). Now, many studies offer positive evidence that the autonomic nervous system can be conditioned instrumentally (*2, 3*). The work of Miller and associates on rats treated with curare shows that the effects of the conditioning do not depend on mediating changes in skeletal behavior but are specific to the responses that are reinforced (*4*).

One promising area of potential application of operant conditioning is in the treatment of disorders mediated autonomically (*5*). Essential hypertension, elevated blood pressure without a demonstrable cause, is a disorder in which the autonomic nervous system may play an important role.

The purpose of this research is to develop a method for the treatment of patients with essential hypertension. This requires instrumentation which can provide an individual with continuous external feedback of his own systolic pressure. Arterial cannulation, the most accurate means of recording blood pressure, is not satisfactory for repetitive laboratory experimentation. Cuff-type intermittent recording procedures are inadequate because they offer insufficient opportunity for feedback and reinforcement.

We developed automatic instrumentation that yields a continuous approximation of a subject's systolic pressure on each successive heartbeat. The essential elements of the recording and reinforcement procedure are shown in Fig. 1. A conventional blood-pressure cuff is wrapped around the upper arm and pumped up to a given pressure by means of a regulated, low-pressure, compressed-air source. The pressure can be held constant or changed in 2-mm steps. The pressure applied to the cuff is measured by a Statham (P23AC) transducer and recorded on one channel of an Offner type R polygraph.

A crystal microphone to detect the Korotkoff sounds (*6*) is mounted in the cuff over the brachial artery, and the output of this microphone is fed into an audio system and displayed as a pulse on a second polygraph channel. When the pressure in the artery is high enough to overcome the occlusive pressure of the cuff, the Korotkoff sound is produced. By setting the cuff at a constant pressure close to the measured systolic pressure of a subject, this system gives information of upward and downward changes in systolic pressure at each successive heartbeat. When Korotkoff sounds occur, the systolic pressure is at or above the cuff pressure; when the sounds disappear, the systolic pressure is below the cuff pressure.

The electrocardiogram is recorded on a third channel of the polygraph, and the R-wave in each successive heart-cycle pattern is detected by an electronic switch. The Korotkoff sounds are also detected automatically by a second switch. Inasmuch as the interval between the R-wave and the Korotkoff sound is approximately 300 msec, the joint occurrence of a heartbeat and the presence, or absence, of the Korotkoff sound in the prescribed time can be automatically determined with appropriate logic modules. Respiration was recorded on a fourth channel by means of a strain gauge belt fastened around the waist.

The operant conditioning of systolic blood pressure was studied in 20 normal male college students between 21 to 27 years of age. The subjects were seated in a semireclining position in a sound- and light-controlled room and were told that (i) the experiment was concerned with the ability of individuals to control certain physiological responses, and (ii) people could often achieve such control when given information about their own responses. Subjects were asked not to move about or tense their muscles during the session and to keep their breathing as regular as possible. Any subject with initial blood pressures above 135 mm systolic or 85 mm diastolic was eliminated from the study.

Subjects were given 25 trials. Each trial period was 65 seconds long, preceded by 11 seconds during which time the cuff was inflated. The beginning of a trial period was signaled to the subject by a blue light. After each trial,

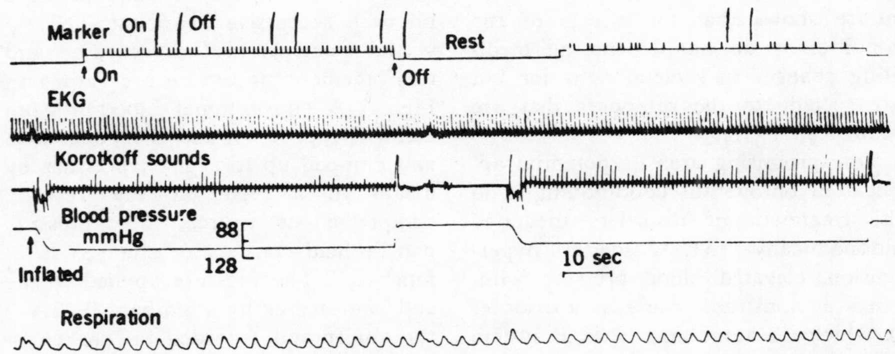

Fig. 1. Segment of polygraph record of a subject in the *down* group, reinforced for lowering his pressure. Two trials are shown. Marker channel indicates onset of blue light (first arrow), feedback (small marks), and slide presentations (large spikes). Rest period begins when blue light is turned off. For this subject in the *down* condition, feedback was administered when Korotkoff sounds were below critical amplitude. Blood-pressure channel shows cuff inflation and deflation. Applied pressure was 116 mm on first trial shown. During this trial there were 76 percent successes. On second trial shown, applied pressure was changed to 114, making task more difficult. Note reduction in success rate to 37 percent on this trial.

the cuff was deflated and there was a rest period of from 20 to 25 seconds (Fig. 1).

Two experimental conditions were studied to control for possible unconditioned effects of the reinforcer. In the *up* condition, ten subjects were reinforced for raising their systolic blood pressure. In the *down* condition, ten subjects were reinforced for lowering their pressure. During each trial, a success was defined by the presence of a Korotkoff sound for *up* subjects, and the absence of this sound for *down* subjects, after each heart beat. This information was fed back to the subject by the programming apparatus, which produced a 100-msec flash of red light and a simultaneous 100-msec tone of moderate intensity for each success (*7*). The reinforcer was a slide of a nude from *Playboy* magazine projected on a screen in front of the subject for 5 seconds after every 20 flashes of light (*8*) (Fig. 1). Each slide was different. Subjects were instructed that the slides offered an incentive for them to try to make the light flash and the tone beep as often as possible.

Five preliminary blank trials were used to obtain a cuff pressure which resulted in Korotkoff sounds on 50 percent of the heartbeats. Thus, all subjects started with exactly the same probability (50 percent) of being above or below this initial cuff pressure and the same probability of receiving feedback and reinforcement. Pairs of subjects were matched on this starting blood-pressure level and assigned at random to each experimental condition.

During each trial, the applied pressure was maintained constant. If more than 75 percent of the heartbeat-Korotkoff sound contingencies were successful on two successive trials, the pressure applied to the cuff was increased by 2 mm for *up* subjects and decreased by 2 mm for *down* subjects on the next trial. This change made the task correspondingly more difficult in each condition (Fig. 1). In the case of success on fewer than 25 percent of the heartbeats, the task was made easier by 2 mm. No change was made if a trial showed 25 to 75 percent successes. These shaping criteria were chosen empirically, and enabled the tracking of the subject's pressure to be rapid and accurate. The purpose of the procedure was to try to maximize the differences between experimental groups and also take into account the possible changes in blood pressure unrelated to the experimental treatment, such as habituation or adaptation to the situation.

The data are cuff pressures on each trial, indicating approximate systolic pressure. The average curves for each condition, grouped for simplicity in blocks of five trials, are shown in Fig. 2. An analysis of variance (*9*) (two

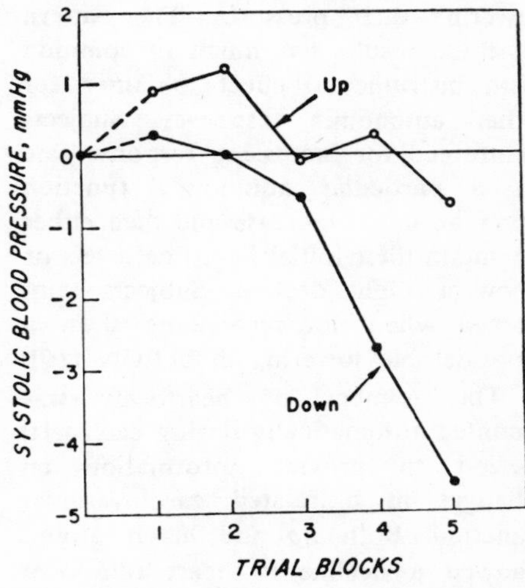

Fig. 2. Average approximate systolic pressures in groups reinforced for increasing (*up*) and decreasing (*down*) blood pressure. Each point is the mean of ten subjects, five trials each. The means were adjusted for slight differences in starting pressure between the two groups (*up*, 120.9 mm; *down*, 120.1 mm) which were set to zero.

experimental treatments, 25 repeated trials) shows that the main effect for conditions is not significant. The main effect for trials and the interaction between trials and treatments are significant beyond the .01 probability level. On the first two trial blocks, in eight of the ten matched pairs, *down* subjects showed a larger decrease in pressure than *up* subjects (see Fig. 2). The greatest differentiation between the two groups appeared during the last two trial blocks. From the third to the fifth trial block, *up* subjects show a decrease of 0.6 mm compared to 4.0 mm for *down* subjects. A *t*-test (matched pairs) of the average difference between these changes is significant beyond $P < .01$. Thus, the *up* group tended to maintain their base line pressure or decrease slightly during the session. The decrease was significantly and consistently more marked in the group reinforced for lowering their pressures. The pattern of these results has much in common with instrumental effects obtained for other autonomic responses. Subjects reinforced for increasing response rate of a particular autonomic function show an initial increase and then either maintain their initial base line levels or show a slight decline. Subjects reinforced when not responding show a considerable lowering of activity (*10*).

The number of heartbeats was counted automatically during each trial period to provide information on changes in a related cardiovascular function. Both *up* and *down* groups showed a decline in heart rate over time. According to the analysis of variance, this trial effect is highly significant ($P < .005$). The main effect for treatments and the interaction between trials and treatments are not significant, indicating no systematic relationship between heart rate and the conditioned blood-pressure changes. Breathing patterns were indistinguishable in the two experimental conditions, and irregularities in breathing appeared with equal frequency in both.

On the basis of these indices of related functions, no particular causal or mediating effect of one system on another accounts for the differentiation of blood pressure observed. Verbal reports elicited from subjects after the session were inconsistent. A few *down* subjects stated that the flashing light probably meant a state of relaxation. Most subjects in both conditions seemed to infer that the experimenter wanted them to get excited. With one or two exceptions, subjects said that they had no control over the physiological response (the flashing light) and no knowledge of what specific bodily function we were trying to condition.

The results of this study indicate that systolic blood pressure can be modified by the use of external feedback and operant reinforcement. The apparatus and techniques described here should prove of value in research on modification of blood pressure in hypertensive patients. The methods also merit consideration in other areas of psychophysiological investigation in which continuous evaluations of systolic blood pressure are desired.

DAVID SHAPIRO, BERNARD TURSKY
ELLIOT GERSHON, MELVIN STERN
Harvard Medical School,
Massachusetts Mental Health Center,
Boston 02115

References and Notes

1. G. A. Kimble, *Hilgard and Marquis' Conditioning and Learning* (Appleton-Century-Crofts, New York, 1961).
2. H. D. Kimmel, *Psychol. Bull.* **67**, 337 (1967).
3. E. S. Katkin and E. N. Murray, *ibid.* **70**, 52 (1968).
4. N. E. Miller, *Proc. N.Y. Acad. Sci.*, in press. L. V. DiCara has reported successful instrumental conditioning of blood pressure in curarized rats, with the use of escape from and avoidance of electric shock as a reward (L. V. DiCara, paper presented at a meeting of the Eastern Psychological Association, Washington, D.C., April 1968).

5. Only one such application has been reported. B. T. Engel achieved marked improvement in patients with atrial fibrillation by providing them external feedback of their cardiac activity and reinforcing slower heart rate (B. T. Engel, paper presented at Pavlovian Society, Princeton, N.J., November 1967).

6. Cuff-type methods depend on the detection of the turbulent flow of blood through the artery under the cuff which is produced when the cuff pressure is slightly lower than the arterial pressure. This turbulence produces an audible sound, the Korotkoff sound.

7. Exteroceptive feedback was used to facilitate the development of control of blood pressure. Augmented sensory feedback has been used extensively in studies of the control of heart rate. See (*3*).

8. These slides have been shown to be an effective operant reinforcer of the galvanic skin response. See G. E. Schwartz and H. J. Johnson, *J. Exp. Psychol.*, in press.

9. This was treated as a two-factor experiment with two levels of one factor (*up, down*) and repeated measures on the second factor (25 trials). For experimental treatments, d.f. = 1/432; for trials and the trial × treatment interaction, d.f. = 24/432. See B. J. Winer, *Statistical Principles in Experimental Design* (McGraw-Hill, New York, 1962), p. 302.

10. D. Shapiro, A. Crider, B. Tursky, *Psychon. Sci.* 1, 147 (1964).

11. Supported by K3-MH-20,476-06, MH-08853-05, and MH-04172-08 (NIMH); also by contract Nonr-1866(43), group psychology branch. We thank Mrs. M. Chartres for her assistance.

19 September 1968; revised 4 December 1968 ■

17

Behavioral Induction of Arterial Hypertension and its Reversal

Herbert Benson, J. Alan Herd, W. H. Morse, and Roger T. Kelleher

ENVIRONMENTAL STIMULI can induce elevations of arterial blood pressure (1). In the preceding paper (2), experiments in the squirrel monkey showed that characteristic increases in blood pressure develop in association with certain patterns of key pressing. In one schedule, the monkeys were required to make a number of key presses after the onset of a light. Completion of this requirement turned off the light and prevented the delivery of noxious stimuli. After appropriate training, whenever the light was turned on, each animal began to key press, and blood pressure increased. When the light was turned off at the completion of the key-pressing requirement, each animal stopped key pressing and blood pressure decreased.

In the present experiments, squirrel monkeys were trained to press a key which turned off a white light and prevented the delivery of noxious stimuli. In order to eliminate the effects of key pressing per se on cardiovascular function, the key was removed after initial training. The schedule was modified so that the white light was turned off only after an increase in blood pressure; thus the rise in blood pressure prevented the delivery of noxious stimuli. This schedule resulted in hypertension. In a later phase of the experiment in the same animals, the blood pressure requirement was reversed; a decrease in blood pressure turned off the white light and prevented noxious stimuli. The level of blood pressure returned toward control values during this phase. Scheduling of stimuli thus induced hypertension and later induced its reversal to control levels of blood pressure.

METHODS

Two hundred seventy-three experiments were performed in three squirrel monkeys (600 to 900 g). The techniques for measuring arterial blood pressure and training the animals to press a key have been described in the preceding paper (2) with the exception that, in the present experiments, the monkeys were maintained in an upright posture by a loose-fitting Lucite neck collar. Each animal was anesthetized with mixtures of halothane and oxygen in order to implant one end of a small polyvinyl chloride catheter (id 0.38 mm and od 0.76 mm) into the aorta of each animal through the right internal or common iliac artery. The other end of the catheter was passed out through the skin to the interscapular region. The catheters were filled with heparin solution (10 mg Na heparin/ml) and kept patent by daily injections of saline (0.9% NaCl) followed by 0.2 ml of the heparin solution. During blood pressure measurements, each animal sat in a restraining chair which was placed within a lightproof, sound-attenuating isolation chamber. The animal could be viewed through a peephole in the chamber. The implanted catheter was connected through tubing to a Statham P23Db pressure strain-gauge transducer, which in turn was connected to a syringe infusion pump. The electrical output of the strain gauge was amplified and recorded on a Grass Instrument Co. polygraph. All pressure values reported were mean arterial blood pressure obtained by electrical filtering. The mean blood pressure of the first 20 min was compared to that of the last 20 min in all sessions. Patency of the arterial catheters was verified at each session by free flow of blood from the open end and also by amplitude of the pressure oscillations.

Control phase. During the 2-week period following surgery and before training began, control measurements of mean arterial blood pressure were made during sessions of 1–4 hr. The isolation chamber was illuminated by a 25-w overhead light. This first 2-week phase was termed "control."

Key-pressing phase. In the next phase of the experiment,

termed "key pressing," the monkeys were trained to press a key under the type of schedule previously described (2). Noxious stimuli (650 v, 60 Hz, through a series resistance, to give a current of 1–10 ma for 200 msec) were delivered in the presence of a white light through two electrodes lying across a shaved portion of the tail. A predetermined number of key presses (usually 30) turned off the white light. If 30 key presses were not made within 30 sec after the onset of the white light, a noxious stimulus was delivered and repeated every 15 sec until either the required number of key presses was made or a total of 10 noxious stimuli had been given. In either case the white light was turned off, leaving the monkey in complete darkness. After 1 min of darkness, the white light reappeared and the cycle repeated.

When stable, rapid key pressing (usually 30 key presses in 15–20 sec) and reproducible blood pressure elevations associated with this key pressing were obtained, the schedule was changed and a second contingency introduced. Under the new schedule, 30 key presses or an increase in mean arterial blood pressure for a specific amount of time turned off the white light. Whenever mean arterial blood pressure rose above some predetermined value (5–10 mm Hg above existing values) a blue panel light was turned on by means of a voltage discriminating circuit attached in parallel with the output of the polygraph driver amplifier. When the mean arterial blood pressure exceeded the predetermined value for a specified interval (1–4 sec), the white light was turned off even if the preset number of key presses had not been made. As long as the mean blood pressure remained above the specified value, the blue light stayed on. The white light stayed off when the blue light was on even after the 1-min time-out period had elapsed. When the mean arterial blood pressure fell below the specified value, the blue light turned off. The cycle then repeated as soon as the regularly scheduled, 1-min timeout had been completed. Thus, whenever the blue light was on, noxious stimuli were never delivered (Fig. 1). This phase was continued until mean arterial blood pressure

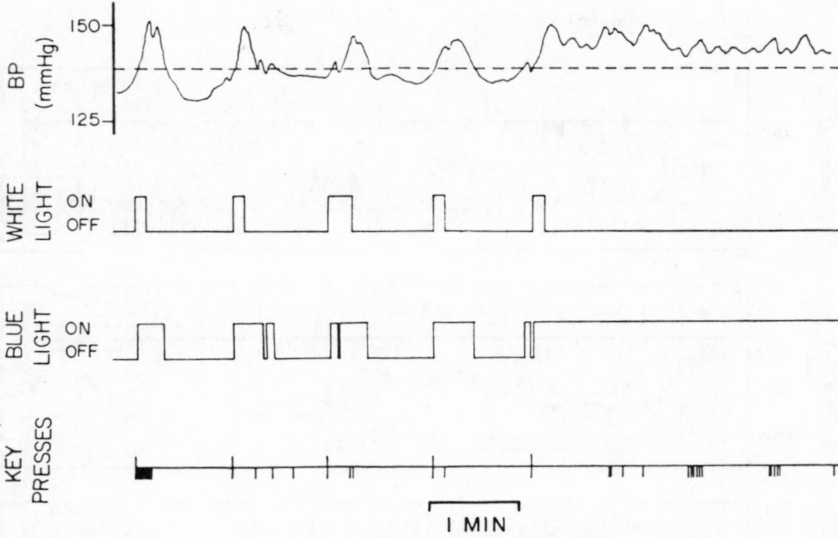

FIG. 1. Representative figure illustrating relation of blood pressure changes to white and blue light sequences and to key pressing. From top to bottom: mean arterial blood pressure (BP); white light indicator; blue light indicator; event marker (vertical stroke across horizontal line indicates start of cycle; stroke below line indicates key press). In first cycle, white light was turned off after 30 key presses; in subsequent cycles, white light was turned off after blood pressure exceeded 139 mm Hg (dashed line) for required period of time. After last cycle, blood pressure remained above 139 mm Hg keeping on blue light, thereby preventing further presentations of white light. (See text.)

TABLE 1. *Mean arterial blood pressure of three monkeys during different phases of blood pressure training*

Phase of Blood Pressure Training	Mean Arterial Blood Pressure, mm Hg±1 SE					
	Monkey 110		Monkey 129		Monkey 144	
	First 20 min	Last 20 min	First 20 min	Last 20 min	First 20 min	Last 20 min
Control	118±5.0	117±6.1	126±1.2	127±2.0	124±1.9	119±7.5
Key pressing	123±1.0	125±1.0	139±1.2	139±1.3	129±1.9	130±1.7
Pressor:						
Early	128±2.5	130±2.4	149±1.4	144±1.8	132±0.5	131±1.8
Late	134±2.3	143±2.2	151±1.0	155±1.5	137±1.1	152±2.7
Depressor:						
Early	137±3.7	137±2.8	152±1.5	148±2.6	147±3.8	144±2.6
Late	133±2.5	114±3.3	151±1.0	138±1.9	144±3.0	133±1.6

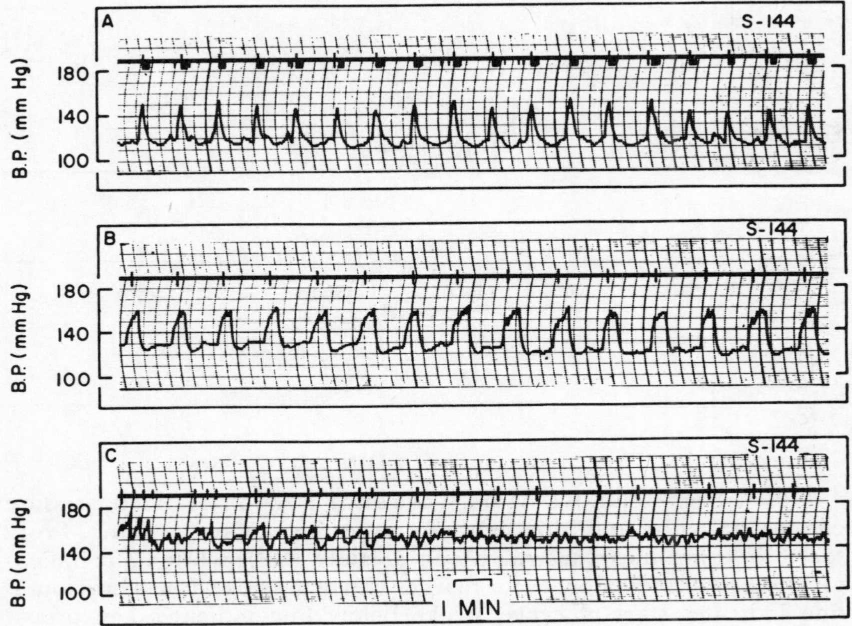

FIG. 2. Characteristic patterning of mean blood pressure (*monkey S-144*) during 3 different phases of experiment in which noxious stimuli were scheduled to occur in presence of a white light. *A*, *B*, and *C*: polygraph records of mean arterial blood pressure in mm Hg (B.P.) with an event marker above blood pressure recording. In event recording, a vertical stroke across line indicates start of a cycle with presentation of white light; a stroke above line, noxious stimulation; a stroke below line, key press. *A*: key-pressing phase. *B:* early pressor phase. *C*: early depressor phase. In *A* and *B*, blood pressure increased with presentation of white light both in presence (*A*) and absence (*B*) of key. In *C*, blood pressure rose slightly, remained unchanged, or fell in presence of white light. When blood pressure rose slightly or remained unchanged in presence of white light (first 5 cycles of record *C*) noxious stimuli were delivered and were followed by transient decreases in mean blood pressure (10–20 mm Hg). When blood pressure fell in presence of white light (cycles 6–14 of *record C*), light was turned off automatically and noxious stimuli were not delivered. In early pressor phase of training (*B*) white light was associated with marked increases in mean blood pressure, whereas in early depressor phase (*C*) same white light was associated with decreases in mean blood pressure.

increased to levels that consistently turned off the white light before completion of the key-pressing requirement.

Pressor phase. The key was removed from the restraining chair in the next phase of the experiment, termed the "pressor phase." As in the key-pressing phase, when blood pressure was above the specified level, the blue light was turned on automatically. After a predetermined time in the presence of the blue light, the white light was turned off, thereby preventing the delivery of noxious stimuli. During successive experimental training sessions, the specified level of blood pressure to be exceeded was gradually increased (10–15 mm Hg above existing values) and the interval of time required for the blood pressure to be above the specified level was gradually lengthened (15–60 sec). Thus each animal was required to elevate its blood pressure to higher and higher levels for increasing periods of time in order to turn off the white light and prevent the delivery of noxious stimuli. After 30–40 min of each session, if the specified level of blood pressure was further increased or left unchanged, the monkey no longer met the requirement, numerous noxious stimuli were administered, and mean arterial blood pressure decreased. Various modifications of the schedule were therefore tried to maintain elevated blood pressures. Empirically, it was noted that lowering the requirement 5–20 mm Hg rather than raising it actually led to more marked and sustained elevation of blood pressure.

Depressor phase. The last phase of the experiment was termed the "depressor phase." After several consecutive training sessions in which mean arterial blood pressure rose and was maintained for 20 min or more at levels of approximately 25 mm Hg higher than pretraining control levels, the schedule was again changed. During one session, the blood pressure requirement was increased to 40–50 mm Hg above existing values for 15–60 sec. As a result, the schedule requirement was seldom met and noxious stimuli were frequently delivered.

In following sessions, the blood pressure requirement was reversed. A decrease in mean arterial blood pressure

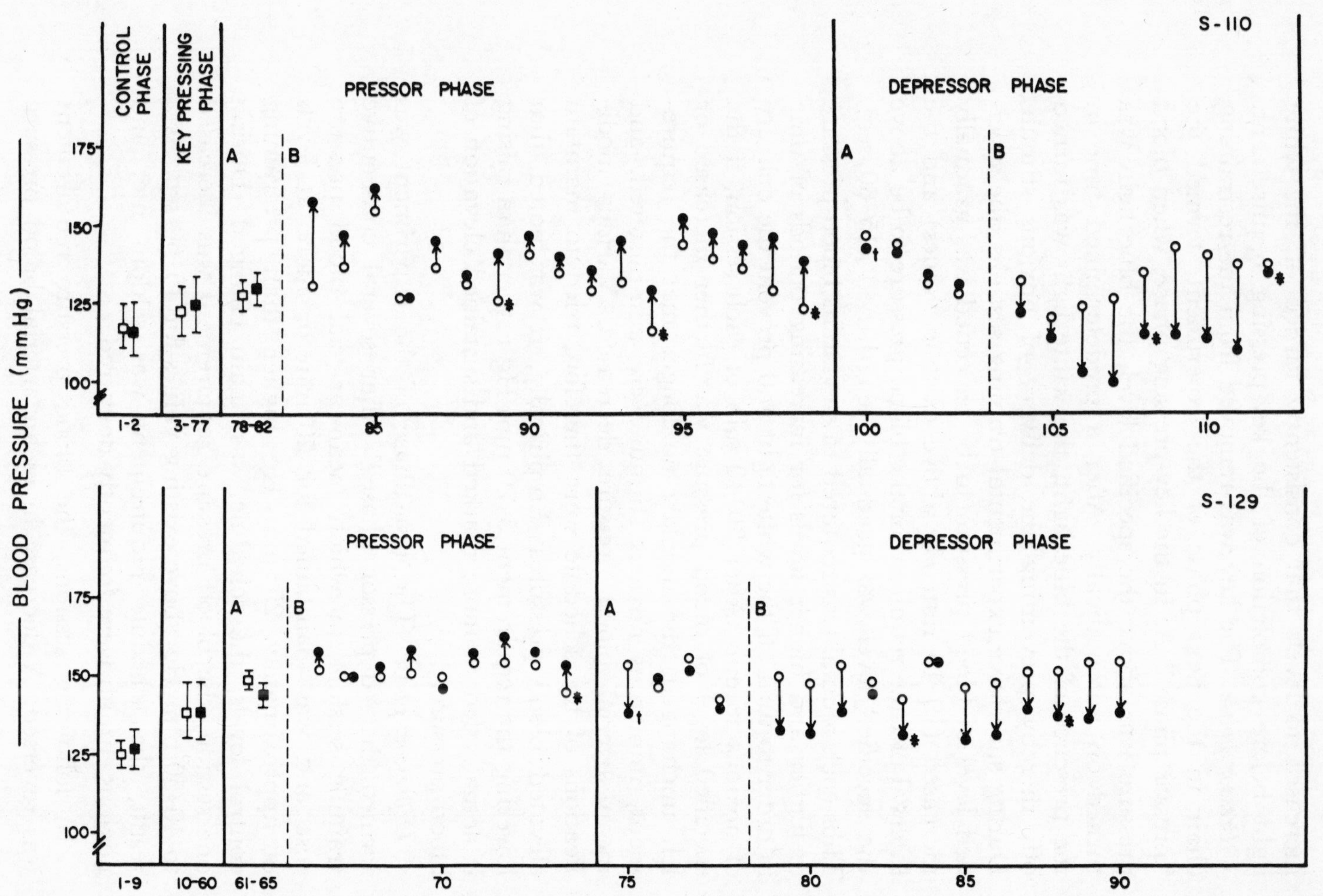

BLOOD PRESSURE (mmHg)
S-110
CONTROL PHASE
KEY PRESSING PHASE
PRESSOR PHASE
DEPRESSOR PHASE
A
B
A
B
175
150
125
100
1-2
3-77
78-82
85
90
95
100
105
110
S-129
PRESSOR PHASE
DEPRESSOR PHASE
A
B
A
B
175
150
125
100
1-9
10-60
61-65
70
75
80
85
90

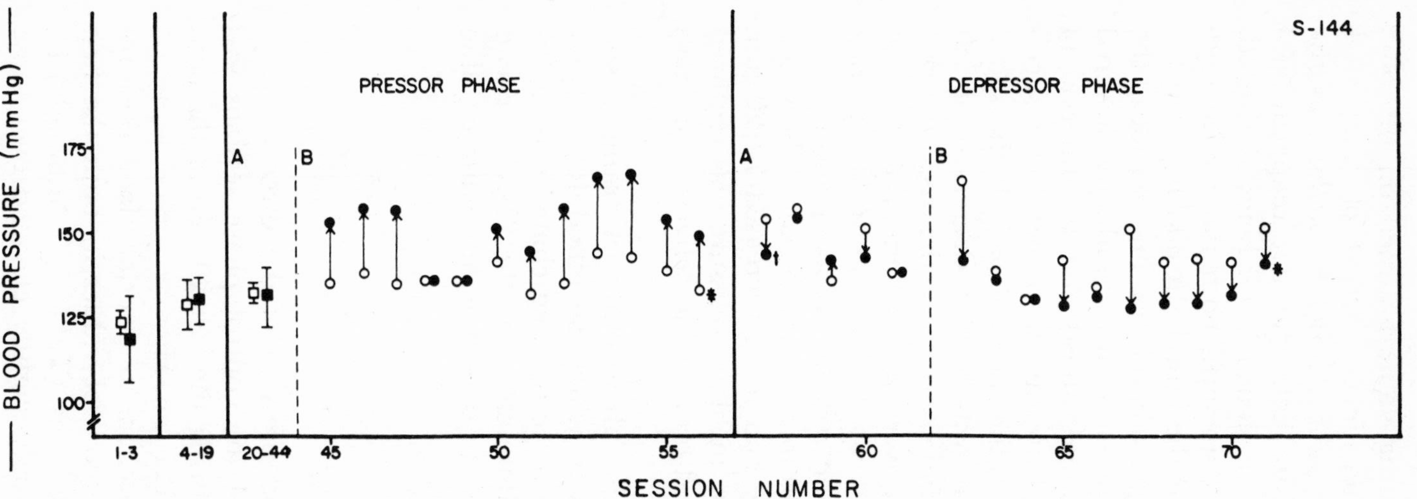

FIG. 3. Changes in mean arterial blood pressure of 3 squirrel monkeys (*S-110*, *S-129*, and *S-144*) during all 273 sessions of training program. Each panel shows mean blood pressure (mm Hg) during control phase, key-pressing phase, early pressor phase, *A*; late pressor phase, *B*; early depressor phase, *A*; and late depressor phase, *B*. Open squares are averages of all mean arterial blood pressures recorded during first 20 min of control, key pressing, and early pressor phases. Closed squares are averages of all mean arterial blood pressures recorded during last 20 min of control, key pressing, and early pressor phases. Short horizontal lines joined by vertical lines through squares represent 1 SD of mean. Open circles are mean arterial blood pressures recorded during first 20 min of individual sessions. Closed circles are mean arterial blood pressures recorded during last 20 min of individual sessions. Asterisks indicate sessions in which no schedule was presented. Daggers indicate sessions in which blood pressure requirement was increased 40–50 mm Hg above existing levels for 15–60 sec (see text). In late pressor phase, mean arterial blood pressures recorded during last 20 min of each session were almost always higher than mean pressures recorded during first 20 min and were significantly higher than those recorded during control phase. In late depressor phase, mean arterial blood pressures recorded during last 20 min of each session were almost always lower than mean pressures recorded during first 20 min and were only slightly different from those recorded during control phase.

(5–10 mm Hg below existing values) turned on the blue light and, after a predetermined interval of time (1–4 sec), turned off the white light. As before, as long as the blue light was on the white light did not reappear and there was no possibility of noxious stimuli being delivered. The cycle did not begin again until both the 1-min time-out period had elapsed and the blue light had been turned off. During successive experimental training sessions, the specified level of blood pressure was gradually lowered (2–10 mm Hg below existing values) and the interval of time required for the blood pressure to be below the specified level was gradually lengthened (2–8 sec). If the specified level of blood pressure was further decreased or left unchanged, the monkey failed to meet the requirement, numerous noxious stimuli were administered, and mean arterial blood pressure increased. Thus, as in the pressor phase, an adjustment in the specified blood pressure level had to be made after 30–40 min of each session. The specified level of blood pressure had to be increased 10–20 mm Hg, i.e., a smaller decrease in blood pressure was required to turn on the blue light and prevent the delivery of noxious stimuli.

Interspersed between the above-outlined training sessions were several sessions in which no schedule was presented. Each animal sat in its restraining chair in the isolation chamber which was illuminated by the 25-w overhead light. No noxious stimuli were delivered, no white or blue lights were presented.

RESULTS

The mean arterial blood pressure of the three monkeys during the first 20 min of the control phase sessions was 123 mm Hg (Table 1). During the last 20 min of the control phase sessions, it was 121 mm Hg.

In the key-pressing phase, mean arterial blood pressure was slightly elevated over control values in both the first and last 20 min of the sessions (Table 1). Although the mean blood pressure was only slightly elevated over those values obtained in the control phase, there was a marked patterning of blood pressure changes associated with the onset of the white light and rapid key pressing (Fig. 2*A*). The blood pressure of each monkey rapidly rose 20–40 mm Hg when the white light came on and each monkey pressed the key. The blood pressure abruptly returned to existing values when the white light was turned off and each monkey stopped key pressing.

In the early stages of the pressor phase, after the key had been removed from the restraining chair, the mean arterial blood pressure was slightly more elevated in both the first and last 20 min of each session (Table 1). The upward patterning of blood pressure continued (Fig. 2*B*). When the white light was on, blood pressure rose 20–40 mm Hg and remained elevated until the requirement was met and the white light turned off; blood pressure then fell rapidly. Thus, similar patterning of blood pressure was produced in absence or presence of the key. During the first session after the key had been removed, each of the monkeys tended to paw at the hole formerly occupied by the key. After a few sessions, this activity ceased and each monkey sat quietly while the white light was on.

In the late pressor phase, the mean arterial pressure of the three monkeys during the first 20 min of each session averaged 141 mm Hg, rising to 150 mm Hg during the last 20 min of each session (Table 1 and Fig. 3). Since the blood pressure progressively increased during each session, the blood pressure requirement was exceeded most of the time, keeping the blue light on. Consequently, the white light appeared less often and only 3.3 noxious stimuli per 2-hr session were administered. Fewer cyclic changes in blood pressure were recorded, but a gradual upward drift of blood pressure occurred (Fig. 4*A*).

In the early stage of the depressor phase, mean arterial blood pressure varied little from the previous pressor phase (Table 1). After the reversal of the blood pressure requirement, the patterns of blood pressure change also reversed. With the onset of the white light, blood pressure usually fell 10–20 mm Hg (Fig. 2*C*) and turned off the white light.

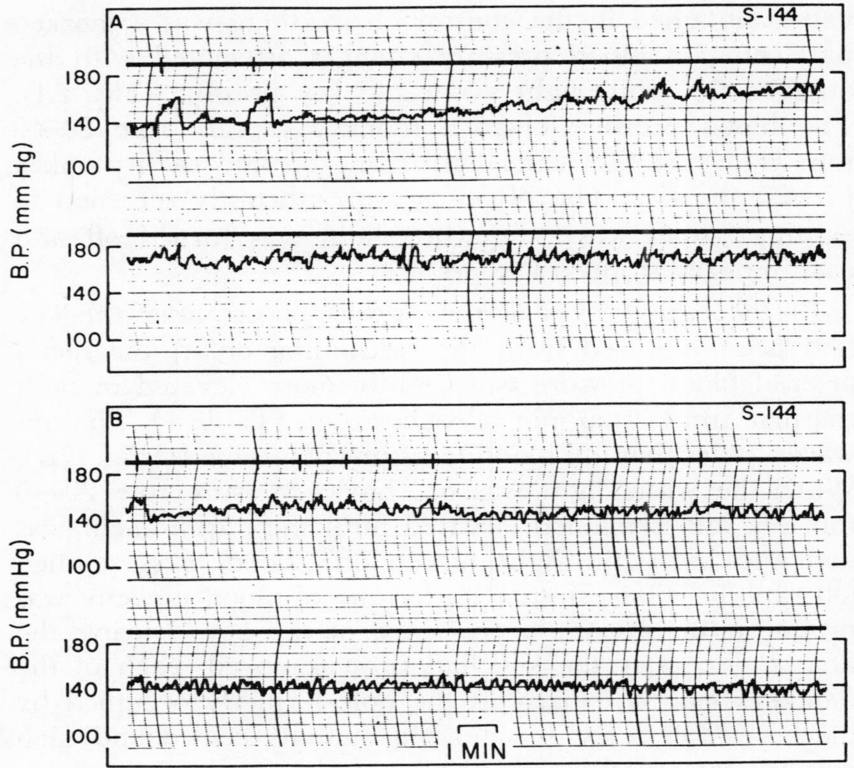

FIG. 4. Progressive changes of mean arterial blood pressure (*monkey S-144*) in late pressor and late depressor phases of experiment. *A* and *B*: polygraph records of mean arterial blood pressure in mm Hg (B.P.) with an event marker above blood pressure recordings. Both *A* and *B* are continuous 40-min records from top left to bottom right. In event recording, a vertical stroke across line indicates start of a cycle with presentation of white light; a stroke above line, noxious stimulation. *A:* late pressor phase. *B:* late depressor phase. In late pressor phase (*A*) a blue light (not represented on record) was turned on whenever blood pressure exceeded a specified level, and in late depressor phase (*B*) same blue light was turned on whenever blood pressure was below a specified level. In *A*, blood pressure increased transiently with two presentations of white light and then remained above level at which blood pressure discriminator was set (140 mm Hg). In *B*, blood pressure decreased transiently in first cycle following presentation of white light and one noxious stimulus. In cycles 2–7 (*B*) blood pressure decreased transiently following presentations of white light without noxious stimuli. During and after cycle 7 (*B*) mean blood pressure remained below level at which blood pressure discriminator was set (155 mm Hg). In late pressor phase (*A*), blood pressure increased progressively in presence of blue light, whereas in late depressor phase (*B*) same blue light was associated with a progressive decrease in blood pressure.

Occasionally blood pressure did not fall until noxious stimuli were delivered. After the white light was turned off, blood pressure returned to previous values. Thus, in contrast to the patterned elevations of blood pressure with the onset of the white light in the pressor phase, patterned decreases were noted in the depressor phase (Fig. 2,*B*, cf. *C*). No progressive changes in blood pressure occurred in this early depressor phase, but in the late depressor phase the mean arterial blood pressure of the three monkeys declined from 143 to 128 mm Hg (Figs. 3 and 4*B*). During this final phase of training, the animals received an average of 33.0 noxious stimuli per 2-hr session, 10 times the number delivered in the pressor phase.

In the several sessions in which no lights or noxious stimuli were presented, mean blood pressure changed in the same direction as in other sessions of that phase (Fig. 3).

DISCUSSION

These experiments demonstrate that mean arterial blood pressure can be made to rise or fall predictably when environmental stimuli are scheduled according to variations in blood pressure. In the preceding paper (2), patterns of change in blood pressure developed in association with patterns of key pressing. The changes in blood pressure cannot be specifically ascribed to key pressing because, as we have shown in the present experiments, patterning of blood pressure persisted even after the key had been removed, i.e., mean arterial blood pressure was directly influenced by the stimulus of the white light. In the pressor phase, the blood pressure could be made to rise reproducibly 20–40 mm Hg when the white light was on. Later, in the presence of the same white light but a reversed blood pressure requirement schedule, the blood pressure could be made to fall reproducibly 10–20 mm Hg. Thus, the patterning of blood pressure changes was under stimulus control of the white light.

As soon as transient elevations of blood pressure were under stimulus control, it was possible to cause progressive and prolonged elevations by appropriate scheduling of stimuli according to blood pressure. Mean arterial blood pressure could be maintained at levels of 150 mm Hg, 30 mm Hg above control levels. Later, under the depressor schedule, blood pressure returned from hypertensive levels toward control values. Hence, behavioral experiments could induce hypertension and its reversal.

The results of the present experiments also show that the changes in blood pressure were not determined by the total number of noxious stimuli, but by the schedule under which they were delivered. As described in the preceding paper (2), the delivery of noxious stimuli alone was not sufficient to cause sustained elevation of mean arterial blood pressure. For example, the animals received the greatest number of noxious stimuli in the early stages of their training, but arterial pressure was not elevated at that time. In the present experiments, the animals received the lowest number of noxious stimuli in the late pressor phase, the time when blood pressure was highest. When an excessive number of stimuli were delivered in the early pressor phase, blood pressure decreased. In order to reverse blood pressure from hypertensive to control levels, 10 times the number of noxious stimuli were delivered. Evidently, environmental stimuli may cause either elevations or depressions of blood pressure depending on the schedule of their presentation.

We thank Dr. A. C. Barger and Dr. P. B. Dews for their encouragement and helpful advice.

Supported by Public Health Service Grants HE 09154 (CV), MH 02094, and MH 07658, and The Medical Foundation, Inc., of Boston.

H. Benson is a Medical Foundation Research Fellow.

J. A. Herd is an Established Investigator of the American Heart Association.

W. H. Morse is the recipient of Research Career Program Award 5-K3-GM-15, 530 from the National Institutes of Health.

R. T. Kelleher is the recipient of Research Career Program Award 5-K3-MH-22, 589 from the National Institute of Mental Health.

Received for publication 19 August 1968.

REFERENCES

1. Cannon, W. B. *Bodily Changes in Pain, Hunger, Fear and Rage. An Account of Recent Researches into the Function of Emotional Excitement.* New York: Appleton, 1929, p. 93.

2. Herd, J. A., W. H. Morse, R. T. Kelleher, and L. G. Jones. Arterial hypertension in the squirrel monkey during behavioral experiments. *Am. J. Physiol.* 217: 24–29, 1969.

18

Operant Conditioning of Increases in Blood Pressure

Lawrence A. Plumlee

ABSTRACT

Four monkeys were presented with 10 sec tones which terminated with shocks. The tones were immediately terminated without shock if the animal's diastolic blood pressure rose above a criterion level and remained high for 1 sec. Termination of the tone was followed by a 5 sec time out. Trials began whenever the pressure dropped below the criterion level. All subjects learned the avoidance task, showing diastolic elevations of up to 60 mm of mercury in response to the tones. A linear relationship was seen between the minimum pressure required for avoidance and the pressure achieved. No change in blood pressure accompanied a second stimulus which was never paired with shock. A fifth control monkey was yoked to one of the experimental monkeys and simultaneously received all shocks and tones as determined by the blood pressure of the experimental animal. The yoked control showed no pressure changes to the tones, but normal pressure elevations to shock.

DESCRIPTORS: Blood pressure, Operant conditioning, Avoidance conditioning, Heart rate. (L. Plumlee)

Several recent publications have shown that autonomic functions may be operantly conditioned. Heart rate (Brenner, 1966; Engel & Chism, 1967; Engel & Hansen, 1966; Frazier, 1966; Hnatiow & Lang, 1965; Miller & DiCara, 1967; Shearn, 1962; Shearn & Clifford, 1964; Trowill, 1967), skin resistance (reviewed by Kimmel, 1967), salivation (Brown & Katz, 1967; Miller & Carmona, 1967), colonic contractions (Miller & Banuazizi, 1968), blood pressure (Miller, 1967), and vasoconstriction (DiCara & Miller, 1968) have been operantly conditioned. Miller & Banuazizi (1968) and DiCara & Miller (1968) have also shown that learned autonomic responses can be specific to the response that is rewarded and do not necessarily reflect arousal state or sympathetic–parasympathetic balance.

The present study was undertaken to determine if the blood pressure of monkeys could be brought under the control of operantly programmed environmental consequences. Cued avoidance trials were programmed to begin whenever the monkey's blood pressure fell below a predetermined trigger level. After 10 sec the CS (tone) terminated with a painful foot shock unless the subject's diastolic pressure increased to above the trigger level and remained there for at least 1 sec. Such a response was rewarded by immediate termination of the tone, avoidance of the shock, and a 5 sec time out.

The assistance of Dr. Joseph V. Brady is gratefully acknowledged.

Address requests for reprints to: Dr. Lawrence A. Plumlee, AR–3, 200 C Street, S.W., Washington, D. C. 20204.

METHOD

Physiological Recording Procedure

Male adult rhesus monkeys, 11–15 pounds, were implanted with chronic intraarterial catheters and placed in primate chairs. The principles of laboratory animal care as promulgated by the National Society for Medical Research were observed. Access to the catheter was denied the monkey by a thoracic plate with chest band (Werdegar, Johnson, & Mason, 1964). The catheterization method previously described (Perez-Cruet, Plumlee, & Newton, 1966) was used, except that a polyvinyl chloride catheter .047 in. × .067 in. was implanted with its unbeveled tip between the renal arteries and the aortic bifurcation. Blood pressures were monitored on Statham gauges P23AA or P23Gb and recorded on a Grass model 5 or Beckman Type R Dynograph which was calibrated daily. Output of the polygraph amplifier operated a minimum diastolic pressure detector (Swinnen, Brown, & Plumlee, 1966) which was used to operate environmental programming equipment. Heart rate and respiration using a circumthoracic belt containing a strain gauge were sometimes recorded simultaneously.

Operant Conditioning Procedure

In four monkeys, a fall in diastolic blood pressure below a given level (the trigger level) started a tone (CS+) in the monkey's booth. The tone could be terminated immediately by a 1 sec period during which no diastolic blood pressures below trigger level occurred (the avoidance response). If the subject failed to hold its blood pressure above trigger level for 1 sec, the tone terminated after 10 sec and was followed instantly by a 100 msec 10 ma shock to the feet. On the first day, small, normally-occurring pressure increases were reinforced by termination of the tone. After the first day, in order to maximize performance, the animals were placed on an automatic adjusting schedule. If an animal received more than three shocks in 15 min, the pressure required for avoidance was lowered. If no shock was received during an hour, the blood pressure level required for avoidance was raised. Training periods varied up to 22 hours per day. In one monkey, a second, higher pitched tone (CS−) was later introduced which could also be terminated by 1 sec of pressures above the trigger level, but which was never followed by shock. This monkey, M.1, was yoked to a control monkey, M.5, which from the beginning of training simultaneously received all shocks and tones as determined by the blood pressure of the experimental animal. Blood pressure of the yoked control was monitored at the same time.

RESULTS

All four subjects learned the avoidance task. Acquisition took from one day to one week, and by the ninth day of training, all animals demonstrated a visible pressure rise to each tone. Since the adjusting schedule changed the pressure requirement according to the number of shocks received, a reduction in the frequency of errors did not occur. Initially, after a few pairings of tone and shock, elevations in pressure lasting several minutes occurred. Later, occasionally no pressor response was seen to the stimuli. With increasing frequency, small rises

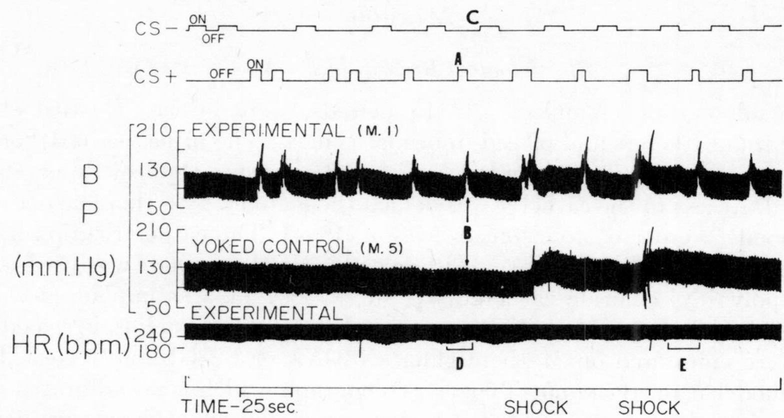

FIG. 1. Discriminated avoidance by diastolic pressure elevation. A. Onset of tone (CS+) which warns that a foot shock will be administered after 10 sec unless the blood pressure of the experimental monkey rises above trigger level and remains there for at least 1 sec; B. Correct blood pressure avoidance response in experimental monkey with no concomitant change in pressure of yoked control; C. Onset of tone (CS−) which has never been paired with shock; D. Cardioacceleration accompanying pressure rise in experimental monkey; E. Heart rate stability accompanying pressure rise in experimental monkey. Response of blood pressure of yoked control to shock illustrates its capacity to show pressure elevations.

and falls in pressure came into phase with tones and intertrial intervals respectively. Then, differences in pressure between these two periods increased. Finally, the time of each elevation shortened.

Figure 1 illustrates typical responses of monkey 1 to randomly presented tones after one month of training. The pressure rises to above trigger level, remains there for a second or two, and then rapidly begins its descent. In Figure 1, the trigger level has been set manually to a much higher pressure requirement than usual in order to illustrate the response to shock. At A, for example, a CS+ presentation is followed by a rapid rise in pressure in the experimental animal at B, with no change in pressure seen in the yoked control. At C, a tone which has never been followed by shock is presented, and no change in pressure in either animal is seen. At D, cardioacceleration accompanies the rise in pressure of the experimental monkey, but at E, a similar pressure change is accompanied by no change in heart rate.

Table 1 summarizes all blood pressure elevations during 30 consecutive minutes of observation. An elevation was defined as increase of diastolic pressure of 12 or more mm of mercury above the pressure baseline during the preceding intertrial interval and occurring in 10 seconds or less. Although M.1 elevated its pressure 12 mm above the baseline of the preceding intertrial interval only 61 trials out of 65, it still achieved perfect avoidance behavior and received no shocks during the 30 minutes. This was possible because the baseline pressure during the intertrial interval preceding the four remaining trials was elevated, making it possible to avoid the shock with an acute elevation of less than 12 mm.

In Figure 2, similar pressure responses to the warning stimulus are seen in three other experimental animals. Table 2 summarizes the responses of all four monkeys

TABLE 1

Comparison of blood pressure changes in experimental and yoked control subjects

Subject Groups	During CS+		During CS−		During Intertrial Interval	
	No. trials	No. elevations >12 mm	No. trials	No. elevations >12 mm	No. trials	No. elevations >12 mm
Experimental (M.1)	65	61	62	0	127	0
Yoked Control (M.5)	65	0	62	1	127	1

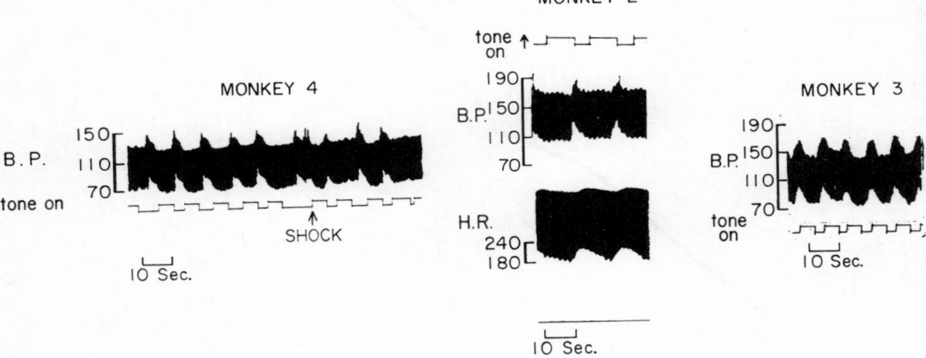

FIG. 2. Shock avoidance by pressure elevation in 3 monkeys

TABLE 2

Avoidance task performance

Animal	Consecutive Trials Observed	Avoidance Response to Tone	No Avoidance Response to Tone	Avoidance Response in Absence of Tone
1	100	100	0	0
2	100	100	0	0
3	84	82	1	1
4	100	100	0	0

during a large number of consecutive trials. Because the response was quite predictable, polygraph records of the performance were not recorded during most of the trials. Thus, it is not possible to present the results of a random selection of trials. In order to reduce the possibility of bias, the data were tabulated from all of a large number of consecutive trials in order to show performance when no selection of individual trials was exercised by the experimenter. The sequences selected were the best performance for each animal in which 100 consecutive trials were available. In the case of M.3, no record of more than 84 consecutive trials existed. Table 2 shows that perfect avoidance performance may be obtained with pressure elevations accompanying each CS+, and no such rises occurring spontaneously. Further, an examination of over 100 consecutive trials of alter-

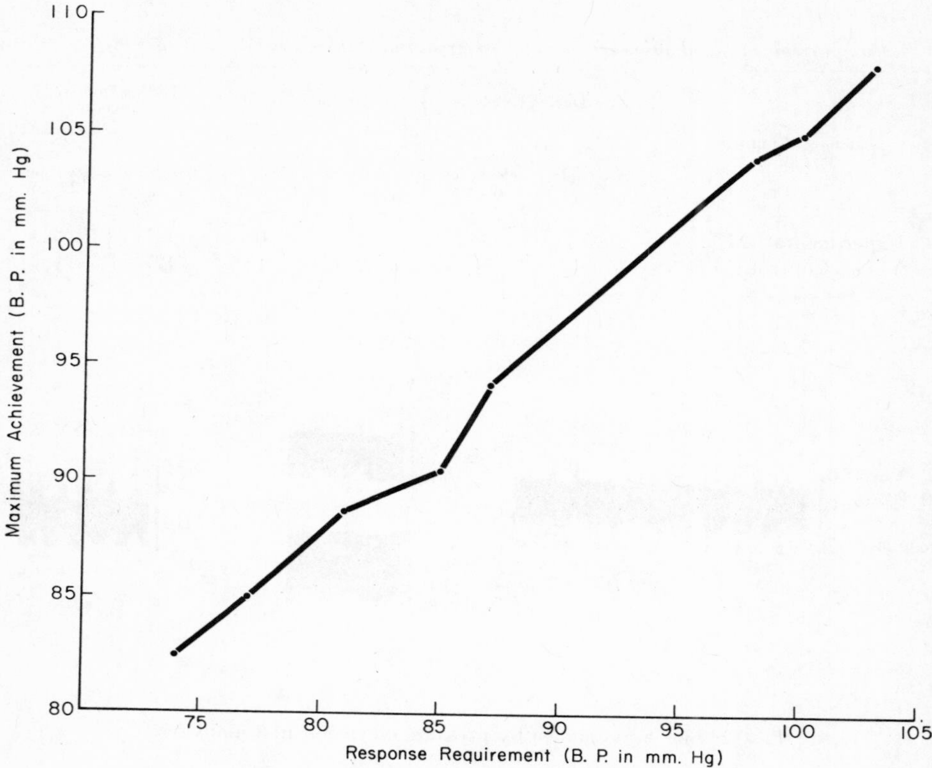

FIG. 3. Relationship between the diastolic pressure required to avoid shock and the actual pressure demonstrated in response to the warning tone. Monkey 3. All points but one are means of 10 determinations.

nating CS+ and CS− showed that M.1 avoided shock on every CS+, but made no pressure responses to any CS−.

Figure 3 is a plot of trigger level (minimum diastolic pressure required for avoidance) as a function of pressure actually achieved during the avoidance trial. To obtain these data, the automatically adjusting schedule was turned off and diastolic pressure level required to terminate the tone and avoid shock was set manually to the levels graphed along the abscissa. These data were acquired on M.3 on two afternoons one week apart. Within this pressure range, when the pressure requirement is low, pressure rises are small; similarly, when the requirement is high, the animal raises its pressure proportionally more. The maximum achievable blood pressure varies from day to day. Thus, the three highest points shown in Figure 3 could not be replicated 8 days later. On the latter occasion, when the response requirement was raised above 90 mm Hg, the animal began to receive frequent shocks and pressure responses to the tone soon extinguished.

Figure 4 illustrates that the heart rate changes which accompany sudden blood pressure increases may be highly variable even when the requirements are similar in the same monkey on the same day. In the upper tracing, large heart rate

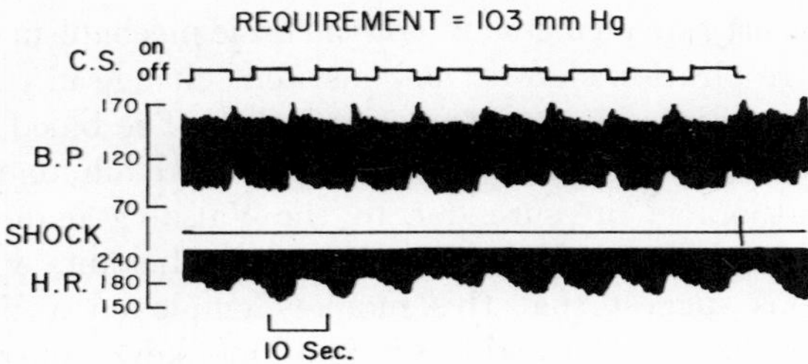

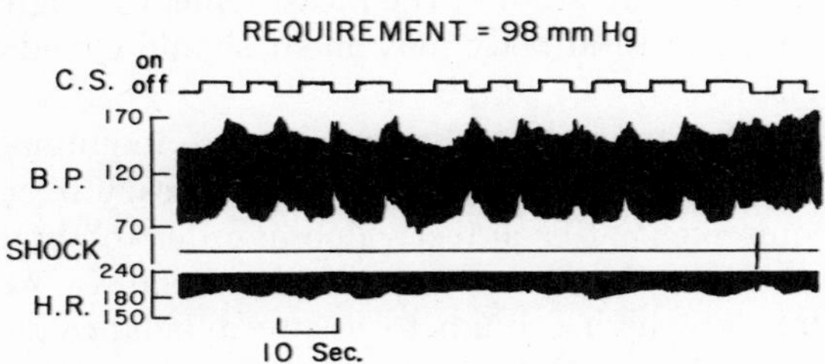

FIG. 4. Two different heart rate patterns accompanying similar blood pressure responses to similar avoidance requirements in same monkey on the same day. Monkey 2.

swings accompany the rises, whereas in the lower tracing, similar blood pressure changes occur with little or no heart rate change. Whether or not heart rate changes accompanied blood pressure elevation did not depend upon the time of day or the duration of training.

DISCUSSION

The ease with which large blood pressure rises can be conditioned on an avoidance design is not surprising in view of the ease with which classically conditioned blood pressure elevations may be secured. The most impressive results are the high level of pressure elevations achieved, the amount of control that may be exerted over height and time of occurrence of the pressure change, the rapidity with which large rises and falls are executed, and the contrast between experimental and yoked control animals. While these avoidance schedules do not produce chronic elevations, it is possible that such schedules may evolve from modifications of these methods. The animals were kept on the schedule for up to 22 hours a day to determine its chronic effects on the blood pressure. We wondered if the repeated transient increases in diastolic pressure every 8–15 seconds would lead to a permanent, self-sustaining hypertension. In M.1, this was carried out for 10 weeks, in M.2 for 7. There was no change in the blood pressure at rest in either animal. Post-mortem histological examination revealed no renal changes attributable to hypertension.

If the stroke volume and peripheral resistance remain constant, an increase in heart rate would lead to an increase in blood pressure. However, rapid, large fluctuations in blood pressure unaccompanied by a simultaneous change in heart rate proves that heart rate changes are not the basic mechanism of these pressure rises. Records of respiration showed no consistent change in rate or depth correlating with pressure increase or return to baseline. The blood pressure records of M.1 shown in Figure 1 are similar to records obtained in humans during sudden increases in intrathoracic pressure due to the Valsalva maneuver (Marshall, Scherger, & Shepherd, 1961; McIntosh, Burnum, Hickam, & Warren, 1954). Visual observations suggest that this monkey employed a Valsalva strategy during some acute elevations. In addition, in all monkeys, contraction of the leg extensor and arm flexor muscles by pushing down on the footrest while pulling on the headplate usually accompanied the rise when the pressure requirement for avoidance was high. In future studies, the measurement of right atrial pressure, tracheal temperature, and total body movement should elucidate the strategies adopted for these rises.

The linear relationship between minimum pressure requirement and achievement shown in Figure 3 illustrates that it is not necessary to set an upper limit on the pressure requirement at each test value in order to obtain a straight line function. If one assumes that pressure elevation requires work expenditure, raising the pressure only high enough to assure avoidance is consistent with efficient behavior. It is also consistent with good stimulus control, since termination of the tone is followed by no further avoidance behavior.

Conditioning studies of autonomic functions may increase our understanding of psychosomatic disease. Many diseases result from chronic hyper-activity of functions under autonomic control. Establishing functional changes seen in disease, or the diseases themselves, by placing subjects in programmed environments, suggests that such diseases are primarily psychosomatic in origin. Variations in the program provide a means of analysis of the pathogenic elements of an organism's relationship with its environment. Further, the specificity of conditioned autonomic responses to the responses which are rewarded strengthens the view that the kind of psychosomatic disease an individual develops in response to stress may be determined, at least in part, by factors other than genetic constitution.

REFERENCES

Brenner, J. Heart rate as an avoidance response. *Psychological Record*, 1966, *16*, 329–336.

Brown, C. C., & Katz, R. A. Operant salivary conditioning in man. *Psychophysiology*, 1967, *4*, 156–160.

DiCara, L. V., & Miller, N. E. Instrumental learning of vasomotor responses by rats: Learning to respond differentially in the two ears. *Science*, 1968, *159*, 1485–1486.

Engel, B. T., & Chism, R. A. Operant conditioning of heart rate speeding. *Psychophysiology*, 1967, *3*, 418–426.

Engel, B. T., & Hansen, S. P. Operant conditioning of heart rate slowing. *Psychophysiology*, 1966, *3*, 176–187.

Frazier, T. W. Avoidance conditioning of heart rate in humans. *Psychophysiology*, 1966, *3*, 188–202.

Hnatiow, M., & Lang, P. J. Learned stabilization of heart rate. *Psychophysiology*, 1965, *1*, 330–336.

Kimmel, H. D. Instrumental conditioning of autonomically mediated behavior. *Psychological Bulletin*, 1967, *67*, 337–345.

Marshall, R. J., Scherger, A., & Shepherd, J. T. Blood pressure during supine exercise in idiopathic orthostatic hypertension. *Circulation*, 1961, *24*, 76–81.

McIntosh, H. D., Burnum, J. F., Hickam, J. B., & Warren, J. V. Circulatory changes produced by the Valsalva maneuver in normal subjects, patients with mitral stenosis, and autonomic nervous system alterations. *Circulation*, 1954, *9*, 511–520.

Miller, N. E. Psychosomatic effects of specific types of training. Paper read at Conference on Experimental Approaches to the Study of Emotional Behavior, at the New York Academy of Sciences, Nov. 16–18, 1967. *Annals of the New York Academy of Sciences*, in press.

Miller, N. E., & Banuazizi, A. Instrumental learning by curarized rats of a specific visceral response, intestinal or cardiac. *Journal of Comparative & Physiological Psychology*, 1968, *65*, 1–7.

Miller, N. E. & Carmona, A. Modification of a visceral response, salivation in thirsty dogs, by instrumental training with water reward. *Journal of Comparative & Physiological Psychology*, 1967, *63*, 1–6.

Miller, N. E., & DiCara, L. Instrumental learning of heart-rate changes in curarized rats: Shaping and specificity to discriminative stimulus. *Journal of Comparative & Physiological Psychology*, 1967, *63*, 12–19.

Perez-Curet, J., Plumlee, L. A., & Newton, J. E. O. Chronic basal blood pressure in unanesthetized dogs using the ring-catheter technique. *Proceedings of the Symposium on Biomedical Engineering*, 1966, *1*, 383–386.

Shearn, D. W. Operant conditioning of heart rate. *Science*, 1962, *137*, 530–531.

Shearn, D. W., & Clifford, G. D. Cardiac adaptation and contingent stimulation. *American Psychologist*, 1964, *19*, 491.

Swinnen, M. E., Brown, C. C., & Plumlee, L. A. A novel peak detector. *Psychophysiology*, 1966, *3*, 98–100.

Trowill, J. A. Instrumental conditioning of the heart rate in the curarized rat. *Journal of Comparative & Physiological Psychology*, 1967, *63*, 7–11.

Werdegar, D., Johnson, D. G., & Mason, J. W. A technique for continuous measurement of arterial blood pressure in unanesthetized monkeys. *Journal of Applied Physiology*, 1964, *19*, 519–521.

19

Instrumental Learning of Vasomotor Responses by Rats: Learning to Respond Differentially in the Two Ears

Leo V. DiCara and Neal E. Miller

In previous studies from this laboratory, we have discussed the importance of the instrumental learning of visceral responses for theories of learning and psychosomatic medicine, and have summarized the literature; we have presented evidence that the instrumental learning of cardiac, gastrointestinal, and renal responses can occur in curarized rats, with the use of either the onset of rewarding brain stimulation or the offset of mildly painful electric shock to the tail as reinforcement (*1*).

DiCara and Miller (*2*) have shown that vasomotor responses in the tail of the rat can be modified by instrumental learning. The purpose of the experiment we now report was to determine whether such vascular learning can be made specific to a given structure, such as one ear. In order to accomplish this, individual measures of vasomotor responses in the ears of the curarized rat were fed into a bridge circuit so that the differences in vasomotor activity between the two ears could be detected. The output of the bridge circuit was then fed in parallel into two pens of a Grass polygraph. One pen, the tip of which was constructed of a brass ballpoint, traced the between-ear difference over the surface of a plate constructed of 25 strips of brass, each 1.7 mm wide, inlaid into a flat piece of Plexiglas and separated from each other by 0.3 mm. When these strips were appropriately connected to programming equipment, reward consisting of electrical stimulation of the brain (ESB) could be delivered whenever the pen tracing the difference between the two ears was on or beyond a specified "reward criterion" strip. The second pen traced an ink record of the exact movements of the first pen. Other pens recorded the separate responses of each ear.

Subjects were 12 male rats of the Sprague-Dawley strain (394 to 558 g), implanted with permanent monopolar electrodes aimed at the medial forebrain bundle with Krieg stereotaxic coordinates of 1.5 mm posterior to bregma, 8.5 mm below the surface of the skull at 1.5 mm lateral to midline. Subsequent histologic examination showed that all of the electrode tips were located in the medial forebrain bundle at the level of the ventromedial nucleus [for details of general method, see (*3*)].

During vasomotor conditioning, the subject lay prone in a harness-supported cloth sling, placed in a soundproof, ventilated enclosure equipped with a loudspeaker delivering a 1000-hz tone at 82 db. The sling was cut so as to allow the subject's forepaws to hang down through the opening and rest on a small platform support. Grass photoelectric plethysmograph transducers were used to measure the vasomotor activity in the ears, forepaws, and tail.

Recordings were taken at a sensitivity of 0.1 mv/cm for the ears, and 1.0 mv/cm for the forepaws and tail. The transducers were rigidly mounted on swivel arms attached to the metal frame of the harness so that the arms could be adjusted in all directions to place the photocells in homotopic positions on the ears (bisecting the central artery) and forepaws. The photocell used to measure vasomotor activity in the tail was placed at the base of the tail, and the tail was elevated about 3 cm to allow excreted boluses to pass without disturbing the position of the photocell. Heart rate was measured with previously implanted stainless steel electrodes, and temperature was measured by a thermistor probe inserted 4 cm into the rectum. A Grass Model 5 polygraph was used for all recordings.

Three to 4 days before vasomotor training, and approximately 2 weeks after electrode implantation, a current ranging from 30 to 100 μa was adjusted for each subject in order to elicit maximum rates of bar pressing for 0.5 second of 60-cycle a-c ESB. Current was then held constant, and on the next 2 days each subject was trained on fixed-interval schedules. During this training, a 1000-hz tone was turned on whenever bar pressing would secure ESB, and turned off during the time-out intervals (lengthened progressively: 5, 10, 20, 30 seconds), at which time the subject was in silence and ESB was not available.

The day after the above training, subjects were injected intraperitoneally with 3 mg of *d*-tubocurarine chloride per kilogram of body weight in a solution containing 3 mg/ml; the animals were fitted with a specially constructed face mask and artificially respirated, with the ratio of inhalation to exhalation being 1:1 (70 cycle/min, and a peak pressure of 20 cm-water). Additional *d*-tubocurarine was infused intraperitoneally at 1.0 mg per kilogram of body weight per hour throughout the experiment.

At the end of a 60-minute habituation period, vasomotor training began. Of the 12 rats used in this experiment, six were rewarded for relatively greater vasodilatation in the right ear (which could also be relatively less vasoconstriction), and the other six were rewarded for relatively greater vasodilatation in the left ear. Assignment to groups was random except for the last two subjects, that had to be assigned to the group rewarded for left-ear dilatation. All subjects had ESB electrodes located on the right side of the brain.

Training consisted of the presentation of trials signaled by the onset of the 1000-hz tone, during which the subject could obtain ESB for making the proper vasomotor response, and time-out or intertrial intervals during which the subject remained in silence and ESB was not available. A total of 300 trials was presented on a variable-interval schedule with a mean intertrial interval of 30 seconds (range: 10 to 50). The fifth trial and every tenth trial thereafter were test trials during which ESB could not be obtained for the first 7.5 seconds of the trial, so that vasomotor performance could be measured without contaminating effects from having the tone turned off or ESB turned on. On the tenth trial and every tenth one thereafter, subjects received so-called blank trials each lasting 7.5 seconds, during which vasomotor activity was recorded, but the subjects remained in silence and could not achieve reward. Throughout training, each block of 20 trials was analyzed, and the criterion level (that is, the difference in vasomotor activity between the two ears required to obtain ESB) was made more difficult if the time to achieve reward fell below an average of 5 seconds and easier if it was above 10 seconds.

The results are presented in Fig. 1.

Since the resistance of the photocell in the photoelectric transducer depends on the amount of light reaching it, changes in vasomotor activity have been expressed as changes in resistance of the photocell. The subjects rewarded for relatively greater vasodilatation in the right as compared to the left ear showed significant increases in dilatation of the right ear ($P < .01$) and significant decreases in vasodilatation of the left ear ($P < .01$), with the difference between the two ears at the end of training being highly significant ($t = 10.12$, $P < .001$). Subjects rewarded for relatively greater vasodilatation in the left as compared to the right ear showed significant vasodilatation of the left ear ($P < .01$) but did not show a significant change in vasomotor activity in the right ear. However, despite the lack of significant vasomotor changes in the left ear in this group, the increase in the difference between the ears during training, as well as the difference at the end of training ($t = 5.18$, $P < .001$), was highly significant.

During training, the between-paw difference in vasomotor response was not significantly correlated with the between-ear difference ($\rho = -.22$), and at the end of training there was no significant difference between the paws ($t = .32$).

These results indicate that the learned differences between the vasomotor responses of the two ears were specific to the ears and not to the entire right or left side of the rat's body. Analyses of heart rate, vasomotor activity in the tail, and temperature did not indicate any significant overall changes either between groups at the end of training or within groups during training; however, both groups showed slight increases in vasoconstriction of the tail and in body temperature. In previous experiments on heart rate, rats learned the discrimination of changing their heart rate dur-

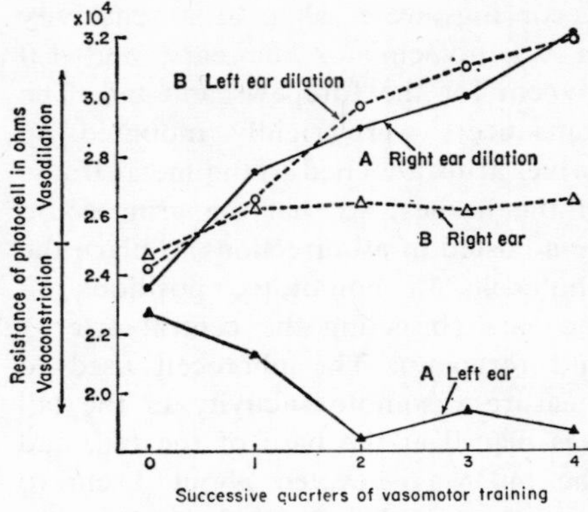

Fig. 1. Learning a difference between the vasomotor responses of the two ears. Group A was rewarded for relatively more dilatation of the right ear; Group B was rewarded for relatively more dilatation of the left ear.

ing the time-in stimulus, when such a response would be rewarded, rather than during the time-out intervals between trials, when it would not. However, analyses of vasomotor responses on blank and test trials in the present experiment did not indicate any evidence of discrimination learning.

The experiment described here adds to an increasing list of successful attempts to secure instrumental learning of visceral responses mediated by the autonomic nervous system, and represents a striking example of the specificity of the learning which can be achieved. Since the vasomotor innervation to the ear of the rat is primarily, if not wholly, sympathetic, the results indicate that the sympathetic nervous system has a greater capacity for specific local activity than usually has been attributed to it. The specificity of visceral learning is further supported by the finding that changes in either heart rate or intestinal contraction can be learned without changes in the other response (*4*). Furthermore, Miller and DiCara (*5*) have secured instrumental learning of changes in urine formation, accompanied by changes in renal blood flow but independent of heart rate and blood pressure, and of vasomotor responses in the tail. The specificity of learning of a variety of visceral responses demonstrated in these experiments makes it difficult to salvage the strong traditional belief that the instrumental learning of visceral responses is impossible by trying, as Black (*6*) has done, to explain the visceral changes produced by training as the indirect effects of somatic learning mediated by the cerebrospinal nervous system.

Finally, if changes in the blood flow to organs other than the kidney can be learned, and if these changes have the specificity that our experiment has demonstrated for the vasomotor responses of the ears, such changes could have significant consequences for psychosomatic medicine.

LEO V. DICARA
NEAL E. MILLER
*Rockefeller University,
New York, 10021*

References and Notes

1. N. E. Miller, *Ann. N.Y. Acad. Sci.* **92**, 830 (1961); ———, *Proc. World Congr. Psychiat. Montreal* **3**, 213 (1963); ——— and A. Carmona, *J. Comp. Physiol. Psychol.* **63**, 1 (1967); L. V. DiCara and N. E. Miller, *ibid.*, in press; ———, in preparation.
2. L. V. DiCara and N. E. Miller, *Commun. Behav. Biol.*, in press.
3. N. E. Miller and L. V. DiCara, *J. Comp. Physiol. Psychol.* **63**, 12 (1967).
4. N. E. Miller and A. Banuazizi, *ibid.*, in press.
5. N. E. Miller and L. V. DiCara, in preparation.
6. A. H. Black, paper presented at the 7th annual meeting of the Psychonomic Society, St. Louis, 1966.
7. Supported by USPHS grant MH 13189 to N.E.M. One of us (L.V.D.) is an advanced research fellow of the American Heart Association.

20 Operant Conditioning of Vasoconstriction

Charles Snyder and Merrill Noble

The question of whether visceral responses can be conditioned by operant techniques has for some time been a controversial matter on both theoretical and empirical grounds. Early empirical attempts failed (e.g., Mowrer, 1938; Skinner, 1938), but during the past few years numerous successful attempts have been reported (Crider, Shapiro, & Tursky, 1966; Fowler & Kimmel, 1962; Kimmel & Hill, 1960; Kimmel & Kimmel, 1963; Miller & Carmona, 1967; Miller & DiCara, 1967; Razran, 1961; Shearn, 1962; Shearn & Clifford, 1964; Trowill, 1967; Van Twyver & Kimmel, 1966).

Since these recent reports apparently refute theoretical arguments that visceral responses cannot directly be instrumentally conditioned (Keller & Schoenfeld, 1950; Kimble, 1961, 1964; Konorski & Miller, 1937; Mowrer, 1947, 1950; Skinner, 1938; Solomon & Wynne, 1954), it is important to note that mediation of the visceral response via instrumentally conditioned skeletal responses appears to have been ruled out for both GSR (Van Twyer & Kimmel, 1966) and heart rate (Miller & Di Cara, 1967; Trowill, 1967).

The status of a third visceral response, vasoconstriction, is less clear; the results of a study of instrumental conditioning of vasomotor changes in the rabbit (Fromer, 1963) failed to provide strong evidence for instrumental conditioning of this response, but data from human *S*s (Snyder & Noble, 1965) suggest that this response can be instrumentally conditioned. In the latter study, however, mediation of the visceral response via certain skeletal responses could not be ruled out. In view of the theoretical and practical importance of the general problem and the large individual differences in patterns of visceral responses (Lacey & Lacey, 1958, 1962),

[1] This research was supported in part by Research Grant HD 00956 from the National Institute of Child Health and Human Development.

[2] Now at the University of Notre Dame.
[3] Now at Pennsylvania State University.

the present study was designed to determine whether or not finger vasoconstriction could be instrumentally conditioned in the absence of mediating skeletal responses.

METHOD

Subjects.—A total of 54 introductory psychology students were assigned to two experimental conditions ($n = 12$), two matched conditions ($n = 12$), and a base-line control ($n = 6$). There were equal numbers of males and females in each group.

Apparatus.—Right index finger volume was recorded from a Yellow Springs-Fels plethysmograph on one channel of an Offner Model R Dynagraph oscillographic recorder at 2.5 mm/sec speed with a Type 9806A ac/dc Input Coupler (set for dc). The electronically controlled heater of the finger cup was modified to dc operation to eliminate 60-cycle interference to EMG recording, and cup water temperature was held to 35° C ± .5°.

Right forearm extensor-flexor EMG was recorded on a second oscillograph channel using a Type 9852 EMG Integrator Input Coupler modified to .25-sec. rise and 1-sec. decay time constants. The EMG electrodes were two standard EKG arm electrodes (32 × 55 mm.) plus a third electrode for ground placed on the forearm near the elbow. Sanborn Redux paste was applied to the electrodes, which were held in place lightly with standard rubber arm straps. Preliminary tests indicated that placement of the two active electrodes for maximal EMG from isometric extension and flexion of the index finger also gave a measureable indication of any movement of muscle tension great enough to produce a finger vasoconstriction or to affect vasomotor tone as indicated by finger pulse volume amplitude.

Abdominal and thoracic respiration were recorded on the third and fourth oscillograph channels with Type 9806A ac/dc Input Couplers (set for dc) from mercury-in-rubber strain gauges around the trunk at the xiphoid and underarm levels. Each gauge operated in a full bridge resistance circuit powered by a mercury cell. The gauges were made up according to Ackner's (1956) description except for the use of pure silver electrodes which are much more stable with mercury than stainless steel.

The reinforcement light was a standard pilot-light assembly turned on by *E*, recorded on the marker channel of the oscillograph, and turned off after 3 sec. by an electronic timer. The light was located on a small sloping panel on the front of *S*'s chair, which also served as a rest for the right arm and plethysmograph cup.

Procedure.—The *S*s were seated in a sound-treated and electrostatically shielded room, and were read the following instructions:

> This is an experiment in learning. We are attempting to condition an involuntary response over which you have no direct control. I will attach pickups to your finger, arm, and trunk to record various activity. In no case will you receive an electric shock or other unpleasant stimulus. All you have to do is watch the signal light that indicates when you are making the correct response. But don't try to make any kind of voluntary response. Just sit in a comfortable position, relax, breath quietly and evenly, and don't make any unnecessary movements. This will take 40 min., so if you get uncomfortable and have to move, go ahead and change position but don't keep moving around any longer than necessary. It is necessary for you to stay awake. If any of the equipment slips or you become uncomfortable you can call me over the intercom, which is on all the time. Do you have any questions?

In the instructions to Group B (base-line control) reference to the signal light and correct response was omitted.

After placement of the electrodes, a 5-min. base-line period was followed by a 25-min. conditioning and a 10-min. extinction period. There were no rest pauses.

All *S*s were treated alike during the base-line period, in which responses were recorded but the reinforcement light was omitted. During conditioning, *S*s in two experimental groups (EC and EN) which were treated alike until extinction, received reinforcement at the peak amplitude of any definite vasoconstriction not preceded by bodily movement; *S*s in Group B were treated just as during the base-line period, while for *S*s in Group C reinforcement was contingent upon 5-sec. vasomotor stability. Each *S* in the latter group received the same number of reinforcements per 5-min. period as did a given *S* in Group EC or EN and in this sense these *S*s were matched with *S*s in Groups EC and EN.

During extinction, the reinforcement light was omitted for all *S*s except those in Group EC, which continued to receive reinforcement at about the same rate as before but now contingent upon 5-sec. vasomotor stability.

Scoring.—Changes in finger volume were scored so as to exclude the respiratory component by measuring the vertical distance between the beginning and end of the change at corresponding points (phase) in the respiratory cycle. In addition, CRs which occurred within 5 sec. after thoracic or abdominal respiratory irregularity were excluded. The cardiac component of the plethysmogram was excluded by measuring along the troughs of the pulse waves of the plethysmogram. Very slight movements of the finger resulted in an abrupt change in both the plethysmogram and the EMG records and vasomotor changes occurring 2–5 sec. after any such movement were also excluded from the main results. Vasoconstrictions which began between 2–5 sec. following the onset of the light (reinforcement) were also excluded. Finally, indexes of both tonic respiratory rate and heart rate were obtained. The former was accomplished by counting the number of respiratory cycles during the first 1 min. during successive 5-min. periods that was free of gross respiratory irregularities. An approximation of heart rate was obtained from the plethysmogram.

The minimum amplitude of vasoconstrictions included in the main results was set individually for each *S* so as to give an operant level during the base-line period of approximately one vasoconstriction free from skeletal muscle artifacts in the plethysmogram, respiration, or EMG records. This criterion which ranged .45 cu. mm.–3 cu. mm. with a mean of 1.52 cu. mm., or a range of 3 mm.–21 mm. of movement on the oscillographic records with a mean of 10.8 mm., permitted *S*s to be equated for initial operant level regardless of recording sensitivity, total finger volume within the plethysmograph cup, and other such variables associated with absolute volume measurements; it also eliminated very small and unreliable volume changes that may not have been vasomotor in nature.

Results and Conclusion

The mean number of vasoconstrictions that met the amplitude criterion

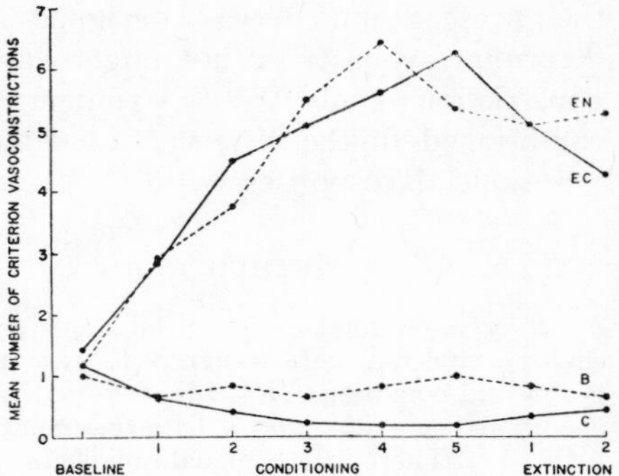

Fig. 1. Mean number of vasoconstrictions for successive 5-min. periods for experimental (EC and EN), matched control (C), and base-line control (B) conditions.

are shown in Fig. 1 for successive 5-min. periods.

As described above, the amplitude criterion was arbitrarily set to provide the same initial base-line operant level, and the actual means obtained during the initial 5-min. base-line period did not differ significantly ($p < .20$). During the conditioning phase there was a marked increase in number of vasoconstrictions for both experimental conditions, a decrease for the matched controls (C) and no essential change for the base-line control (B). During the final two 5-min. acquisition periods, and during extinction, each *S* in Groups EC and EN gave more criterion vasoconstrictions than did his matched control. Since there was no overlap between these distributions, statistical tests of the differences between experimental and control groups did not seem appropriate.

The data of Group C included a number of scores of zero and were not included in the analyses of variance described below. However, the overall difference between Groups B and C during the conditioning phase was significant, $t\ (28) = 2.71$, $p < .02$, but the apparent increase in performance of

Group C during extinction was not significant ($p > .20$).

The results of an analysis of variance of Groups EC and EN during conditioning showed a significant practice effect, F (4, 80) = 12.64, $p < .001$. All other main effects and interactions were nonsignificant ($p > .20$). Extended-trend analysis of variance (Grant, 1956) revealed that only the linear, F (1, 80) = 44.30, $p < .001$, and quadratic, F (1, 80) = 5.90, $p < .05$, components of the overall practice effect were significant. The linear component accounted for approximately 88%, and the quadratic component for approximately 12%, of the variance.

Similar analyses which included the final 5-min. conditioning period together with the 10-min. extinction period also were performed. The tendency for Group EC performance to decline relative to that of Group EN during extinction proved to be reliable as evidenced by a significant linear component of the EC + EN × Extinction interaction, F (1, 40) = 4.15, $p < .05$. That linear component accounted for 96% of the variance. Further analyses showed that there was a significant downward linear trend for Group EC, F (1, 40) = 9.03, $p < .01$, but not for Group EN ($F < 1.0$).

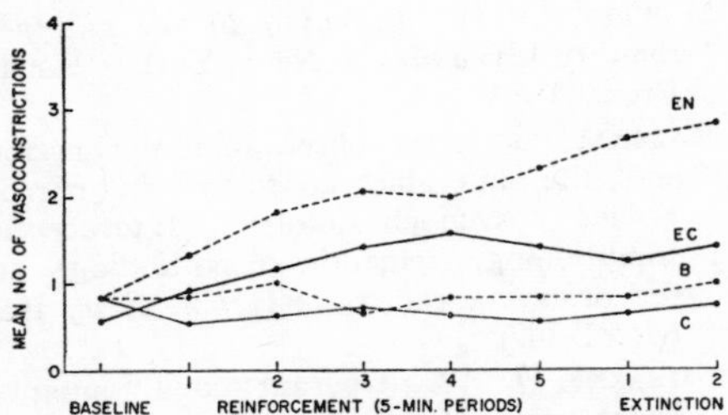

Fig. 2. Respiratory-related vasoconstrictions of experimental (EC and EN), matched control (C), and base-line control (B) conditions.

The decline in performance of Group EC during extinction together with the decline in performance of Group C during the acquisition phase suggests that these conditions resulted in learning not to respond. This effect is small, however, compared to the effects of conditioning for Groups EC and EN during acquisition and, moreover, was not found in an earlier study (Snyder & Noble, 1965). Consequently, any conclusions at this point must be highly tentative.

Respiratory irregularities.—As stated earlier, vasoconstrictions associated with respiratory irregularity were excluded from the main results. It is of interest, however, to note that, as shown in Fig. 2, vasoconstrictions which followed within 2–5 sec. a respiratory irregularity increased for the experimental conditions, but not for the control conditions. The binomial sign test showed that differences among groups were not significant at the beginning of acquisition ($p > .30$), but that the difference between either experimental group and the control *S*s was significant by the end of training ($p < .05$), indicating that respiratory irregularity also became conditioned. The tendency for the performance of Group EN to be greater than that of Group EC was not reliable ($p > .30$). The failure to find any decrease in mean number of vasoconstrictions for the experimental conditions during extinction is not surprising in view of the fact that only those respiratory irregularities which were followed within 2–5 sec. by a criterion vasoconstriction were reinforced. This represents a partial reinforcement situation, and resistance to extinction would be expected.

EMG control.—No instances of reflexive vasoconstrictions to phasic

EMG activity were scored because large bursts of EMG activity threw the plethysmograph off scale and single spikes had no effect upon vasomotor activity. Continuous, moderate EMG activity was found to correlate with obvious irregularity in the plethysmogram; in fact, the plethysmograph employed was more sensitive to muscle artifact than the EMG recording.

Respiratory rate and heart rate.—There were no significant differences among conditions either for mean respiratory rate or for mean heart rate.

In conclusion, the results indicate that like GSR and heart rate, vasoconstriction can be instrumentally conditioned, and that this was independent of gross bodily movement, muscle tension, heart rate, respiratory irregularity, and minute finger movements. The results thus provide additional evidence which is contrary both to theories which state that ANS responses can be learned only by classical conditioning, and theories which assume that the learning of somatic and visceral responses are mediated by functionally different neuroanatomical systems.

REFERENCES

Ackner, B. A simple method of recording respiration. *J. psychosom. Res.*, 1956, **1**, 144–146.

Crider, A., Shapiro, D., & Tursky, B. Reinforcement of spontaneous electrodermal activity. *J. comp. physiol. Psychol.*, 1966, **61**, 20–27.

Fowler, R. L., & Kimmel, H. D. Operant conditioning of the GSR. *J. exp. Psychol.*, 1962, **63**, 563–567.

Fromer, R. Conditioned vasomotor responses in the rabbit. *J. comp. physiol. Psychol.*, 1963, **56**, 1050–1055.

Grant, D. A. Analysis of variance tests in the analysis and comparison of curves. *Psychol. Bull.*, 1956, **53**, 141–154.

Keller, F. L., & Schoenfeld, W. N. *Principles of psychology.* New York: Appleton-Century-Crofts, 1950.

Kimble, G. A. *Conditioning and learning* (2nd ed.) New York: Appleton-Century-Crofts, 1961.

Kimble, G. A. Categories of learning and the problem of definition: Comments on Professor Grant's paper. In A. W. Melton (Ed.), *Categories of human learning.* New York: Academic Press, 1964.

Kimmel, E., & Kimmel, H. D. A replication of operant conditioning of the GSR. *J. exp. Psychol.*, 1963, **65**, 212–213.

Kimmel, H. D., & Hill, F. A. Operant conditioning of the GSR. *Psychol. Rep.*, 1960, **7**, 555–562.

Konorski, J., & Miller, S. Further remarks on two types of conditioned reflex. *J. gen. Psychol.*, 1937, **17**, 405–407.

Lacey, J. I., & Lacey, B. C. Verification and extension of the principle of autonomic response stereotypy. *Amer. J. Psychol.*, 1958, **71**, 50–73.

Lacey, J. I., & Lacey, B. C. The law of initial value in the longitudinal study of autonomic constitution: Reproducibility of autonomic responses and response patterns over a four-year interval. *Ann. N. Y. Acad. Sci.*, 1962, **98**, 1257–1290.

Miller, N. E., & Carmona, A. Modification of a visceral response, salivation in thirsty dogs, by instrumental training with water reward. *J. comp. physiol. Psychol.*, 1967, **63**, 1–6.

Miller, N. E., & Di Cara, L. Instrumental learning of heart rate changes in curarized rats. *J. comp. physiol. Psychol.*, 1967, **63**, 12–19.

Mowrer, O. H. Preparatory set (expectancy)—a determinant in motivation and learning. *Psychol. Rev.*, 1938, **45**, 62–91.

Mowrer, O. H. On the dual nature of learning—a reinterpretation of "conditioning" and "problem solving." *Harv. educ. Rev.*, 1947, **17**, 102–148.

Mowrer, O. H. *Learning theory and personality dynamics.* New York: Ronald Press, 1950.

Razran, G. The observable unconscious and the inferable conscious in current Soviet psychophysiology: Interoceptive conditioning, semantic conditioning, and the orienting reflex. *Psychol. Rev.*, 1961, **68**, **81–147**.

Shearn, D. W. Operant conditioning of heart rate. *Science*, 1962, **137**, 530–531.

Shearn, D. W., & Clifford, G. D. Cardiac adaptation and contingent stimulation. *Amer. Psychologist*, 1964, **19**, 491. (Abstract)

SKINNER, B. F. *The behavior of organisms: An experimental analysis.* New York: Appleton-Century, 1938.

SNYDER, C., & NOBLE, M. Operant conditioning of vasoconstriction. Paper presented at the meeting of the Midwestern Psychological Association, Chicago, April 1965.

SOLOMON, R. L., & WYNNE, L. S. Traumatic avoidance learning: The principles of anxiety conservation and partial irreversibility. *Psychol. Rev.,* 1954, **61,** 353–385.

TROWILL, J. A. Instrumental conditioning of the heart rate in the curarized rat. *J. comp. physiol. Psychol.,* 1967, **63,** 7–11.

VAN TWYVER, H. B., & KIMMEL, H. D. Operant conditioning of the GSR with concomitant measurement of two somatic variables. *J. exp. Psychol.,* 1966, **72,** 841–846.

(Received June 3, 1967)

IV

ELECTRODERMAL ACTIVITY

21

Differentiation of an Autonomic Response through Operant Reinforcement

David Shapiro, Andrew B. Crider, and Bernard Tursky

Abstract

Making reinforcement contingent upon the emission of a skin potential response maintains a stable rate of responding in contrast to a declining rate in a non-contingently reinforced control group. The effect is independent of changes in related response systems.

Attempts to condition visceral responses with operant techniques have produced varying results. Negative findings (Skinner, 1938, p. 112; Mowrer, 1938; Mandler et al., 1962) are consistent with two factor learning theories which limit autonomic conditioning to classical techniques. Recent studies suggest that autonomic responses can be modified with operant techniques (Shearn, 1962; Lisina, 1961; Fowler & Kimmel, 1962; Kimmel & Kimmel, 1963). The extent to which such conditioning occurs independently of concomitant or mediating changes in other autonomic and somatic functions has not been systematically investigated.

This study attempted to condition a discrete autonomic response with operant techniques while monitoring the activity of related response systems. A spontaneous change of short duration in skin potential was selected as the response to be conditioned. Skin potential and heart rate levels were measured to determine the effect of such conditioning procedures on the basal activity of these autonomic variables. Respiration records were taken as a check on the possible mediating influence of irregularities or changes in rate of breathing.

Skin potential responses were recorded from electrodes placed on an active sweating area on the thenar eminence of the right palm and an inactive area on the dorsal aspect of the right forearm. A response was defined as any change of 0.5 mV or greater as recorded with an R-C coupled amplifier having an input time constant of 1 sec.

Skin potential level was recorded from the same electrodes by means of a DC coupled amplifier at a reduced gain to study gradual change over time. Similarly, a cardiotachometer was used to measure continuous changes in heart rate from standard electrocardiograph electrodes placed bilaterally on the lower chest. Respiration cycle was recorded with a strain gage fixed around the waist. These measures were recorded on an Offner Type R dynagraph.

The subjects were seated in a lounge chair in a soundproofed, temperature-controlled room. They were told that the purpose of the experiment was to study the effectiveness of various devices for measuring thought processes and were asked to think actively about emotional experiences. They were also instructed that each time the recording apparatus detected their emotional thinking they would hear a tone and thereby earn five cents.

Eighteen student nurses were divided into experimental and control groups by roughly matching each experimental subject with a control on the basis of the number of skin responses emitted during an initial 5-min. resting period. This was followed immediately by a 30-min. period during which the experimental subjects were reinforced with a 70-db tone each time a response occurred. Since the tone in itself generally evoked skin potential activity, no response occurring within 10 sec. of a tone was reinforced. Each control subject received the same number of tones as her experimental match

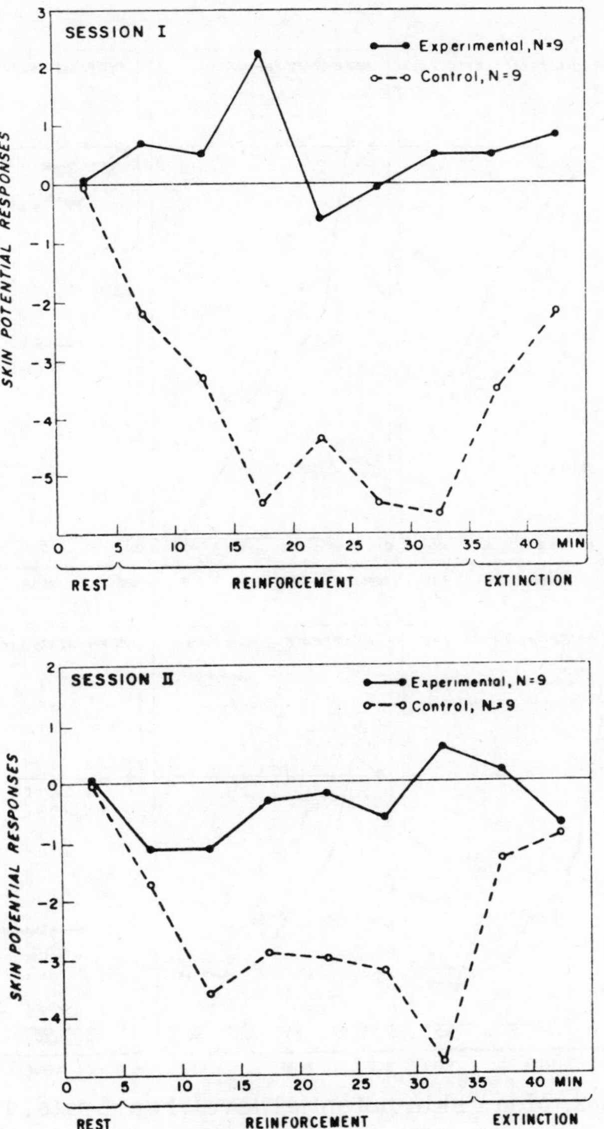

Fig. 1. Mean skin potential responses per 5-min. interval adjusted for resting period responses.

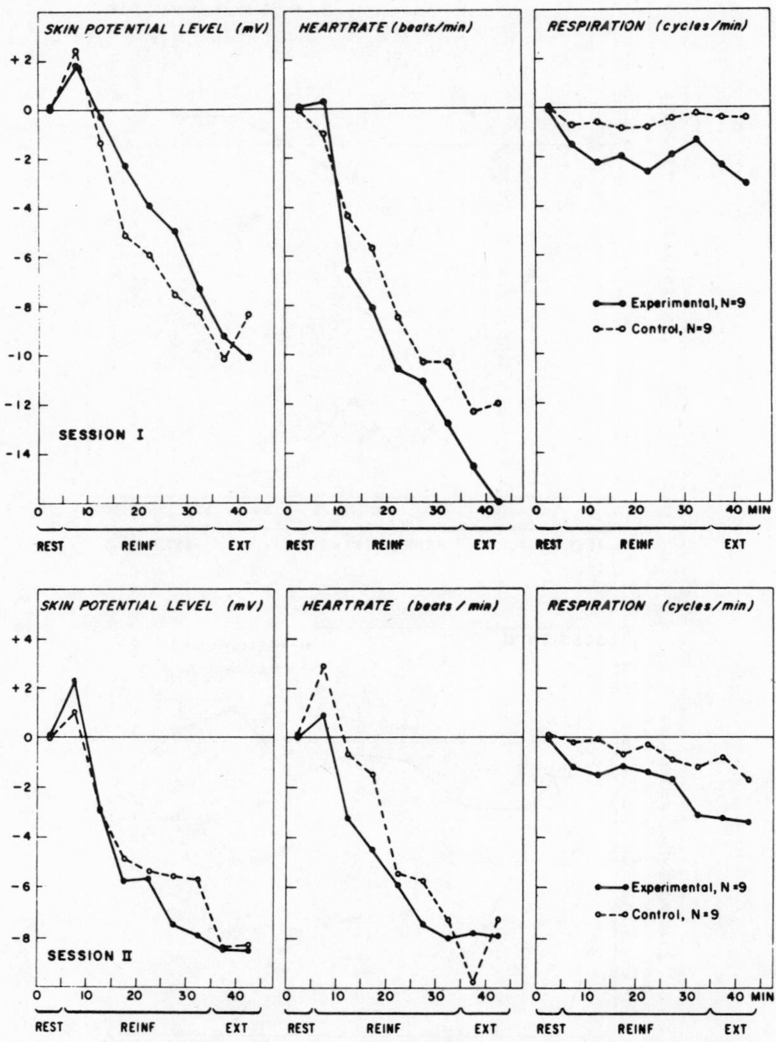

Fig. 2. Mean skin potential level, heart rate, and respiration per 5-min. interval adjusted for resting period levels.

but at points of no skin response activity and not less than 10 sec. following an emitted response. The reinforcement period was followed by a 10-min. extinction period during which no tones were sounded.

The experiment was repeated in exactly the same manner for each subject after an interval of several days.

The results for both sessions were analyzed in terms of the number of skin responses spontaneously emitted in each 5-min. interval of reinforcement and extinction, equated for baseline by subtracting out the number of resting period responses (Fig. 1). During each session the experimental group maintained a stable rate of response, while the control group declined over time. Although the effect became less pronounced, the results of the second session parallel the first. For each session a group by time-interval analysis of variance for correlated observations was carried out with the reinforcement period data. A significant group by time-interval interaction for both session I ($P < .001$) and session II ($P < .05$) confirms differential responsiveness over time. There was in addition a significant overall difference between groups in session I ($P < .025$). During extinction there was a marked increase in rate of response in the control group with no change apparent among the experimental subjects. The extinction data suggest that skin potential responses occurred as a component of an orienting reaction to the absence of tones in the final period. While this is apparent in the control group it is not seen among the experimental subjects who were already responding at a high level.

The above results are in complete accord with those reported by Fowler & Kimmel (1962) and Kimmel & Kimmel (1963).

Skin potential level and cardiotachometer records were sampled every 20 sec. and analyzed in terms of 5-min. time intervals adjusted for the resting period activity. Respiration records were analyzed by counting the number of inspiration-expiration cycles per min. during each interval, also adjusted (Fig. 2). No resting, reinforcement, or extinction period differences in any of these variables were found during either session; both groups showed similar trends over time. There were slightly more skin responses associated with breathing irregularities in the experimental than in the control group. When all such responses are discarded, the overall group effect of session I and the interaction effect of session II remain significant ($P < .025$ and $P < .05$ respectively).

All subjects were interviewed after each session. When tape recordings of the interviews were rated blindly along a 5-point scale for the subjects' perceived association between emotional thoughts and the occurrence of a tone, experimental and control group subjects were alike in finding only a moderate relationship between the two events. The control group did not report becoming less involved in the thinking task than did the experimental subjects.

Our results indicate that operant reinforcement facilitates skin response activity independently of observable changes in respiration pattern, skin potential and heart rate levels, and ideation. While we did not note any apparent differences in gross movement between the two groups, more systematic investigation of the skeletal muscle components of such conditioning would be of value. It seems unlikely, however, that muscular activity would affect one response system independently of the others measured in this study. Operant reinforcement provides a technique for investigating the extent to which an autonomic response can be differentiated from other autonomic and somatic behavior.

References

FOWLER, R. L., & KIMMEL, H. D. Operant conditioning of the GSR. *J. exp. Psychol.*, 1962, 63, 563-567.

KIMMEL, ELLEN, & KIMMEL, H. D. A replication of operant conditioning of the GSR. *J. exp. Psychol.*, 1963, 65, 212-213.

LISINA, M. I. In G. Razran. The observable unconscious and the inferable conscious in current Soviet psycho-physiology: interoceptive conditioning, semantic conditioning, and the orienting reflex. *Psychol. Rev.*, 1961, 68, 81-148.

MANDLER, G., PREVEN, D. W., & KUHLMAN, C. K. Effects of operant reinforcement on the GSR. *J. exp. Anal. Beh.*, 1962, 5, 317-321.

MOWRER, O. H. Preparatory set (expectancy): a determinant in motivation and learning. *Psychol. Rev.*, 1938, 45, 62-91.

SHEARN, D. W. Operant conditioning of heart rate. *Science*, 1962, 137, 530-531.

SKINNER, B. F. Behavior of organisms. New York: Appleton-Century-Crofts, 1938.

Note

1. Supported by Research Career Development Award K3-MH-20, 476-01, NIMH; ONR Contract Nonr-1866(43), Group Psychology Branch; and Mental Health Research Training Program, NIMH. Dr. Eric Kandel offered many helpful suggestions during the preparation of this report.

22 Reinforcement of Spontaneous Electrodermal Activity

Andrew B. Crider, David Shapiro, and Bernard Tursky

It is generally presumed that autonomic responses are inherently noninstrumental and therefore incapable of modification with operant procedures (Kimble, 1961; Morgan, 1965). Further, both Skinner (1938) and Mowrer (1938) have reported unsuccessful attempts to condition autonomic activity instrumentally. Recently, however, several laboratories have independently reported successful conditioning of electrodermal responses with contingent stimulation techniques in human *S*s (Fowler & Kimmel, 1962; R. J. Johnson, 1963; Kimmel & Kimmel, 1963; Shapiro, Crider, & Tursky, 1964). The typical effect is a maintenance of the prereinforcement operant level in contrast to a declining rate of response in a control situation. Similar results have been reported by Mandler, Preven, and Kuhlman (1962) and Rice (1964), but interpreted as demonstrating phenomena other than strict operant conditioning. With regard to other autonomic responses, Shearn (1962) demonstrated that the frequency of temporary heart rate accelerations could be increased by making shock avoidance contingent upon their occurrence, while Razran (1961) cites a Soviet experiment in which peripheral blood volume changes were modified when vasodilation produced escape from shock.

This report deals with a series of studies based on the contingent stimulation of spontaneous skin potential responses recorded from the palmar surface of the hands in human *S*s. The effects of this procedure on the skin potential response, as well as its effects on related response systems, will be reviewed.

Method

Spontaneous autonomic activity refers to fluctuations in autonomic effector activity which occur in the resting individual and at points uncorrelated in time with specifically imposed exteroceptive stimuli (L. C. Johnson, 1963; Lacey & Lacey, 1958; Wilson & Dykman, 1960). Such responses are termed spontaneous solely in the sense that their stimulus sources are unknown or unimportant for the conduct of a given investigation. The definitional similarity of spontaneous autonomic activity and so-called emitted skeletal responses is apparent; in the following experiments spontaneous skin potential responses were regarded as potentially modifiable autonomic operants.

Spontaneous skin potential responses recorded from the footpads of cats have been studied with neurophysiological techniques by Wang (1964), who isolated various excitatory and inhibitory centers for such activity at spinal, subcortical, and neocortical CNS levels. With respect to a null reference electrode, palmar skin potential waves can be negative going, positive going, or diphasic. The peripheral mechanisms subserving these two components are not well understood. While both are abolished with atropinization (Wilcott, 1964) or pe-

[1] This research was supported by the National Institute of Mental Health and by the Office of Naval Research as follows: NIMH Research Career Development Award K3-MH-20,476-02, Mental Health Research Training Program 5T1-MH-7084-05, Research Grants MH-08853-01 and 04172-05; ONR Contract Nonr-1866(43), Group Psychology Branch.

ripheral sympathetic section (Rothman, 1954), there is some evidence that both epidermal cells and sweat glands are involved in their production (Edelberg & Wright, 1964; Shaver, Brusilow, & Cooke, 1962).

Apparatus

Palmar skin potential activity was recorded on two separate channels of an ink-writing eight-channel Offner Type R dynagraph. Skin potential level was measured as the dc potential difference between two nonpolarizing silver-silver chloride sponge electrodes (O'Connell & Tursky, 1960), one of which was placed on the thenar eminence of the right palm and the other on the dorsal aspect of the right forearm approximately 20 cm. distant. Skin potential responses were recorded on an R-C coupled channel having an input time constant of 1 sec. and a gain setting 10 times that of the dc channel. With a pen excursion of 1 mv/cm, this combination of R-C coupling and increased sensitivity made it possible to recognize rapid changes from base line as small as .1 mv.

Continuous recordings were also made of respiration and heart rate during each experiment. Respiration cycle was measured by means of a strain gauge respirometer fixed around the waist. A cardiotachometer measured beat-by-beat variation in heart rate via standard EKG electrodes placed bilaterally on the upper arms.

Experiment 5 utilized a measure of gross body activity, recorded via a transducer consisting of a crystal phonograph cartridge whose stylus was replaced by a 3-in. section of piano wire terminating in a lead weight. It was fixed to the springs of *S*'s chair and mechanically damped to exclude normal building vibrations. The output of this device was amplified and quantified by means of a voltage comparator integrator described elsewhere (Tursky, 1964). A portion of a typical polygraph record is reproduced in Figure 1.

Procedure

Except where otherwise noted, all *S*s were either female college students or student nurses. The *S*s were seated in a lounge chair in a sound-attenuated, temperature-controlled room maintained at 72° F. and separated by a one-way mirror from the recording chamber. After placement of the various recording devices by a female assistant, standard instructions were read by one of the *E*s, followed by a 5-min. period during which resting level recordings were made of all channels. The start of the experiment was signaled by a long tone.

The *S*s received the following instructions, designed to maintain alert wakefulness during the course of the session:

> We are interested in studying the effectiveness of various devices for measuring thought processes. What we are going to ask you to do is simple, but requires cooperation on your part. We have attached recording devices to various parts of your body. We would like you to recall and actively think about situations in your life in which you have felt especially emotional.
>
> Because we have found that it is not necessarily easy for a person to sustain recalled emotional experiences over an extended period, we are going to provide you with some incentive. Whenever our recordings detect your emotional thinking, you will hear a tone like this (tone). Each tone is worth five cents. This money will accumulate during the time you are here and will be paid to you as a bonus after the session. You may be able to earn as much as $3.00 extra. The session will last less than 1 hr. Are there any questions?
>
> Please try not to move around a great deal. These recording devices are very sensitive and tend to be disrupted if you move.
>
> After I leave the room, there will be a short rest period. When you hear a long tone, you can start thinking about emotional experiences.

The reinforcement criterion was the production of a skin potential response of at least .5-mv. amplitude, regardless of direction. In terms of the recording sensitivity utilized in the majority of these experiments, this represented a pen deflection which could be easily observed and reliably reinforced. An amplitude distribution of 360 skin potential fluctuations recorded from the R-C coupled channel, representing 40 consecutive responses from the resting records of each of nine *S*s, showed that responses of .5 mv. or greater accounted for some 40% of all observed changes in this group.

Effects of Reinforcement on Response Rate

Experiment 1[2]

Procedure. Eighteen *S*s were divided into a contingent reinforcement (CR) experimental group and a noncontingent reinforcement (NCR) control group by approximately matching each CR *S* with a control on the basis of the number of criterion skin potential responses emitted during the 5-min. resting interval. The rest period was immediately followed by a 30-min. period during which each CR *S* was reinforced with a 70-db. tone of 1.5-sec. duration for each occurrence of a criterion response. Since the tone itself generally elicited skin potential activity, no response occurring within 10 sec. of a tone was reinforced. Each NCR *S* received the same number of tones as her CR match spread over the 30 min., but at points of no skin potential change and not less than 10 sec. following any emitted response.

Results. Subtracting the number of rest-

[2] A preliminary report of this experiment was published by Shapiro et al., 1964.

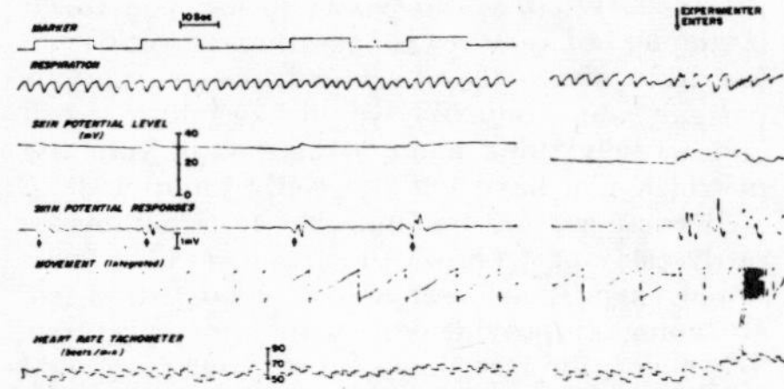

Fig. 1. Portion of typical polygraph record. (Panel on left was recorded during a conditioning experiment. Marker indicates onset of 1.5-sec. tone followed by "time-out" period. Arrows indicate reinforced skin potential responses. Polarity of this channel is in opposite direction from skin potential level. Panel on right illustrates reaction to presence of *E* at end of session.)

ing interval responses from the number of criterion responses emitted during successive 5-min. intervals, save for the 10-sec. periods following reinforcement, adjusted for slight group differences not adequately controlled by the matching procedure. As seen in Figure 2, the CR group maintained a mean rate of response at or above the resting level, while the NCR controls declined over time. These differences were statistically analyzed by a groups-by-time intervals analysis of variance for correlated observations. Both the overall difference between the CR and NCR groups ($p < .025$) and the groups-by-time intervals interaction ($p < .001$) were significant.

Experiment 2

This experiment involved a similar demonstration of the CR-NCR effect, but substituted an intraindividual comparison for the interindividual control group method of the first experiment.

Procedure. Two groups of 10 *S*s each were approximately matched in terms of resting period response rate. Following the resting interval, each group received 20 min. of contingent reinforcement and 20 min. of noncontingent reinforcement in the same session. A CR-NCR sequence in one group was balanced with a NCR-CR sequence in the second group. Noncontingent reinforcement was administered at a rate equivalent to the mean number of responses per minute emitted by the CR *S*s in the first experiment.

Results. In Figure 3 the data were plotted in terms of the number of criterion responses emitted during successive 5-min. intervals adjusted for resting level.

Of the CR-NCR *S*s, eight emitted fewer criterion responses during the second 20 min. than during the first, one *S* increased slightly, and one remained constant. In the NCR-CR comparison, eight *S*s showed an increase in responsiveness from the first to second 20-min. period, one declined slightly, and one remained constant. Thus, CR produced more responses than NCR in 16 out of 18 comparisons ($p < .002$, two-tailed binomial test). It is noticeable that both CR and NCR effects were influenced by fatigue-like processes occurring over time; in the second 20-min. period, both CR and NCR levels are less elevated than those of the first 20-min. period. There were no significant differences in the mean number of tones per interval administered in the four conditions.

Conditioning as a function of operant level. It is known that *S*s differ reliably in the frequency of spontaneous electrodermal responses emitted under resting conditions, (e.g., Lacey & Lacey, 1958). Conditioning

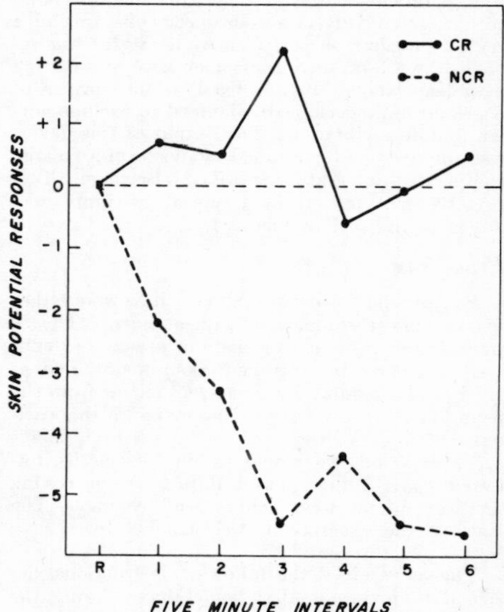

Fig. 2. Mean spontaneous skin potential responses per 5-min. interval for 9 CR *S*s and 9 NCR *S*s, adjusted for resting rate.

curves as a function of such operant-level differences were examined by combining the data of the first 25 equivalent min. of the above two experiments, yielding CR and NCR groups of 19 *S*s each. Each group was ranked in terms of the number of criterion resting interval responses. A homogeneous subset of five *S*s was drawn from the upper half of the CR distribution and matched with a similarly homogeneous set of five *S*s from the upper half of the NCR distribution. The same procedure was used to select two groups of five *S*s from the lower half of each distribution. The plots of the performances of these four groups appear in Figure 4. While both high- and low-operant-level groups appear equally conditionable, their respective curves vary in accordance with a differential regression towards a common mean of some seven responses per interval.

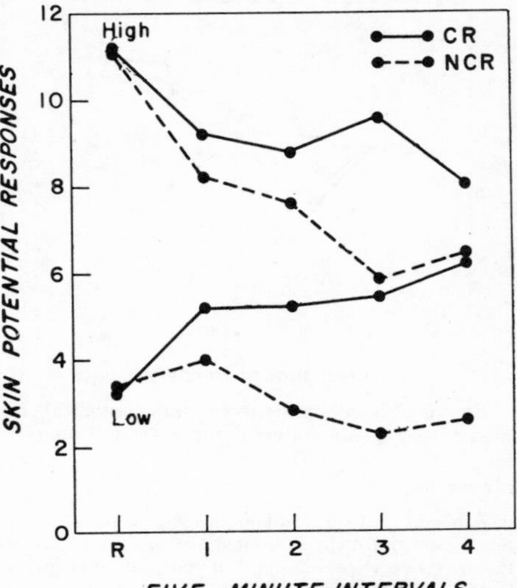

FIG. 4. Mean spontaneous skin potential responses per 5-min. interval for 5 *S*s in high- and in low-operant level subgroups.

Discussion. It is apparent that contingent reinforcement of electrodermal responses increases the probability of their occurrence when evaluated against a noncontingent reinforcement control. The differential effect of these two schedules is observable both between CR and NCR groups and within groups serving as their own controls. While the effect appears in both high- and low-operant-level groups, the former *S*s decline in response frequency towards an apparent "ceiling." Whether this limit reflects the inherent constraints of the effector system or simply the reinforcement schedule employed deserves further investigation. More traditional learning curves, reminiscent of those generated in classical GSR conditioning studies, are produced by low-operant-level *S*s, or by those *S*s adapted to low rates of activity with NCR prior to being placed on a CR schedule, as in the NCR-CR group of Experiment 2. These results thus fully confirm and extend the findings of other investigations of operant electrodermal conditioning previously referred to, despite the rather wide variations among studies in recording techniques, response criteria, instructions, and reinforcing stimuli.

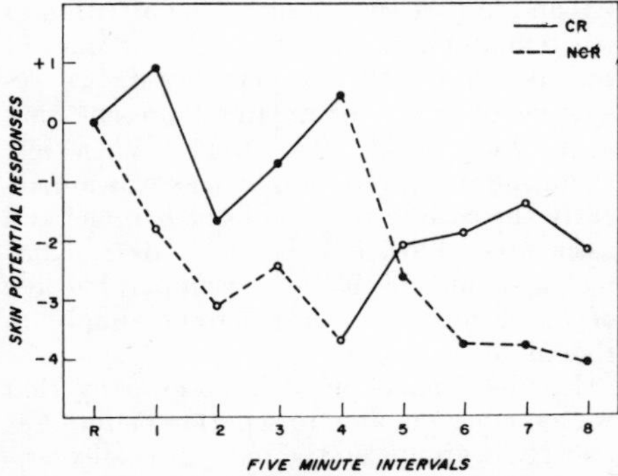

FIG. 3. Mean spontaneous skin potential responses per 5-min. interval for 10 CR-NCR *S*s and 10 NCR-CR *S*s, adjusted for resting rate.

Some Effects of Instructional Set

The above effects occurred under conditions of active ideation induced by preliminary instructions. The following two studies were designed to evaluate the contribution of this set to the CR-NCR differences obtained. Experiment 3 examined the course of spontaneous skin potential response rate under a nonreinforcement, activating set condition; Experiment 4 examined the effects on conditionability of neutral or nonactivating instructions.

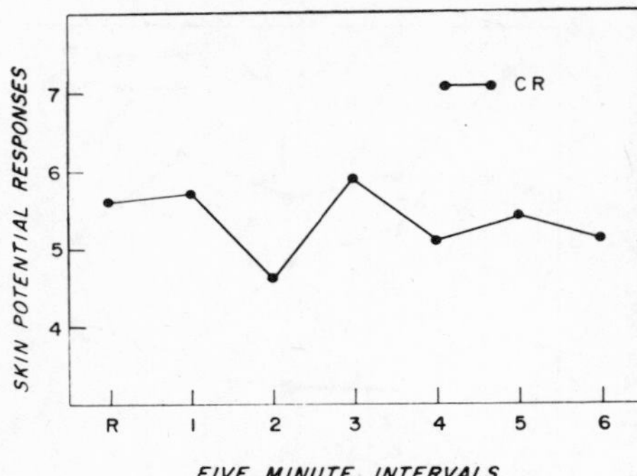

FIG. 5. Mean spontaneous skin potential responses per 5-min. interval for 7 "conditioners."

Experiment 3

Procedure. Five student nurses were administered essentially the same instructions read to *S*s in the first two experiments, but without mention of feedback in the form of tones. Instead, they were told that their bonus would depend on the number of instances of "emotional thinking" observed in their records during the course of the session.

Results. Criterion responses were counted in the manner employed in the previous studies, with a 10-sec. lapse after each tallied response. The mean resting interval rate for these *S*s was 3.4 responses, declining to a low of 2.6 responses in the third 5-min. interval and rising slightly to 3.0 responses in the fourth interval. This operant level and time trend was indistinguishable from that shown by the low operant, NCR *S*s described in Figure 4.

Experiment 4

Procedure. After attachment of the various recording devices, 15 *S*s were told simply that the purpose of the study was to investigate the effectiveness of various devices for registering the effect of sensory stimulation. They were further instructed to sit quietly, remain alert, and expect to hear tones from time to time. After a 5-min. resting interval, tones were made contingent on the production of criterion skin potential responses.

Results. Of the total, seven *S*s produced no criterion responses during the resting interval or for the succeeding 15 min., after which the session was discontinued. One *S* produced several reinforced responses in the first 15 min. but none thereafter. The mean interval-by-interval rate of the remaining seven *S*s is presented in Figure 5. This curve is comparable to that of the CR groups in the first two experiments and is taken as evidence of a conditioning effect in this group.

Discussion. While Experiment 3 would profit from the addition of a high-operant-level group, the similarity in time trends of the five *S*s of this experiment and the low-operant NCR group of Figure 4 suggests a near equivalence of the two conditions. Since comparable adaptational trends are also seen when experimental stimuli are programed randomly with respect to electrodermal activity (L. C. Johnson, 1963), it appears that the major variable responsible for the effects observed in the first two experiments is the CR schedule. While it is possible that an NCR schedule has associative consequences not apparent in our data, there is at present no justification for distinguishing NCR from either random stimulus or stimulus absent conditions.

Experiment 4 demonstrates that while pseudorequests to human *S*s to produce emotional thoughts are not requisite to the conditioning effect, they occasion the production of reinforcible operant-level responses in a greater percentage of *S*s than do more neutral instructions. The specific set does not seem to interact with the enhancement of response frequency with CR, however, since the great majority of *S*s in this experiment who showed any degree of operant-level activity generated a mean curve comparable to those produced with CR in the first two studies. The attrition of the remaining *S*s is partially an artifact of the response criterion employed, as responses of .5 mv. or greater represent only a portion of all fluctuations. When less stringent criteria are applied, as was apparently the case in the studies of Kimmel and associates (Fowler & Kimmel, 1962; Kimmel & Kimmel, 1963), no comparable loss of *S*s occurs under instructions simply to remain alert.

In this connection, it is noteworthy that we have been unable to separate *S*s into CR and NCR groups on the basis of postexperimental interviews. Following Experiment 1, tape recordings of each interview were rated blindly along a 5-point scale for *S*s' perceived association between recalled emotional thoughts and the occurrence of a tone. The mean ratings for both CR and NCR groups were similar, with the average *S* reporting an inconsistent temporal relationship between the two events. Neither did the interviews yield any evidence that the NCR *S*s became less involved in the thinking task over time than the CR *S*s.

Concomitant Changes in Other Autonomic and Skeletal Measures

Figure 6 presents mean skin potential level, heart rate, and respiration rate data collected on both CR and NCR groups in Experiment 1. Skin potential level and cardiotachometer records were sampled every 20 sec. and averaged over 5-min. intervals. Respiration records were analyzed by computing the mean number of inspiration-expiration cycles per minute for each interval. The three variables were adjusted for small group differences in resting levels by subtracting this value from each succeeding interval.

It can be seen that the conditioning of skin potential responses in the CR group occurred against a background of declining activity in skin potential level and heart rate, with slight changes in respiration rate. These trends were in no case significantly different from the pattern of adaptation of the NCR group. Corresponding trends were also obtained in the analysis of these three variables in Experiment 2, indicating that the modification of skin potential response frequency with CR was independent of concomitant changes in the activity levels of other autonomic and quasi-autonomic effectors.

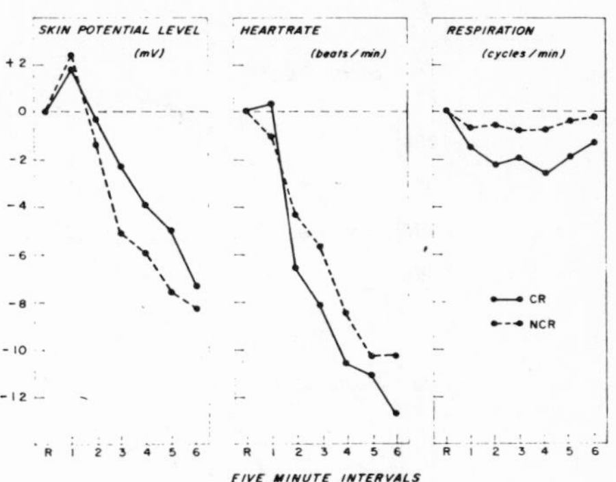

Fig. 6. Mean skin potential level, heart rate, and respiration rate per 5-min. interval for CR and NCR groups of Experiment 1, adjusted for resting period levels.

Experiment 5

Procedure. This experiment was performed to obtain data on the time course of gross skeletal activity during skin potential response conditioning. The movement transducer described above produced a sensitive, objective, and continuous record of postural adjustments, many of which would have been unobservable by eye (see Figure 1). In a rough calibration, performed on a female *S* of average weight and stature, reading aloud produced a 100% increase in registered activity over the resting rate, while active fidgeting produced a 1200% increment.

Skin potential responses were observed with a gain setting twice that employed in the previous experiments, and the reinforcement criterion was adjusted from *S* to *S* to match individual differences in modal amplitude changes during rest. Reinforcements were administered by means of an electronic switch that was triggered by the criterion voltage output from the galvanometer.

Following the resting interval *S*s received 5 min. of NCR, followed by 30 min. of CR. From a group of several *S*s run under these conditions, the data of three individuals who showed the greatest enhancement of skin potential response frequency were selected for presentation in Figure 7: *S*s 502 and 494 (female) were reinforced for responses of ⅛ mv. or greater and *S* 476 (male) was reinforced for responses of at least ¼ mv. Criterion skin potential responses were counted in the usual manner. The movement variable was analyzed by counting the number of integrated "ramps" produced during each interval. For purposes of comparability, both variables were expressed as percentages of the resting interval. The means and ranges of interval-by-interval skeletal activity from a control group of 9 *S*s who received 30 min. of randomly presented tones are shown in the fourth panel of the figure.

Results. In none of the three cases was there any systematic correspondence between changes in skin potential response frequency and changes in amount of skeletal activity. The movement trends were more or less idiosyncratic from *S* to *S*, were of small magnitude compared with the large increases in skin potential responding, and in no interval exceeded the range of the random reinforcement control group.

Discussion. Our observations give little support to the notion proposed by Mandler

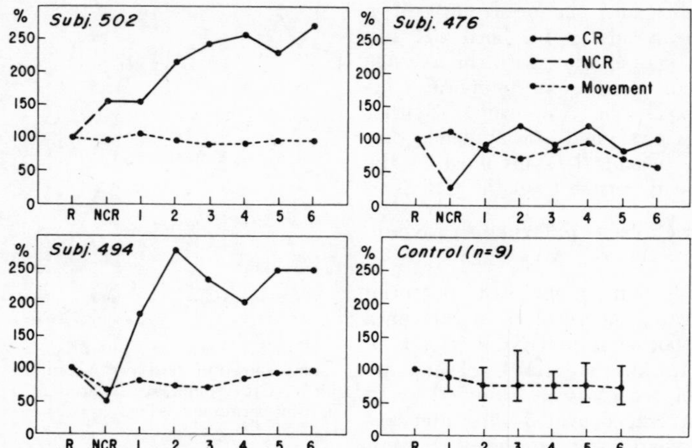

FIG. 7. Spontaneous skin potential responses and gross skeletal activity per 5-min. intervals of rest, NCR, and CR for three cases, plus means and ranges of gross skeletal activity for random reinforcement control group.

et al. (1962) that operant electrodermal conditioning is attributable to a nonspecific autonomic arousal brought about by the simple introduction of reinforcing stimuli. An NCR control leaves little doubt that the effect is specific to pairing response with reinforcement, and the data of Figure 6 indicate that the activity levels of other autonomic variables are unaffected by contingently reinforcing skin potential changes. While no attempt was made to correlate transient changes in heart rate or respiration pattern with the presence-absence of skin potential responses, these data raise the interesting possibility that certain response characteristics of any given autonomic effector can be differentiated from the activity of other effectors depending on the reinforcement contingency chosen by *E*.

Figure 7 indicates that gross skeletal activity also is uncorrelated with changes in conditioned skin potential activity. In a replication of the studies of Kimmel and his associates (Fowler & Kimmel, 1962; Kimmel & Kimmel, 1963), Rice (1964) found an increase in forearm EMG changes in the early stages of operant conditioning of finger-to-finger skin resistance responses. EMG-accompanied GSRs declined abruptly after 12 min. of reinforcement, however, with no corresponding decrease in the frequency of responses not accompanied by EMG changes. Rice was also able to show a conditioning effect in a group of high-operant-level *S*s by reinforcing only those responses not accompanied by EMG changes.

General Discussion

Operant conditioning is traditionally defined as the strengthening of a spontaneously occurring behavioral event through reinforcement. The application of operant procedures and terminology to the study of ANS activity thus requires a suspension of judgment as to qualitative distinctions between autonomic and skeletal responses and their mechanisms of modification. The limited research in this area does, moreover, point to a fruitful extension of experimental behavior analysis to autonomic responses.

Any such behavioral analysis would profit from a finer study of the physiological correlates of spontaneous autonomic activity. There is little doubt that in the intact organism electrodermal responses are regulated by higher CNS structures, with important centers at hypothalamic and striatal levels, and that the elicited electrodermal response is "long-circuited" through these centers (Wang, 1964). Whether spontaneous electrodermal fluctuations are produced by the unsupported activity of these centers, or are to some degree under the control of afferent inputs from interoceptors, has not been determined.

There is no suggestion, either in our data or in the comments of other investigators, that operantly conditioned electrodermal responses are produced by afferent feedback from gross skeletal activity. There are, however, any number of covert responses—slight respiratory irregularities, muscle twitches, and presumably even visceral activity—which are capable, under conditions of novelty, of acting as stimuli for an electrodermal orienting response. Reinforcement of this reaction would tend to inhibit its adaptation in the manner of Sokolov's (1963) conditioned orienting reflex. Sokolov has also shown that conditioned orienting entails a reduction of sensory thresholds, which would mean that as conditioning progresses other relatively minor covert responses would also tend to evoke electrodermal activity. Thus operant reinforcement of electrodermal activity may result in a modified feedback loop, in which covert response-produced stimuli feeding into sudomotor centers elicit an electrodermal response, which is followed by a reinforcing stimulus. Reinforcement, in turn, would serve to enhance the reactivity of these centers to other, formerly subthreshold, peripheral inputs.

REFERENCES

Edelberg, R., & Wright, D. J. Two galvanic skin response effector organs and their stimulus specificity. *Psychophysiology,* 1964, **1**, 39–47.

Fowler, R. L., & Kimmel, H. D. Operant conditioning of the GSR. *J. exp. Psychol.,* 1962, **63**, 563–567.

Johnson, L. C. Some attributes of spontaneous autonomic activity. *J. comp. physiol. Psychol.,* 1963, **56**, 415–422.

Johnson, R. J. Operant reinforcement of an autonomic response. *Dissertation Abstr.,* 1963, **24**, 1255–1256. (Abstract)

Kimble, G. A. *Hilgard and Marquis' conditioning and learning.* (2nd ed.) New York: Appleton-Century-Crofts, 1961.

Kimmel, E., & Kimmel, H. D. A replication of operant conditioning of the GSR. *J. exp. Psychol.,* 1963, **65**, 212–213.

Lacey, J. I., & Lacey, B. C. The relationship of resting autonomic activity to motor impulsivity. In H. C. Solomon, S. Cobb, & W. Penfield (Eds.), *The brain and human behavior.* Vol. 36. Baltimore: Williams & Wilkins, 1958. Pp. 144–209.

Mandler, G., Preven, D. W., & Kuhlman, C. Effects of operant reinforcement on the GSR. *J. exp. Anal. Behav.,* 1962, **5**, 317–321.

Morgan, C. T. *Physiological psychology.* New York: McGraw-Hill, 1965.

Mowrer, O. H. Preparatory set (expectancy): A determinant in motivation and learning. *Psychol. Rev.,* 1938, **45**, 62–91.

O'Connell, D. N., & Tursky, B. Silver-silver chloride sponge electrodes for skin potential recording. *Amer. J. Psychol.,* 1960, **73**, 302.

Razran, G. The observable unconscious and the inferable conscious in current Soviet psychophysiology: Interoceptive conditioning, semantic conditioning, and the orienting reflex. *Psychol. Rev.,* 1961, **68**, 81–147.

Rice, D. G. Operant GSR conditioning and associated electromyogram responses. Unpublished doctoral dissertation, University of Wisconsin, 1964.

Rothman, S. *Physiology and biochemistry of the skin.* Chicago: Univer. Chicago Press, 1954.

Shapiro, D., Crider, A., & Tursky, B. Differentiation of an autonomic response through operant reinforcement. *Psychon. Sci.,* 1964, **1**, 147–148.

Shaver, B. A., Jr., Brusilow, S. W., & Cooke, R. E. Origin of the galvanic skin response. *Proc. Soc. Exp. Biol. Med.,* 1962, **110**, 559.

Shearn, D. W. Operant conditioning of heart rate. *Science,* 1962, **137**, 530–531.

Skinner, B. F. *Behavior of organisms.* New York: Appleton-Century, 1938.

Sokolov, Ye. N. *Perception and the conditioned reflex.* Oxford: Pergamon Press, 1963.

Tursky, B. Integrators as measuring devices of bioelectric output. *Clin. Pharmacol. Therap.,* 1964, **5**, 887–892.

Wang, G. H. *The neural control of sweating.* Madison: Univer. Wisconsin Press, 1964.

Wilcott, R. C. The partial independence of skin potential and skin resistance from sweating. *Psychophysiology,* 1964, **1**, 55–66.

Wilson, J. W. D., & Dykman, R. A. Background autonomic activity in medical students. *J. comp. physiol. Psychol.,* 1960, **53**, 405–411.

(Received May 6, 1965)

23 Operant Electrodermal Conditioning under Partial Curarization

Lee Birk, Andrew B. Crider, David Shapiro, and Bernard Tursky

It has been shown that unelicited electrodermal responses can be operantly conditioned by reinforcing their occurrence with a variety of stimuli (Crider, Shapiro, & Tursky, 1966; Fowler & Kimmel, 1962; Johnson, 1963). The following demonstration was designed to determine whether such an effect can be obtained during a drug-induced minimization of skeletal activity. A volunteer *S* (L. B.) was administered an immobilizing dose of d-tubocurarine prior to a reinforcement session and the results compared with a previous undrugged session.

Method

To establish a stable baseline prior to the drug session, *S* was run for six sessions over a 3-wk. period under the following procedure (described more fully in Crider et al., 1966). Skin potential responses were recorded through an R-C coupled amplifier having an input time constant of 1 sec. via an active electrode placed on the right palm and a reference electrode placed on the right forearm. Following an operant level determination, *S* was continuously reinforced with a 70-db. tone of 1.5-sec. duration for any negative-going skin potential fluctuation of 0.1 mv. or greater. Reinforcements were administered by means of an electronic switch triggered by the criterion voltage change. Continuous recordings were also made of skin potential level, heart rate, and respiratory rate.

The curarization session took place 20 days after the sixth control session. As in previous sessions, *S* was seated in a semireclining position in a lounge chair. Following placement of the recording devices, an anesthesiologist[2] started a continuous intravenous infusion of 5% dextrose and water. The *S* was allowed 10 min. to adapt to the infusion procedure, after which the administration of d-tubocurarine was begun via the already running infusion. A total of six separate doses was infused over a period of 11 min., totaling 16.5 mg. or about 0.25 mg/kg. Administration ceased when *S*, whose vertebral column formed an angle of 45° with the horizontal, was unable to lift his head from the chair. At this criterion, *S* was also unable to lift any extremity, open his eyes, swallow, or utter intelligible words, although slight lateral rolling head movements were possible with great effort. He was able to breathe throughout the session without marked subjective distress.

After the anesthesiologist left the experimental chamber, a 4-min. operant level (OL) determination was made, followed by a 16-min. continuous reinforcement period. The assumption that this 20-min. session would coincide with maximal drug effect was corroborated by *S*'s postexperimental verbal report (see Goodman & Gilman, 1956, p. 605).

Criterion skin potential responses were tallied for each minute and averaged across 4-min. intervals. Skin potential level and cardiotachometer records were sampled every 20 sec. and averaged over each interval. Respiration was analyzed by averaging inspiration-expiration cycles per minute for each interval.

Results

Figure 1 presents the conditioning curves generated under curare and in the fifth precurare session, chosen for comparison because of its nearly equivalent operant level. Both show a similarly heightened frequency of response with contingent reinforcement comparable to that observed in studies previously referred to. This contrasts with a declining rate of response observed under noncontingent reinforcement schedules. It appears, therefore, that the level of curarization employed here had no marked effect on the conditioning phenomenon.

From Figure 2 it can be seen that the curare session was characterized by declining trends in respiratory rate, skin potential level, and heart rate. These trends are also in accord with

[1] This research was supported by the National Institute of Mental Health, as follows: NIMH Research Career Development Award K3-MH-20,476-02; Mental Health Research Training Program 5T1-MH-7084-05; Research Grants MH-08853-01 and 94172-05.

[2] Leroy Vandam, to whom we extend our thanks.

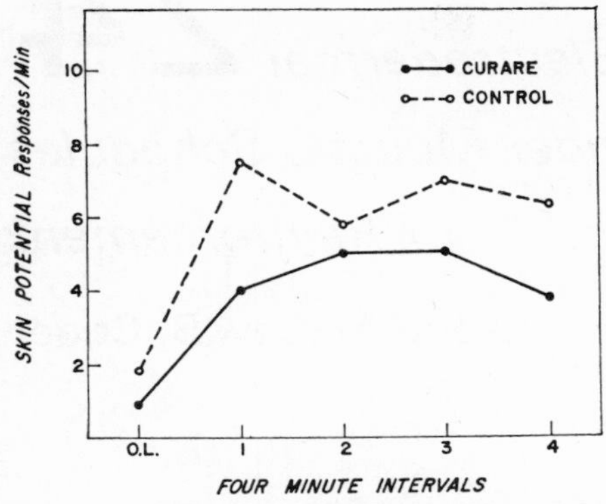

FIG. 1. Comparison of skin potential responses during curare and control sessions.

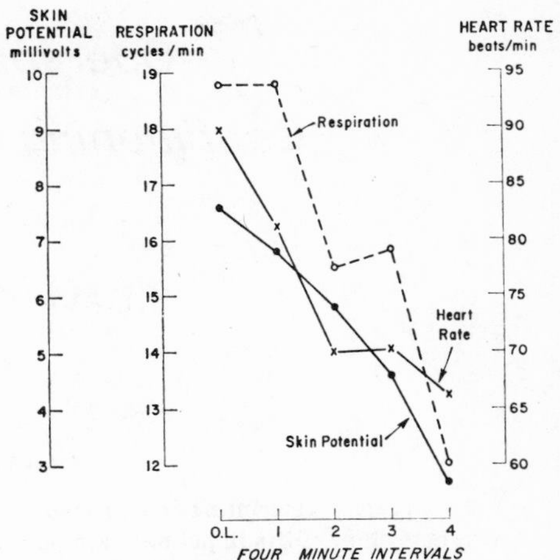

FIG. 2. Time trends of respiration rate, heart rate, and skin potential level during curare session.

previous findings that operant reinforcement facilitates electrodermal activity independently of observable changes in the activity levels of these variables.

DISCUSSION

While negative evidence from a single *S* is always preliminary, these results at least limit the easy universality of notions which find autonomic conditioning a derivative of a behavioral response (e.g., Smith, 1964). They are also consistent with other studies reporting a lack of correlation between operantly conditioned electrodermal activity and continuous measures of somatic activity (Crider et al., 1966; Rice, in press; Van Twyver & Kimmel, in press). Although the level of curarization employed was not sufficient to completely block the neuromuscular junction, there is no doubt that it precluded gross motor responses as contributory factors to the conditioning effect. This does not, of course, rule out relatively subtle interactions between the electrodermal response and covert muscular activity, ongoing autonomic functioning, or respiratory irregularities and patternings. Whether these variables are to be classified as "mediating" phenomena cannot be decided until such questions are phrased in terms of known physiological mechanisms or testable hypotheses.

REFERENCES

CRIDER, A., SHAPIRO, D., & TURSKY, B. Reinforcement of spontaneous electrodermal activity. *J. comp. physiol. Psychol.*, 1966, **61**, 20–27.

FOWLER, R. L., & KIMMEL, H. D. Operant conditioning of the GSR. *J. exp. Psychol.*, 1962, **63**, 563–567.

GOODMAN, L. S., & GILMAN, A. *The pharmacological basis of therapeutics.* (2nd ed.) New York: Macmillan, 1956.

JOHNSON, R. J. Operant reinforcement of an autonomic response. *Dissert. Abstr.*, 1963, **24**(3), 1255–1256.

RICE, D. G. Operant GSR conditioning and associated electromyogram responses. *J. exp. Psychol.*, 1966, in press.

SMITH, K. Curare drugs and total paralysis. *Psychol. Rev.*, 1964, **71**, 77–79.

VAN TWYVER, H. B., & KIMMEL, H. D. Operant conditioning of the GSR with concomitant measurement of two somatic variables. *J. exp. Psychol.*, in press.

(Received December 27, 1965)

24 Operant Electrodermal Conditioning under Multiple Schedules of Reinforcement

David Shapiro and Andrew B. Crider

ABSTRACT

An operant discrimination procedure was used to determine whether human subjects show shifts in palmar skin potential response rate under varying schedules of reinforcement. In Experiment I, a monetary reinforcer was either given or subtracted on a short fixed ratio schedule during alternating 5-min periods. In Experiment II, the schedules alternated between 10-min periods in which either a variable number of responses or long interresponse times were reinforced. Each experiment consisted of seven subjects tested over several sessions. The results indicated differences in response rates consonant with changes in the scheduling of reinforcement. Concurrent recordings of respiration, heart rate, and skin potential level showed that the reinforcement effects were generally specific to the skin potential response variable.

DESCRIPTORS: Operant conditioning, Electrodermal activity, Skin potential response (SPR), Reinforcement.

Contrary to generally held views (Kimble, 1961; Morgan, 1965), recent evidence suggests that autonomic activity can be modified through operant conditioning. When a reinforcing event is made contingent upon the occurrence of unelicited fluctuations in heart rate, peripheral blood flow, or palmar electrodermal activity, higher rates of responding are observed than in various noncontingent reinforcement conditions (Crider, Shapiro, and Tursky, 1966; Engel and Hansen, 1966; Fowler and Kimmel, 1962; Lisina, 1965; Miller and DiCara, 1967; Shearn, 1962; Trowill, 1967). In order to extend the analogy between such results and the more traditional operant conditioning of skeletal responses, a finer control of the frequency characteristics of autonomic operants would be desirable. In the experiments reported here, the rate of emission of electrodermal responses in human subjects was studied under two multiple schedules of reinforcement. The aim was to determine whether changes in the scheduling of

This paper was read at the Society for Psychophysiological Research meetings in Denver, Colorado, October, 1966. The investigation was supported by National Institute of Mental Health Research Grants MH-08853-03, MH-04172-06, and K3-MH-20,476-04 (Career Development Award); also by Office of Naval Research Contract Nonr-1866(43), Group Psychology Branch.

We thank Mr. Bernard Tursky for his generous assistance in facilitating this reasearch. Drs. William Morse and Roger Kelleher contributed many helpful suggestions.

Address requests for reprints to: Dr. David Shapiro, Massachusetts Mental Health Center, 74 Fenwood Road, Boston, Massachusetts 02115.

intermittent reinforcement correlated with an external signal would result in adaptive shifts in response rate.

Method

Skin potential responses (SPRs) were recorded as the potential difference between an active palmar electrode and a forearm reference electrode via an R-C coupled amplifier having an input time constant of 1 sec. Individual differences in directionality of skin potential responsiveness were taken into account by defining a response as either a positive or a negative going fluctuation from the zero baseline of this channel. The criterion amplitude varied from 0.10 to 0.25 mv from subject to subject depending on the modal amplitude of unelicited responses observed under pre-experimental resting conditions. In order to avoid reinforcing both waves of any diphasic response, a 3-sec dead period was interposed after any criterion fluctuation. Each response was fed back to the subject in the form of a 200 msec tone that was automatically triggered by an electronic switch on the basis of the galvanometer voltage output. Reinforcement schedules were programmed by relay equipment. Ten cents was given as the reinforcer, administered by changing a slide indicating the running monetary total on a screen in front of the subject. Criterion responses and reinforcements were registered on a Gerbrands cumulative recorder. In addition, continuous recordings were made of skin potential level, heart rate, and respiration, as described in Crider *et al.* (1966). The subject was isolated from the recording, stimulus, and programming equipment in a sound attenuated, temperature controlled room.

The subjects were 14 male university student volunteers selected because of the electrodermal lability noted in their records from previous experiments. They were fully informed of the problem being investigated, including the response measure and the contingencies of reinforcement. Data were obtained in weekly sessions. The experiment was terminated either by the development of a consistent discrimination between two schedules or repeated failure over several sessions.

Experiment I

This experiment required a discrimination between reinforcement (S+) and punishment (S−) conditions, as indicated by turning on or off a small light placed in view of the subject. Subjects earned 10 cents for every three criterion responses produced during S+ periods and lost 10 cents for every three responses during S− periods.[1] Each session consisted of an initial 5-min period in which a resting, or operant level, response frequency was determined, followed by three 5-min S+ and two 5-min S− periods in alternating sequence.

Table 1 presents the results from the seven subjects run under these conditions, showing the mean rate of SPRs per minute during the last two sessions in which each subject participated. Five subjects (A-E) showed evidence of a successful

[1] Specifically, the S− condition was a response cost procedure (Weiner, 1962). Pilot work had shown this to be more effective in producing response suppression than simple nonreinforcement.

TABLE 1

Skin potential responses per min during operant level (O. L.), reinforcement (S+), and punishment (S−) periods averaged over final two sessions

Subject	Sessions	O. L.	S+	S−	S+	S−	S+
A	3–4	5.4	10.9	2.8	7.5	3.5	7.4
B	8–9	2.4	5.5	2.5	6.1	1.1	3.4
C	3–4	2.4	6.3	3.4	4.3	2.4	3.9
D	3–4	7.0	9.3	5.7	8.8	5.5	6.6
E	2–3	8.4	10.1	6.5	9.8	4.5	8.7
F	7–8	3.8	7.1	5.1	4.1	3.4	3.5
G	6–7	5.1	9.7	7.3	6.0	5.6	4.3
Means		4.9	8.4	4.8	6.6	3.7	5.4

discrimination between S+ and S−, superimposed on a generally declining rate of response over time. Subjects F and G did not form a consistent discrimination. The mean response rate during S+ periods for the five successful subjects was approximately 140% that of the initial operant level. The mean rate during S− periods was only 80% as great. On the average, the gain or loss of ten cents (the reinforcer) occasioned a SPR within six seconds 58% of the time. These "elicited" responses accounted for 21% of all SPRs occurring in S+ and 17% of all responses occurring in S−.

Figure 1 shows cumulative recordings of the SPR operant obtained in subjects A and B over four consecutive sessions during which the discrimination was formed (sessions 1–4 for A and 5–8 for B). Despite individual differences in rate and stability of response, the development of the discrimination is evident in both subjects. These two subjects manifested no clearcut differential activity between S+ and S− periods in heart rate, respiration rate, or skin potential level. Subjects C, D, and E tended to have higher respiration and heart rates during S+ periods. On the average, all seven subjects maintained initial (operant) heart rate and skin potential levels over time, regardless of the reinforcement condition. This contrasts with the declining trends in these functions found in previous experiments employing simple continuous reinforcement (Shapiro, Crider, and Tursky, 1964). A similar lack of habituation in autonomic levels is seen in difficult Pavlovian discriminations (Johnson, 1963).

Experiment II

In this experiment, periods of variable ratio (VR) reinforcement were alternated with a schedule in which long interresponse times were reinforced (differential reinforcement of low rates—DRL). Under VR3, reinforcers followed from one to five responses in random order, while under DRL20″ reinforcement was made contingent on responses following a pause of at least 20 sec. Each schedule was accompanied by a different state of the signal light. The schedules were in force for 10-min periods that alternated in ABA or BAB sequence in consecutive sessions.

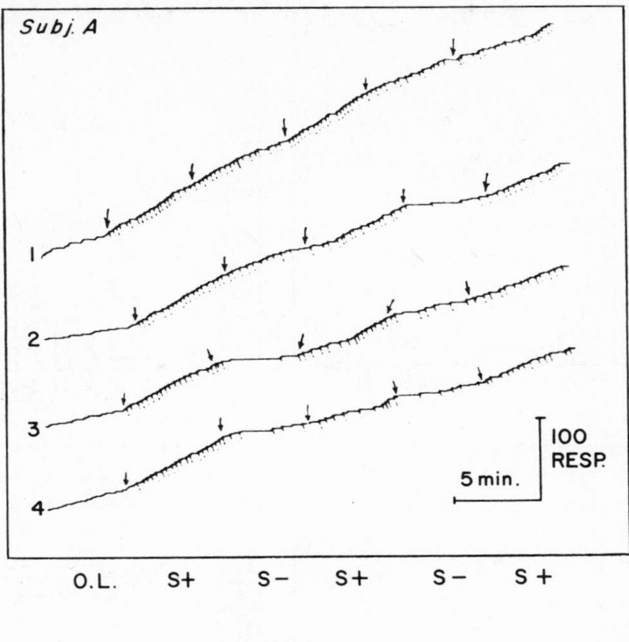

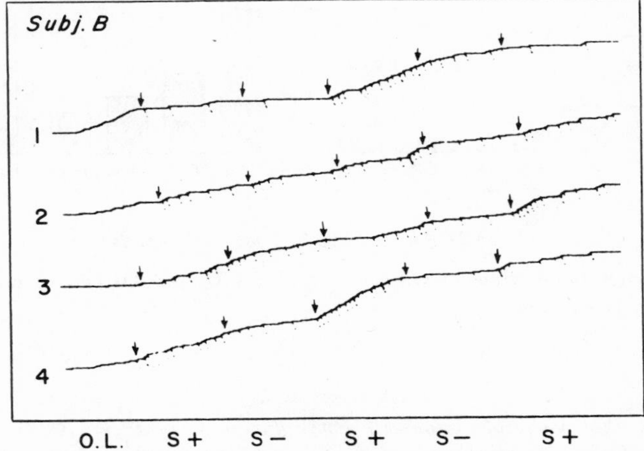

FIG. 1. Development of discrimination between S+ and S− conditions in two subjects over four sessions. Pips indicate addition or subtraction of monetary reinforcer following three emitted responses; arrows indicate schedule change.

Table 2 gives the mean response rates under VR and DRL for each of the seven subjects in their last two sessions, i.e., three 10-min periods of each schedule. Higher response rates were observed in VR than in DRL in all subjects, as would be anticipated (Ferster and Skinner, 1957). Mean rate under VR3 was 200% that of the operant level and under DRL20″ 133% of operant level. On the average, the reinforcer was followed by SPR occurrence within six seconds

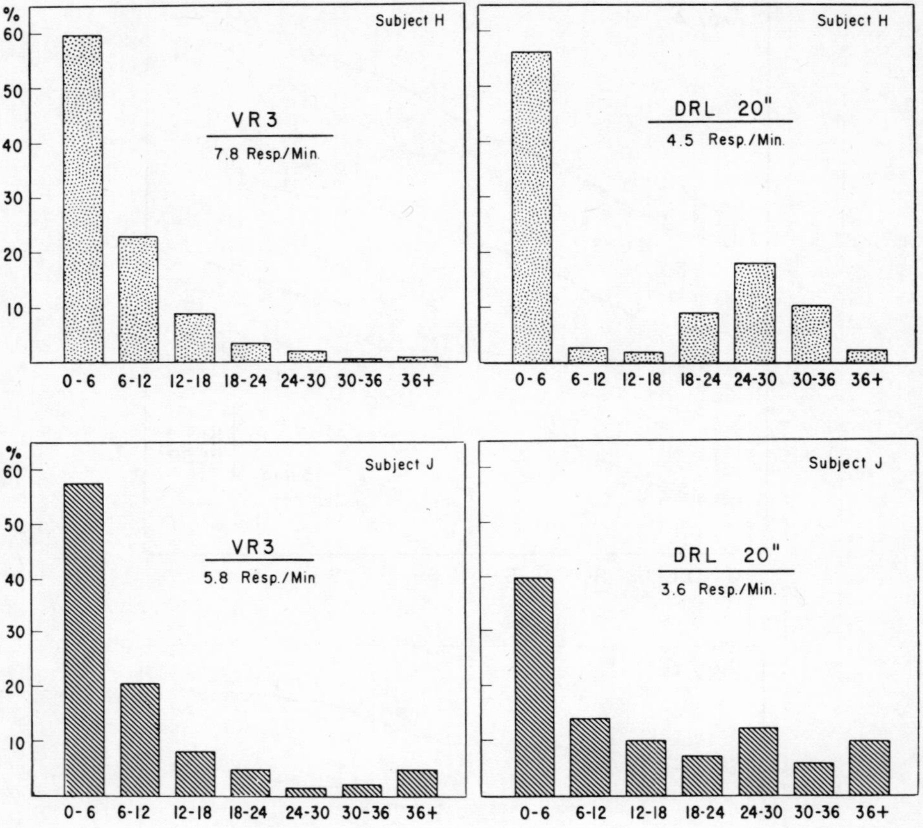

FIG. 2. Interresponse time distributions under VR3 and DRL20″ conditions for two subjects over several sessions.

TABLE 2

Skin potential responses per min during operant level (O. L.), variable ratio reinforcement (VR3), and differential reinforcement of low rates (DRL20″) conditions in final two sessions

Subject	Sessions	O. L.	VR3	DRL20″
H	3–4	1.4	8.0	4.9
J	6–7	0.6	5.3	3.8
K	7–8	6.3	6.1	4.5
L	3–4	1.9	4.2	3.7
M	7–8	3.3	8.3	5.2
N	2–3	1.6	5.4	3.3
O	5–6	6.1	5.1	2.5
Means		3.0	6.1	4.0

57% of the time under VR and 64% of the time under DRL. As in the first experiment, such short latency responses accounted for approximately one-fifth of all responses (18%) in each condition. For purposes of comparing the differential effectiveness of VR and DRL, the interresponse time distributions of two subjects are shown in Figure 2, representing the last three sessions for subject H and the last four sessions for subject J. The higher relative frequency of short interresponse times under VR3 contrasts with the shifting of the distribution to longer times under DRL20″. In both schedules, about half the SPRs occurred within six seconds of a previous SPR, suggesting that volleys of closely spaced responses were not uncommon.

There was little evidence of any systematic correspondence between the two different schedules and the other response measures averaged across subjects. Although respiration rate, skin potential level, and heart rate tended to be slightly higher in VR than in DRL, only the mean difference of 3.2 beats per minute in heart rate was significant.

Discussion

The rate of emission of electrodermal responses is apparently subject to a good deal of environmental control via the manipulation of reinforcement contingencies. The first experiment demonstrates that a successful operant discrimination can be formed between reinforcement and punishment conditions. In the second experiment, differential response patternings were obtained as a function of the scheduling of intermittent reinforcement. Both experiments suggest that conditioned variations in response rate are specific to the reinforced response and do not necessarily depend on corresponding variations in activity of other autonomic functions. On the basis of evidence obtained with curarized subjects (Birk, Crider, Shapiro, and Tursky, 1966; Trowill, 1967; Miller and DiCara, 1967), mediation of operant autonomic conditioning by a skeletal operant also appears unnecessary to the results shown here.

These experiments are adaptations of operant procedures to the problem of instrumental conditioning of autonomic activity in human subjects. They differ from animal studies and other human operant research in the support given subjects through instructions concerning the exact contingencies of reinforcement. These instructions were thought to be justified in light of the a priori supposed difficulty in bringing electrodermal activity under any degree of fine control. The interaction between such instructions and the reinforcement procedures employed in producing positive results is an interesting problem. Simply instructing subjects to produce electrodermal responses without feedback does not lead to sustained modifications of response rates (Crider *et al.*, 1966; Stern and Kaplan, 1966). Moreover, Johnson and Schwartz (in press) found no difference between informed and noninformed groups in the amount of electrodermal response suppression produced with contingent punishment. On the other hand, contingent reinforcement of electrodermal responses with only minimal instructions to subjects to stay alert produces successful conditioning (Fowler and Kimmel, 1962; Rice, 1966). The gradual development over sessions of the discriminations reported here (Figure 1) also suggests a reinforcement or training, as opposed to a purely cognitive, interpretation of the phenomenon.

This is not to say that some degree of cognitive involvement may not be a concomitant of operant electrodermal conditioning. In post experimental interviews, the subjects of the present studies reported a good deal of problem solving activity. Although none claimed to be able to predict the exact occurrence of a feedback tone or a reinforcer, each worked out an idiosyncratic mental strategy to meet the demands of the various schedules. Especially difficult was learning to inhibit responding under punishment conditions or to delay it under DRL. Knowledge of the exact contingencies of reinforcement is probably not necessary to obtain the effect, however. As an illustration, Gavalas (1967) obtained successful conditioning by reinforcing subjects for producing electrodermal responses while leading them to believe that the reinforcer was contingent on behavior in a card sorting task being performed at the same time.

Subjects in psychological experiments respond to the experience by actively producing hypotheses concerning the exact nature of the study (Orne, 1962). By failing to provide instructions, the experimenter does not do away with such problem solving, but simply increases its variability. Insofar as this activity represents an alerted state of the subject, it is probably a necessary, though not sufficient, condition for successful operant electrodermal conditioning. Investigation of its relationship to conditioned autonomic behavior, however, will have to await the development of methodologies more powerful than the post-hoc verbal report. In any event, the use of operant techniques in the study of autonomic processes in human subjects promises to clarify the role of environmental contingencies in the acquisition and maintenance of both normal and pathological functioning.

REFERENCES

Birk, L., Crider, A., Shapiro, D., & Tursky, B. Operant conditioning of electrodermal activity under partial curarization. *J. comp. physiol. Psychol.*, 1966, *62:* 965–966.

Crider, A., Shapiro, D., & Tursky, B. Reinforcement of spontaneous electrodermal activity. *J. comp. physiol. Psychol.*, 1966, *61:* 20–27.

Engel, B. T., & Hansen, S. P. Operant conditioning of heart rate slowing. *Psychophysiol.*, 1966, *3:* 176–187.

Ferster, C. B., & Skinner, B. F. *Schedules of reinforcement.* New York: Appleton-Century-Crofts, 1957.

Fowler, R. L. & Kimmel, H. D. Operant conditioning of the GSR. *J. exp. Psychol.*, 1962, *63:* 573–577.

Gavalas, Rochelle, J. Operant reinforcement of an autonomic response: Two studies. *J. exp. Anal. Behav.*, 1967, *10:* 119–130.

Johnson, H. J. Decision making, conflict and physiological arousal. *J. abnorm. soc. Psychol.*, 1963, *67:* 114–124.

Johnson, H. J., & Schwartz, G. E. Suppression of GSR activity through operant reinforcement. *J. exp. Psychol.*, in press.

Kimble, G. A. *Hilgard and Marquis' conditioning and learning.* (2nd ed.) New York: Appleton-Century-Crofts, 1961.

Lisina, M. I. The role of orientation in the transformation of involuntary reactions into voluntary ones. In L. G. Voronin, A. N. Leontiev, A. R. Luria, E. N. Sokolov, & O. S. Vinogradova (Eds.) *Orienting reflex and exploratory behavior.* Washington: American Psychological Association, 1965. Pp. 450–456.

Miller, N. E., & DiCara, L. Instrumental learning of heart rate changes in curarized rats. *J. comp. physiol. Psychol.*, 1967, *63:* 12–19.

Morgan, C. T. *Physiological psychology*. New York: Mc-Graw-Hill, 1965.

Orne, M. T. On the social psychology of the psychological experiment: With particular reference to demand characteristics and their implications. *Amer. Psychol.*, 1962, *17:* 776–783.

Rice, D. G. Operant conditioning and associated electromyogram responses. *J. exp. Psychol.*, 1966, *71:* 908–912.

Shapiro, D., Crider, A., & Tursky, B. Differentiation of an autonomic response through operant reinforcement. *Psychon. Sci.*, 1964, *1:* 147–148.

Shearn, D. W. Operant conditioning of heart rate. *Science*, 1962, *137:* 530–531.

Stern, R. M. & Kaplan, B. E. Voluntary control of the galvanic skin response. Paper read at the Society for Psychophysiological Research, Denver, Colorado, October, 1966.

Trowill, J. A. Instrumental conditioning of the heart rate in the curarized rat. *J. comp. physiol. Psychol.*, 1967, *63:* 7–11.

Weiner, H. Some effects of response cost upon human operant behavior. *J. exp. Anal. Behav.*, 1963, *6*, 193–195.

25

Suppression of GSR Activity through Operant Reinforcement

Harold J. Johnson and Gary E. Schwartz

Despite early failures to demonstrate operant conditioning of autonomic activity (Mowrer, 1938; Skinner, 1938), there have been a number of recent reports indicating success in this endeavor. A number of studies have reported modification of spontaneous GSR activity using positive reinforcement (Crider, Shapiro, & Tursky, 1966; Fowler & Kimmel, 1962; Kimmel & Kimmel, 1963; Mandler, Previn, & Kuhlman, 1962; Rice, 1966). On the other hand, there is only a single study reporting the suppression of spontaneous GSR activity through operant conditioning (Senter & Hummel, 1965). Grings (1965) also reports suppression of GSRs through punishment, but the GSRs were elicited rather than spontaneous.

In the Senter and Hummel study, electric shock was presented to experimental *S*s contingent upon the emission of spontaneous GSR activity. A yoked control group was used and all *S*s were run for eight experimental sessions. The only data presented in the study is the number of periods in which spontaneous emissions occurred during adaptation and test periods for each session. In each session, the experimental *S*s showed a decreased rate of responding during test periods while the control group showed an increased rate of responding. No data were presented concerning trial-by-trial changes during conditioning and extinction, or the amount of activity in a period.

The single report of suppression of spontaneous GSR responding is in some ways encouraging and is highly relevant to other fields, where suppression of autonomic responses is of primary concern (Eysenck, 1960; Wolpe, 1958). However, it is difficult to evaluate the effectiveness of the procedure and compare it with the positive reinforcement studies because of the nature of the data presented. The positive reinforcement studies have generally accomplished conditioning and extinction in a single session and data is often presented in terms of changes over rather brief time periods. It is the aim of this study to more fully investigate the suppression of GSR responding through presentation of a negative reinforcer.

[1] This research was supported in part by Grant MH 10398 from the National Institutes of Health. The authors gratefully acknowledge the suggestions and criticisms of Joseph J. Campos and F. Elaine Donelson. The experiment was conducted at Cornell University.

Further, it has been implied (Grings, 1965; Kimmel & Kimmel, 1963) that operant conditioning of the GSR is accomplished even when *S*s have no knowledge of the contingency between reinforcement and response. Indeed, *S*s who have expressed awareness of response contingencies have been dropped from some studies (Kimmel & Kimmel, 1963). Considering the confusion which has developed in other areas regarding the awareness issue (Eriksen, 1962) this issue bears close investigation. Instructions to *S*s were manipulated and a postexperimental interview was used in an attempt to evaluate this problem.

Method

Design.—The experiment used a simple 2 × 2 factorial design. One variable was the operant reinforcement procedure. This consisted of a contingent (C) group which received a negative reinforcer contingent upon emission of spontaneous GSRs and noncontingent group (NC) which received the same number of reinforcers per minute during periods of no spontaneous GSR activity.

The second variable was a variation in preexperimental instructions to *S*s. One group of *S*s (I group) received instructions which indicated that the reinforcer was dependent upon their activity. The second group (NI group) received no information regarding the reinforcement contingency. The detailed instructions for the two groups were:

> I group: Specifically, your task in this experiment will be simply to try and limit the number of sounds coming through the earphones. The number of sounds that will be presented will depend upon your physiological responses, and not upon any gross movement. We do not know how people will do this, so I cannot tell you exactly what to do. All I can say is that it depends on you, things that happen inside you. Outside things such as noises of the machine, pipes rattling, or movements which you make have nothing to do with how often the sounds will be presented. The fewer number of sounds that you hear, the better you are doing.

> NI group: Specifically, your task in this experiment will be simply to sit quietly and listen to the sounds coming through the earphones. I have to emphasize this point, because the better you follow these instructions, the better your record will be.

Subjects.—Thirty-two female undergraduates from Cornell University served as *S*s. They were volunteers paid for their participation in the experiment.

Procedure.—Upon entering the laboratory, *S*s were assigned to one of the four experimental conditions (NI-C, NI-NC, I-C, I-NC) on the basis of their order of arrival for the experiment. The *S*s were then seated in a reclining chair in an electrically shielded room and the physiological recording devices were attached.

All *S*s received a set of general instructions explaining that the physiological recordings were being obtained to determine how people respond to sounds which would be presented during the experiment. They were also told not to move during the experiment and to remain attentive. Following these general instructions half the *S*s were then given the I instructions and half were given the NI instructions. The *E* then left the room telling *S*s that the experiment would last approximately 30 min. and would end when he returned.

After preliminary equipment adjustments, a 4-min. adaptation period began. During this period the base rate of unelicited GSR responding was determined for each *S*. A spontaneous or unelicited GSR was defined as a change in skin resistance of greater than 500 ohms which did not occur within 6 sec. following any observable event which could give rise to a GSR. The following were considered relevant events: coughs, sighs, deep breaths, gross skeletal movement, outside noises, abrupt increases or decreases in HR, and 25 mv. changes in forearm EMG potential. The criterion for operant responding was set at a minimum of two unelicited GSRs during the 4-min. adaptation period. The *S*s who showed fewer than two such responses were dropped from the experiment.

The conditioning period followed next and was begun by the presentation of a reinforcer to all *S*s without regard to GSR activity. Thereafter, the C *S*s received the negative reinforcement each time an unelicited GSR occurred. The running of a C *S* was followed by an NC *S* who received the reinforcement schedule of her predecessor. The conditioning period lasted 16 min.

Immediately after the conditioning period, the extinction period began and lasted for 8 min. During this period no reinforcements were presented to either group.

At the end of extinction *S*s were interviewed with structured questions to gather information about their feelings and actions during the experiment.

Apparatus.—The physiological responses were recorded on a Grass Model 5 D polygraph. Two Grass Model 5P1 preamplifiers were used to record GSR activity. One channel was set at very high sensitivity to detect 500-ohm changes in skin resistance. The second used a lower sensitivity. The use of two channels aided *E* in making the distinction between elicited and unelicited GSRs as the high sensitivity recording frequently exceeded the limits of the recorder. The GSR electrodes were zinc, 8 mm. in diameter, and were enclosed in a Plexiglas housing which successfully eliminated electrode movement artifacts. The electrodes were placed on the second and fourth fingers of the right hand using a zinc-sulfate paste (Lykken, 1959).

Heart rate was obtained from a photoelectric plethysmograph attached to *S*'s right ear. Muscular activity was obtained from a right forearm muscle group using silver cup electrodes placed 1 in. apart. These responses were also recorded on the polygraph.

The negative reinforcer was a 700-cps square-wave tone generated by a RCA WA-44C audio generator. This signal was amplified and presented to *S*s through Lafayette Model F767 earphones. The duration of the tone was 3 sec. with an intensity of 95 db. measured at the earphones. Pilot work had indicated that this combination of frequency, intensity, and duration resulted in a sound consistently judged by *S*s as unpleasant. Furthermore, GSRs to the sound showed little adaptation.

Quantification of data.—The number of unelicited GSRs per 4-min. time samples was computed for each *S*. This resulted in seven scores; one for adaptation, four for conditioning, and two for extinction. These raw scores were then converted into two different types of derived scores which had been used in previous studies. Simple difference scores were obtained by subtracting the adaptation scores from the conditioning and extinction scores. Percentage transformations were obtained by adding unity to every raw score, dividing each score from the conditioning and extinction periods by the adaptation score, taking square roots and multiplying by 100 (Fowler & Kimmel, 1962). The other physiological variables were only used by *E* in making the judgment about spontaneity of electrodermal activity.

Results

Adaptation period.—The mean number of GSRs for each group for each time period is presented in Table 1. Analysis of variance was performed on the raw data for the adaptation period to determine if there were any differ-

TABLE 1

Number of GSRs per Time Sample

	Adaptation	Conditioning				Extinction	
	1	1	2	3	4	1	2
I-C							
$\bar{X}$	10.00	8.50	4.00	2.38	2.38	3.75	2.38
σ	(4.21)	(2.96)	(2.24)	(2.23)	(2.00)	(4.24)	(3.46)
NI-C							
$\bar{X}$	15.50	6.62	4.25	3.75	3.00	6.00	3.50
σ	(11.08)	(3.53)	(4.55)	(3.23)	(3.04)	(8.20)	(4.87)
I-NC							
$\bar{X}$	14.38	17.25	13.88	13.62	11.38	17.75	17.25
σ	(8.07)	(12.76)	(12.62)	(9.95)	(8.31)	(10.43)	(10.11)
NI-NC							
$\bar{X}$	10.25	11.82	8.50	7.88	8.75	13.00	11.25
σ	(7.63)	(6.62)	(4.85)	(4.86)	(6.44)	(9.35)	(8.48)

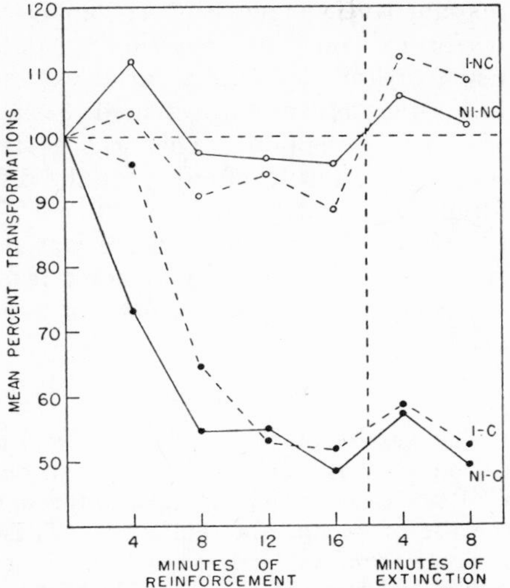

FIG. 1. Mean percentage transformations of spontaneous GSRs during reinforcement and extinction.

ences between the various treatment groups during this period. This analysis yielded nonsignificant *F*s for both the main effects and the interaction.

Conditioning.—Parallel analyses of variance were performed on both the simple difference scores and the percentage transformations with C and I as the between variables and time (T) as the within variable. The C effect was highly significant,[2] F (1, 28) = 31.43, $p < .001$. The C × T interaction was significant for the percentage scores only, F (3, 84) = 3.03, $p < .05$, while the I effect was not significant. These data are presented graphically in Fig. 1. Here it can be seen that the C groups show a substantial decrease in spontaneous GSR activity across the 16 min. of reinforcement.

The above analyses, because they are based on derived scores, were unable to detect any interaction between C and I from the adaptation period to the first 4 min. of reinforcement. In order to assess such changes within the first 4 min., t tests were performed. The difference in number of responses between C and NC groups across instructions for the first 4 min. was significant, t (30) = 2.71, $p < .02$. Comparing C and NC groups for the NI condition alone yielded significant effect, t (14) = 3.98, $p < .002$, but this same effect was not significant when evaluated for the I condition alone. These findings indicate that the contingency variable differentially affects the number of responses in the I and NI groups and is suggestive of an interaction between I and C during the first 4 min. of reinforcement (See Fig. 1).

Extinction.—The analysis of variance for the two 4-min. extinction periods again yielded a highly significant C effect, F (1, 28) = 53.72, $p < .001$. None of the other *F*s in this analysis was significant. It is interesting to note the increases in all groups from the last 4 min. of reinforcement to the first 4 min. of extinction. This rise in rate of response was tested by an analysis of variance which included the last 4 min. of reinforcement and the first 4 min. of extinction. This analysis yielded the expected effect for C. The T effect was also significant, F (2, 56) = 5.34, $p < .01$, indicating that the rise in the first period of extinction is a significant effect.

Postexperimental interview.—Each *S* reported the tone as being unpleasant and the average rating for all *S*s was 7.94 on a 10-point scale going from very pleasant to very unpleasant. There were no differences between the various treatment groups on this rating. However, seven C *S*s and six NC *S*s did indicate that the tone became less

[2] In this and subsequent comparisons the *F*s are those for the percentage of transformations. The levels of significance for the difference score analyses are comparable unless otherwise indicated.

unpleasant as the experiment progressed.

Only 10 of the 32 *S*s gave a positive answer when asked if they felt they were doing something which determined whether or not they received a tone. Surprisingly, these *S*s were almost evenly split between the four conditions. This distribution suggests that neither the I or C conditions were very effective in eliciting a feeling of control.

When *S*s were questioned further regarding possible hypotheses, even though they may have responded negatively to the initial question, the I *S*s produced an average of 5.0 hypotheses and the NI *S*s 4.4. This lack of difference between the groups is surprising considering the instructions given to the I *S*s. However, it probably reflects the comments of many of the I *S*s who said nothing worked and that they gave up trying to influence the tone. The number of hypotheses offered was also evenly split between the C and NC conditions.

Of all the hypotheses offered by *S*s, only one categorization appeared with any sizeable frequency. About 60% of the *S*s made some comments which involved a calm-tense dimension, but there were again no significant differences among the treatment groups.

High and low operant Ss.—In order to gain some further insight into the process which seems to be occurring, treatment groups were split into high and low responders during adaptation. A median split was performed and *S*s were grouped as high and low responders in the C and NC conditions. The function generated was very similar to that presented in Fig. 1 with operant level substituted for I-NI (low operant for I). That is, the NC *S*s tended to show only slight decrements in response and the C *S*s manifested suppression of response regardless of operant level. The results of analyses performed on the conditioning data using operant level as a variable were the same as those presented earlier with one exception. This was a significant operant level effect for the difference score analyses, $F\ (1, 2) = 9.37$, $p < .001$. Interestingly, there was no significance attributed to operant level when percentage transformations were used.

Discussion

The results of this study support the conclusion that suppression of spontaneous GSR activity can be accomplished through the use of an operant reinforcer. These findings confirm and extend those of Senter and Hummel (1965). The changes observed during extinction are especially interesting. First there was a significant initial rise in responsiveness of all groups from the last 4 min. of reinforcement to the first 4 min. of extinction. This rise can probably be attributed to an orienting or perceptual disparity response resulting from a change in reinforcement conditions. It should also be noted that even after 8 min. of extinction, the groups do not begin to converge. This finding of virtually no extinction after 8 min. does not correspond to earlier findings using positive reinforcement (Fowler & Kimmel, 1962; Kimmel & Kimmel, 1963). This may reflect a difference in the strength of the reinforcer used in this study as compared to those used in the positive reinforcement studies.

One of the aims of this study was to determine the effectiveness of instructions upon the conditioning and concomitantly to gain some insight into the question of the *S*s' awareness of reinforcement contingencies. By and large, the instructional manipulation and postexperimental inquiry were not very revealing. The instruction effects were not significant in any overall analysis, but there was suggestion of differential changes occurring as a result of instructions and contingency within the first 4 min. of conditioning. The instructions seemed to inhibit rather than facilitate conditioning during this period.

The interview used in this study did not result in any consistent information concerning *S*s' expectancies, but more systematic inquiry is needed before one concludes that such learning occurs without awareness. Similarly, the question of what is being conditioned is not resolved. Although care was taken to eliminate obvious artifacts, the possibility still exists that other highly correlated responses were, in fact, being manipulated.

REFERENCES

Crider, A., Shapiro, D., & Tursky, B. Reinforcement of spontaneous electrodermal activity. *J. comp. physiol. Psychol.,* 1966, **61,** 20–27.

Eriksen, C. W. (Ed.) *Behavior and awareness: A symposium of research and interpretation.* Durham: Duke Univer. Press, 1962.

Eysenck, H. J. (Ed.), *Behavior therapy and the neuroses.* London: Pergamon, 1960.

Fowler, R. L., & Kimmel, H. D. Operant conditioning of the GSR. *J. exp. Psychol.,* 1962, **63,** 563–567.

Grings, W. W. Verbal-perceptual factors in the conditioning of autonomic responses. In W. F. Prokasy (Ed.), *Classical conditioning: A symposium.* New York: Appleton, 1965. Pp. 71–89.

Kimmel, E., & Kimmel, H. D. A replication of operant conditioning of the GSR. *J. exp. Psychol.,* 1963, **65,** 212–213.

Lykken, D. T. Properties of electrodes used in electrodermal measurement. *J. comp. physiol. Psychol.,* 1959, **52,** 629–634.

Mandler, G., Previn, D. W., & Kuhlman, C. K. Effects of operant reinforcement on the GSR. *J. exp. Anal. Behav.,* 1962, **5,** 317–321.

Mowrer, H. H. Preparatory set (expectancy)—a determinant in motivation and learning. *Psychol. Rev.,* 1938, **45,** 62–91.

Rice, D. G. Operant conditioning and associated electromyogram responses. *J. exp. Psychol.,* 1966, **71,** 908–912.

Senter, R. J., & Hummel, W. F. Suppression of an autonomic response through operant conditioning. *Psychol. Rec.,* 1965, **15,** 1–5.

Skinner, B. F. *The behavior of organisms.* New York: Appleton-Century-Crofts, 1938.

Wolpe, J. *Psychotherapy by reciprocal inhibition.* Stanford: Stanford Univer. Press, 1958.

(Received September 12, 1966)

26
Ability of Actors to Control Their GSRS and Express Emotions

Robert M. Stern and Nancy L. Lewis

ABSTRACT

Ability to control GSRs through the use of ideational stimulation was studied in 26 professional actors. During one 10-min period, "Respond," *S*s had to make as many GSRs as possible; during the other period, "Rest," *S*s were instructed to inhibit their GSRs. All *S*s received continuous visual feedback of their responses during both periods. No relationship was found between ability to control GSRs and emotional expression as measured by the ratings of directors. However, as predicted, method actors performed at a significantly higher level than non-method actors. An additional finding was that those actors who normally experience sweating as their primary response to stress performed at a higher level than non-sweaters.

DESCRIPTORS: GSR, Emotions, Voluntary control of ANS, Actors. (R. Stern)

In a recent study, Stern and Kaplan (1967) showed that the ability to produce and inhibit galvanic skin responses (GSRs) through the use of ideational stimuli varied considerably among college students. One personality variable that was suspected of playing a role in determining this ability was creativity, i.e., the more creative an individual, the easier it will be for him to conjure up the appropriate imagery to produce a GSR. A second variable that was thought to be related to this ability was the amount of practice an individual had had in deciding what class of events was for him emotional and then imagining a specific scene of sufficient intensity to produce a GSR.

The purpose of the present investigation was to study the ability of professional actors to voluntarily control GSRs and to relate this to the following variables:

(1) creativity, as measured by success in expressing emotions on the stage,

(2) the technique used by the individual actor to express emotions on the stage, and

(3) the predominant physiological change that the actor experiences in real-life emotional situations.

Of primary concern in this study was the procedure used by the actor to express emotions on the stage, i.e., *method* actors versus *non-method* actors, since it was felt that this would reveal information about the importance of

This research was supported by grant MH 13062-01, National Institute of Mental Health of the U.S. Public Health Service and a NSF Undergraduate Research Participation grant to Penn State. The cooperation of the Theatre Arts Department, and in particular J. Peter Shapiro, is gratefully acknowledged.

Address requests for reprints to: Dr. Robert M. Stern, Department of Psychology, Pennsylvania State University, University Park, Pennsylvania, 16802.

practice. The *method* school of acting, which was first popularized by Stanislavsky (1950) and is today supported by leading directors such as Lee Strasberg (Hethmon, 1965), teaches aspiring actors to express emotions through the use of personal emotional memory. The more traditional school of acting states that emotions are expressed in the theatre by means of sterotyped facial expressions and voice tones which the actor learns from his teacher (Dean, 1941). *Method* actors, on the other hand, are taught to make use of their memory of specific sensory experiences that were connected with personal emotional situations, which were similar to the emotion that they are trying to recreate. It was, therefore, hypothesized that *method* actors as compared with *non-method* actors would show a higher degree of voluntary control over their GSRs. No direct evidence was available at the onset of the study to support the hypothesis; however, pilot studies had shown that the ability to control GSRs improves with practice and *method* actors are individuals who have repeatedly practiced the relevant technique.

A serendipitous finding of a previous study (Stern and Kaplan, 1967) indicated that for the college students used as subjects, those who displayed the highest degree of control over their GSRs usually were individuals who reported that their predominant physiological change during real-life emotional situations was sweating. This finding was considered to be particularly interesting both because of the relationship of GSRs to sweat-gland activity and because this was the first evidence of a constitutional variable which correlated with the ability to control GSRs. It was one of the purposes of this study to see if this relationship prevailed for actors as well.

METHOD

Subjects

The *S*s were 15 male and 11 female professional actors who appeared in productions of the Pennsylvania State Festival Theatre during the summer of 1966. The actors and actresses ranged in age from 17 to 51 with a median of 22. They were paid $3.00 for their participation; none had ever taken part in a GSR experiment.

Apparatus

GSRs were recorded on a Stoelting Dermograph, set at maximum sensitivity. The finger electrodes used were 1 in. square pieces of silver; Sanborn Redux was applied before the electrodes were attached to *S*. A microammeter was wired in parallel with the Dermograph and provided the *S*s with continuous visual feedback of their own responses. Sound-deadening ear cushions were worn by *S* to keep out any noises.

Procedure

Experimental sessions were conducted in the women's dressing rooms of the two theatres in which the actors were appearing. The *S* was seated at a dressing table and shielded from the rest of the room by a folding screen. The meter was placed in front of *S*, the electrodes were attached, instructions were given and then the ear cushions were put on.

The experiment consisted of two 10 min periods, "Rest" and "Respond," the order being reversed for half the *S*s. Prior to the Rest period, *S*s were instructed to try and make the needle on the meter go to the right (an increase in skin resistance) by relaxing. Before the Respond period, *S*s were told to make as many deflections of the needle to the left as they could by thinking about emotional situations. All *S*s were cautioned to avoid movements and irregular breathing.

Post-session Questionnaire

Following each session, *S* was given a written questionnaire to complete with several items, only the following two of which are considered relevant to this report.

1. "What basic technique do you use on the stage to portray a particular emotion?"
2. "When you are anxious about something, how does this feeling usually manifest itself? (sweating, blushing, heart palpitations, irregular breathing, stomach difficulties, etc.)"

Directors' Ratings

Three directors rated each actor on a scale of 1–5 on their ability to express emotions on the stage.

Quantification of GSR Data

A response was scored as one GSR if it was a drop in resistance of approximately 125 ohms or greater. (Unfortunately a degree of non-linearity in the sensitivity of the GSR unit produced some variability.) A GSR was considered completed if the record leveled off for two sec or turned back. Following a needle reset or *S* movement, no GSRs were counted for five sec.

Results and Discussion

Fig. 1 shows the data for 21 of the *S*s divided into *method* actors and *non-method* actors on the basis of their answers to the post-session question which was concerned with the technique they use in portraying emotions on the stage. The curves in this graph and those in Fig. 2 were smoothed by the use of a 3-point-running-average over minutes of the session. The answers of five *S*s were such that it was not possible to place them in either category. As can be seen in the figure, both groups emitted GSRs at a higher rate during Respond than during Rest. Both groups also showed an adaptation effect during the 10 min Respond period, something which was not found in the earlier study (Stern and Kaplan, 1967). A comparison of the relative ability of *method* and *non-method* actors to control their GSRs can be seen by looking at the difference between the Respond and Rest functions for each group in Fig. 1. As hypothesized, the *method* actors displayed better voluntary control of their GSRs than *non-method* actors. A *t* test calculated using the difference score for each *S* between the total number of GSRs made during Respond and during Rest indicated a

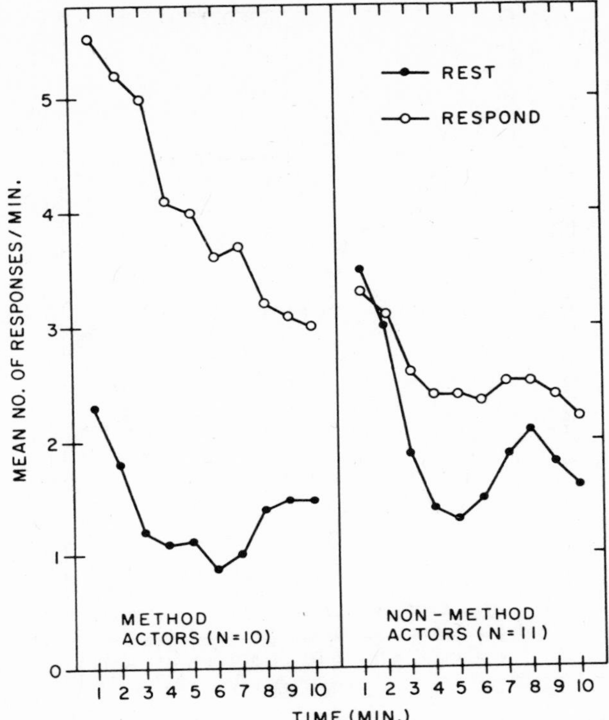

FIG. 1. Mean number of GSRs emitted by *method* actors and *non-method* actors.

highly significant difference between the groups ($t = 3.40$, $df = 19$; $p < .01$). Unfortunately, a direct comparison between the *non-method* actors and college students was not possible since the college students were run on a Fels Dermohmmeter, whereas all of the actors were run on the Stoelting portable GSR unit which turned out to be quite non-linear. However, an admittedly superficial comparison of the *non-method* actors and the college students run in the previous study indicated that if any differences exist, the college students showed a greater difference between GSRs emitted during Respond and during Rest than the *non-method* actors.

These two findings—better GSR control by *method* actors than *non-method* actors, and no better control by *non-method* actors than college students—support the hypothesis that relevant practice and not a general creativity factor is primarily responsible for these results. The possibility exists, however, that the more creative young actor tends to become a *method* actor. As an indirect check on this, the three directors of the actors used in this study were asked to rate on a scale from 1—5 each actor on his ability to portray emotions on the stage. The mean ratings were correlated against the ability to control GSRs with the result being that no systematic relationship was found. Further analysis revealed that some *non-method* actors and some *method* actors received very high ratings for portraying emotions on the stage, indicating that, according

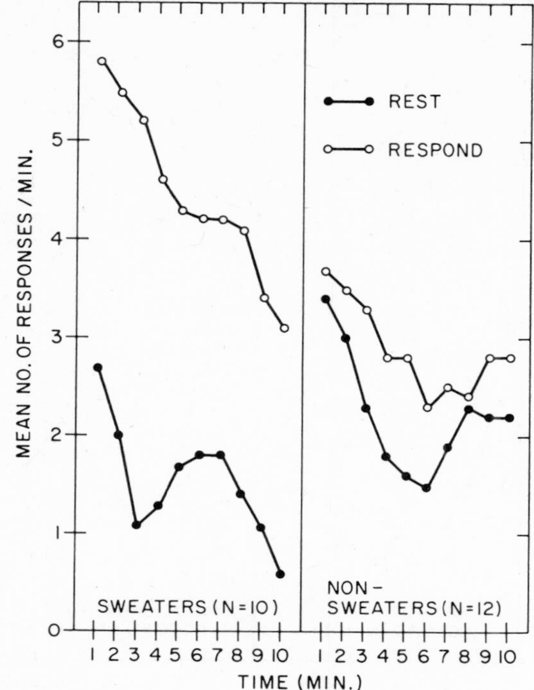

FIG. 2. Mean number of GSRs emitted by actors whose predominant physiological response to stress is sweating and the mean number emitted by those who do not sweat.

to the three judges used, either technique can lead to success. The high ratings received by some of the *non-method* actors were the cause of the nonsignificant correlation between ability on the stage and ability to control GSRs. In other words, these individuals received high ratings from the directors, but scored low on ability to voluntarily control their GSRs.

Fig. 2 shows the GSR data for 22 of the same actors divided into those who reported that they experience sweating as one of their primary manifestations of anxiety in a real-life situation and those who do not experience sweating. It can be seen in this graph that, as predicted, those *S*s who reported sweating showed better voluntary control of their GSRs than *S*s who do not sweat. A t test calculated as described above indicated that the difference between the groups was significant ($t = 2.95$; $df = 20$; $p < .01$).

This finding, namely, that by simply asking a person his predominant physiological change in a stress situation, you can predict how well he can voluntarily control his GSRs, has appeared in both studies from this laboratory. It was surprising to find that the verbal report of *S*s would be so strongly related to any physiological measure. A comparison of Fig. 1 with Fig. 2 reveals that whether a person reported that he sweats when anxious or not is as powerful a variable as whether or not he is a *method* actor.

In brief, the results of this study indicate that the ability to control GSRs is related to the amount of practice an individual has had and his primary physiological response to stress, but not to creativity as measured by success in expressing emotions on the stage.

REFERENCES

Dean, A. *Fundamentals of play directing*. New York: Farrar & Rinehart, 1941.

Hethmon, R. H. *Strasberg at the actors studio*. New York: Viking, 1965.

Stanislavsky, K. *Stanislavsky on the art of the stage*. London: Farber & Farber, 1950.

Stern, R. M. & Kaplan, B. E. Galvanic skin response: Voluntary control and externalization. *J. Psychosom. Res.*, 1967, *10*: 345–353.

V

SALIVATION, URINE FORMATION, AND GASTRIC MOTILITY

27 Operant Salivary Conditioning in Man

Clinton C. Brown and Ruth A. Katz

ABSTRACT

Operant conditioning of the human parotid salivary response was attempted in order to bypass some of the difficulties encountered in classical conditioning procedures. Spontaneous, unstimulated parotid saliva was collected from two groups of human subjects under two reward conditions. One group was rewarded for salivary peaks, defined as an increase in on-going salivary rate to a criterion of 3 drops/5 sec; the other was rewarded only at times of non-peaking or minimal salivation. A significant increase in the number of peaks was found in the "peak reward" group, but there was no change in salivary rate in the "non-peak reward" group. The data demonstrate salivary conditioning in human subjects by means of operant reward technique.

DESCRIPTORS: Operant reward, Salivation, Parotid, Conditioning.

Despite the ease with which classical salivary conditioning has been demonstrated in animals, attempts at salivary conditioning in human subjects have been either fruitless or questionable in nature.

This paradox has been variously explained as the result of: 1) gross interspecies differences, particularly in the innervation of the salivary glands, or 2) a significant difference in the role of parotid secretion in human and animal digestive processes.

Despite intensive efforts over the past 18 months to make the stimuli and the experimental situation conducive to classical conditioning, this laboratory has not been able to produce, in normal subjects, statistically significant evidence of salivary conditioning. By this is meant a measurable increase in the rate of salivation elicited by the conditional stimulus and noted either in the interval between the conditional (CS) and unconditional stimulus (US) or during unreinforced trials with the positive stimulus. If a less stringent criterion is used, salivary conditioning has been accomplished (Brown and Katz, 1967). B. W. Feather has since succeeded in salivary conditioning with humans (personal communication). Both studies used an acid stimulus rather than a food stimulus.

We have, however, recently completed two studies which demonstrate that

We wish to express our appreciation to A. M. I. Wagman, Ph.D. for editorial assistance and to C. T. Smith for technical assistance.

Supported by NIH Grant No. MH 10731–01 and administered by Friends of Psychiatric Research, Inc., 52 Wade Avenue, Baltimore, Maryland 21228.

Address requests for reprints to: C. C. Brown, Psychophysiology Laboratory, Spring Grove State Hospital, Baltimore, Maryland 21228.

a different kind of salivary conditioning in the human subject may be demonstrated by means of operant techniques. The phenomenon appears to be closely related to what Brown (1966) has called control conditioning and other investigators (Shearn, 1962; Miller and Carmona, 1966) have reported for heart rate in man and salivation in dogs. The operant reward technique was adopted on the hunch that the acid commonly used as US in classical conditioning, while reflexly creating salivary flow, might be so noxious as to produce a masking defense reflex (Sokolov, 1963). We therefore substituted positive (rewarding) stimulation in an operant procedure.

METHOD

Eighteen randomly selected male and female subjects ranging in age from 15–28 were arbitrarily divided into two groups of nine *S*s each. They were paid $8.00 for serving as subjects on five consecutive days at the same time of day. On the first day the subject was told that the study was concerned with salivation, that nothing unpleasant would happen and that no shocks or stimuli would be given. Parotid collection cups were applied bilaterally, and integrated salivary flow recorded from both glands on an Offner polygraph (Sutherland and Katz, 1961). Continuous skin potential (SP) and digital photoplethysmograph records were also taken. The subject was instructed to relax in a comfortable chair in a pleasant room while listening to soft music played for the duration of the one-hour test. Each subject was asked to press a button to signal swallowing. This procedure was followed for both groups for Days 1–3.

On Day 4, a 5″ x 7″ electric sign and a counter were placed in the subject's field of view. The sign was made of translucent plastic bearing the word "BONUS" in 3″ black letters which were visible briefly only when illuminated. The counter was set at zero at the beginning of each test. These two additional items were pointed out to the subject. He was then told that on Days 4 and 5, in addition to the fee he was to receive for serving as a subject, he would receive a 10 cent bonus each time the sign lit up and that the counter would keep track of the number of bonuses he earned. He was advised that it was impossible for him to determine the reason for the administration of the bonus and that consequently he could not control the number received.

For one group (peak-reward) the experimenter administered a bonus each time that a spontaneous increase in salivary rate occurred. This peak had to meet the criterion of a deflection approximately equal to 3 drops in 5 seconds.

The other group (non-peak reward) also received bonuses but never during a period of peak salivation. The criterion for reward was the absence of peaking rather than a specified low rate of salivation. The number of rewards given to each subject in this group on each of the reward days equalled the mean number of rewards given to the peak-reward group on that day (13 on Day 4 and 26 on Day 5).

Each one hour test record was scored for the number of drops of saliva, the number of peaks reaching the criterion, and the number of swallows signalled by the subject. Within-subject comparisons by use of t tests for correlated means were made within each group using the averages of the scores obtained on each measure for each S under reward and non-reward conditions.

A period of 20 sec following each signalled swallow was selected for inspection to detect any salivary peaking which could possibly be attributed to the preceding swallow. This interval is approximately double the time required for an acid stimulated salivary response to reach maximum amplitude (Brown and Katz, 1966). If swallowing per se can serve as a stimulus for salivation, one may reasonably expect the response to become visible in approximately the same length of time required for other stimuli to elicit it.

Results

The peak-reward group, as shown in Fig. 1 and Table 1 revealed a significant increase in peaks and swallows under reward conditions. The number of drops also increased but the increase did not quite attain significance due to marked inter-subject variability.

When reward was not contingent on salivary peaking, but on its absence (non-peak reward group), no significant change in peaks was seen. There was also, as shown in Fig. 2 and Table 1, no change in the other measures.

For both groups, under reward and non-reward conditions, the percentage of swallows which was followed within 20 sec on the record by a salivary peak was found to be at most 10%. Thus, 90% of all recorded swallows had no effect on subsequent spontaneous salivation of the subject.

Post test interviews revealed that only one or two *S*s were aware that the bonus was given for salivation although they knew that salivation was being measured. The presence of other transducers may have deceived the subjects.

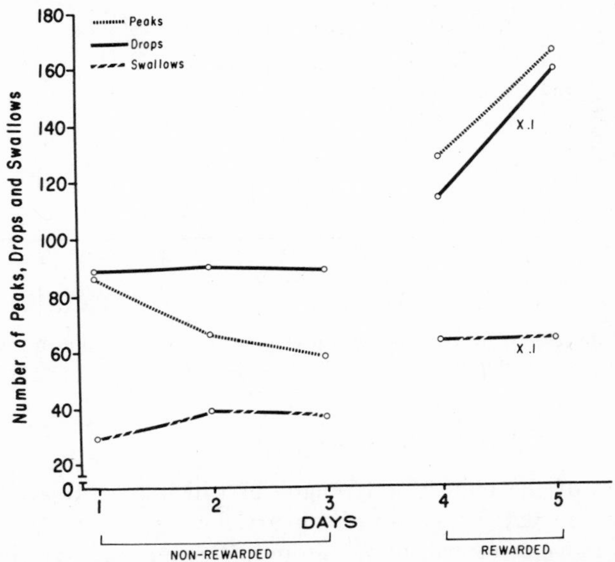

Fig. 1. Comparison of salivation measures under reward and non-reward conditions for the peak reward group. Measures of drops and swallows have been reduced by a factor of 0.1.

TABLE 1

Comparison of peaks, drops, and swallows under conditions of reward and non-reward for the peak-reward group and the non-peak reward group

Peak-reward group	Days 1–3 Mean	Days 4–5 Mean	% Increase	t
Drops	891	1356	52	1.827
Peaks	70	146	108	2.1801*
Total swallows	350	635	81	2.131*
Coincident swallows	29/350 = .08	65/635 = .10	2	
Non-peak reward group				
Drops	947	1104	16	1.053
Peaks	109	118	8	.200
Total swallows	794	904	14	1.246
Coincident swallows	77/794 = .10	87/904 = .10	0	

* Significant at .05 level.

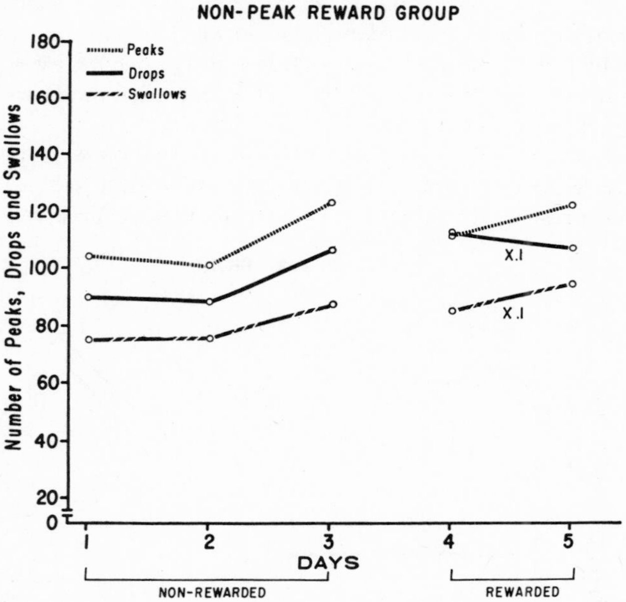

FIG. 2. Comparison of salivation measures under reward and non-reward conditions for the non-peak reward group. Measures of drops and swallows have been reduced by a factor of 0.1.

Discussion

Movements of the mouth and tongue of which the subject was not aware may well have modified the rate of salivary flow (Feather, 1966) and in future studies such movements will be monitored. We can only say conclusively that under both reward conditions our subjects did not increase the percentage of coincident swallows, i.e. those followed by salivary peaks, even though the total number of swallows which was signalled was significantly increased for the peak-reward group. It seems reasonable to conclude that it was the increase in salivary peaks which elicited the increase in the number of swallows rather than vice-versa.

This study indicates that salivary secretion in the human subject may be brought under stimulus control by means of operant reward techniques without evidence of mediation by swallowing.

The relative ease with which salivation was conditioned using an operant paradigm suggests that involuntary, autonomically mediated systems may be experimentally controlled by procedures previously thought to be limited in their effective application to voluntary systems.

REFERENCES

Brown, C. C. Telemetry and Psychophysiology. Presented at the Association for the Advancement of Medical Instrumentation (AAMI), Boston, July 1966.

Brown, C. C., & Katz, R. A. Psychophysiologic aspects of parotid salivation in man: I. Stimulus intensity and salivary output. *Psychophysiology*, 1966, *3:* 273–279.

Brown, C. C., & Katz, R. A. Psychophysiologic aspects of parotid salivation in man: II. Effects of varying inter-stimulus intervals on salivary conditioning. *Psychophysiology*, 1967, *4:* 99–103.

Feather, B. W. Effects of concurrent motor activity on the unconditioned salivary reflex. *Psychophysiology*, 1966, *2:* 338–343.

Miller, N., & Carmona, A. Modification of a visceral response, salivation in thirsty dogs, by instrumental training with water reward. Presented at EPA meetings, April 14–19, 1966.

Shearn, D. Operant conditioning of heart rate. *Science*, 1962, *137:* 530–531.

Sokolov, Ye. N. *Perception and the conditioned reflex.* (trans. by S. W. Waydenfeld), New York: Pergamon Press, 1963. p. 14.

Sutherland, G. F. & Katz, R. A. Apparatus for study of salivary conditional reflex in man. *J. appl. Physiol.*, 1961, *16:* 740–741.

28 Modification of a Visceral Response, Salivation in Thirsty Dogs, by Instrumental Training with Water Reward

Neal E. Miller and Alfreda Carmona

Thirsty dogs rewarded by water for bursts of spontaneous salivation showed progressive increases in salivation, while other dogs, rewarded for brief periods without salivation, showed progressive decreases. No obvious motor responses were involved, but dogs rewarded for decreasing salivation appeared more drowsy than those rewarded for increasing it. Implications for learning theory and psychosomatic medicine are mentioned.

The traditional view of most learning theorists who have written specifically on the problem is that, while skeletal responses mediated by the somatic nervous system can be learned as instrumental responses rewarded by the law of effect, visceral responses mediated by the autonomic nervous system can be learned only by classical conditioning (Keller & Schoenfeld, 1950; Kimble, 1961; Konorski & Miller, 1937; Mowrer, 1947, 1950; Skinner, 1938; Solomon & Wynne, 1954).

In sharply challenging this position, Miller (1961, 1963a, 1963b, 1964) has emphasized that this issue has fundamental implications not only for theories of learning and its neurophysiological basis, but also for practical problems of psychosomatic symptoms and of individual differences in patterns of autonomic responses. While the reinforcement for classical conditioning must be a stimulus eliciting a UR similar to the one to be learned by conditioning, any reward can reinforce any response in instrumental learning. Thus if visceral responses can be modified only by classical conditioning, they can be learned and maintained only in situations involving reinforcing stimuli with the unconditioned ability to elicit that particular visceral symptom; whereas if they can be learned as instrumental responses, rewarded by the law of effect, their initial learning and subsequent performance can be reinforced by any one of a wide variety of rewards (Dollard & Miller, 1950).

There is an extensive literature on the modification of many visceral responses by classical conditioning (Bykov, 1957; Kimble, 1961; Pavlov, 1927). For a long time the experimental evidence on the possibility of instrumental learning of visceral responses has been limited to incidental mention of the negative results of two unpublished studies (see Mowrer, 1938, and Skinner, 1938). Very recently, however, there have been a number of reports of the successful modification of the galvanic skin potentials in humans by instrumental techniques (Crider, Shapiro, & Tursky, 1966; Fowler & Kimmel, 1962; Johnson, 1963; Kimmel & Kimmel, 1963; Snyder & Noble, 1965). Furthermore, Shearn (1962) and Shearn and Clifford (1964) have reported modifying the heart rate in humans and in rabbits by instrumental techniques. But there are a large number of ways in which the GSR and heart rate can be affected via skeletal responses, such as breathing; the possibility of such mediational responses probably is much greater with human than with animal *S*s. For this reason we thought it desirable to work with dogs and to use a visceral response, salivation, which has been extensively employed in classical conditioning. It seemed worthwhile to determine whether this response could be modified in any way by instrumental training procedures before going on to the difficult attempt to eliminate all possibilities of mediation via skeletal responses.

[1] Presented in part at the 1966 meetings of the Eastern Psychological Association and at the National Academy of Sciences (Miller, 1966); supported by Grant MH00647 from the National Institute of Mental Health.

[2] Both now at Rockefeller University.

The purpose of the present study was to see whether it is possible to change the rate of spontaneous salivation of mildly thirsty dogs by using water to reward some of them when they increase their rate and others when they decrease their rate. Preliminary tests showed that the water did not elicit any appreciable change in salivation; furthermore, the effects of any classical conditioning produced by any unconditioned effect of water were controlled by the procedure of rewarding different groups for changes in opposite directions.

Experiment 1

Method

Subjects. The *S*s in this experiment were 10 naive mongrel dogs of both sexes purchased through the Yale animal care facilities. Their weights were 18–30 lb.

Apparatus. We used the method of recording salivation from the parotid duct developed by Sheffield (1957).[3] Under Nembutal anesthesia, the parotid duct was cannulated with polyethylene PE50 tubing which was brought outside through a hole in the cheek and fastened to a parallel stainless steel plate on the inside of *S*'s cheek. A stainless steel tube, which made an airtight and watertight seal with the flanged terminal of the cannula, was threaded into the outer Plexiglas plate and served as the exit for the salivary flow and the spigot to which the recording device was attached.

The recording was made by connecting a polyethylene tubing (PE240) between the stainless steel tube in the outer Plexiglas plate and a closed water system whose lower end was placed about 18 in. below the level of the floor of the experimental cage. The closed system was filled with a mixture of 50% water and 50% ethyl alcohol. The saliva entering in the upper end of the system allowed fluid to drip out of the lower end. Thus changes in the viscosity of the saliva did not affect the output. The lower end of this system was a No. 27 hypodermic needle, from which each emerging drop touched an upright needle and momentarily closed a circuit to an electronic relay (Grason-Stadler Drinkometer) which activated a counter and an ink-writing recorder.

The experimental chamber was a $3\frac{1}{2} \times 3\frac{1}{2} \times 5\frac{1}{2}$ ft. wooden one with glass wool between the walls. A blower unit provided ventilation and white noise. This cage was only relatively soundproof. The clicks of the relays energized by each drop of liquid were always present. A one-way mirror allowed *E* to observe *S* during the experiment.

Procedure. During the first week after operation *S*s were habituated to the experimental situation. On each session, *S* was placed in the experimental chamber and the working condition of the cannula was tested by giving *S* a pellet of food, after which the door was closed. Twenty-four hours after cannulation, only salivation in response to one pellet of food was recorded. For the rest of the experiment, *S*s were tested after 16 hr. of water deprivation. Each session lasted 30 min. after the initial test for the condition of the cannula. For the next 5 days, *S*s were run twice a day in habituation sessions involving only the recording of spontaneous salivation. The *S*s salivating between 20 and 40 drops/min (a middle range from which either increases or decreases were possible) were kept and randomly assigned to one of two groups of 5 dogs each; 9 dogs had to be discarded in the process of securing 10 *S*s.

After habituation, the sessions were reduced to one a day and the instrumental training procedure began. Group 1 was first given 10 sessions during which nonsalivating was rewarded. After this the cannulas were removed and *S*s allowed to rest for 15 days before cannulas were implanted again in the opposite cheeks. Then *S*s were given 10 sessions of reward for the opposite response of salivating. For Group 2 the procedures were exactly the same, except that salivating was rewarded first and nonsalivating second.

In all cases the reward was the delivery of 20 ml. of water when *E* closed a switch. After each reward there was a 30-sec. time-out period (not marked by any cue given to *S*) during which no reward could be earned. The initial criterion for rewarding nonsalivation was a pause of 2 sec. between drops. This criterion was progressively increased to a pause of 30 sec. between drops. The initial criterion for rewarding salivation was 2 drops per 10 sec. This criterion was progressively increased to 10 drops per 10 sec.

In preliminary work on other dogs we had found that, if a CS preceded the delivery of water, the *S*'s orientation response to this CS and the water dish temporarily inhibited all salivation. Therefore, no CS was used. Between rewards *S* sat with its head just above the dish ready to drink as soon as water was delivered.

For each of the 10 days of training, the total number of drops of salivation per day was recorded for each *S*. The slope of the curve for drops per day on successive days of training was calculated for each *S* according to the procedure described by Edwards (1962).

Results

There was a general tendency for the slopes to be negative, indicating decreased

[3] More detailed description available in mimeographed manuscript from F. D. Sheffield, Department of Psychology, Yale University, New Haven, Connecticut.

salivation as training progressed; but during the first period of training, the average negative slope of −2.7 for Group 1 was greater than that of −0.08 for Group 2. Thus the difference was in the expected direction. Furthermore, there was no overlap between the two groups—all five *S*s in Group 1 had negative slopes, and in Group 2 two *S*s had smaller negative slopes, while the other three had positive ones. Such a perfect dichotomy between the two groups would be expected by chance only 1 time out of 512.[4]

During the second period of training, when each group was rewarded for the opposite response, the difference between them reversed, but was small (−0.07 vs. −0.12) and unreliable. Finally, when each *S*'s score during reward for decreasing was subtracted from its score during reward for increasing, the net difference for 8 out of 10 *S*s was in the expected direction, and the mean +1.3, also in the expected direction, was reliably different from zero ($p < .05$, $df = 9$).

Although the foregoing overall effects were reasonably reliable, the differences were small and the slopes for individual *S*s were unreliable.

Experiment 2

The purpose of this experiment was to see whether more extensive training would produce more clear-cut instrumental learning of salivation in individual dogs. For those dogs whose salivary ducts remained in good condition, a secondary purpose was to see whether the effect of the initial training could be reversed by rewarding the opposite response.

Method

Subjects. We used 6 naive mongrel dogs, 1 female and 5 males, purchased from the Department of Animal Care at Yale University.

Procedure. After being habituated to the experimental situation for 2 wk., *S*s were implanted with cannulas in the parotid gland and 24 hr. later were put on a schedule of 16 hr. of water deprivation preceding each day's preliminary tests. During 7–12 sessions of 1 hr. each, spontaneous salivation was recorded and, if *S*s were salivating less than 10 or more than 35 drops/min, they were discarded. We had to give preliminary tests to 11 dogs in order to get 6 within this range. After this test period, *S*s were assigned to two matched groups which were then assigned by the flip of a coin to the two conditions of treatment.

The *S*s in each group received the same amount of food every day 1–2 hr. before the experimental session. They were run 45 min. a day for 5 days a week. Before the experimental session there was 15 min. of habituation during which the door of the experimental chamber was left open. During weekends *S*s had access to water ad lib, but during the rest of the week received water only during the experimental sessions. An exception occurred with *S*s rewarded for speeding up after being rewarded for slowing down; in those cases, because of the small number of rewards received in the experimental situation, *S*s had to be given water after the experimental sessions.

In order to flush out the cannula with more profuse salivation than food elicits, and also to determine that it was functional, we squirted 5–10 ml. of a mixture of 50% ethyl alcohol and water into *S*'s mouth each day while *S* was standing in the experimental chamber before the experimental session began. Then we followed this mixture with a pellet of food to remove the aversive taste. Whenever *S* salivated less than 100 drops/min in this test, we took the cannula off. We did the same as soon as signs of infections were suspected. After the wound had healed, we implanted the cannula in the same cheek and resumed training. To prevent local infections, we daily applied 2–3 ml. of sulfodiazol around the wound.

The groups were randomly assigned to the following treatments:

Group 1, fast first. Three dogs were rewarded for speeding up during 37, 40, and 39 days, respectively. After a rest period of about 3 mo., we found that one *S* had been infected with distemper and had to be discarded. The other two were run next for 40 days with reward for slow salivation. After a rest period of 2–3 mo., these same *S*s were run for 40 days, with reward now for fast salivation.

Group 2, slow first. Three *S*s were rewarded for slowing down for 40 days. After 3 mo. of rest, one had to be discarded because the salivary gland did not work well. The other two were run for 40 days with reward for fast salivation, and then finally run for 30 days with reward for slowing down. All *S*s were run in the same apparatus, and as a further control *S*s getting opposite training were run on the same day. The criterion for giving the reward was similar to that in Experiment 1, but, with *S*s rewarded for decreases in salivation, the initial criterion of a 2-sec. interval without salivation was progressively increased up to 60 sec. in the final stages. During training for increases in salivation, we initially rewarded every burst of spontaneous salivation of 1 drop or more in a 5-sec. period and progressively increased this

[4] All probabilities reported in this paper are for two-tailed tests.

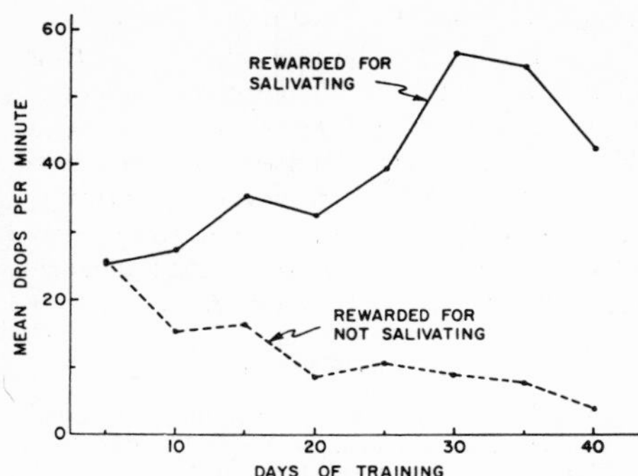

FIG. 1. Mean curves of instrumental learning by three thirsty dogs rewarded with water for increases or decreases in spontaneous salivation.

criterion to 7 drops in 5 sec. in the last stages of the experiment.

The slopes of the curves for amount of salivation on successive days of training were calculated for each *S* in the way described for Experiment 1.

Results

The results of the first period of training are presented in Figure 1. In Group 1 the dogs rewarded for fast salivation increased with slopes of +.91, +.59, +.29, each reliably different from zero ($p < .001$, .016, .013; $df = 35$, 38, 37, respectively). In Group 2 the dogs rewarded for not salivating decreased with slopes of −.50, −.90, −.45, each reliably different from zero beyond the .001 level ($df = 38$). During the early part of this training, *S*s in Group 1 received more rewards than those in Group 2; but during the last 10 days of training, the criterion for Group 1 had been increased so that the two groups did not differ in amount of water received as a reward.

When the conditions of reward were reversed during the second period of training, the signs of the slopes reversed. The slopes in Group 1 were −.59 and −.92, both reliable at the .001 level; those in Group 2 were +.07 and +.16, reliable at the .007 and .052 levels, respectively.

When the conditions of reward were reversed yet again during the third period of training, one *S* in Group 1 continued to decrease its salivation, showing a slope of −.17, reliable ($p = .02$) in the wrong direction, while the other dog showed a slope of +.02 in the correct direction, but not statistically significant. In Group 2 both *S*s showed successful reversals with slopes of −.58 and −1.2 in the expected direction which were highly reliable ($p < .002$, .001).

When the slopes of individual *S*s during different conditions of reward were compared (i.e., *S* 1 fast minus *S* 1 slow for first vs. second training periods, and separately for third vs. second periods), the algebraic differences of all eight possible comparisons are in the direction to be expected on the basis of the rate rewarded, with six significant at the .001 level, one at the .01 level, and one not significant.

Discussion

The results clearly show that the autonomically innervated visceral response of salivation can be modified by an instrumental training procedure. From the point of view of the theory and neurophysiology of learning, it is important to know whether the visceral response was directly modified or whether the change was mediated by the learning of some skeletal response which in turn had an unlearned tendency to affect the visceral response. We did not observe any obvious skeletal responses, such as chewing or panting, regularly associated with the amount of salivation, but more subtle ones may have occurred.

In an attempt to get more accurate data on the skeletal response most likely to affect salivation, we recorded the frequency of respiration by a strain-gauge pneumograph in 5 cases of reward (in different stages of Experiment 2) for increases in salivation and 4 cases of reward for decreases. As supplementary data we also recorded the number of heartbeats by taking EKGs. The records for these responses and for salivation were scored during the 5–10, 20–25, and 35–40 min. periods of each day. The daily averages were used to compute separate correlations between these measures for each type of training on each dog.

Salivation on daily tests was positively correlated with number of breaths on daily tests for all 9 cases, indicating a definite relationship between rate of breathing and amount of salivation. Similarly, the number of heartbeats was correlated with the amount of salivation for 7 out of the 9 cases. There was a general tendency for the number of breaths to decrease as training progressed, but no appreciable difference between the groups in this respect.

In order to determine how much of the effects of training on salivation could be accounted for on the basis of any effects on breathing, partial correlations were calculated. With the effects of breathing partialed out, correlations between days of training and amount of salivation were positive for the 5 cases rewarded for increasing and negative for 3 of the 4 rewarded for decreasing, a difference in the expected direction reliable at a little better than the 10% level. When the effects of both heart rate and breathing were partialed out, the correlations between training and salivation were larger (i.e., more positive) for all of the cases rewarded for increased salivation than for any of them rewarded for decreased salivation, a difference reliable beyond the 1% level by Fisher's exact test. These results indicate that the difference in salivation between the cases rewarded for increasing and for decreasing salivation probably cannot be attributed solely to differences in rate of breathing.

We started to proceed to the next step of more conclusively ruling out the effects of breathing and other skeletal responses by paralyzing them with curare and maintaining the dogs on artificial respiration, but found that this drug elicited continuous copious salivation, without the pauses and bursts necessary for rewarding increases or decreases. Furthermore, the saliva was viscous and likely to clog the cannula.

A second question is whether the learned response is relatively specific to the salivation that was rewarded, or is a part of some more general pattern, such as arousal. We observed that *S*s rewarded for increasing salivation tended to be especially alert; the EEG records taken on two of them were characterized by many movement artifacts and a typical arousal pattern of fast, low-voltage activity. The *S*s rewarded for not salivating tended to be more drowsy, sitting quietly in front of the water dish, sometimes appearing to take short naps. The EEG records on two of them showed much more slow-wave, high-voltage activity. These observations suggest that the salivation may have been part of a larger pattern of general arousal vs. relaxation.

REFERENCES

Bykov, K. M. *The cerebral cortex and the internal organs.* (Ed. & trans. by W. H. Gantt) New York: Chemical Publishing, 1957.

Crider, A., Shapiro, D., & Tursky, B. Reinforcement of spontaneous electrodermal activity. *J. comp. physiol. Psychol.,* 1966, **61,** 20–27.

Dollard, J., & Miller, N. E. *Personality and psychotherapy.* New York: McGraw-Hill, 1950.

Edwards, A. L. *Statistical methods for the behavioral sciences.* New York: Holt, Rinehart & Winston, 1962.

Fowler, R. L., & Kimmel, H. D. Operant conditioning of the GSR. *J. exp. Psychol.,* 1962, **63,** 563–567.

Johnson, R. J. Operant reinforcement of an autonomic response. *Dissert. Abstr.,* 1963, **24,** 1255–1256. (Abstract)

Keller, F. S., & Schoenfeld, W. N. *Principles of psychology.* New York: Appleton-Century-Crofts, 1950.

Kimble, G. A. *Conditioning and learning.* (2nd ed.) New York: Appleton-Century-Crofts, 1961.

Kimmel, E., & Kimmel, H. D. A replication of operant conditioning of the GSR. *J. exp. Psychol.,* 1963, **65,** 212–213.

Konorski, J., & Miller, S. Further remarks on two types of conditioned reflex. *J. gen. Psychol.,* 1937, **17,** 405–407.

Miller, N. E. Integration of neurophysiological and behavioral research. *Ann. N. Y. Acad. Sci.,* 1961, **92,** 830–839.

Miller, N. E. Animal experiments on emotionally-induced ulcers. *Proc. World Congr. Psychiat., June 4–10, 1961, Montreal,* 1963, **3,** 213–219. (a)

Miller, N. E. Some reflections on the law of effect produce a new alternative to drive reduction. In M. E. Jones (Ed.), *Nebraska symposium on motivation: 1963.* Lincoln: University of Nebraska Press, 1963. Pp. 89–91. (b)

Miller, N. E. Some implications of modern behavior theory for personality change and psychotherapy. In D. Byrne & P. Worchel (Eds.), *Personality change.* New York: Wiley, 1964. Pp. 149–175.

MILLER, N. E. Extending the domain of learning. *Science,* 1966, **152,** 676. (Abstract)

MOWRER, O. H. Preparatory set (expectancy)—a determinant in motivation and learning. *Psychol. Rev.,* 1938, **45,** 62–91.

MOWRER, O. H. On the dual nature of learning—a reinterpretation of "conditioning" and "problem solving". *Harv. educ. Rev.,* 1947, **17,** 102–148.

MOWRER, O. H. *Learning theory and personality dynamics.* New York: Ronald, 1950.

PAVLOV, I. P. *Conditioned reflexes.* (Trans. by S. V. Anrep) London: Oxford University Press, 1927.

SHEARN, D. W. Operant conditioning of heart rate. *Science,* 1962, **137,** 530–531.

SHEARN, D. W., & CLIFFORD, G. D. Cardiac adaptation and contingent stimulation. *Amer. Psychologist,* 1964, **19,** 491. (Abstract)

SHEFFIELD, F. D. Salivary conditioning in dogs. *Yearbk. Amer. Philos. Soc.,* 1957, 284–287.

SKINNER, B. F. *Behavior of organisms.* New York: Appleton-Century, 1938.

SNYDER, C., & NOBLE, M. Operant conditioning of vasoconstriction. Paper read at the Midwestern Psychological Association, Chicago, April 1965.

SOLOMON, R. L., & WYNNE, L. S. Traumatic avoidance learning: The principles of anxiety conservation and partial irreversibility. *Psychol. Rev.,* 1954, **61,** 353–385.

(Received March 11, 1966)

29

Instrumental Learning of Urine Formation by Rats; Changes in Renal Blood Flow

Neal E. Miller and Leo V. DiCara

MILLER, NEAL E., AND LEO V. DICARA. *Instrumental learning of urine formation by rats; changes in renal blood flow.* Am. J. Physiol. 215(3): 677–683. 1968.—Rats paralyzed by curare and maintained on artificial respiration were rewarded by electrical stimulation of the medial forebrain bundle for changes in the rate of urine formation. Each of the seven rats rewarded for increases learned increases, and each of the seven rewarded for decreases learned decreases. The changes in the opposite directions were each highly reliable ($P < .001$). The two groups differed in rate of glomerular filtration and renal blood flow, but not in blood pressure, heart rate, core temperature, or vasomotor activity.

learning, visceral; learning, of renal blood flow; conditioning, operant; psychosomatic medicine; kidney function; renal circulation; autonomic nervous system; cardiovascular; renal physiology; glomerular filtration

---

THE STRONG TRADITIONAL BELIEF has been that the autonomic nervous system, mediating glandular and visceral responses, is limited to a more primitive type of learning, classical conditioning, whereas the cerebrospinal system, mediating somatic responses, is the only one that can be modified by a higher type of learning called instrumental learning, trial-and-error learning, type II conditioning, or operant conditioning. (See Konorski (10); Skinner (20); and Kimble's summary (9).)

The question of whether or not the two peripheral divisions of the nervous system are functionally different in this way has fundamental significance for the neurophysiology of learning, for the normal functioning of these systems in maintaining homeostasis, for their malfunction in the production of psychosomatic symptoms, and for the possibility of using instrumental learning in the therapy of such symptoms.

The belief that the autonomic nervous system is inferior with respect to learning has been challenged theoretically by Miller (11–14). Recent experiments have shown that either increases or decreases, respectively, can be produced, by rewarding them, in the following autonomically mediated visceral responses: salivation (17), heart rate (5, 19, 21), blood pressure (6), and peripheral vasomotor responses (7). Furthermore, the specificity of visceral learning has been shown by the fact that rewarding changes in intestinal contractions produces such changes without affecting heart rate, while rewarding changes in heart rate produces such changes without affecting intestinal contractions (16), and that differential vasomotor responses can be elicited from the two ears (8). All but the first of the foregoing studies have ruled out the indirect effects of overt skeletal responses by paralyzing them with curare.

In classical conditioning, the reinforcement must be a stimulus that has the unconditioned ability to elicit the specific response to be conditioned. The learning consists merely of attaching the same response to a new stimulus, the conditioned one. In instrumental learning the reinforcement, usually called a reward, does not need to be able to elicit the response to be learned; it has the property of strengthening any immediately preceding response. Thus, a given response can be reinforced by a variety of rewards, or a given reward can reinforce a variety of responses. This greater flexibility has given rise to the notion that instrumental learning is a different and higher type than classical conditioning.

A number of studies summarized by Bykov (2) and Corson (3) have shown that increases in the rate of urine formation can be classically conditioned by using as the unconditioned stimulus water loads that normally produce diuresis, and that decreases can be conditioned by using as the unconditioned stimulus a noxious stimulus such as an electric shock. The purpose of the present study was to see whether or not the volume of urine formed, and attendant changes in glomerular filtration and renal blood flow, could be increased or decreased, respectively, by instrumental learning.

In addition to its relevance to renal function and to the plasticity of the autonomic nervous system, evidence for instrumental learning of urine formation would help to rule out one of the arguments that has been used by those continuing to advocate the traditional inferiority of the autonomic nervous system. These advocates (cf. 1) maintain that the subjects may be learning to send out impulses from the motor cortex to peripheral muscles that are paralyzed by curare, and that these impulses may be innately or classically conditioned to autonomic changes and thus indirectly mediate them. While such a view has some plausibility for changes in heart rate, it becomes less plausible the greater the variety of autonomically mediated responses that can be demonstrated to be affected in a specific way by instrumental training. The formation of urine appears to be less likely than cardiovascular responses to be affected by impulses to somatic motor activity.

METHODS

The subjects were 14 male, albino Sprague-Dawley rats, weighing between 570 and 640 g, which were successfully prepared with electrodes and catheters. Their bladders were chronically catheterized in order to eliminate bladder contractions. They were paralyzed by curare in order to eliminate bodily movement, and maintained on artificial respiration. Since they were paralyzed, they had to be rewarded either by escape from electrical shock (which was not chosen because of a marked unconditioned response to the shock) or by direct electrical stimulation of the lateral hypothalamus, which was found to produce a much smaller unconditioned response. In order to rule out any unconditioned or cumulative effects of the electrical stimulation of the brain or any effects of any other aspects of any of the procedures used in the experiment, half of the rats selected at random were rewarded for increased rates of urine formation and the other half for decreased rates. The first small changes in the desired direction were promptly rewarded; as the rats learned, they were gradually shaped to make larger responses by progressively increasing the difficulty of the criterion. Urine was collected in 10-ml samples, and the rate of flow was monitored by a counter of tiny drops which activated a device for automatically delivering the reward. In order to secure a rapid enough flow so that small changes in rate could be measured promptly, the rats had to be preloaded with Ringer solution via a chronic catheter into the jugular vein. To maintain the water load constant in both of the groups throughout the experiment, 0.5 ml of Ringer solution was infused automatically after every 0.5 ml of urine collected. Training consisted of 220 rewarded trials, administered over a period of approximately 3 hr.

After it became clear that the first animals were learning, additional measures of blood pressure, peripheral vasoconstriction, and temperature were taken on the remaining 10 rats. After further evidence of successful learning, glomerular filtration was measured by using inulin-^{14}C, and the rate of renal blood flow by using tritiated *p*-aminohippuric acid (PAH) for the last eight rats, four rewarded for increases and four rewarded for decreases. The details of the method follow.

Electrode Implantation and Pretesting

Electrodes were implanted by the procedure described by Miller et al. (18) and illustrated in Fig. 7B of that article. A pair of enameled, stainless steel, surgical needles (Anchor no. 1843) with straight tapered points bare for 0.5 mm at the tip were used as monopolar electrodes aimed at the medial forebrain bundle of the lateral hypothalamus at the A-P level of the ventromedial nucleus with Krieg stereotaxic coordinates of 1.5 mm posterior to bregma, 8.5 mm below the skull top at point of entry, and 1.5 mm lateral to the midline. Subsequent histology revealed that all electrodes were closely grouped within 0.7 mm^3 rostrocaudal at the middle one-third of the ventromedial nucleus, laterally in the region between the fornix and internal capsule and at a depth just dorsal to the ventromedial nucleus above the fornix column at this level. A stainless steel screw on the skull was used as an indifferent electrode.

Two weeks after implantation, the electrodes were tested for reward effects according to the procedure described by Miller and DiCara (19). After shaping to press a bar delivering 0.5 sec of 60-cycle a-c electrical stimulation of the brain (ESB), each rat was further trained on *day 1* for two 30-min sessions separated by a 30-min rest, with each bar press rewarded. Rats pressing fewer than 500 times during the first of these sessions were discarded. The current for individual rats was adjusted within a range of 30–100 μa to secure optimal reward without motor side effects, and kept constant thereafter. On *day 2*, rats received 10 min of training with each bar press rewarded, and then 10 min under each of four fixed-interval schedules with reward turned off after each delivery for 5, 10, 20, and 30 sec, respectively, during the successive training conditions. Throughout the experiment, a 60-w light and a 1,000 cycle, 82 db tone, serving as time-in stimuli, were on whenever pressing the bar would deliver reward, and were off whenever it would not, i.e., during the time-out intervals. *Day 3* repeated the 50-min sequence given on *day 2.*

Implantation of Indwelling Catheters and ECG Electrodes

Two days after the end of the preceding tests and training, the rats were anesthetized with 40 mg/kg of Nembutal. According to a procedure described by Davis and Weeks (4), a catheter consisting of PE polyethylene tubing connected to a length of Vivosil tubing (i.d. .012, o.d. .025 mm) was implanted via the jugular vein into the right auricle. The polyethylene part of the tube was threaded under the skin and emerged from the back of the neck. The catheter was filled with Ringer solution and sealed by a stainless steel pin.

Next, a 7-cm incision was made on the rat's abdomen, and the bladder exposed. A small incision was made in the bladder, and the slightly flared end of a catheter consisting of 6.5 cm of polyethylene tubing (i.d. 2.0 mm) was inserted approximately 5 mm, to the point where the ureters enter. Surgical silk was then wrapped round the bladder wall to push it snugly against the catheter. Into the other end of the polyethylene catheter had been thrust a 2.5-cm, snugly fitting, stainless steel tube, approximately 0.5 mm from the end of which a piece of 000 stainless steel (Surgaloy) wire had been wrapped and soldered. The ends of the wire were then threaded through the sides of the abdominal wall at the top of the incision, drawn together, and tied so that the stainless steel tubing protruded from the abdomen. The abdominal wall and skin were then sutured together. Surgaloy wires for ECG recording were stitched subcutaneously in the middle of the back at shoulder level and on the chest over the sternum.

Procedures Just Prior to Training

Preliminary water load. Two days after the preceding operation, the experiment began. In order to make sure that the urine flow was normal, and to start all animals from the same initial condition of adequate hydration, Ringer solution was infused by the venous catheter for 2 hr at the rate of 5 ml/hr, while the rat was in a small holding cage over a funnel. Then urine was collected for an additional hour without further infusion, during which time the rat presumably equilibrated his water load.

Curarization and respiration. The rats were injected ip with 3.0 mg/kg of *d*-tubocurarine chloride (Squibb) in a solution containing 3 mg/ml. As soon as the rat showed signs of difficulty in breathing, it was fitted with a face mask made from the mouthpiece of a rubber balloon, the lower lip of which was secured by placing it behind the upper incisors and the upper portion of which fitted snugly over the top of the snout. The balloon end of the mask was connected via a rubber stopper with a hole drilled through its middle to the tube from a small-animal respirator (E&M Instrument Co., model VE5 QG). The rat was respirated at a 1:1 I/E ratio, 70 cycles/min with a peak reading of 20 cm water pressure. In previous experiments, such respiration had been found to maintain the rat's heart at a stable and normal level for up to 5 hr. After the lateral abdomen had been locally anesthetized by 0.15 ml of 2% Xylocaine hydrochloride, a small 30-gauge needle was inserted ip, and additional *d*-tubocurarine was infused at the rate of 1 mg/kg per hr until the end of the experiment.

Experimental chamber and drop counter. The subject was placed on its left side on a folded towel in a soundproof, ventilated, picnic lunchbox (Grason-Stadler). The stainless steel tube from the bladder cannula was attached to a polyethylene tube filled with water, leading directly down to a drop counter consisting of a 27-

gauge hypodermic needle pointing down directly at the tip of a surgical needle mounted so that the gap between them could be adjusted with a micrometer. As each drop bridged the gap, it activated an electronic relay operating a counter, was pierced, and ran down the needle. The gap was adjusted so that each drop averaged 25 μliters.

The surgical needle below was supported in the middle by a downward-sloping metal bar, so that each drop ran down to the lower end of the needle and dripped into a collection chamber. Each time 0.5 ml was collected, this chamber automatically emptied into a test tube below and activated a pump to infuse 0.5 ml of Ringer solution into the venous catheter at the rate of 15 ml/hr. Afterwards, a carousel automatically rotated a new test tube into position below the collection chamber. The purpose of these infusions was to maintain the water load constant throughout the experiment for the rats in the two groups. Since each infusion came after a sample of urine was collected, any differences between the two groups in the time before the next infusion could be only an effect of differential rates of preceding urine formation and not a cause of that differential rate. No constant error in this procedure could produce opposite effects on the two groups and any variable errors (including barely conceivable effects of positive feedback) would be expected to be distributed by chance.

At a point 5 mm from where the tube to the drop counter fitted over the stainless steel tube emerging from the rat, a needle from a small hypodermic syringe filled with air was inserted into this tube. By this means, a small bubble of air could be injected into the column of fluid to serve as a marker.

Cardiovascular and temperature measurements. The ECG picked up by the electrodes on the sternum and back was recorded by a Grass model 5 polygraph, which activated the heart-beat counting device described by Miller and DiCara (19). Vasomotor responses from the middle of the tail were recorded by a photoelectric transducer (Grass, model PTTI) connected to the polygraph. Temperature was recorded by a thermistor probe inserted 4 cm up the rectum. Blood pressure was recorded by a cuff and microphone (Carter Electronics Lab, model SM-65) placed at the base of the tail.

Infusion of water load and labeled tracers. The tube from the venous catheter was connected to the common end of a Y tube through one arm of which Ringer solution could be infused at 10 ml/hr. This infusion was started 30 min after connections had been made to the rat and the door of the soundproof chamber had been closed. The other end of the Y tube, sealed off by a tight clamp when not in use, was connected to a screw-driven tuberculin syringe for infusing labeled tracers. Ten minutes after the infusion of Ringer solution was started, 0.5 ml of a mixture containing 33 μc of tritiated PAH and 16.8 μc of inulin-^{14}C was infused over a period of 30 min.

Ten minutes after the end of the infusion of isotopes, the following measuring procedures were started. Urine output for the next 10 min was measured. Marker air bubbles were inserted into the column of urine where it left the rat at the beginning of this period, at the end of 5 min, and at the end of the 10 min. Subsequently, the first drop of urine following each bubble was collected on a slide. At the end of the 5th min, the infusion of Ringer solution was discontinued for 90 sec, and a drop of blood was withdrawn via that arm of the Y tube. From each of the drops of urine or blood, a 20-μl sample was obtained via a micropipette to which was added 0.5 ml of 5% trichloroacetic acid, after which it was placed under refrigeration. Later, the samples were centrifuged for 10 min at 800 $\times$ *g*. A 100-μl aliquot was added to 15 ml of Bray's solution and counts for ^{14}C and ^{3}H were made on a Packard Tri-Carb scintillation counter. Clearance and flow rates were then calculated.

Randomization and additional pretraining measures. To make sure that the experimenter could not unconsciously introduce any bias, all of the preceding adjustments were made before the random assignment of a rat to the increase or decrease group, except that the last rat had to be assigned to the increase group in order to make the numbers come out even. In order to determine an appropriate initial setting of the criterion for reward, the drop-count scores for 3 min were recorded. The vasoconstriction score was noted, as was the temperature. Three measures of systolic blood pressure were taken and averaged.

Instrumental Training Trials

All subjects were given a total of 240 trials over a period of approximately 3 hr, which were presented with unpredictable intertrial intervals ranging from 20 to 50 sec with a mean of 30 sec. On each regular training trial, the light and tone came on, indicating that the reward was available, and as soon as the rat exceeded the criterion—which for half of the rats was to form urine fast enough so that two drops were counted in less than a specified time, and for the other half was to form urine slowly enough so that a specified time elapsed before two drops were counted—the reward was delivered and the tone and light were turned off. On the basis of the immediately preceding sample, this criterion was set so that reward would be achieved, on the average, 15 sec after the light and tone indicated the beginning of a trial. Every 20 regular trials, the criterion was adjusted, if necessary; if the rat was averaging less than 10 sec to achieve the reward, the criterion was made harder; if he averaged more than 20 sec, it was made easier.

In order to secure a measure of urine response to the time-in stimulus that was uncontaminated by delivery of reward or turning off the light and tone, the 5th trial, and every 10th one thereafter, was a special test trial. During the first 15 sec of such trials, the reward circuit was disconnected while drops were being counted on a special counter. At the end of that time, the special counter was turned off, and the reward circuit was turned on. Thus, from the rat's point of view, it was a regular trial during which he happened to have to work a little longer in order to achieve the reward.

In order to be able to compare performance when the time-in stimulus (tone and light) was present and reward could be achieved with that under the time-out condition when reward could not be achieved, and thus to find out whether the rat had learned to discriminate between these two conditions, the 10th trial, and every 10th one thereafter, was a blank trial. During blank trials, the special drop counter was actuated for 15 sec without either the time-in stimuli or the reward circuit being turned on. For the rat this merely meant a longer than usual intertrial interval. In total, then, the rat received 200 regular rewarded training trials, 20 test trials with the reward slightly delayed, and 20 blank trials, which were not training trials at all.

The volume of urine formed during a 10-min period shortly before training had already been collected. Similar samples were taken during the last 10 min of each quarter of training.

The same procedure of infusing radioisotopes and collecting samples that was used prior to training was again started 40 min before the end of training. Immediately after the end of formal training, trials were continued while temperature, vasoconstriction scores, and blood pressure were determined.

In order to provide an additional control for the effect of keeping the water load constant by replacing with Ringer solution for the urine that was formed, the last four rats, two in each group, were run for an additional 2 hr during which training conditions remained the same but the infusion of Ringer solution was discontinued.

Experiment on Unconditioned Effects

Rewarding half of the rats, selected at random, in the main experiment for increasing and the other half for decreasing their rates of urine formation controlled for any unconditioned or progressive effects of the electrical stimulation of the lateral hypothalamus used as a reward. Nevertheless, in order to determine the degree to which such effects might be present, four additional rats were run through the procedures that were used in training, except that *a*) isotope tracer procedures were omitted, *b*) the rewarding electrical stimulation was always given 15 sec after the time-in stimulus was turned on regardless of the rate of urine formation by the rat, and *c*) after 120 trials (90 min) with this procedure, the electrical stimulation was turned off for the next 120 trials, after which it was turned on again for 120 more trials and off for the final 120 trials. During all these trials, the time-in signal went off at the end of 15 sec.

RESULTS AND DISCUSSION

Unconditioned Effects

In the experiment on unconditioned effects, the urine production dropped from a rate of 3.7 drops/min during the 15 sec that the light and tone were on immediately before the electrical stimulation of the brain to a rate of 2.7 for the 15 sec immediately afterwards ($t = 4.5$, df = 3, $P < .001$). Since this difference was not present during the periods without electrical stimulation of the brain ($t = 0.2$), it must have been produced by this stimulation rather than by the tone and light. This moderate unconditioned effect could have been an inhibition of urine formation or a dilatation of the tubules.

Since half of the rats in the main experiment were rewarded for increases and the other half for decreases, this unconditioned decrease could not have accounted for the results of that experiment.

The same amount of urine was formed during the 90-min blocks with and those without stimulation (t = 0.08). Therefore, we can conclude that there were no appreciable cumulative effects of such stimulation Since there were no progressive long-term trends throughout the course of the 6 hr, either in the average amounts (t = 0.8) or those for the individual rats, we can also conclude that the procedure of infusing Ringer solution at the same rate that urine was formed kept the animals in equilibrium at a moderately high rate of urine formation but did not have either any progressive or any positive-feedback effects that would distort the results of the main experiment.

Main Experiment

Before training, the two randomly selected groups showed no differences that approached statistical reliability in any of the measures that were taken. The average scores for the increase and decrease groups were, respectively: for urine formation during the 3 hr before curarization, 0.0079 and 0.0085 ml/min per 100 g body weight (t = .04); for rate of 100% reinforced bar pressing in 30 min, 659 and 772 (t = .09); and for strength of current used to induce bar pressing, 71 and 64 μa (t = 0.4). During the first five trials of training for the increase and decrease groups, respectively, the average number of drops of urine during the 15 sec before the onset of electrical stimulation was 4.3 and 4.8 (t = 1.0) and the decrements produced by stimulation in the next 15 sec of urine formation were 1.2 and 1.4. Each of these decrements was reliable ($P < .01$), but the difference in the decrements in the two groups did not approach reliability (t = .09). Both groups received

TABLE 1. *Summary of changes during training*

Rat No.	Urine Formation, ml/min per 100 g		Heart Rate, beats/min		Blood Pressure, mm Hg		Body Temperature Rectal, °F		Peripheral Vasomotor Activity, mv		Glomerular Filtration, ml/min per 100 g		Renal Blood Flow, ml/min per 100 g		Osmolarity, mosmols	
	B	E	B	E	B	E	B	E	B	E	B	E	B	E	B	E
Increase rats																
340	.023	.030	450	522											569	386
334	.014	.019	270	486											500	541
388	.016	.029	360	318	136	134	98.3	104.0	11.7	10.9					488	668
370	.018	.030	360	384	135	124	97.9	103.5	12.1	10.1	.56	.88	3.51	6.36	415	358
389	.007	.016	390	372	126	130	98.1	102.3	14.3	12.9	.35	.51	1.10	2.13	410	641
394	.026	.044	354	330	141	146	98.0	103.4	10.7	10.3	.70	1.24	1.60	5.23	666	565
401	.012	.026	378	474	130	124	98.3	104.0	14.5	12.8	.40	1.12	1.96	3.15	432	701
Mean	.017	.028	367	412	136	132	98.1	103.4	12.6	11.4	.67	.94	2.04	4.22	497	571
Decrease rats																
367	.018	.007	390	383											435	840
345	.015	.003	402	480											452	903
366	.012	.005	360	384	157	152	98.5	104.0	14.1	13.2					587	2,815
393	.015	.006	348	300	135	145	98.6	103.6	12.0	11.3	.55	.06	3.33	.74	455	500
402	.030	.009	396	426	134	152	98.0	104.5	12.1	10.9	.73	.32	2.37	1.67	236	790
397	.027	.008	391	408	143	158	99.3	104.2	12.3	11.7	.64	.14	1.52	1.25	330	627
392	.020	.003	325	348	120	120	98.6	103.8	13.1	12.0	.49	.13	2.69	1.37	508	775
Mean	.019	.006	373	390	138	145	98.6	104.0	12.7	11.8	.60	.16	2.48	1.26	429	1,036

B: before training; E: end of training.

the same number of brain stimulations as rewards during training; that the rate was approximately constant is indicated by the fact that the average number of rewards during the last 30 min was 36 for the increase and 38 for the decrease groups ($t = 0.6$). In all of the foregoing measures, the low t scores indicate great overlapping of the two groups before training. Finally, there were no reliable correlations between the amount of change during training and any of the following: amount of current used for electrical stimulation, rate of bar pressing elicited by this stimulation, or amount of transient reduction in urine flow induced by this stimulation.

The results of the most pertinent measures at the beginning and end of training are presented in Table 1. Figure 1 shows in more detail the results of the two types of training on the amount of urine collected during the 10-min samples. It can be seen that the rats rewarded for increases learned to increase the rate of urine formation, while those rewarded for decreases learned to decrease it. Each of these changes in opposite directions was highly reliable ($t > 6.6$, df = 6, $P < .001$). Changes in the rewarded direction occurred in each one of the 14 rats. Table 1 shows that, whereas there was great overlap between the two groups in the rate of urine formation before training, there was a great separation at the end of training, with the lowest rate of any rat rewarded for an increase being .016 and the highest rate for any rat rewarded for a decrease being .009 ml/min per 100 g.

Similar results were secured in terms of the count of drops during test trials and blank trials. Scores on both types of trials increased approximately equally for rats rewarded for increases and decreased approximately equally for rats rewarded for decreases.

Although the training produced an over-all change in the rate of urine formation, there was no reliable difference (which would have indicated a discrimination) between the urine formation during the time-in signal indicating that reward could be achieved as compared with that during time-out conditions when reward could not be achieved. In a previous experiment on heart rate (19), there was little tendency for any discrimination to develop during the period while the criterion was being shifted to shape the learning of a larger response, but evidence for discrimination did develop during additional trials while the criterion was held constant at the difficult level. Such additional trials were not given in this experiment. Furthermore, our trials, averaging 15 sec, probably were too short in relationship to the various lags in the system—crossing the membrane, filling the tubules, etc.—to provide favorable conditions for learning a discriminatory response to the time-in stimulus pattern.

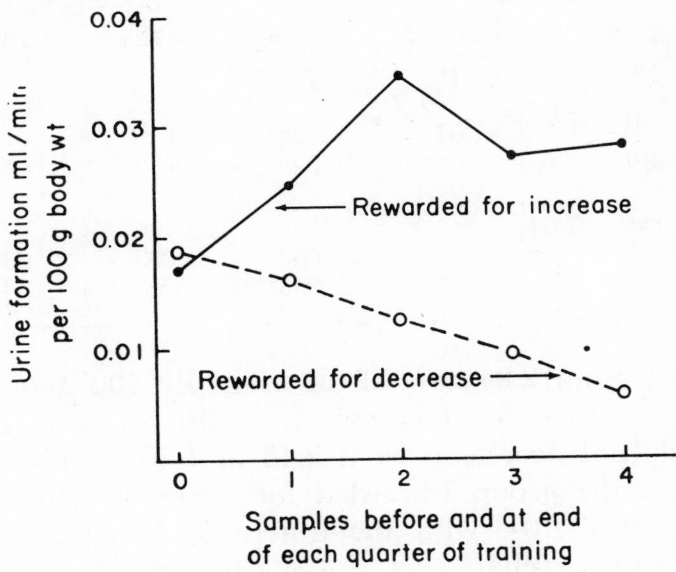

FIG. 1. Instrumentally learned changes in rate of urine formation.

Looking at the other measures, the group rewarded for increases in urine formation increased its rate of glomerular filtration from 0.67 to 0.96 ml/min per 100 g (t = 3.6, df = 3, P < .05), while the group rewarded for decreases decreased filtration from 0.60 to 0.16 ml/min per 100 g (t = 13.9, df = 3, P < .001). The difference between the two groups in this measure was highly reliable (P < .001). This change in filtration rate was clearly not the result of any general changes in blood pressure, since neither group showed a consistent change and the slight and unreliable mean differences that were observed—a greater increase in the decrease group—happened to be in the wrong direction to produce the observed changes in filtration. Similarly, there were no appreciable changes in heart rate.

The main cause for the difference in glomerular filtration appears to have been the renal blood flow, which

TABLE 2. *Urine formation during additional 2 hr without infusion of Ringer solution, ml/min per 100 g*

Rat No.	30 Min	60 Min	90 Min	120 Min
Increase rats				
394	.043	.035	.022	.017
401	.029	.028	.013	.010
Decrease rats				
397	.006	.005	.003	.002
392	.005	.003	.003	.001

increased from 2.04 to 4.22 ml/min per 100 g in the group rewarded for increases (t = 3.6, df = 3, P < .05), while it decreased from 2.48 to 1.26 ml/min per 100 g in the group rewarded for decreases (t = 2.4, df = 3, P < .10), with the difference between them being highly reliable (t = 4.3, df = 6, P < .001). These different vasomotor changes produced in the kidney by instrumental learning procedures were not completely general, since all of the animals showed increased vasoconstriction of the tail and the correlation between this increase and that in the rate of renal blood flow was insignificant (ρ = .02). There was no difference in the amount of vasoconstriction shown by the two groups (t = .05).

Finally, the group rewarded for decreases showed an increase in osmolarity from 419 to 1,036 (t = 2.2, df = 6, $P < .07$), but this slightly more than doubling in osmolarity, which is only of borderline statistical significance, could be accounted for by the fact that glomerular filtration was reduced to almost one-fourth. Thus, the increased osmolarity cannot be said with any certainty to represent an increase in the secretion of ADH.

The group rewarded for increased urine formation did not show the reduction in osmolarity that might be expected if the increase had been produced by an inhibition of the secretion of ADH. There was no consistently reliable change in osmolarity (t = 1.2, df = 6, $P > .30$). This result is analogous to the one frequently secured when the changes in the rate of urine formation are secured by classical conditioning. Conditioned increases produced by the use of water loads as an unconditioned stimulus frequently are not accompanied by any reduction in osmolarity, and even may be accompanied by an increase in it (3).

All of the 10 animals in which it was measured showed an increase in body temperature during the experiment, from a mean of 98.4 to 103.7, something that would be expected by chance less than one time in 500. There was no difference between the two groups in this respect (t = 0.7, df = 8, $P > .60$). This increase appears to have been an effect of the electrical stimulation of the brain used as a reward, since in the experiment on unconditioned effects a similar increase was observed during both periods when ESB was administered, while a return toward normal appeared during both periods when it was turned off.

The results of the rats given 2 hr of additional training without any infusion of Ringer solution are presented in Table 2. The two rats rewarded for increase continued to form urine at a high rate for an hour, after which dehydration began to decrease their rate. Nevertheless, during each of the four 30-min periods measured, each of them formed at least four times as much urine as did either of the two rats rewarded for decreases. These results show that the procedure of infusing Ringer solution to hold the water load constant is not essential for maintaining a difference between the rats rewarded for increases and those rewarded for decreases in the rate of urine formation; they serve as an additional control for the effects of the infusion.

As soon as the procedure of infusing the Ringer solution was stopped, the continued difference in rates of urine formation produced by the different rates rewarded in the two groups began to produce a difference in their water balances—maintaining the high water load of the rats rewarded for a slow rate and rapidly reducing that of the rats rewarded for a high rate. In spite of the fact that this difference worked against the effects of reward, the latter were able to maintain the differences between the two groups for at least 2 hr. Thus the results show, as would be expected from those in the main experiment, that learned modifications of kidney function can produce deviations from the normal homeostatic regulation of water balance. In the present experiment, changes in water balance were produced by reinforcing an internal glandular response, that of the kidney; in a previous experiment, changes in water balance had been produced by reinforcing external skeletal responses, water seeking, and consumption (15). Thus, both the internal, physiological and the external, behavioral mechanisms for maintaining water balance are subject to modification by learning.

Dr. Eric Windhager, Dept. of Physiology, Cornell Medical School, gave helpful guidance in regard to the radioisotope clearance techniques.

Work on this study was supported by Public Health Service Grant MH 13189 from the National Institute of Mental Health to N. E. Miller and by an Advancedes Rearch Fellowship from the American Heart Association to L. V. DiCara. A preliminary report was presented at the 1967 Meeting of the Psychonomic Society by L. V. DiCara.

Received for publication 8 February 1968.

REFERENCES

1. Black, A. H. *The Operant Conditioning of Heart Rate in Curarized Dogs: Some Problems of Interpretation.* Paper presented at the Annual Meeting of the Psychonomic Society, 7th, St. Louis, 1966.
2. Bykov, K. M. *The Cerebral Cortex and the Internal Organs,* translated and edited by W. H. Gantt. New York: Chemical Publishing Co., 1957.
3. Corson, S. A. Conditioning of water and electrolyte excretion. *Assoc. Res. Nervous Mental Disease* 43: 140–199, 1966.
4. Davis, J. D., and J. R. Weeks. Chronic intravenous cannulas for rats. *J. Appl. Physiol.* 19: 340–341, 1964.
5. DiCara, L. V., and N. E. Miller. Changes in heart rate instrumentally learned by curarized rats as avoidance responses. *J. Comp. Physiol. Psychol.* 65: 8–12, 1968.
6. DiCara, L. V., and N. E. Miller. Instrumental learning of systolic blood pressure by curarized rats: dissociation of cardiac and vascular changes. *Psychosom. Med.* In press.
7. DiCara, L. V., and N. E. Miller. Instrumental learning of peripheral vasomotor responses by the curarized rat. *Commun. Behav. Biol.* 1: 209–212, 1968.
8. DiCara, L. V., and N. E. Miller. Instrumental learning of vasomotor responses by rats: learning to respond differentially in the two ears. *Science* 159: 1485–1486, 1968.
9. Kimble, G. A. *Hilgard and Marquis' Conditioning and Learning.* New York: Appleton-Century-Crofts, 1961.
10. Konorski, J., and S. Miller. Further remarks on two types of conditioned reflex. *J. Gen. Psychol.* 17: 405–407, 1937.
11. Miller, N. E. Integration of neurophysiological and behavioral research. *Ann. N.Y. Acad. Sci.* 92: 830–839, 1961.
12. Miller, N. E. Animal experiments on emotionally-induced ulcers. *Proc. World Congr. Psychiatry, 3rd, Montreal, 1961* 3: 213–219, 1963.

13. MILLER, N. E. Some reflections on the law of effect produce a new alternative to drive reduction. In: *Nebraska Symposium on Motivation*, edited by M. R. Jones. Lincoln, Nebraska: Univ. of Nebraska Press, 1963, p. 65–112.
14. MILLER, N. E. Experiments relevant to learning theory and psychopathology. *Proc. Intern. Congr. Psychol., 18th, Moscow, 1966*. In press.
15. MILLER, N. E. Behavioral and physiological techniques: rationale and experimental designs for combining their use. In: *Handbook of Physiology. Alimentary Canal*. Washington, D.C.: Am. Physiol. Soc., 1967, vol. I, chapt. 4, p. 51–61, Fig. 1.
16. MILLER, N. E., AND A. BANUAZIZI. Instrumental learning by curarized rats of a specific visceral response, intestinal or cardiac. *J. Comp. Physiol. Psychol.* 65: 1–7, 1968.
17. MILLER, N. E., AND A. CARMONA. Modification of a visceral response, salivation in thirsty dogs, by instrumental training with water reward. *J. Comp. Physiol. Psychol.* 63: 1–6, 1967.
18. MILLER, N. E., E. E. COONS, M. LEWIS, AND D. D. JENSEN. Electrode holders in chronic preparations. B. A simple technique for use with the rat. In: *Electrical Stimulation of the Brain*, edited by D. Sheer. Austin, Texas: Univ. of Texas Press, 1961, p. 51–54.
19. MILLER, N. E., AND L. V. DICARA. Instrumental learning of heart-rate changes in curarized rats: shaping, and specificity to discriminative stimulus. *J. Comp. Physiol. Psychol.* 63: 2–19, 1967.
20. SKINNER, B. F. *The Behavior of Organisms*. New York: Appleton-Century, 1938.
21. TROWILL, J. A. Instrumental conditioning of the heart rate in the curarized rat. *J. Comp. Physiol. Psychol.* 63: 7–11, 1967.

30 Role of Striated Muscle in Urination

Jack Lapides, Robert B. Sweet and Louis W. Lewis

EDITORIAL NOTE BY NEAL E. MILLER

Excellent evidence that people can learn significant voluntary control of an automatically mediated response has been around for some time but came to my attention only recently because it was published in a journal not usually read by psychologists. There has been some debate about whether skeletal muscles play a role in the control of urination. The following study, in which the skeletal muscles of human subjects were paralyzed by curare or succinylcholine, clearly shows that human subjects can control urination solely via the autonomic nervous system. It is obvious that for this particular visceral response both immediate feedback and strong rewards and punishments are available to provide good conditions for learning. It is conceivable that rats can learn a considerable variety of visceral responses while people can learn only one, but that seems improbable.

Micturition is a complicated act involving smooth as well as striated muscles; the function of each type of muscle has been a controversial subject from the earliest experiments on urination[1] to the present era. One group of observers[2, 3] believes that micturition can be started as well as stopped by direct cortical control over bladder smooth muscle. The other school which includes Galen, contends that activity of striated muscle is necessary for the voiding of urine. Recent indirect evidence obtained by fluoroscopy and cinefluorography have led some investigators[4] to state that the voluntary mechanism for starting or stopping urination is mediated *entirely* through striated muscles. Hinman and co-workers,[5] also on the basis of cinefluorography, disagree with this conclusion.

In an effort to resolve this controversy we have conducted experiments[6] in humans to determine the part played by the striated perineal muscles in micturition. It was found that paralysis of the perineal musculature obtained by bilateral pudendal nerve block did not affect initiation of urination nor did it produce incontinence except in one patient who had undergone prostatectomy; these observations are in agreement with previous work.[7, 8] However, the amount of time necessary to stop urination during pudendal block was increased from a normal of 1–2 seconds to 8–10 seconds; this finding has not been noted previously.

Since some research workers feel that the abdominal and diaphragmatic musculature forms a part of the voluntary mechanism necessary for control of micturition, it was decided to carry the investigation a bit further and produce complete flaccid paralysis of all of the skeletal muscle of the human body. Under these circumstances it was hoped that an unequivocal answer could be obtained relative to the function of striated muscle in urination.

METHOD

The muscle relaxants D-tubocurarine and succinylcholine were used in separate trials to produce flaccid paralysis of the striated muscle. One female and 15

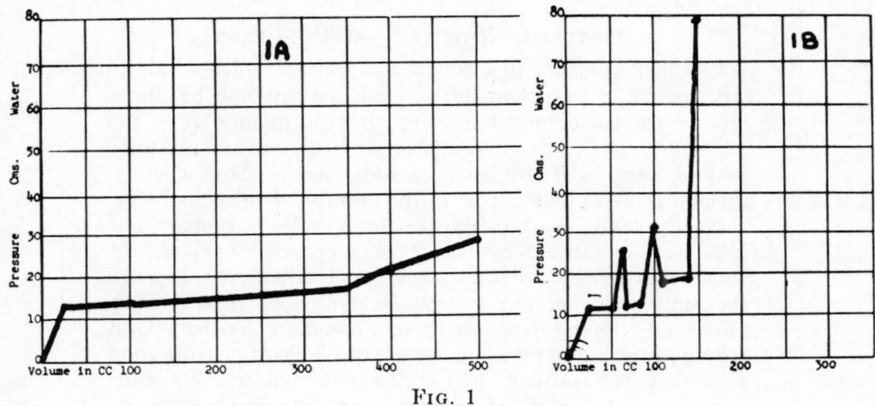

Fig. 1

male patients served as subjects; they were fully conscious throughout the experiments.

1) Since the effects of the skeletal muscle relaxants on the human autonomic nervous system are not known definitely at the present, it was considered necessary to ascertain these effects first. Nine male patients with neurogenic bladders exhibiting uninhibited or uncontrolled contractions were selected as subjects for this part of the experiment. Four of the patients exhibited reflex neurogenic bladders[9] associated with traumatic transverse myelitis while five subjects demonstrated uninhibited neurogenic bladders resulting from multiple sclerosis and cerebrovascular accidents. Figure 1, *A* demonstrates the cystometrograph of a normal bladder while figure 1, *B* illustrates the cystometrograph of a bladder with uncontrolled contractions.

Each patient was catheterized and several control cystometrographs obtained. A muscle relaxant was administered then intravenously until flaccidity of all skeletal muscle was evident. The bulbocavernosus reflex as described by Bors[10] was used to test for flaccidity of the perineal musculature. Artificial respiration (intermittent manual compression of rebreathing bag) was necessary in all cases because of diaphragm and intercostal muscle paralysis. Cystometrographs were repeated during the period of complete flaccid paralysis. Four of the patients were administered both succinylcholine and D-tubocurarine on different occasions while the remaining five were given only succinylcholine.

2) When the experiments on the neurogenic bladders had been completed and evaluated, attention was directed toward the seven patients with normal bladders. For several days prior to the experimental run, each patient was instructed to practice emptying the bladder in the supine position; this was done in order to accustom the patient to voiding in the posture assumed during complete flaccid paralysis. On the day of the experimental run the patient was placed in the supine position, catheterized and the bladder filled with physiological saline. In three of the patients the amount of fluid instilled was 300 ml.; in the four other patients the amount instilled was governed by the patient's sensation of fullness and varied with the patient. After the bladder was filled with fluid, the catheter was removed and the patient requested to initiate urination and then to stop it. After the performance times were noted, the patient was paralyzed completely with succinylcholine or D-tubocurarine as previously described. When complete skeletal flaccidity was present, the patient was catheterized again, the bladder filled with the same amount of fluid used in the control micturition and the catheter withdrawn. The patient was asked to initiate and then to inhibit micturition. The time intervals taken to carry out the commands were noted. Several patients were placed in the semi-upright position before being requested to void in order to observe continence.

RESULTS

1) Amounts of succinylcholine and D-tubocurarine necessary to produce flaccidity of all striated muscle were insufficient to abolish or significantly impair the uncontrolled contractions of the patients with neurogenic bladders. Cystometrographs from a typical experiment are illustrated in figure 2. A is the control cystometrograph and B is the cystometrograph obtained during succinylcholine administration. Observe that the two graphs are essentially identical. Muscle relaxants did not impair transmission of nerve impulses over the parasympathetic pathways to the bladder.

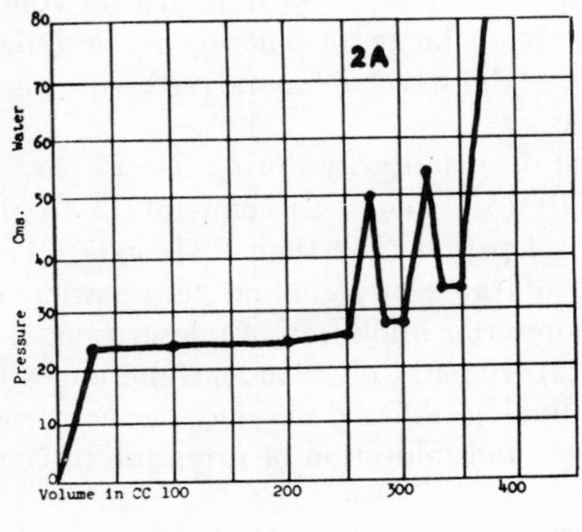

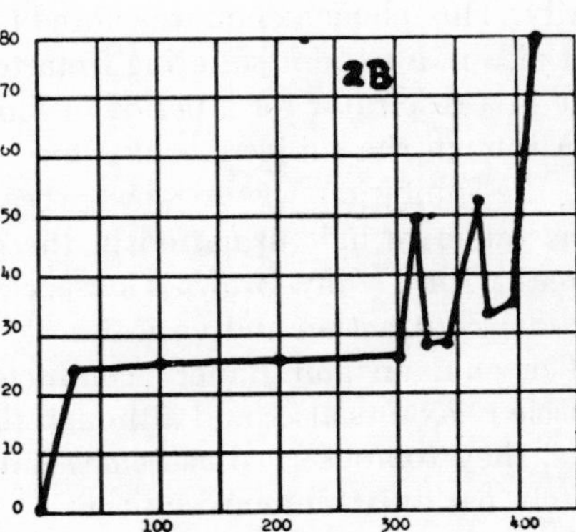

FIG. 2

2) Of the patients with normal bladders five were able to start micturition within two seconds after command while the remaining two required five seconds to initiate urination. All of the normal patients could stop urination within two seconds after request.

After complete flaccid paralysis with one or the other of the muscle relaxants all of the patients could initiate urination within their control time interval, i.e., 2–5 seconds. However, none of the patients could inhibit urination within the control time of 2 seconds but required 8–15 seconds for cessation of micturition. Measurement of voided fluid in these patients indicated that inhibition of urination had been accomplished before 75 per cent of the instilled fluid volume had been evacuated. It was observed that after the bladders had been filled with saline, leaking of fluid from the urethral meatus did not occur either in the supine or semi-erect positions.

DISCUSSION

In 1933 Denny-Brown and Robertson[2] studied micturition in three normal patients by means of endoscopy, cystometry and sphincterometry. They concluded that micturition is initiated by a voluntary contraction of the vesical smooth muscle itself. The activity of such striated muscles as perineal, abdominal and diaphragmatic during initiation of micturition, were considered associated movements and not directly connected with the starting of urination. These investigators observed that the vesical neck or sphincter opened only after bladder muscle had started to contract. The external urethral sphincter in turn dilated following detrusor contraction and opening of the vesical neck. In no instance was the striated external sphincter observed to open before the bladder had started to contract.

They noted also, that with abrupt inhibition of urination, the external urethral sphincter constricted rapidly followed by a gradual cessation of vesical muscle contraction and slow closure of the vesical neck.

The observations made during the present investigation support the findings of Denny-Brown and Robertson in every respect. With all skeletal muscle paralyzed normal patients were observed to start and stop micturition voluntarily. This phenomenon was noted not only in the acute experiments described but also in a patient suffering from tetanus who was kept paralyzed with anectine in a respirator for a period of four days.[11]

Although the subjects could stop urination voluntarily during flaccid paralysis, the inhibition was a slow process (10–15 seconds) as compared with the time taken to halt urination in the control period (less than 2 seconds). This agrees with Denny-Brown's observation of the slow, gradual termination of bladder contraction and vesical neck closure after inhibition of micturition.

Dortenmann and Bauer[12] conducted experiments on bladder function with muscle relaxants also; and although their findings differed in many respects from ours, they too observed *voluntary* initiation and inhibition of urination in completely paralyzed human subjects.

Incontinence of urine was not observed during paralysis of striated muscles. This observation confirms our findings with pudendal block and is in accord with previous experiments[6] which indicated that urinary continence can be maintained by either the vesical neck or external urethral sphincter. Both must be injured to result in a typical "dripping type" of urinary incontinence.

CONCLUSIONS

Micturition can be initiated and terminated *voluntarily* without the use of any striated muscle in the human body.

Striated muscle is necessary for the sudden, rapid inhibition of urination.

Urinary continence can be maintained by the vesical neck (internal vesical sphincter) alone in the absence of functioning striated perineal musculature.

Accepted for publication September 27, 1956.

[1] Mosso, A. and Pellacani, P.: Sur les fonctions de la vessie. Arch. Ital. de Biol., **1:** 97, 1882.

[2] Denny-Brown, D. and Robertson, E. G.: On the physiology of micturition. Brain, **56:** 149, 1933.

[3] Le Gros Clark, F.: Some remarks on the anatomy and physiology of the urinary bladder and of the sphincters of the rectum. J. Anat. and Physiol., **17:** 442, 1883.

[4] Muellner, S. R. and Fleischner, F. G.: Normal and abnormal micturition: A study of bladder behavior by means of the fluoroscope. J. Urol., **61:** 233, 1949.

[5] Hinman, F., Jr., Miller, G. M., Nickel, E., Steinbach, H. L. and Miller, E. R.. Normal micturition. Calif. Med., **82:** 6, 1955.

[6] Lapides, J., Gray, H. O. and Rawling, J. C.: Function of striated muscles in control of urination: (1) Effect of pudendal block. Surg. Forum, **6:** 611, 1955.

[7] Langworthy, O. R., Kolb, L. C. and Lewis, L. G.: Physiology of Micturition. Baltimore: The Williams and Wilkins Co., 1940.

[8] Emmett, J. L., Daut, R. V. and Dunn, J. H.: Role of the external urethral sphincter in the normal bladder and cord bladder. J. Urol., **59:** 439, 1948.

[9] Nesbit, R. M., Lapides, J. and Baum, W. C.: Fundamentals of Urology. Ann Arbor: J. W. Edwards, 1953.

[10] Bors, E., Comarr, A. E. and Moulton, S. H.: The role of nerve blocks in management of traumatic cord bladders. J. Urol. **63:** 653, 1950.

[11] Vial, B. A.: Succinylcholine and artificial respiration in the treatment of tetanus: Report of a case. U. Mich. Med. Bull., **21:** 123, 1955.

[12] Dortenmann, S. and Bauer, K. M.: Untersuchungen über den Einfluss der quergestreiften Muskulatur auf die Blassenfunktionen mit Hilfe eines Muskelrelaxans. Die Medizinische, **15:** 528, 1955.

31

Homeostasis and Reward: T-Maze Learning Induced by Manipulating Antidiuretic Hormone

Neal E. Miller, Leo V. DiCara, and George Wolf

MILLER, NEAL E., LEO V. DICARA, AND GEORGE WOLF. *Homeostasis and learning: T-maze learning induced by manipulating antidiuretic hormone.* Am. J. Physiol. 215(3): 684–686. 1968.—Albino rats were injected with antidiuretic hormone (ADH) if they chose one arm of a T maze and with isotonic saline vehicle if they chose the other one. Normal rats preloaded with H_2O learned to choose the vehicle side where their kidneys could excrete the excess water more promptly, but rats with diabetes insipidus preloaded with hypertonic NaCl learned to choose the opposite side where ADH helped them to excrete excess NaCl. Normal controls that were not preloaded showed no learning. Thus, an excess of H_2O or NaCl can function as a drive, and the return to normal levels produced by an internal glandular response (instead of an external consumatory one) can function as a reward. We advance the hypothesis that drives and rewards induced in this way can produce visceral learning that may serve to maintain homeostasis.

kidney function; drive; reinforcement; motivation; diabetes insipidus; electrolyte balance; conditioning, operant; reward

---

RECENT WORK from this laboratory (see 4 for references) has shown that glandular and visceral responses are subject to instrumental learning, in which the reward does not immediately elicit the specific response to be learned (as does the unconditioned stimulus in classical conditioning) but instead strengthens any immediately preceding response. This evidence refutes the strong traditional belief that the autonomic nervous system is more primitive than the cerebrospinal one, in that responses mediated by the autonomic are subject only to classical conditioning, while those mediated by the cerebrospinal are subject to instrumental learning.

Under the conditions that were present during mammalian evolution, the functional utility of instrumental learning by the cerebrospinal nervous system is obvious. The skeletal responses mediated by it operate on the external environment so that there is survival value in the ability to learn responses that bring rewards such as food, water, or escape from pain. The fact that most of the responses mediated by the autonomic nervous system do not have such direct action on the external environment was one of the reasons for believing that they were not subject to instrumental learning. Is the recently demonstrated instrumental learning ability of the autonomic nervous system something with no normal physiological function, a mere accidental by-product of the survival value of cerebrospinal learning, or does instrumental learning of autonomically mediated responses have some physiological function, such as helping to maintain the constancy of the internal environment?

Perhaps homeostatic regulation is not limited to innate mechanisms but can be achieved also via the instrumental learning of visceral responses. But, in order for instrumental visceral learning to function homeostatically in this way, a deviation from the optimal level would have to function as a drive to motivate such learning, and a change toward the optimal level would have to function as a reward to reinforce the learning of the visceral response that restored the optimal level.

When a mammal has less than the optimal level of water, this deficiency serves as a drive (thirst) to motivate learning, and the overt consumatory response of drinking functions as a reward to reinforce the learning of the responses that lead to the water that restores the optimal level (3). The purpose of the present experiment was to determine *a*) whether other deviations from homeostasis, for example, instead of a deficiency, an excess of H_2O, will serve as a drive to motivate learning, and *b*) whether the correction of these deviations by an internal glandular, instead of an external consumatory, response can serve as a reward to reinforce learning.

The general design of the experiment was to give one group of rats a water load immediately before each learning trial in a T maze. The stem of the maze had two longitudinally running dowel sticks as a floor. Standing on these sticks was slightly uncomfortable, so that the rats were motivated to move off of them into the arms of the T. If they turned in one direction, they received an injection of antidiuretic hormone (ADH), which delayed the renal excretion of the water. If they turned in the other direction, they received an injection of the isotonic saline vehicle, which did not interfere with excretion of the excess water. A second group of rats had diabetes insipidus produced by lesions in the median eminence. These rats received a load of hypertonic saline immediately before being run in the maze and received the same injections, of ADH if they chose one side and vehicle if they chose the other. But, for these rats the homeostatic function of the injections was reversed. While the ADH interfered with the return to homeostasis by the water-loaded rats, it aided the return to homeostasis by the salt-loaded rats with diabetes insipidus because it allowed them to reabsorb water so that they could excrete urine of higher NaCl concentration (2). In addition to using these two groups, a third group, consisting of normal rats without any preloading, was run in the maze to control for any rewarding or aversive effects that the injection of ADH might have independently of its homeostatic effects.

METHODS

The subjects were 24 male, adult (400–500 g) albino Sprague-Dawley rats on ad lib. food and water. These were divided into three groups of eight animals each: *A*) normal rats preloaded via a stomach tube with 20 ml of distilled water at body temperature 2 min before being run in the T maze; *B*) normal rats which did not receive any preload; *C*) rats with diabetes insipidus, which received a load of 20 ml of 0.33 M NaCl via a stomach tube 2 min before being run in the T maze. Diabetes insipidus was produced by a lesion in the median eminence made by passing a d-c current of 1 ma for 20 sec through the tip of an electrode with the stereotaxic coordinates of 2 mm posterior to bregma on the midline and 0.25 mm above the base of the brain. The skull was positioned flat in the stereotaxic instrument. Only rats with a daily intake of more than 200 ml were used.

The T maze consisted of a starting box 10 inches long, 6 inches wide, and 15 inches high, the floor of which consisted of two dowel sticks 0.75 inches in diameter spaced symmetrically and running longitudinally. A guillotine door opened from the starting box to a 6-inch-square solid section with black floor and walls. Each arm of this T had the same dimensions as the starting box, with a guillotine door at its entrance. In order to make them as distinctive as possible, the right arm had black-and-white striped walls and a floor of 0.5-inch wire mesh, under the far end of which was a beaker containing mothballs. The left arm had grey walls and a floor of fine window screen, under the far end of which was a beaker containing ethyl alcohol.

Each rat was given one trial per day in the T maze, being placed in the start box with its head facing away from the T junction. Immediately after it entered one of the goal boxes, the guillotine door was closed and the rat was given an ip injection of 1 ml of either ADH (Pitressin, 200 milliunits) or of the isotonic saline vehicle and replaced in the same goal box for 90 min. The following measures were taken: latency of response from placement in start box to entering end box; amount of urine excreted during 90 min in the end box; daily water consumption; and water consumption during the first 15 min in the home cage after the 90-min stay in the maze.

For half of each group, ADH was given in the left-hand goal box and the vehicle in the right; for the other half, these conditions were reversed. On odd-numbered days, the rats had a free choice in the maze; their performance on these days was used to measure learning. In order to give the animals equal experience on both sides and an equal number of injections of each kind, on even-numbered days, each rat was forced to the side opposite to the one chosen on the preceding day. All rats were given a total of 60 training trials, 30 of which were free choice. After this, the conditions of training were reversed for an additional 60 trials, with the ADH now given on the side where vehicle previously had been injected, and vice versa. Since some of the animals with diabetes insipidus began to get sick and die during the latter part of the second 60 trials, the reversal phase of training was discontinued for this group. The rats in it all seemed to be in good health, however, at the end of the first 60 days. While any unnoticed decline in their vigor might have reduced their speeds, it could not have produced, under the balanced conditions of this experiment, any tendency for them to choose predominantly a specific arm of the T maze.

RESULTS

During the 90 min in the end box, the ADH-injected rats excreted less urine than the vehicle-injected ones, the means being, respectively, 1.5 and 11.8 ml for normals preloaded with H_2O, 5.8 and 18.9 ml for diabetic rats preloaded with 0.33 M saline, and 0.8 and 1.2 ml for normals not preloaded. The first two of these differences were reliable beyond the 0.1 % level and the third beyond the 5 % level. The injections had no appreciable effect on daily water consumption, but the rats with diabetes insipidus drank less water after the ADH injection than

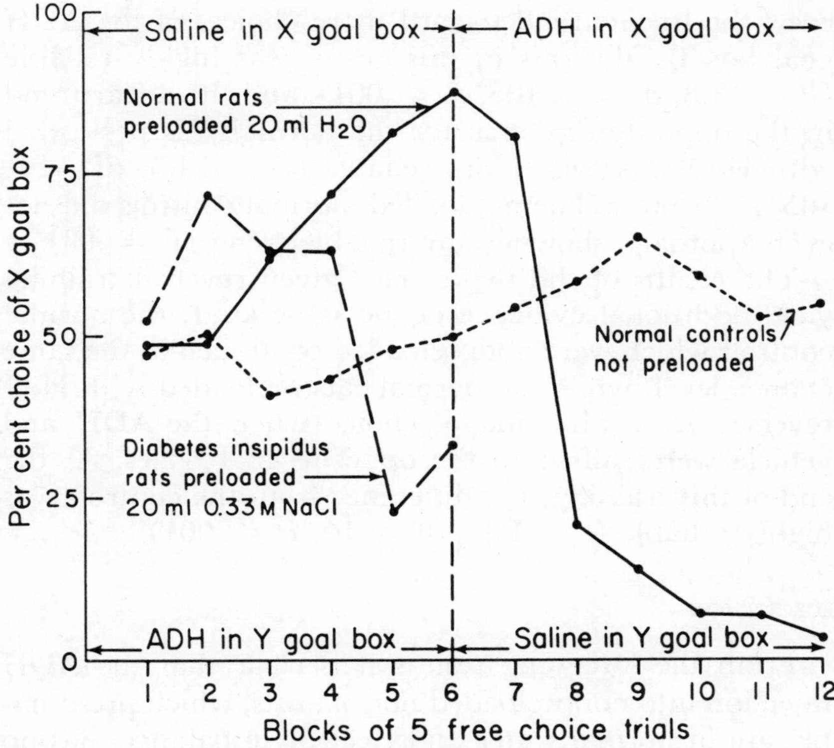

FIG. 1. Learning in a T maze rewarded by manipulating levels of antidiuretic hormone.

after the vehicle one (6.4 vs. 10.3 ml; $t = 6.2$, df $= 7$, $P < .001$).

The choices during training in the T maze are presented in Fig. 1. During the first stage of training, it can be seen that the normal rats preloaded with water learned to go away from the ADH side of the T maze, while the diabetic rats preloaded with 0.33 M NaCl learned to go toward the ADH side. The normal controls which were not preloaded showed no learning. Although the curve for the rats with diabetes insipidus preloaded with NaCl solution appears to go up before it goes down, this initial reversal is not statistically reliable, since a trend analysis of variance (5) showed that the quadratic trend, indicating curvilinearity, had an F of only 2.2 whereas one of 3.9 is needed for a reliability at the 5% level. However, the linear trend toward more choices of the ADH goal box by the rats in this group was highly reliable ($F = 13.8$, df $= 1, 105$, $P < .001$), and the linear trend in the opposite direction for the normal rats preloaded with H_2O was even more reliable ($F = 24.5$, df $= 1, 105$, $P < .001$). The nonloaded, normal controls did not even approach showing any reliable trend ($F = .03$).

The results of the two groups given reversal training yield additional evidence of the same kind: the normal controls which were not preloaded continued at the same chance level, while the normal rats preloaded with H_2O reversed their direction of choice when the ADH and vehicle were shifted to the opposite goal boxes. At the end of this training, the difference from the controls was highly reliable ($t = 4.37$, df $= 14$, $P < .001$).

DISCUSSION

From the foregoing results it is clear that the ADH injection into nonpreloaded normal rats, which presumably are in normal water and electrolyte balance, has no appreciable aversive or rewarding effect. When the rats are preloaded so that they have an excess of either H_2O or NaCl, however, they are motivated to learn, and they choose the side, which is avoiding the ADH for the H_2O-loaded group and approaching the ADH for the NaCl-loaded group, where they receive the injection that allows their kidneys to restore normal balance most rapidly.

Previous experiments have shown that a deficit in either water or salt can function as a drive to motivate learning and that, under these conditions, the consumatory response of drinking water or saline, respectively, that corrects these deficits, functions as a reward to reinforce learning (1, 3). The present experiment is novel in that *a*) it is an excess rather than a deficit of H_2O that functions as a drive, and *b*) the return to normal balance is effected by an internal, glandular rather than by an external, consumatory response. Generalizing these results, we advance the hypothesis that, at least in cases where homeostasis is mediated via the central nervous system, deviations in any direction, if large enough, can function as a drive and the prompt restoration to normal levels by any means can function as a reward.

Combining the results of the present experiment, demonstrating the rewarding effects of return to homeostasis, with our previously cited other experiments demonstrating that glandular and visceral responses can be instrumentally learned, we would expect the animal to learn glandular and visceral responses which promptly restore any deviation to the proper homeostatic level. Whether or not this theoretically possible learning actually will occur will depend on whether the innate homeostatic mechanisms control the level closely enough so that deviations large enough to function as a drive are allowed to occur. Even if the innate controls should be accurate enough to preclude learning in most cases, there still remains the possibility that abnormal circumstances, such as pathology, can interfere with innate control. In this case, we are led to expect that visceral learning rewarded by a return to homeostasis might be available as a supplementary mechanism. It remains for further research to establish how important such theoretically possible learning actually is in maintaining normal homeostasis.

Work on this study was supported by National Institute of Mental Health Research Grant MH 13189 to N. E. Miller.

L. V. DiCara is an Advanced Research Fellow of the American Heart Association.

G. Wolf is an Established Investigator of the American Heart Association.

REFERENCES

1. Denton, D. A. Salt appetite. In: *Handbook of Physiology. Alimentary Canal.* Washington, D.C.: Am. Physiol. Soc., 1967, vol. I chapt. 31, p. 433–459.
2. Fusco, M., R. L. Malvin, and P. Churchill. Alterations in fluid, electrolyte, and energy balance in rats with median eminence lesions. *Endocrinology* 79: 301–308, 1966.
3. Miller, N. E. Behavioral and physiological techniques: rationale and experimental designs for combining their use. In: *Handbook of Physiology. Alimentary Canal.* Washington, D.C.: Am. Physiol. Soc., 1967, vol. I, chapt. 4, p. 51–61.
4. Miller, N. E., and L. V. DiCara. Instrumental learning of urine formation by rats; changes in renal blood flow. Am. J. Physiol. 215: 677–683, 1968.
5. Winer, B. J. *Statistical Principles in Experimental Design.* New York: McGraw-Hill, 1962.

Received for publication 8 February 1968.

32

Instrumental Learning by Curarized Rats of a Specific Visceral Response, Intestinal or Cardiac

Neal E. Miller and Ali Banuazizi

When deeply curarized rats, maintained on artificial respiration, were rewarded by electrical stimulation of the medial forebrain bundle for relaxation of the large intestine, spontaneous intestinal contraction decreased; but if subsequently rewarded for intestinal contraction, it increased to above base-line level. When rewards stopped, the learned response extinguished. Deeply curarized rats rewarded for increased or decreased intestinal contraction showed progressive changes in the appropriate direction but heart rate did not change; conversely, rats rewarded for high or low rates of heartbeat, respectively, learned to change rates appropriately but showed no changes in intestinal contraction. Thus, instrumental learning of 2 visceral responses can be specific to the rewarded response, a fact which rules out several alternative interpretations.

The traditional view, as expressed in Kimble's (1961) authoritative summary, has been that "for autonomically mediated behavior, the evidence points unequivocally to the conclusion that such responses can be modified by classical, but not instrumental, training methods [p. 100]." This view has been sharply challenged in theoretical articles by Miller (1951, 1961, 1963a, 1963b) and by experimental studies from his laboratory (Miller & Carmona, 1967; Miller & DiCara, 1967; Trowill, 1967). In the last two of these studies, increases or decreases, respectively, in heart rate were secured and the mediation of these changes by instrumentally learned overt skeletal responses was ruled out by completely paralyzing such responses with curare. In studies that have not used this essential control, and hence that may be interpreted as showing indirect effects of the learning of skeletal responses, other investigators recently have also reported the modification by instrumental training procedures of heart rate, as well as of galvanic skin responses and vasodilatation, while a few investigators have reported failure to achieve such modification.[2]

In view of the crucial implications of experiments on the instrumental learning of visceral responses for theories of learning and for theories of psychosomatic symptoms, it seemed highly desirable to investigate the problem further. It is conceivable that the changes in heart rate observed in the preceding studies with curarized rats were the result of generalized changes in the level of arousal, of the type observed by Miller and Carmona (1967) while using instrumental training to change the rate of spontaneous salivation. It is also conceivable that *S*s were learning to send out from the cortex motor impulses which would have elicited struggling if the neuromuscular junction had not been paralyzed by curare, and that these centrally initiated impulses either had an innate or a classically conditioned effect on heart rate. The foregoing kinds of interpretation, preserving the traditional view that the instrumental learning of visceral responses is impossible, will be much less plausible if it is found that a second type of visceral response can be modified by instrumental training and if the learned

[1] This study was supported by National Institute of Mental Health Research Grant MH 13189.

[2] Papers reporting positive results, in addition to those cited in our previous report (Miller & DiCara, 1967) are: Birk, Crider, Shapiro, and Tursky (1966), Crider, Shapiro, and Tursky (1966), Engel and Chism (in press), Engel and Hansen (1966), Greene (1966), Greene and Nielsen (1966), Hnatiow and Lang (1965), Kimmel and Baxter (1964), Kimmel, Sternthal, and Strub (1966), Lisina, cited by Razran, (1961, p. 121), Shapiro, Crider, and Tursky (1964), and Van Twyver and Kimmel (1966). Negative findings have been reported by Harwood (1962) and Mandler, Preven, and Kuhlman (1962).

changes in it are found to be independent of those in heart rate.

The purpose of the present study was to determine whether curarized rats can learn a different type of autonomically controlled visceral response, namely, the contraction or relaxation of the large intestine. If such learning is possible, are the effects on intestinal contraction and heart rate (*a*) highly positively correlated (expected if they are indiscriminate ones of general arousal or by-products of the rate at which impulses to struggle are being sent out from the motor cortex); (*b*) highly negatively correlated (expected if the training is primarily changing the balance between sympathetic and reciprocally innervated parasympathetic arousal); or (*c*) uncorrelated (expected if the effects of training are specific to the particular type of visceral response rewarded)?

Experiment 1

The purpose of this exploratory experiment was to determine whether or not intestinal contractions could be modified by instrumental training.

Method

The general procedure of using electrical stimulation of the brain (ESB) in the medial forebrain bundle to reward the deeply curarized rat has been described by Trowill (1967) and by Miller and DiCara (1967).

Subjects. The *S*s were seven adult male albino Sprague-Dawley rats, weighing 450–600 gm., with chronic monopolar electrodes, at least one of which yielded a reward effect, aimed bilaterally at the medial forebrain bundle.

Apparatus. For preliminary testing of reward effects a Grason-Stadler bar 3 in. above floor level in a 12 × 12 × 18 in. box delivered 0.5 sec. of 60-cps ac stimulation regulated by a micropot used as a voltage divider and delivered from the stimulating electrode to an indifferent one on the skull. An 820 K ohm resistor in series with *S* minimized the effects of fluctuations in brain resistance. During intestinal training *S* lay on a folded towel in a soundproof, ventilated picnic lunch box (Grason-Stadler) and was artificially respirated by means of a mask made from the mouthpiece of a rubber balloon, the lower lip of which was secured by placing it behind the upper incisors and the upper portion of which was secured by pulling it tightly over the top of *S*'s mouth. The balloon end of the mask was connected via a rubber stopper with a hole drilled through its middle to a short Y junction connected to the inspiration and expiration tubes from a small-animal artificial respirator (A. & M. Instrument Co., Model V5KG).

To monitor intestinal motility, a small balloon tied to the end of a 7-cm. dull-point gauge No. 19 hypodermic needle was inserted through the rectum into the large intestine approximately 4 cm. beyond the anal sphincter and filled with 2 ml. of water; the needle connected via a short length of polyethylene tubing to a pressure transducer (Statham Laboratories, Model P23Dc) yielding an electrical output in proportion to the pressure which was amplified by a dc amplifier and fed in parallel into two pens of a polygraph (Grass, Model 5D). The first pen had a ball point, which made contact with a surface of a plate constructed of five strips of brass, each 10 mm. wide, inlaid into a piece of Plexiglas and separated from each other by approximately 1 mm. By connecting these strips appropriately to electronic relays, the amount of time that the excursions of the pen were beyond a specified recording level could be timed and a reward could be delivered when the pen was beyond a specified reward criterion level. Two Plexiglas strips prevented this pen from going beyond the limits of the commutator.[3] The second pen traced the record of intestinal motility, and a third pen indicated when rewards were delivered.

Procedure for implantation and pretesting. Enameled, stainless-steel surgical needles (Anchor brand, No. 1843), with straight, tapered points bare for 0.5 mm. at the tip, were implanted by the procedure described and illustrated by Miller, Coons, Lewis, and Jensen (1961, Figure 7B). They were aimed at the medial forebrain bundle at the A-P level of the ventromedial nucleus with Krieg's stereotactic coordinates of 1.5 mm. posterior to bregma, 8.5 mm below the skull top at point of entry, and 1.7 mm. and 1.3 mm. to the right or left of the midline, respectively.

A 30–80-μa. current was adjusted initially to elicit alertness and sniffing and then maximum bar pressing without undue side effects from each individual *S*. Then *S*s were shaped to press the bar to secure 0.5 sec. of 60-cps ac ESB. Current was held constant throughout the experiment. After shaping on Day 1, each *S* was trained for two 30-min. sessions separated by 30-min. rest with each bar press rewarded; *S*s pressing fewer than 500 times in the first of these 30-min. sessions were discarded.

Procedure for intestinal training. On a subsequent day *S*s were initially injected intraperitoneally with 3 mg/kg of *d*-tubocurarine chloride, and every hour subsequently until the end of the experiment with a supplementary dose of 1 mg/kg. As soon as *S*s showed signs of difficulty in breathing, they were fitted with a face mask and respirated at 1:1 I/E ratio, 70 breaths per minute, with a peak reading of 20 cm. of water pressure. At the end of a 30-min. habituation period, the circuits

[3] This commutator was designed by Gordon Ball and built by Nigel Cox.

were connected to reward a given type of response, namely relaxation for 3 and contraction for 4 *S*s, and the apparatus automatically delivered 0.5 sec. of rewarding ESB once every second as long as the pen was beyond the specified criterion in the proper direction. In order to prevent too frequent ESB, whenever *S* had achieved 5 rewards in 5 sec. the reward circuit was turned off for 3–4 sec. At the end of 60-min. training for a reward for one type of response, the reward circuit was switched to the opposite criterion and the opposite type of response was automatically rewarded for 120 min. Then all rewards were turned off for a 60-min. extinction period. After this recording was discontinued, *S*s were maintained on artificial respiration for 2–5 hr., until they were able to breathe by themselves.

Results

The *S*s remained deeply curarized throughout the experiment; the first signs of slight muscle twitching or of fluctuations in the respirometer pressure indicating breathing movements did not occur for at least 1 hr. after the end of recording.

Figure 1 shows typical samples of the record of the first *S* run. Even after a given type of training, contractions were episodic, so that not all samples of record as short as those shown were representative of the overall effect. That the records shown in Figure 1 are indeed representative is demonstrated by Figure 2, which shows the results of scoring the entire record to yield the amount of time above the contraction criterion during 30-min. periods throughout the experiment. The difference between the number of seconds above the scoring criterion for 10 1-min. periods at the end of habituation and similar samples of performance at the end of reward for relaxation was reliable ($p < .01$). Similar t tests showed that the performance at the end of reward for contraction was reliably above that at the end of habituation ($p < .01$) and that at the end of the nonrewarded extinction period ($p < .05$).

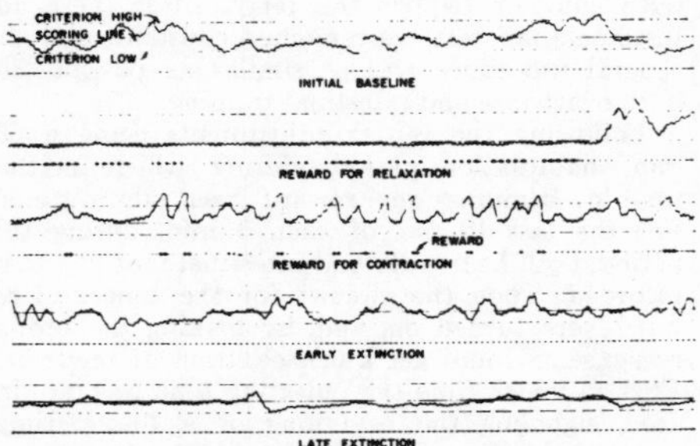

FIG. 1. Record of the instrumental learning of an intestinal response by a curarized rat. (The top record shows spontaneous activity during habituation, which carries the record above and below the scoring criterion. Reward for relaxation lowers base line and reduces number of spontaneous contractions; subsequent reward for contraction raises base line and increases contractions; nonrewarded extinction produces return to initial base line in about 20 min. Each sample of record is for 2 min.)

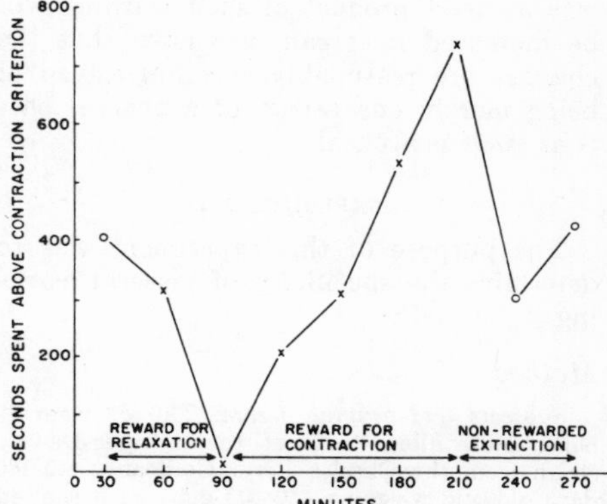

FIG. 2. Instrumental learning of intestinal relaxation and then contraction.

All *S*s showed similar results. For the 2 *S*s rewarded first for relaxtion, the decline was reliable beyond the 1% level; for the 4 *S*s rewarded first for contraction, the increase was reliable beyond the 1% level for 3 and beyond the 5% level for 1 *S*. All of the foregoing showed reversal learning and extinction.

Initially, the ESB used as a reward sometimes had some effect on intestinal contraction, increasing it in certain *S*s and decreasing it in others; but this effect disappeared during the first few minutes of training. Since the same ESB could be used to reinforce opposite responses in the same *S*, it is hard to explain this learning on the basis of any lingering small unconditioned effects of the ESB. Thus it seems highly probable that the changes in intestinal contractions were indeed the product of instrumental learning; the probability that it was a direct product of such learning will be increased if it can be shown that the changes are reasonably specific instead of being merely one aspect of a general pattern, such as arousal.

Experiment 2

The purpose of this experiment was to determine the specificity of visceral learning.

Method

Subjects and general design. The *S*s were 12 naive male albino rats of the Sprague-Dawley strain from the Charles River Company, 140–160 days old and weighing 450–600 gm.; each showed satisfactory reward effects to stimulation via electrodes of the types described in Experiment 1. Both heart rate and intestinal contractions were recorded for each *S,* but 6 were rewarded for changes in intestinal contraction, while 6 were rewarded for changes in heart rate (3 for increases and 3 for decreases of each response). All *S*s were deeply curarized throughout the visceral training.

Apparatus. For preliminary testing of reward effects, the apparatus was the same as that used in Experiment 1, except that the stimulation during this test and subsequent training was produced by a Tektronix power supply, wave generator, and pulse generator set up to produce dc square-wave pulses of negative polarity, 1 msec. in width with the interval between pulses adjusted at 5–10 msec. and the amplitude at 10–40 v., in order to maximize reward effects for individual *S*s. A 10 K ohm resistor was in series with *S*. Once a setting was selected during preliminary training, it was maintained throughout the experiment. The duration of the pulse train was 0.3 sec.

For monitoring intestinal activity, the apparatus was identical to that in Experiment 1, except that an Offner dynograph was substituted for the Grass polygraph and that a more detailed objective score was secured by connecting each strip of the commutator to a separate counter activated at 5 pps whenever the pen was on the strip leading to that particular counter.[4]

For recording heart rate, Beckman bipotential skin electrodes placed on shaved areas over the sternum, on the abdomen, and on the middle of the back (used as a ground) were connected directly to the amplifiers of an Offner dynograph, Type RB. The output of the amplifier to the dynograph pen was used to activate a 32-ohm relay which in turn was fed into a Grason-Stadler counter. The circuit for automatically delivering reward when the heart rate was beyond a specified criterion has been described by Trowill (1967). A predetermining counter counted heartbeats during a precisely timed interval (which had to be varied somewhat from an average of 1 sec. in order to achieve a finely graded series of criteria). For rewarding fast rates, the reward circuit was activated whenever the beats during the interval exceeded a specified number; for rewarding slow rates, reward was delivered whenever the beats fell below a specified number. Two such counters operated alternately, allowing the inactive one to reset.

Pretesting. At least 7 days after the electrodes were implanted, *S*s were shaped to press the bar, then given 1 hr. of training with each press reinforced. Rats pressing fewer than 500 times during the first 30 min. of this hour were discarded. For the following hour they were trained on a schedule (FI 10 sec.) with 10 sec. of time-out during which no rewards were delivered after each rewarded bar press. The following day they were given an additional hour of similar training. Throughout this and subsequent training, whenever pressing the bar would deliver a reward, a 600-cps, 2-pps tone was used as a time-in stimulus; this tone was off during the time-out interval.

Visceral training. Approximately 3 hr. prior to the experiment, *S*s were brought into the experimental room, deprived of food and water; they were curarized and artificially respirated in the way described in Experiment 1. For intestinal recording the same sensitivity setting (200 μv/cm) was always used, but the balancing circuit was adjusted so that the typical base-line resting level of the pen between episodes of contraction was on the second strip of the commutator. This level was easily identified, although occasional, atypical relaxations carried the pen down to the first strip. At the same time the EKG amplifiers were adjusted so that only the spike component of the QRS complex tripped the relay. After these adjustments, all settings remained constant, and the general procedure was as similar as possible for both heart-rate and intestinal training.

Following the initial adjustments came a 30-min. habituation period, during which neither time-in stimulus nor reward was administered. For the last 10 sec. of each 3 min. during this period, both heart rate and intestinal motility were recorded. Then the circuits for the tone and reward were turned on, and, by making the proper response, *S* could get a 0.3-sec. train of rewarding ESB, at which time the pulsating tone was terminated, signaling the beginning of a 10-sec. time-out period before the next trial.

The *S*s were randomly distributed among the four experimental conditions, with reward administered (*a*) when heart rate was faster than the criterion, (*b*) when heart rate was slower than the criterion, (*c*) when intestinal contractions were above the criterion (i.e., on or above the third strip), and (*d*) when intestinal contractions were below the criterion. The initial criterion for heart rate was selected, on the basis of performance during habituation, as the level which *S* achieved on

[4] The authors thank A. R. Wagner for the use of his laboratory space and some of his equipment purchased with funds from National Science Foundation Grant GB 3523.

approximately half of the brief intervals during which heart rate was counted. Whenever *S* improved enough so that it met the criterion in 2 sec. or less on 6 consecutive trials, it was changed to a harder one; whenever *S* failed twice in successive 10-sec. periods to meet the criterion it was changed to an easier one, but the need for such remedial retraining was quite rare. Changes were in 2% steps. No such shaping was necessary for the intestinal training.

Throughout training, every tenth trial was a "test" during which the reward circuit was turned off for the first 10 sec. of the time-in stimulus while the responses were measured in exactly the same way as they had been during habituation. This gave the same sample of measurement throughout habituation and training, avoiding any selective or interferring factors that might have been introduced if the trial had been terminated and the ESB administered as soon as the criterion was met. At the end of the 10-sec. recording interval, the response counters were turned off and the reward circuit was turned on, so that *S* was, immediately thereafter, rewarded if it had already exceeded the criterion or as soon as it met it. In short, test trials involved an unsignaled 10-sec. delay in the possibility for reward, a procedure which was designed to produce less interference with learning than the use of nonrewarded test trials. A total of 500 trials were given during training, which lasted for approximately 4 hr.

The heart-rate score was the number of heart-beats during each 10-sec. period. An integrated intestinal contraction score was computed by multiplying the number of seconds that *S* spent on any commutator strip by the number of the strip, with the one indicating complete relaxation numbered 1 and the top one 5.

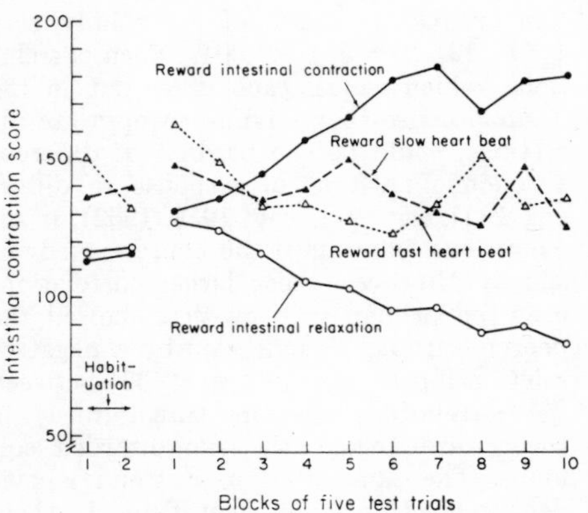

FIG. 3. When increases or decreases, respectively, in spontaneous intestinal contraction were rewarded, changes in the appropriate direction were produced; when heart rate was rewarded, intestinal contractions were unaffected.

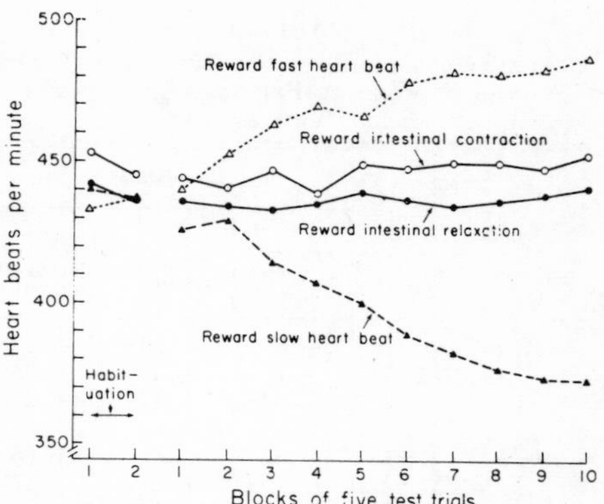

FIG. 4. When increases or decreases in heart rate were rewarded, rates changed in the appropriate direction; when intestinal contraction was rewarded, no changes in heart rate were produced.

Results and Discussion

The results for intestinal motility are presented in Figure 3 and for heart rate in Figure 4. Intestinal contraction increased when rewarded, decreased when relaxation was rewarded, and remained virtually unchanged when either increased or decreased heart rate was rewarded. Similarly, heart rate increased when a fast rate was rewarded, decreased when a slow rate was rewarded, and remained virtually unchanged when either intestinal contraction or relaxation was rewarded.

The foregoing results are in line with the suggestive evidence for specificity shown in the ingenious experiments by Birk, Crider, Shapiro, and Tursky (1966), Crider, Shapiro, and Tursky (1966), and Shapiro, Crider, and Tursky (1964). In addition to not having the control of complete curarization, those experiments did not use a balanced design; since only one response was rewarded, failure of the unrewarded response to show a reliable change could be attributed to innate unmodifiability or to lack of a sufficiently reliable measurement of that response. In the balanced design of the present experiment, such interpretations are ruled out by the reliable changes produced in each response when it is rewarded.

Results for individual *Ss* are presented

TABLE 1

Correlations of Each *S*'s Test Scores with the Number of Previous Training Trials

Group	*S*	Correlation[a]		
		Rewarded responses	Non-rewarded responses	Heart rate and intestinal contractions
IC+	1	+.62	+.23	+.32
	2	+.76	+.32	+.15
	3	+.32	+.43	+.21
IC−	4	−.82	−.07	+.05
	5	−.79	+.01	+.17
	6	−.58	+.40	−.20
HR+	7	+.75	−.22	−.19
	8	+.79	−.28	−.29
	9	+.78	+.05	+.06
HR−	10	−.96	−.09	+.07
	11	−.76	−.26	+.26
	12	−.92	−.10	+.05
M (*z*)		+.78	+.05[c]	+.06
(*t*)[b]		8.7	0.7	1.0

[a] $p = .05$ for $r = .27$; $p = .01$ for $r = .35$; $p = .001$ for $r = .44$.

[b] $df = 11$.

[c] Computed on the assumption that nonrewarded response should change in direction predicted for rewarded response.

in Table 1. The amount of learning by each *S* is shown by the size of the correlation between the number of each test trial during training and the score achieved on that trial. The positive correlation for each *S* rewarded for increases showed that the scores of each of them increased during training; the negative correlation for each *S* rewarded for decreases showed that the scores of each of them decreased during training. Eleven of the correlations indicating learning by each *S* were reliable beyond the .001 level and the twelfth at the .05 level. The average correlation in the predicted direction, computed by changing the signs of the correlations for rats rewarded for decreases and using the *z* score transformation (McNemar, 1962), was +.78 and highly reliable.

If one assumes that both the rewarded and nonrewarded responses reflect the effect of the same general factor (such as level of activation or rate of impulses to struggle sent out from the motor cortex), the predicted effect of reward should be in the same direction for both. Therefore the average for the nonrewarded response, shown at the bottom of Column 2 in Table 1, was computed in the way just described. The opposite assumption of a reciprocal relationship between sympathetic and parasympathetic responses predicts the same-sized coefficient but with the opposite sign. It can be seen that the average correlation (+ .05) was not significantly different from zero and hence did not support either hypothesis. The contrast between this extremely low coefficient and the much higher one for the rewarded response showed that on the average the learning of these visceral responses was highly specific. When the correlations for individual *S*s are inspected, the considerably larger size of the one for the rewarded response indicates for almost every *S* a high degree of specificity, especially when one remembers that the percentage of variance accounted for is the square of the correlation.

It should be noted that there are somewhat more of the larger (but still low) correlations for nonrewarded responses than would be expected in a random distribution around a mean of essentially zero ($\chi^2 = 22$, $df = 5$, $p < .001$). Such correlations, which cancel each other out in the mean because they deviate in opposite directions, could be the product of different autonomic patterns of response in different *S*s (Lacey & Lacey, 1958, 1962) or superstitious learning of the nonrewarded response. However, these larger correlations occurred primarily in *S*s that showed the poorer learning, as indicated by a negative relationship ($r = -.68$, $p < .02$) between the correlations (*Z*-score transformed) in the rewarded and the nonrewarded columns. The same kind of a trend is evident in the third column of Table 1, where the larger correlations between heart rate and intestinal contractions tended to occur in *S*s that showed poorer learning ($r = -.59$, $p < .05$). The hypothesis that both the intestinal and cardiac responses were mediated by the learning of some general reaction, such as arousal, struggling, or a shift in the sympathetic-parasympathetic balance, would predict exactly the opposite results.

In the learning of skeletal responses, poorer learning is frequently associated with less specificity. For example, during early trials of leg-lifting conditioned as an avoidance response, the dog will struggle and move all four legs, although, after good learning, it will make only the precise lifting movement required of one leg. Thus the correlation of poorer learning with less specificity in the present experiment is an example of similarity between skeletal and visceral learning.

REFERENCES

Birk, L., Crider, A., Shapiro, D., & Tursky, B. Operant electrodermal conditioning under partial curarization. *J. comp. physiol. Psychol.*, 1966, **62,** 165–166.

Crider, A., Shapiro, D., & Tursky, B. Reinforcement of spontaneous electrodermal activity. *J. comp. physiol. Psychol.*, 1966, **61,** 20–27.

Engel, B. T., & Chism, R. A. Operant conditioning of heart rate speeding. *Psychophysiology,* 1967, **3,** 418–426.

Engel, B. T., & Hansen, S. P. Operant conditioning of heart rate slowing. *Psychophysiology,* 1966, **3,** 176–187.

Greene, W. A. Operant conditioning of the GSR using partial reinforcement. *Psychol. Rep.*, 1966, **19,** 571–578.

Greene, W. A., & Nielsen, T. C. Operant GSR conditioning of high and low autonomic perceivers. *Psychon. Sci.*, 1966, **6,** 359–360.

Harwood, C. W. Operant heart rate conditioning. *Psychol. Rec.*, 1962, **12,** 279–284.

Hnatiow, M., & Lang, P. J. Learned stabilization of heart rate. *Psychophysiology,* 1965, **1,** 330–336.

Kimble, G. A. *Hilgard and Marquis' conditioning and learning.* (2nd Ed.) New York: Appleton-Century-Crofts, 1961.

Kimmel, H. D., & Baxter, R. Avoidance conditioning of the GSR. *J. exp. Psychol.*, 1964, **68,** 482–485.

Kimmel, H. D., Sternthal, H. S., & Strub, H. Two replications of avoidance conditioning of the GSR. *J. exp. Psychol.*, 1966, **72,** 151–153.

Lacey, J. I., & Lacey, B. C. Verification and extension of the principle of autonomic response stereotypy. *Amer. J. Psychol.*, 1958, **71,** 50–73.

Lacey, J. I., & Lacey, B. C. The law of initial value in the longitudinal study of autonomic constitution: Reproducibility of autonomic responses and response patterns over a four-year interval. *Ann. N. Y. Acad. Sci.*, 1962, **98,** 1257–1290.

Mandler, G., Preven, D. W., & Kuhlman, D. K. Effects of operant reinforcement on the GSR. *J. exp. Anal. Behav.*, 1962, **5,** 317–321.

McNemar, Q. *Psychological statistics.* New York: Wiley, 1962.

Miller, N. E. Comments on multiple-process conceptions of learning. *Psychol. Rev.*, 1951, **58,** 375–381.

Miller, N. E. Integration of neurophysiological and behavioral research. *Ann. N. Y. Acad. Sci.*, 1961, **92,** 830–839.

Miller, N. E. Animal experiments on emotionally-induced ulcers. *Proc. World Congr. Psychiat., June 4–10, 1961, Montreal,* 1963, **3,** 213–219. (a)

Miller, N. E. Some reflections on the law of effect produce a new alternative to drive reduction. *Nebraska Symposium on Motivation: 1963.* Lincoln: University of Nebraska Press, 1963. Pp. 65–112. (b)

Miller, N. E., & Carmona, A. Modification of a visceral response, salivation in thirsty dogs, by instrumental training with water reward. *J. comp. physiol. Psychol.*, 1967, **63,** 1–6.

Miller, N. E., Coons, E. E., Lewis, M., & Jensen, D. D. Electrode holders in chronic preparations. B. A simple technique for use with the rat. In D. E. Sheer (Ed.), *Electrical stimulation of the brain.* Austin: University of Texas Press, 1961. Pp. 51–54.

Miller, N. E., & DiCara, L. Instrumental learning of heart-rate changes in curarized rats: Shaping, and specificity to discriminative stimulus. *J. comp. physiol. Psychol.*, 1967, **63,** 12–19.

Razran, G. The observable unconscious and the inferable conscious in current Soviet psychophysiology: Introceptive conditioning, semantic conditioning, and the orienting reflex. *Psychol. Rev.*, 1961, **68,** 81–147.

Shapiro, D., Crider, A. B., & Tursky, B. Differentiation of an autonomic response through operant reinforcement. *Psychon. Sci.*, 1964, **1,** 147–148.

Trowill, J. A. Instrumental conditioning of the heart rate in the curarized rat. *J. comp. physiol. Psychol.*, 1967, **63,** 7–11.

Van Twyver, H. B., & Kimmel, H. D. Operant conditioning of the GSR with concomitant measurement of two somatic variables. *J. exp. Psychol.*, 1966, **72,** 841–846.

(Received April 6, 1967)

33

The Interpretation of Gastric Motility: Apparent Bias in the Reports of Hunger by Obese Persons

Albert Stunkard, and Charles Koch

This paper describes the correlation between the presence of gastric motility and reports of hunger in four groups of persons: obese and nonobese men and women. Current clinical conceptions of hunger are based on reports that the experience of hunger occurs primarily during contractions of the empty stomach. We were therefore surprised to find that a number of obese women showed an apparent "denial of hunger" in the presence of gastric motility.[15] An extension of this study to men, using an improved method of data analysis, reveals that denial of hunger is a special case of a more general bias of obese subjects in associating hunger with gastric motility. Obese persons seem particularly prone to one extreme or the other, with women tending to denial of hunger and men to "exaggeration of hunger."

Submitted for publication Feb 25, 1964.

Professor of Psychiatry, University of Pennsylvania (Dr. Stunkard). Resident in Psychiatry, University of Pennsylvania (Dr. Koch).

Supported in part by Grant M-3684, National Institute of Mental Health, National Institutes of Health, United States Public Health Service.

Method

The female subjects were 17 obese and 18 nonobese women from the General Medical Clinic of the Hospital of the University of Pennsylvania. The median per cent overweight of the obese women, as calculated from standard height-weight tables,[12] was 62, with a range from 26 to 182. The median age of the obese women was 30, whereas that of the nonobese women was 34, a difference which did not approach statistical significance.

Twenty obese and 19 nonobese men were studied. Because of difficulty in obtaining a sufficient number of obese men, we were forced to use three referral sources, each contributing approximately one third of the obese and nonobese groups. The sources were the General Medical Clinic, the Psychiatric Outpatient Department, and University students. The responses of subjects from the different sources do not appear different. The median per cent overweight of the obese men was 43.5, with a range from 21 to 297. Their median age was 26 years, whereas that of the nonobese men was 28.5 years, a difference which did not approach statistical significance. A summary of the information on each subject is presented in the tables. Note that both groups of women were somewhat older on the average than the men, and that their per cent overweight was greater and less variable. All subjects were in good health and were free of any condition believed to influence food intake (diabetes, hypoglycemia, etc), gastric motility (peptic ulcer, gastritis, etc), or the ability to tolerate the experimental procedure (congestive failure, pulmonary insufficiency, etc). Control subjects in each group were not only of normal weight, but reported no past history of overweight or underweight. There was no apparent difference between the obese and control groups in regard to ethnic background, educational level or socioeconomic status. In an effort to obviate the effects of prior learning, no subjects were used who had previously undergone gastric intubation.

The following experimental procedure was used. The subject arrived at the laboratory at 9 AM after an overnight fast. A gastric balloon attached to a Levin tube was inserted into the stomach and was inflated to a pressure of 15 cm of water and a volume of approximately 90 cc. The tube was withdrawn until resistance was encountered at the cardia, and then the tube was anchored at the nose with adhesive tape. Gastric contractions were recorded on a kymograph by means of a water manometer. The test lasted four hours.

"MOTILITY"

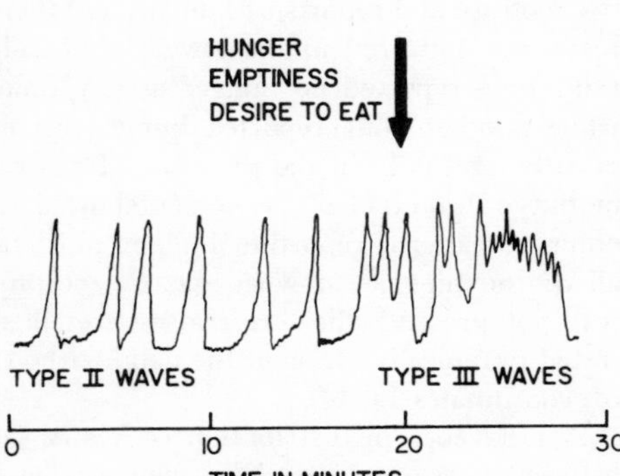

Fig 1.—Gastric pressure tracing during the presence of gastric motility. Type II and type III waves are shown and the subject's report of presence of hunger, emptiness, and desire to eat is entered directly on the motility tracing.

"NO MOTILITY"

NO HUNGER
NO EMPTINESS
NO DESIRE TO EAT

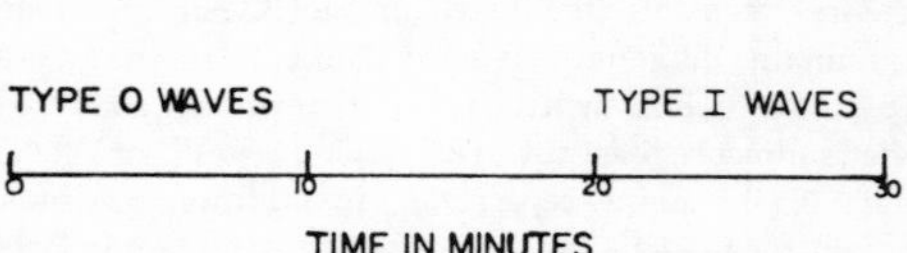

TIME IN MINUTES

Fig 2.—Gastric pressure tracing during absence of gastric motility. No motility is called type O waves, and motility of low amplitude is called type I. Subject's report of absence of hunger, emptiness and desire to eat is recorded directly on the motility tracing.

The kymograph tracing was interpreted according to the original description of Cannon[1] and of Carlson[2] as standardized by Code.[3] According to these authors, gastric motility can be divided into four basic patterns, of which we scored types 0 and I as "no motility," and types II and III as "motility."

In an effort to delineate the components of the hunger experience, 200 obese and nonobese persons were asked what they meant when they said they were hungry. The only replies regularly associated with a report of "hunger" referred to sensations of "emptiness" in the upper abdomen, or a desire to eat, or both. Accordingly, in the studies to be described, questions as to "hunger," "emptiness," and "desire to eat" were always asked together. Since in these studies the three replies were usually associated, and since no pattern of nonassociation was discerned, only results of the reports of "hunger" will be described.

At 15-minute intervals during the measurement of gastric motility each subject was asked whether or not he was hungry at that time. Information about gastric motility and reports of hunger was thus collected 17 times. This information is summarized in the following manner. For each subject we plot on the vertical axis the probability of hunger responses during gastric motility, and on the horizontal axis the probability of hunger responses in the absence of gastric motility. Such a plot of the so-called "hits" of the vertical axis against the so-called "false alarms" of the horizontal axis reduces each subject's two-by-two table relating "hunger" and "no hunger" to "motility" and "no motility" to a single point on the graph. The relatively small number of responses for each subject results, of course, in a large variability about each of these points. The standard deviation of responses in the left upper quadrant, for example, is typically of the order of ±0.15, although the variability decreases to zero as the responses approach the 0.0,0.0 and 1.0,1.0 coordinates.

The usefulness of such representation has been described in the recent literature on signal detection theory[8,16] and is illustrated in Fig 3. Any point on the main diagonal (labeled "bias") represents a response of no sensitivity, one determined solely by the subject's bias towards saying "yes" or "no." The 0.5,0.5 point represents, in addition, no bias, which means random responses to a randomly presented stimulus. Perfect sensitivity (with no bias) is represented by coordinates 0.0 (horizontal) and 1.0 (vertical), and increasing sensitivity is measured by proximity to this point. This system of analysis makes it possible to represent the two independent factors which enter into any psychophysical judgment—physiological sensitivity to the stimulus and psychological bias as to the criteria used in making the judgment.

Results

Gastric Motility As a Function of Obesity.—The gastric motility of the obese subjects did not appear to differ from that of the

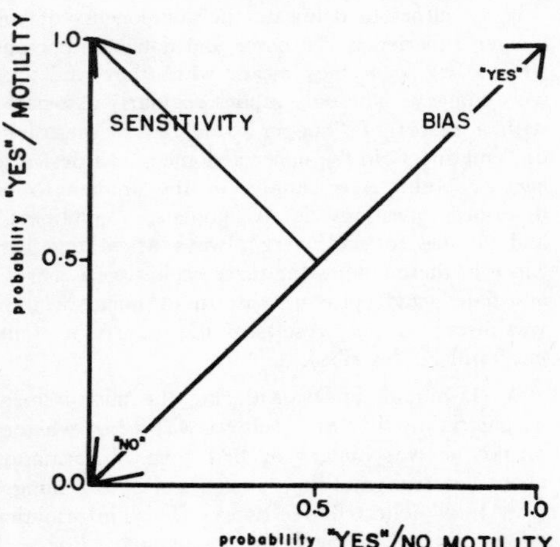

Fig 3.—Analysis of hunger responses according to signal detection model. The vertical axis represents the probability of the subject's saying "yes" in the presence of gastric motility (a "hit"), and the horizontal axis the probability of his saying "yes" in its absence (a "false alarm"). These axes define the parameters of "sensitivity" and "bias" which are illustrated in this figure and described more fully in the text.

control subjects. Although nonobese men showed a higher median per cent of time during which motility was present (55 vs 42.5 for the obese men), this difference did not approach statistical significance, and among the women the results were reversed (46.5 for the nonobese vs 49 for the obese).

Among the women there was a slight tendency for increasing age to be associated with decreasing time during which gastric motility was present ($r=0.347$, $P=<0.05$), but there was no similar trend among the men. In neither sex was any relation found between per cent overweight and gastric motility. Accordingly, it appears that differences in reports of hunger between obese and nonobese groups were not due to differences in gastric motility.

Gastric Motility and Reports of Hunger As a Function of Obesity.—The most important finding of this study is the difference between obese and nonobese groups in their correlation of gastric motility with reports of hunger. Nonobese women showed a high correlation of presence of gastric motility with reports of hunger, and, conversely, between absence of gastric motility and reports of no hunger. This finding is illustrated graphically in Fig 4, which shows that the responses of most of the nonobese women were in the left upper quadrant. Most of the obese women, on the other hand, failed to show a comparable correlation between gastric motility and reports of hunger, and their responses clustered in the lower left quadrant. Five reported no hunger at any time, while another four reported hunger infrequently, and only in the presence of gastric motility. Two other obese women showed an opposite pattern, reporting hunger most or all of the time, even when gastric motility was not present; their responses are illustrated graphically at, or in the neighborhood of, coordinates 1.0,1.0.

A difference in distribution of responses between obese and nonobese men was also found (Fig 5). The nonobese men did not show as high a degree of association between gastric motility and reports of hunger as did the nonobese women. Despite an apparently

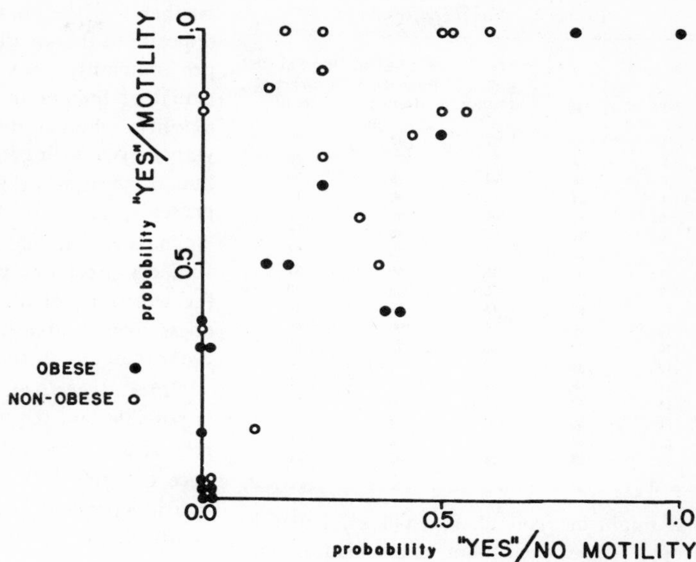

Fig 4.—Hunger responses of women. Each point represents the results of one test. "Hits" are plotted on the vertical axis and "false alarms" on the horizontal axis. Nonobese women appear to show moderate sensitivity with little bias. Many obese women showed marked bias, most commonly of a "nay-saying" type.

lessened sensitivity, however, their responses showed no consistent pattern of bias. Such bias was a prominent feature of the responses of the obese men. Seven reported hunger most or all of the time, even when gastric motility was not present. Three obese men, on the other hand, showed the pattern characteristic of the modal obese woman, rarely or never reporting hunger, even in the presence of gastric motility. These extremely biased responses of many obese persons were infrequent among nonobese men and women.

A strictly quantitative assessment of the sensitivity and bias involved in the distributions shown in Figs 4 and 5 requires assumptions which are not warranted by the available data. The clustering of responses of many of the obese subjects at the 0.0,0.0 and the 1.0,1.0 coordinates, however, suggests that a rough comparison of the degrees of

TABLE 1.—*Obese Women*

No.	Age	Height, Ft, In (Cm)	Weight, Lb (Kg)	% Overweight	% Motility	% "Yes" During Motility	% "Yes" Absence of Motility	Night-Eating Syndrome
1	27	5,4 (162.6)	280 (127.0)	119	25	50	18	+
2	28	5,2.5 (158.8)	185 (83.9)	62	23	0	0	+
3	53	5,0 (12.7)	196 (88.9)	69	28	40	42	0
4	45	5,0.5 (153.7)	200 (90.7)	71	37	40	38	0
5	39	5,2 (157.5)	196 (88.9)	62	43	50	14	0
6	44	5,0 (152.4)	197 (89.4)	70	55	0	0	+
7	55	5,5 (165.1)	243 (110.2)	86	49	14	0	0
8	48	5,1 (154.9)	332 (150.6)	181	19	100	100	0
9	50	4,11 (149.9)	216 (98.0)	90	21	0	0	+
10	23	5,3 (160.0)	160 (72.6)	29	58	38	0	0
11	30	5,0 (152.4)	175 (79.4)	51	60	0	0	+
12	23	5,3.5 (161.3)	182 (82.6)	46	22	67	25	0
13	24	5,1 (154.9)	180 (81.6)	53	55	0	0	+
14	24	5,6.5 (168.9)	176 (79.8)	29	54	100	78	+
15	27	5,4 (162.6)	200 (90.7)	56	69	33	0	0
16	29	5,2 (157.5)	182 (82.6)	50	68	78	50	0
17	27	5,9 (175.3)	295 (133.8)	102	55	33	0	+

TABLE 2.—*Nonobese Women*

No.	Age	% Motility	% "Yes" in Presence of Motility	% "Yes" in Absence of Motility
1	20	77	92	25
2	53	57	100	50
3	23	34	83	55
4	34	19	100	17
5	56	64	88	0
6	34	95	86	0
7	21	62	73	25
8	37	48	50	37
9	33	38	100	50
10	23	45	88	14
11	20	54	100	60
12	32	55	78	43
13	54	17	0	0
14	33	39	60	33
15	60	26	100	25
16	39	37	14	11
17	59	40	37	0
18	38	75	83	50

bias might be fruitful. If a perpendicular is dropped from each point to the main diagonal, the distance from this intersection to the 0.5,0.5 point which indicates no bias can be determined. Such values provide an approximate estimate of the degree of bias, irrespective of direction. A two-tailed *t* test based on this criterion reveals that obese men and women show significantly more bias than their nonobese counterparts (for women, $P > 0.05 < 0.10$; for men, $P < 0.005$). Note further that the obese women exhibit a bias opposite to that of the obese men; theirs was predominantly "nay-saying," resulting in a denial of hunger in the presence of gastric motility, whereas the obese men tended towards "yea-saying," resulting in reports of hunger even when gastric motility was not present.

The marked bias of many of the obese subjects precludes meaningful estimates of the sensitivity of the obese groups. The nonobese women appear to show greater sensitivity than any of the other three groups.

Other Aspects of Reports of Hunger.—It is possible that the obese subjects' biases are related to neurotic distortions of the experience of hunger. To assess this hypothesis, two independent psychological measures of neurosis were collected: presence of the "night-eating syndrome" [14] and scores on the Taylor Test for Manifest Anxiety.[17] The latter test was of particular interest because of recent evidence that it may be primarily a measure of the effort to convey an impression of social undesirability.[17] As such, it might be expected to throw light upon the defensive functions of the hunger responses of the obese persons.

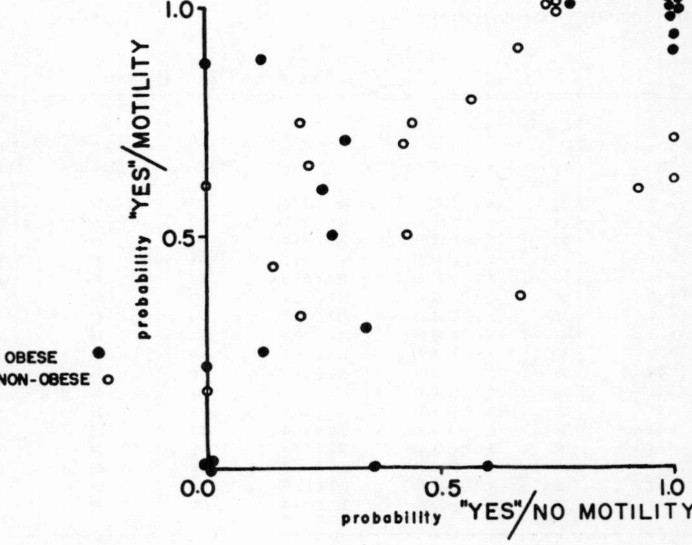

Fig 5.—Hunger responses of men. Each point represents the results of one test. "Hits" are plotted on the vertical axis and "false alarms" on the horizontal axis. Obese subjects again show bias more commonly than do the nonobese, although among men this was more commonly of a "yea-saying" type. Note the curious response of subject No. 16, at coordinates 0.6 (horizontal), 0.0 (vertical).

TABLE 3.—*Obese Men*

No.	Source	Age	Height, Ft, In (Cm)	Weight, Lb (Kg)	% Overweight	% Motility	% "Yes" During Motility	% "Yes" in Absence of Motility
1	Medicine	19	6,2 (188.0)	340 (154.2)	91	30	100	100
2	"	39	5,7 (170.2)	258 (117.0)	77	36	100	100
3	"	42	6,2 (188.0)	272 (123.4)	53	36	0	0
4	"	38	5,8 (172.7)	217 (98.4)	44	30	60	25
5	"	26	5,11.5 (181.6)	235 (106.6)	43	45	25	12
6	"	48	5,9 (175.3)	210 (95.2)	36	51	90	100
7	"	53	5,11 (180.3)	220 (99.8)	35	50	100	78
8	"	32	5,8 (172.7)	195 (88.5)	29	33	0	0
9	"	22	5,1 (154.9)	155 (70.3)	24	41	0	0
10	School	22	5,6 (167.6)	227 (102.9)	60	35	50	27
11	"	21	6,0 (182.9)	265 (120.2)	59	66	22	0
12	"	22	5,9 (175.3)	235 (106.6)	52	34	0	36
13	"	24	5,10 (177.8)	217 (98.4)	37	80	30	34
14	"	26	5,11 (180.3)	200 (90.7)	25	52	89	12
15	"	23	5,11.5 (181.6)	198 (89.8)	21	72	100	100
16	Psych.	26	5,8 (172.7)	600 (272.2)	297	44	0	60
17	"	45	5,5 (165.1)	230 (104.3)	66	78	93	100
18	"	29	5,10 (177.8)	238 (107.9)	51	28	100	100
19	"	23	5,11 (180.3)	230 (104.3)	41	56	88	0
20	"	41	5,10.5 (180.1)	200 (90.7)	26	39	71	30

The night-eating syndrome, a pattern of food intake consisting of morning anorexia, evening hyperphagia, and insomnia, was previously shown to occur in some obese persons during periods of emotional stress. The denial of hunger of some obese women during our tests, which were all conducted in the morning, suggested that they might be suffering from the night-eating syndrome. This expectation was confirmed: eight suffered from the night-eating syndrome, and these women show a more marked bias than most of the other nine (Fig 6). The responses of five of them, for example, cluster at the point 0.0,0.0 indicating a maximal bias.

TABLE 4.—*Nonobese Men*

No.	Source	Age	% Motility	% "Yes" During Motility	% "Yes" in Absence of Motility
1	Medicine	27	31	60	92
2	"	70	63	91	67
3	"	65	45	50	43
4	"	33	80	71	100
5	"	38	60	17	0
6	"	42	49	80	57
7	"	36	60	70	42
8	School	25	19	66	22
9	"	25	29	75	20
10	"	21	72	37	67
11	"	19	70	33	20
12	"	35	55	100	75
13	"	25	48	62	0
14	"	25	37	75	44
15	"	28	53	100	75
16	Psych.	27	60	44	14
17	"	28	83	100	100
18	"	29	45	100	75
19	"	34	61	62	100

No significant relationship was found between the hunger responses and scores on the Taylor Test for Manifest Anxiety.

Mention should be made of the response of obese subject No. 16, at coordinates 0.6 (horizontal), 0.0 (vertical). Such responses in the region below the main diagonal are of considerable theoretical interest. They represent not a lack of sensitivity, but an anomalous response of a person who, precisely because of his sensitivity, is able rather consistently to report no hunger in the presence of gastric motility, and hunger in its absence. Other aspects of the subject who produced this curious record are of interest for the light they cast upon it. He is a 600-lb (272.2 kg) man who was diagnosed as suffering from a borderline schizophrenic state and who, some time after he had been tested, underwent a year of psychiatric inpatient treatment which brought to light a pervasive negativism.

Comment

The nature and significance of the gastric pressure changes measured in these studies deserve comment. Pressure changes recorded

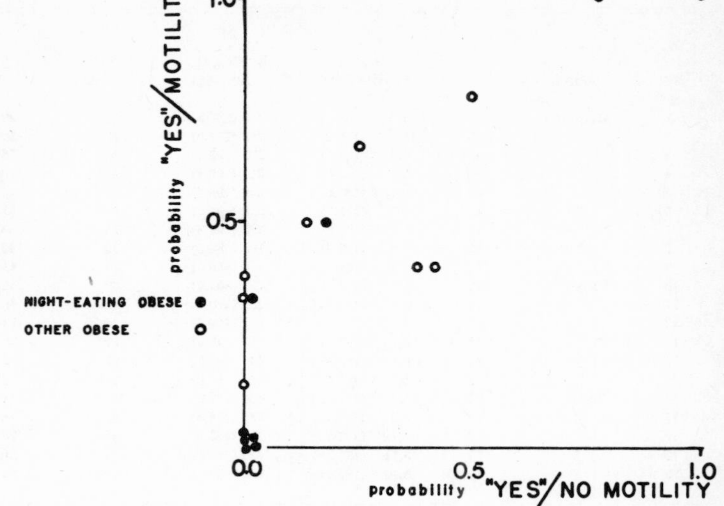

Fig 6.—Hunger responses of obese women. Those suffering from the "night-eating syndrome" show a more marked bias than the others.

by an intragastric balloon were first described in 1912 by Cannon, who called them "gastric hunger contractions," and ascribed them a major role in the regulation of appetite and hunger. Early workers viewed them as the determinants of not only the experience of hunger, but also of the regulation of food intake, through their putative influence in converting changes in the internal milieu into an experience which motivated food-seeking behavior. They have been allotted a more modest role since the demonstration that gastrically denervated animals can regulate food intake without difficulty,[8] and that the removal of the entire stomach in man does not abolish hunger sensations or the ability to regulate food intake.[9] Indeed, it has been doubted that intragastric pressure changes represent anything more than artifacts of the presence of a foreign body in the stomach. Investigators using the electrogastrograph [4] and a small pressure-sensitive intragastric telemetering capsule [13] have both reported that large, abrupt changes in gastric activity, or pressure, were associated almost entirely with the presence of an inflated balloon in the stomach. A later unpublished attempt to confirm these electrogastrographic studies was unsuccessful.[10] Studies from this laboratory in which intragastric pressure was measured by a small (1.8 mm in diameter) polyethylene tube indicate that pressure changes similar to those found with a balloon present also occur in its absence, although the balloon may somewhat increase the frequency and alter the character of the pressure changes.

Another technical problem is suggested by the apparently greater sensitivity of women than men among the nonobese subjects. (The sensitivity of the obese groups is masked by the extent of their bias.) This apparently greater sensitivity of the nonobese women may be no more than an artifact which was removed when more effective controls against cuing of the subjects were introduced at the beginning of the study of men. These influences of our measurement techniques suggest caution in generalizing the findings of this study.

The most important finding of this study was the difference in the association of gastric motility and reports of hunger between obese and nonobese persons. What is the nature of this finding and what are its implications?

Our data suggest that this difference represents a bias of many obese persons that may interfere with the normal associations between hunger and gastric motility which were made by the nonobese controls. Presumably this is a learned response, and one which may serve as a form of internal and external communication. That obese persons might readily learn such methods of communicating about so significant and so conflict-ridden a subject as hunger seems entirely reasonable. The difference in direction of the bias is of interest in this regard, because it parallels the reports of food intake of obese persons. It is a clinical axiom that obese women frequently underestimate their food intake. In a recent study of obese men,[11] on the other hand, it was reported that they rarely underestimated their food intake, and indeed, often boasted about it in a jovial, stereotyped manner.

If both reports of hunger and estimates of food intake serve as forms of communication, this difference in bias may result from the difference in the opportunities available to obese men and women to conceal their overeating. It sems reasonable for an obese housewife, who can overeat in the solitude of her kitchen, to underestimate food intake. For an obese man, who must often do his overeating in public, such deception may not be feasible, and so he may take refuge in a jovial, stereotyped assertion of his appetite.

The idea that the major determinant of the responses of obese people is a learned bias and not a lessened sensitivity is one that can be tested. If it is true, it should be possible to teach such persons to overcome their bias and so to utilize their now-wasted sensitivity. The accompanying report describes an example of just such an occurrence: an obese man was taught to overcome a strong naysaying bias and to alter markedly his perceptions of gastric motility.[5]

How do these findings bear upon our understanding of disorders in the regulation of food intake? This regulation is achieved by adjusting current intake to both prior food intake and changing levels of energy expenditure. In most people these adjustments seem to occur largely unconsciously and with remarkable sensitivity to the appropriate cues. If, however, a person has difficulty appreciating these cues, and the present study suggests that some obese people may not appreciate sensations related to hunger, then automatic, unconscious, regulation of food intake may not be possible. Instead of its being a casual gratification of a biological desire, eating may of necessity become a matter of conscious and often desperate choice at meal after meal. Such is the testimony of many obese persons—it has been years since they were able to trust their senses as to how much to eat. Every meal requires careful and often conflict-ridden decisions based on previous caloric intake, previous caloric output and current caloric alternatives. The regulation so laboriously accomplished by such means compares unfavorably with that achieved automatically by the wisdom of the body.

Summary

The presence of gastric motility was correlated with reports of "hunger" in groups of obese and nonobese men and women. Nonobese women frequently reported "hunger" in the presence of gastric motility and "no hunger" in its absence, while nonobese men showed a more random association of these variables. Obese subjects, on the other hand, showed strikingly different patterns. Many obese women reported "hunger" infrequently, even in the presence of gastric motility, whereas many obese men reported "hunger" frequently, even in the absence of gastric motility. Analysis of these findings by a signal detection model suggests that they arose from a high degree of bias on the part of many obese subjects. For these subjects, the need to report "hunger" or "no hunger" was a more important determinant of their response than was their sensitivity to gastric motility. Such faulty perception of visceral cues may play a part in the impaired regulations of food intake of some obese persons.

Albert Stunkard, MD, University of Pennsylvania, Department of Psychiatry, Philadelphia, Pa 19104.

REFERENCES

1. Cannon, W. B., and Washburn, A. L.: An Explanation of Hunger, Amer J Physiol 29:441-454, 1912.

2. Carlson, A. J.: The Control of Hunger in Health and Disease, Chicago: University of Chicago Press, 1916.

3. Code, C. F.; Hightower, W. C.; and Morlock, C. G.: Motility of the Alimentary Canal in Man: Review of Recent Studies, Amer J Med 13:328-351, 1952.

4. Davis, R. C.; Garafolo, L.; and Kveim, K.: Conditions Associated With Gastro-Intestinal Activity, J Comp Physiol Psychol 52:466-475, 1959.

5. Griggs, R. C., and Stunkard, A. J.: The Interpretation of Gastric Motility: II. Sensitivity and Bias in the Perception of Gastric Motility, in this issue.

6. Grossman, M. I.; Cummins, G. M.; and Ivy, A. C.: The Effect of Insulin on Food Intake After Vagotomy and Sympathectomy, Amer J Physiol 149:100-102, 1947.

7. Jackson, D. N., and Messick, S.: "Response Styles and the Assessment of Psychopathology," in Measurement in Personality and Cognition, edited by Samuel Messick and John Ross, New York: Wiley, 1961.

8. Luce, R. D.: "Detection and Recognition," in Handbook of Mathematical Psychology, edited by R. D. Luce, R. R. Bush, and E. Galanter, New York: Wiley, 1963, vol 1.

9. MacDonald, R. M.; Inglefinger, F. J.; and Belding, H. W.: Late Effects of Total Gastrectomy in Man, New Eng J Med 237:887-896, 1947.

10. Machella, T. E.: Personal communication to the author.

11. Mendelson, M.; Weinberg, N.; and Stunkard, A. J.: Obesity in Men: A Clinical Study of 25 Cases, Ann Intern Med 54:660-671, 1961.

12. Metropolitan Life Insurance Company: Statist Bull Metrop Life Insur Co 24:6, 1943.

13. Penick, S. B.; Smith, G. P.; and Hinkle, L. E.: Gastric Motility in Hungry Men Recorded by a Transensor Capsule, Clin Res 10:192, 1962.

14. Stunkard, A. J.; Grace, W. J.; and Wolff, H. G.: The Night-Eating Syndrome: A Pattern of Food Intake Among Certain Obese Patients, Amer J Med 19:78-85, 1955.

15. Stunkard, A. J.: Obesity and the Denial of Hunger, Psychosom Med 21:281-289, 1959.

16. Swets, J. A.; Tanner, W. P., Jr.; and Birdsall, T. G.: Decision Processes in Perception, Psychol Rev 68:301-340, 1961.

17. Taylor, J. A.: A Personality Scale of Manifest Anxiety, J Abnorm Soc Psychol 48:285-290, 1953.

34

The Interpretation of Gastric Motility: Sensitivity and Bias in the Perception of Gastric Motility

Robert C. Griggs and Albert Stunkard

This paper describes a study of factors that determine the perception of gastric motility. Our interest in this phenomenon was aroused by the results of an earlier investigation of the correlation of gastric motility and reports of hunger,[7] which confirmed the Cannon-Carlson thesis that persons of normal weight tend to report "hunger" in the presence of gastric motility, as measured by an intragastric balloon, and "no hunger" in its absence.[3,4] This study revealed, further, that many obese persons do not make this association: their reports of hunger tend to be determined not by the presence or absence of gastric motility, but by a response bias. Many obese men reported hunger most of the time, whereas many obese women reported no hunger most of the time.

These findings raise the question whether some obese persons are unable to perceive gastric motility. Such an incapacity, which could result from a defective neural apparatus, from faulty learning, or from both, might account for their impaired regulation of food intake. The method of data analysis described by signal detection theory suggests a means to distinguish between the structural and learning alternatives, as well as to explore the significance of the deviant responses previously observed among obese subjects. This method, which was described briefly in the foregoing paper,[7] and extensively elsewhere,[5,9] makes it possible to disentangle the two independent factors that enter into any psychophysical judgment—physiological sensitivity to the stimulus and psychological bias in the criterion used in making the judgment. It makes clear that the extreme bias manifested by many obese subjects precludes an accurate assessment of their sensitivity. Such an assessment could be made, however, if the bias of the obese subjects could be reduced by reinforcement of their responses in a direction opposite to that of their usual tendency. Furthermore, appropriate patterns of reinforcement in normal subjects should generate the kinds of biased responses that have been observed in obese subjects. Finally, adding additional reinforcement might increase the sensitivity of both obese and nonobese subjects, if such sensitivity can be learned.

The studies reported below were carried out in an effort to influence the perception of gastric motility of two male graduate students in their middle twenties, one obese and one nonobese.

Method

The obese subject had a lifelong history of difficulty in controlling his body weight, and he suffered periodically from a mild "night-eating syndrome."[6] He was selected both because of these factors and because of a pronounced bias in his reports of hunger, which had been discovered in previous studies. This bias was of a "nay-saying" type which resulted in a "denial of hunger," a pattern shown previously to be more characteristic of obese women than of obese men,[7] and particularly common in those manifesting the night-eating syndrome. The nonobese subject was of normal weight and he had no history of difficulty in the regulation of food intake or body weight.

Each experiment lasted four hours, beginning at 8 AM following an overnight fast. Subjects lay on a cot and did not eat, drink, or smoke during the procedure.

Gastric motility was inferred from pressure changes recorded from an open-tipped catheter located in the gastric antrum. The catheter, a 40% barium-polyvinyl plastic tube with an outside diameter of 1.8 mm, was positioned in the stomach so that its tip was located 4 or 5 cm below the point where reversal of the respiratory pressure indicated the diaphragm. Pressure changes were recorded on a polygraph * via a pressure transducer † attached to the catheter. Patency of the catheter lumen was maintained by a constant infusion of water at a rate of 0.19 cc/minute delivered by an infusion pump.‡ Respiratory and extraneous movements were measured by a pneumograph attached around the upper abdomen and connected by an air column to a strain gauge whose output was recorded on the polygraph. Any extragastric pressure changes resulting from these sources could thus be observed and discounted whenever they occurred.

Subjects were asked to report two types of events, the presence of hunger and the occurrence of gastric contractions. As in the previous study,[7] they were asked whether or not they were hungry every 15 minutes for a total of 17 times. The experimenter, who observed the pressure record, intermittently signaled the subject to report whether or not he felt a contraction. The experimenter scheduled his requests for reports in a random manner so that half of them occurred when the subject was in fact showing a pressure elevation and half when he was not. About 60 such inquiries were made in each experimental period.

Three types of experiments were carried out. In the first, the spontaneous ability of the subject to perceive individual contractions was determined. In the second, an attempt was made to increase this ability by informing him after each response about its accuracy and by paying him money for correct answers and fining him for incorrect ones according to the symmetrical payoff matrix shown in Table 1. In the third type of experiments, efforts were made to alter the pattern of response by using asymmetric patterns of payment, as shown in the lower two payoff matrices in Table 1. All experiments of each condition were completed before experiments of the next condition were begun.

A few comments on these payoff matrices are in order. The top one of Table 1 is symmetrical and it was designed to increase the subject's sensitivity by making correct responses valuable and incorrect ones expensive. The two lower ones illustrate prototypes of the matrices used to try to reproduce the extreme biases such as those exhibited by some obese persons in their reports of hunger. The lower right asymmetrical matrix is designed to encourage a nay-saying bias, that is, to increase the frequency with which the subject reports "no contraction." Such an asymmetrical matrix rewards a correct "no" in the absence of a contraction more highly than a correct "yes" in the presence of a contraction, and it penalizes errors in a similar manner, so that a "no" in the presence of a contraction is less costly than a "yes" in its absence. The lower left asymmetrical matrix is designed to induce a "yea-saying" bias by an analogous method.

Data on the detection of contractions are presented in a form frequently used in signal detection studies: The estimated probability (ie, relative frequency) that the subject said "yes" in the presence of a contraction (a "hit") is plotted along the ordinate, and the estimated probability that he said "yes" in the absence of a contraction (a "false alarm") along the abscissa. The rationale for this method is described briefly in the foregoing paper,[7] and extensively elsewhere.[5,9]

Two differences in experimental design between this and the previous study deserve emphasis. First, an open-tipped catheter was substituted for the intragastric balloon to record pressure changes. This change was desirable because the Levin tube to which the balloon was attached produced so much discomfort that repeated testing of the same person was essentially impossible. The open-tipped catheter is easily tolerated, and it aroused no complaints of discomfort. The balloon method also has the disadvantage that gastric contractions around the balloon tend to pull it further into the stomach, resulting in a tug on the tube in the nasopharynx that could thereby serve as an extragastric signal for the occurrence of contractions. Such signals are not transmitted by the open-tipped catheter. A third

TABLE 1.—*Pay-Off Matrices* *

	Symmetrical	
	"Yes"	"No"
Contraction	+10¢	−10¢
No contraction	−10¢	+10¢

	Yea-Saying			Nay-Saying	
	"Yes"	"No"		"Yes"	"No"
Contr.	+15¢	−15¢	Contr.	+5¢	−5¢
No contr.	−5¢	+5¢	No contr.	−15¢	+15¢

* The patterns of reinforcement utilized in the present study are illustrated in the payoff matrices above. The symmetrical matrix at the top of the table rewards and punishes the subject so as to increase the accuracy of his responses. The asymmetrical matrices at the bottom of the table, in addition, differentially reward responses with a "yea-saying" and a "nay-saying" bias. Similar matrices with an asymmetry as great as 2-18 were also used in the present study.

* Offner Electronics Co. Dynograph, Type R.

† Statham P23BB strain gauge pressure transducer.

‡ Harvard Electronics Co. Model 660-910.

advantage, for this study, of the open-tipped catheter is its smaller time lag in the recording of gastric motility. Finally, this method measures over-all changes in the intragastric pressure rather than changes in localized segments of the stomach. A study comparing the two methods revealed that they both recorded pressure changes of approximately the same type, indicating that results obtained by the two methods can be legitimately compared.[8]

The second difference between the two investigations is in the character of the responses studied. The first study focused only upon reports of hunger or no hunger, whereas the present one also assessed the ability of subjects to recognize individual gastric contractions. This latter measure has at least two advantages: a greater number of responses are available for study (approximately 60 per experiment versus 17), and the task is more clearly defined. The subjects said that it was far easier to report the presence or absence of a gastric sensation than to make a judgment of so ambiguous a state as "hunger."

Results

Gastric Motility and Reports of Hunger. In the first condition, comprising experiments 1 and 2, we assessed the ability of the subject to associate reports of hunger with gastric motility. As in the previous study, every 15 minutes each subject was asked if he felt hungry; he was also asked to detect gastric contractions in the absence of information feedback. The hunger responses of the two subjects under this condition exhibited the patterns previously shown to be characteristic of obese and nonobese persons, with the former showing biased and the latter unbiased responses. The bias of the obese man, however, was of a nay-saying type, resulting in the denial of hunger, seen more commonly among obese women than among obese men. In other words, he tended to report no hunger during gastric motility as well as during gastric quiescence. Table 2 shows that the obese man associated hunger with gastric motility only 10% of the time, whereas the nonobese man associated hunger with motility 72% of the time. Both subjects reported hunger about 10% of the time during quiescence. It appears therefore that feelings of hunger in this obese man are uncorrelated with the contractions, whereas they are strongly correlated in the nonobese subject.

Assessment of the subjects' ability to associate reports of hunger with gastric motility continued throughout the experiments involving feedback of information about contractions. No change from base-line conditions was observed.

Gastric Motility and Recognition of Contractions: The Obese Subject.—The obese subject was tested under seven conditions for a total of 15 experiments. The results are shown in Fig 1, where the number beside each point indicates the experimental condition. The surrounding ellipse represents chance variation within one standard deviation, based upon the assumption of only binominal variability.

Point No. 1 is the average of the results of the first two experiments, which were conducted to determine the subject's spontaneous level, prior to information feedback. The initial nay-saying bias is apparent.

Point No. 2 is the average of the three experiments with information feedback and a symmetrical payoff matrix. Interestingly, the major change is a marked decrease in the initial nay-saying bias. An increase in sensitivity may also have occurred.

Point No. 3 is the average of the results of two experiments with the asymmetrical 5-15-cent payoff matrix designed to restore the subject's original nay-saying bias (Table 1, lower right). This maneuver almost exactly reproduces the subject's original pattern of responses.

Point No. 4 is the average of the results of two experiments with the 5-15-cent asymmetrical payoff matrix designed to establish

TABLE 2.—*Comparison of Reports of Hunger Obese and Nonobese* *

	"Hunger" During Motility	"Hunger" During No Motility
	%	
Obese	10	9
Nonobese	72	13

* Comparison of the reports of hunger by the obese and the nonobese subject during the no-feedback conditions of the first two experiments. "Denial of hunger" by the obese subject is clearly shown.

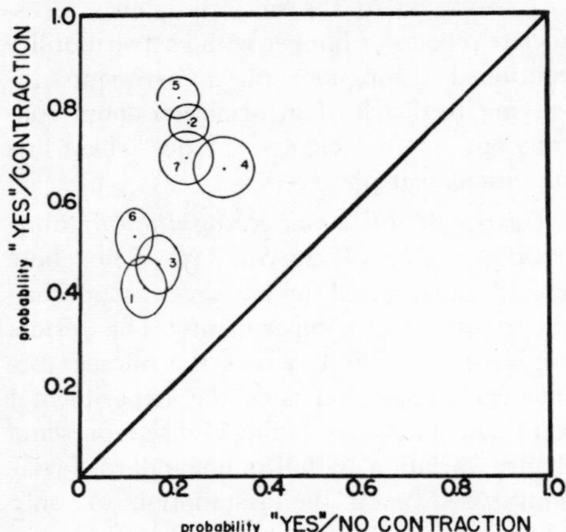

Fig 1.—Effect of information feedback with varying payoff matrices on perception of gastric motility by the obese subject. The ordinate is the probability of saying "yes" in the presence of a contraction (a "hit"); the abscissa is the probability of saying "yes" in the absence of a contraction (a "false alarm"). The conditions which generated the seven points are described in the text.

a yea-saying bias. Surprisingly, a yea-saying bias was not produced, and the subject's responses returned to the relatively unbiased behavior produced by a symmetrical pattern of reinforcement.

Point No. 5 is the average of the results of two experiments using an even more asymmetrical payoff matrix in an effort to produce a yea-saying bias. The 5-15-cent matrix was replaced by a 2-18-cent matrix which, it was hoped, would heighten the motivation to respond "yes." The results were unexpected: no increase in yea-saying was achieved, although there may have been an increase in sensitivity.

Point No. 6 is the average of the results of two experiments which used an extreme (2-18-cent) payoff matrix designed to produce a nay-saying bias. The effect of this matrix did not differ from that of the 5-15-cent matrix used in condition 3. It is interesting that both of these asymmetrical matrices resulted in a response very similar to the subject's spontaneous level.

Point No. 7 is the average of the results of two experiments without feedback that replicated condition 1, one year after the completion of the experiments described above. The subject did not show his former nay-saying bias, and it appears that the learning achieved through the use of information feedback with payoffs has persisted for a period of a year.

Gastric Motility and Recognition of Contractions: The Nonobese Subject.—The results of the 12 experiments with the nonobese subject are shown in Fig 2. In contrast to the obese subject, he was more sensitive and his responses were less easily influenced.

Point No. 1 shows the average of the results of the first four experiments where no information feedback was used. His sensitivity appears greater and his bias less than that of the obese subject, even after the latter had had the benefit of information feedback.

Point No. 2 is the average of the results of three experiments in which information was fed back and the payoffs were symmetrical. No significant change from the no-feedback condition resulted.

Point No. 3 is the average of the results of three experiments with a nay-saying payoff matrix. A slight nay-saying bias may have developed, but this subject appeared less responsive to the asymmetry in payoffs than was the obese subject.

Point No. 4 is the average of the results of two experiments with a yea-saying payoff matrix. No such bias resulted, and this point does not differ significantly from No. 3, obtained with a nay-saying matrix.

Detection of Contractions as a Function of Their Amplitude.—The ability of the subjects to detect gastric contractions as a function of their amplitude was examined. Because the total number of contractions was small, they were divided into only two classes —those with an amplitude of 10-20 cm of water, which we call "weak," and those with an amplitude greater than 20 cm of water, which we call "strong." The subjects were far more successful in detecting strong contractions than they were in detecting weak ones. Fig 3 shows that the probability of detecting strong contractions was greater than that of weak ones for each subject under

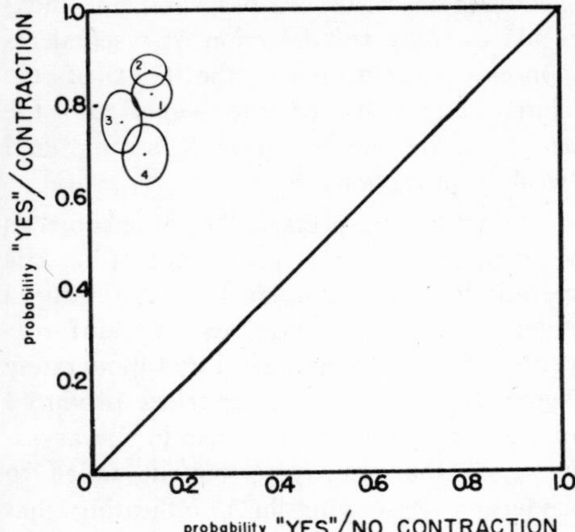

Fig 2.—Effect of information feedback with varying payoff matrices on perception of gastric motility by the nonobese subject. The coordinates are as described in Fig 1. The conditions which generated the four points are described in the text.

each condition. The samples from which these probabilities were determined ranged from 14 to 41 contractions with a mean of 23 for the weak contractions, and from 27 to 61 with a mean of 40 for the strong contractions.

The Obese Subject's Interpretation of His Experience.—What is being biased in these experiments—the subject's perceptions or just his responses? If the former, he would actually perceive a change in the amount of gastric motility under different conditions of reinforcement; if the latter, he would employ a conscious strategy designed to maximize rewards. A spontaneous remark by the obese subject favors the perceptual hypothesis. He had taken some pride in his achievement on the experiments with symmetrical payoffs, and he seemed puzzled by his failure to detect contractions during the first experiment with a nay-saying payoff matrix. Toward the end of the experiment, which actually contained a higher than average amount of gastric motility, he said in a perplexed manner, "It's amazing. I absolutely try to give the correct answer but just can't. Everything seems 'no' to me." If this remark itself was not part of a conscious strategy, it suggests that the obese man experienced a real change in his perception.

Long-Term Effects of the Procedure.—Both subjects reported that the experimental

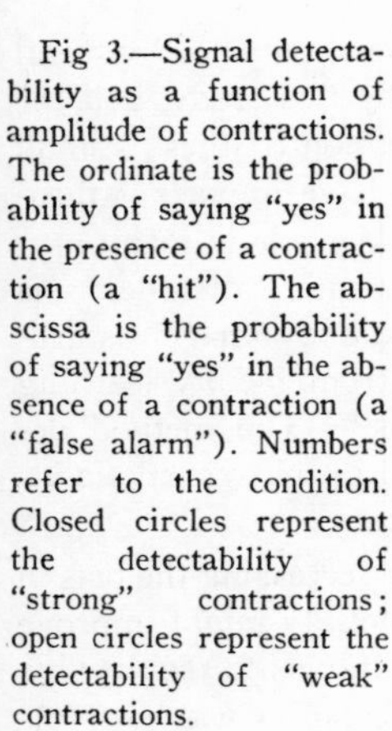

Fig 3.—Signal detectability as a function of amplitude of contractions. The ordinate is the probability of saying "yes" in the presence of a contraction (a "hit"). The abscissa is the probability of saying "yes" in the absence of a contraction (a "false alarm"). Numbers refer to the condition. Closed circles represent the detectability of "strong" contractions; open circles represent the detectability of "weak" contractions.

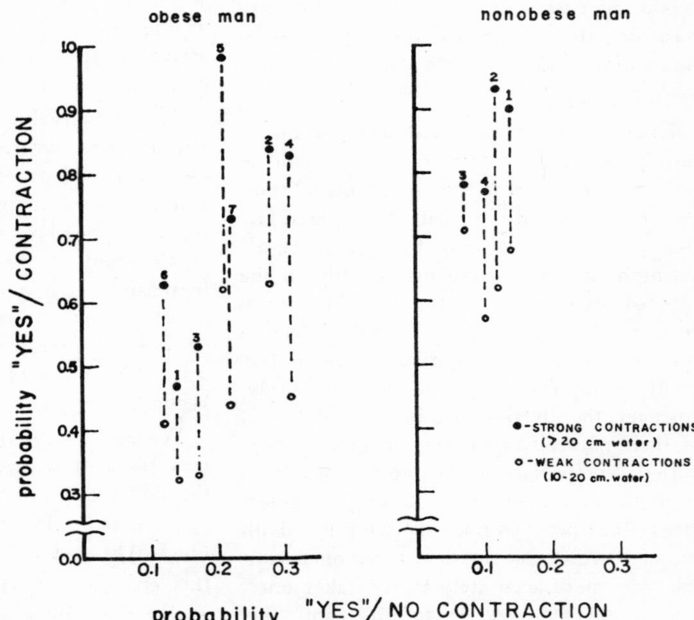

procedure produced a heightened awareness of visceral sensations which was still present a year later. This awareness was coupled with what they believed to be a far greater capacity for discrimination of gastric sensations. A year later the obese subject reported that this awareness was present in undiminished intensity; the nonobese subject reported that it had decreased. It is of interest that the obese subject felt that these changes had had no effect upon his experience of hunger, his regulation of food intake, or his night-eating syndrome. "Contractions don't have much to do with when or how much I eat." As before his participation in these studies, his food intake was only rarely prompted by feelings of hunger, and he found it as difficult as ever to control overeating. Despite continuing efforts at weight control, he gained three more pounds during the year following the studies.

Comment

This study provides a clear-cut answer to a major question raised by the previous one. One obese person, at least, was able to learn to overcome a strong nay-saying bias and to demonstrate a high degree of sensitivity in reports of gastric motility. He thereby demonstrated that his apparently faulty perception had not been wholly due to an imperfect neural apparatus.

Experiments in the signal detection literature have shown that, just as bias can be altered by changes in the conditions of reinforcement, sensitivity can be altered by changes in the strength of the signal. Signal strength can be manipulated readily in the case of exteroceptive stimuli, which have been used exclusively in signal detection studies to date. Interoceptive stimuli are less easily manipulated. In the present study, however, the division of stimuli on the basis of their naturally occurring magnitude permitted a comparison of the effect of the signal strength on the subject's capacity to detect them. It is perhaps not surprising that both subjects were able to detect stronger contractions more accurately than weaker ones. The regularity and magnitude of this phenomenon, however, bespeaks an unexpected capacity for detection of visceral responses, particularly in the light of the relative ambiguity of the signal and the crudity of the dichotomy into "strong" and "weak" contractions.

At least three aspects of the obese subject's performance deserve comment: (1) His operant level was similar to that produced under nay-saying conditions of reinforcement; (2) both symmetrical and moderately asymmetrical yea-saying matrices produced unbiased results; (3) increase in the asymmetry of the yea-saying matrix failed to produce a yea-saying bias, indicating that asymmetry of the payoff matrices does not necessarily produce a perceptible asymmetry in response. The possibility of making a false alarm, even under conditions favoring yea-saying, is apparently not as acceptable as the possibility of failing to identify a contraction under conditions favoring nay-saying. It is as if the motivation for saying "yes" were different from that for saying "no," so that differential payments of relatively small amounts of money proved less motivating than these long-standing biases. The subject's resistance to the influence of a yea-saying matrix provides evidence, over and above his spontaneous level, of the strength of his habitual nay-saying bias.

The remarkable sensitivity to visceral stimuli demonstrated by this study, and the ease with which perception of these stimuli could be manipulated, have important theoretical implications. A key role in many psychotic and psychosomatic disorders is increasingly ascribed to body image disturbances arising in part from faculty learning of visceral perceptions.[1,2] The method described here should aid in the systematic investigation of such events.

We had thought that decreasing the bias in perception of gastric motility might improve the ability of the obese subject to regulate his food intake. This expectation was not fulfilled. Although his bias was decreased and this change was part of his conscious experience, no apparent effect upon his ability to control his overeating was observed. There are at least two implications of this negative finding. It would appear to provide still further confirmation of the relative unimportance of the stomach in the regulation of food intake. At the same time, the dissociation between detection of contractions and reports of hunger may imply that only when visceral perceptions are integrated into a more complex motivational state of "hunger" do favorable changes in perception favorably alter the regulation of food intake.

The responses of the nonobese man are also worthy of comment. He was, even without training, more sensitive to his gastric motility than the obese man. Furthermore, it was not possible to influence materially his responses in the experimental situation. Two questions arise. Was his greater sensitivity related to his more effective regulation of food intake? Did this greater sensitivity make him less susceptible to influence under the conditions of the experiment? A survey of persons varying both in their sensitivity to gastric motility and in their ability to regulate food intake should answer these questions and provide useful information about the part played by the perception of gastric motility in the regulation of food intake.

Summary

A signal detection method of analysis was used to assess the ability of an obese and a nonobese male subject to detect gastric motility as measured by an open-tipped catheter. Subjects were selected to exemplify previously demonstrated response patterns of obese and nonobese persons in that the obese man tended to deny "hunger" both during gastric motility and quiescence, whereas the nonobese man usually associated "hunger" with motility.

Both subjects could perceive contractions, and their sensitivity was found to increase as a function of higher amplitude of contraction. The nonobese man was more sensitive than the obese man, and it was not possible to influence materially his responses in the experimental situation. In contrast, the obese man, who initially showed a strong "nay-saying" bias, manifested by the frequent failure to report contractions when they were present, could be influenced to overcome the bias by use of information feedback with appropriate patterns of reinforcement.

By overcoming a strong "nay-saying" bias and demonstrating a high degree of sensitivity in reports of gastric motility the obese subject showed that his apparently faulty perception was not due wholly to an imperfect neural apparatus. These findings suggest that bias in reports of gastric motility can be learned and they imply that the disordered hunger feelings of some obese persons have resulted from past experiences which fostered the development of bias.

Albert Stunkard, MD, University of Pennsylvania, Department of Psychiatry, Philadelphia, Pa 19104.

REFERENCES

1. Bruch, H.: Falsification of Bodily Needs and Body Concept in Schizophrenia, Arch Gen Psychiat (Chicago) 6:18-24, 1962.

2. Bruch, H.: Perceptual and Conceptual Disturbances in Anorexia Nervosa, Psychosom Med 24:187-194, 1962.

3. Cannon, W. B., and Washburn, A. L.: An Explanation of Hunger, Amer J Physiol 29:441-454, 1912.

4. Carlson, A. J.: The Control of Hunger in Health and Disease, Chicago: University of Chicago Press, 1916.

5. Luce, R. D.: "Detection and Recognition," in Handbook of Mathematical Psychology, edited by R. D. Luce, R. R. Bush, and E. Galanter, New York: Wiley & Sons, Inc., 1963, vol 1.

6. Stunkard, A. J.; Grace, W. J.; and Wolff, H. G.: The Night-Eating Syndrome: A Pattern of Food Intake among Certain Obese Patients, Amer J Med 19:78-85, 1955.

7. Stunkard, A. J., and Koch, C.: The Interpretation of Gastric Motility: I. Apparent Bias in the Reports of Hunger by Obese Persons, this issue, p 74.

8. Stunkard, A. J., and Reinhardt, A.: The Effect of an Intragastric Balloon on Gastric Motility in Man, in preparation.

9. Swets, J. A.; Tanner, W. P., Jr., and Birdsall, T. G.: Decision Processes in Perception, Psychol Rev 68:301-340, 1961.

VI

ELECTROENCEPHALOGRAPHIC ACTIVITY

35

Conditioned Discrimination of the EEG Alpha Rhythm in Humans

Joe Kamiya

ABSTRACT

Combining methods of behavioral analysis and physiological recording in a study of "introspection," a conditioned discrimination training procedure was used to determine whether there were private stimulus concomitants of the EEG alpha burst. With monopolar recording of the EEG (ear to left occiput, scalp electrodes), and with the subject lying relaxed with eyes closed, E sounded a single ding of a bell aperiodically about five times a minute, randomly scheduling the dings to coincide half the time with the occurrence of an EEG alpha burst, and half the time with the absence of alpha. The subject was instructed that with each ding in the training trials he was to try to guess either "yes" or "no," depending on "how he felt" at the time of the bell. The "yes" responses to dings with alpha and the "no" responses to dings without the alpha were reinforced by being called "correct" by the experimenter. Results:

In about 50 to 500 trials, six subjects learned to a criterion of 80% correct responses. As a control over the possibility that the bell ding sounded differently on the two types of trials, in a second procedure the subject were asked to make guesses out loud as to when their "yes" or "no" states occurred, without the aid of the bell. The subjects could do that also. One subject could not offer any verbal description of why he performed so accurately. Five felt that it was related to what they were thinking about or to their efforts to imagine visual scenes. In some of the subjects the discrimination appears possible only if they try deliberately to alternate their states of mind between visual imagination and relaxed inattention in a regular pace so that they can more easily discern their state as required by the experimenter. In a third type of procedure, it is found that the alpha burst can be produced by the subject, upon command from the experimenter, who asks S to produce the states accompanying their verbal responses in the discrimination procedure.

Preliminary analyses of cardiac and respiratory activity, eye movement and muscle tension indicate that these are not stimulus sources accompanying changes in the EEG.

36

Operant Control of the EEG Alpha Rhythm and Some of its Reported Effects on Consciousness

Joe Kamiya

The history of psychology seems to be a history, or a prehistory, of the mind-body problem. Modern psychology began by dealing with one aspect of this problem, epistomology, the question of how we come to know the outside world. The early psychologists, working in the 1870s, administered sensory stimuli to *S*s and were concerned with seeing what kinds of responses the *S*s gave under specific verbal instructions, and then trying to develop what they considered a psycho-physical relationship. But this wasn't satisfactory because there was too much disagreement among investigators concerning the theoretical structures. There were debates, for instance, as to whether thoughts could be thoughts without containing images. The whole picture of experimentation in learning and often in physiological psychology also was concerned with the question of what it is that makes the organism, human or the animal, interact with the world in the way it does. The concept of mind was with us long before any more sophisticated, philosophical, scientific, logical outlook had been developed, but as the psychologists, especially after the behaviorist revolution, began to think about this problem, they began to use concepts like mind not in the older, introspectively oriented sorts of ways but rather as hypothetical constructs. The hypothetical constructs, such as images, dreams, hopes, fears, etc., stood or fell according to how *useful* they were in ordering the data and making valid predictions.

The problem for the scientist of private experience—i.e., that you can sit there and all of a sudden become aware of the way the seat fits your pants—is the kind of problem to which I like to think psychology ought to be addressing itself. But it's a very difficult problem, and many of the attempts psychology has made in the past to become more "scientific" seem to me to have been detours which have left the undergraduate psychology student, after he has finished his first year of courses and has gotten properly brainwashed with the behavioristic routine, unsatisfied because he still *knew* that he had dreams, that he had these funny sensations inside, that he could let his mind wander, etc., etc. One of the chief difficulties is that it is very very difficult when an individual says, "I had a dream last night," to know whether in fact he is telling the truth. This emerges as a problem because we have no independent way of indexing the occurrence of a dream, or the occurrence of an image, or the occurrence of a hope, etc.

By and large, psychology is still dealing with the uttered word as the primary source of datum to make these inferences of a hypothetical sort about what went on behind those words, how those words emerged.

As psychology has been struggling with the problem of the usefulness of introspective reports, other disciplines, those of a more biological orientation, have been dealing with the relationship between behavior and brain function. This work has had important implications for psychology. It is now possible to index the activity of a variety of events within the brain: we can put electrodes on the scalp and observe patterns of electrical activity that reflect brain activity. It is the possibility of correlating the electrical activity of the brain with *S*s' reports of their experience and their behavior which has intrigued me as a psychologist, for I feel it is a more solid approach than building hypothetical mental mechanisms on the basis of verbal reports alone.

Suppose we assumed that a particular brain process that we could identify was associated with some particular subjective state. Then we may ask: "Can people be trained to discern the comings or goings of brain rhythms, say the EEG alpha rhythm, just by using the standard learning procedures that have been developed for use with rats and pigeons?" What we did in our laboratory in Chicago several years ago was the following: We attached EEG electrodes on the scalp (occipital-to-ear lobe tracing) in order to detect the EEG alpha rhythm. Our experimental question was, if we set up the proper discrimination procedure, could we get individuals to say "A" when the alpha rhythm was present and to say "B" when the rhythm was absent? The *S* was told that from time to time he would hear a bell ring once; when he heard it, he was to make a guess as to whether at that time he was in brain wave state A or brain wave state B. And as soon as he made his response we told him if he was correct. You will recognize this as a pretty straightforward discrimination learning problem. The stimuli to be discriminated were not in the outside world, as is the conventional case with most experimental psychology, but were instead located inside the *S*, and were presumably correlated with the comings and goings of the alpha rhythm.

After a *S* had been given his task, he began to call out his guesses. Usually in the first hour or so there was essentially chance performance; in the second hour the *S* very often began to get about 60% correct. In the third hour, *S*s typically got about 75 or 80% correct. Some *S*s became 100% correct in naming their brain wave state. Thus *S*s had learned to make correct discriminative responses to something that was going on inside their brains.

At a simple-minded level of thinking about correctness, one can say that the *S* here had learned to read his own brain, or his mind, if you will. In so far as that is true, one can argue that he has learned to make some kind of an introspective response. But that's perhaps primarily a pedantic lession in academic psychology. What may interest you more is what the *S*s reported about how they carried out this task. Any particular *S*, even one who got nearly 100% correct, was not necessarily able to articulate in English just *how* he was able to do this. He may have felt that he should say "A" sometimes and "B" at other times, but not be able to give us any reasons, but this was the extreme case. Many *S*s, however, did offer various kinds of verbal explanations on the basis for their discriminative responses. One thing that needs to be mentioned before discussing these reports, however, is this: any *S*s who were sort of mid-way in the learning process, that is when they were getting more than 50% correct, but not yet 100% correct, were quite unable to tell us how they were able to do it. They would offer various verbalizations, but then immediately qualify them by saying they weren't all sure, and so on. I find this rather interesting; it suggests that we have succeeded in teaching individuals to make internal discriminations or perceptions about themselves whose dimensions are so unfamiliar that they are unable to give a clear-cut verbal description.

Now let me report an important finding that we came upon quite by accident. We said to the *S*, "Look, obviously you've learned how to make this discrimination quite well. Let's see if you can produce those states that you have been calling 'A' and 'B' upon our command." The *S* was told that if he heard the bell ring twice, he was to enter into that state that he'd been calling "A"; and as soon as he heard the bell ring once, he was to change into the other state that he'd been calling "B." We found that having successfully gone through the discrimination training, most *S*s seemed to have at the same time acquired the skills necessary for the control of the brain wave states that they had been discriminating. I think there may be other explanations possible to account for this finding, but they are not this striking, and certainly you do not see results like this in an untrained subject.

When I moved from Chicago to San Francisco, I asked myself the natural question, "Can people be trained to control their brain waves without this prior discrimination training?" The answer seems to be yes. The procedure I used is quite simple. I set up an electronic device which would turn on a sine-wave tone in the *S*'s room whenever the alpha rhythm was present. The tone would disappear as soon as the alpha rhythm would disappear. The *S* was told: "Hear that tone? That's turned on by your brain wave." And after waiting for a few minutes for the *S* to settle down, I said: "Now let's see if you can learn how to control the percent of time that the tone is present. First we'll have you try and keep the tone *on* as much as you possibly can, and then we'll have you try to keep the tone *off* as much of the time as you can."

I did various kinds of experiments along these lines; sometimes I just trained people to suppress the alpha rhythm without any training for enhancing it. I found generally speaking, that most *S*s had the ability to learn how to produce more alpha or to suppress it, although the extent to which they were able to suppress it is a little bit in question. Figure 1 illustrates the results of suppression training only. On the ordinate I have scaled the percentage of time that alpha was absent, i.e., suppressed. Consider only the section of the graph under Phase A for now. As you can see, for six of the seven individual curves of performance, there is a general upward trend. The lowest curve depicts the one exception, an *S* who seems to be going the opposite direction. We ran this individual several more trials with an added factor, discussed below, and he soon joined the group of individuals back up at the other end of the graph.

By this time I had found, especially for suppressing the alpha rhythm, a fair degree of agreement by the different *S*s on their reports as to *how* it was done, especially as the *S*s became more and more proficient in the task. I would often hear that visual imagery was the answer, that all they would have to do was to conjure up an image of a person's face, hold it, and look at it very *carefully*, so much so that they could actually see the features of the person's face. If the *S* was able to fixate something like this, or if he fixated the spots that float around in the visual field, the visual phosphenes, or if he engaged in other kinds of visual imagery activity, it seemed to be effective in turning alpha off.

Because of this agreement among the successful alpha-suppressors as to the value of visual imagery in suppressing the alpha rhythm, I decided to

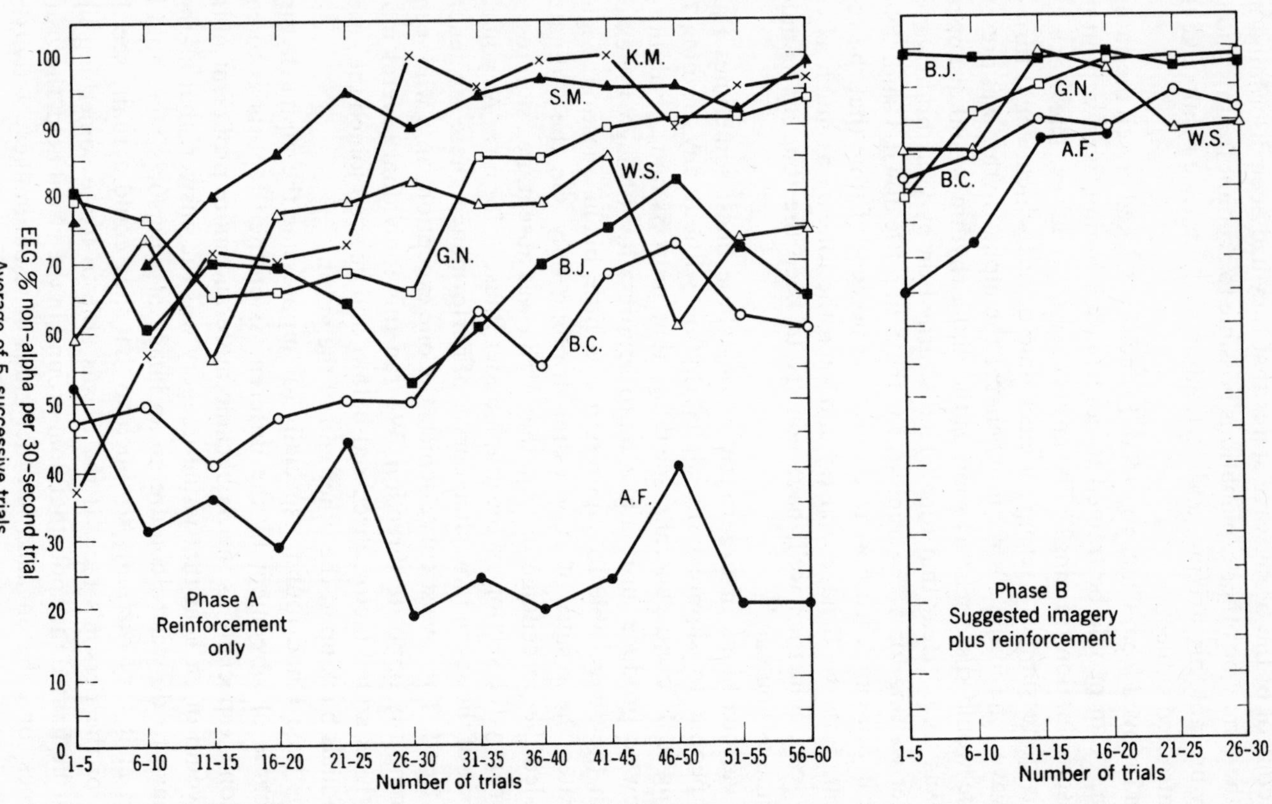

EEG % non-alpha per 30-second trial
Average of 5 successive trials
100
90
80
70
60
50
40
30
20
10
0
1-5
6-10
11-15
16-20
21-25
26-30
31-35
36-40
41-45
46-50
51-55
56-60
Number of trials
K.M.
S.M.
W.S.
G.N.
B.J.
B.C.
A.F.
Phase A
Reinforcement
only
Phase B
Suggested imagery
plus reinforcement

Figure 1

see what would happen with our less successful *S*s if they were instructed to try suppressing the alpha rhythm by using intense visual imagery. Most of the *S*s found it very helpful. In Figure 1, the portion labeled Phase B, you can see that the least successful *S*, who was only reaching about 20% non-alpha time, went up to 80%, and some of my other poor *S*s reached 100% non-alpha time on many trials. The *S*s typically reported, "Yes, that did help very much"; they *kind of* thought so, but weren't entirely sure that it did, and were trying other ways of alpha suppression they had hypothesized would be successful even after the instructions to use visualization. This manner of checking hypotheses seems to be quite essential if the *S* is to learn.

This area is full of all kinds of superstitious learning. Partial success brought on by some practically correct attempts inside the *S*'s head often are so rewarding to *S*s that they persist in these and plateau at a certain level. In order to get themselves up to a higher level of performance they have to be willing to shake loose from that partially correct hypothesis and find one that has more elements of correctness in it.

I want now to discuss the results of an experiment which attempts to control for one important question that is raised by this suppression-only experimental design. That question is this: it might be possible that simply as a result of the number of trials or the number of sessions of a repeated task that an *S* would get less and less alpha, i.e., that it is simply a shift in the *S*'s baseline EEG and has nothing to do with the tone or the reinforcement at all. I thought the best way to control for that would be to use one and the same *S* and train him, on opposite blocks of five trials each, to turn alpha on or to turn it off. So I gave five trials, each trial 30 seconds long in which the *S* was to keep alpha on, and after those five trials were over he was then instructed to rest for a while while "we tuned up our machine." Unbeknownst to the *S*, I recorded the number of seconds of alpha he would have gotten had the tone been switched into his room, even though it was not. This was an attempt to get some kind of a baseline of percent alpha time. Then, I asked him to suppress alpha, that is to keep the tone off. What we have then are alternating blocks of three kinds of trials: enhance alpha, rest, suppress alpha.

Figure 2 indicates the sort of results obtained from this kind of experiment, which I have replicated several times. The ordinate is scaled in percentage time alpha was present, rather than the time it was absent, as Figure 1 was. The three curves represent the average performances of a group of seven *S*s under the three conditions, enhance, rest, suppress. Considering the Suppress Alpha condition first (labeled "low alpha" in Figure 2), there is clearly a downward trend, as would be predicted, although it is not as pronounced as the curve in Figure 1. The increase

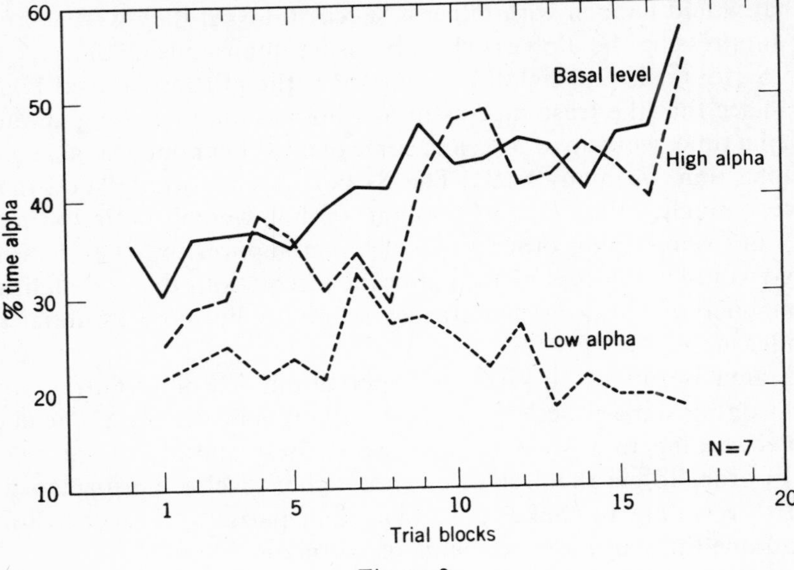

Figure 2

in the curve for the Enhance Alpha condition is much more striking, and it is clear that there is a very significant aifference in percent alpha time between the enhance and suppress conditions for these *S*s. Thus it's not merely a function of alpha decreasing with the passage of time; there does seem to be some measure of volitional control. Now, consider the third curve (labeled "basal level") which represents the percent alpha score *S*s got in the rest periods when they weren't getting the tone: as a matter of fact they were specifically told *not* to try producing tones, they were told that the tones would be switched off and just merely to wait until we got our machine aligned. This curve shows a generally upward trend, yet if this were really *a baseline* sort of thing one would have expected it to have stayed level. My interpretation of this upward trend is that the *S*s do not remain the same after having learned these two kinds of tasks; the experimental tasks apparently set them into certain *preferred* modes of waiting, and the preferred mode is the higher alpha state. There will be additional evidence for this later, but let me cite one important piece right now. If the *S*s are asked which of the two tasks—enhance or suppress—they prefer, they almost invariably say they greatly prefer learning how to turn alpha *on*. A good many of them will say suppressing alpha is easy, all you have to do is conjure up an image, but it's not particularly fun or pleasant, and it's much more interesting to keep oneself in the high alpha state.

Let me report a few verbalizations about the high alpha state, besides its general pleasantness, that have been reported by *S*s. First, they all seem to say that this seems to have to do with some kind of *relaxation* of the mental apparatus, not necessarily relaxation in the motor system, but a kind of a general calming-down of the mind. Second, it is a state in which it is good not to be thinking too much about the outside world, or how the experiment is going, or even how you're doing, but one in which you just sort of listen for the tone and just let it carry you along. Some of them have said in this context that it's one of those things where you stop being *critical* about anything, including the experiment. I will elaborate the description of the high alpha state in a moment, but it suffices to say now that I feel these reported characteristics of the high alpha state support my hypothesis that during rest periods the *S*s tend to continue to produce high alpha, and thus the "baseline" goes up.

Another indirect bit of evidence that supports this interpretation of the increase of percent alpha time in the baseline condition is that with *S*s who are given only suppression training, their percent alpha time in baseline conditions tends to decrease. That is, they seem to continue doing what they have been trained to do. But when they have been trained for both enhancement and suppression, they prefer enhancement, and baselines go up.

Now let me describe a little bit more fully what *S*s say about this high alpha state, and what kinds of people seem to be especially good at producing it. Tentatively, I have found that there are certain kinds of individuals that do much better at learning how to control alpha, especially learning how to increase alpha; and these individuals so far seem to be those who have had some interest in and practice of what I shall loosely call "meditation." It doesn't necessarily have to be of the Zen or Yoga school, or any other formal school of meditation. If the individual has had a long history of introspection on his own, he seems to be especially good at enhancing the alpha rhythm. Also he is likely to be an individual who uses words like *images, dreams, wants*, and *feelings*. It have come to the conclusion that there are a large number of people who really don't know exactly what you are talking about when you talk about images and feelings. To such people the words describe something that *somebody else* might have; but these people do not seem to have any degree of sensitivity to such things themselves. These people do not do well in my experiments, they do not gain a high degree of control over their alpha rhythm.

Let me mention some other characteristics of people who are good at learning alpha control, although these are impressions that I don't consider well proven yet. Psychotherapists, of the type who lay great store by such techniques as sensitivity training and the other sorts of growth techniques used at such places as Esalen Institute, seem to be good at learning alpha control. People who look you in the eye and feel at ease in close interpersonal relationships, who are good at intuitively sensing the way you feel, are also good at this. Also I notice I generally tend to have more positive liking for the individual who subsequently turns out to learn alpha control more readily. I find this especially true of females, for some reason!

I mentioned that people who have practiced meditation are good *S*s. I have been carrying on further experiments with these meditators, as I have come to call them, and have been pushing them to see how high they could go in percent time of alpha, and whether their testimony changes over a period of many sessions. I have run one or two of these *S*s on sort of "marathon" runs, at one time attempting to do it on a 20-hour basis. I've been finding that these *S*s do seem to continue to improve their skills by doing this over and over for as much as 20 and 30 hours total time. One *S* plateaued after about 50 hours and didn't improve much beyond that; but this individual started at about 15% alpha and got himself up to 85% alpha. That's a rather substantial change; certainly if you look at the EEG record before and after, it would look like the EEG records of two different people.

I think it's rather intriguing that those who practice meditation, are skillful at this task and it makes me listen with more interest to reports about the effects of meditation. For example, the ineffability of the meditative state, so often stressed in mystical writings is similar to statements many of my *S*s make, e.g., "In this experiment, you keep asking me to describe this darned alpha state. I can't do it; it has a certain feel about it, sure, but really, it's best left undescribed; when I try to analyse what it is I don't do well."

The high alpha state is a desirable thing to my *S*s. It used to be that to motivate *S*s to participate in these experiments I payed them a certain flat rate per hour and threw in a bonus on some occasions. But now it's gotten to be the case that they're almost ready to pay *me* to serve as *S*s, especially if I say I will let them turn on alpha for an extended period of time! I no longer pay *S*s, and I have a list a mile long from various people who call me on the telephone or who write me from New York, and other places all over the country to ask if they can come over and serve as subjects! I think that perhaps this is somewhat indicative, although certainly not conclusively so, that the state itself is a nice one to be in. I have tried it myself for only a couple of hours, so I can't give you my own introspective reports on this. Before one can be sure that this is something that people really will strive for, one needs to see that this state works as a motivator in the conventional sorts of ways that are available for testing such a concept.

Finally let me briefly describe the latest experiment in this area. The experiments all have to do with changing the *amplitude* and *frequency of occurrence* of the alpha rhythm, the extent to which there were bursts of alpha rhythm of sufficient amplitude to trigger our tone device. But, is it possible to train *S*s to vary the *frequency* of the alpha rhythm? My engineer devised a cycle period analyser and thus had a system whereby each EEG alpha wave cycle was compared against a standard duration: if the wave was *longer* in duration than the standard, the *S* heard a *high*-pitched click, while if the wave was *shorter* than the standard, he heard a *low*-pitched click. I chose a standard duration for each *S* in such a fashion that approximately half of his clicks would be high clicks and approximately half would be low clicks. I now told the *S* to try to increase the predominance of one kind of click over the other. Again we used the *S* as his own control, alternating blocks of trials in which he was to raise alpha frequency with blocks in which he was to decrease alpha frequency.

Figure 3 presents the mean performance of ten *S*s on alpha *frequency* control. I have plotted the mean percent difference between the conditions when they were trying to produce high clicks in predominance on the one hand, and the condition in which they were trying to produce low

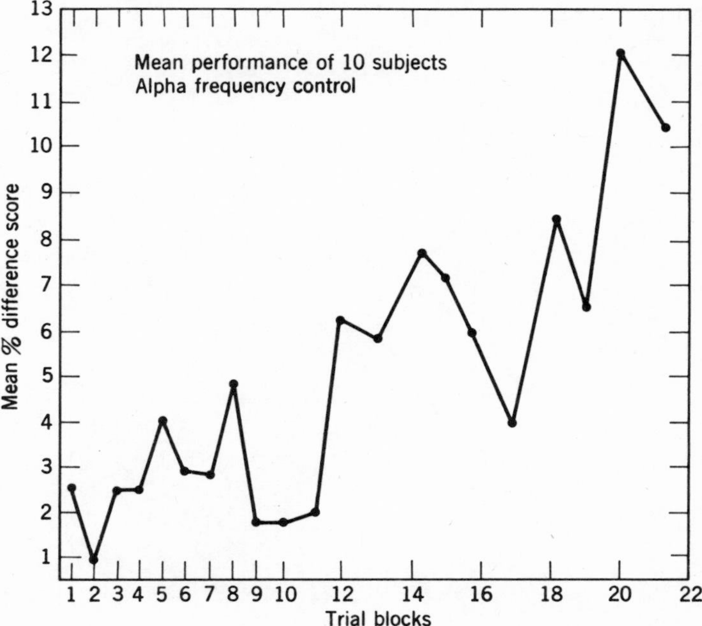

Figure 3

clicks on the other. As you can see, the difference between those two experimental conditions grows as a function of trial blocks and it's a rather striking change. One can actually see the difference in the raw EEG record in some cases, although I have yet to run the magnetic tape data through the computer to see exactly how much the frequency shifts. I can say that for some *S*s the shift is as large as two cycles per second.

I do not know, and nobody really knows for sure what the significance of a change alpha frequency is. However, we can note that when individuals are drowsy, their alpha frequency tends to slow up a little. When they are asleep and in a stage 1 EEG state, with the eyes moving, there is some slowed alpha activity mixed in with the irregular theta rhythms. Recently I ran across a report that some Japanese investigators had been looking at the EEGs of Zen masters, and found that as they were doing their meditative exercises, their alpha rhythm became more abundant and tended to be slower. It's not just drowsiness that will bring this slowing about; as a matter of fact, any marked degree of drowsiness is almost always accompanied by a drop in the control that my *S*s have over their alpha rhythm.

I shall refrain from speculating on where this work is going to lead. Clearly we have a rather strong effect, but many questions remain to be answered about it. My laboratory will be busy for many years to come working in this area.

37

Discrimination of Two Sleep Stages by Human Subjects

Judith S. Antrobus and John S. Antrobus

ABSTRACT

Three adult females were awakened from EEG stage 1 REM and stage 2 of sleep and instructed to guess the stage of sleep from which they had been aroused. One subject obtained 27 correct discriminations (guesses) out of 29 awakenings after 84 training trials (awakenings). A second subject gave no indication of learning the discrimination in 65 trials; a third discriminated the two sleep stages significantly better than chance in 138 trials. It was suggested that the first subject successfully formed a *stage 1 REM/2* concept that matched the occurrence of the sleep stages better than did the concept *dreaming*. Several problems in concept formation of internal states during sleep were discussed.

DESCRIPTORS: EEG, Sleep, Dreaming, Discrimination. (John Antrobus)

Ever since developing measures of the electrical activity of the nervous system, scientists have been searching for electrophysiological patterns which coincide in time with an individual's phenomenal states, or conversely, for classes of private experience which match commonly observed electrophysiological patterns. The strength of the association between the physiological event and words used to describe or name a phenomenal event (word label), cannot be stronger than the strength of the association between the original phenomenal event and its verbal label. Useful word labels can be learned only when the public—the verbal community—can observe the event to be labelled.

It is particularly remarkable, therefore, that a strong association should be discovered between physiologically-defined states of sleep and reported *dreaming* (Aserinsky and Kleitman, 1955; Dement and Kleitman, 1957). The accuracy of matching the dreaming/non-dreaming classification with EEG-defined sleep stages 1 REM and NREM[1] has ranged between 64 and 80% in experi-

This research was supported by research grants MH 06733 to the Research Center for Mental Health, New York University and MH 10956 to J. L. Singer and J. S. Antrobus from the National Institute of Mental Health, United States Public Health Service.

The authors express thanks to Drs. A. A. Arkin, J. E. Barmack, D. R. Goodenough, R. R. Holt, J. Kamiya and J. L. Singer for their helpful comments of earlier drafts of the paper.

Judith S. Antrobus now at The City College of The City University of New York.

Address requests for reprints to: John S. Antrobus, City College, City Univ. of New York, New York, N. Y. 10031.

[1] REM is the abbreviation for *rapid eye movements* which occur exclusively during emergent stage 1 sleep as defined by EEG criteria. Emergent stage 1 is commonly called stage 1 REM. Emergent stage 1 is distinguished from non-emergent stage 1 in that the former is

ments where subjects were awakened from the two sleep states and asked to describe their mentation prior to being aroused (Murray, 1965). Phi co-efficient indices of the association range as high as 0.76, with modal values between 0.5 and 0.6; no other single verbal classification is more strongly associated with REM/NREM sleep.

Given the rather haphazard way in which one learns to label the experience *dreaming*, it is reasonable to assume that a subject might learn to make a more accurate discrimination of the private experiences associated with REM versus NREM sleep, if he were trained by a more systematic discrimination procedure.

The experimental design is similar to that of a concept formation task where the subject is presented with a sequence of objects, randomly sampled from two classes of objects. He is reinforced for correctly guessing the class to which the object belongs (e.g., cubic or spherical). Instead, however, of being presented with public stimuli, the subjects in the present study are directed by the experimenter to respond to their own self-generated private states at different points in time. In the present experiment subjects were repeatedly awakened from two different EEG-defined sleep stages, asked to guess which sleep stage they had been in, and then informed whether or not they were correct.

The experimental design was suggested by Kamiya's (1962) work on the discrimination of EEG-alpha. The experimenter rang a bell about five times a min. Every time the bell rang the subject was to say "Yes" or "No" depending upon "how he felt" at the time of the bell. If he said "Yes" when the bell was accompanied by EEG-alpha or "No" when alpha was absent, the experimenter replied, "Correct." Kamiya was able to train six waking subjects to discriminate the presence or absence of their own EEG alpha with 100% accuracy in 50 to 500 trials. Because of high cost of sleep research and the limited number of awakenings possible each night, the present investigators had to be content with a more modest number of trials.

Method

Subjects

Three unmarried nurses, ages 23 to 25, working on night shift were selected as subjects so that they could sleep in the laboratory during the day. All were classified as high recallers of dreams—recall from at least three nights/wk (Antrobus, Dement and Fisher, 1964) on the basis of a questionnaire and a daily sleep log filled out by the subject for two wks prior to her being selected for the experiment. Subjects were told the purpose of the experiment: to see if anyone could tell the difference between two kinds of sleep. None of the subjects had read any of the sleep articles in popular magazines and they were asked to refrain from doing so. They were told not to discuss their ideas about the study with the experimenter in order to avoid getting any inadvertent cues from the experimenter about the two sleep conditions. The sleep log items relative to dreaming were imbedded among other items about a variety of aspects

preceded by a period of stage 2, 3 or 4 sleep. Non-emergent stage 1 sleep generally occurs shortly after falling asleep. NREM sleep refers to any of the other stages, non-emergent stage 1, stage 2, 3 or 4.

about sleep. It is assumed that the sleep log had little or no effect on the subjects' performance in the experiment. It did, however, insure that they had some ability to attend to and recall their sleeping experience, thereby giving the experiment a reasonable chance of success.

Procedure

Standard procedures were used to measure parietal EEG and horizontal REMs during the night. Recordings were made on 4 channels of an Offner type R polygraph at 10 mm/sec.

Subjects were awakened during stage 1 REM and stage 2 according to the following criteria:

A subject was awakened only if she had been in a given sleep stage for at least 6 min. The 6-min interval allowed the experimenter to be absolutely certain of his scoring and insured that the sleep stage was not a transient one. If a body movement or muscle artifact occurred, the awakening was made at least 2 min after the last observed disturbance. In addition to the usual criteria for identifying stage 2, it was required that the immediate 1-2 min interval prior to an awakening include EEG spindles (12-14 cps sinusoidal envelopes). All stage 1 REM awakenings were made in the presence of an ongoing sequence of rapid eye movements. In summary, all awakenings were made during unambiguous, clear-cut intervals of stage 1 REM and stage 2 awakenings was quasi-random.

The subject was told that she would be awakened from 2 different types of sleep: "type A" (1 REM) and "type B" (stage 2). She was told that she would be awakened several times during the night from both types A and B sleep. The subject was awakened either by a buzzer or by hearing her name called over an intercom. She was asked "What type of sleep were you in?" If her guess was correct the experimenter replied, "Yes, that is correct. You were in ——— sleep. Now go back to sleep." If her guess was wrong, the experimenter replied, "No, that was incorrect. You are in ——— sleep. Now go back to sleep." At the end of each sleep session the subject wrote her impressions of the two types of sleep. In order to prevent the experimenter from giving the subject any cues which might assist her in learning the discrimination, the experimenter did not read or discuss the subject's written impressions in the presence of the subject. Every possible effort was made to keep the subject's cooperation and motivation at a high key. In order to limit the fatigue from multiple awakenings, subjects were never scheduled more than three sessions consecutively.

Results

Subject Z. J. was awakened 56 times during stage 1 REM and 57 times during stage 2 over 16 "nights" for an average of seven awakenings/night. Her discrimination improved to the point where during the final four nights, she made 27 correct discriminations out of 29 awakenings (see Fig. 1), a level of accuracy that far exceeds ($p < 0.00001$) that expected by random guessing. To our knowledge, such a sharp discrimination has never been made by any subject whose REM and NREM experiences were compared through the tra-

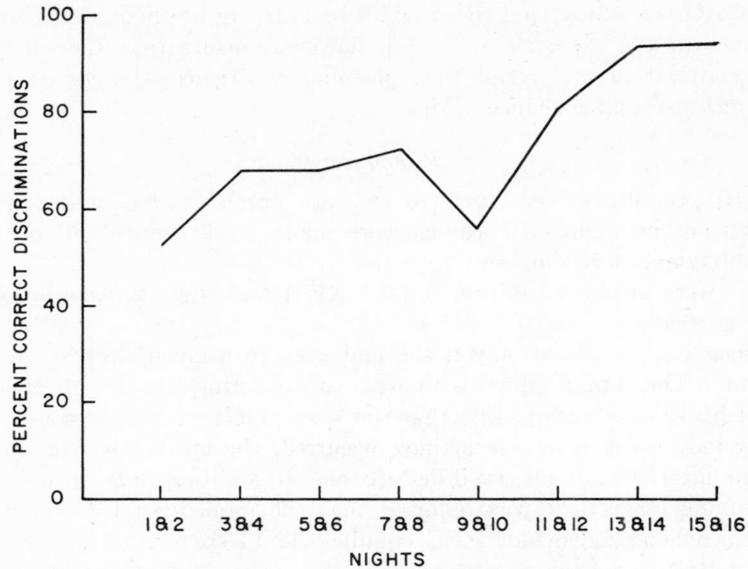

FIG. 1. Percentage of correct discriminations between stage 1 REM and stage 2 sleep averaged over successive pairs of nights.

ditional method of matching the subject's verbal descriptions of his introspections with the sleep stages from which they were obtained. For *Z. J.* the 1 REM and stage 2 experiences became almost perfectly distinguishable. Her achievement demonstrates, therefore, that by means of a discrimination training technique, a person can be trained to distinguish the private stimuli from which his 1 REM and stage 2 experiences develop.

An examination of the two errors made by *Z. J.* indicates that her discrimination was perfect when she was permitted a sufficiently long interval within a sleep stage before being awakened. Her two errors were made when she was awakened from only seven min of stage 2 sleep, which had been preceded by over 30 min of 1 REM sleep. These were the most extreme in the entire study. It is not clear whether the errors can be attributed to the short stage 2 interval or the long preceding 1 REM period. It seems plausible that the private stimuli upon which the subject bases her discrimination, lag in time behind the more discrete EEG changes from one sleep stage to another.

Z. J.'s retrospective reports suggest that her discrimination did not rest on one single variable. "I am almost positive that if you have a good sound dream, no matter how agitating it is, it is type A sleep (stage 1 REM). . . The sounder . . . deeper . . . more relaxed . . . sleep is type A and the restless sleep is type B (stage 2; see Rechtschaffen, Verdone and Wheaton, 1963). With respect to differences in mentation during the two stages she volunteered, ". . . you are having type A sleep when you have a vivid dream . . . very clear . . . like a motion picture [with] a lot of action." By contrast, "I am not sure if you can call the . . . vague . . . dreams you have in type B sleep 'dreams,' since you are on the verge of waking, they might just be things you are drowsily thinking about." *Z. J.* said that the difference between A and B was sometimes quite difficult to detect, particularly if she had a "restless disturbing dream, while sleeping soundly" or if she was very tired. Excessive fatigue seemed to contaminate the usefulness of "depth of sleep" as a discriminative cue. In addition to the visual element, which is a well established discriminator between 1 REM and NREM sleep, *Z. J.*'s reports emphasize the non-cognitive differences. She reports that excitement and agitation, which define the highest pitch of wakefulness coincide, in time, with the "deepest," and most "relaxed" part of her sleep. It is small wonder that the association of these two seemingly incompatible variables during stage 1 REM sleep should handicap the subject's discrimination of 1 REM and stage 2 sleep.

The relation of rapid eye movements to the stage 1 REM experience is far from clear in the experimental literature. Rapid eye movements have been associated with visual scanning of the dreaming imagery (Roffwarg, Dement, Muzio and Fisher, 1962), the suppression of imagery (Antrobus, Antrobus and Singer, 1964; Antrobus, Dement and Fisher, 1964), attempts to recapture the imagery (Kubie, 1962), and general activity of the dream, not necessarily visual (Oswald, 1962). If moving imagery occurs only during rapid eye movements and if moving imagery is essential to the subject's identification of stage 1 REM sleep, then her discrimination should break down if her stage 1 REM awakenings are confined to intervals of ocular *quiescence* within stage 1 REM, again, compared to stage 2. During four additional sessions, *Z. J.* was awakened six times in stage 2 and fifteen times following intervals of at least 30 sec of ocular *quiescence* within stage 1 REM. *Z. J.* maintained the accuracy of her discrimination by identifying 14 out of the 15 stage 1 REM awakenings and 5 out of 6 stage 2 awakenings. So that she would not have an opportunity to learn a new basis for discriminating 1 REM from stage 2 sleep, *Z. J.* was not informed whether her judgments were correct. In her retrospective reports, *Z. J.* reported no change in her criteria. It appears that the private experiences which define stage 1 REM do not coincide exclusively with the rapid eye movements. That is, if moving visual imagery is confined exclusively to the intervals which include rapid eye movements, then moving visual imagery is not the basis of the subject's discrimination. The discrimination method might be usefully extended to determine if subjects can discriminate rapid eye movement and ocular quiescent intervals within stage 1 REM sleep.

Subject C. G. slept in the laboratory for 11 "nights" during which time she was awakened 32 times from 1 REM sleep and 33 times from stage 2 for an average of six awakenings/night. A change in her nursing schedule obliged her to leave the experiment prematurely. At no time, however, did she show any evidence of learning the discrimination between 1 REM and stage 2 sleep; she guessed the sleep stage correctly on only 34 out of 65 trials. There are several plausible explanations for her lack of success. Her sleep was more disturbed than *Z. J.*'s, as evidenced in the fewer occasions that permitted awakenings under the established criteria: 5.9/night compared with 7.1 for *Z. J.* As do many night workers, *C. G.* was accustomed to dividing her sleep into two intervals, often separated by a meal. Although her sleep improved through the experiment, her basic sleep cycle was clearly abnormal. Indirect evidence from several studies indicates that the correspondence of EEG behavior and mentation may be drastically altered under abnormal conditions. For example, sleep deprived subjects may show EEG delta activity, typical of stage 4 sleep when, in fact, they are carrying on a conversation and walking about a room. As already mentioned, *Z. J.* had found discrimination difficult when she was sleep deprived. Secondly, subject *C. G.* was an extreme intellectualizer and seemed relatively insensitive to the class of subjective cues that *Z. J.* found useful in making her discrimination. In response to her inability to sleep and to the frustration of the discrimination task, she increasingly adopted a rigid intellectual style: "So to correct this error in organization and approach to [the] problem, I here forthwith commence my new method of analysis . . . scientific method . . . I identify the problem . . ." In order to handle her frustration, *C. G.* became progressively intropunitive: ". . . annoyed with myself when I'm wrong . . . exasperated . . . Am I really contributing something toward this project or am I just taking up time and space?" Against such opposition, the experimenter found it impossible to arouse the high level of motivation and rapport that had existed with subject *Z. J.*

Subject N. U. slept for 22 sessions with an average of 6.4 awakenings/night, 60 in 1 REM sleep and 78 in stage 2. On the very first sessions she proposed that A sleep was "not just dreaming, but actually being part of the dream, and B was not dreaming." By the final session she was certain that, "A is a deeper sleep, one in which I am probably dreaming. B is a lighter sleep. I have often felt as if I was awake when I was called and the answer was B." Her discrimination over all sessions was significantly better than chance (χ^2 = 8.2; $p < 0.01$) but nowhere near perfect. Two factors appear to have handicapped her performance. The quality of her sleep varied greatly from session to session so that a given sleep stage was most likely experienced quite differently on consecutive sessions. On approximately one-half of the sessions she reported extreme fatigue; in one instance she had not slept for 27 hrs. She developed a respiratory infection during the last half of the experiment and, unknown to the experimenters, took a moderate dose of *Chlortrimetron* (sessions 15 through 19) which made her "very drowsy." On the first session after giving up the antihistamine she made eight correct discriminations out of eight awakenings. On the 19th session her sleep deteriorated with frequent interruptions marked by coughing and her discrimination fell off. *N. U.* became discouraged and the experiment, therefore, was terminated. The number of trials for both *C. G.* and *N. U.* was relatively small even for a concept formation experiment carried out in the waking state (Kamiya, 1962). It is entirely possible that if the trials had been continued under more favorable conditions, the subjects may have been as successful as *Z. J.*

Discussion

The experiment indicates that it is possible for a person to learn to discriminate stage 1 REM from stage 2 with perfect accuracy. In so learning the discrimination, the subject has formed a new concept which we must call by the public operations that define it: a *stage* 1 *REM*/2 concept. The introspective reports of *Z. J.* indicate that the concept of *dreaming* played an important role in attaining the *stage* 1 *REM*/2 concept. That an experimental procedure, so radically different from the verbal report technique of the original investigators should independently emphasize the distinguishing characteristics of dreaming and vivid visual imagery, substantially confirms the initial claims of Aserinsky and Kleitman (1955) and Dement and Kleitman (1957).

Dreaming, however, does not tell the whole story. Not only is dreaming often reported outside stage 1 REM, but Monroe, Rechtschaffen, Foulkes and Jensen (1965) have demonstrated that judges can clearly distinguish stages 1 REM from NREM dreams. Moreover, both the accuracy of *Z. J.*'s performance and her introspective reports suggest that in addition to dreaming, she successfully used other cues to make her discriminations. Like the judges who sorted verbal reports into REM and NREM categories, *Z. J.* also used these various cues in combination. For example, a long visual, hallucinatory, emotional sequence might indicate 1 REM sleep but a short visual image, or a long "thinking" sequence with vague imagery, or visual images with little associated affect might indicate NREM sleep.

The *stage* 1 *REM*/2 concept, as developed in this study, should be distinguished from the criteria used by judges to sort a pair of verbal reports into REM/NREM categories (Monroe *et al.*, 1965). Judges were required, in effect, to decide which of the two reports in each pair was the most REM-like. Their judgment was *comparative* in contrast to the more difficult *categorical* judgment asked of the subjects in this study. Monroe's judges used *both* reports in the pair to make one judgment. The concept formation task which engaged the present subjects required that they develop a fixed, stable boundary criterion and judge each consecutive event as belonging to one category or the other.

The notion of a *stage* 1 *REM*/2 concept assumes a substantial measure of stability in the association of EEG and experience. Although the effects on REM/NREM reports of time of night and night-to-night changes in physiological states such as fatigue have not been studied systematically, available evidence from a number of sources suggests that the effects are sizable. Subjects in the present study performed poorly when fatigued. Pivik and Foulkes (1966) found that 1 REM reports are more dream-like when subjects are deprived of REM sleep. REM awakenings are progressively more vivid and emotional as time of night increases (Verdone, 1965). Judges sorting verbal reports made their best discriminations when reports were matched for night and time of night (Monroe *et al.*, 1965). Many drugs are known to affect the sleep EEG although their effect on the associated verbal reports has not been reported. It is possible, therefore, that a subject might learn, or begin to learn, to discriminate REM from NREM sleep but lose the discrimination under altered sleep conditions. Assessment of the effects of these conditions on the experience of the sleeper is necessary to determine the stability of any concept learned by discriminating different physiologically-defined classes of sleep. It has been observed that night workers commonly have a disturbed sleep cycle with many spontaneous awakenings and that the disturbance is greatest for those who change back and forth between night and day shifts. *Z. J.*, who had worked a night shift for over three years, slept very well in the lab and made a much better discrimination than the other subjects who had worked night duty for less than one month and slept poorly in the lab. Regular night sleepers, with more stable sleep patterns, might be more successful in discriminating sleep states than the night workers employed in the present study.

REFERENCES

Antrobus, J. S., Antrobus, Judith S. & Singer, J. S. Eye movements accompanying daydreaming, visual imagery, and thought suppression. *J. of ab. and soc. Psychol.*, 1964, *69:* 244–252.

Antrobus, Judith, Dement, W. & Fisher, C. Patterns of dreaming and dream recall. *J. of ab. and soc. Psychol.*, 1964, *69:* 341–344.

Aserinsky, E. & Kleitman, N. Two types of ocular motility during sleep. *J. of ap. Physiol.*, 1955, *8:* 1–10.

Dement, W., & Kleitman, N. The relation of eye movements during sleep to dream activity: An objective method for the study of dreaming. *J. of exp. Psychol.*, 1957, *53:* 339–346.

Kamiya, J. Behavioral and physiological concomitants of dreaming. Progress Report. National Institute of Health Grants M-2116 and M-5069. Feb., 1962.

Kubie, L. S. The concept of dream deprivation: A critical analysis. *Psychosom. Med.*, 1962, *34:* 62–65.

Monroe, L. J., Rechtschaffen, A., Foulkes, D., & Jensen, Judith. Discriminability of REM and NREM reports. *J. of pers. and soc. Psychol.*, 1965, *2:* 456–460.

Murray, E. J. *Sleep, dreams and arousal.* New York: Appleton-Century-Crofts, 1965.

Oswald, I. *Sleeping and waking: Physiology and psychology.* New York: American Elsevier, 1962.

Pivik, T., & Foulkes, D. "Dream deprivation": Effects on dream content. *Sci.*, 1966, *153:* 1282–1284.

Rechtschaffen, A., Verdone, D., & Wheaton, J. Reports of mental activity during sleep. *Can. Psychiat. Assoc. J.*, 1963, *8:* 409–414.

Roffwarg, H., Dement, W., Muzio, J., & Fisher, C. Dream imagery: Relationship to rapid eye movements of sleep. *Arch. of gen. Psych.*, 1962, *7:* 235–258.

Verdone, P. Temporal reference of manifest dream content. *Percep. and mot. skills*, 1965, *20:* 1253–1268.

38 Occipital Alpha Rhythm Eye Position and Lens Accommodation

Edmond M. Dewan

People can be taught to control voluntarily their own alpha rhythms. This can be used to send messages in Morse code when an electroencephalogram pattern is used as part of a computer programme. Such procedures may help to explain the mechanisms by which the alpha rhythm is "blocked" or "unblocked".

THE "blocking" or attenuation of occipital alpha rhythms has been associated with "attention" since the discovery of electroencephalography. Many conditions favour the appearance of these rhythms such as eye closure, patternless visual fields if eyes are open and, of primary importance, a lack of ocular fixation. On the other hand, such things as pattern vision, ocular fixation, intense concentration on mental tasks or on non-visual stimuli and the perception of surprising, alerting, or affective stimuli favour the "abolition" of alpha activity. This much was shown in 1934 by Adrian and Matthews[1] and has been confirmed many times since then. On the other hand, interpretation of these facts has been quite varied.

Berger's original interpretation was that alpha "blocking" results from attention in general. When Adrian and Matthews subsequently found that alpha activity was generated in the occipital lobe rather than from the entire brain, and that it was "blocked" far more effectively by visual than by non-visual activity, they advanced the idea that non-visual "blocking" was caused by a spread of desynchronizing activity from other areas to the occipital lobe. They also explained the effects of eye opening and closure on the alpha rhythms by the presence or absence of visual attention. A similar explanation was used for alpha blocking by the deliberate accommodation and convergence of the eyes (or "looking") in complete darkness or with eyes closed. This view seems to be generally accepted today[2], although there have been additional variations. For example, Cobb[3] has proposed that pattern vision may cause the blocking response by the series of momentary stimuli resulting from repeated fixation and accommodation. These would act only momentarily, as in the case of unexpected stimuli, but would have a continuous effect because of rapid repetition. It is significant that he found that this happened when something had to be kept in focus, as Adrian and Matthews noted the subjective sensation of accommodation and convergence when they attempted to block activity in the absence of visual stimulation, that is, in darkness or with eyes closed.

More recent experiments have raised additional questions regarding the role of attention. For example, Mundy-Castle[4] noted that the blocking response was not in one to one correspondence with either attention or visualization and that, as was also noted earlier[1], considerable mental activity and attention are possible without this response. In addition some workers[5] have reported that the blocking response does not correlate significantly with measures of efficacy of perception or attentiveness. Of especial interest is the observation of Oswald[6] that in some cases during intense auditory alertness the alpha rhythms can appear, but in this situation they were accompanied by "loss of ocular fixation and accommodation". He therefore proposed that the blocking response represents an "increase of specific visual alertness which may be but one component of general arousal". Still more recently, Mulholland and Evans as well as myself[7–10] independently noted that eye position can play an important part in alpha rhythm activity, the extreme upward position tending greatly to enhance it. As will be shown here, one can entertain the hypothesis that this effect may be caused by a tendency of the eyes to defocus and relax convergence when in the maximum upward position. This would be consistent with Oswald's observation that the presence of alpha activity is accompanied by absence of fixation and accommodation as well as Adrian's remark that alpha activity is never present during fixation.

The main purpose of this article is to show that it is possible voluntarily to control one's alpha activity by the manipulation of oculomotor configuration and accommodation with accuracy sufficient to send Morse code to a computer and to have the latter type out the corresponding letters automatically on a teleprinter, and to re-examine the "attention hypothesis" in the light of this demonstration.

The procedure was as follows. A 'Grass model 7 Polygraph' was used to record occipital electroencephalogram activity with one pair of transcortical bipolar paste electrodes over the left and right occipital areas. This signal was passed through a 10 c/s band pass filter sufficiently broad to pass the alpha activity of all the subjects studied. This filtered signal was fed to a Schmitt trigger which generated a pulse each time the signal voltage exceeded a threshold so adjusted that during the presence of significant alpha activity there was a pulse for each wave crest. This output was fed to a 'LINC' computer which was programmed to ignore "drop ins" (spurious pulse trains) or "drop outs" (short interruptions of pulse trains) of the duration indicated in Table 1. It was also programmed to decide on the basis of duration whether or not a dot or dash was being sent, to display this information acoustically and/or visually, to decide on the basis of a pause duration when a character had been completed, and to translate the Morse character into its alphabetical equivalent. It then displayed the letter on a cathode ray tube in the subject's field of vision, deleted the complete letter (assumed incorrect), whenever the subject sent a sufficiently prolonged dash, or caused the teleprinter to type out the letter on completion of the succeeding one.

The subject was placed in a screened acoustically isolated room and provided with the following types of feedback: (1) he could hear the Schmitt trigger output through a loudspeaker; (2) he was informed of the registration of dots and dashes through an intercom or, in the case of subject C. H., he was given this information automatically by audio-signals triggered by the computer; (3) he could check his results by viewing the cathode ray tube through a window. (As expected, it was found that this could be done without spurious generation of alpha activity.)

The subject was instructed to control his alpha activity by the following technique. For maximum voltage he was to turn his eyes to the extreme upward position consistent with comfort and avoid fixation or convergence and accommodation. For minimum or "blocked" activity he was either to open his eyes and fixate them on a nearby object (C. H. and S. M.) or else to keep his eyes closed and fixate them on an imaginary point located almost immediately in front of his lids (E. D.).

Table 1

Subject	Sec/ correct letter	No. of errors	No. of letters	Dot-dash dividing time in 1/9 sec	Inter-character interval in 1/9 sec	Drop-out drop-in time in 1/9 sec
C.H.	49	16	26	26	56	4
	35	13	26	26	56	4
S.M.	48	9	15	34	50	6
	34·6	10	26	20	60	6
E.D.	33·6	4	26	34	50	6

The observations and results were as follows: Figs. 1*a* and 1*b* show electroencephalograph recordings of subject E. D. while sending Morse code with eyes closed. Fig. 1*c* shows the effect of the 10 c/s filter on the signal in Fig. 1*b*. Fig. 2 shows a word typed out using the deleted feature of the programme to remove errors before print out. Fig. 3 shows the print out for three subjects attempting to send the alphabet without corrections. Table 1 gives the corresponding times used, number of errors, and computer parameter settings. As can be seen, 35 sec is about the minimum average time needed for each correct letter. Of interest is the "drop in–drop out time" (on the order of 0.5 sec) which gives a measure of the duration during which alpha activity might spontaneously disappear during eye elevation or spontaneously appear during eye fixation. Because this parameter was not very carefully adjusted to optimize performance and because it was used for both the "drop in" and "drop out" durations, it can be considered as being only a rather crude measure of a stochastic component superposed on the voluntary control.

It was further observed that preliminary practice was necessary, external influences could be distracting, and that the subjects reported occasional difficulties in remembering the number of dots and dashes just sent and this accounted for some of the errors in performance. Acoustic feedback had to be low in volume and the subject had to become habituated to it in order to avoid an uncontrolled blocking response; several hours of this task proved to be very fatiguing. Finally, when I myself was a subject, I found that merely turning the closed eyes to an extreme elevation did not result in alpha activity if I focused and converged my eyes in that position. Fixation or accommodation and convergence had to be avoided to prevent the blocking response. As indicated, this result was incorporated in the technique used by the subjects in this experiment. It would be difficult to interpret these results if one were to assume that the alpha rhythm is blocked or "unblocked" exclusively by some sort of attention mechanism; because, during the experiment, the subject was in a state of constant attention. He was forced to listen attentively to the acoustic feedback and to keep track of

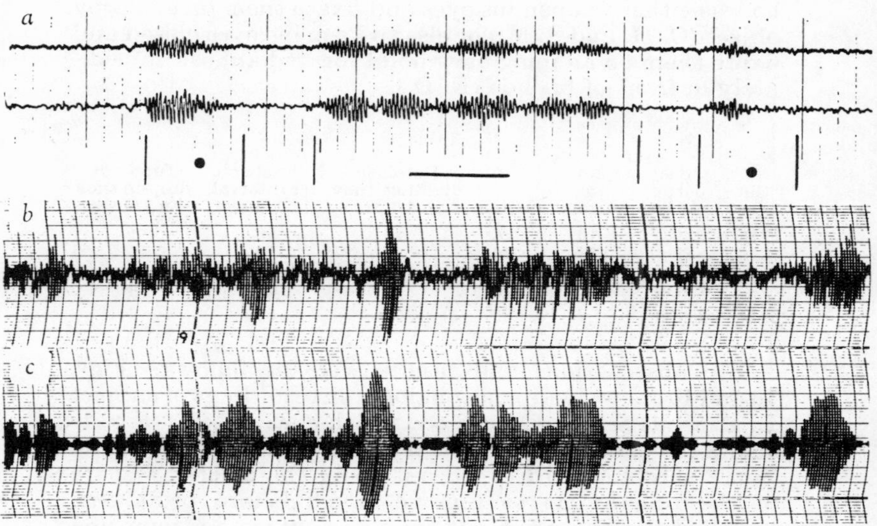

Fig. 1. Electroencephalograph records during the "transmission" of Morse code by voluntary control of alpha activity (subject E. D.); *a*, record for "sending" the letter *R* (·— ·); *b*, record for "sending" the letter *K* (—·—); *c*, record for signal in (*b*) after filtering with 10 c/s band pass filter.

X

CYBERNETICS

EI

Fig. 2. Teleprint output of computer during the use of the delete feature of the programme. In this way errors are not typed out; however, this is at the expense of time.

(*a*)

ABCDEF GAHIJKA

(*b*)

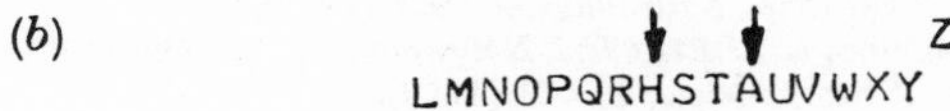

Z

LMNOPQRHSTAUVWXY

ABBCIDEFGHIEJKLMNOPEDQRSTTUXUTVWKXYZ

(*c*)

Fig. 3. Typed output from three subjects sending the alphabet; *a*, subject C. H.; *b*, subject S. M.; *c*, subject E. D. Note that the sending is far from perfect. In *a* a 5 was sent instead of an *H*. This is an example of how subjects tend to forget the number of dots being sent (*H* being four dots and 5 being five dots). Note also that the letter *Z* was inserted by hand because only "previous" letters are typed out (see text).

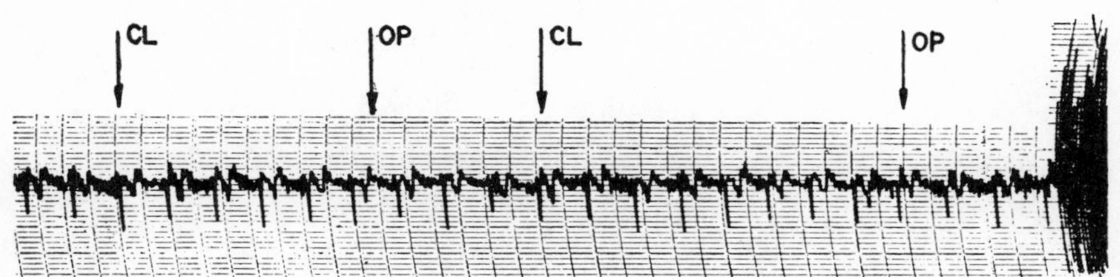

Fig. 4. Voltage from back of neck during eye opening and closure (and hence alpha rhythm "off" and "on"). The same settings were used as Fig. 1*b*. Notice the increase in signal at the end when neck muscles were deliberately contracted.

what he was "sending." Subjectively, he was not consciously manipulating his attention but, instead, his accommodation, convergence and eye elevation. Although *cyclo*mydril, a drug causing pupilary dilation and preventing accommodation, seemed to have no effect on the ability of one subject (E. D.) to send Morse code, it would be interesting to know the effects of other drugs such as curare. If curare prevented to sending of Morse code (and yet allowed alpha activity) one would be forced to consider seriously the possibility that alpha rhythms are strongly dependent on peripheral ocular activity. If the contrary were found, a more exclusively central mechanism of the type currently assumed would probably be involved and, as a bonus, one would have a possible "channel of communication" from curarized patients. In any event, it is clear that the common denominator for the blocking response of the alpha activity, if one exists, is not yet known.

I thank S. Michaels for devising the 'LINC' programme which made this experiment possible and for several valuable suggestions. I also thank B. Farley and T. Mulholland for discussions.

[1] Adrian, E. D., and Matthews, B. H. C., *Brain*, **57**, 355 (1934) (see also Durup, G., and Fessard, *Année Psychol.*, **36**, 1 (1935)).

[2] Penfield, W., and Jasper, H. H., *Epilepsy and the Functional Anatomy of the Human Brain*, 189 (Little, Brown, 1954).

[3] Hill, J. D. N., and Parr, G., *Electroencephalography*, 238 (Macmillan, 1963).

[4] Mundy-Castle, A. C., *EEG Clin. Neurophysiol.*, **9**, 643 (1957).

[5] Fraisse, P., and Voillaume, C., *EEG Clin. Neurophysiol.*, **13**, 485 (1961).

[6] Oswald, I., *EEG Clin. Neurophysiol.*, **11**, 601 (1959).

[7] Mulholland, T., and Evans, C. R., *Nature*, **207**, 36 (1965).

[8] Mulholland, T., and Evans, C. R., *Nature*, **211**, 1278 (1966).

[9] Dewan, E. M., *Air Force Cambridge Res. Labs.*, 64–910.

[10] Dewan, E. M., *Proc. Symp. on Biomed. Eng.*, **1**, 349 (1966).

39 Feedback Electroencephalography

Thomas Mulholland

* *

The brief history of electroencephalography contains many examples of experimental methods whereby presentation of a stimulus to the subject is controlled by the subject's EEG. In 1934, Adrian reported listening to the alpha rhythm recorded from himself and presented through a loud speaker. He then tried to correlate the subjective impression of looking or not with the absence or presence of the alpha rhythm (Adrian & Matthews, 1934). Other control loops permitted triggering of the stimulus at some phase of the alpha rhythm or generated stimulus frequencies which "followed" brain rhythm fre-quencies (Shipton, 1949; Walter & Walter, 1949; Corriol & Gastaut, 1950; Hewlett, 1951; Turton, 1952). The feedback loop described here which I developed with S. Runnals (Mulholland & Runnals, 1961) was based on the *suppression* of the alpha rhythm by the controlled stimulus. That feature, combined with the diversity of its applications, distinguished this stimulus-EEG loop from those reported previously.

Since 1961 interest in feedback methods and in particular feedback EEG has quickened as the unique advantages of the method have become apparent. Many studies using feedback stimuli controlled by occipital alpha have been reported (Mulholland & Runnals, 1962a, b, 1963, 1964; Kamiya, 1962, 1967, 1968; Mulholland, 1964, 1967; Runnals & Mulholland, 1964, 1965; Dewan, 1964, 1967; Mulholland & Evans, 1965, 1966; Mulholland & Davis, 1966; Bundzen, 1966; Hart, 1967; Nowlis, 1968). Feedback methods bring more than simply increased efficiency or convenience. Feedback changes the dynamics of the system being studied. This point is overlooked by those who see no new principles in feedback EEG and who wrongly assume that the effect of repeated stimulation is essentially the same whether the stimulus is controlled by feedback from the EEG or not. Figure 1 shows a typical arrangement of the apparatus. In the following discussion feedback loop refers to the external path from the EEG to the stimulus to the subject.

SPECIAL USEFULNESS OF FEEDBACK METHODS

1. Nulling a closed internal loop. In this application an external path is so arranged that the effect of an internal biological feedback control loop is cancelled or nulled. In the work of Stark (1959) and Young (1962) this utilization of an external feedback path is prominent. Young connected the measured eye position to target position, permitting a control of the target position by the eye position. By adjusting the parameters of the external

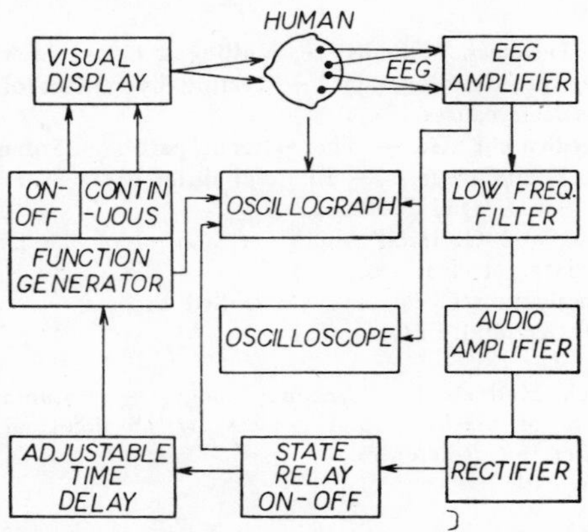

Figure 1. Flow diagram of the external path from EEG to stimulus display

path he was able to eliminate the effective visual feedback. The eye tracking control system was then "open-loop" permitting an analysis of its response to various input functions.

2. Environmental clamping of a biological control system. Here the response of the system under consideration is fed back to the input. If the gain of the feedback element in the internal loop is increased, instability oscillations may occur. The frequency of these can be measured, and discontinuous or continuous oscillations observed, from which inferences about the system may be drawn (Stark, Kupfer and Young, 1965). Also, such "clamping" decreases the variability of the response. Other functions can be introduced into the system and their effect identified and evaluated. The loop described here is an example of this "clamping" application.

3. Testing hypotheses of causality. Usually, in conformity with traditional methods, the experimenter does not permit any feedback from observed response or output (effect) on hypothesized stimulus or input (cause). *A* is varied in a controlled way and the effect on *B* is observed. However for certain situations a mutually dependent variation may show more efficiently a causal link between *A* and *B*. In this case the experimenter connects the effect *B* to hypothesized cause *A* using positive feedback. Especially in the case where the probability that *B* varies in conjunction with *A*, is less than 1, then the positive feedback test may reveal the causal link more efficiently than the "open-loop" method. If there is in fact a causal link, then the response

measured at *B* will vary according to a function given by the system transfer function. This change of magnitude may be an easily detected and unambiguous phenomenon, and for this reason may be valuable to the experimenter. If there is no causal link, then connecting *B* to *A* will not result in a shift of the magnitude of *B*. To my knowledge no explicit application of feedback method to test hypotheses about causal links has been reported.

4. Testing hypotheses of the difference between causes. If *A* and *B* are assumed to have the same effect on *C* then connecting *C* back to *A* or to *B* by an external feedback path should produce the same change in *C*. However if *A*-were, say, opposite from *B* in its effect on *C*, then connecting *C* to *A* could result in positive feedback while connecting *C* to *B*, negative feedback. The difference in the system response to positive or negative feedback may provide phenomena which were unambiguous and thereby would assist the experimenter in his determination of the differences between *A* and *B* in relation to *C*. This principle was applied in a test of Sokolov's (1963) hypothesis that increase or decrease of illumination had the same effect on EEG activation, or alpha "blocking" by connecting the EEG alpha back to two kinds of stimuli: an increase, and a decrease of illumination (Mulholland & Runnals, 1964a).

STANDARDIZATION OF FEEDBACK EEG

We have tried to control the relevant variables in order to standardize the method. In this we have been partially successful. However, we have greatly reduced the variability associated with lack of control of: the stimulus conditions; orienting responses to acoustical or uncontrolled visual stimuli; the contingency between the EEG alpha and the presentation of the stimulus. By recording eye position we can verify the presence of the large ocular excursions which are sometimes associated with increased occurrence of alpha (Mulholland & Evans, 1965, 1966; Dewan, 1967; Fenwick & Walker, 1967; Kris, 1968). The sound-reduced electrically shielded room in which the subject is tested is also temperature-humidity controlled and ventilated.

We recognize that the anxiety of the subjects needs to be minimized. Persons coming to the laboratory are never referred to by other than their name, never as "subjects". Each person is asked to be a participant in the study and their importance is constantly affirmed by staff. While the subject is being prepared for recording, the experimenters attempt to put him at ease by engaging in light conversation, by explaining each step in the placement of electrodes and by giving the subject an opportunity to ask questions. This concern for the person who volunteers requires planning and time but the resulting reduction of anxiety in the recording room is proof of its effectiveness. Success is in terms of an absence of movement and a speedy onset of occipital alpha. Only one of more than 1200 persons tested discontinued the experiment due to fear in the situation.

Recordings of the electroencephalogram (EEG) and electro-oculogram (EOG) are obtained with the person comfortably seated in a very quiet audiometric test room (Industrial Acoustics). The interior of the room is 8 × 10 ft., painted flat black. Rated noise reduction, outside to inside, is greater than 79 db in the range 300—6900 Hz. There is some low frequency noise produced by the air conditioning system which was measured at 30—40 db inside the room. This is barely audible to those who can "hear" it at all. When overhead lights are off there is no light in the room.

Controlled visual stimuli are presented through a window. Because of diffusion, reflection and refraction, there is a penumbra around the projected visual stimulus and some loss of definition. The former is not serious and can be controlled by masks. At the intensity levels used in these studies, subjects report that the visual stimuli are clear with well-defined contours.

All visual stimuli are projected from behind, over the subject's head, onto a minimal-glare screen located 5 ft. in front of the subject. The maximum intensity of the projected light with no filters or transparencies in the light path in 7.2 ft. candles measured at the screen with a Weston light meter with visual correction filters. The spectrum of light projected is not evaluated. Standard 300 or 500 watt projection lamps are used with a high quality projector (Lietz, Pradovit F.). Colored stimuli are produced using Kodak Wratten Filters; patterned stimuli, with appropriate 35 mm transparencies.

Silver disk electrodes are attached to the scalp with electrolytic paste (Grass). The electrodes are usually placed over the occipital and parietal regions on both sides, from O_1-P_3 and O_2-P_4 in the International EEG nomenclature. The ground electrode is placed over the mastoid. The EOG electrodes are biopotential skin electrodes, silver/silver chlordie (Beckman) which are either non- or very slowly polarizing. These are attached to the outer and inner corners of each eye and above and below the left eye. The relation between angle of eye movement and recorded DC shift is not routinely calibrated since head position usually is not fixed in these studies, and only large excursions of the eyes are of interest. However before each experiment the DC level for straight ahead, extreme left, right, up and down positions are recorded. For EEG the Grass Model 5P5 preamplifier and 5E driver amplifier are used; for EOG the Grass Model 5P1 low level DC preamplifier and Model 5E driver are used.

The external path. The amplified EEG is led from the Grass 5E driver amplifier to the input of a continuously adjustable band pass filter (Krohn-Hite Model 330M). The attenuation slope is a nominal 24 db per octave. Maximum attenuation greater than 80 db; pass-band gain is zero db $\pm$ 1 db. The band pass of the filter is $\pm$ 1 cps, the resting alpha frequency at center. Filter output is fed to an input stage of an audio amplifier (Scott Model 299-B). Output from the audio amplifier is rectified (International Rectifier Model J29B1). Output of the rectifier is put into the coil of a relay (Potter-Brumfield B523D-6V). This EEG "state" relay controls the coil circuit for a time delay relay (Potter-Brumfield CDD38 30003). Time delays are adjustable in the range 0.1 to 10 seconds. This relay controls the stimulus function.

Stimulus functions. The ON-OFF discontinuous functions are produced by electrically controlled shutters in the various visual display, tachistoscopic and projection apparatus. The continuous function generator (Philbrick) provides various controlled voltage functions which are fed to the input of a servo-system in a visual display (Barkley & Dexter) designed to provide controlled variation of stimulus intensity without a change of color. The servo--system controls the angular rotation and position of two neutral density circular filters (Kodak) which are placed in the light path of a 35 mm projector (Lietz, Pradovit, F.) with an 85 mm lens. The projector lamp is powered by direct current. The angle of rotation of the n. d. filters is a function of input voltage and since the two filters counter-rotate, straddling the film plane, they permit a smooth change of projected light with no intensity gradient across the light field. By the use of fixed neutral density filters in the slide magazine in conjunction with the variable filters, a wide range of illumination functions can be employed. The filters cause no serious diffraction and images can be projected through them.

Each circular filter has a density range from 0.05 to 1.5 thus the combined density will be from 0.1 to 3.0 using 1.3" aperture. This yields a maximum transmission along the optic axis of about 75%. The light beam after passing through the filters can be adjusted for height, direction, and elevation by adjustable frontsilvered mirrors.

The servo is linked to the filters by a precision differential with minimum backlash, driven by a power servo amplifier-motor (Diehl). This combination of amplifier, motor and filters has a nearly flat frequency response

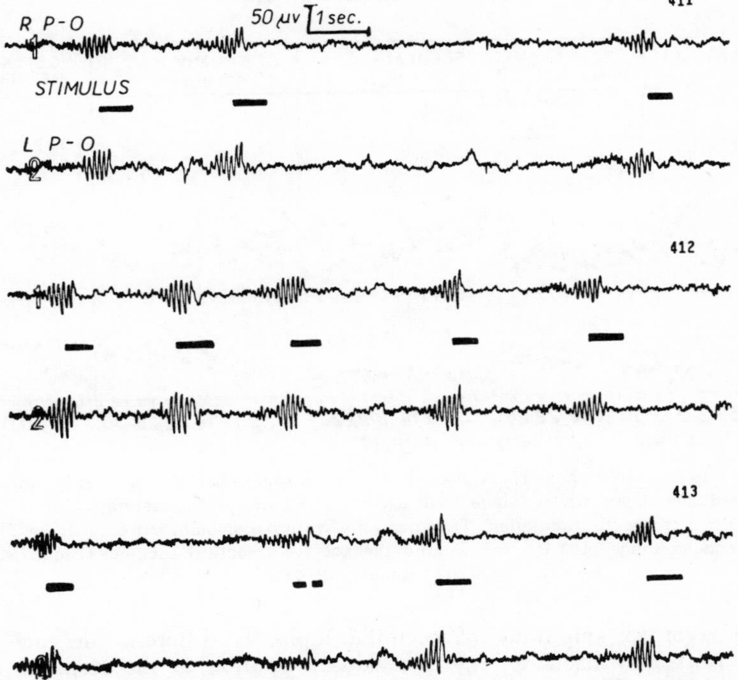

Figure 2. Parietal-occipital EEG during feedback visual stimulation. Stimuli presented more than 100 times previous to this excerpt. Eyes open

to 1 cps. Currently we are using various linear voltage functions which yield a logarithmic change in illumination when either alpha or no-alpha is present in the EEG. Studies using continuous stimulus functions will not be reported here since they are not completed at this time.

In order to reduce variation due to the external path, all apparatus parameters are maintained at fixed values. Calibration values are as follows. With 10 Hz input into the Grass Model 5P5 EEG preamplifier which produces $\pm$ 1 cm peak-to-peak deflection on the Grass oscillograph, output from the 5E driver amplifier is between 8.6 and 8.9 rms volts. Output from the filter, band pass 9—11 Hz is near 14 volts p-p. Output from the audio amplifier is 12 volts p-p. Output from the rectifier is approximately 6V p-p with a + 6 volt DC component. These values are maintained throughout the experiments.

In general alpha was defined as a signal recorded from parietal-occipital or occipital electrodes 8—13 cps; $>$25% maximum "resting" amplitude of the

8—13 cps signal recorded with eyes closed; . 2— . 4 second duration. This combination of values caused the state relay to go ON (alpha); otherwise it was OFF (no-alpha). As the amplitude of alpha decreased from 25% to a smaller value, the probality of state relay ON decreased. This transition from a probability 1 to zero was steep and presented no problems for our s udies. Since the relay is OFF when signal frequencies are less than 8 and greater than 13 Hz the behavior of the *S* must be monitored in order to prevent recording during drowsiness or sleep. We require *S* to press a key at quasi-regular intervals or to press the key in response to the feedback stimulus.

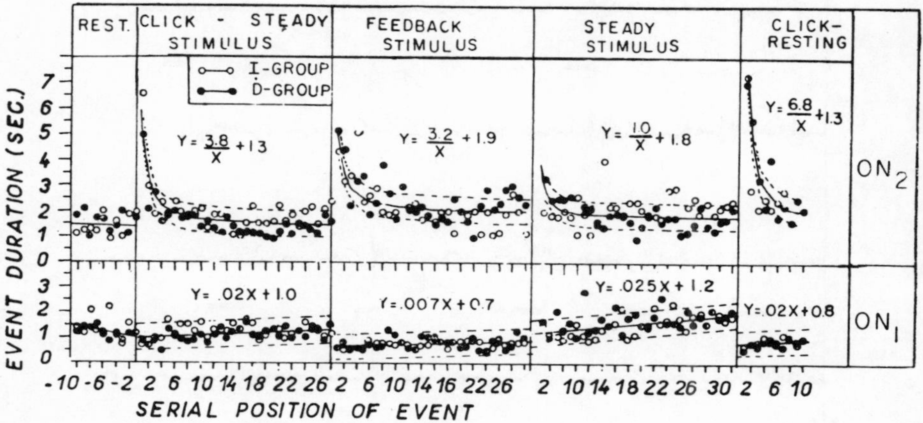

Figure 3. Alpha (ON_1) and No-Alpha (ON_2) mean durations during steady and feedback visual stimulation. Open circles, alpha feedback is an increase of stimulation; filled circles, alpha feedback is a decrease of stimulation. The equations are for a smooth, fitted function. The dashed curves provide a range of ± 0.5 sec. around the average smoothed fuction. (From: Mulholland & Runnals, 1964)

The recorded amplitude of occipital alpha is different for each subject and this causes by itself, a large individual variation in the response to feedback stimulation. We adjust the attenuation of the EEG at the Grass Preamplifier so that the maximum alpha amplitude produces nearly the same voltage at the filter input. For some subjects this means that the recorded EEG is attenuated, for others it is increased to "normalize" the alpha amplitude. This procedure does not eliminate individual differences but does reduce them considerably. Other parameters of the loop are maintained at fixed values so that the "state" relay in ON when alpha (as defined before) is present and OFF when alpha is absent. At the present time there is no definite agreement on the standardization of feedback EEG method, but it is clear that without standardization, results from different laboratories cannot be compared meaningfully.

THE EFFECTS OF FEEDBACK STIMULATION

The best description of the effects of feedback stimulation would be provided by a quantitative analysis, using the method and theory of control systems expressed in mathematical terms. However, no such description has yet been provided for feedback EEG and this lack, an impediment to the development of feedback methods in EEG, is a challenge to control scientists and cyberneticians. The description of the effects of feedback on the EEG which I shall present here is in terms of temporal changes in the duration of alpha events (△t alpha) and no-alpha (△t no-alpha) with and without an external feedback path. As such it is a precursor to the kind of analysis suggested above, illustrating the utility of the method and the diversity of problems to which it can be applied.

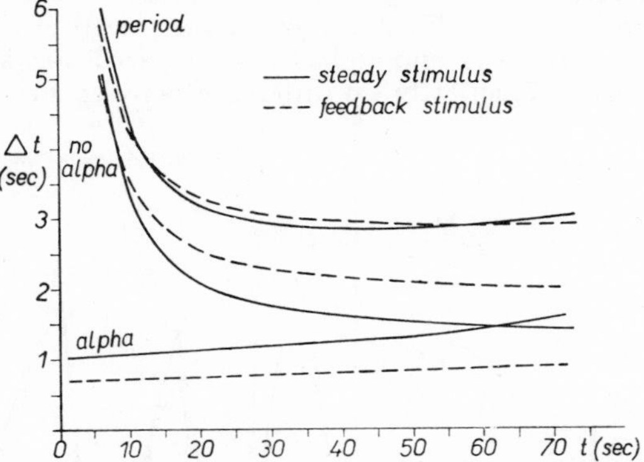

Figure 4. Smoothed Δt alpha, Δt no-alpha as functions of t. Replotted from data in Figure 3. Feedback and steady stimulation have similar effects. (From: Mulholland, 1967)

When alpha causes a stimulus *ON* and no-alpha, stimulus *OFF*, the EEG is a series of relatively brief alpha events and, initially, longer no-alpha durations.

These latter are due to the activation of the EEG by the feedback stimulus. The stimulus *ON* and *OFF* durations (corrected for delays) can be an index of alpha and no-alpha (activation). In Figure 2, the effects of feedback visual stimulation on the P-O EEG is shown.

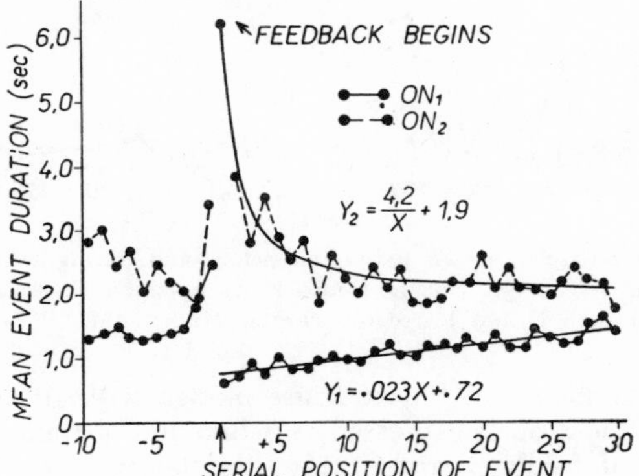

Figure 5. Mean Δt alpha, Δt no-alpha functions of serial number of events, before and during complex feedback stimulation. Equations for smoothed functions. Compare with Figure 3. (From: Mulholland & Runnals, 1964b)

The EEG also alternates between alpha and no-alpha following onset of a steady light (Bagchi, 1937). During steady stimulation the average time series of alpha and no-alpha durations can be described with the same class of functions as for feedback stimulation. However the particular values of the parameters are different. In Figure 3, the EEG response to steady and feedback stimulation is shown (Mulholland & Runnals, 1964a).

During feedback the temporal series of alpha durations (Δt alpha) and no-alpha duration (Δt no-alpha) are different; average Δt alpha is a linear function of the number of prior alpha events while the average Δt no-alpha is approximated by a hyperbolic function. When the period of the average alpha, no-alpha cycle (Δt alpha + Δt no-alpha) is compared for steady and feedback stimulation, there is practically no difference between these two conditions.

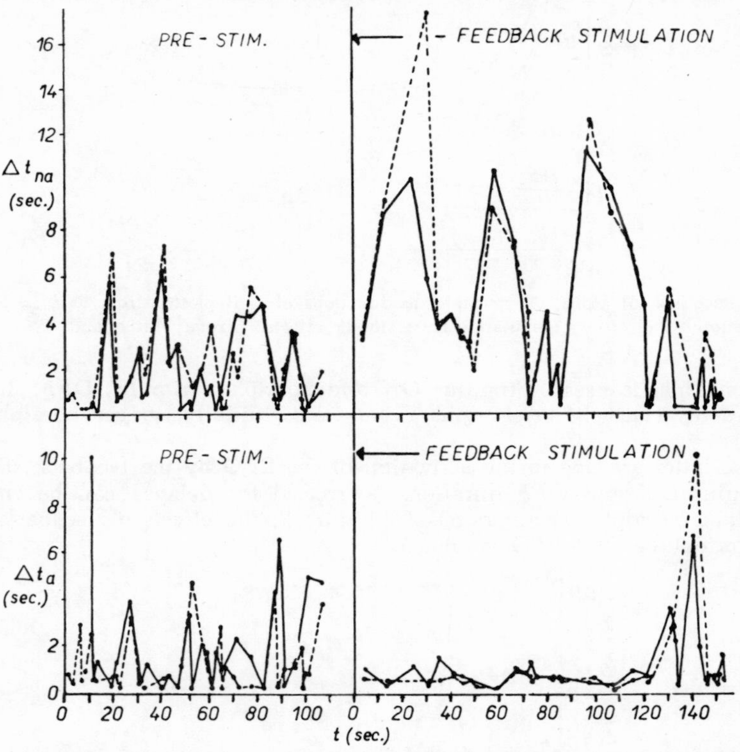

Figure 6. Δt alpha, Δt no-alpha for an individual before and during feedback. Solid lines are R P-O; dashed, L O-P recordings. During feedback Δt no-alpha shows an oscillatory decline for both L and R records (From: Mulholland, 1967)

In Figure 4, the data shown in Figure 3 are plotted with Δt alpha, Δt no-alpha and (Δt alpha + Δt no-alpha) expressed as a function of time (t). The similarity between the effects of feedback and steady stimulation is obvious. In Figure 5 (Mulholland & Runnals, 1964b) the change in EEG following onset of a complex feedback visual-auditory stimulus in shown. Here alpha caused a light ON, sound OFF; "no-alpha" caused light OFF, sound ON. Note that the value of the fitted function for average Δt no-alpha at asymptote is 1.9 sec. which is the same as that during feedback visual stimulation and shown in the previous figures.

The distribution of Δt alpha and Δt no-alpha is not Gaussian but more like a Poisson distribution. Long alpha and no-alpha durations are unlikely events. This means that the increased duration of no-alpha, activation durations following stimulation reflects a transitory disturbance. The system tends to a minimum duration of activation or desynchronization.

The average functions Δt alpha and Δt no-alpha as functions of N stimulations mask individual differences. In Figure 6 a series of Δt alpha and Δt no--alpha from an individual bilateral record is shown. The durations of no-alpha during stimulation vary in a non-random way as shown by Morrell and Morrell (1962) and confirmed by Bundzen (1965) and myself (Mulholland, 1964). If the individual EEG alpha records are "normalized" as described before, the

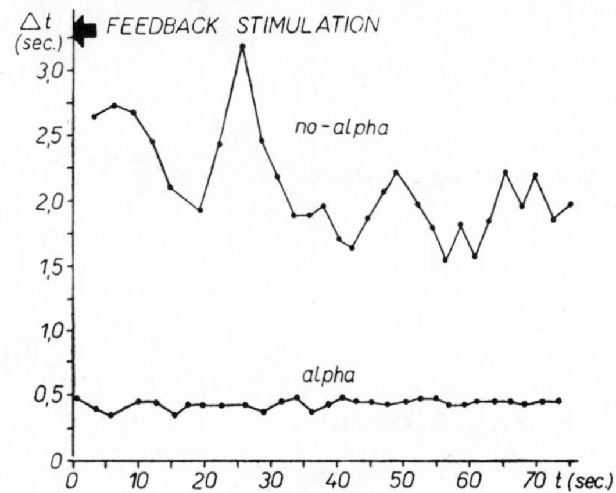

Figure 7. Mean Δt alpha, Δt no-alpha during feedback visual stimulation. EEG was "normalized" at the input. Δt no-alpha shows characteristic oscillatory decrease (From: Mulholland, 1967)

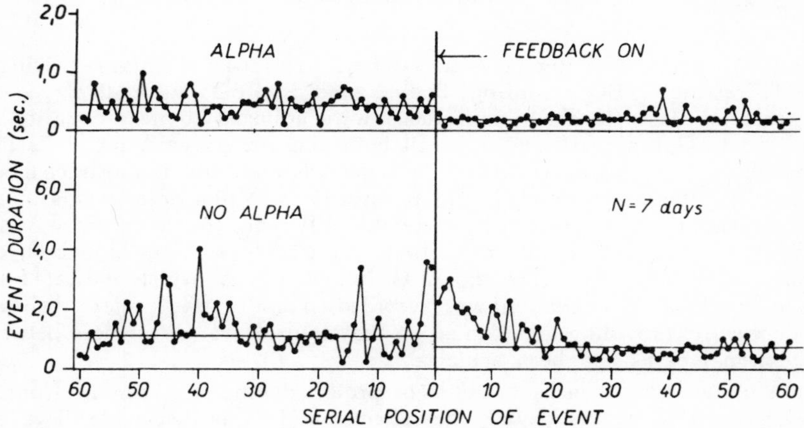

Figure 8. Δt alpha, Δt no-alpha function of serial number of events before and during feedback. Each point average of 7 days recording from single subjects. Feedback stabilizes Δt alpha and Δt no-alpha. (From: Mulholland & Runnals, 1962b)

individual variation is reduced and the average data do typify the response of individuals (Mulholland, 1967). See Figure 7.

In my opinion, the feedback method is the most reliable and controlled way for studying the habituation of the EEG response to stimuli. In addition, after a stable response has been obtained, the effect of other variables on the

EEG can be tested while the variation of the EEG is reduced by feedback. The stabilization of the alpha-activation cycle during feedback is seen clearly in individual recordings. Figures 8 and 9 show average results from one subject who was tested on seven different days. The variation of the averages is less during the feedback.

B i l a t e r a l d i f f e r e n c e s. We examined differences between left and right parietal-occipital EEG during feedback control of the stimulus. The feedback path was duplicated for each side. By means of a switch the feedback stimulus could be controlled by either the left or right P-O EEG.

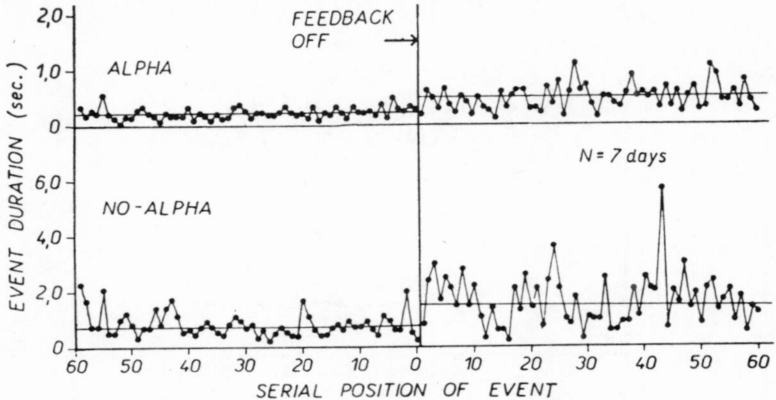

Figure 9. Same as Figure 8, except during and after feedback stimulation. Feedback stabilizes Δt alpha, Δt no-alpha (From: Mulholland, 1967)

Control over mismatching between external paths was maintained. During the initial "resting" EEG recording, both L and R EEGs were displayed on a dual beam oscilloscope. Their amplitudes were adjusted to near equality by adjusting the EEG input attenuation until both pen writers gave p-p deflection, of 1.0 cm during resting alpha. The oscilloscope was then put in substract mode, displaying the difference between L and R recordings. Minor adjustments in the amplitude attenuation were made to reduce this difference to a minimum. At the beginning of the experiment, the EEG from o n e side was put into both external paths and the behavior of the EEG "state" relays, which indicated the occurrence of alpha or activation was recorded. These "state" relays behaved almost identically, providing a confirmation that control over differences between the two external paths had been achieved.

The stimulus was a large bright spot projected onto a screen in front of the subject. Even with eyes closed, the reflected light was detectable. Two experiments, one with eyes closed the other with eyes open, were conducted. In the first, six volunteers, in the second, eight were tested.

The first condition, in darkness, ended with the accumulation of about thirty alpha — activation intervals. The next four were feedback conditions: "right feedback" where the recording from R P-O leads was put into the external path controlling the stimulus; the alternate condition using L P-O recording was "left feedback". Two runs of each feedback condition were continued until about 30 alpha — no-alpha durations were obtained for each. A recording in darkness completed the series.

In general, both L P-O and R P-O recordings were similar under the various steady stimulation and feedback stimulation conditions. However they were not as similar as were the recordings of the same EEG put in both channels. Some individuals had an EEG which was bilaterally symmetrical while others showed occasional gross asymmetries e.g., a well-defined alpha burst occurring on one side with no evidence of alpha on the other.

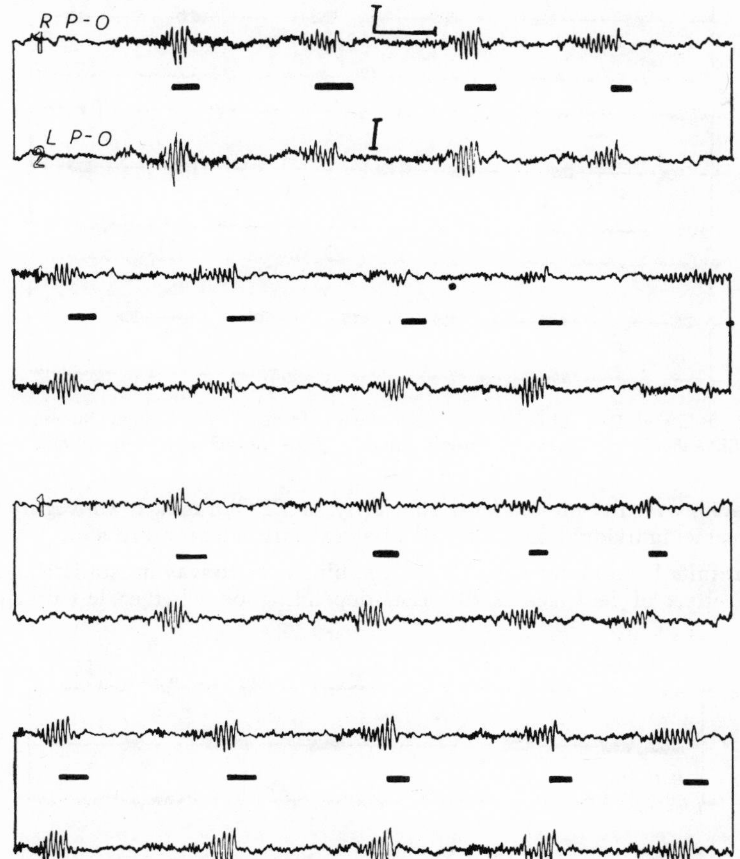

Figure 10. P-O bilateral EEG for one subject showing transitory bilateral asynchrony during feedback. More than 100 stimulations occurred before this excerpt. Disturbance of synchrony begins at stimulus 6, continues through stimulus 13. Eyes open

Graphs of Δt^{on}, Δt^{off} as a function of t were prepared for each subject for both L and R recondings. In order to estimate the correlation between these two graphs a perpendicular was drawn from each recorded point. The intersection of this with the curve for the opposite side was recorded. These numbers were then analyzed using Pearson product-moment correlation (r). Correlations for individuals were transformed to z values, which were combined to give a mean z which was transformed to *r*. Table I presents the overall *r* between L and R sides for each condition. Mean durations and standard deviations of durations were not significantly different for the right or left sides. Correlation coefficients were always greater than 0.86 for matching tests while correlation between L and R recordings were between 0.30 and 0.65. These results were consistent

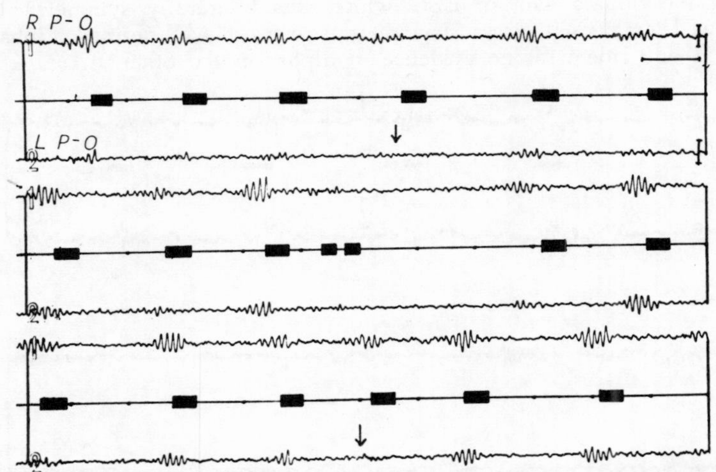

Figure 11. Bilateral P-O EEG from a man during feedback stimulation. Subsequent clinical EEG borderline normal. Arrows indicate gross asynchrony and asymmetry

with previous results on bilateral similarity of EEG tracings. In Figures 10, 11 and 12 some individual examples of bilateral mismatching are shown.

If results from many subjects are combined, there was no statistical evidence that the effect of feedback is different depending on whether left or right P-O

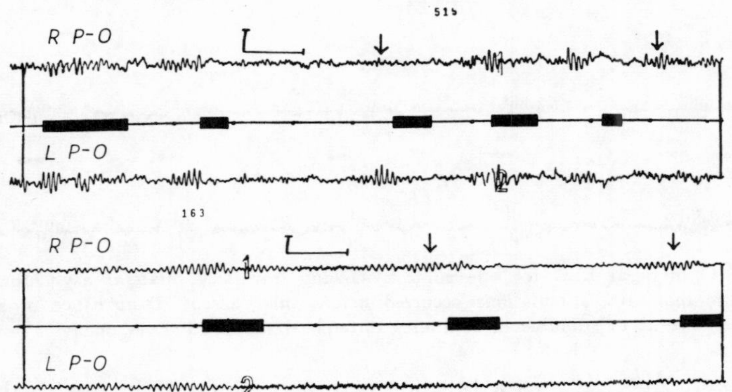

Figure 12. Bilateral P-O EEG from two normal women during visual feedback stimulation. Eyes open. Arrows show gross asynchrony and asymmetry

EEG is controlling the stimulus. However some individual records showed a large difference between "left feedback" and "right feedback" conditions. The significance of these individual results can be evaluated only by further research on this topic. We conclude that if a reproducible difference between the response of the left and right cortex recording under feedback conditions occurs,

Table 1. Correlation (r) coefficients between left and right Parietal-Occipital recordings for Δt alpha and Δt no-alpha (State relay ON, OFF durations)

Eyes Open	Match-ing	Dark-ness	R drive	L drive	R drive	L drive	Dark-ness	Match-ing
OFF (no alpha)	.96	.57	.55	.64	.56	.46	.63	.90
ON (alpha)	.87	.60	.32	.35	.38	.47	.37	.93
ON + OFF	.90	.61	.46	.59	.43	.45	.47	.93
Eyes Closed								
OFF (no alpha)	.95	.44	.49	.57	.59	.52	.49	.97
ON (alpha)	.96	.32	.36	.38	.31	.33	.56	.98
ON + OFF	.98	.40	.46	.45	.44	.41	.29	.96

in the absence of a definite asynchrony or asymmetry under resting conditions it would show an atypicality of orienting to visual stimuli worthy of more comprehensive evaluation (Runnals & Mulholland, 1964). However, the utility of this method for EEG diagnosis of brain dysfunction or pathology awaits further research.

F e e d b a c k a s a n i n c r e a s e o r d e c r e a s e o f s t i m u l a t i o n. When the apparatus is arranged so that alpha causes an increase, and no-alpha a decrease of stimulation, this is called Loop 1. When alpha causes a decrease, and no-alpha an increase of stimulation, the arrangement is called Loop 2. For adults receiving a visual stimulus with eyes closed the EEG responses for

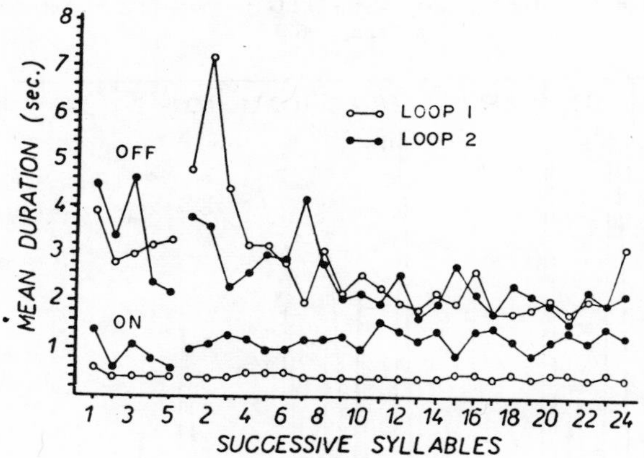

Figure 13. Mean Δt alpha, Δt no-alpha for each of 25 different nonsense syllables. Each point based on response to six stimuli by seven young adults. After habituation of response to first syllable (1—5) Δt no-alpha increases for new syllable. Successive different syllables produce less response. Loop 1 is similar to Loop 2 for Δt no-alpha, Loop 2 always produces longer Δt alpha than Loop 1

Loop 1 and Loop 2 are similar especially if the change of stimulation is from some intermediate value, e.g., when the brightness of light increases with alpha, decreases with activation (Mulholland & Runnals, 1964). See Figure 3. However when the change of stimulation is more complex, the difference between Loop 1 and Loop 2 is more evident. In Figure 13 the average response to successive nonsense syllables is shown. The first thirty presentations are the same word. After that 24 different words are presented six times each. The greatest difference between Loop 1 and Loop 2 occurs for the initial syllable. However as a di-

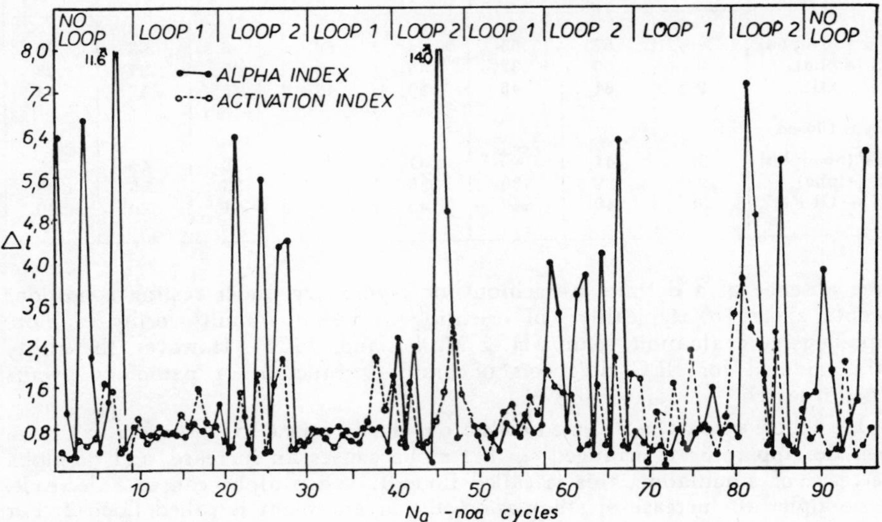

Figure 14. Δt alpha, Δt no-alpha from L P-O recordings during two kinds of feedback stimulation for 7 year old boy. During Loop 1 the EEG is more stable. See text (From: Mulholland, 1967)

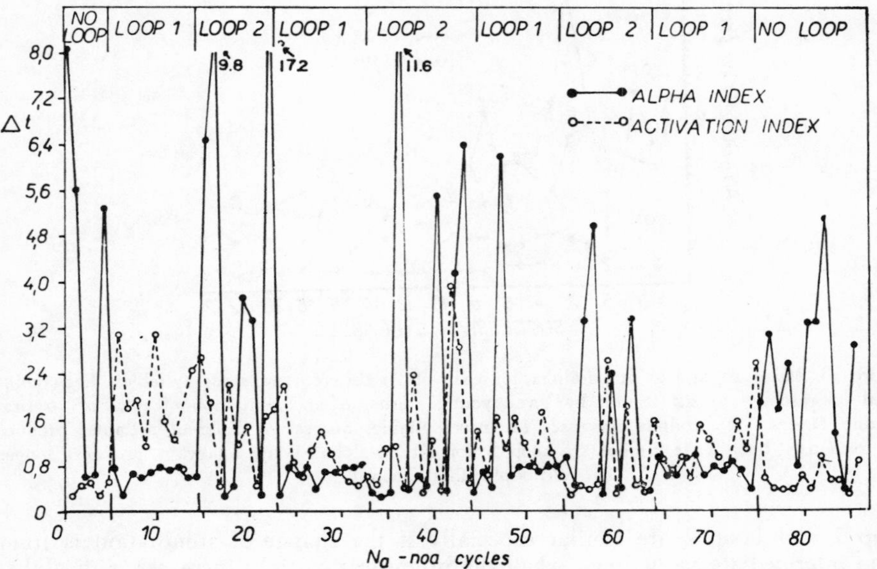

Figure 15. Δt alpha, Δt no-alpha from L P-O recordings during two kinds of feedback. During Loop 1 EEG is less variable. Note initial response to feedback during Loop 1. Subject is an eleven year old boy (From: Mulholland, 1967)

minished response occurs to each successive different stimulus, the activation intervals obtained with Loop 1 and 2 become more similar. The Δt alpha for Loop 2 is longer than Δt alpha for Loop 1 and the variability is greater. This experiment and others (Mulholland & Davis, 1966) also show that the EEG component of the orienting response can exhibit habituation to classes of stimuli as suggested by Sokolov (1963).

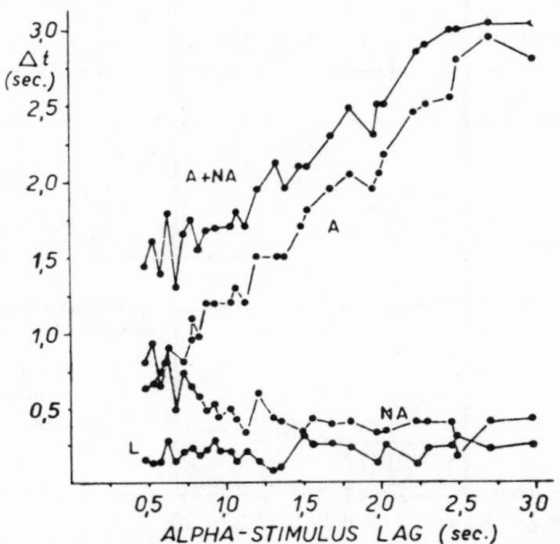

Figure 16. Mean Δt alpha, Δt no-alpha, Δt alpha + Δt no-alpha as a function of the lag between alpha and feedback stimulation. Subject normal woman, 63 years of age

In contrast to young adults, children younger than 12 years show a definite difference between Loop 1 and Loop 2 with patterned visual stimuli. In Figures 14 and 15 the effects on Δt alpha and Δt no-alpha of the two kinds of feedback are shown for two children. In general the response series is much more stable and Δt alpha, Δt no-alpha are briefer during Loop 1 compared to Loop 2. This difference between adults and children implies a developmental change in the response to feedback which awaits further research.

Delayed feedback. Changing the values of any of the parameters of the external path would be expected to change the response of the system during feedback. The time delay between the onset of alpha and the onset of stimulation has a particularly clear effect. In Figure 16 the average Δt alpha and Δt no-alpha to a brief flash triggered by alpha is shown. As the time delay between alpha and stimuli increases the duration of Δt alpha increases; Δt no-alpha decreases. As longer time delays occur, the stimulus occurs less often. Under these conditions alpha occurs in longer bursts and the response to the stimulus is reduced. The latency of activation (L) increases slightly as the delay increases. The duration of alpha increases with increased feedback delay up to a limit given by the period of the spontaneous alternation between alpha and no-alpha or minimal alpha.

VISUAL CONTROL SYSTEMS AND OCCIPITAL ALPHA

Control of the electroencephalogram by the subject. Kamiya (1962) has independently developed a method for training *Ss* to voluntarily facilitate or inhibit their own alpha rhythm. Essential to his method

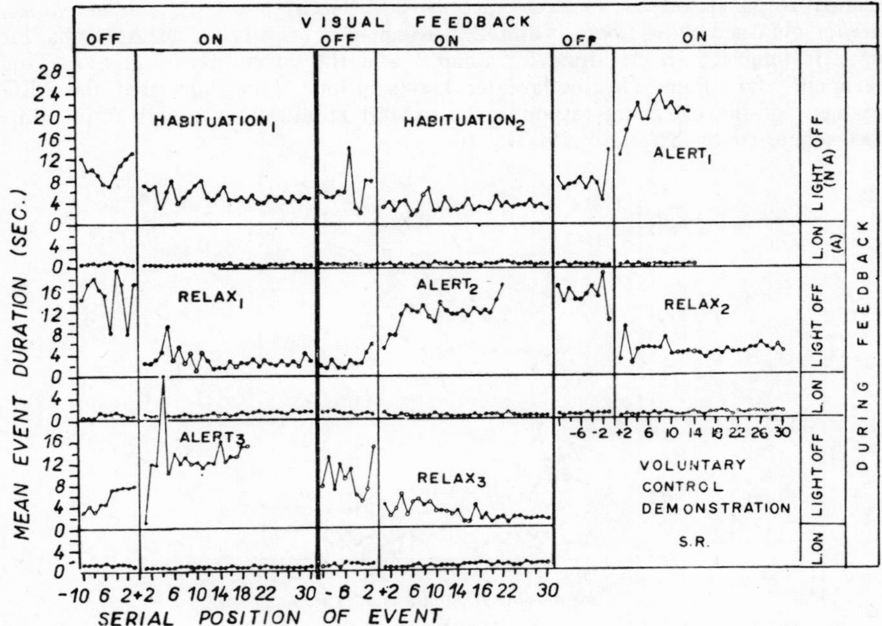

Figure 17. Δt alpha and Δt no-alpha before and during feedback. First two runs were for habituation. Then subject tried alternately "to be alert" or "to be relaxed". For Δt no-alpha there was a definite effect of the subjective attempt to be relaxed or alert. Normal woman

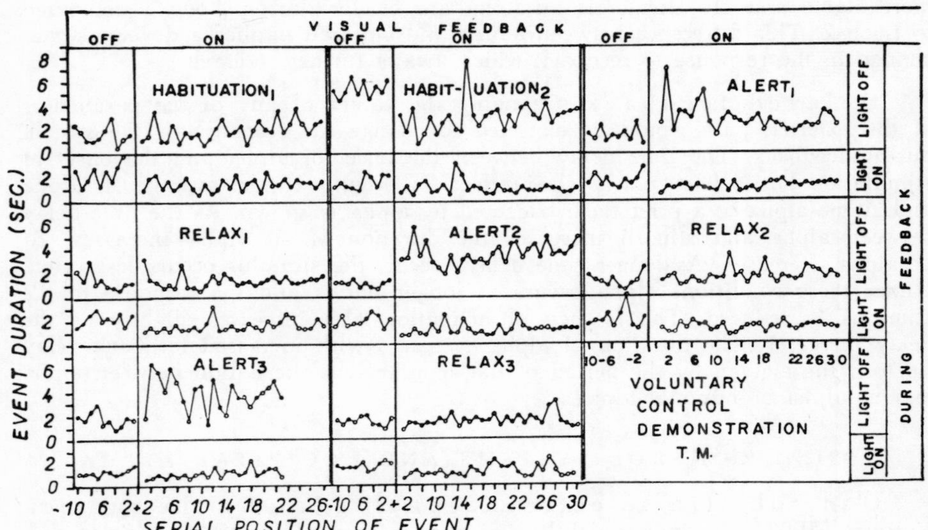

Figure 18. Same as Figure 17. Normal man

is informational feedback which "tells" the subject when alpha is occurring or not. A number of researchers (Runnals & Mulholland, 1965; Mulholland & Evans, 1966; Dewan, 1967; Hart, 1967; Nowlis, 1968) have also reported that *Ss* can learn to control their EEG if they receive feedback. In Figures 17 and 18 two examples of such control of the feedback EEG by the subject is shown, "Alert" is trying to be alert, "Relax" is to not be alert. The difference between the two conditions is clear, Δt no-alpha increases when the subject is trying to

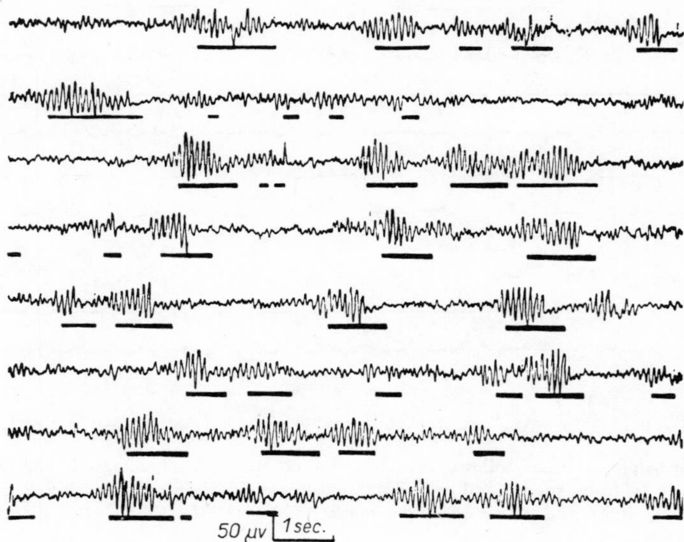

Figure 19. Rapid alternation of "alpha" and "no-alpha" as subject attempted to keep meter pointer at midpoint. Needle moved right of dial when alpha occurred; left when alpha was not present. Heavy bars indicate periods of "alpha". (From: Mulholland & Evans, 1966)

be "alert". The function being controled voluntarily by the subject when he produces changes in the EEG is not known. For reasons discussed below. I believe that the subject learns to control oculomotor functions which in turn can facilitate or inhibit the occurrence of alpha in the occipital EEG.

Both Adrian and Matthews (1934), Durup & Fessard (1935) and Adrian (1943) suggested that ocular and oculomotor functions were involved in alpha occurrence and non-occurrence. Adrian (1943) stated that alpha did not occur during attention and fixation and appeared when attention and fixation were lost. Durup & Fessard's (1935) description of changes of alpha occurrence concomitant with changes of fixation, convergence and accommodation illustrate this point. Oswald's (1959) note that alpha occurs during intense auditory attention when accompanied by a loss of ocular fixation and accommodation is further evidence that oculomotor functions are involved in the alpha-activation cycle.

Figure 19 presents results obtained when a trained subject viewed a meter pointer which moved to the right when alpha occurred and to the left when activation occurred. This P-O EEG excerpt was obtained during a time when he was trying to keep the pointer at the center zero position, which involved a rapid alternation between EEG alpha and activation. For the initial stages of this experiment the subject was unable to control the meter pointer because he

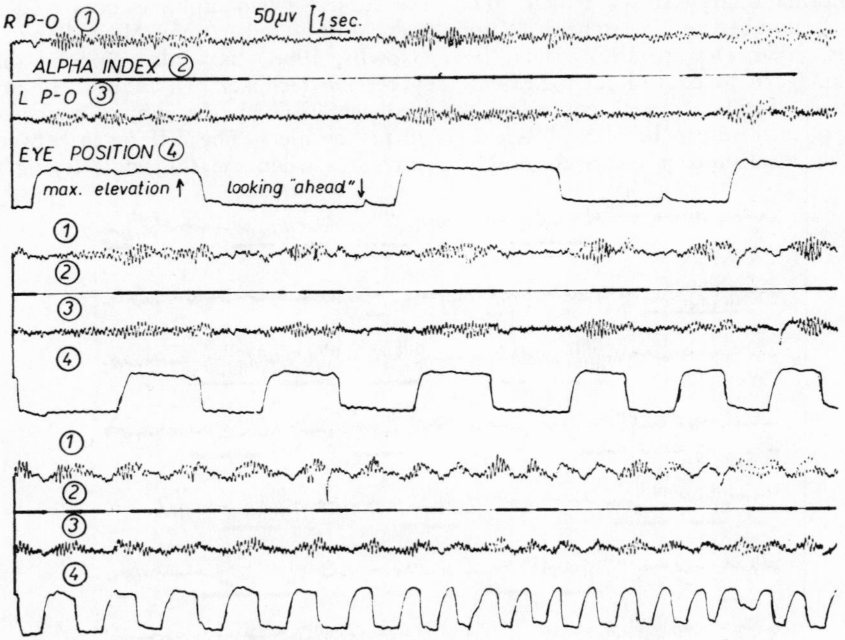

Figure 20. The alpha rhythm follows the elevation of the eyes. Tracings 1 and 3 are P-O EEG; Tracing 4 is from super- and suborbital electrodes for left eye. Tracing 2 is auditory stimulus (heavy line) and a one sec. time mark. Subject TBM (From: Mulholland, 1967)

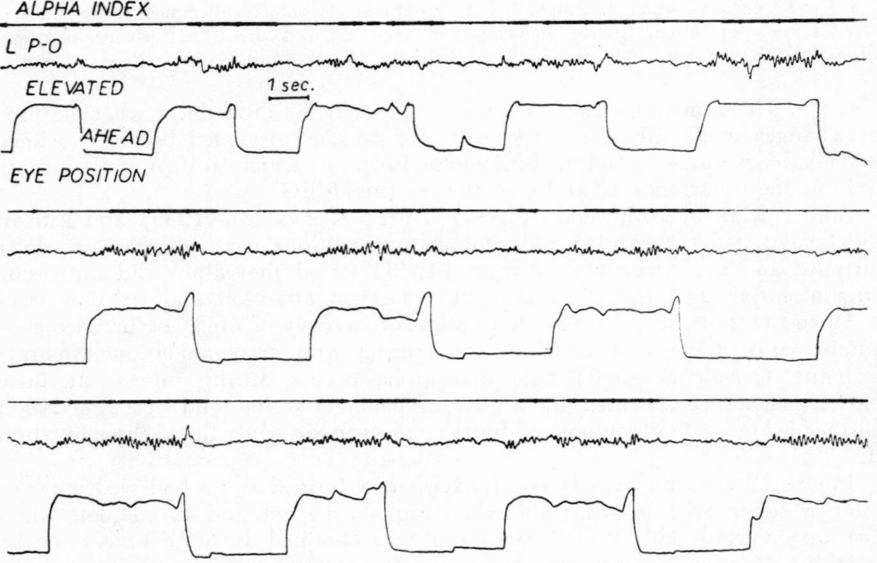

Figure 21. Same as Figure 20 except no auditory feedback. Subject young man (From: Mulholland, 1967)

could not produce alpha "at will". When the pointer moved to the end of the scale he "brought it back" by closing his eyes, opening them to observe the movement of the pointer. After awhile he needed only to "get the pointer going" by briefly closing his eyes, then opening them. Later on he did not close his eyes. In Figure 19 eyes were open throughout (Mulholland & Evans, 1966).

The performance of this subject is reminiscent of that reported by Durup & Fessard (1935). Their subject was able to alternately increase or decrease the amount of EEG alpha by shifting his "attention" alternately from visual to auditory stimuli. They reported that their subject controlled the occurrence of EEG alpha by regulating visual accommodation, convergence and fixation. Likewise, the subject in our studies reported that when attempting to move the meter pointer to the right, he would look at a distant plane with indefinite fixation producing an indistinct, blurred image of the pointer. He was able to judge its approximate position. Under these conditions alpha occured. To be "alert" he reported that he focussed on the meter pointer, tracking it with a clear distinct visual image.

Mulholland & Evans, (1965, 1966) and Dewan (1966, 1967) observed that the occipital alpha rhythm is facilitated when the eyes are elevated maximally, a maneuver which can be uncomfortable requiring continuing effort. Though not all subjects show this effect it is reliable in those who do. Moreover, this facilitation can occur with eyes open or closed, with or without visual stimulation and while inspecting and reporting on a visual after-image. The facilitation of the alpha rhythm by both vertical and lateral deviation of the eyes has been confirmed by Kris (1966, 1968), and Fenwick and Walker (1967). Kris reports an angular deviation greater than 30° is required to facilitate alpha.

In Figure 20 an example of this effect is shown (Mulholland, 1967). The subject, myself, was in darkness with eyes open. As the eyes were raised and lowered the close following by the EEG alpha occurred. When alpha had been recorded from the right P-O deviation for at least 0.5 seconds, a radio program was presented over earphones providing high information auditory feedback. Despite this, alpha was not blocked as long as the eyes were elevated. When eye positions were alternating most rapidly, each period of eye elevation was associated with an alpha "burst". In Figure 21 the results for a naive subject in darkness, eyes open and no feedback stimulation are shown. When his eyes were elevated, alpha occurred.

Accommodation, Fixation and Target Clearness. In order to analyze the variables responsible for facilitation of alpha when the eyes are elevated we made use of the fact that when one eye views a near target, the other eye turns inward, even if it is occluded. When accommodation is relaxed, giving an unclear, blurred target the non-viewing eye turns outward. These accommodative vergence movements are different from the associative tracking movements which occur when both eyes move together in pursuit of a moving target (Stark, Kupfer and Young, 1965).

The clearness of the target was indicated by pressing a key, ON means blurred; OFF means clear. The EEG was recorded using a Grass Model 5 polygraph. The position of the eyes was evaluated by means of electrodes placed in the inner and outer canthi of each eye. The DC signal recorded from these gave a good indication of the occurrence and direction of movement. It was not intended to give a precise measure of the angular deviation.

The general plan was to evaluate the EEG and eye movements under conditions of viewing a stationary target and tracking a moving target. In the former (1) the accommodative vergence movements were evaluated which gave an index of lens accommodation. In the latter (2) the target was moving and vergence and tracking movements were evaluated in relation to the EEG.

In a dark room, subjects were seated at a viewing box so arranged that the target was visible only to the right eye. Head movements were controlled by having the subject bite on a rigid mouthpiece. The target, a black cross on an orange background illuminated from behind was about 0.80 mm diameter and 115 mm from the viewing eye when it was pointed "straight ahead". In experiment 2 the target moved back and forth laterally, 0.6—1.0 Hz, amplitude 25 cm. The target attached to a cam-driven rod which was audible to the subject.

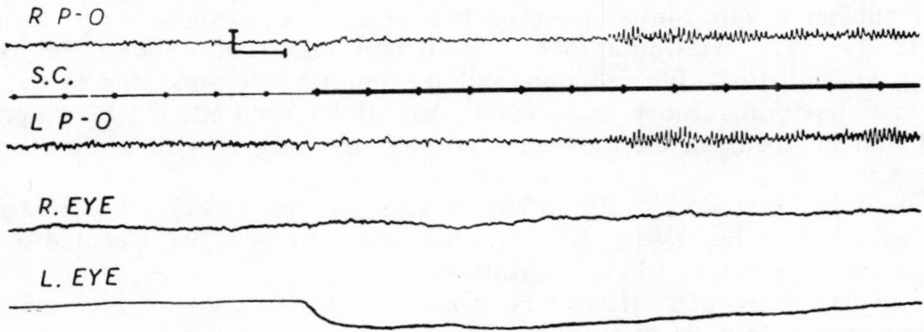

Figure 22. Following accommodative divergence of left eye, alpha occurs after a delay. Tracing 1, R P-O EEG; 2, index of target clearness, thick line equals blurred target; 3, is L P-O; 4, DC EOG, lateral movements for R eye; 5, same for L eye. Down is eye movement toward the left; upward toward the right

Experiment 1. Five subjects were studied while they focussed the target (Focus) and when they did not focus resulting in a blurred target (Blur). "Focus" means the left, non-viewing eye was turned inward, the viewing eye not moving and straight ahead and the subject reported a clear target. "Blur" means that the left eye turned outward, right eye not moving and subject reported an unclear, blurred target. "Focus" and "Blur" trials were alternated in irregular sequence.

Figure 22 illustrates the EEG change when the subject switched from "Focus" to "Blur". In all qualitative features it is like the results obtained for every subject and every trial where alpha was recorded: synchronous activity occurred with higher amplitude and for longer periods during the "Blur" trials compared to the "Focus" ones; there was always a delay from 1 to more than 5 seconds between the onset of accommodative divergence and report of blurred target. These were not coincident with onset of alpha. The accommodation hypothesis was not confirmed.

Experiment 2. This was designed to test the hypothesis that the delay discussed above was a period during which the feedback gain of the eye position control system was decreasing. According to this hypothesis when alpha occurred, the eye position was not controlled by feedback from the target. If eye position control were important for the blocking of alpha, then when the eyes were tracking the target with either accommodative divergence or convergence, and target clear or blurred, minimal alpha should occur. When eye tracking stopped alpha should occur, even though the target were still moving and visible.

Four subjects were then tested in three oculomotor conditions: (a) focussing and tracking the target (FT) in the so-called "pursuit mode"; (b) blurring the target while tracking it (BT); (c) blurring the target and not tracking it (BNT).

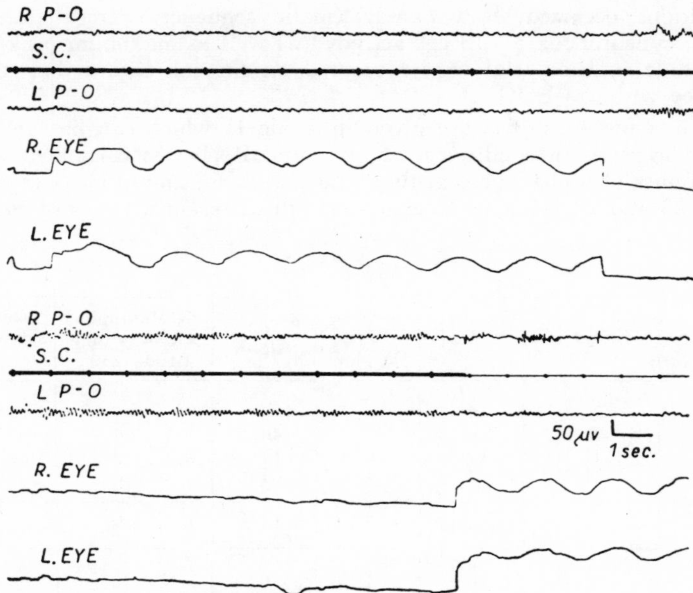

Figure 23. Alpha occurs in the occipital EEG after the cessation of pursuit tracking and report of blurred target. With onset of pursuit tracking and convergence of the left eye with a report of focussed target alpha is blocked. Tracings as in Figure 22

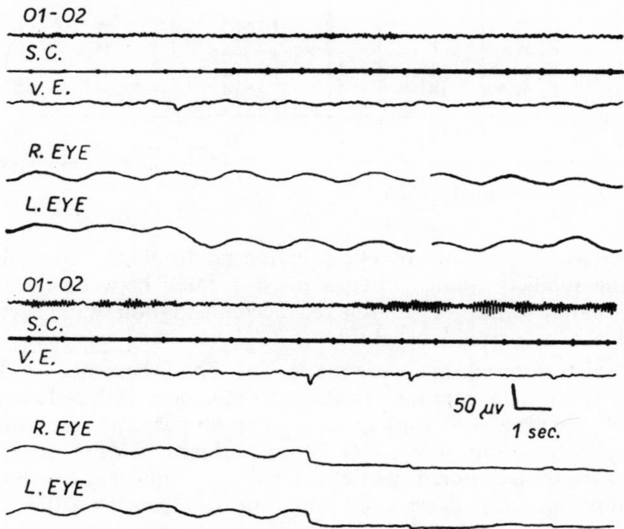

Figure 24. During pursuit tracking, L eye converged and focussed target, minimal alpha occurs. With pursuit tracking, L eye diverged, target blurred, more alpha is seen; with cessation of tracking, L eye diverged and target blurred, much more alpha is recorded. Tracing 1, O_1-O_2 EEG; 2, target clearness; 3, vertical eye movements; 4, right eye lateral movements; 5, left eye lateral movements

These conditions occurred in a non-systematic sequence. Percent-time alpha (identifiable, synchronous 8—13 cps activity) as well as maximum peak-to-peak alpha amplitude in each trial were measured on the ink record and averaged and correlated with the EOG.

The results for 4 subjects are given in Table II which can be summarized as follows: The series of conditions FT, BT, and BNT was correlated with increasing amounts of alpha (percent-time and maximum amplitude) respectively. See Figures 23 and 24. This series also seems to represent a series of conditions

Table 2

Conditions	Subjects	N Trials	N Trials With Alpha	% Time Alpha	$\overline{X}$ Maximum Alpha Amplitude (μv)	Range of Alpha delay (Sec)
Focus-Track	B. P.	10	0	0	—	—
	B. K.	8	8	40	9.6	0.8— 2.7
	E. D.	12	6	17	8.0	6.8—19.0
	E. M.*	3	1	04	—	—
	$\overline{X}$	8.2	3.7	15.0	8.8	
Blur-Track	B. P.	10	1	0	—	—
	B. K.	3	3	51	10	0.0— 6.9
	E. D.	10	8	31	14	0.6—15.2
	E. M.*	17	9	03	36	0.8—17.0
	$\overline{X}$	10.0	5.2	21.2	20.0	
Blur-No Track	B. P.	9	9	55	29	1.8— 6.6
	B. K.	6	6	52	19	0.0— 6.0
	E. D.	11	11	64	36	0.6—11.0
	E. M.*	18	17	14	50	0.2—12.8
	$\overline{X}$	11.0	10.7	46.2	33.5	

* — Psychiatric patient receiving Librium

which have decreasing amounts of eye positioning feedback. Note, however, in Table II that the subjects studied often showed a delay between the onset of the change of oculomotor condition to relaxed accommodation with accommodative divergence and the occurrence of alpha.

Four additional observations were made: (a) The occurrence of alpha during BNT trials increased with repeated trials in terms of a reduced delay between the cessation of tracking and alpha, an increased percent-time alpha and increased maximum alpha amplitude; (b) In one subject (EMD) the FT, BT, and BNT conditions were monitored while the subject registered by hand key his subjective attempt to "pay close attention", or to "pay as little attention as possible consistent with the task". With 12 trials per condition the subject showed no effect of "attention" and merely reproduced results similar to Table II; (c) For two subjects, the target was turned off during FT leaving the subject in complete darkness. In both cases alpha promptly appeared in less than one second with increased amplitude provided that no "tracking" movements occurred; (d) One subject was instructed to continue tracking an imaginary target.

Smooth tracking movements lasted for about 1 second and were followed by increasingly irregular lateral undulatory and saccadic movements accompanied by increasing amounts of alpha. Another subject shifted from pursuit tracking (FT) to a succession of saccadic movements. Alpha occurred abundantly during the sequence of saccadic movements.

THE OCCIPITAL ALPHA – ACTIVATION CYCLE AS A CONTROLLED PROCESS

The relationship between visual stimulation and occipital alpha can be better understood as a relationship between occipital alpha and the visual control systems which use feedback from target intensity, blur and location to control eye movement and position, to optimize the reception of visual information. Accordingly, the temporal series of alpha and no-alpha durations should be associated with a temporal series of changes in the visual control systems alternating between feedback control (activation) and loss of feedback control (alpha). The succession rate and duration of accurate fixations, the visual field scanning or sampling function may be a controlled function. From this point of view the temporal series of Δt alpha and Δt no-alpha may reflect a biological control system which oscillates between limits which are time varying; which can be modified by an external feedback path; which recovers to near its original values when disturbed by an impulse or step-change of stimulation; which in response to redundant stimulation, exhibits a shift toward an asymptote which is assumed to be an optimum for the conditions; and which shifts to a new optimum with a change in stimulation.

In Figure 25 a schema for the temporal variation of Δt alpha and Δt no-alpha is shown. An alpha control process is assumed which varies between limits associated with stability of the triad of accommodation (pupillary diameter, vergence movements and lens accommodation) and pursuit tracking movements at the upper limits, and instability at the lower limits. These limits are themselves time varying. In addition a threshold function is described which determines the state of the EEG. In this particular example, the response following a step increase of stimulation is shown.

This kind of schema permits a quantitative description of the various changes of Δt alpha and Δt no-alpha which have been described in experiments: the alternation between alpha and no-alpha; the initial increase of the period of the alpha, no-alpha cycle following a step increase of stimulation; the subsequent oscillatory decrease in the period of the alpha, no-alpha cycle as stimulation continues; the asymptote value which is approached as stimulation continues; the different time series of Δt alpha and Δt no-alpha.

It is clear that transient or intermittent or feedback stimulation will have effects depending upon the transfer function of the system as well as the parameters of the stimulus. If the alpha, no-alpha cycle is connected to a stimulus by an external feedback path, the system behavior will reflect the complex resultant of the external feedback parameters interacting with the system. It is also clear that Δt alpha and Δt no-alpha are neither useful nor meaningful measures unless the time series information is preserved.

THE ACTIVATION OF THE OCCIPITAL EEG

In the usual interpretation of the "blocking" of the occipital alpha or occipital EEG desynchronization, emphasis is given to the contribution of afferent, sensory inflow especially along collaterals to the brain stem reticular formation passing upward to the thalamus and projected from there to cortex. This conceptualization does not provide for the possibility of desynchronization of the occipital EEG by efferent, oculomotor processes. It is interesting at this point that Jasper and Penfield (1949) reported that desynchronization of the ECoG recorded from the precentral gyrus occurs with the intention to make a movement. The so-called mμ rhythm is desynchronized by movement of the hand or foot.

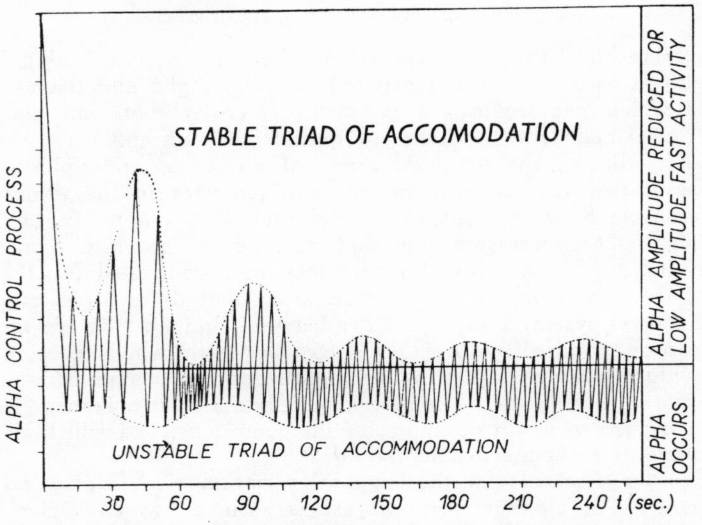

Figure 25. Hypothetical schema for temporal changes of Δt alpha, Δt no-alpha following step--change of stimulation

In the experiments on pursuit tracking and occipital alpha presented before, the cessation of pursuit tracking with relaxed accommodation producing a blurred visual image of the target, was associated with abundant alpha, even though the target was visible and moving. When pursuit tracking movement began with relaxed accommodation, target blurred but visible and moving, alpha was attenuated or desynchronized. This shows that an oculomotor process, in this case a tracking movement, can produce cortical activation or desynchronization of the occipital EEG. On this point the work of M. Bender and his associates is especially important. They have shown that "electric stimulation or lesion within the brain stem invariably causes alternations in eye movements but it is not generally appreciated that the reticular activation system and the oculomotor pathway are located in the same zone of the brain stem tegmentum" (Bender & Shanzer, 1964).

There is probably a large amount of cortex given over to control of eye movement since (1) there is widespread representation of eye movement control processes in brain (Bender, 1967) and (2) there is such a high nerve--to-muscle fiber ratio for the etraocular muscles. Following this line of speculation, the greatest number of desynchronized neurons occur when there is both a large afferent inflow and an extensive oculomotor control process, e.g. when tracking a moving target. As these two processes decrease the process of desynchronization becomes less extensive and briefer.

Previous observations on the occipital EEG may be interpreted by means of an afferent-efferent, sensory-oculomotor theory of desynchronization of the occipital EEG. When the eyes are closed, there is a lack of stimulation and eye positioning control since there is no feedback from the extra-ocular muscles, and alpha desynchronization does not occur. However a voluntary positioning of the eyes can occur in the dark resulting in a transitory stabilization of eye position and a transitory blocking of alpha. In my opinion this is the basis for alpha "blocking" when the subject "tries to see" in the dark. Eye position in the absence of feedback control soon becomes unstable and desynchronization does not continue.

When the eyes are opened in a patterned visual environment both afferent processes and eye position control are maximum. Desynchronization is widespread and long lasting. If the environment is not patterned, as in a Ganzfeld, eye positioning control is much less efficient and briefer desynchronization or activation occurs (Avant, 1965).

When the eye positioning control system is not operative, when feedback gain of the eye position control systems is greatly reduced, desynchronization of occipital EEG is less likely. This is perhaps the case, for some subjects, when eyes are elevated. Those subjects who show practically no desynchronization when the eyes are elevated, may not show desynchronization even when the stimulus increases by intense illumination of the eyes when they are elevated. To me this shows that, in the absence of an eye positioning control using feedback, a visual stimulus may have little effect on the synchronous occipital EEG.

In the case where the subject is "paying attention" or orienting to a stimulus, it is clear that eye position control is present, and so is a desynchronized or activated occipital EEG. However if the eye positioning control using feedback is reduced or absent then "paying attention" has little effect. "Paying attention" to a stabilized afterimage with eyes elevated did not abolish alpha in two subjects tested (Mulholland & Evans, 1966). In the experiments with E. Dewan reported before, when he subjectively "paid attention" to the moving target but did not track it, there was no evident effect of this subjective process on the occipital EEG. Studies on the blocking of alpha with mental calculation or visual imagery, or studies on the learned voluntary control of the occipital alpha need to record eye movements before the relevance of these psychological processes can be assumed (Mulholland, 1967).

The desynchronization of the occipital EEG by tactual and auditory stimuli may reflect the simultaneous eye positioning movements which occur as part of the orienting response (Zikmund, 1965). In experiments on this problem, visual stimuli are usually absent or minimal and without pattern in order to avoid uncontrolled responses to visual stimuli. Under these conditions the visual control system must operate with reduced or no feedback from a visual target. Desynchronization is less likely since controlled eye positioning occurs less often.

In the case of saccadic movements, since vision is suppressed during the movement (Latour, 1962), target information is reduced and the eye positioning system is deprived of an error signal. Under these conditions desynchronization would not occur for a brief time after onset of a saccade, provided that another movement, which can desynchronize the EEG, does not occur too soon.

The perception of visual stimuli should be different depending on whether the stimulus is presented during desynchronization or synchronization of the occipital EEG. However the difference would be that expected when the eye positioning system is controlled by feedback from a visual target compared to when is not. More research on this topic is needed.

The two factor, sensory-oculomotor hypothesis is relevalent only to the desynchronization of the occipital EEG. I would suggest that the occipital desynchronization process be considered to be not merely the obverse or negative of the synchronization process. Whatever the synchronizing process is, it is more probable if sensory-oculomotor processes occurring in the occipital cortex are reduced, but the reduction of excitation in the occipital cortex does not by itself account for synchrony. Some new principle is required (Brumlík, Richeson and Arbit, 1967; Dewan, 1967), and we await conclusive neurophysiological studies on this point. This line of speculation implies that there exists a state of occipital cortex which is neither a desynchronized state, nor a synchronized state. Such a state is perhaps unstable and goes to a state of synchrony if no other process intervenes.

PROSPECTS FOR FEEDBACK METHODS IN PSYCHOPHYSIOLOGY

Volutary control of physiological processes. Feedback via an external path permits the subject to receive information about processes which are not represented in subjective consciousness. For instance we are not aware of our brain rhythms. We cannot learn to control the occurrence of alpha without feedback. However with feedback stimulation which informs the subject when alpha is OFF or ON a subject can learn to control his alpha rhythm (Kamiya, 1962; Runnals & Mulholland, 1965; Mulholland & Evans, 1966; Dewan, 1967; Hart, 1967). Control of physiological processes as seen in the highly skilled practice of Yoga or in the trance state of the mystic requires years of training. The voluntary control of these processes may be more efficiently learned using feedback training permitting an increase in the range of psychological experiences associated with physiological changes for more people. Similarly, subjects can learn to control their GSR well enough to send Morse code (Simonov & Velueva, 1966) if they receive GSR feedback. The possibility of new and powerful kinds of training is evident here. The subject can control the occurrence of a behavior or response which before was an unconscious reflex, by learning a voluntary response. In some cases where the reflex process is not operating properly, a learned voluntary process may compensate for the defect. For instance, a child who has reading difficulties due to inaccurate reflex fixations and movements might be trained first under "open--loop" conditions to voluntarily position the eyes and then be trained with "closed-loop" feedback to impose a voluntary control upon his eye movements during reading. If subjects could be trained to voluntarily maintain the physiological state associated with increased alertness, then students or school children might be better able to orient to stimuli in a way which would make their learning process more efficient. This kind of training might be useful in the treatment of neurotic symptoms whereby the subject learns to control not only the final response but the physiological precursors to this response which are not in awareness but which are brought into awareness by feedback. More research on these topics is required.

Control of machines and devices by physiological processes. The process of learning to voluntarily control the occurrence of physiological reflexes should be distinguished from regulation of physiological processes by instrumental conditioning. In the former, the subject must be able to initiate the response in relation to a self-generated cue, i.e., he decides when he will control the response voluntarily, and this involves a further control of the schedule of response occurrence. Voluntary control of physiological processes permit one to use such processes to send messages given the appropriate apparatus (Dewan, 1964, 1967; Simonov, 1966). Once voluntary control is achieved any machine or device can be connected to the controlled response and be controlled by it. This could include prostheses as suggested by Prof. Benes of Charles University in Prague (Hammond, 1966).

Another application involves controlling the information received by the subject as a function of his orienting. For instance a teaching machine might be controlled by the EEG or other components of the orienting response. If the subject were maximally orienting, information would be received at the maximum level; if orienting decreased information would be less; if orienting were absent, a warning or waking signal would be given. Kuprianovitch (1965) reported a method for controlling the level of sleep as defined by the EEG, using a visual feedback stimulus. When the sleep EEG was within a particular controlled range, an auditory signal was automatically presented to test the hypothesis that learning could occur during sleep.

Diagnosis. The use of feedback EEG and feedback EOG may be useful in the diagnosis of neurological conditions which result, not in an abnormal clinical EEG nor abnormal ocular reflexes *per se,* but in a defect in the organization of the orienting response. The habituation and dishabituation of the electroencephalographic and electro-oculographic components of the orienting responses using various feedback configurations and stimuli which range from simple to complex symbolic patterns could be studied in a standardized way and and compared to norms. Bundzen (1966) has applied feedback EEG to studies of psychasthenia.

Group processes. Studies of interpersonal and group processes might achieve a new dimension using feedback methods. If the physiological response for one person were information to another, and his physiological response were information to the first, the behavior of both might be modified. For instance if heart beat were indicated by a flashing light worn say on the lapel and a man and woman were introduced, the man's light might flash more quickly. The woman perceiving this might be slightly embarrassed, and her light would begin flashing more quickly. The man would perceive this, and his light would flash more, etc. etc. Soon both lights may be flashing at a fast rate. I would hope that both would laugh when this occurred, restoring equilibrium! It would be of great interest do discover if such procedures could teach people to have a much greater sensitivity to each other's behavior which reflects emotional states. If a small group of persons were controlling, via their brain rhythms, a group reinforcement system, so arranged that a reward occurred only if alpha was present in the occipital recordings of all members of the group, one might induce a learning process whereby the alpha would occur coincidently among members of the group more often. With this may come a behavioral similarity as well.

Art and Theatre. There is a possible application of feedback from physiological processes to the stimulating environment which, though not scientific, might be of interest to artists involved in producing novel sensory patterns to evoke more intense and novel psychological and emotional responses. The currrent use of stroboscopic, multi-colored illumination, with unusual and rhythmic auditory stimuli combined with movement and dance illustrate this interest. This can be extended now to permit the artist to control the environment by his own physiological responses. As he learns to control these by feedback training, he may then attempt to achieve a control of the stimulating environment which would result in patterns of light, sound and movement which would be judged to be artistic and aesthetically valuable which, in turn, affect his physiological response. Feedback need not reduce the creative process, nor restrict the response of the individual. It may in fact, add a new means of achieving control over physiological processes by higher nervous functions.

REFERENCES

Adrian, E. D.: The dominance of vision. Ophthalmological Society, U. K., 1943, 63 : : 194—207. — **Adrian, E. D. & Matthews, B. H. C.**: The Berger rhythm: potential changes from the occipital lobes in man. Brain, 1934, 57 : 355—385. — **Avant, L. L.**: Vision in the Ganzfeld. Psychol. Bull., 1965, 64 : : 246—258. — **Bagchi, B. K.**: The adaptation and variability of response of the human brain rhythm. J. Psychol., 1937, 3 : 463—485. — **Bender, M. B.**: The oculomotor system and the alpha rhythm. Paper presented to the Conference on Attention in Neurophysiology, Teddington, England. October 4, 1967. To be published in: C. R. Evans and T. Mulholland (Eds.). Proceedings of the Conference on Attention in Neurophysiology (in preparation). — **Bender, M. B., Shanzer, S.**: Oculomotor pathways defined by electric stimulation and lesions in the brain stem of monkey. In: M. B. Bender (Ed.). The Oculomotor System, New York, Hoeber 1964, 81—140. — **Brumlik, J., Richeson, W. B., Arbit, J.**: The origin of certain electrical cerebral rhythms. Brain Research, 1967, 3 : 227—247. — **Bundzen, P. V.**: Autoregulation of functional state of the brain: an investigation using photostimulation with feedback. Fed. Proc. Translation Supplement, July-August 1966, 25 : No. 4, T551—T554. Translated from: Fiziol. Zh. (Moskow), 1965, 51: No. 8, 936. — **Corriol, J. et Gastaut, H.**: Le circuit déclancheur électronique de H. Shipton et W. Grey Walter. Son application en électroencéphalographie clinique. Rev. neurol., 1950, 82 : 608—613. — **Dewan, E. M.**: Communication by electroencephalography. Air Force Cambridge Research Laboratories, Special Report No. 12, 1964, 1——7. — 11. **Dewan, E. M.**: Communication by voluntary control of the electroencephalogram. Proc. Symp. Biomedical Engineering, Marquette University, June, 1966. — **Dewan, E. M.**: Occipital alpha rhythm, eye position and lens accommodation. Nature, 1967, 214 : 975——977. — **Durup, G., Fessard, A.**: L'eléctrencephalogramme de l'homme. Observations psychophysiologiques relatives a l'action des stimuli visuels et auditifs. L' Ann. psychol., 1935, 36 : 1—35. — **Fenwick, P. B. C., Walker, S.**: The effect of eye position on the alpha rhythm. Paper read at the Conference on Attention in Neurophysiology, Teddington, England, October 4, 1967. To be published in: C. R. Evans and T. Mulholland (Eds.). Proceedings of the Conference on Attention in Neurophysiology (in preparation). — **Hammond, P. H.**: The control of artifical limbs. Discovery, March, 1966, 21—25. — **Hart, J. T.**: Autocontrol of EEG alpha. Paper presented at the Seventh Annual Meeting of the Society for Psychophysiological Research, San Diego, October 20—22, 1967. — **Hewlett, M. G. T.**: An electronic trigger mechanism. Electroenceph. clin. Neurophysiol., 1951, 3 : : 513—516. — **Jasper, H. H., Penfield, W.**: Electrocorticograms in man: Effect of voluntary movement upon electrical activity of the precentral gyrus. Arch. Psychiat., 1949, 183 : : 163—174. — **Kamiya, J.**: Conditional discrimination of the EEG alpha rhythm in humans. Paper presented at the meeting of the Western Psychological Association, San Francisco, 1962. — **Kamiya, J.**: EEG operant conditioning and the study of states of consciousness. In: D. X. Freedman (Chm.) Laboratory studies of altered psychological states. Symposium at American Psychological Association, Washington, D. C., Sept. 4, 1967. — **Kamiya, J.**: Conscious control of brain waves. Psychology Today, 1968, 1 : 56—60. — **Kris, E. C.**: Electrophysiological sensitive measures of alertness. NASA ER-2372, Sept. 1966. National Aeronautics and Space Administration, Electronics Research Center, Cambridge, Massachusetts, 1966. — **Kris, E. C.**: EEG and EOG measurement while learning to position the eyes. Electroenceph. clin. Neurophysiol., 1968, 24 : 189. — **Kuprianovich, L.**: The process of instruction during sleep can be regulated. Technika-Molodezhi (Technology to youth) No. 11, Moscow, 1965 : 26—28. — **Latour, P. L.**: Visual threshold during eye movements. Vision Res., 1962. 2 : 261—262. — **Lombroso, S. T.**: Electrophysiological observations during voluntary saccades, Discussion presented to the Conference on

Attention in Neurophysiology. Teddington, England, October 3, 1967. To be published in: C. R. Evans and T. Mulholland (Eds.), Proceedings of the Conference on Attention in Neurophysiology (in preparation). — **Morrell, L., Morell, F.**: Non-random oscillation in the response-duration curve of electrographic activation. Electroenceph. clin. Neurophysiol., 1962, 14 : 724—730. — **Mulholland, T.**: The electroencephalogram as an experimental tool in the study of internal attention gradients. Trans. N. Y., Acad. Sci., 1962, Ser. II, 24, 6 : 664—669. — **Mulholland, T.**: Variations in the response-duration curve of successive cortical activation by a feedback stimulus. Electroenceph. clin. Neurophysiol., 1964 16 : 394—395. — **Mulholland, T.**: The concept of attention and the electroencephalographic alpha rhythm. Paper read at the Conference on Attention in Neurophysiology, Teddington, England, October 3, 1967. To be published in: C. R. Evans and T. Mulhollands (Eds.), Proceedings of the Conference on Attention in Neurophysiology (in preparation). — **Mulholland, T., Davis, E.**: Electroencephalographic activation: nonspecific habituation by verbal stimuli. Science 1966, 152 : : 1104—1106. — **Mulholland, T., Evans, C. R.**: An unexpected artefact in the human electroencephalogram concerning the alpha rhythm and the position of the eyes. Nature, 1965, 207 : 36—37. — **Mulholland, T., Evans, C. R.**: Oculomotor function and the alpha-activation cycle. Nature 1966, 211 : 1278-1279 — **Mulholland, T., Runnals, S.**: A brain response-sensory stimulus feedback system. Digest of Papers, 1961 International Conference on Medical Electronics. L. E. Flory (Chm.), Princeton, New Jersey, 1961, 166. — **Mulholland, T., Runnals, S.**: Evaluation of attention and alertness with a stimulus brain feedback loop. Electroenceph. clin. Neurophysiol., 1962, 14 : 847—852 (a). — **Mulholland, T., Runnals, S.**: A stimulus-brain feedback system for evaluation of alertness. J. Psychol., 1962, 54 : 69—83 (b). — **Mulholland, T., Runnals, S.**: The effect of voluntarily directed attention on successive cortical activation responses. J. Psychol., 1963, 55 : 427—436. — **Mulholland, T., Runnals, S.**: Cortical activation during steady and changing stimulation. Electroenceph. clin. Neurophysiol., 1964, 17 : : 371—375. — **Mulholland, T., Runnals, S.**: Cortical activation by alternate visual and auditory stimuli. Cortex 1964b; 1 : 225-232. — **Nowlis, D. P.**: Early observations on a system providing EEG alpha feedback. Hawthorne House Research Memorandum 78, 1968. — **Oswald, I.**: The human alpha rhythm and visual alertness. Electroenceph. clin. Neurophysiol., 1959, 11 : 601. — **Runnals, S., Mulholland, T.**: A method for the study of bilateral asymmetry of cortical activation. Amer. J. EEG Technol., 1964, 4 : 15—18. — **Runnals, S., Mulholland, T.**: Selected demonstrations of voluntary regulation of cortical activation. Bedford Research, 1965, 11 : 26. — **Shipton, H. W.**: An electronic trigger circuit as an aid to physiological research. J. Brit. Inst. Radio Engrs., 1949, 4 : 374—383. — **Simonov, P. V., Valueva, M. N.**: On physiological mechanism of conditioned switching in man. EPEBAH, 1966, 387—400. — **Sokolov, Y. N.**: Perception and the conditioned reflex. Pergamon, New York, 1963. Trans. from Vospriyatiye Uslovnyl Refleks. Moscow, Univ. Press Moscow 1958. — **Stark, L.**: Stability, oscillations and noise in the human pupil servomechanism. Proc. IRE, 1959, 47 : 1925— —1939. — **Stark, L., Kupfer, C., Young, L.**: Physiology of the visual control system. NASA Contract Report 238, National Aeronautics and Space Administration, Washington, D. C., June, 1965, 31. — **Turton, E. C.**: An electronic trigger used to assist in the EEG diagnosis of epilepsy. Electroenceph. clin. Neurophysiol., 1952, 4 : 83—91. — **Walter, V. J., Walter, W. G.**: The central effect of rhythmic sensory stimulation Electroenceph. clin. Neurophysiol., 1949, 1 : 57—86. — **Young, L. R.**: A sampled data model for eye tracking movements, Sc. D. Thesis, Massachusetts Institute of Technology, Cambridge, Mass., 1962. — **Zikmund, V.**: Eye movements during orienting reaction and during some kinds of mental activity. Mechanisms of the Orienting Reaction in Man, International Colloquium, Smolenice Bratislava, Sept., 1965. SAV, Bratislava, 1967, 93—99. — **Zikmund, V.**: The time course of the oculomotor component of orienting reaction. Paper read at the Conference on Attention in Neurophysiology, Teddington, England, October 5, 1967. To be published in: C. R. Evans and T. Mulholland (Eds.), Proceedings of the Conference on Attention in Neurophysiology (in preparation).

T. M., Perception Laboratory Veterans Administration Hospital Bedford, Massachusetts, USA.

40

EEG Correlates of Sleep: Evidence for Separate Forebrain Substrates

Maurice B. Sterman and Wanda Wyrwicka

The discovery of electrical activity from the surface of the brain provided a means for an objective investigation of a behavioral phenomenon which had puzzled man since the dawn of history, sleep. Yet, paradoxically, the development of the EEG in this pursuit has been meager in succeeding years which have seen ever-increasing sophistication in the use of brain stimulation and lesion procedures, evoked neuroelectric activity, and unit discharge recordings. Many neurophysiologists today tend still to record from the surface of the cerebral hemispheres as if they were a homogeneous neural mass, paying slight attention to the fact that somewhat different kinds of electrical activity may be observed from different cortical areas. Moreover, the terms 'synchronized' and 'desynchronized' express the accepted level of differentiation of EEG patterns which one encounters in much of the current animal literature.

A greater resolution of the EEG has been accomplished by investigators interested in the neural substrates of learning and reinforcement. Localized EEG recordings obtained from cats engaged in instrumental performance have disclosed a number of discrete EEG slow wave phenomena associated with equally distinctive behaviors. Out of these findings has emerged a growing realization that the EEG may provide much more than just an electrophysiological correlate of sleep.

EEG patterns during wakefulness

Two distinctive EEG rhythms seen in the alert cat have been thoroughly investigated in our laboratory over the past few years. Both of these phenomena can be related to behavioral inhibition and, as will be shown, to sleep. Moreover, there is good reason to believe that the neural mediation of these rhythms involves the forebrain structures of interest to this group.

Adult cats prepared surgically with pairs of small stainless steel screws placed into the skull over specific cortical gyri were placed in a behavior chamber and attached by way of a cable, slipring assembly and counterweight system to a Grass VI polygraph. After habituation to the chamber and recording cable they were trained to depress a lever in order to obtain fortified liquid food from an adjacent feeding

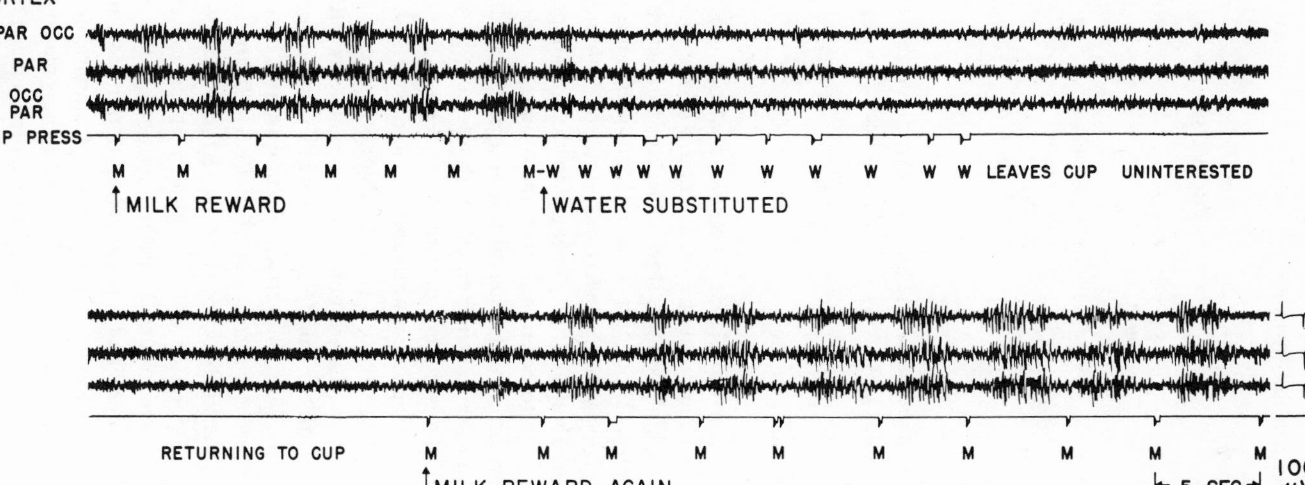

Fig. 1. Examples of post-reinforcement synchronization (PRS) obtained during consumption of a milk reward. When water is substituted as the reward, the food-deprived animal continues pressing and drinking for some time, but does not develop this slow wave rhythm. Upon returning to the lever and finding milk once again, the animal proceeds with his work and gradually demonstrates prominent and sustained PRS. (From Clemente *et al.*[4].)

TABLE I

COMPARISON OF POST-REINFORCEMENT EEG SYNCHRONIZATION WITH VARIATION OF REWARD SUBSTANCE IN THE CAT

(A) SC No. 62, food deprived		*(B) SC No. 60, water deprived*	
Reward substance	*Mean % PRS in first 25 responses*	*Reward substance*	*Mean % PRS in first 25 responses*
Complete liquid food	89 ± 5	Fortified milk	68 ± 14
Chicken broth	78 ± 6	Dextrose (10%)	57 ± 12
Fortified milk	67 ± 8	Water	51 ± 16
Water	0	Quinine (0.8%)	17 ± 6

apparatus. Other experiments required the animal to withhold this response during a given tone or light. Continuous EEG recordings were collected during the performance of these tasks.

Post-reinforcement synchronization

If one records electrocortical activity from the posterior areas of the cerebral cortex in a cat during the consumption of a small amount of wet food he will observe a brief shift in the EEG to high voltage, slow (synchronized) activity. When we first observed this phenomenon we thought it to be most peculiar, since this type of EEG activity is generally associated with the behavioral states of drowsiness and sleep and was not expected from an animal actively engaged in the consumption of milk. The instrumental situation proved to be an important factor in this observation, since the desynchronized EEG during one instrumental response and preceding the next provided a clear demarcation of this phenomenon[4]. With reference to the objective *event* of reinforcement, we termed this rhythm 'post-reinforcement synchronization', or PRS. The PRS is specific to the appropriate reward and develops progressively as the animal becomes relaxed in the feeding situation (Fig. 1). Quantitative evaluation[11] indicated that the percentage of instrumental responses accompanied by PRS remains stable for a given substance, but differs among substances in relation to their desirability or appropriateness (Table I). Moreover, it is important to note that a novel stimulus presented during the occurrence of an episode of PRS will abolish the slow wave activity without necessarily disturbing the animal's performance[4]. We have observed, also, that PRS activity is suppressed in a very hungry or otherwise aroused animal.

More recent studies have localized this activity on the dorsal cortical surface to the middle and posterior lateral and suprasylvian gyri in the cat, and determined its frequency range among different animals at 4–12 c/sec[7]. Comparisons of EEG patterns during various behavioral states have shown that PRS activity may develop also during feeding *ad libitum*, and that this activity is identical in frequency and topography to the slow wave EEG pattern seen in drowsiness or at sleep onset (Fig. 2).

In attempting to determine, specifically, the behavioral significance of the PRS

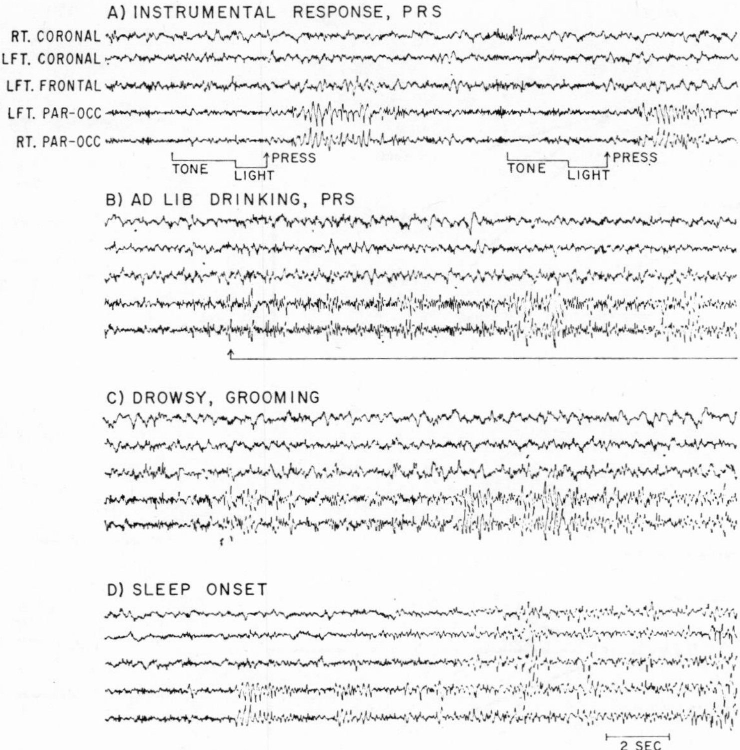

Fig. 2. Localized EEG recordings obtained from the same animal under several different circumstances. Note the similarity in configuration of the slow wave rhythms from dorsal posterior cortex present in each case.

pattern, we have employed a novel conditioning procedure. If an untrained hungry animal is placed in the behavior chamber, and the apparatus is arranged to provide food whenever this rhythm appears spontaneously, its behavioral correlate can be actively elicited. This is achieved by passing the PRS rhythm through a series of finely tuned frequency filters to determine the frequency of peak voltage for a given animal. Whenever this frequency occurs at a level significantly above background voltage a relay is closed which, in turn, operates the feeder apparatus (Fig. 3). Under these circumstances, 4 animals demonstrated a characteristic development of conditioned behavioral and EEG responses (Fig. 4). At first the animal merely became quiet, often sitting in front of the feeder. Within a dozen or so trials he reclined regularly after drinking the reward, and sometimes failed to approach the feeder in response to its operation. Finally, within 30–50 reinforcements, the animal disregarded the feeder entirely and responded to its operation by demonstrating drowsy behavior and generalized EEG synchronization. If a novel stimulus was presented at this point the animal would once more approach the feeder and consume the reward. Under control conditions animals will normally consume 200 or more reinforcements in a

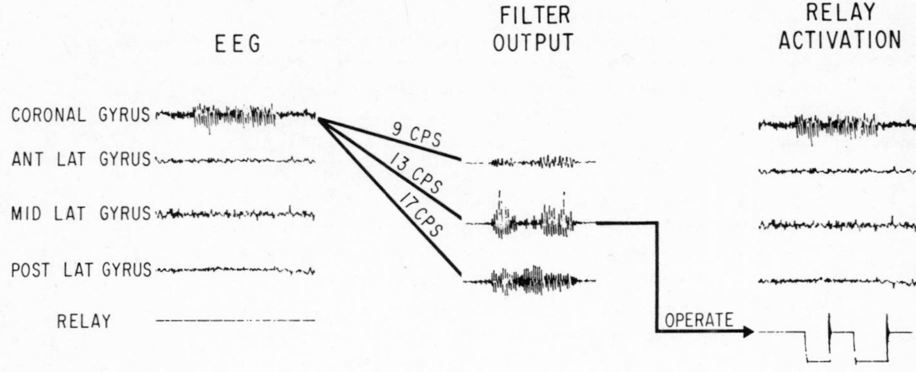

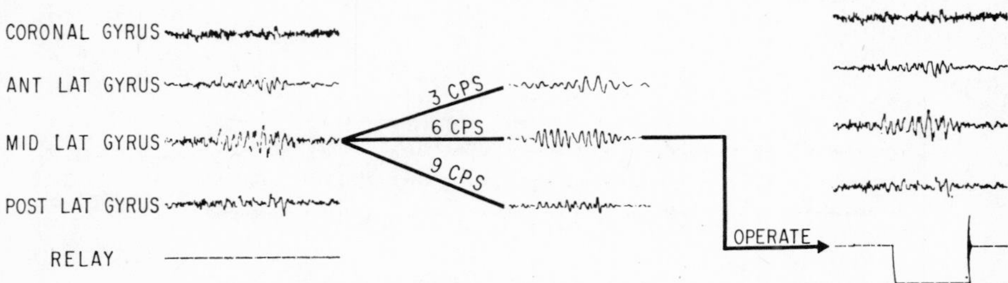

Fig. 3. Frequency analysis of post-reinforcement synchronization (bottom traces) showing how peak frequency filter is used to operate a relay switch. Traces at top show a similar analysis of the sensorimotor rhythm from coronal gyrus, which is discussed later in the text.

given session. We have interpreted these findings as indicating that drowsiness and satiety, the central nervous system substrates of which we feel are one and the same, were here conditioned by making the presentation of food contingent upon the occurrence of their common EEG correlate[13]. Furthermore, this conditioning reflected the development of an active inhibitory response with regard to behavior. We thus propose that the PRS be associated functionally with an active process related to the reduction of drive.

The sensorimotor rhythm

A different, and equally discrete EEG rhythm has been recorded from the sensorimotor cortex of animals during various behaviors whose common feature is the withholding of some class of somatomotor response[1,5,7,8]. In our experience, this sensorimotor rhythm (SMR), as we have labeled it, consists of a 12–20 c/sec burst or train of synchronized activity, of a somewhat lower voltage than the PRS, and

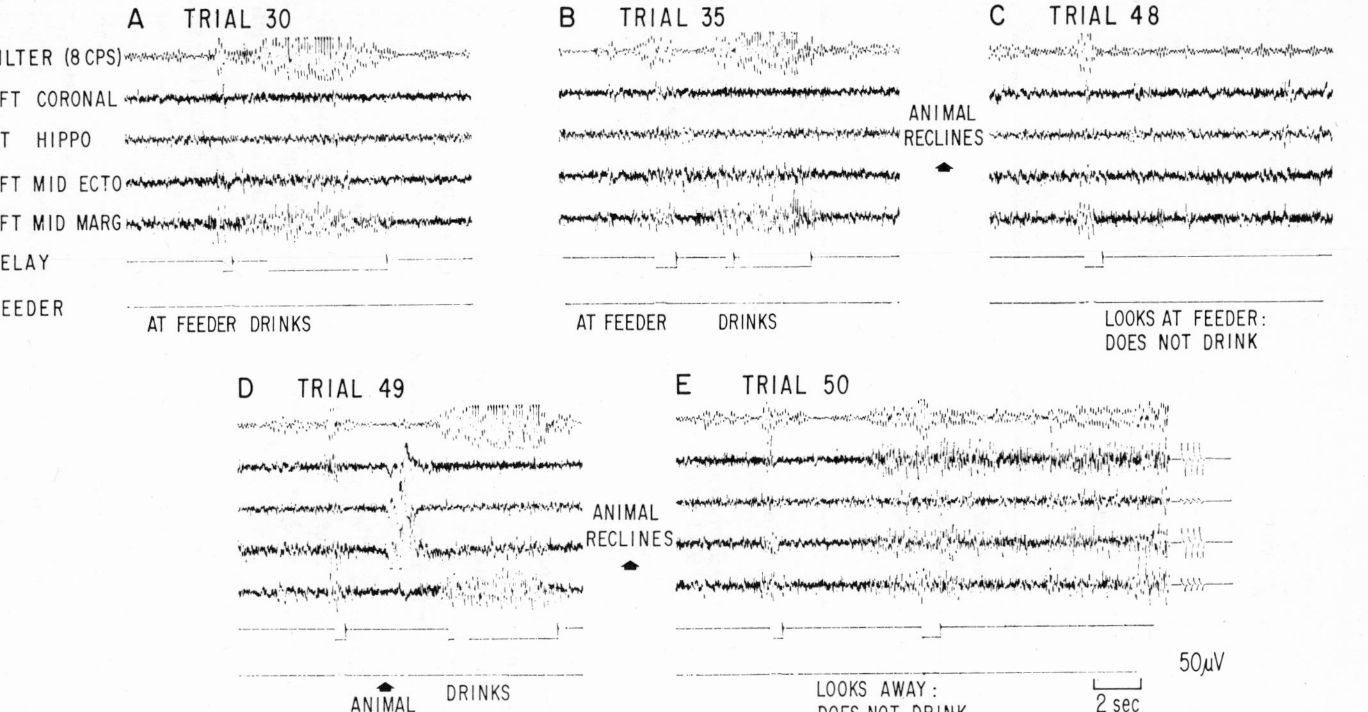

Fig. 4. EEG and behavioral changes associated with automatic reinforcement of spontaneous PRS-like rhythms from lateral (also called marginal) gyrus. A 12 sec delay circuit is activated during drinking so that corresponding PRS does not continually activate feeder. Animal first becomes quiet, then reclines frequently between trials and occasionally fails to respond. Eventually (in 50 trials or less) the animal disregards the food and responds to the feeder click with an EEG pattern of light sleep. The primary PRS frequency in this animal was 8 c/sec, as is shown in tuned filter response (top trace).

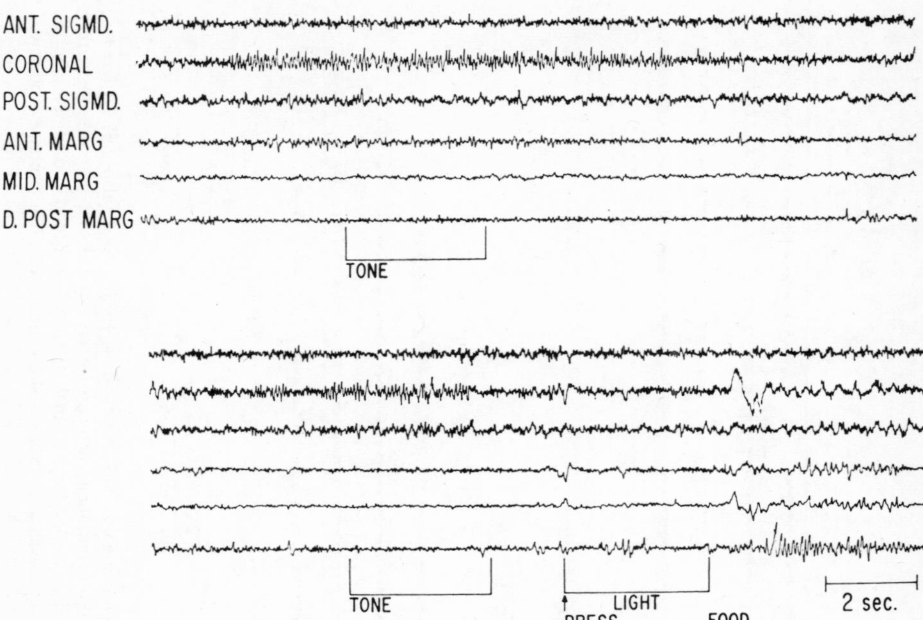

Fig. 5. The sensorimotor rhythm from coronal gyrus is shown here elicited in relation to a negative conditioned stimulus. The animal has learned not to press the lever during sounding of the tone. A lever-press after the tone activates a light which is followed by food. Note localized PRS from lateral gyrus during milk consumption in bottom traces, and delay of lever pressing in relation to prolonged SMR in top record.

localized instead to the coronal gyrus of the cat[7]. The SMR is observed in response to a negative conditioned stimulus (Fig. 5), in the extinction of a learned alimentary response, and, spontaneously, in an alert but quiet animal. This rhythm is quite obvious if electrodes are properly placed over coronal gyrus, since it typically arises from the characteristic pattern of low voltage fast activity seen in the motor cortex of a waking animal.

If the presentation of food to a hungry cat was made contingent upon the occurrence of this rhythm, a characteristic pattern of stereotyped behavior developed in association with a conditioned SMR (Fig. 6). The procedure was identical to that described above for auto-conditioning of the PRS; however, in this case SMR activity operated the feeder mechanism. Each animal became very alert, but gradually assumed a stereotyped motionless stance. The characteristic postures differed from one animal to the next, but a consistent feature among the 4 animals studied to date was an almost intense cessation of movement. When these postures were assumed and maintained the SMR always followed. Animals trained in this manner were able to obtain 150–200 reinforcements within a 2 h session. Other behavioral tests, such as conditioning of control EEG patterns and extinction, confirmed the fact that the SMR had become a conditioned EEG response. Behaviorally, this EEG response was clearly correlated with volitional somatomotor inhibition.

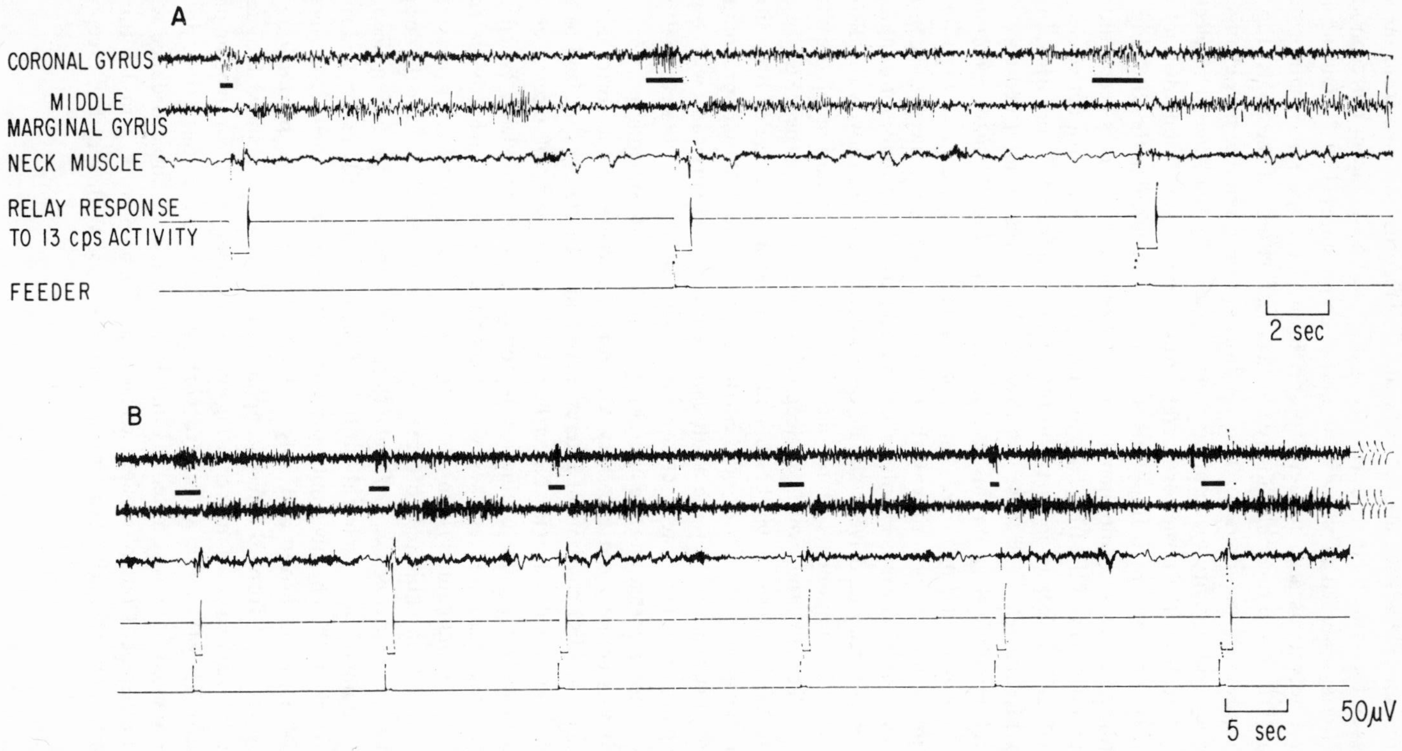

Fig. 6. Example of conditioned sensorimotor rhythm in the cat. In A, bursts of the SMR are seen to operate the relay and feeder at regular intervals in the trained animal. A stereotyped alert but motionless behavior developed in association with the recurrent appearance of this rhythm. B shows the same phenomenon recorded at a slower paper speed.

Comparison of the PRS and SMR with EEG records obtained from the same cortical sites in the same animals during sleep disclosed a significant correspondence between patterns observed in the waking and sleeping states (Fig. 7). During sleep, spindle burst activity is seen to develop specifically over the sensorimotor cortex (Fig. 7C and D), and to resemble closely the electrical pattern of the SMR (Fig. 7B). As sleep progresses the configuration and amplitude of sleep spindles may change, but their frequency and primary localization remain essentially the same. The electrical activity recorded from the posterior lateral gyrus during sleep is quite distinct from this. It consists of periodic high amplitude, slow activity identical to the PRS (Fig. 7A), together with an increasing component of large 2–4 c/sec slow waves as sleep continues. It appears that in terms of the frequency and topography of these distinctive electrical patterns, the EEG of sleep is, in part, constituted of both the SMR and PRS. In view of the behavioral correlates of these two rhythms which were established above, can we not therefore define sleep on this level as *the volitional inhibition of somatomotor activity and the reduction of drive*? This conclusion was subjected to experimental evaluation in our most recent study. If the EEG pattern of sleep reflects the contributions of two separate neural mechanisms, one associated with somatomotor inhibition and the other with reduced drive, it should be possible to alter selectively the distribution of these two components of the sleep EEG by manipulation of the factors antecedent to sleep. In other words, if sleep is initiated by satiety the PRS-like component of its EEG pattern should be enhanced, whereas if it is initiated by volitional decision in the absence of satiety the SMR-like component should predominate.

To test this hypothesis, three adult cats were prepared surgically with bipolar electrodes (pairs of stainless steel screws) placed specifically over the coronal gyrus on the sensorimotor cortex and the lateral gyrus on the posteromedial portion of the cerebrum. It was actually not necessary to record elsewhere on the cortex in these animals, since the EEG rhythms of interest here had been clearly localized to these areas in previous investigations. Nevertheless, adjacent pairs of electrodes were included to provide a similar evaluation of topography during sleep. When these animals recovered from surgery they were habituated to the previously described behavior chamber and cable system and overtrained in the task of lever-pressing for milk. They were then subjected to three distinct experimental treatments designed to establish *satiation*, *extinction*, *or fatigue* as specific conditions preceding sleep. For satiation, the animals were prefed about half of their normal daily ration and simply allowed to work in the chamber until they were completely satisfied. In extinction, they were 24 h food-deprived and only allowed 50 normal trials before food was permanently removed from the feeder apparatus. Finally, to achieve fatigue the animals were placed in a constantly rotating activity wheel for 8 h prior to being placed in the chamber. They then were allowed to work in the chamber as long as they desired. Each of these conditions was tested 3 times in every animal, in accordance with a randomized schedule. Control sessions consisting of a brief work period within the chamber were alternated with experimental test sessions. With very few exceptions,

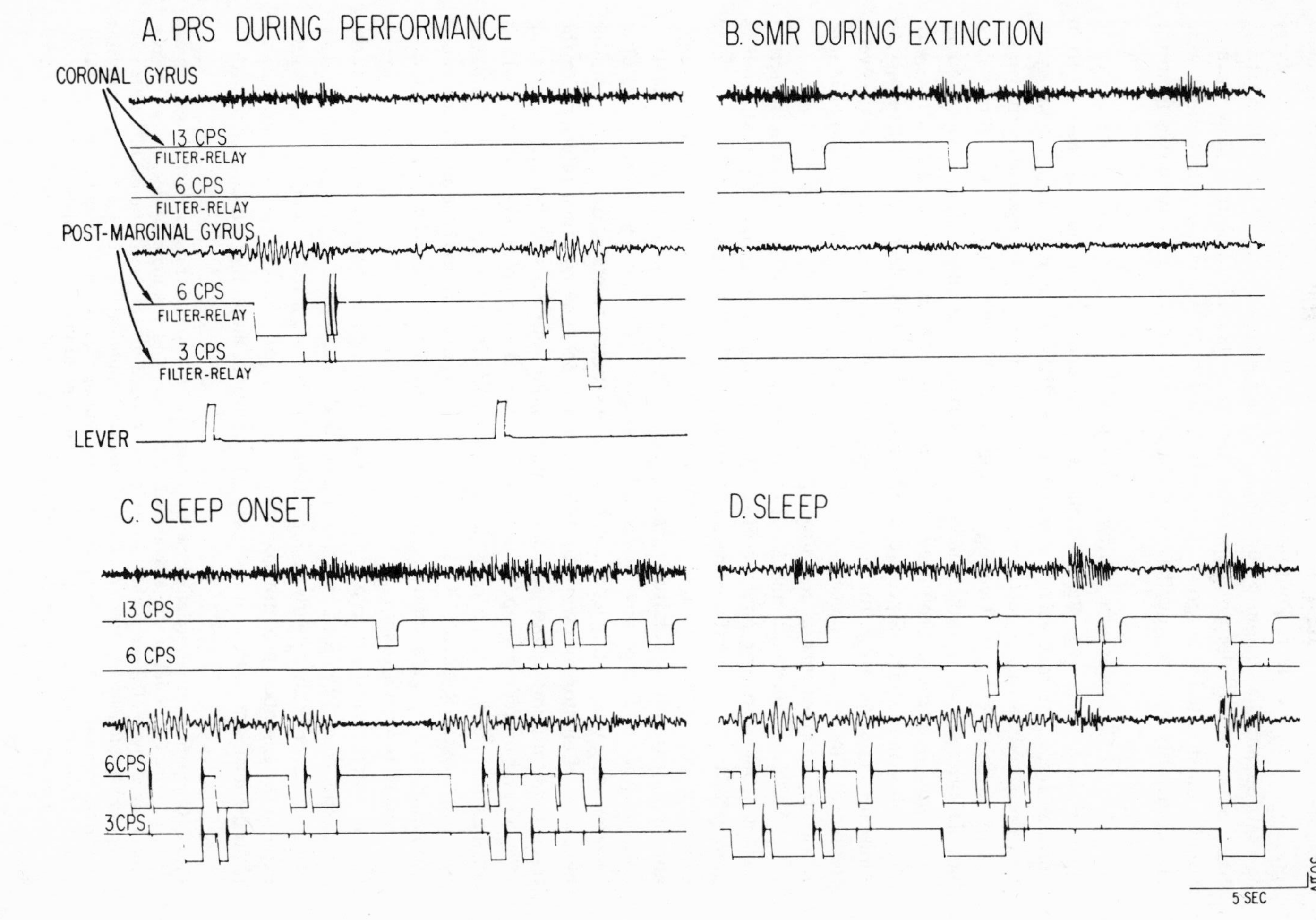
A. PRS DURING PERFORMANCE
B. SMR DURING EXTINCTION
CORONAL GYRUS
13 CPS
FILTER-RELAY
6 CPS
FILTER-RELAY
POST-MARGINAL GYRUS
6 CPS
FILTER-RELAY
3 CPS
FILTER-RELAY
LEVER
C. SLEEP ONSET
D. SLEEP
13 CPS
6 CPS
6CPS
3CPS
50µV
5 SEC

these treatments led more or less rapidly to the occurrence of sleep following lever-pressing in the chamber. Continuous EEG records were obtained throughout every test session, and the sessions were terminated after at least 10 min of EEG sleep activity had been recorded. Sleep onset was defined as a maintained shift to slow wave or spindle burst activity progressing uninterrupted to a pattern in which both were simultaneously present, together with the addition of 3 c/sec slow waves from the lateral gyrus.

The EEG records were analyzed automatically by a filter relay apparatus, which quantified, in 5 sec units, the independent occurrence of spindle burst activity from coronal gyrus, the independent occurrence of PRS-like activity from lateral gyrus, and the simultaneous occurrence of both of these rhythms, together with 3 c/sec activity from lateral gyrus (see Fig. 7C and D for examples). During the 10 min sleep samples brief periods of wakefulness were observed under all conditions. These tended to be most frequent or prolonged with sleep following extinction, and least so with sleep following fatigue. Since we were interested only in sleep, it was necessary to omit these episodes from analysis. The various components of the sleep EEG were, therefore, converted to percentages of total sleep time, in order that comparisons related to the experimental conditions preceding sleep could be achieved.

Analysis of behavior and EEG rhythms during the interval between the termination of lever-pressing and the onset of sleep yielded some interesting differences among the results of the three different treatments. The fatigued animal went quickly to sleep after a variable number of responses, and the EEG showed a correspondingly abrupt shift from low voltage fast activity in both cortical areas to spindle bursts in coronal gyrus and large slow waves in lateral gyrus. The latency of sleep onset after lever-pressing was usually less than 2 min. The satiated animal showed more variable behavior, often stopping work to groom for a short time and then returning to the lever. It was not unusual to see a substantial amount of PRS activity during grooming (Fig. 2) and preceding sleep. Several animals would stop grooming and stare with a fixed gaze at the feeder, during which time the EEG from the coronal gyrus displayed long trains of SMR activity. This was terminated by a return to the lever, by continued grooming, or by the assumption of a sleep posture. Sleep onset was usually rapid once the animal stopped grooming or lever-pressing. Finally, in extinction, animals showed the most stereotyped presleep behavior of all. Although sleep onset was usually variable, taking from 10 min to over 1 h, the animals characteristically alternated periods of frantic lever-pressing with periods of equally intense lever avoidance. They would explore the chamber, groom vigorously, and eventually squat down at some distance from the lever and stare at it intensely. On occasion, an animal would recline with its

Fig. 7. Comparison of post-reinforcement synchronization and sensorimotor rhythm with activity from the same electrodes in the same animal at the onset and at a more advanced stage of sleep. Frequency analysis of these rhythms (see Fig. 3) provided for the precise tuning of a filter relay system which automatically registered their occurrence. Additional filter relay channels are shown at other frequency values found to be predominant also during sleep. Both visual and automatic analysis of these records indicates that the two rhythms are present during sleep. In addition, 6 c/sec activity appears over coronal gyrus and 3 c/sec activity becomes increasingly prevalent from lateral (marginal) gyrus.

back to the lever and specifically avoid looking at it. During periods of motor quiescence, the EEG tracing from coronal gyrus displayed frequent and prolonged bursts of the SMR, while the lateral gyrus showed only occasional episodes of slow wave activity. The contrast in amount of SMR activity between extinction and satiation is demonstrated in Fig. 8. The dominant and unique feature of the EEG during extinction was a significant increase in SMR activity in relation to the suppression of motor activity.

In the transition from the EEG patterns of wakefulness to those of sleep, there was an essentially complete dissociation between rhythmic activity from the coronal and lateral gyri. Prolonged trains of 12–20 c/sec coronal discharge were accompanied by high voltage 4–12 c/sec slow waves from lateral gyrus, and the waxing and waning of one did not occur in any obvious relationship to comparable fluctuations in the other (Fig. 7). The absolute frequency of these distinctive rhythms varied between animals within the ranges noted above; however, the modal values were 13 c/sec for coronal gyrus and 6 c/sec for lateral gyrus, which corresponded exactly to modal values for the SMR and PRS, respectively. The configuration, frequency and localization of rhythmic activity from the coronal gyrus was identical to the SMR for a given animal. Although slow wave activity was more diffusely present over the cerebral hemispheres, and particularly so in classical sensory projection areas, peak amplitudes developed at the middle and posterior lateral and suprasylvian gyri at PRS frequencies. There was a more or less gradual shift to increasing 3 c/sec activity from dorsal cortex as sleep prevailed. Eventually some bursting activity appeared in posterior cortical areas, but large slow waves were only rarely observed over motor cortex. Whenever

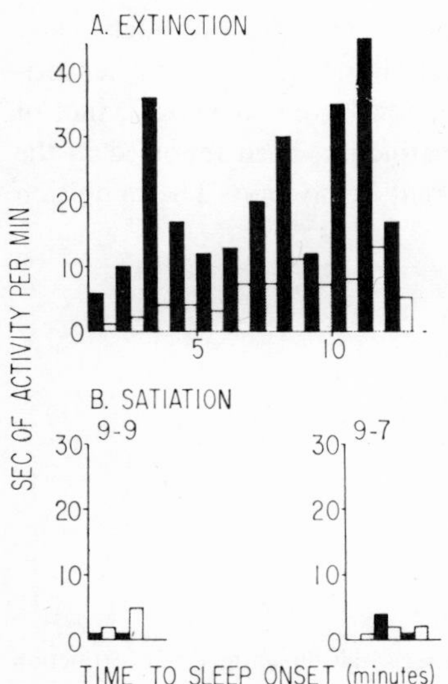

Fig. 8. Comparison of the postpress occurrence of coronal gyrus sensorimotor rhythm (black) and lateral gyrus post-reinforcement type activity (white) after presleep treatments of extinction and satiation. The unique feature of the EEG during extinction is a predominance of the sensorimotor rhythm during periods of motor quiescence.

the animal moved towards a 'lighter' stage of sleep the dissociation again became complete.

Analysis of electrocortical activity during sleep indicated that the three presleep treatments produced significantly different sleep patterns. These differences are exemplified in Fig. 9, which provides the mean distribution of independent and combined activity from sensorimotor and dorsal cortical leads, obtained with three replications of each treatment in one of the animals. With fatigue the predominant pattern was combined activity with relatively little independent spindle or slow wave discharge. Large slow waves (3 c/sec) from lateral gyrus were present and sustained very shortly after the onset of sleep. In contrast to this, with extinction the distribution is essentially equally divided between the three possible configurations. This treatment was unique in evoking large amounts of independent spindle burst activity. Sleep in the satiated animal was characterized by independent lateral gyrus slow wave activity alternating with combined activity. There was very little independent coronal spindle activity in this condition. Similar trends were apparent in the other animals tested.

COMMENTS AND CONCLUSIONS

A frequency analysis of the sleep EEG in cats was performed as early as 1953 by Hess *et al.*[6]. In this study cortical electrodes were placed 10 mm apart, over frontal, parietal and occipital areas, and bipolar recordings obtained. A succession of dominant, but overlapping frequencies were described through the various stages of sleep, as defined at that time. The placement of recording electrodes, however, made it impossible to discriminate distinct rhythms and led to an unavoidable 'smearing', which masked the kind of distinctions noted in the present study. Although many investigators have alluded to the localization of spindle burst activity over sensorimotor areas and slow wave activity over parietal cortex, few have taken cognizance of this important fact and only recently has a similar distinction been reported in the waking animal and correlated with significantly different behaviors[7]. The definition

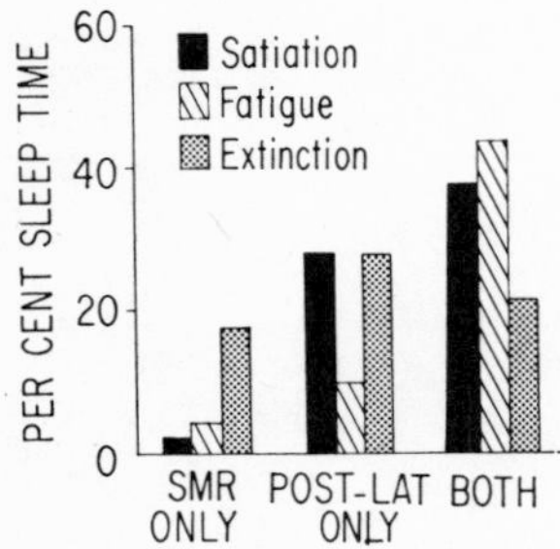

Fig. 9. Distribution of EEG components during sleep following satiation, fatigue and extinction treatments, as described in text. 'SMR only' refers to the independent occurrence of coronal gyrus spindle burst activity, 'Post-lat. only' to independent lateral gyrus slow wave activity and 'Both' to the simultaneous occurrence of these rhythms. At least 5 sec of sustained or recurrent activity was required of each component before it could be registered in the automatic cumulative analysis. Same animal as in Fig. 8.

of EEG rhythms based upon topographical distinctions and behavioral correlates has allowed us to separate, meaningfully, the components of the 'synchronized' EEG and to manipulate their occurrence predictably by the appropriate alteration of antecedent behavioral conditions.

Presumably, these distinctive rhythms have their origins in separate thalamo-cortical feedback circuits. In his illuminating presentation Dr. Schlag has described to us specific projections in the cat from ventrolateral and dorsolateral thalamus to sensorimotor and parietal cortex, respectively. His findings regarding the appearance of augmenting and recruiting responses in these and other areas suggest that they constitute separate functional units of thalamo-cortical interaction. Velasco and Lindsley[14] have shown that orbital cortex lesions abolish spontaneous and induced sensorimotor spindle bursts and Skinner has shown us how this same effect can be obtained more efficiently by lesion or temporary blockade of the inferior thalamic peduncle, a fiber system projecting from midline thalamus to the orbital gyrus in the cat. Finally, our research group has been working for some time on the potent somato-motor inhibition produced by orbito-cortex stimulation[2], and Drs. Sauerland and Nakamura have specified the pathway and mechanism involved. The mechanisms subserving cerebrally initiated somatomotor inhibition and giving rise to the SMR appear, therefore, to involve an interrelationship between motor cortex, thalamus and the orbital cortex. One might speculate, regarding the SMR, that the relatively constant proprioceptive feedback which may be characteristic of general somato-motor inhibition provides conditions suitable for oscillation in the thalamo-cortical circuits involved.

Neurophysiological evidence is somewhat less comprehensive with regard to the mediation of PRS activity. In addition to the evidence provided here by Dr. Schlag, it has been well-demonstrated that the posterior lateral and suprasylvian gyri receive direct projections from the nucleus lateralis posterior, the pulvinar, and the lateral geniculate nucleus in the cat[9]. In fact, the cortical region from which the PRS is recorded in the cat corresponds to the primary and secondary visual projection areas in this species. We are presently exploring the possibility that this type of EEG activity may be characteristic of all sensory projection areas during conditions of reduced drive. In previous experiments we have observed that bilateral destruction of the nucleus of the diagonal band of Broca, or of this fiber pathway itself, leads to a progressive reduction in PRS activity (Fig. 10). Animals continue to work effectively in the instrumental situation, but show evidence of increased food drive and locomotor hyperactivity[12]. Stimulation of these same structures results in the termination of feeding behavior, and, as we have stressed in the past, elicits the behavioral and EEG manifestations of drowsiness and light sleep[10]. These effects appear to be mediated both directly via limbic pathways to the cortex and hypothalamic pathways to the thalamus, and indirectly through connections with midbrain structures which modulate thalamo-cortical functions[3].

The successful dissociation of discrete components of EEG activity achieved here by manipulation of presleep conditions suggests that sleep does, in fact, reflect EEG activity characteristic of the operation of at least two independent forebrain

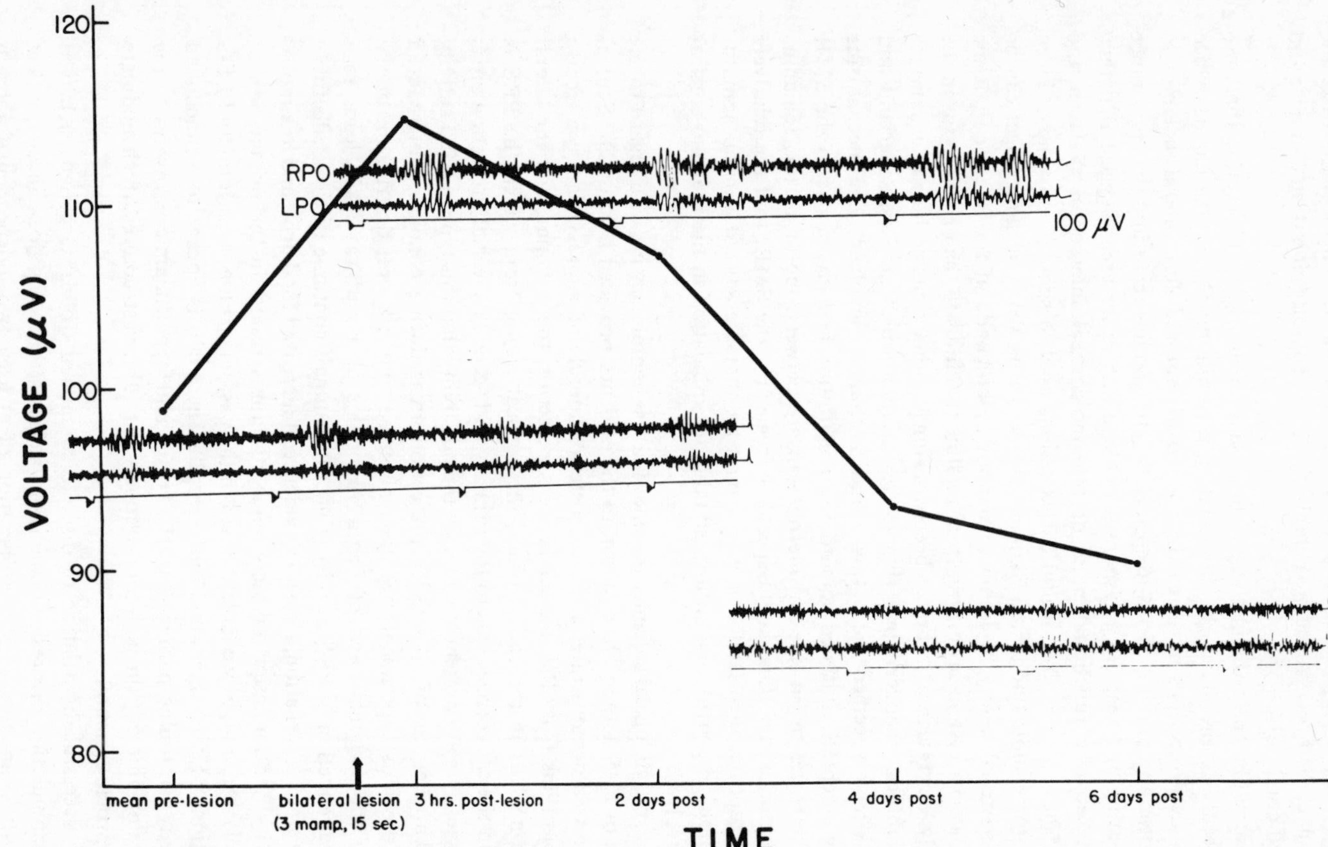

Fig. 10. Effect of bilateral destruction of the nucleus of the diagonal band of Broca upon postreinforcement synchronization. Mean voltage changes over time are shown by this curve in addition to representative samples of actual recordings obtained during the various stages. Effect was statistically reliable and sustained for over 5 weeks.

inhibitory mechanisms, one associated with the inhibition of somatomotor function and the other with the subsidence and suppression of basic drives. Although it must be remembered that the investigation of sleep states was preliminary and involved only a few animals, we can nevertheless speculate as to the role of these mechanisms in establishing the sleep states described. Fatigue resulted in an overwhelming predominance of simultaneous spindle burst and slow wave activity from their respective cortical fields. The influence upon individual neural generators was general, reflecting, perhaps, a primitive regulation of forebrain mechanisms by peripheral influences integrated in the brain stem and serving the purpose of restoration. In addition to this kind of an inhibitory influence, satiation acted also to enhance the forebrain substrates of slow wave activity. If we relate these substrates to the structures which appear to modulate the PRS, we may classify them as paleocortical. Older parts of the forebrain, such as the limbic system and hypothalamus, can be ascribed the function of initiating and terminating 'innate' behaviors related to the homeostatic needs of the organism. Finally, extinction was effective in dissociating the two forebrain generators and was distinguished by its specific evocation of rhythmic sensorimotor activity. The SMR is a manifestation of influences originating in the neocortex, and thus reflects the highest level of inhibitory integration. We may attribute this to the most recent of inhibitory mechanisms, which is dependent upon the development and differentiation of a neocortex. Such a classification provides a useful model for a continued assault upon the related problems of inhibition and sleep.

SUMMARY

During experiments designed to localize certain EEG phenomena associated with specific aspects of behavior, we realized the general similarity of these rhythms to several components of the EEG pattern observed during slow wave sleep (SS). A systematic temporal and topographic analysis of these rhythms was thus performed. Cats with pairs of small electrodes placed over every cortical gyrus were trained to press a lever in order to receive food. During the consumption of this reward a 4–12 c/sec slow wave pattern was observed over the posterior lateral and adjacent posterior suprasylvian gyri. We have previously termed this phenomenon 'post-reinforcement synchronization' (PRS), and related it to a quantum-like process of 'drive' reduction. The animals were subsequently trained to withhold responses during a discriminated negative stimulus and a delay stimulus presented between trials. With learned inhibition of response, a 12–20 c/sec slow wave rhythm appeared discretely over the coronal and anterior and posterior sygmoid gyri. We have termed this pattern the 'sensory-motor rhythm' (SMR). These animals were allowed also to sleep in the behavior chamber while EEG recordings were obtained from the same cortical electrodes. Manifestations of both the PRS and SMR were clearly present in the transition from wakefulness through the various stages of SS. We believe that the PRS and SMR rhythms reflect the activity of two independent neural mechanisms operating during wakefulness in association with drive reduction and learned motor inhibition, respectively. The fact that these same rhythms are present in an intensified form during SS suggests the active participation of the same two neural mechanisms in establishing the behavioral and EEG characteristics of this sleep state. By manipulation of these tangible factors as experimental variables antecedent to sleep we hoped to demonstrate (a) that the above-stated proposal is correct, and (b) that sleep, therefore, is not a single entity, but may take many forms depending upon the different physiological mechanisms mediating the needs of the animal.

Three conditions were established prior to sleep observations. Animals were totally *satiated* by prefeeding and unlimited instrumental response at the chamber feeding mechanism, they were kept hungry by allowing a few responses and then instituting an acute *extinction*, or they were *fatigued* by 8 h of constant activity. These manipulations did, in fact, produce distinctive patterns of sleep onset and of sleep itself. With satiation and fatigue sleep onset was rapid, with no distinctive EEG characteristic. After extinction, sleep onset was variable and preceded by significantly enhanced SMR activity. Sleep itself was characterized by enhanced spindle (coronal gyrus) activity in extinction, slow wave (lateral gyrus) activity in satiation, and combined activity in fatigue. These differences agreed, generally, with those predicted by our hypothesis and allow us now to seek several separate neural substrates involved in the mechanism of sleep.

ACKNOWLEDGEMENTS

The authors gratefully acknowledge the technical assistance provided by Mr. Sidney Ross.

This investigation was supported by the Veterans Administration and U. S. Public Health Service, N. I. H. Grant No. 10083.

REFERENCES

1 Anokhin, P. K., Electroencephalographic analysis of cortico-subcortical relations in positive and negative conditioned reactions, *Ann. N. Y. Acad. Sci.*, 92 (1961) 899–938.

2 Clemente, C. D., Chase, M. H., Knauss, T. K., Sauerland, E. K., and Sterman, M. B., Inhibition of a monosynaptic reflex by electrical stimulation of the basal forebrain or orbital gyrus in the cat, *Experientia (Basel)*, 22 (1966) 844–849.

3 Clemente, C. D., and Sterman, M. B., Basal forebrain mechanisms for internal inhibition and sleep, *Ass. Res. nerv. Dis. Proc.*, in press.

4 Clemente, C. D., Sterman, M. B., and Wyrwicka, W., Post-reinforcement EEG synchronization during alimentary behavior, *Electroenceph. clin. Neurophysiol.*, 16 (1964) 355–365.

5 Gastaut, H., Some aspects of neurophysiological basis of conditioned reflexes and behavior. In G. E. W. Wolstenholme and C. M. O'Connor (Eds.), *Ciba Foundation Symposium on the Neurological Basis of Behavior*, Churchill, London, 1958, pp. 255–272.

6 Hess, Jr., R., Koella, W. P., and Akert, K., Cortical and subcortical recordings in natural and artificially induced sleep in cats, *Electroenceph. clin. Neurophysiol.*, 5 (1953) 75–90.

7 Roth, S., Sterman, M. B., and Clemente, C. D., EEG correlates of reinforcement, internal inhibition, and sleep, *Electroenceph. clin. Neurophysiol.*, in press.

8 Rougeul, A., Observations électroencéphalographiques du conditionnement instrumental alimentaire chez le chat, *J. Physiol. (Paris)*, 50 (1958) 494–496.

9 Sprague, J. M., Visual, acoustic and somesthetic deficits in the cat after cortical and midbrain lesions. In D. P. Purpura and M. Yahr (Eds.), *The Thalamus*, Columbia Univ. Press, New York, 1966, pp. 391–414.

10 Sterman, M. B., and Clemente, C. D., Forebrain inhibitory mechanisms: sleep patterns induced by basal forebrain stimulation in the behaving cat, *Exp. Neurol.*, 6 (1962) 103–117.
11 Sterman, M. B., Roth, S., and Clemente, C. D., EEG correlates of instrumental performance, *Fed. Proc.*, 22 (1963) 399.
12 Sterman, M. B., Roth, S., and Clemente, C. D., The effect of brain lesions on EEG correlates of alimentary behavior, *Physiologist*, 6 (1963) 280.
13 Sterman, M. B., Wyrwicka, W., and Roth, S., Electrophysiological correlates and neural substrates of alimentary behavior in the cat. In *Neural Regulation of Food and Water Intake*, N.Y. Acad. Sci. Symp., 1967 (in press).
14 Velasco, M., and Lindsley, D. B., Role of orbital cortex in regulation of thalamocortical electrical activity, *Science*, 149 (1965) 1375–1377.

DISCUSSION

Dr. Scheibel: Dr. Sterman, we have some data that suggest that what you call PRS occurs in the newborn cat. 2 years ago, we implanted kittens within 48 h of birth and then followed them from the viewpoint of their behavior and EEG. We were particularly interested in what psychiatrists call 'patterns of gratification', and we noticed that when a kitten was kept away from its mother for several hours and then put in its box it would run directly for its mother and suckle. Usually, during the first 30 sec to 1 min of frantic suckling, low voltage fast activity was present, but when the kitten began to relax, slow potentials similar to yours began to appear. The kitten slowed down and seemed to enjoy it more, and we called this 'satiety' or 'gratification' suckling.

Dr. Trachtenberg: Dr. Sterman, this post-reinforcement synchronization that you describe is really a 'reward rhythm', isn't it?

Dr. Sterman: We prefer to relate it simply to the event of reinforcement. The term post-reinforcement synchronization is an objective one; it refers to the fact that the rhythm occurs after the reinforcement. If you use the term 'reward' I think you are making a subjective interpretation, and this is something we have tried specifically to avoid.

Dr. Trachtenberg: Well, but what interests me is the way this synchronization is related to the reward. Before receiving the reward, the animal is anticipating it and is excited. Having received the reward, it is relieved from excitation. The spindles appearing, then, may be due, not to an active inhibition but to a release.

Dr. Sterman: I don't think so. In fact there is evidence to the contrary, namely that lesions in the basal forebrain area abolish the PRS and that stimulation of this area induces a PRS-like pattern. This seems to be an active process. When you use the term release, you implicitly suggest something passive. I think that this reintroduces an old source of semantic confusion.

Dr. Villablanca: Dr. Sterman, do you know if stimulation or destruction of the so-called hypothalamic satiety and feeding centers described by Anand and Brobeck[1] has any EEG effects?

Dr. Sterman: Yes, EEG synchronization has been induced by stimulation of the hypothalamic ventromedial nucleus in cats[5] and in rabbits[7]. With lesions in this region, animals become hyperactive and vicious. Dr. McGinty, in our laboratory, is now systematically studying the effects of lesions in the basal forebrain area. Lesions placed medially anywhere within this general area elicit, to a greater or lesser degree, such effects as hyperphagia, hyperactivity, obstinate progression, hypersensitivity, and significant decreases in sleep.

Dr. Villablanca: By what mechanism does the afferent information relative to feeding finally lead to cortical synchronization and sleep? I have observed[9] that tube feeding may promote sudden episodes of paradoxical sleep in chronic decerebrate cats.

Dr. Sterman: Your observations are probably related to influences upon caudal brain stem mechanisms. As for the thalamo-cortical mechanisms we have been discussing here, it may be that EEG synchronization occurs whenever they are left to function autonomously. There is some evidence for this kind of organization. Andersson *et al.* found that deafferentation of a given sensory relay nucleus of the thalamus results in a synchronization of the electrical activity within its specific cortical projection area[2]. Satiety may correspond to the active suppression of afferent impulses which normally elicit or guide innate behaviors such as feeding.

Dr. Hall: Dr. Sawyer, are these EEG rhythms observed following food reinforcement similar to the postcoital rhythms that you have described?

Dr. Sawyer: Yes, probably. It may be interesting to point out that a lesion in the area that Dr. Sterman mentioned makes the female cat permanently anestrous; conversely, implants of estrogen there will bring on permanent estrous behavior.

Dr. Buchwald: Several years ago[3], we made some observations comparable to those that Dr. Sterman has described in his presentation. But we noticed that when the animals were placed in the dark, there was very little EEG synchronization of any kind though the animals responded at about the same rate as they would in the light.

Dr. Sterman: How long did you follow these animals?

Dr. Buchwald: For a number of days. After a few days, the EEG synchronization, indeed, reappeared in the dark.

Dr. Sterman: Doesn't this suggest that the novel aspect of being in the dark perhaps excited the animals? I should have stressed the fact that if the animal is very hungry or in any way overly excited, the PRS is depressed. It only appears when the animal relaxes and concerns himself solely with bar-pressing. I think one can imply, therefore, that we are dealing with some competition between excitatory and inhibitory influences.

Dr. Clemente: This brings forward the more general question of the value of EEG patterns as correlates of behavior.

Dr. Buchwald: The situation of animals placed in the dark is a case of definite dissociation between behavior and EEG rhythms.

Dr. Schlag: So is the cryogenic blockade just described by Skinner and Lindsley.

Dr. Sterman: Yes, and there are other cases: immature animals, brain lesion and pharmacological effects. This does not reflect upon the fact that the EEG is a valuable indicator of the behavioral state of the normal, mature animal. Other important physiological indicators are irregular in infants and become abnormal as a result of lesions and drugs; however, we do not question their validity under usual circumstances.

Dr. Schlag: I would go one step further and say that there may also be a good correlation between the EEG and single unit activity. Some time ago, we measured the intensity of neuronal population activity in the cerebral cortex and mesencephalic reticular formation and compared it to the cortical EEG. Under specific conditions, the correlation was very high indeed—so high that we could predict the evolution of EEG patterns by simply observing the level of neuronal population activity[8]. Now, direct correlations between neuronal population activity and some behavioral states have been pointed out[6].

Dr. Demetrescu: Thalamo-cortical evoked potentials are also useful indices of the state of the animals; maybe these are even more informative than the EEG itself.

Dr. Clemente: It seems to me that there are at least three important areas where stimulation results in the inhibition of ongoing behavior. One is the orbitofrontal cortex, another is the medullary reticular formation, and the third is the anterior lobe of the cerebellum. Now we know that stimulation of the orbitofrontal cortex can synchronize the EEG, and Magnes *et al.*[4] have shown that stimulation of the caudal part of the reticular formation synchronizes the EEG. But what about the cerebellum? I do not know of a similar study concerned with the cerebellum and I think that this would be an interesting subject for study.

1 ANAND, B. K., AND BROBECK, G. R., Hypothalamic control of food intake in rats and cats, *Yale J. Biol. Med.*, 24 (1951) 123–140.
2 ANDERSSON, S. A., NORRSELL, V., AND WOLPOW, E. R., Cortical synchronization and desynchronization via spinal pathways, *Acta physiol. scand.*, 61 (1964) 144–158.

3 BUCHWALD, N. A., HORVATH, E. E., WYERS, E. J., AND WAKEFIELD, C., Electroencephalogram rhythms correlated with milk reinforcement in cats, *Nature (Lond.)*, 201 (1964) 830–831.
4 MAGNES, J., MORUZZI, G., AND POMPEIANO, O., Synchronization of the EEG produced by low-frequency electrical stimulation of the region of the solitary tract, *Arch. ital. Biol.*, 99 (1961) 33–67.
5 MORGANE, J. P., The function of the limbic forebrain-limbic midbrain system in the regulation of food and water intake. In *Neural Regulation of Food and Water Intake*, N.Y. Acad. Sci. Symp., 1967 (in press).
6 PODVOLL, E. M., AND GOODMAN, S. J., Averaged neural electrical activity and arousal, *Science*, 155 (1967) 223–225.
7 KAWAKAMI, M., AND SAWYER, C. H., Induction of behavioral and electroencephalographic changes in the rabbit by hormone administration or brain stimulation, *Endocrinology*, 65 (1959) 631–643.
8 SCHLAG, J., AND BALVIN, R., Background activity in the cerebral cortex and reticular formation in relation with the electroencephalogram, *Exp. Neurol.*, 8 (1963) 203–219.
9 VILLABLANCA, J., Behavioral and polygraphic study of 'sleep' and 'wakefulness' in chronic decerebrate cats, *Electroenceph. clin. Neurophysiol.*, 21 (1966) 562–577.

41

Instrumental Conditioning of Sensorimotor Cortex EEG Spindles in the Waking Cat

Wanda Wyrwicka and Maurice B. Sterman

EXPERIMENTS WITH INSTRUMENTAL conditioning have shown that essentially any behavioral act may be conditioned. Recently, it has become evident that not only somatomotor reactions, but also autonomic responses [9, 10, 15] and electrocortical activity [4, 5] can be successfully regulated through instrumental conditioning. With regard to the latter, the most distinctive changes which can be detected in EEG recordings are those associated with the behavioral transition from wakefulness to sleep. Low-voltage, high-frequency (desynchronized) activity which characterises the alert animal is progressively replaced by a sequence of high-voltage, low-frequency (synchronized) patterns indicative of the several stages of "slow-wave" sleep. Classical conditioning of both spontaneous [12] and electrically-induced [16] EEG slow-wave patterns has been demonstrated in cats. Using a shock-avoidance procedure, Izquierdo *et al.* [4] were able to establish a conditioned EEG desynchronization to tone in sleeping cats, who showed no visible signs of an associated behavioral awakening. Kamiya [5] has obtained evidence for instrumental conditioning of the EEG alpha rhythm in man by using verbal reinforcement.

Our laboratory has been interested in patterns of localized EEG slow-wave activity which can be recorded from the cortex of the awake cat in relation to certain specific behaviors. One of these patterns, which we have termed the "sensorimotor rhythm" or SMR, can be detected over coronal and

---

[1] This investigation was supported by the Veterans Administration and by P.H.S. grant No. MH-10083. Additional support for preliminary studies, carried out at the Nencki Institute of Experimental Biology, Warsaw, Poland, was provided for W.W. by the Foundation's, Fund for Research in Psychiatry, New Haven, Connecticut.

peri-cruciate cortex during the extinction of an instrumental response and in association with negative conditioned responses [11, 13]. This rhythm consists of a 12–20 cps burst of synchronized EEG activity over motor cortex in an awake animal whose EEG from other cortical areas is clearly desynchronized. The specific details of cortical localization, electrical characteristics and other salient features of this rhythm have been published elsewhere [11]. In the present investigation we attempted to utilize this EEG rhythm as an instrumental response and to determine whether or not this response could be subjected to modification through instrumental conditioning procedures. Moreover, we were interested in establishing more clearly the functional significance of the SMR by reference to its direct behavioral correlate.

METHOD

Each of four adult cats (3 males and 1 female, weighing 3–4.5 kg) were surgically prepared with $\frac{1}{8}$-in. jeweler's screws threaded into the skull in pairs over coronal gyrus bilaterally, and in an anterior-posterior sequence, 2 mm apart, over lateral (also termed marginal) gyrus, on one side. After recovery from surgery the animals were observed individually in a sound-attenuated behavior chamber supplied with a one-way viewing window and a relay-operated, dip-type liquid feeder apparatus. They were attached by a 20-connector cable, slip-ring assembly and counterweight mechanism to a Grass VI electroencephalograph.

Training sessions were carried out three times a week on alternate days. The animals were food-deprived for approximately 22 hr when brought to the experimental chamber. They received condensed milk during experimental testing, supplemented by dry meat or fish 1–2 hr after its termination. The animal initially learned to obtain liquid food from the feeder apparatus through manual random presentations of 0.5 cc milk in the feeder cup. In order to achieve the objectives outlined above it was necessary to establish an instrumental conditioning situation, utilizing the SMR as the response to be reinforced. This was accomplished in the following manner. Spontaneous SMR activity, which is during which milk reinforcement was presented randomly. In the fourth session visual detection of an episode of SMR activity lasting for at least one-half second was utilized as a criterion for reinforcement by contiguous presentation of milk. The discrete nature of this rhythm and the contrasting low-voltage pattern of desynchronization in its absence made this criterion quite reliable. After one such "shaping"

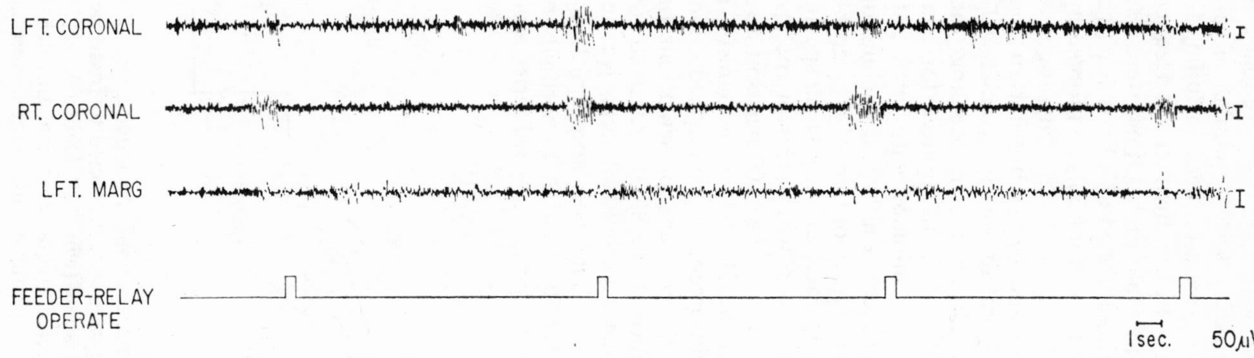

FIG. 1. Example of recurrent conditioned SMR activity recorded bilaterally from sensorimotor cortex in the cat. In this animal milk reinforcement was presented for SMR from right coronal gyrus. Asymmetry of SMR activity from the two hemispheres is usually observed, and may be attributed to differences in electrode placement or individual differences among the animals. Note the absence of spindle activity from marginal gyrus during coronal SMR discharge.

session, reinforcement of the SMR was accomplished automatically. Sensorimotor EEG signals were fed into a series of solid-state amplifiers with twin "T" feedback networks. These networks were tuned to select a number of very narrow frequency bands and to reject all others (including harmonics of selected frequencies). In this manner the frequency components of SMR activity from each animal were determined. Peak frequencies were established as 13 cps for two animals "easily" detected visually, was initially reinforced by manual operation of the feeder device. The rhythm was usually visualized over both left and right sensorimotor areas simultaneously, but was seldom absolutely symmetrical (Fig. 1). Recordings from the electrode pair yielding the highest voltage of SMR activity were thus utilized for conditioning tests. The experiment was initiated with three test sessions and 17 cps for the two others. Calibrated attenuation circuits were used to adjust the voltage output of a given filter. This output was fed to an integrator circuit, which specified that a signal containing at least 4–6 waveforms of the appropriate peak SMR frequency, at a voltage 100 per cent above background level, would activate the feeder mechanism (Fig. 2).

With automatic reinforcement established, conditioning sessions consisted of 50 automatic presentations of milk in relation to the occurrence of the SMR. After 20 consecutive conditioning sessions the experimental series was terminated by an extinction session. This was accomplished by withholding reinforcement for 30 min after 15 normally-reinforced trials. After a two-day interval, a control series was initiated

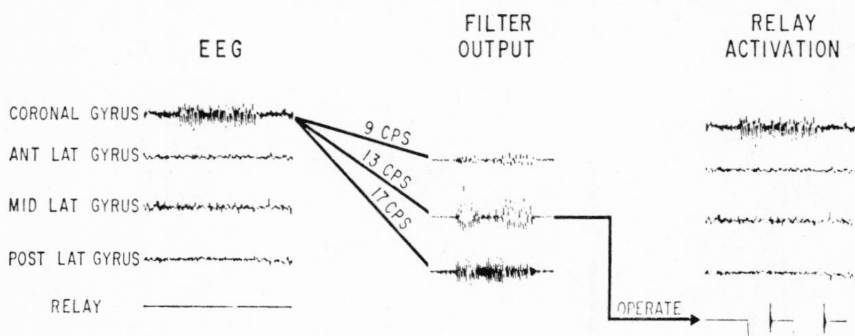

FIG. 2. Samples of spontaneous spindle activity recorded from the coronal gyrus of waking cats were subjected to frequency analysis as shown above. Solid state twin "T" feedback networks then passed the precisely-selected frequency signal through adjustable voltage attenuator and integrator circuits, which were adjusted to establish certain criteria for the operation of a final relay circuit. During training sessions this relay activated the chamber feeder apparatus.

in which milk was presented manually only in relation to a desynchronized pattern visually detected in the coronal gyrus (also utilizing the absence of automatically-registered SMR activity). This series was terminated by an extinction session identical to that described above.

Following another two-day interval, a third series was carried out in which the previously-conditioned SMR was again reinforced during 20 consecutive sessions, followed by an extinction session.

Throughout each of these series visual observations were carried out simultaneously with the recording of EEG activity, and of the relay discharges associated with the occurrence of the SMR (see Fig. 2). Quantification of the SMR for evaluation of experimental effects was achieved by measurement of the frequency and duration of relay closure. During extinction sessions the relay system continued to indicate the presence of the SMR, but was no longer connected to the feeder mechanism of the chamber.

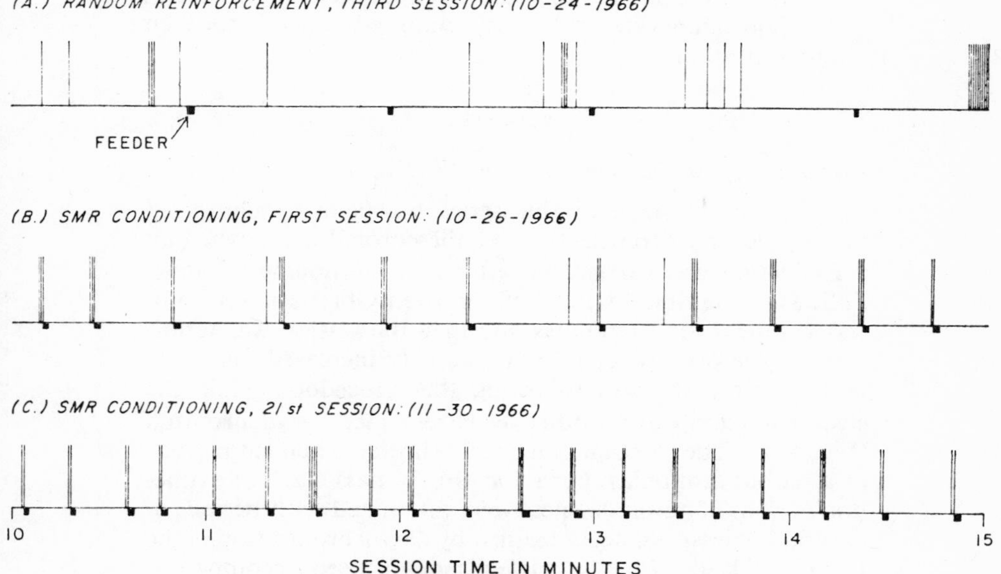

FIG. 3. The charts shown above indicate the time distribution of 0.25 sec epochs of SMR activity during (A) initial random milk reinforcement, (B) first training session, and (C) last training period before extinction test. Each vertical line represents and individual epoch, and adjacent lines indicate continuing SMR activity. These data samples were taken, in each instance, during a continuous five-min period, ten min after the initiation of a given session.

animal appeared to "freeze" briefly during normal activities, such as walking or searching (Fig. 3-*A*). In the first training session the cats usually began to assume a definite motionless posture between reinforcements and to display the SMR regularly (Fig. 3-*B*). The previously random behavior was replaced by a systematic alternation between feeding and the assumption of motionless stereotyped postures. Each animal developed its own specific posture. One cat (SW-1) characteristically arched its back while its eyes were fixed straight ahead. Two other animals (SW-2, SW-3) went to a fixed location at the front of the chamber and stood "at attention." The fourth cat (SW-4) assumed a half-sitting posture at the wall opposite to the feeder. Other observations suggest that a decrease in muscle tonus and a suppression of respiratory activity accompany the occurrence of SMR activity. With the development of such behavior the SMR, which previously had appeared only randomly or in relation to the onset of sleep [13], showed not only an increase in percentage occurrence, but also a systematic temporal distribution, recurring periodically in relation to these various stereotyped postures (Fig. 3-*C*). When milk was delivered upon the occurrence of the SMR the animals immediately approached the feeder. After consuming the milk they returned quickly to their original positions.

RESULTS

Conditioning of the Sensorimotor Rhythm

In the early stage of this series bursts of synchronized activity occurred transiently over the coronal gyrus when an

Extinction was carried out after 20 conditioning sessions. Following the first 15 trials of the twenty-first session, milk was withheld for 30 minutes and then the session was terminated. The occurrence of SMR activity increased markedly in the first 5 minutes following this procedure, while the electrical activity from other leads remained desynchronized (Fig. 4-*A*). The cats simultaneously showed extended periods of alert but motionless behavior. In the next stage of extinction periods of motionlessness were alternated with periods of general restlessness, accompanied by desynchronization of the EEG in all leads. Finally, brief episodes of sleep, accompanied by synchronized patterns in all leads could be observed.

Conditioning of Desynchronized Sensorimotor Area Electrical Activity

In this experimental series milk reinforcement was made contingent upon the occurrence of a desynchronized (low-voltage, high-frequency) electrical pattern from the sensorimotor cortical area. The previous conditioned SMR response faded in the first training session and was replaced by a consistently desynchronized pattern. The behavior of the cats changed also. They became constantly active, searching the chamber, circling and performing frequent discrete

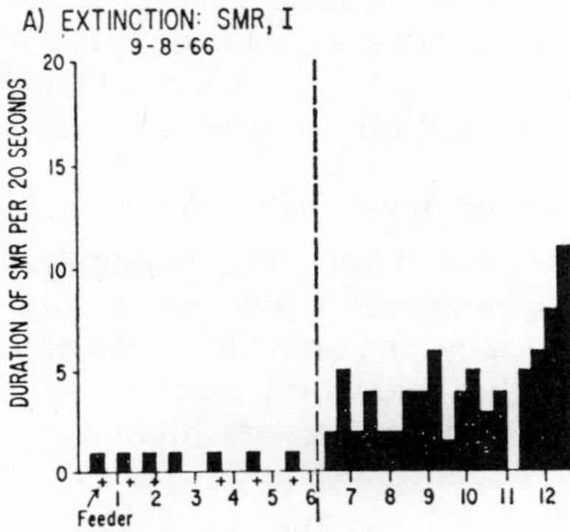

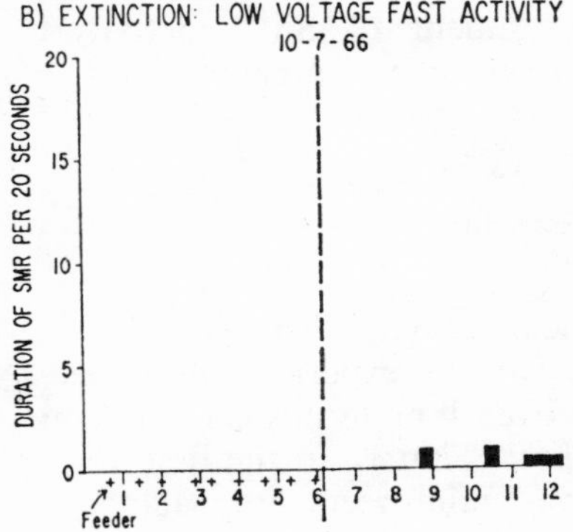

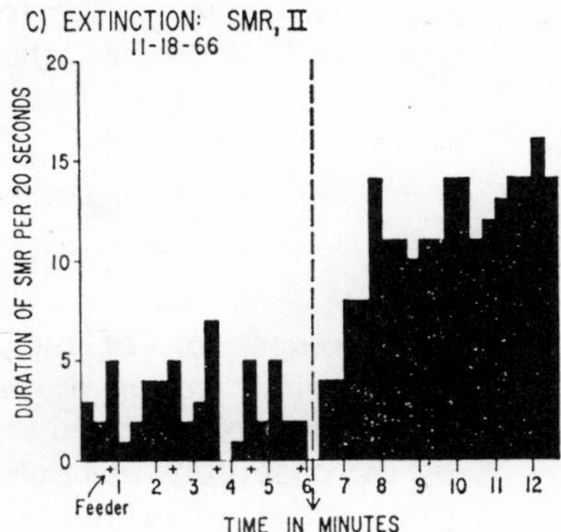

FIG. 4. These graphs indicate the course of events during the initial period of extinction in each training series for one cat. The broken vertical line shows the point at which food was withdrawn in each instance. The absolute amount of SMR activity in sequential 20-sec samples is shown on the ordinate for a period six minutes before and after food withdrawal. Plus marks indicate operation of the feeder prior to the beginning of extinction. In series A and C, SMR activity was the reinforced sensorimotor cortex pattern. In series B, low-voltage fast activity from the same leads was reinforced. Note the specific increase in the occurrence of SMR activity during the initial period of extinction in A and C, and its contrasting absence in B. The enhancement of SMR activity during extinction was usually greater after reconditioning than after the original training series.

spontaneous movements. The previously-observed quiescent postures were absent from their behavior.

Once more an extinction session terminated this experimental series. As can be seen in Fig. 4-*B*, the initial period of extinction was characterized by an absence of the SMR, in contrast to its enhanced status under these circumstances in the first series. The behavior of the cats was also different; they showed an intensification of the restless activity described above. After some 15 min the cats tended to become less active and frequently reclined. During these quiet episodes the SMR was observed to reappear.

Reconditioning of the Sensorimotor Rhythm

For this last series the filter relay system was used again to provide milk reinforcement in association with the occurrence of the SMR. Initially, the animals displayed desynchronized cortical activity and behavioral restlessness. However, they gradually returned to postures resembling those displayed in the first SMR conditioning series, concurrently with the reappearance of related, and systematically-generated, periodic SMR activity over the coronal gyrus. The extinction session produced results similar to those observed in the first series (Fig. 4-*C*). However, the enhancement of SMR activity during the initial minutes of extinction was greater in this series.

DISCUSSION

A comparison of the electrical activity displayed from sensorimotor cortex during the various experimental series described here indicated that the overall pattern was modified by the conditions of reinforcement. That is, in the first and third series when the SMR was reinforced, this activity increased in its rate of occurrence and became systematically distributed in time. Moreover, the SMR was considerably enhanced during the initial period of extinction. Conversely, in the second series when milk was presented in association with the absence of SMR activity, a desynchronized EEG pattern dominated the sensorimotor cortex recordings, and no increase in SMR activity accompanied the initial period of extinction. These findings are in accord with the results usually obtained during the training and extinction of typical instrumental conditioned reactions [6, 7]. From these facts we may conclude that the discrete SMR reinforced by milk in our experiments became an instrumental conditioned response.

Behaviorally, conditioning of the SMR leads to a recurrent

alternation between idiosyncratic motionless postures and feeding reactions. This observation may lead some to conclude that the corresponding SMR was simply associated with the development of these stereotyped postures. However, the postures assumed by the animals differed from one another and were sometimes variable in a given animal in the course of training. Moreover, the appearance of a given posture was necessary but not sufficient for the occurrence of SMR activity. The latter typically developed several seconds after the stereotyped posture was assumed. Ultimately, the behavioral feature common to all animals in every session was an absolute suppression of phasic-motor activity in an alert animal. This suggests, instead, the development of a "central state" of inhibition which is reflected both by the suppression of phasic-motor activity and by the change in the electrical activity from sensorimotor cortex. The general nature of this central influence is indicated, also, by the observation of a suppression of respiratory activity in relation to the occurrence of a conditioned SMR.

The SMR resembles the activity induced in sensorimotor cortex by stimulation of the basal ganglia or non-specific thalamus, manipulations which are known also to induce and facilitate motor inhibition [2, 3]. Extrapyramidal feedback, moreover, has been attributed an important role in motor inhibition by Massion [8], who demonstrated a filtering process in ventrolateral thalamus which modulates cerebellar input to sensorimotor cortex. Sensory feedback also may influence thalamo-cortical motor mechanisms, as suggested by the findings of Andersson and Wolpow [1]. Transection of the primary somatosensory projection pathways in the spinal cord of cats resulted in the appearance of slow-wave activity, specifically in somatosensory cortex. They suggest that the elimination of afferent input from specific receptor systems can alter the physiological organization of corresponding cortical projection areas. Presumably, therefore, extrapyramidal, proprioceptive and interoceptive stimuli can provide sufficient feedback information for the CNS to define this inhibitory state and to activate the pathways required for its production in order to obtain food. We are presently attempting to determine the subcortical structures involved in the generation of the SMR, with the hope that this will shed light, also, upon the central mechanisms of behavioral inhibition. Findings and implications in this regard have been discussed elsewhere [11, 13, 14].

REFERENCES

1. Andersson, S. A. and E. R. Wolpow. Localized slow wave

activity in the somatosensory cortex of the cat. *Acta Physiol. Scand.* **61**: 130–140, 1964.

2. Buchwald, N. A., F. E. Horvath, A. Soltysik and C. Romero-Sierra. Inhibitory responses to basal ganglia stimulation. *Bol. Inst. Estud. Méd. Biol.* **22**: 363–377, 1964.
3. Buser, P., A. Rougeul and C. Perret. Caudate and thalamic influences on conditioned motor responses in the cat. *Boln. Inst. Estud. méd. biol. Univ. nal. Méx.* **22**: 293–307, 1964.
4. Izquierdo, I., W. Wyrwicka, G. Sierra and J. P. Segundo. Etablissement d'un reflexe de trace pendant le sommeil naturel chez le chat. *Actualités Neurophysiol.* **6**: 277–296, 1965.
5. Kamiya, J. Conditioned discrimination of the EEG alpha rhythm in humans. Presented at Western Psychological Association, San Francisco, 1962.
6. Kimble, G. A. *Hilgard and Marquis: Conditioning and Learning.* New York: Appleton-Century-Crafts, 1961.
7. Konorski, J. *Conditioned Reflexes and Neuron Organization.* Cambridge: University Press, 1948, p. 217.
8. Massion, J. M. Etude d'une structure motrice thalamique, le noyau ventrolateral et de sa regulation par les afferences sensorielles. Doctoral Thesis, Univ. of Paris, Paris, France, 1968.
9. Miller, N. E. and A. Carmona. Modification of a visceral response, salivation in thirsty dogs, by instrumental training with water reward. *J. comp. physiol. Psychol.* **63**: 1–6, 1967.
10. Miller, N. E. and L. DiCara. Instrumental learning of heart rate changes in curarized rats. *J. comp. physiol. Psychol.* **63**: 12–19, 1967.
11. Roth, S. R., M. B. Sterman and C. D. Clemente. Comparison of EEG correlates of reinforcement, internal inhibition, and sleep. *Electroenceph. clin. Neurophysiol.* **23**: 509–520, 1967.
12. Rowland, V. Electrographic responses in sleeping conditioned animals. In: *Ciba Foundation Symposium, Nature of Sleep*, edited by G. E. W. Wolstenholme and M. O'Connor. Boston: Little, Brown, 1961, pp. 284–306.
13. Sterman, M. B. and W. Wyrwicka. EEG correlates of sleep: Evidence for separate forebrain substrates. *Brain Res.* **6**: 143–163, 1967.
14. Sterman, M. B. Central mechanisms of forebrain-induced inhibitory states in the cat. *Electroenceph. clin. Neurophysiol.*, in press.
15. Trowill, J. A. Instrumental conditioning of the heart rate in the curarized rat. *J. comp. physiol. Psychol.* **63**: 7–11, 1967.
16. Wyrwicka, W., M. B. Sterman and C. D. Clemente. Conditioning of induced electroencephalographic sleep patterns in the cat. *Science N.Y.* **137**: 616–618, 1962.

42 Operant Control of Neural Events in Humans

Joel P. Rosenfeld, Alan P. Rudell, and Stephen S. Fox

Abstract. *Human subjects were trained by traditional methods of instrumental conditioning to change the amplitude of a late component of the auditory evoked potential with and without oscilloscopic feedback of their performance.*

Fox and Rudell (*1*) trained hungry cats to change the amplitude of a late component of their visual evoked response by reinforcing them with milk whenever the response reached a specified amplitude. We have now trained human subjects in a similar task; the experimental design was somewhat modified to satisfy conditions created by the use of human subjects.

The aforementioned workers and others have aptly described the general aims of the operant control approach and the disadvantages of earlier attempts (*2*) to decode brain waves via the demonstration of neural correlates of behavior. This report, as part of the operant control program, aims to specify brain wave components as potential information carriers as demonstrated by their ability to yield to operant control.

Human subjects were used in the hope of getting at the mechanisms responsible for the operant control of evoked potential components. We expected that, by asking successful subjects how they were able to "control their brain waves," we might obtain suggestive information. Furthermore, we wished to confirm our belief that the operant conditioning of neural events is a general enough phenomenon to be reliably observed in humans as well as in cats. There are advantages in the use of humans; application of scalp electrodes can obviate long hours of surgical placement required in animals less inclined to restraint, and human subjects can be instructed quickly before and interviewed easily after a session.

Our experiments were under the control of a PDP-8 computer (Digital Equipment Corporation). One hundred stimuli (tonal pips) were presented every 4 seconds, and the evoked response was averaged. We selected for each subject a negative-going peak at about 200 msec (*3*) as the criterion component. The computer's next operation was the presentation of a second hundred stimuli, after which it calculated and stored the mean difference between the voltage of the average responses 200 msec before the stimulus (base line) was given and the voltage during the 20 msec selected earlier as the criterion. During training, the computer would reinforce a subject (with money) for increasing the calculated mean difference by 1 standard deviation. Differences rather than pure criterion amplitudes were evaluated to ensure that artifacts of long duration would not be rewarded. In the third phase, stimuli were presented as before, and a running record of reinforcements

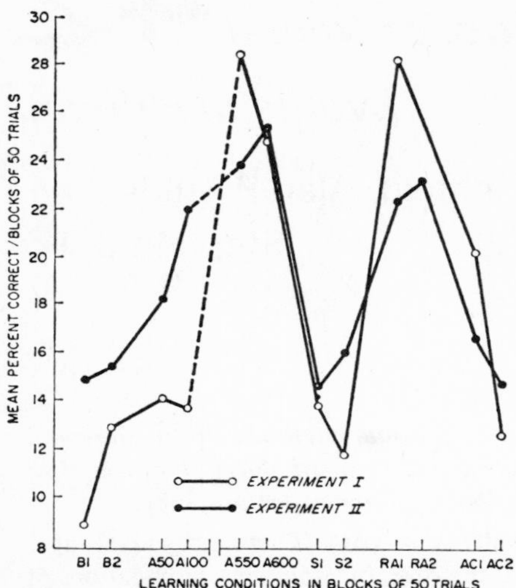

Fig. 1. Percentage of responses reaching criterion as a function of experimental conditions.

was kept. Before the subjects were given instructions, a last set of 100 samples was given so that we could determine the subject's "chance level" of performance. (This last, nonreinforced set is referred to later as the B1 and B2 blocks or *base*.) The subjects were then told to find some "state of mind which would change their brain waves" and that a signal light would inform them of a success, each worth 10 cents. (All subjects were told to be as still as possible.) The subjects in experiment 1 could watch their own performance on an oscilloscope. Subjects in experiment 2 sat in dim light.

Acquisition trials followed. (A100 hereafter refers to acquisition trials 50 through 100). The term *final acquisition* means the last hundred acquisition trials. After A600, the trial block including trials 550 and 600, subjects were told to suppress making correct responses; the signal light would continue flashing, but it now meant the loss of 10 cents. (These trials, 100 in all, are called later S1 and S2 or *suppress.*) Then 100 more acquisition trials were run (later called *reacquisition* or blocks RA1 and RA2). Finally, trials to control for artifact were run (hereafter, AC1 and AC2 or *artifact*). In AC1 and AC2, the tone stimulus was removed altogether (in experiment 1) or attenuated (in experiment 2) to a level whose evoked component in the criterion segment was, according to our pilot data, essentially absent—down by 6 μv. We reasoned that a subject generating a myogenic artifact does not need a stimulus-evoked wave to change but only knowledge of where to generate the artifact. The display provided that knowledge in experiment 1, and the attenuated stimulus did the same in experiment 2. In view of recent data (*4*), our special concern for controlling artifacts seems justified.

In all cases, silver–silver chloride electrodes were fastened with collodion to vertex and mastoid (ground). The raw signal was amplified (80,000 = gain) by a Grass preamplifier set to pass signals between 1.5 and 100 hz. The amplifier signal went to A–D channels of the PDP-8 and also, in experiment 2, to the A–D channels of the Mnemotron CAT 400B. The tone, triggered by the PDP-8, was a 20-msec train of 0.1-msec square pulses. The burst was amplified with an audio preamplifier whose output was delivered to Koss earphones. The PDP-8 computer displayed only single sweeps after B1; the CAT was used to collect average evoked responses during the five training conditions.

The epoch of the PDP-8 was 500 msec, beginning 200 msec before stimulus. The long base line was to insure a sample large enough to be averaged to a straight line. The CAT, also with a 500-msec epoch, began sampling 50

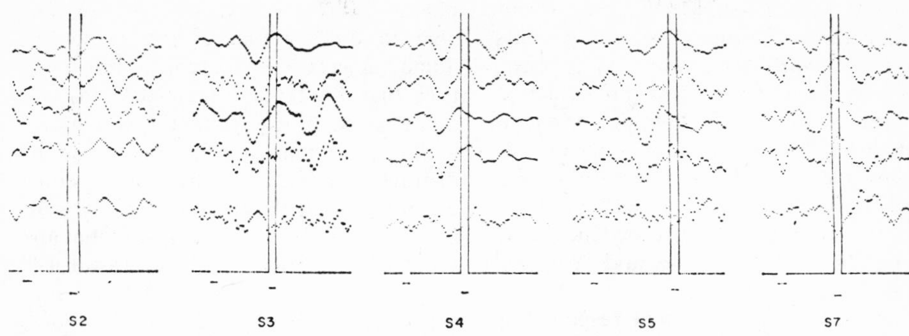

Fig. 2. Average evoked potentials in five experimental conditions. Top trace, B1 and B2; next, A550 and A600; middle, S1 and S2; next, R1 and R2; bottom, AC1 and AC2. First pulse shows stimulus location. Height is arbitrary. Second pulse shows criterion segment; its height represents 12.5 μv.

msec before the stimulus so as to continue sampling after the PDP-8 epoch ended, in order to allow further monitoring for artifact. This meant sacrificing the first 150 msec sampled by the PDP-8. The delayed CAT sweep was achieved with Tektronix waveform and pulse generators.

The percentages of correct responses as a function of learning condition are elevated in the appropriate conditions (Fig. 1). The raw B1 and B2 scores added to the S1 and S2 scores were compared (in a within-subject one-tailed *t*-test) to the sum of the A550, A600, R1, and R2 scores of experiment 2. (For experiment 1, the sum of B1, B2, S1, and S2 scores was compared with the sum of A500, A550, A600, and R1 scores, since no R2 trials were run.) Conditions were pooled to control for any changes occurring over time.

The differences in scores were significant in both experiments. In experiment 1, $t = 2.55$; d.f. = 4, and $.025 < P < .05$; in experiment 2, $t = 4.69$, d.f. = 6, and $.0005 < P < .005$. In view of the decrease in scores (Fig. 1) on the artifact control blocks, the significant differences obtained testify to the successful control of responding attained by subjects. From the *artifact* traces of Fig. 2, it would seem unlikely that a large artifact occurred systematically which could account for the elevated amplitudes obtained by successful responses. During the criterion segments, these traces do not show any negative-going peaks.

Differences in mean amplitudes of the criterion segments between *base* and *suppress* (pooled) and *acquisition* and *reacquisition* (pooled) approached significance ($t = 1.71$, d.f. = 4, and $.05 < P < .10$). Amplitude data were collected only in experiment 2. Absolute amplitudes rather than peak-to-peak differences were tested, inasmuch as the average amplitude of arbitrarily selected 20-msec samples of voltage before stimulus occurrence across learning conditions and across subjects is a flat, straight line, the major difference between neighboring conditions being 0.37 μv.) Records of the averaged evoked potentials of experiment 2 for subjects 2 through 5 and 7 are shown in Fig. 2. There is no uniform pattern of response. Subject 2 seems to generate a new component in the critical segment. Subjects 4 and 5 seem to change the latency of an existing component, whereas subject 7 seems to increase the amplitude of all components in *final acquisition* compared to *base*. Subject 3's records show no evidence of reacquisition although performance measured in number of correct responses does show an increase in R1 and R2 over B1, B2, S1, and S2. Subjects 3 and 5 demonstrate increased variability in acquisition and reaquisition.

There was no consistent response to the question, "What did you do to get rewarded?" The responses were of three types: (i) imagined sights such as "I imagined seeing a pin stick me in the head each time I heard the tone"; (ii) imagined sounds such as, "I heard a [second] tone"; and (iii) special attention to various aspects of the stimulus, for example, "I tried not to hear just one tone but an *on* part and an *off* part." Of the 12 subjects in experiments 1 and 2, four responses were of type (i), four were of type (ii), and two were of type (iii), with two subjects reporting that they had to change their strategy from time to time, sometimes imagining a sound, sometimes a sight, with a variety of specific imagery.

The operant control demonstrated here is far from large. These subjects at best did not exceed 30 percent successful responses when chance success was about 16 percent.

It should be noted that the averaged amplitude increases in *acquisition* and *reacquisition* are not apparent although the increases in success scores (Fig. 1) are. Such results agree with the small absolute size of the effect of reinforcement and may be explained if one assumes that, during acquisition trials, a subject may show an increase in criterion responses while missing criterion on unsuccesful trials by a wider margin than during unsuccessful trials of *base* and *suppress*.

Our data do not offer any simple explanation of the operant control phenomenon. Clearly, subjects are not able to quickly perfect the response. Some subjects volunteered the information that immediate feedback on the oscilloscope in experiment 1 was more a hindrance than a help. They said it distracted them from the intense concentration that they needed to do well in the task. If some simple behavior regularly resulted in the rewarded neural event, subjects watching the oscilloscope should have been more quickly able to discover the right technique. There seems, however, to be little difference in the scores of the two experiments. The variety of verbal reports and the various types of changes seen in the average evoked potentials in the criterion segment and elsewhere argue against the idea that subjects can learn a simple motor response whose somasthetic feedback or efferent command generates the rewarded amplitude change. In view of the controls for systematic movements, such an interpretation becomes even less tenable. The use of earphones as the vehicle of stimulus presentation makes it unlikely that learned changes in receptor orientation are the simple explanation of the phenomena. This general kind of interpretation might, however, be successfully revised to account for the phenomenon by basing it upon the notion that subjects can learn to attend (or not to attend) to the stimulus, behavior whose neural correlate could be an enhanced component (*5*). Yet Fox and Rudell reported two successfully conditioned voltage changes of opposite direction. It seems unlikely that control of attention could be mastered with such specificity.

The lack of uniformity in verbal and neural responses makes it difficult to propose a specific mechanism for the operant control reported, even if we have eliminated notions involving a regularly occurring neural correlate (efferent or afferent) of a movement.

It is likely that subjects are learning to generate some internal state which may mediate an altered evoked potential by either increasing the overall excitability of many neuron populations, or by increasing the size of a particular population so that when the population is activated, its greater effective size yields an enhanced voltage (*6*). The verbal reports of the subjects suggest that behavior they call "imagining" can bring the relevant state about.

JOEL PETER ROSENFELD
ALAN P. RUDELL, STEPHEN S. FOX
Department of Psychology,
University of Iowa, Iowa City 52240

References and Notes

1. S. Fox and A. Rudell, *Science* **162**, 1299 (1968).
2. E. R. John, *Annu. Rev. Physiol.* **23**, 451 (1961); F. Morrell, *Physiol. Rev.* **41**, 443 (1961); H. Yoshii and N. Ogura, *Med. J. Osaka Univ.* **11**, 1 (1960); D. S. Runchkin, J. Villegas, E. R. John, *Ann. N.Y. Acad. Sci.* **115**, 799 (1964); J. Villegas, *ibid.* **122**, 362 (1964).
3. G. E. Chatrian, M. C. Peterson, J. A. Lazarte, *Electroencephalogr. Clin. Neurophysiol.* **24**, 458 (1968); G. E. Celesia, R. J. Broughton, T. Rasmussen, C. Bronch, *ibid.*, p. 458.
4. R. G. Bickford, J. L. Jacobson, D. T. R. Cody, *Ann. N.Y. Acad. Sci.* **112**, 204 (1964).
5. R. Hernandez-Péon, A. Scherrer, M. Jouvet, *Science* **123**, 331 (1956); E. Garcia-Austt, *Electroencephalogr. Clin. Neurophysiol.* **24**, suppl. 101 (1963); P. Spong, M. Haider, D. B. Lindsley, *Science* **148**, 395 (1965).
6. S. S. Fox and J. H. O'Brien, *Science* **147**, 888 (1965); S. S. Fox and R. J. Norman, *ibid.* **159**, 1257 (1968); E. R. John, *Mechanisms of Memory* (Academic Press, New York, 1967), p. 228.
7. Supported by PHS grant MH 11834 to S.S.F.

6 February 1969; revised 5 May 1969 ■

VII

ELECTROMYOGRAPHIC ACTIVITY

43 Escape and Avoidance Conditioning in Human Subjects without Their Observation of the Response

Ralph F. Hefferline, Brian Keenan, and Richard A. Harford

Abstract. An invisibly small thumb-twitch increased in rate of occurrence when it served, via electromyographic amplification, to terminate or postpone aversive noise stimulation. Subjects remained ignorant of their behavior and its effect. Their cumulative response curves resembled those obtained in similar work with animals. Other subjects, informed of the effective response, could not produce it deliberately in a size small enough to qualify for reinforcement.

When the human subject has "voluntary control" of the response to be conditioned, experimental results are in general less predictable and reproducible than those obtained from animals. This is commonly attributed to "self instruction"—that is, to variables experimentally uncontrolled. In the study reported here this problem was circumvented by working with a response so small as to preclude a history of strengthening through discriminable effect upon the environment—in fact, so small as to occur unnoticed by the subject.

The electromyographic setup employed was a modification of that previously reported (*1*). The subject sat in a shielded enclosure in a reclining chair. Recording electrodes were attached to the palmar base of the left thumb and to the medial edge of the left hand. Three additional sets of dummy electrodes were applied in some instances, to suggest that a comprehensive study of body tensions was being conducted. Muscle-action potentials across the left hand were amplified by a factor of 1 million and rectified, and their average momentary values were displayed on a meter. They were also permanently recorded by an Esterline-Angus recording milliammeter.

Twenty-four adults served as subjects. Records from 12 were ruined by apparatus failure, excessive artifact, or failure of the subject to sit still. Results are reported from eight men and four women ranging in age from 18 to 50 and divided into four groups of three each.

Group 1, with four sets of electrodes attached, were told that the study concerned the effects on body tension of noise superimposed on music. Their task was to listen through earphones and, otherwise, do nothing. Group 2, also with all electrodes attached, were told that a specific response, so small as to be invisible, would temporarily turn off the noise or, when the noise was not present, postpone its onset. Their task was to discover and make

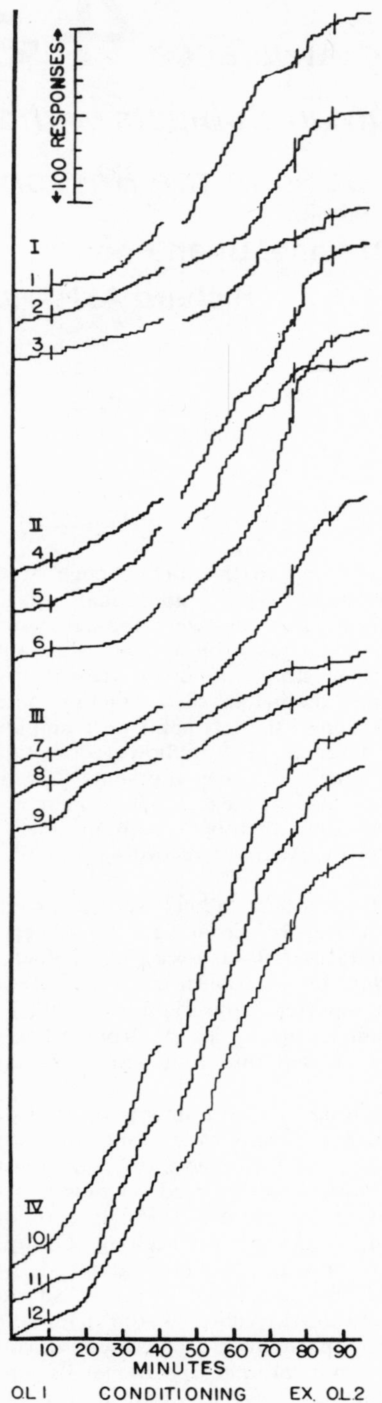

use of the response. Group 3 (with recording electrodes only) were informed that the effective response was a tiny twitch of the left thumb. Group 4 were given the same information as group 3 but had, in addition, a meter before them during the first half-hour of conditioning, which provided a potential basis for them to use the visual presentation of their response as a "crutch" for proprioceptive observation of the response.

Experimental procedure was identical for all groups. While the subject relaxed and listened to tape-recorded music through earphones, the experimenter watched the meter on his panel for 5 to 10 minutes to select for later reinforcement a response of a size occurring not more than once in 1 or 2 minutes. It was a ballistic swing of the pointer up and back over a few scale divisions. This represented, for a particular subject, a momentary voltage increment at the electrode of 1, 2, or 3 μv.

After the operant level for this response had been recorded for 10 minutes (*OL 1* in Fig. 1), conditioning was begun by superimposing on the music an aversively loud, 60-cycle hum. Whenever the experimenter saw on the meter an instance of the selected response, he pressed a key. This turned off the noise for 15 seconds or, when it was already off, postponed noise resumption for 15 seconds. [This type of avoidance schedule, mentioned in 1950 (*2*), has been extensively employed by Sidman in animal work (*3*).]

After an hour of conditioning, with a 5-minute intermission at the half-hour point, 10 minutes of extinction occurred during which the subject's response was ineffective in terminating continuously present noise. During final 10 minutes of music only, the extent of

Fig. 1. Cumulative response curves for adult human subjects in a situation where an invisibly small and unnoticed thumb-twitch either terminated or postponed noise stimulation. *OL 1* and *2*, initial and terminal operant level determinations, respectively; *ex.*, extinction.

recovery of the original operant level was recorded.

Figure 1 presents cumulative response curves for each subject. Conditioning is clearly indicated by the positive acceleration in the rate of responding for all subjects except subjects 2 and 3 in group 3. These two kept so busy producing voluntary thumb-twitches that the small, reinforceable type of response had little opportunity to occur.

When interviewed later, all members of group 1 still believed that they had been passive victims with respect to the onset and duration of noise, and all seemed astounded to learn that they themselves had been in control. Subjects 1 and 2 of group 2 reported that they early gave up searching for an effective response and thus, in effect, transferred themselves to group 1. Subject 3 of group 2 professed to have discovered an effective response sequence, which consisted of subtle rowing movements with both hands, infinitesimal wriggles of both ankles, a slight displacement of the jaw to the left, breathing out—and then waiting. Subject 1 of group 3 gave evidence of conditioning perhaps because he misconstrued the instructions. Instead of making the response a quick contraction, he spent his time very gradually increasing pressure on an imaginary switch button. This may have kept deliberate activity at a level low enough for the correct response to break through and be reinforced.

Group 4 subjects, provided with their own meter, obtained many more reinforcements than the others, an effect which continued through the second half-hour of conditioning, with the meter removed. While the meter did not enable them to achieve direct control of the discrete response, it seems to have provided a basis for rapid responding within a range which included the reinforced size. This showed on the meter as rapid oscillation.

The technique employed in this study (*4*) offers possibilities for investigating human behavior, in a sense, at the animal level. Research now in progress is concerned with attempts to clarify the circumstances under which the human subject may come to discriminate verbally—that is, to become conscious of—his small responses.

RALPH F. HEFFERLINE
BRIAN KEENAN
RICHARD A. HARFORD
Department of Psychology, Columbia University, New York

References and Notes

1. R. F. Hefferline, *Trans. N.Y. Acad. Sci.* **20**, 739 (1958).
2. ———, *Genet. Psychol. Monographs* **42**, 231 (1950).
3. M. Sidman, *Science* **118**, 157 (1953).
4. This investigation was supported by research grant M-2961, National Institutes of Health, U.S. Public Health Service, and by grants from the Higgins Fund and the Columbia University Council for Research in the Social Sciences.

44

Feedback of Speech Muscle Activity during Silent Reading: Rapid Extinction

Curtis D. Hardyck, Lewis F. Petrinovich, and Delbert W. Ellsworth

Abstract. *Surface electromyograms of the laryngeal muscles were made while subjects read silently. Those who showed an increase in electrical activity over that at relaxation were provided with auditory feedback of the muscle activity. This treatment resulted in immediate and long-lasting cessation of the subvocalization. This method should prove valuable in treating some reading problems.*

The phenomenon of subvocal speech has been of great interest to educators concerned with the teaching of reading. It has, however, received little systematic study, with the exception of the work by Sokolov and by Edfeldt (*1*). Subvocalization is considered one of the most difficult problems to overcome in increasing reading speed. An individual who subvocalizes to any great extent is limited to a top reading speed of approximately 150 words per minute—a maximum attainable while reading aloud. We use the term subvocalization to include a wide range of activity, from inaudible articulations and vocalizations to audible whispering while reading. If subvocal activity includes movements of the lips and jaw, some corrective measures are possible. However, if the activity is limited to the vocal musculature, eliminating the response becomes more complex, especially since individuals are often not aware they are subvocalizing.

The initial study of subvocalization required the examination of several subjects with strong subvocalization patterns to determine whether subvocalization during silent reading could be detected by surface electromyograms recorded from the throat. A successful technique was developed with mesh electrodes placed over the thyroid cartilage. An ink-writing oscillograph was used to record the electromyogram (EMG). At maximum sensitivity of the oscillograph unit, the electrical activity of the vocal muscles can be detected while the subject is reading (if subvocalization is present), in contrast to a minimum signal (approximately 3 μv) obtained when the subject is relaxed, and to an extremely strong signal (approximately 1 mv) obtained when the subject speaks during normal conversation. To determine the presence or absence of subvocalization, the subject first selects reading material which he reads for 30 minutes, and during this time an oscillograph record of the EMG is obtained.

The presence of subvocalization is determined by asking the subject to stop reading, then to begin reading, and then to stop reading. Each time, the changes in the EMG record are noted. The presence of subvocalization can be detected quite reliably, there being a large increase in action potentials when

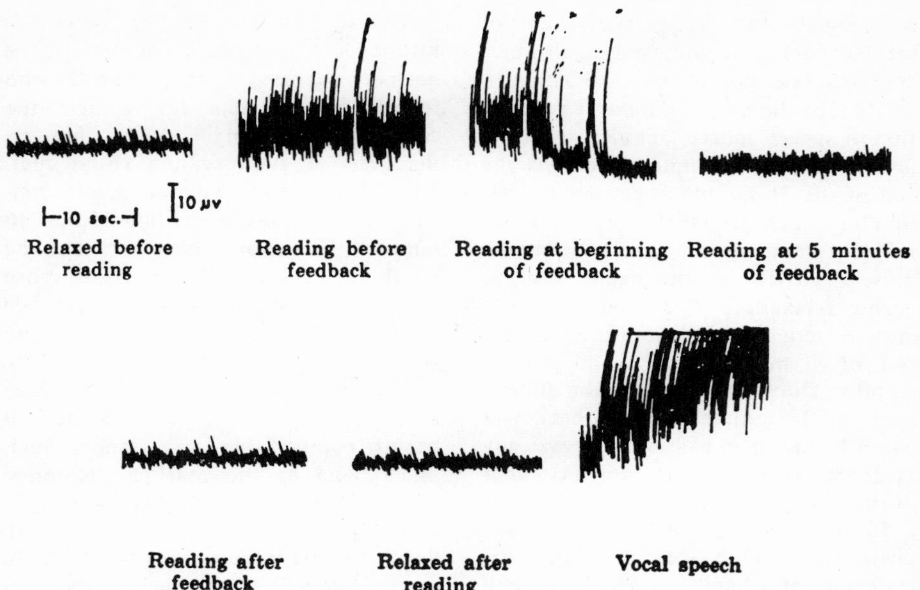

Fig. 1. Electromyograms recorded from the laryngeal muscles. This record is typical of those obtained from the 17 subjects who received feedback treatment.

reading begins and an immediate cessation of this activity when reading stops.

Treatment of subvocalization is done in the following manner. The subject is asked if he is aware that he subvocalizes or reads aloud to himself. A brief discussion is held with the subject, informing him of his response tendencies while reading. The feedback technique is then introduced. The subject is told that he will be able to hear the activity of his vocal muscles as he reads, and that this will help him to eliminate the problem. The manner in which signals emanating from the vocal muscles are detected is explained to him. The subject is then given earphones to wear, and is asked to remain relaxed. Feedback is introduced by channeling the output of the oscillograph amplifier to an audio-amplifier, and then to earphones. When the subject is relaxed, the audio-circuit is opened and the subject is asked to swallow. The swallow results in an immediate burst of static in the earphones. The subject is then requested to experiment with the sound to satisfy himself that he can control it (stop it and start it) by such actions as talking, swallowing, turning his head, clenching his jaw muscles, and so forth. The subject is allowed to continue experimenting with the feedback, and with its control, until he states that is is able to control its presence or absence. The subject then begins reading while attempting to keep the EMG feedback to a minimum, that is, to maintain silence in the earphones.

A total of 50 college students from a reading improvement class were tested; it was found that 17 subvocalized. All subjects who subvocalized were treated in the manner described above. Originally, it was planned to administer the feedback treatment over several sessions to determine the number of feedback treatments necessary to establish normal reading patterns. This was found to be unnecessary. The feedback treatment was remarkably effective (Fig. 1). In all cases one session of the feedback was sufficient to produce complete cessation of subvocalization.

Most of the subjects showed a reduction of speech muscle activity to resting levels within a 5-minute period. At the end of the 30-minute experimental session in which feedback was given, all subjects were able to read with speech muscle activity at the same level as during relaxation. The level of speech muscle activity was monitored at the end of 1 month, and again after 3 months. During these tests, the subject read for 30 minutes; no feedback was used. None of the subjects gave any evidence of subvocalization in either of these tests.

In many ways, this is a surprising phenomenon. One does not expect the extinction of a habit—especially a habit which presumably had existed during the entire time the subject had been reading—to occur so quickly and easily. However, if the overlearned response of subvocalization is placed in conflict with a second, even more strongly overlearned response, extinction should be quite rapid. Such a second response is the ability to make a fine motor adjustment of the speech musculature on the basis of auditory cues. Exactly this response is involved under conditions of feedback of laryngeal EMG activity. The subject is required to make fine motor adjustments on the basis of auditory cues. Attempts to reduce the speech muscle activity by instructions alone were not successful. The subjects were not aware of their subvocal activity even when told they were subvocalizing, and were unable to reduce it without the feedback.

This ability to make fine motor adjustments of restricted muscle groups has been reported by Basmajian (*2*) who found that subjects can control the contractions of single motor units on the basis of auditory and visual cues. His subjects also achieved stable control of the muscles within 15 to 30 minutes. He reports, however, that the aural feedback in all subjects is more useful than visual display on a cathode ray tube monitor; the visual display served a subsidiary purpose (*2*).

The motor-auditory feedback loop as a cue produced by response might well be of considerable importance. Such phenomena as the marked disruption of speech under conditions of delayed auditory feedback provide ample evidence for the importance of this cue to normal speech. Consequently, the evocation of this overlearned response-produced cue may result in the rapid extinction of subvocalization under the conditions of auditory feedback.

CURTIS D. HARDYCK
LEWIS F. PETRINOVICH
DELBERT W. ELLSWORTH
Institute of Human Learning, University of California, Berkeley

References and Notes

1. A. W. Edfeldt, *Silent Speech and Silent Reading* (University of Chicago Press, Chicago, 1960); A. N. Sokolov, in *Psikhologiches-Kaya Nauka V SSSR, I* (Akad. Pedag. Nauk RSFSR, Moscow, 1959), p. 488 [translation in *Psychological Science in the USSR, I* (U.S. Joint Publications Res. Serv. No. 11466, Washington, D.C., 1961), p. 669].
2. J. V. Basmajian, *Science* **141**, 440 (1963).
3. Aided by HEW Office of Education contract 5-1126.

45 Control and Training of Individual Motor Units

J. V. Basmajian

Abstract. *Experiments clearly demonstrate that with the help of auditory and visual cues man can single out motor units and control their isolated contractions. Experiments on the training of this control, interpreted as the training of descending pathways to single anterior horn cells, provide a new glimpse of the fineness of conscious motor controls. After training, subjects can recall into activity different single motor units by an effort of will while inhibiting the activity of neighbors. Some learn such exquisite control that they soon can produce rhythms of contraction in one unit, imitating drum rolls, etc. The quality of control over individual anterior horn cells may determine rates of learning.*

It is a commonplace observation that very gentle contractions of skeletal muscles recruit only a few motor units and that, on relaxation, human beings can promptly repress all neuromuscular activity in large areas under voluntary control (*1*). However, little attention has been paid to the fine voluntary control of individual motor units. In 1960 Harrison and Mortensen (*2*) reported that subjects were able to maintain isolated activity of several different motor units in the tibialis anterior as recorded from surface electrodes and confirmed by needle electrodes. The implications of this finding led to an intensive systematic investigation with special indwelling electrodes.

By definition, a motor unit includes a spinal anterior horn cell, its axon, and all the muscle fibers on which the terminal branches of the axon end (Fig. 1). This motor unit "fires" when an impulse reaches the muscle fibers, the response being a brief twitch. The electrical potential accompanying the twitch is now well documented. The twitch frequency has an upper limit of about 50 per second. With indwelling electrodes, individual motor units are identifiable by their individual shapes; these remain relatively constant unless the electrodes are shifted.

The subjects of these experiments were provided with two modalities of "proprioception" that they normally lack, namely, they heard their motor unit potentials and saw them on monitors. The subjects were 16 normal persons ranging in age from 20 to 55. All but five were under 24 and only one was female.

The main muscle tested in all subjects was the right abductor pollicis brevis (Fig. 2). In two subjects the tibialis anterior was also tested; in another, the biceps brachii and the extensor digitorum longus were tested on other occasions. The recording and monitoring apparatus is illustrated in Fig. 2.

The indwelling electrodes used have already been described in detail (*3*). They are nylon-insulated Karma alloy wires 0.025 mm in diameter, which are introduced into the muscle as a pair by means of a hypodermic needle that is immediately withdrawn. In the case of a small muscle like the abductor pollicis brevis, the activity of all its motor units are probably detected while the fascial coat of the muscle isolates the pick-up to this muscle alone.

After placement and connection of the electrodes, the subjects spent 5 to 10 minutes becoming familiar with the response of the electromyograph to a range of movements and postures. They were invariably amazed at the responsiveness to even the slightest effort. Then they began learning how to maintain very slight contractions, which were apparent to themselves only through the response of the apparatus. This led to increasingly more demanding effort involving many procedures intended to reveal both their natural talent in controlling individual motor units and their skill in learning and retaining tricks with such units. Individual units were identified by the characteristics of their potentials which show considerable difference on the oscilloscope and, to a lesser extent, on the loudspeaker. Film recordings of potentials were made for confirmation (Fig. 3).

Generally, experiments on one muscle were limited to about half a day. Within 15 to 30 minutes all subjects had achieved notably better willful control over gentle contractions. In this time almost all had learned to relax the whole muscle instantaneously on command and to recruit the activity of a single motor unit, keeping it active for as many minutes as desired. A few had difficulty maintaining the activity of such a unit, or in recruiting more units. No relationship was obvious to age, manual dexterity, or anything that might have been invoked as an underlying explanation of the differences in performance. Two of the apparently most dexterous persons performed only moderately well. The youngest persons were among both the worst and the best performers. General personality traits did not seem to matter.

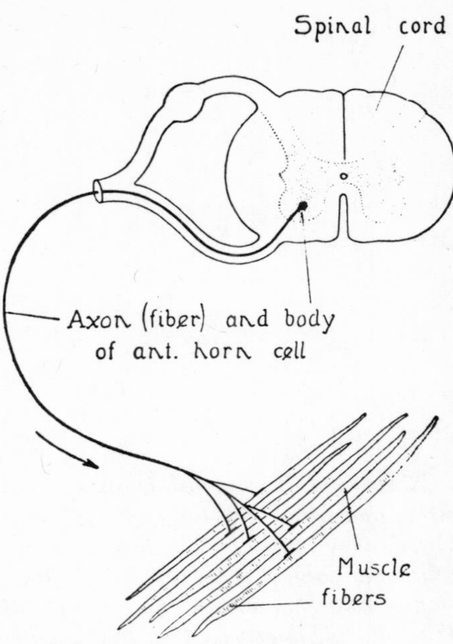

Fig. 1. Diagram of a motor unit of skeletal muscle.

After about 30 minutes the subject was required to learn how to repress the first motor unit he had become familiar with and to recruit another one. Most subjects were able to do this and gain mastery of the new unit in a matter of minutes; only one subject required more than 15 minutes. More than half of the subjects could repeat the performance with a third new unit within a few minutes. A few subjects could recruit a fourth or a fifth isolated unit. The next problem facing a subject was to recruit, unerringly and in isolation, the several units over which he had gained control.

Here there was a considerable variation in skill. About one in four could respond easily to the command for isolated contractions of any of three units. About half the subjects displayed much less skill in this regard, even after several hours and even though they may have learned other bizarre tricks. Several subjects had particular difficulty in recruiting the asked-for units. They groped around in their conscious efforts to find them and sometimes, it seemed, only succeeded by accident.

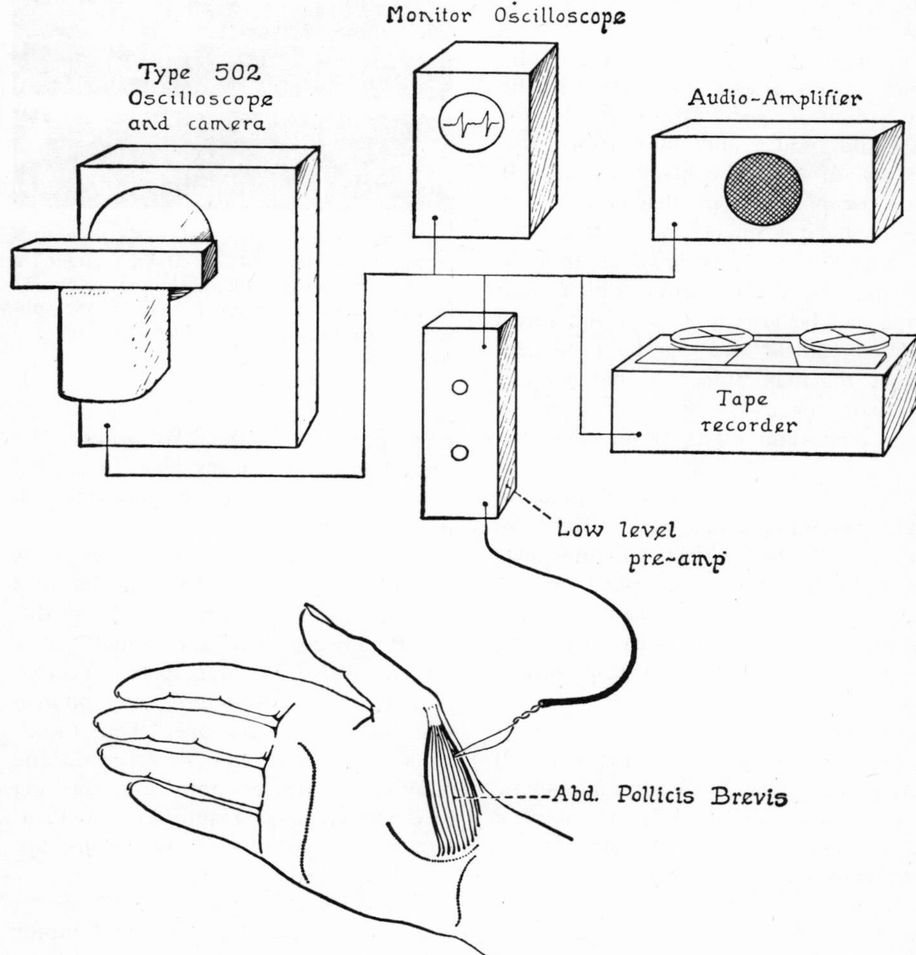

Fig. 2. Technique of recording from abductor pollicis brevis.

The subjects with the finest control were then trained to learn various tricks. Several were tested for their powers of recalling specific units into activity in the *absence* of the aural and visual feedbacks which were so important to most of the subjects. Three subjects could recall units voluntarily under these handicaps, but they were unable to explain how they could do it.

Other tests showed that in all subjects the aural feedback is more useful than the visual display on the cathode ray tube monitor. The latter served only a subsidiary purpose.

After 60 to 90 minutes, most of the subjects were tested and trained in the production of specific rhythms. Almost all could reduce and increase the frequency of a well-controlled unit. It soon became apparent that motor units do not have a single characteristic frequency. Rather, they have an individual maximum rate below which their firing can be greatly slowed and single isolated contractions can be produced. Above the maximum rate that is characteristic for a specific unit, overflow takes place and other motor units are recruited.

Subjects learned to control units so that they could produce various rhythms. Almost all the subjects in the later experiments who were asked to try these (10 of 11) succeeded. Various gallop rhythms, drum-beat rhythms, doublets, and roll effects were produced and recorded (*4*).

The experiments reported above suggest that pathways from the cerebral cortex can be made to stimulate single anterior horn cells while neighboring anterior horn cells remain dormant or are depressed. Although the skills learned in these experiments depended on aural and visual feedbacks from muscles, the controls are learned so quickly, are so exquisite, and are so well retained after the feedbacks are eliminated in some subjects, that one must not dismiss them as tricks. The underlying mechanisms seem to involve active suppression of neighboring anterior horn cells.

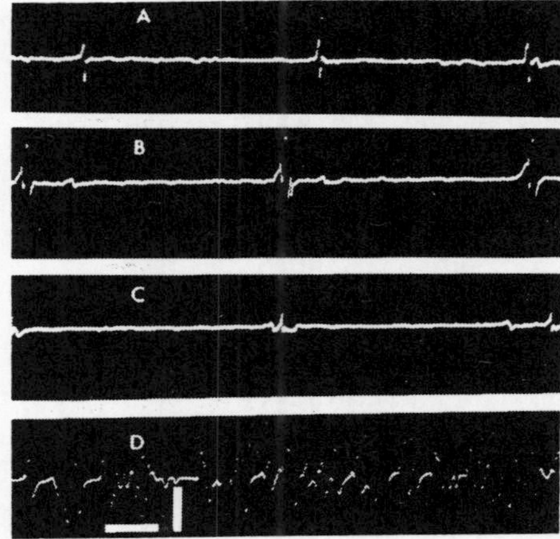

Fig. 3. Electromyograms of potentials from motor units, *A*, *B*, and *C*, and of a weak-to-moderate contraction (tracing *D*) in the abductor pollicis brevis of one subject. Calibrations lines: 25 msec and 200 μv.

A number of obvious problems emerge from the differences in the rates of learning of motor unit skills by different subjects. New but limited studies by Harrison (*5*) suggest that accomplished athletes have no better control than other subjects over their motor units. Future studies to ascertain the relation of rates of motor unit learning to dexterity, special abilities, and techniques of teaching motor skills are called for.

The extremely fine ability to adjust the rate of firing of individual motor units is a novel concept. Above a characteristic frequency, which varies from cell to cell, overflow to neighbors occurs. Detailed studies of these characteristics should expose some of the underlying control mechanism in the spinal cord (*6*).

J. V. BASMAJIAN
Department of Anatomy, Queen's University, Kingston, Ontario

References and Notes

1. J. V. Basmajian, *Muscles Alive: Their Functions Revealed by Electromyography* (Williams and Wilkins, Baltimore, 1962), p. 7.
2. V. F. Harrison and O. A. Mortensen (abstract), *Anat. Record,* **136**, 207 (1960); **144**, 109 (1962).
3. J. V. Basmajian and G. Stecko, *J. Appl. Physiol.* **17**, 849 (1962).
4. Excerpts of tape recordings were played to the annual meeting of the American Association of Anatomists, April 1963, as part of a paper of mine.
5. V. F. Harrison (abstract), *Anat. Record,* **145**, 237 (1963).
6. Supported by grants from the Muscular Dystrophy Association of Canada and the Medical Research Council of Canada. Glenn Shine provided technical assistance.

46

Conscious Control and Training of Motor Units and Motor Neurons

J. V. Basmajian

Studies of neuromuscular and spinal-cord function have been growing increasingly complex in recent years without offering clearer answers to many fundamental problems. Especially confusing and fragmentary are theories on the influence of various cortical and subcortical areas on spinal motor neurons and motor units in man. It is therefore refreshing to be able to use and advocate a technique that not only proves to be quite simple but also promises to reveal considerable fundamental information. Ironically, the technique is only a modification of ordinary electromyography. This modification consists of regarding electromyographic potentials not for their own intrinsic value but as the direct mirroring of the activity of spinal motor neurons. Thus the group of muscle fibres in a motor unit is considered only as a convenient transducer that reveals the function of the nerve cell.

Perhaps the ultimate irony is that in their classic paper establishing the modern era of electromyography in 1929, Adrian and

Bronk suggested that "... The electrical responses in the individual muscle fibres should give just as accurate a measure of the nerve fibre frequency as the record made from the nerve itself." Even earlier, Gasser and Newcomer (1921) had shown that "the electromyogram is a fairly accurate copy of the electroneurogram." Perhaps as a reflection of the general turning away from man as an experimental animal in favour of more exotic beasts and preparations, no real use of these early conclusions has been made until recently. In fact, the implications in Gasser and Newcomer's work did not lead to any systematic use of electromyography for studying the behaviour of individual spinal motor neurons in any species even though the action potential of a motor unit picked up by direct electromyography reflects the activity of its spinal motor neuron.

No great progress was made until 1928–29 when Adrian and Bronk published two classic papers on the impulses in single fibres of motor nerves in experimental animals and man. Their method consisted of cutting through all but one of the active fibres of various nerves and recording the action currents from that one fibre. They also succeeded in making records directly from the muscles supplied by such nerves. Somewhat incidentally, Adrian and Bronk introduced the use of concentric needle electrodes with which the activity of muscle fibres in normal human muscles could be recorded. Meanwhile Sherrington (1929) and his colleagues had crystallized their definition of a motor unit as "an individual motor nerve together with the bunch of muscle-fibres it activates." (Universally, later workers have also included in their definition the cell body of the neuron from which the nerve fibre arises.)

Although in subsequent years the concentric needle electrode was seized upon for extensive use, until the Second World War only a handful of papers appeared on the characteristics of action potentials from single motor units in voluntary contraction. In 1934, Olive Smith reported her observations on individual motor unit potentials, their general behaviour and their frequencies. She showed that normally there is no proper or inherent rhythm acting as a limiting factor in the activity of muscle fibres; rather, the muscle fibres in a normal motor unit simply respond to each impulse they receive. Confirming earlier work of Denny-Brown (1929) she set at rest the false hypothesis of Forbes (1922) that the muscle fibres or motor units were fatiguable at the frequencies they were called upon to reproduce by their nerve impulses.

Forbes had also suggested that normal sustained contraction requires rotation of activity among quickly fatiguing muscle fibres. Smith proved that such a rotation need not occur and that an increase in contraction of a whole muscle involves both increase in frequency of impulses in the individual unit and an accession of new units which are independent in their rhythms. The frequencies ranged from 5 to 7 per second to 19 to 20 per second, although "highly irregular discharge may occur at threshold both during the onset of a contraction and during the last part of relaxation." Finally, she proved that tonic contraction of motor units in normal mammalian skeletal muscle fibre, the existence of which was widely debated, does not exist. Two generations later, there are people in muscle research still not aware of her definitive studies.

Lindsley (1935), working in the same physiology laboratory as Smith, determined the ultimate range of motor unit frequencies during normal voluntary contractions. Although others must have been aware of the phenomenon, he seems to have been the first to emphasize that at rest "subjects can relax a muscle so completely that . . . no active units are found. Relaxation sometimes requires conscious effort and in some cases special training." In none of his subjects was "the complete relaxation of a muscle difficult." Since then, this finding has been confirmed and refined by hundreds of investigators, using much more sophisticated apparatus and techniques than those available in the early 30's.

Lindsley also reported that individual motor units usually began to respond regularly at frequencies of 5 to 10 per second during the weakest voluntary contractions possible and some could be fired as slow as 3 per second. The upper limit of frequencies was usually about 20 to 30 per second but occasionally was as high as 50 per second. Earlier, Adrian and Bronk (1928, 1929) had found the same upper limit of about 50 per second for the nerve impulses in single fibres of the phrenic nerve and from the diaphragm of the same preparations.

Gilson and Mills (1940, 1941), recording from single motor units under voluntary control, reported that discrete, slight and brief voluntary efforts may call upon only a single potential (i.e., a single twitch) of a motor unit being recorded. Twenty years later, Harrison and Mortensen (1962) showed that by means of surface and needle electrodes action potentials of single motor units could be identified and followed during slight voluntary contractions in tibialis anterior. Subjects provided with auditory and visual cues could produce "single, double and quadruple contractions of single motor units . . ." and in one case, ". . . the subject was able to demonstrate predetermined patterns of contraction in four of the six isolated motor units."

Using special indwelling fine-wire electrodes (p. 32), I had no difficulty in confirming these findings (Basmajian, 1963), and on this basis I was able to elaborate techniques for studying the fine control of the spinal motor neurons, especially their training, and the effects of volition. Later, my colleagues and I further developed and described our system of testing and of motor unit training. We demonstrated the existence of a very fine conscious control of pathways to single spinal motor neurons (Basmajian, Baeza and Fabrigar, 1965). Not only can human subjects fire single neurons with no overflow (or perhaps more correctly, with an active suppression or inhibition of neighbours), but they can also produce deliberate changes in the rate of firing. Most persons can do this if they are provided with aural (and visual) cues from their muscles.

Following the implantation of fine-wire electrodes and routine testing, a subject needs only to be given general instructions. He is asked to make contractions of the muscle under study while listening to and seeing the motor unit potentials on the monitors (fig. 52). A period of up to 15 minutes is sufficient to familiarize him with the response of the apparatus to a range of movements and postures.

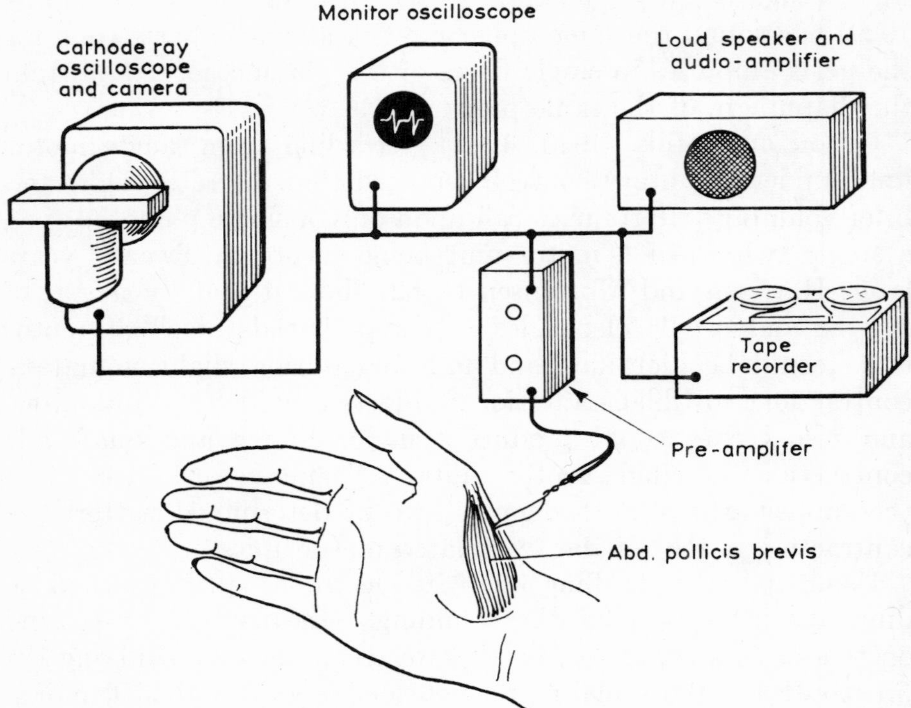

FIG. 52. Diagram of arrangement of monitors and recording apparatus for motor unit training. (From Basmajian, 1963b.)

Subjects are invariably amazed at the responsiveness of the loudspeaker and cathode-ray tube to their slightest efforts, and they accept these as a new form of "proprioception" without difficulty. It is not necessary for subjects to have any knowledge of electromyography. After getting a general explanation they need only to concentrate their attention on the obvious response of the electromyograph. With encouragement and guidance, even the most naive subject is soon able to maintain various levels of activity in a muscle on the sensory basis provided by the monitors. Indeed, most of the procedures he carries out involve such gentle contractions that his only awareness of them is through the apparatus. Following a period of orientation, the subject can be put through a series of tests for many hours.

Several basic tests are employed. Since people show a considerable difference in their responses, adoption of a set routine has proved to be impossible. In general, however, they are required to perform a series of tasks. The first is to isolate and maintain the regular firing of a single motor unit from among the 100 or so a person can recruit and display with the technique described. When he has learned to suppress all the neighbouring motor units completely, he is asked to put the unit under control through a series of tricks including speeding up its rate of firing, slowing it down, turning it "off" and "on" in various set patterns and in response to commands.

After acquiring good control of the first motor unit, a subject is asked to isolate a second with which he then learns the same tricks; then a third, and so on. His next task is to recruit, unerringly and in isolation, the several units over which he has gained the best control.

Many subjects then can be tested at greater length on any special skills revealed in the earlier part of their testing (for example, either an especially fine control of, or an ability to play tricks with, a single unit). Finally, the best performers are tested on their ability to maintain the activity of specific motor-unit potentials in the absence of either one or both of the visual and auditory feedbacks. That is, the monitors are turned off and the subject must try to maintain or recall a well-learned unit without the artificial "proprioception" provided earlier.

Any skeletal muscle may be selected. The ones we have used most often are the abductor pollicis brevis, tibialis anterior, biceps brachii and the extensors of the forearm.

Ability to Isolate Motor Units

Almost all subjects are able to produce well-isolated contractions of at least one motor unit, turning it off and on without any interference from neighbouring units. Only a few people fail completely to perform this basic trick. Analysis of poor and very poor performers reveals no common characteristic that separates them from better performers.

Most people are able to isolate and master one or two units

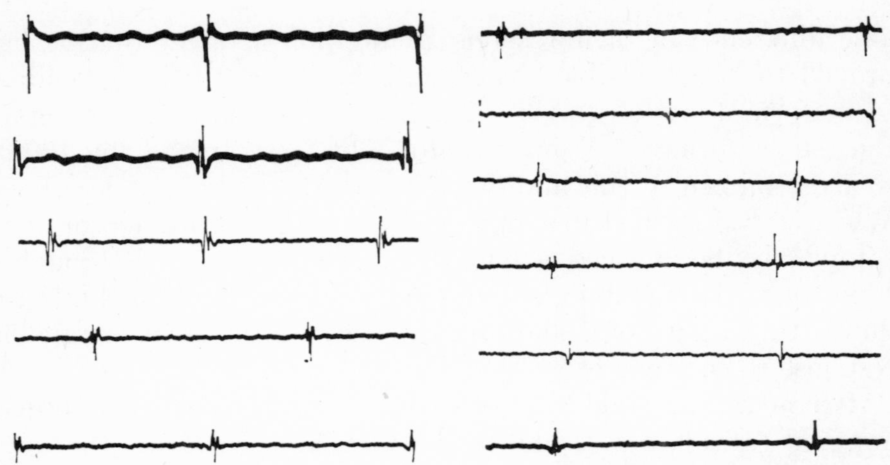

Fig. 53. Eleven different motor units isolated by a subject in quick succession in his abductor pollicis brevis. (From Basmajian, Baeza and Fabrigar, 1965.)

readily; some can isolate and master three units, four units and even six units or more (fig. 53). This last level of control is of the highest order, for the subject must be able to give an instant response to an order to produce contractions of a specified unit without interfering activity of neighbours; he also must be able to turn the unit "off" and "on" at will.

Control of Firing-Rates and Special Rhythms

Once a person has gained control of a spinal motor neuron, it is possible for him to learn to vary its rate of firing. This rate can be deliberately changed in immediate response to a command. The lowest limit of the range of frequencies is zero, i.e., one can start from neuromuscular silence and then give single isolated contractions at regular rates as low as one per second and at increasingly faster rates. When the more able subjects are asked to produce special repetitive rhythms and imitations of drum beats, almost all are successful (some strikingly so) in producing subtle shades and coloring of internal rhythms. When tape-recorded and replayed, these rhythms provide striking proof of the fineness of the control.

Individual motor units appear to have upper limits to their rates beyond which they cannot be fired in isolation; that is, overflow occurs and neighbours are recruited. These maximum frequencies range from 9 to 25 per second (when the maximum rates are carefully recorded with an electronic digital spike-counter). Almost all lie in the range of 9 to 16 per second. However, one must not infer that individual motor units are restricted to these rates when many units are recruited. Indeed, the upper limit of 50 per second generally accepted for human muscle is probably correct, with perhaps some slightly higher rates in other species.

Reliance on Visual or Aural Feedback

Some persons can be trained to gain control of isolated motor units to a level where, with both visual and aural cues shut off, they can recall any one of three favorite units on command and in any sequence. They can keep such units firing without any conscious awareness other than the assurance (after the fact) that they have succeeded. In spite of considerable introspection, they cannot explain their success except to state they "thought about" a motor unit as they had seen and heard it previously. This type of training probably underlies ordinary motor skills.

Variables Which Might Affect Performance

We find no personal characteristics that reveal reasons for the quality of performance (Basmajian, Baeza and Fabrigar, 1965). The best performers are found at different ages, among both sexes, and among both the manually skilled and unskilled, the educated and uneducated, and the bright and the dull personalities. Some "nervous" persons do not perform well—but neither do some very calm persons.

Carlsöö and Edfeldt (1963) also investigated the voluntary control over individual motor units. They concluded that: "Proprioception can be assisted greatly by exeroceptive auxiliary stimuli in achieving motor precision." Nevertheless, Wagman, Pierce and Burger (1965), using both our technique and a technique of recording devised by Pierce and Wagman (1964), emphasize the rôle of proprioception. They stress their finding that subjects believe that certain positions of a joint must be either held or imagined for success in activating desired motor units in isolation.

We have recently completed an investigation into various factors which affect motor unit training and control (Simard and Basmajian, 1967; Basmajian and Simard, 1967; Simard, Basmajian and Janda, 1968). We find that moving a neighbouring joint while a motor unit is firing is a distracting influence but most subjects can keep right on doing it in spite of the distraction. We tend to agree with Wagman and his colleagues who believe that subjects require our form of motor unit training before they can fire isolated specific motor units with the limb or joints in varying positions. Their subjects reported that "activation depended on recall of the original position and contraction effort necessary for activation." This apparently is a form of proprioceptive memory and almost certainly is integrated in the spinal cord.

Our observations were based on trained units in the tibialis anterior of 32 young adults. They showed that motor unit activity under conscious control can be easily maintained despite the distraction produced by voluntary movements elsewhere in the body (head and neck, upper limbs and contralateral limb). The control of isolation and the control of the easiest and fastest frequencies of discharge of a single motor unit were not affected by those movements (fig. 54).

Turning to the effect of movements of the same limb, we found that in some persons a motor unit can be trained to remain active in isolation at different positions of a "proximal" (i.e., hip or knee), "crossed" (ankle), and "distal" joints of a limb (fig. 55). This is a step beyond Wagman, Pierce and Burger (1965) who observed that a small change in position brings different motor units into action. Consequently they noted the important influence of the sense of position on the motor response. The present investigation shows that in order to maintain or

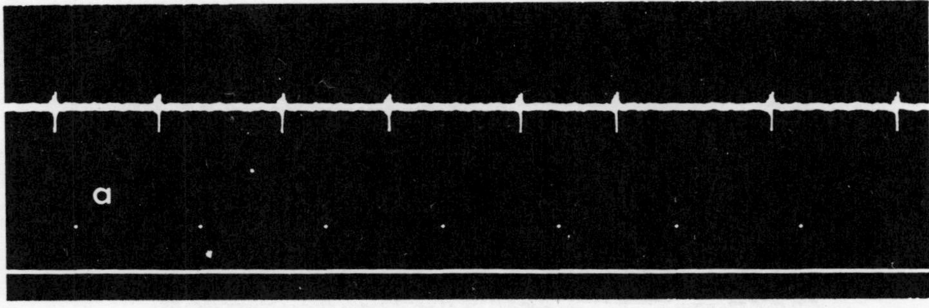

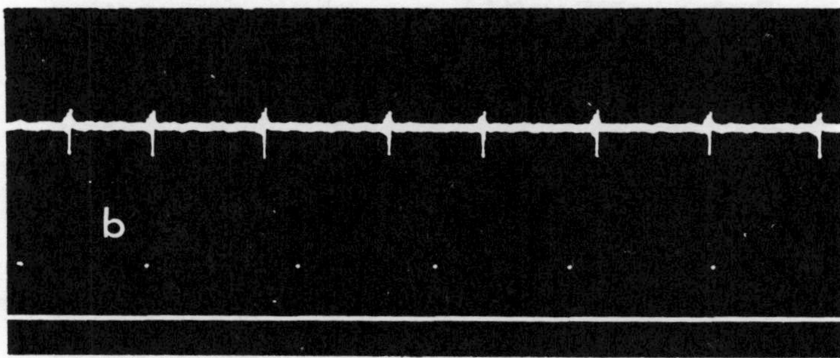

FIG. 54. Sample EMG of (*a*) the easiest, and (*b*) the fastest frequency of discharge of a motor unit in the right tibialis anterior during movements of the contralateral limb (time mark: 10-msec intervals). (From Basmajian and Simard, 1966.)

recall a motor unit at different positions, the subject must keep the motor unit active during the performance of the movements; and, therefore, preliminary training is undeniably necessary.

The observation that trained motor units can be activated at different positions of a joint is related to the work of Boyd and Roberts (1953). They suggested that there are slowly adaptive end organs of proprioception, which are active during movements of a limb. They observed that the common sustained discharge of the end organs in movements lasted for several seconds after attainment of a new position. This might explain why a trained single motor unit's activity can be maintained during movements.

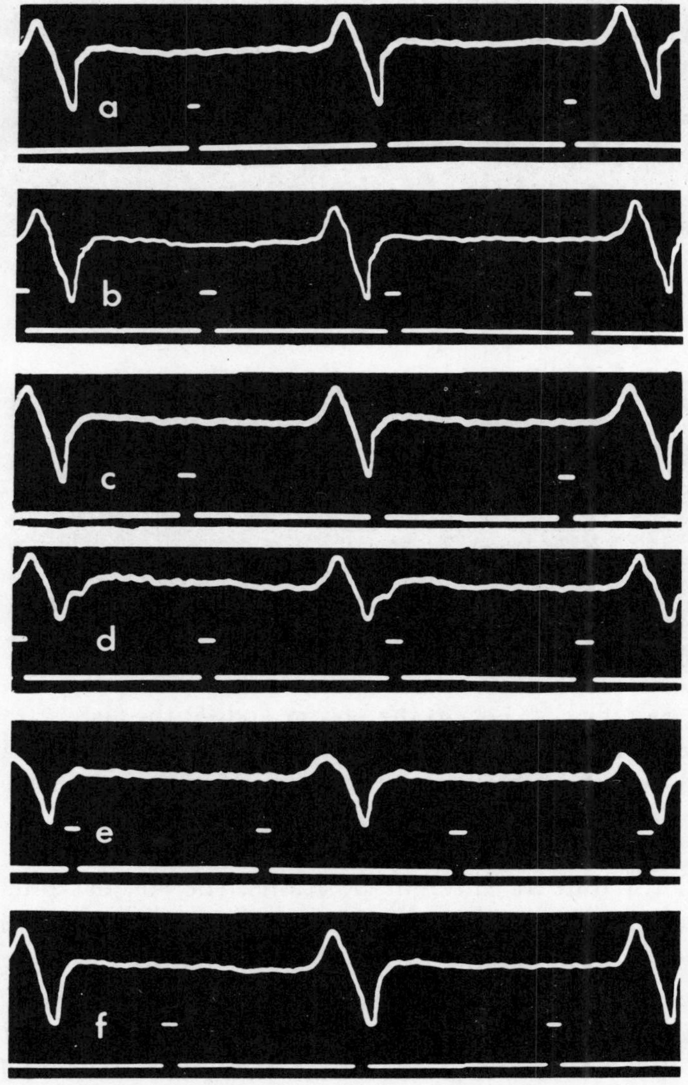

Fig. 55. EMG of a controlled motor unit potential at different "held" positions of the right lower limb: *a*, neutral; *b*, in lateral rotation at the hip; *c*, in medial rotation at the hip; *d*, dorsiflexion of the ankle; *e*, plantarflexion; *f*, toes extended. (Calibration: 100 μV, 100-msec intervals.) (From Basmajian and Simard, 1966.)

The control of the maintenance of a single motor unit activity during "proximal," "crossed" and "distal" joint movements in the same limb has been proved here to be possible providing that the technique of assistance offered by the trainer is adequate. The control over the discharge of a motor unit during proximal and distal joint movements requires a great concentration on the motor activity. But when one considers the same control during a "crossed" joint movement, there are even greater difficulties for obvious reasons.

The Level of Activity of Synergistic Muscles

The problem of what happens to the synergistic muscles at the "hold" position or during movements of a limb has been taken into consideration only in a preliminary way. The level of activity appears to be individualistic. Active inhibition of synergists is learned only after training of the motor unit in the prime mover is well established.

Practical Applications

Many applications are emerging for the use of motor unit training, e.g., in the control of myoelectric prostheses and orthoses, in neurological studies and in psychology. Very recently a novel application for treatment ot reading disorders has been reported (see p. 324).

47

Transfer Following Operant Conditioning in the Curarized Dog

Abraham H. Black

Abstract. *Dogs were trained to press a pedal to avoid shock. They were then operantly reinforced for making or for refraining from making a series of electromyographic responses while almost completely curarized. Tests after recovery from curarization showed that operant conditioning under curare influenced the original pedal-press response.*

Curare-like drugs have been employed to control skeletal responding in many types of conditioning experiments. Black, Carlson, and Solomon (*1*), Black and Lang (*2*), Black (*3*), and Church and LoLordo *et al.* (*4*) used curare to prevent skeletal responding in research on classical (*5*) heart-rate conditioning. Trowill (*6*) and Di Cara (*7*) used it for the same purpose in research on the operant conditioning of heart rate. Black (*8*), Solomon and Turner (*9*), Leaf (*10*), and Overmier (*11*) studied the effects of prior classical conditioning in curarized subjects on subsequent operant responding in the normal state. In experiments such as these, the assumption has been made that operant reinforcement of skeletal activity does not take place under curare. There are, however, a number of ways in which operant reinforcement that influences skeletal behavior might occur in curarized subjects. If curarization is incomplete, vestigial skeletal responses might be operantly reinforced. Even if curarization is complete, central correlates of overt responses might be reinforced by operant procedures. In each of these cases, the effects of operant reinforcement under curare could transfer to overt skeletal behavior in the normal state. There is, however, so little evidence on operant reinforcement in curarized subjects and its transfer, that one does not know whether this hypothesis is plausible or not. The experiment described here was carried out in order to provide such evidence.

Our first aim was to demonstrate that almost completely curarized subjects could be operantly trained either to increase the rate of or to refrain from making electromyographic responses (*12*). Our second and main goal was to determine the effects of such operant training on subsequent skeletal behavior in the normal state.

In the first phase of the experiment, operant conditioning procedures were employed to train dogs to avoid an intense shock. The avoidance response was pressing a pedal with the left foreleg in the presence of a conditioned stimulus (CS). The CS was a 75-db white noise. A potentiometer measured vertical leg movements. Electromyographic responses were recorded from two locations on the left forelimb by the use of Grass needle electrodes. One location was over the deltoid muscle and the other was over the extensor carpi radialis and the extensor digitorum muscles.

On each trial, if the dog pressed the pedal during the CS-shock interval (15 seconds), the CS was terminated and shock avoided. If the dog failed to press during the CS-shock interval, a series of 4- to 10-ma shocks of one-tenth second duration were presented at 5-second intervals until a pedal-press was made. Fifty trials a day were given. Each dog was trained to a criterion of 20 consecutive avoidances, and then given 50 additional training trials on the next day.

Twenty-four hours later, each dog was given ten more training trials, and then the second phase of the experiment began. The dogs were curarized with doses of *d*-tubocurarine chloride *(13)* which varied from 7.8 to 30.0 mg, with a mean of 14.7 mg. Curare injections and artificial respiration were continued during the course of the experiment. The dogs were maintained at a level of curarization such that electromyographic responding in the left foreleg occurred, but little or no leg movement was present. Of those trials under curare on which electromyographic activity was observed, the median amplitude of vertical movement was 0 mm and the range was 0 to 50 mm. The scores were clustered at the lower end of the distributions; for example, 95 percent of the observations were less than 10 mm. The effects of curare were further assessed by comparing

Table 1. Results for the dogs trained to make electromyographic responses under curare (group R) and for the dogs trained to refrain from electromyographic activity under curare (group NR).

Item	Median number		Mann-Whitney U test
	Group R	Group NR	
Original avoidance training in the normal state			
Shock trials to criterion	7.5	7.5	$P(U \leq 3.0) = 0.880$
Training in the curarized state			
Trials to criterion	24.0	23.0	$P(U \leq 20) = 0.800$
Shock trials to criterion	4.5	8.0	$P(U \leq 17) = 0.130$
Shock per shock trial	2.4	4.6	$P(U \leq 5) = 0.002$
Total shocks (median)	12.0	51.5	$P(U \leq 5) = 0.002$
Test in the normal state			
Pedal-presses during first 10 seconds of the CS	10.0	3.0	$P(U \leq 3.5) = 0.001$
Pedal-presses on trial No. 1	35.5	4.0	$P(U \leq 10.0) = 0.010$
Trials to five consecutive failures to respond to the CS	50+	17.0	$P(U \leq 14.5) = 0.037$

maximum electromyographic responses and vertical leg movements on the last conditioning trial in the normal state and on the first conditioning trial under curare. (Measurements were taken before the presentation of shock, if shock occurred on a given trial.) The median of the maximum leg movements was 185 mm in the normal state and 2.5 mm under curare; the median of the maximum electromyographic responses was 2.59 mv in the normal state and 0.17 mv under curare.

The treatments under curare were designed to operantly reinforce a high rate of electromyographic activity in one group of dogs (group R) and a zero rate of such activity in a second group of dogs (group NR). The nine dogs in group R were reinforced for making a series of seven electromyographic responses to the CS. On each trial, a series of seven electromyographic responses during a 7-second CS-shock interval resulted in CS termination and avoidance of shock. If the dog failed to make seven electromyographic responses during the CS-shock interval, a series of shocks was presented. The shocks were programmed 7 seconds apart and were delayed for 7 seconds by an electromyographic re sponse. The series of shocks and the CS were terminated by the seventh electromyographic response. Eight other dogs, in group NR, were reinforced for a period of 7 seconds *without* an electromyographic response to the CS. On each trial, 7 seconds of electromyographic inactivity resulted in CS termination and avoidance of shock. If an electromyographic response occurred in the presence of the CS, it was followed by a brief shock. The CS and shocks could be terminated only by 7 seconds of muscular inactivity. Dogs in both groups were trained until they met a criterion of 20 consecutive trials without a shock, or until 130 trials had passed.

In the third and final phase of the experiment, I tested the transfer from training under curare to pedal-pressing in the normal state. On this test, which took place 2 days after recovery from curarization, each dog was given a series of 50 extinction trials without shock. On each extinction trial, the CS terminated only after 10 seconds had passed without a pedal press. This special extinction procedure was employed in order to reduce the amount of time required for between-group differences to emerge. In previous work, I found that the response was extremely difficult to extinguish when we employed the usual extinction procedures in which the pedal-press produced CS termination. Also, this special extinction procedure permitted us to obtain measures of pedal-pressing before the data were contaminated by the reinforcing effects of CS termination.

The results are shown in Table 1. The number of shock trials before the acquisition criterion was reached is an index of speed of conditioning of the pedal-press response. There was no significant difference between the two groups on this measure.

During the curare phase of the experiment, 15 of the 17 dogs that were trained met the criterion of 20 consecutive trials without a shock. The two dogs that did not meet the criterion were in group R. (One reached 19 trials without a shock, and the other reached four; the latter dog was discarded from the experiment.) These results indicate that operant procedures can be used to alter the probability of electromyographic responses under curare.

There was no difference between the groups in number of trials required to meet the training criterion under curare. However, the dogs trained to make an electromyographic response required

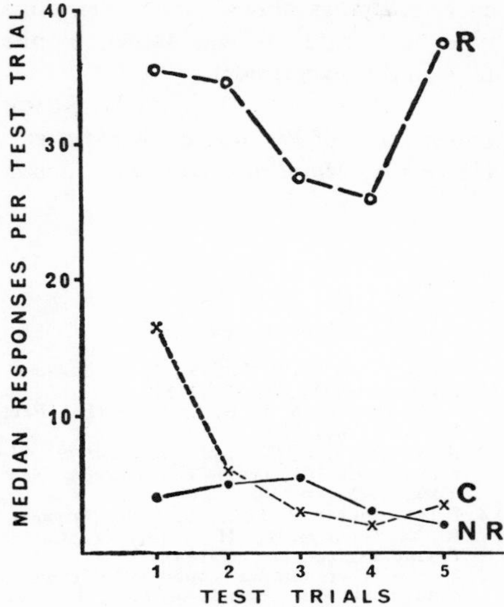

Fig. 1. Pedal-presses for each of the first five test trials for dogs trained to make electromyographic responses (R), dogs trained to refrain from electromyographic responding (NR), and control dogs (C).

significantly fewer shocks per trial and significantly fewer total shocks to reach the criterion. It took less shock to train the dogs to make a series of electromyographic responses than to train them to refrain from electromyographic activity under curare.

On the test after recovery from curarization, the dogs that were trained to make electromyographic responses under curare displayed significantly greater rates of pedal-pressing than the dogs that were trained to refrain from electromyographic activity on all the measures shown in Table 1. Reinforcing electromyographic activity under curare led to a high level of pedal-pressing in the normal state, and reinforcing electromyographic inactivity led to a low level of pedal-pressing in the normal state.

There is some question concerning the interpretation of these test data because of the differential exposure to shock under curare. One might argue, for example, that more occurrences of shock under curare or more pairings of CS and shock under curare result in fewer pedal-presses in the normal state. These possibilities were ruled out by four control groups which received different numbers of classical pairings of CS and shock under curare: 0, 4, 12, or 36 pairings. (In addition, all control dogs received 20 presentations of the CS alone in order to equate them with the experimental dogs, since the latter received 20 presentations of CS alone on the criterion trials under curare.) There was no significant difference among these control groups on any of the test measures after curarization.

In Fig. 1 the R, NR, and control groups are compared. (Data for the control groups were combined for this comparison, since there were no differences among them.) Data are shown for the median number of pedal-presses per trial for the first five test trials. The operant conditioning procedures under curare that increased electromyographic responding produced a significant increase in pedal-pressing relative to the control-group dogs on all five test trials. The operant procedures under curare that decreased electromyographic responding produced a significant decrease in pedal-pressing relative to the control-group dogs on the first test trial. Thus, transfer from operant conditioning under curare is different from that produced by classical conditioning procedures.

The present results lead one to question the curare procedure as a control for operant reinforcement that is related to overt skeletal responding. For example, in experiments on the operant conditioning of heart rate under curare, it may very well be that electromyographic responses were actually conditioned, and that these led to reflexive changes in heart rate. Similarly, demonstrations of how prior classical conditioning in curarized subjects affects subsequent operant behavior in the normal state may in fact be demonstrations of the effects of prior operant conditioning. This is not to say that all the results of training procedures in curarized subjects can be accounted for by operant conditioning; rather, the point is simply that relevant operant conditioning may be occurring inadvertently during experiments on classical conditioning.

These conclusions, of course, hold only for subjects that are not completely curarized. Would not the completely curarized preparation avoid the difficulties described above? While the present experiment provides no direct answer to this question, the data from the operant conditioning of refraining from electromyographic activity are relevant. Two major sources of reinforcement were employed: (i) avoidance of shock and CS termination following a period of no electromyographic activity, and (ii) the occurrence of shock following electromyographic activity. It seems reasonable to assume that at least the former type of reinforcement could occur in completely curarized subjects. If this is so, then the completely curarized preparation would not avoid the difficulties described above. Before we can decide on the adequacy of the completely curarized preparation unequivocally, however, we must determine how much electromyographic activity, if any, is necessary for operant conditioning and transfer to occur. It may be that no electromyographic activity is necessary, and that the transfer could be mediated by the reinforcement of central nervous system events associated with movement (*14*). If this is the case, then the operant reinforcement of motoneuron electrical activity in completely curarized dogs should show the same type of transfer as was shown by the dogs in this experiment.

A. H. Black

Department of Psychology, McMaster University, Hamilton, Ontario, Canada

References and Notes

1. A. H. Black, N. J. Carlson, R. L. Solomon, *Psychol. Monog.* **76**, 29 (1962).
2. A. H. Black and W. H. Lang, *Psychol. Rev.* **71**, 80 (1964).
3. A. H. Black, in *Classical Conditioning*, W. Prokasy, Ed. (Appleton-Century-Crofts, New York, 1965), p. 20.
4. R. M. Church V. LoLordo, J. B. Overmier, R. L. Solomon, L. H. Turner, *J. Comp. Physiol Psychol.* **62**, 1 (1966).
5. In this report classical conditioning refers to conditioning that depends on the pairing of two stimuli—the conditioned stimulus and unconditioned stimulus. Operant conditioning refers to conditioning that depends on the pairing of a response and a reinforcing stimulus.
6. J. A. Trowill, paper presented at the meetings of the Eastern Psychological Association, New York, 1966.
7. L. V. Di Cara, paper presented at the meetings of the Eastern Psychological Association, New York, 1966.
8. A. H. Black, *J. Comp. Physiol. Psychol.* **51**, 519 (1958).
9. R. L. Solomon and L. H. Turner, *Psychol. Rev.* **69**, 202 (1962).
10. R. C. Leaf, *J. Comp. Physiol. Psychol.* **58**, 446 (1964).
11. J. B. Overmier, *ibid.* **62**, 15 (1966).
12. R. F. Hefferline, B. Keenan, and R. A. Harford [*Science* **130**, 1338 (1959)] have demonstrated that human subjects could be operantly trained to make extremely small electromyographic responses. It should be pointed out also that training a subject to refrain from electromyographic activity is not identical with training a subject to "hold still" in the normal state. The latter could involve a cessation of electromyographic activity as in relaxation, or a high level of electromyographic activity as in isometric contractions where movement does not occur because one set of muscles is pitted against another.
13. The *d*-tubocurarine chloride was provided by E. R. Squibb and Sons of Canada Ltd.
14. While the central nervous system events that might be involved cannot be specified, it does seem that feedback from the pedal-press is not a crucial factor, since Gorska and Jankowska [*Acta. Biol. Exp. Polish Acad. Sci.* **21**, 219 (1961)] and Taub, Bacon, and Berman [*J. Comp. Physiol. Psychol.* **59**, 275 (1965)] have demonstrated operant conditioning after deafferentation. The present results are consistent with theirs in that operant conditioning in curarized dogs had a clear-cut effect on subsequent pedal-pressing even though the afferent feedback was very

different under curare from that occurring during pedal-pressing.

15. Supported by research grant MY-2741 from the National Institute of Mental Health, USPHS, by research grant No. 81 from the Ontario Mental Health Foundation, and by research grant APY-42 from the National Research Council of Canada. I thank C. Batenchuk and A. Dalton for their assistance in carrying out the experiment.

48 Feedback Technique for Deep Relaxation

Elmer F. Green, E. Dale Walters, Alyce M. Green and Gardner Murphy

Deep relaxation of striate muscle is of importance in many clinical and research projects. In some therapeutic procedures patients are trained to significantly reduce their tension levels, as, for example, in the process of overcoming anxiety (Haugen, Dixon, & Dickel, 1963; Jacobson, 1938), in desensitization treatment (Wolpe & Lazarus, 1966), and in meditative and depth-dimensional psychotherapy (Schultz & Luthe, 1959; Luthe, 1965). In the latter some patients experience altered states of consciousness while in the deeply relaxed state.

A general problem of relaxation procedures is that it often takes days, or weeks, before a subject or patient can relax to a satisfactory degree. Many projects might be aided by the use of a simple feedback method we have developed through which *extremely low* tension levels, tending toward zero, can be voluntarily achieved by some subjects in a few minutes. For example, seven out of twenty-one subjects were able to achieve intermittent neuromuscular silence within twenty minutes and maintain it for the duration of an experiment, thirty minutes or more. Before using the feedback system, we were unable to approach the zero level of tension quickly in any subject.

Before discussing the feedback technique in detail, we should note that although many researchers know that a *zero* muscular tension level can be achieved by an experimental subject and can be recorded from "skin" electrodes, others are unaware of it or believe that it is not possible. In regard to zero muscular tension, Basmajian (1962, pp. 41–44) notes that, ". . . the usual definition of 'tone' should be modified to state that the general tone of a muscle is determined both by the passive elasticity or turgor of muscular (and fibrous) tissues and by the active (though not continuous) contraction of muscle in response to the reaction of the nervous system to stimuli. Thus, at complete rest, a muscle has not lost its tone even though there is no neuromuscular activity in it. . . . In no nor-

This study was supported in part by Grant MH 14439, National Institute of Mental Health.

Address requests for reprints to: Elmer E. Green, Research Department, The Menninger Foundation, Topeka, Kansas 66601.

mal muscle at complete rest has there been any sign of neuromuscular activity, even with multiple [needle] electrodes. . . . Where did the false concept of continuous neuromuscular activity during rest originate? Chiefly it seems to be a widespread misinterpretation of Sherrington's *postural tonus*. There is no denying that any muscle that is helping to hold the subject upright shows various degrees of activity . . . [but] the human upright position allows many of the limb muscles to relax completely."

In order to help subjects to voluntarily relax dorsal forearm muscle groups (external digitorum communis and external carpi ulnaris) quickly down to low levels, in an operant self-conditioning paradigm, we use a visual feedback circuit employing a meter driven by the pen-motor voltage of a polygraph channel, which in turn is driven by the rectified voltage from a high-gain electromyographic (EMG) circuit, as shown schematically in Fig. 1. Single motor unit (SMU) firing, shown in Figs. 2 and 3, is clearly revealed near the zero level of muscle tension.

Information can be fed back easily through either the visual or auditory modality, but whenever it is necessary for the subject to remain alert although deeply relaxed, we use the meter so the subject must keep his eyes open. It is usually necessary, at the start of a deep-relaxation session, to use low gain on the polygraph because subjects cannot otherwise get a less-than-top reading on the meter. When muscle tension decreases (so that the meter needle drops toward the low end of the scale) the polygraph gain is increased by the experimenter and the subject is informed that the meter sensitivity is being increased. Usually within a few minutes the gain has been increased to a level high enough to reveal SMU firing on the strip-chart writeout. Although SMU firing reportedly can reach rates as high as 50 per second, the rates in our subjects gradually dropped to about six or seven per second during a few minutes of relaxation. The next decrease in many of our subjects, after reaching this relatively low rate, was suddenly to zero, as shown in Fig. 2. In other words, while approaching zero firing there seemed to be a "natural" minimum rate, or reflex-threshold rate, which appeared in the absence of an effort by the subject to produce specific firing rates lower than this spontaneous value. This finding suggests the existence of a CNS "gate circuit" that controls a neural reflex oscillator whose minimum rate is normally quite constant. According to this hypothesis, voluntary single firings could result from short gate-openings controlled by the CNS. A neural oscillator-and-gate circuit would also account for the relative ease with which some of Basmajian's subjects learned to produce unique firing patterns such as "gallop rhythms, drum-beat rhythms, doublets, and roll effects" (Basmajian, 1963).

In a study of covert (subliminal) muscle tension responses during mental imagery, several subjects reported "body-image" changes or illusions. After about twenty minutes of feedback training, with instructions to "try to get the meter needle down to zero," five out of seven subjects made statements such as "My arm feels like a bag of cement," ". . . like a ton of lead," ". . . as if it is moving away from me," "I had to look at it [the hand and arm] to see if it was still in the same place," ". . . to see if I had moved it," etc. It was sometimes necessary for the subject to adjust his arms in order to find the optimal posture for the achievement of a zero level of firing. In naive subjects in a proper setting (reclining chair, quiet

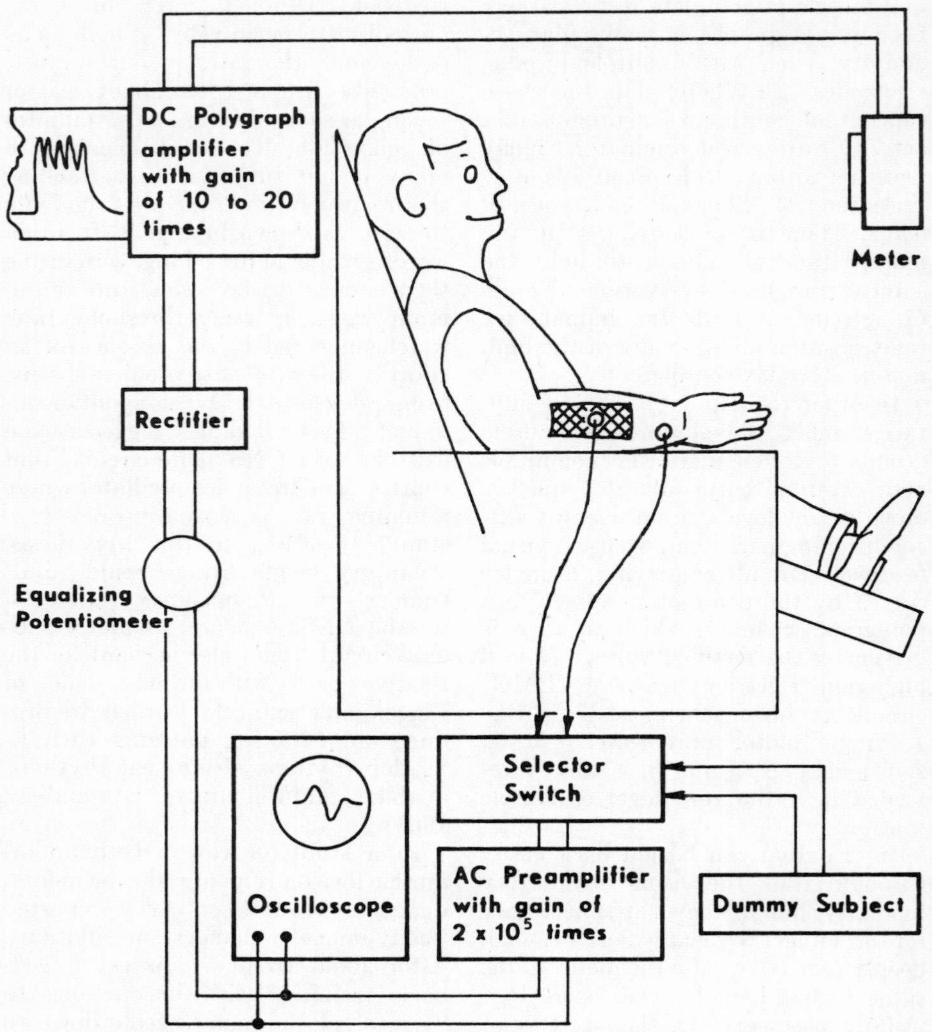

FIG. 1. Block diagram of EMG feedback system. Viewing distance of the meter is about four feet. The large "active" forearm electrode is referenced to a small disc electrode placed over the bony styloid process. A ground electrode on the subject is sometimes needed to reduce 60 Hz noise. The dummy subject is used for calibration of the system. It is an electrical stimulus generator whose biphasic rectangular pulses are "rounded off" with a choke and capacitor so as to produce trains of pulses very similar in appearance to single motor unit firing as seen on the oscilloscope.

room, dim lights, etc.) relaxation, seemed to spread over a large part of the body. We have also found that a normal subject can learn, with a little instruction and practice, to dissociate his right forearm from the rest of his muscular system to the extent that he can tense his abdomen, left arm, leg muscles, or neck muscles, without causing any observable increase of tension in the right arm.

When EMG signals are used for feed-

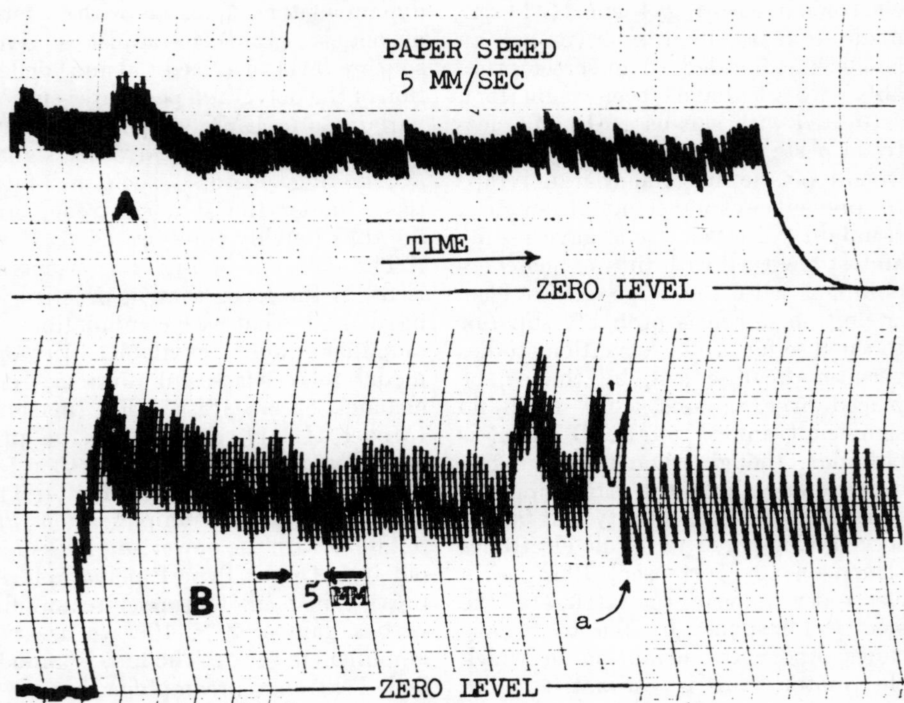

FIG. 2. Sawtooth patterns from Beckman polygraph using the "After" circuit of Fig. 5. (A) Showing a long burst of reflex-type single motor unit firing, 6 to 7 per second, with sudden and total cessation at the end. (B) Zero level is followed by reflex-type single motor unit firing. Paper speed is changed at point "a" from 5 to 10 mm/sec. Amplitude scales were arbitrarily selected, but in both (A) and (B) the SMU firing was accompanied by the oscilloscope pattern of Fig. 3. Delay in rise from zero level probably results from the time constant of the rectifier.

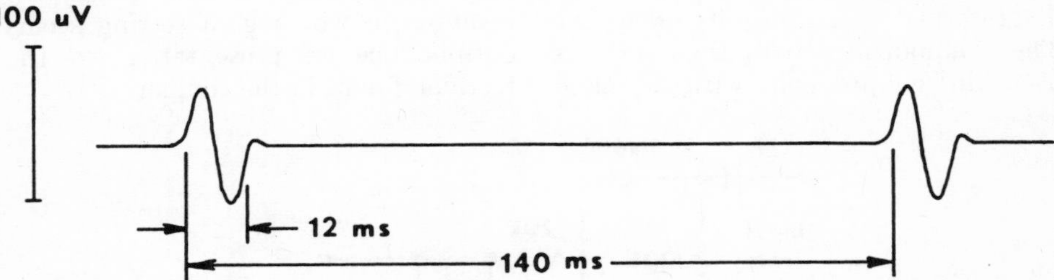

FIG. 3. Drawing of oscilloscopic trace of two single motor unit firings obtained with a skin electrode, very similar to the photographic records obtained by Harrison and Mortensen (1962) in which traces from needle electrodes and skin electrodes were compared.

back of relaxation levels, the surface electrode has a great advantage over the needle electrode because it picks up signals from a very large area. In addition, the skin electrode is less alarming to the subject. In working with forearm muscles, we have taped to the skin (in a few subjects) a rectangular screen electrode measuring 4 cm $\times$ 14 cm made from platinum mesh (though an anodized, chlorided, silver screen probably would have been superior), saturated with salt-based Redux electrode paste and covered with a lightweight plastic material (Saran Wrap) to prevent evaporation. Although a standard "disc" electrode gave results almost identical with those which were obtained with the large screen electrode, the screen is probably superior to the disc for use in relaxation studies because the disc can, by chance, be placed over an area from which no significant amount of EMG signal is obtained, though when moved a few centimeters along the same muscle, considerable SMU activity can be observed. Compared to needle electrodes (Harrison & Mortensen, 1962), skin electrodes appear to attenuate the electrical response by ten to twenty times, under the condition in which the motor unit is estimated (from muscle configuration) to lie within one to one and one-half centimeters of the skin surface.

High-gain detection of SMU firing from skin electrodes is not easy to achieve with standard polygraph equipment. We have found, however, that the general circuit configuration and voltage gains shown in the block diagram of Fig. 1 are quite adequate. The "equalizing potentiometer" is used in conjunction with a hand dynamometer (squeezed with a force of one kilogram) to adjust a given squeeze-level to a fixed (50 mm) deflection of the polygraph pen. Inter-subject variations in EMG amplitude due to electrode placement and other signal-attenuating factors, such as fatty tissue under the skin electrode, etc., are thus roughly equilibrated.

The diagram of Fig. 1 is "generalized" in the sense that in practice we have used whatever combinations of amplifiers were available. In one moderately successful system, one channel of a Grass EEG machine (Model IIID) was used as the preamplifier, and the signal was then fed into the rectifier shown in Fig. 4. The final stage of DC amplification was obtained with the power amplifier of an old-style Offner Type P polygraph. At present, instead of a single preamplifier with a gain of 2×10^5, we use two amplifiers in series. The first is a modified Tektronix preamplifier having a *nominal* voltage gain of 1000, with a lower bandpass cutoff frequency (3 db point, voltage ratio) at 110 Hz. The Tektronix preamplifier is followed by a modified Beckman EMG input-coupler (Type 9852, of a Beckman Type R polygraph) and by its associated preamplifier in which the signal is further amplified by 200 to 1000 times, depending on which gain setting seems optimal, before presentation to the rectifier circuit in the coupler.

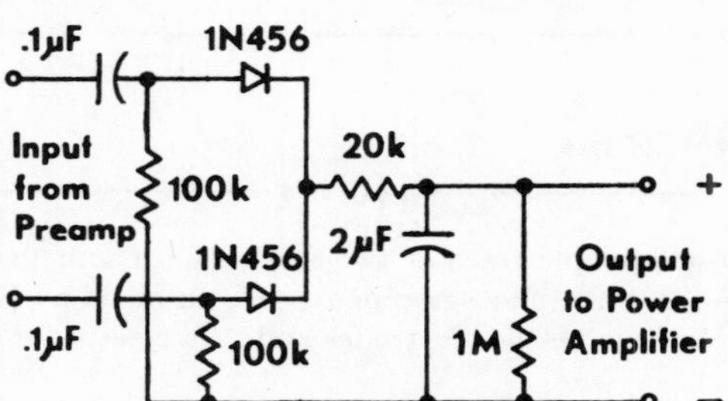

FIG. 4. Rectifier circuit useful for observation of single motor unit firing. Input through stand-off capacitors makes possible the use of preamplifiers having high-voltage DC outputs. The IN456 diodes have a high back resistance, making necessary the use of a bleeder resistor across the 2.0 μf output capacitor.

It is especially important to have control of the voltage gain in the preamplifier stage so that the fixed threshold voltage of the rectifier diodes, about 0.4 volt, can be used as described below, as an optionally chosen and variable threshold, for detection of EMG signals. For instance, if the total gain in the preamplifier stage is set unusually low, perhaps at 5×10^4, then any EMG signal with amplitude less than 8 μv will not be detected because it will not rise above the 0.4 volt diode threshold after amplification. If the signal continuously increases in amplitude above 8 μv, the amplitude of the DC output will increase in a nonlinear manner, typical of the "toe" section of a sigmoid-type curve. On the other hand, if the total preamplifier gain is set unusually high, perhaps at 2×10^6, then every signal with amplitude greater than 0.2 μv will be detected. In this latter case, electrical noise from various non-striate tissues, random noise from the preamplifier stages themselves, 60 Hz noise from the surroundings, and EMG signals from muscle bundles located much deeper than those with which we are concerned, all contribute to a fluctuating and "grassy" baseline and result in an unstable feedback meter at the zero level, after all trace of SMU firing has ceased. After examining the records from a few subjects, we chose to use a total gain in the preamplifier stages of about 2×10^5, about 400 in the Tektronix preamplifier and 500 in the Beckman preamplifier. Thus any noise or signal less than 2.0 μv at the input to the first preamplifier stage was not observed. Possibly this gain setting would not be satisfactory for use with deep-bodied muscles, but it appears to be near optimum when used with the muscles of the dorsal forearm. It is possible to summarize here by saying that when the preamplifier stage had a total gain of about 2×10^5, the EMG signal which reached the rectifier (usually mistakenly called an "integrator") during SMU firing was significantly larger than the voltage threshold of the rectifier diodes. DC amplification of about twenty times, after rectification, was usually sufficient to clearly show SMU firing.

It should be noted that if the active and reference electrodes are placed close together, as is advocated for the detection of very local signals, bipolar common-mode rejection in the preamplifier will cause the EMG signals to be significantly reduced in amplitude. A monopolar arrangement such as we use, however, in which the reference electrode is relatively far from the active site and in contact with a body region which is electrically quite quiet

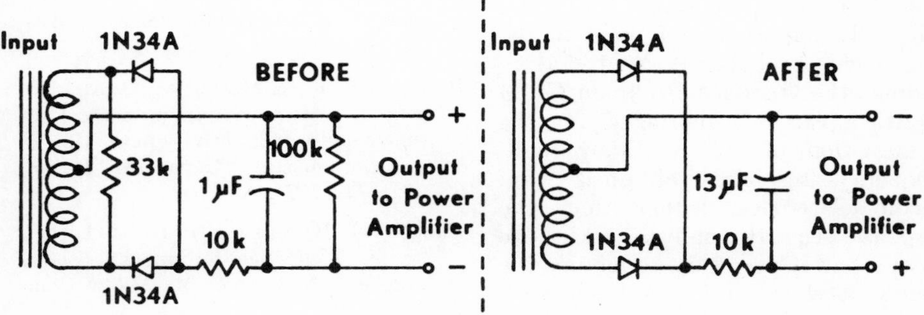

Fig. 5. Modification of rectifier in EMG Input Coupler Type 9852 (Beckman polygraph), for observation of single motor unit firing. A bleeder resistor is not needed across the output in the "After" diagram because the IN34A diodes have a low back resistance. Diode reversal changes the pen motor direction.

(the bony styloid process of the dorsal wrist), allows the EMG signal to reach its maximum amplitude.

Most of the SMU firings, in the muscle groups with which we work, appear to have an oscillatory frequency (not firing rate) between 95 Hz and 75 Hz, at which frequencies the preamplifier stage has voltage attenuations of 3.5 and 4.5 db, respectively. In other words, the SMU oscillatory frequency band is located just below the 110 Hz shoulder in the filter roll-off curve of the preamplifier. This location for the SMU frequency band is useful because at high muscle tension levels the voltage *gain* (not output) of the preamplifier is found to be significantly lower than the gain at close-to-zero (SMU) levels, resulting from the fact that the tensed muscle produces an EMG signal in which most of the large amplitude components are found at frequencies *below* 75 Hz. The overall effect of this use of nonlinear filter characteristics is to attenuate those signals which are produced by erratic bursts of high muscle tension, which normally tend to overdrive the meter, while at the same time there is allowed adequate sensitivity of the feedback meter at low tension levels. In future work with EMG feedback, it is intended to use an improved filter with a steeper cutoff slope (for additional attenuation of the signal resulting from high muscle tension) and a more optimally located cutoff point in the filter curve, probably at about 95 Hz.

Since the standard Beckman EMG input-coupler is constructed for the examination of the relatively high-frequency responses obtained from needle electrodes, rather than the response seen through the skin, we have found it necessary to modify the coupler as shown in Fig. 5. Each tooth of the sawtooth pattern in Fig. 2 resulted from one of the SMU responses seen on the monitoring oscilloscope, Fig. 3, and was essentially identical with the SMU response found by Harrison and Mortensen (1962) with skin electrodes, though our figure is hand-drawn from visual observation rather than photographed. The SMU firing in Figs. 2 and 3 was from one of us (E.D.W.) using the large screen electrode. The forearm was almost totally relaxed after fifteen minutes of practice with the feedback meter. Occasional trains of SMU firing appeared interspersed with periods of neuromuscular silence. By proper placement of a small disc electrode instead of the large screen electrode, with E.D.W. again as subject, essentially identical results (with those shown in Figs. 2 and 3) were obtained.

Elmer E. Green

E. Dale Walters

Alyce M. Green

Gardner Murphy

Research Department
The Menninger Foundation
Topeka, Kansas

REFERENCES

Basmajian, J. V. *Muscles alive: Their functions revealed by electromyography.* Baltimore: Williams & Wilkins, 1962.

Basmajian, J. V. Control and training of individual motor units. *Science*, 1963, *141*, 440–441.

Harrison, V. F., & Mortensen, O. A. Identification and voluntary control of single motor unit activity in the tibialis anterior muscle. *Anatomical Record*, 1962, *144*, 109–116.

Haugen, G. B., Dixon, H. H., & Dickel, H. A. *A therapy for anxiety tension reactions.* New York: MacMillan Company, 1963.

Jacobson, E. *Progressive relaxation.* Chicago: University of Chicago Press, 1938.

Luthe, W. (Ed.) *Autogenic training: Correla-*

tions psychosomaticae. New York: Grune & Stratton, 1965.

Schultz, J. H., & Luthe, W. *Autogenic training: A psychophysiologic approach in psychotherapy*. New York: Grune & Stratton, 1959.

Wolpe, J., & Lazarus, A. A. *Behavior therapy techniques*. Oxford: Pergamon Press, 1966.

VIII

METHODOLOGY: WHAT IS CONDITIONING

49

Instrumental Conditioning of Autonomically Mediated Behavior

H. D. Kimmel

The distinction between classical and instrumental conditioning was first brought into clear focus by Skinner (1935, 1937, 1938). His identification of two types of learning (Type S or classical and Type R or instrumental) was based upon the notion that there are two different kinds of behavior, *respondent* and *operant,* and that each lends itself to modification by a particular procedure. Thus, respondent behavior is *elicited* reliably by some well-identified stimulus or stimuli and is characterized by the kind of learning studied by Pavlov. Operant behavior, in contrast, is *emitted* in the absence of any identifiable stimulation and is modified in the fashion of Thorndike by reinforcement which follows its occurrence.

Accepting a suggestion apparently first made by Schlosberg (1937), Skinner considered the possibility that the two types of conditioning could be related to the somatic-autonomic division of the nervous system and tentatively concluded that they could. Although, as he observed, children seem to learn to cry "real tears" in relation to their consequences,

> Glands and smooth muscles do not naturally produce the kinds of consequences involved in operant reinforcement, and *when we arrange such consequences experimentally, operant conditioning does not take place* [italics added]. We may reinforce a man with food whenever he "turns red," but we cannot in this way condition him to blush . . . [Skinner, 1953, p. 114].

Skinner's belief that autonomic responses could only be conditioned classically, however, was based upon meager factual evidence. Mowrer (1938) attempted to make the GSR instrumental in a shock-avoidance situation without success. Skinner and Delabarre (Skinner, 1938) tried to condition vasoconstriction using a positive reinforcer, but obtained inconclusive results. Careful search of the literature of the 1920s, 1930s, and 1940s reveals not one *systematic* experiment to support Skinner's conclusion. Several independent explorations of the possibility of instrumental modification of autonomically mediated responses were done in the late 1950s and early 1960s and their results published in the last few years. These studies have used both operant reward and avoidance procedures and have dealt with smooth muscle responses such as heart rate and vasomotor behavior as well as glandular responses such as salivation and the GSR. This paper summarizes and reviews this work.

Operant Reward Training

Mandler, Preven, and Kuhlman (1962) attempted to influence the rate of emission of spontaneous GSRs by rewarding the response with a light which signalled the earning of money. They reported increased emission of GSRs for some of their *Ss*, although the overall average number of responses declined over the several days of experimentation. Compared with control periods during which reinforcement was not given, more responses occurred in the experimental periods in spite of the general decrease in rate of responding. These investigators, however, were inclined to

[1] Done under grants MH-06060-03 and 12262-02 from the National Institute of Mental Health.

interpret their results as agreeing with Skinner.

Kimmel and Hill (1960) employed a yoked control procedure to study the instrumental conditionability of the GSR with odors as reinforcement. Each control *S* was matched to an instrumental *S* in the number of reinforcements received during each minute of acquisition, with the only difference between the two *S*s being that the instrumental *S* received the odor immediately following the emission of a response while the control *S* received the odor only at times when he was not responding. Both pleasant and unpleasant odors were used for different groups of *S*s. No difference in rate of responding was found between the instrumental and yoked control *S*s during the reinforcement period, but when the odors were omitted during extinction, the instrumental *S*s showed a brief increase in response frequency and the control *S*s showed a decrease, regardless of which odor had been used. Using the same type of yoked control procedure, Fowler and Kimmel (1962) reinforced *S*s who were seated in a dark soundproof room by presenting a brief dim light following emission of spontaneous GSRs. The light had no instructed significance; it merely interrupted the total darkness of the situation. After a few minutes during which both groups showed a decline in rate of responding, the instrumental group began to increase in frequency of emission of GSRs while the control group continued to decrease. At the end of the reinforcement period and for the first 2 minutes of extinction, the difference in number of responses emitted per minute between the two groups was statistically significant. During the acquisition period, the interaction between experimental conditions and minutes of training was also significant. The shift from olfactory to visual reinforcement, with its attendant reduction in delay of reinforcement, may have been the main reason for the improved result.

The essential aspects of the Fowler and Kimmel (1962) study were replicated by Kimmel and Kimmel (1963). Modifications in method that were intended to clarify the earlier observations were: (*a*) the habituation period prior to reinforcement was increased to permit fuller acclimatization of the *S* to the experimental situation (response frequency during reinforcement and extinction in these studies was expressed relative to prereinforcement measures); (*b*) the duration of the light reinforcement was reduced from 1.0 to .1 second; and (*c*) the duration of the extinction period was extended to permit further analysis of response frequency trends following the termination of reinforcement. The results of the Kimmel and Kimmel (1963) study, shown in Figure 1, provided more convincing evidence of operant conditioning of an autonomically mediated response than had been obtained previously. Not only did the instrumental *S*s make significantly more responses than their yoked controls during reinforcement and extinction, but the instrumental *S*s' response frequency during acquisition and most of extinction was higher than it had been during the last 5 minutes of habituation prior to reinforcement. During extinction, the response frequency trends of the two groups were opposite, an observation that further supports the conclusion that learning had been achieved during reinforcement. In the last minute of extinction, both groups' response rates equalled 100% of their initial rates. The fact that the control group showed an increase in responding during extinction suggests the possibility that these *S*s, rather than participating in a kind of "placebo" control treatment, in which their responsiveness gradually declined during acquisition, were, in fact, instrumentally conditioned to *not* respond. Thus, during extinction, this tendency toward nonresponding gradually disappeared and responding, accordingly, increased.

Further support for both the last-mentioned point and general confirmation of the phenomenon of instrumental conditioning of the GSR with visual reinforcement has been obtained in a study by Greene (1966). In the Greene experiment, a "true" yoked control procedure was used, in which the instrumental and control *S*s were run simultaneously in adjoining soundproof chambers. With simultaneous data collection from instrumental and control *S*s, the control *S* received the visual reinforcer whenever the instrumental *S* emitted a GSR; in this way, the light occasionally occurred following an emitted response, and,

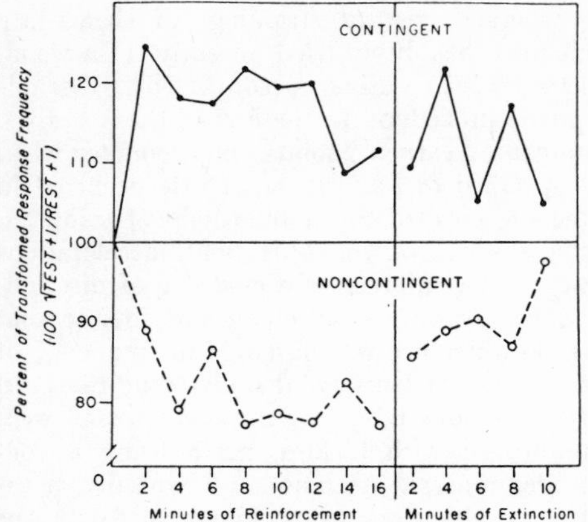

FIG. 1. Relative frequency (percent) of spontaneous GSRs during acquisition and extinction under conditions of operant reward training (visual reinforcement). (The *contingent* group received reinforcement following emitted GSRs while the *noncontingent* group received the reinforcement at times of nonresponding.)

more often, at varying times between emitted responses. Figure 2 shows the overall pattern of Greene's results during reinforcement and extinction and indicates that the control group's decline in responding during reinforcement was not as marked as was found by Kimmel and Kimmel (1963), presumably because of the less consistent reinforcement of nonresponding. The difference in response frequency between the instrumental and control groups was statistically significant during acquisition but fell short of significance during extinction.

Findings of a similar nature have been described by Shapiro, Crider, and Tursky (1964) in a study of emitted skin potential responses (the "endosomatic" GSR) by *S*s who were instructed to "think emotional thoughts." Their experimental group was matched to its control before conditioning on the basis of emitted responses. A tone was delivered contingent upon emitted skin po-

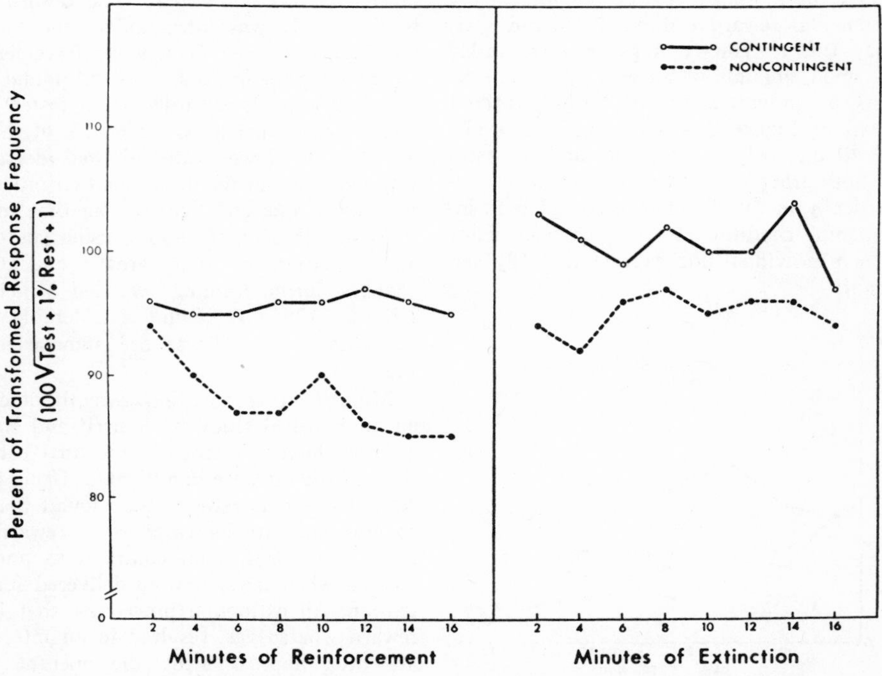

FIG. 2. Relative frequency (percent) of spontaneous GSRs during acquisition and extinction under conditions of operant reward training (visual reinforcement). (The visual reinforcement was delivered to yoked pairs of *S*s simultaneously when the *contingent* member responded.)

tential responses in the experimental condition, while the controls received the same number of reinforcements at times when responses were not occurring. Each occurrence of the tone signalled that *S* had earned $.05. The experimental *S*s emitted a greater number of responses both during reinforcement and extinction than did the control *S*s, and, during the first experimental session, the experimental *S*s were responding at a frequency level above that shown in the preliminary period of no stimulation. A second session several days later showed that the differences between the two groups during reinforcement was again significant, although both groups had fallen to a lower level of responding in relation to the prereinforcement period. On both sessions the control *S*s' response rate increased substantially during extinction, as had previously been reported by Fowler and Kimmel (1962) and Kimmel and Kimmel (1963).

Another glandular response, salivation, has recently been shown to be modifiable by instrumental reward training (Miller & Carmona, 1967). Thirsty dogs were rewarded with small amounts of water for spontaneous bursts of salivation or for pauses without salivation. Figure 3 summarizes the results over 40 days of such training and indicates that both groups' response rates changed systematically in the direction expected with instrumental conditioning. The changes made by each individual dog were statistically significant.

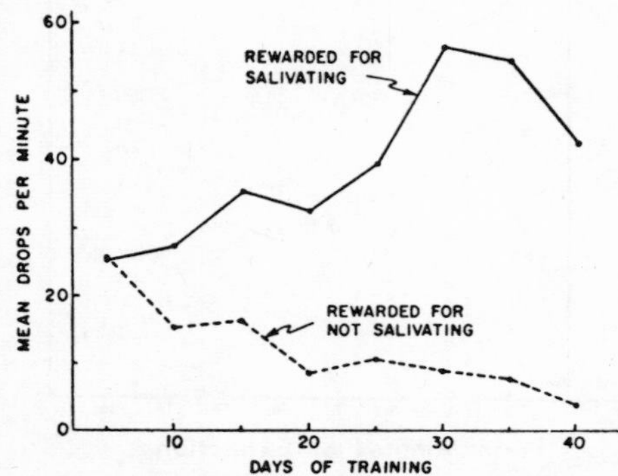

FIG. 3. Average rate of salivation of thirsty dogs rewarded with water for salivating or for not salivating (from Miller & Carmona, 1967).

Operant reward training of heart-rate changes has been tried recently (Harwood, 1959, 1962), using a controlled (discriminated) procedure. In the first of these studies, human *S*s earned "points" on a counter when a specified *increase* in heart rate occurred in the presence of S_D, a blue light. During S_Δ, the absence of the blue light, accelerations were not reinforced. No evidence of discriminative operant conditioning was found; that is, no difference was observed in frequency of accelerations between the two conditions. In the second study, heart-rate *decreases* were reinforced with nickles, again using a controlled operant technique. Essentially negative results were again found, although the possibility was suggested that decelerations lasting for only one heartbeat may have been more frequent in the presence of S_D. More recently, Trowill (1967) has found both increases and decreases in heart rate of curarized rats to be modifiable by instrumental reward training, although, again, the amount of change was quite small. The reward used in this study was intracranial shock, which was shown to be effective reinforcement of bar-pressing beforehand. Trowill found that 15 of 19 animals rewarded for a faster heart rate showed increases, while 15 of 17 rewarded for slower rates showed decreases. Improvements in the discrimination procedure (adding a tone and light during time in sessions, lengthening time-outs, using successive approximation to require greater and greater changes during training) enabled Miller and DiCara (1967) to obtain considerably more dramatic results. These are summarized in Figure 4.

Miller (in press) has also recently described an unpublished study by himself and Banuazizi in which spontaneous intestinal contractions in the rat were investigated. Using intracranial shock as reward, they found that the spontaneous rate increased when reward was made contingent upon contractions and decreased when rewards were delivered for distention. An extinction procedure, that is, no reward whatsoever, resulted in an intermediate rate, approximating the operant level. Thus these results are quite similar to those previously mentioned in which the GSR of human *S*s was employed.

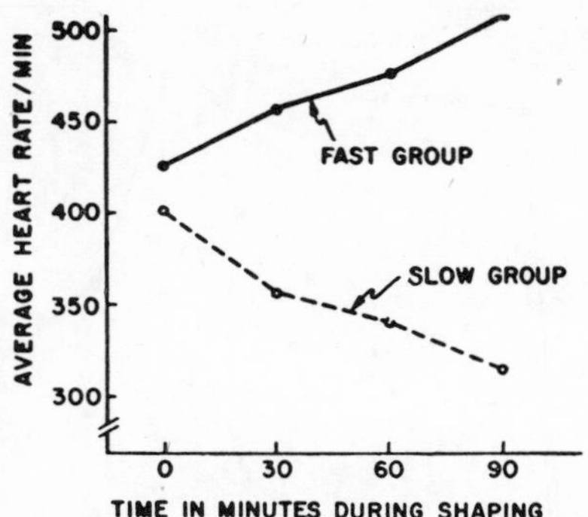

FIG. 4. Average heart rate of curarized rats rewarded with intracranial shock for increases or decreases in rate (from Miller & DiCara, 1967).

Avoidance Training

The first study of instrumental avoidance conditioning of an autonomically mediated response with somewhat positive results was done by Shearn (1960). He attempted to condition heart-rate *acceleration* by making the criterion response effective in shock avoidance. His results indicated that the avoidance procedure led to significantly more accelerations than were observed during a control procedure, although no significant reduction in the number of shocks received was achieved. Respiratory changes were also observed, but these would have been more likely to facilitate deceleration rather than acceleration.

Another smooth muscle response recently studied, the vasomotor reflex, may also be instrumentally conditionable, according to Razran's (1961) description of an experiment by Lisina. Five *S*s were stimulated with a painful electric shock, which normally produces vasoconstriction. If vasodilation occurred, the shock was terminated. Only if the *S*s were allowed to observe their own photophethysmographs during the experiment did they learn to vasodilate and escape the shock.

An instrumental avoidance procedure was recently used in a GSR conditioning study by Kimmel and Baxter (1964). In that experiment, a 1,000 cps tone of 40 db. was the CS and an electric shock to the fingertips was the UCS. The tone had a duration of 1.0 second and the CS-UCS interval was 5.0 seconds so that anticipatory CRs could occur. Yoked pairs of *S*s were run; whenever the avoidance member of a pair made an anticipatory GSR that was judged by *E* to be of criterion magnitude, both *S*s avoided the shock on that trial. The control treatment was thus similar to a partial-reinforcement classical conditioning procedure, since omission of the shock was in no way contingent upon the control *S*'s behavior. Figure 5 presents the results of that study, showing that the avoidance procedure led to significantly larger GSRs than occurred in the control condition. If nothing but classical conditioning were occurring, this difference would not be expected. Figure 6 presents the results of two replications (Kimmel, Sternthal, & Strub, 1966) of the Kimmel and Baxter (1964) avoidance GSR study. In both of the replications, an automatic electronic device was substituted for the experimenter to determine whether or not a particular response was to be reinforced by shock-avoidance, and an extinction series was added. In the first replication, a circular red light was the CS; and in the second replication, the CS was a 1,000 cps tone of 40 db. In both replications, the CS had a duration of 1.0 second and the CS-UCS interval was 5.0 seconds.

In neither replication of the Kimmel and Baxter (1964) study were the differences between the avoidance and control *S*s statistically significant, although the observed differences were in the same direction as had been obtained earlier. In both studies, the apparent differences during extinction were greater and more consistent than they were in acquisition. The combined probability for the two extinction differences was significant at

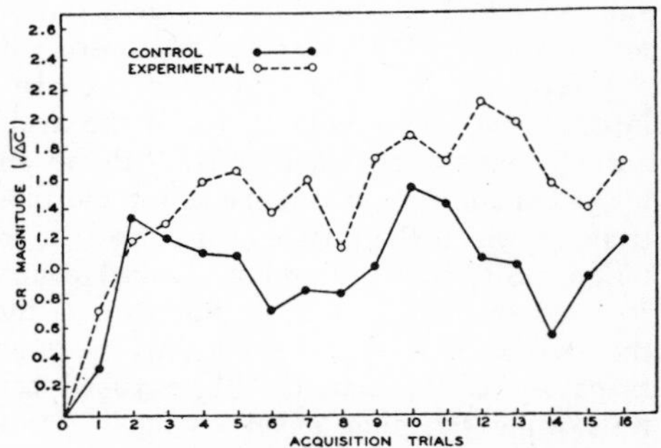

FIG. 5. Average magnitude of anticipatory GSRs of avoidance (experimental) and yoked partial-reinforcement classical (control) groups.

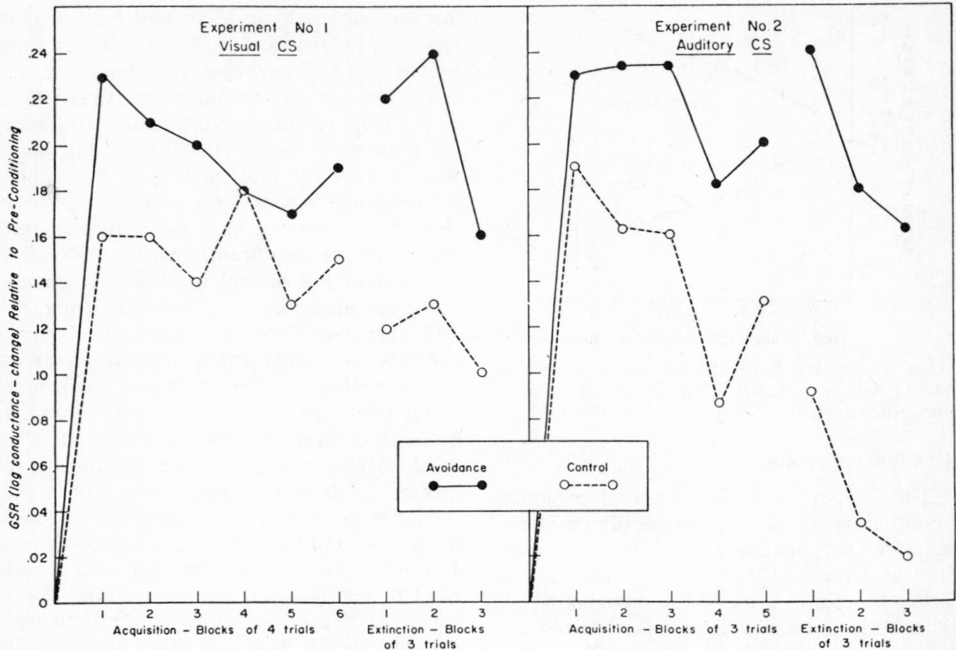

FIG. 6. Average magnitude of anticipatory GSRs of avoidance and yoked control groups in two replications of avoidance GSR conditioning.

$p < .025$. Since substantial variability in the operation of the automatic reinforcement device was noticeable during these experiments, their results seemed to be in reasonable accord with the earlier findings.

Grings (1965) has described an instrumental avoidance study using the GSR, and his results are consonant with Kimmel and Baxter's. In addition to an avoidance group and its yoked control, Grings ran a punishment group and another yoked control. All *S*s were run with a 5.0-second CS-UCS interval. On those trials on which the avoidance *S* made a criterion GSR, both he and his yoked control avoided the shock. On those trials on which the punishment *S* made a criterion GSR, both he and his yoked control received the shock. The results showed that the avoidance *S*s made significantly *more* anticipatory GSRs than did their yoked controls while the punishment *S*s made significantly *fewer* anticipatory responses than did their yoked controls. Conditioned suppression (i.e., punishment) of the rate of occurrence of spontaneous GSRs using contingent shock has also been reported by Senter and Hummel (1965).

Methodological Problems

Two different classes of methodological difficulties have beset workers in this area. The first involves inadvertent confounding that may be associated with the yoked control designs that have frequently been used. In studies of reward training of the emitted GSR, for example, each time that the experimental *S* emits a criterion response and receives a reinforcement, the reinforcement may *elicit* another response. To eliminate contamination of the data from this source, Kimmel and Hill (1960), Fowler and Kimmel (1962), Kimmel and Kimmel (1963), Trowill (1967), and Miller and DiCara (1967) employed a "grace," or time-out period following each reinforcement, during which a response would neither be reinforced nor counted. These grace periods occurred at times of low probability of unelicited responding in the *S* rewarded for responding, since he had just emitted a response. But, the control *S*s received the reinforcements at times when they were not responding. Thus, these time-out periods for the latter *S*s occurred at times of high probability of unelicited responding. Any tendency for the experimental *S* in the Kimmel studies to increase in frequency of unelicited responding during acquisition would be accompanied by a reduction in response frequency in the controls, since they were matched in number of reinforcements per minute. The burden then falls on the extinction data to establish that a difference between the groups actually exists, since the acquisition data suffer from confounding. Only in the Fowler and Kimmel (1962) and Kimmel and Kimmel (1963) extinction data were there significant differences in extinction. Miller and Carmona (1967) found that their water reward did not elicit a salivary response, so this was not a problem in their study.

Another type of confounding may occur in yoked control studies of avoidance conditioning, as Church (1964) has pointed out. As he has noted, when the avoidance and control *S*s are mismatched in classical conditionability, an asymmetrical relationship may exist in which the avoidance *S* benefits from extra shocks when he is the poorer conditioner but the control *S* does not benefit from extra shocks when he is the poorer conditioner. Recently, Kimmel and Sternthal (1967) attempted to replicate the Kimmel and Baxter (1964) study using pairs of *S*s who were matched in classical conditionability and GSR responsivity. The results tended to support Church's argument to some extent, since the previously observed difference in GSR magnitude between avoidance and control *S*s was greatly reduced and just missed reaching statistical significance. No difference between experimental and control groups was found during extinction. As Kimmel and Sternthal pointed out, however, support for Church's argument does not necessarily imply that confounding was the *entire* basis of the previously observed effect, since the acquisition difference did not vanish completely. The experimental procedures needed to achieve good matching on classical conditionability may have interfered with the acquisition of an avoidance response. The point is well-taken, nonetheless, that yoked control procedures may result in better results than would otherwise occur.

The second type of methodological problem stems from the possibility that the reported instances of instrumental conditioning of the various autonomically mediated responses studied may be artifactual consequences of some somatic response which has been conditioned instrumentally. While this argument is quite reasonable, it should be noted that somatic responses that might produce deceleration of the heart rate, such as an abrupt abatement of respiration (a gasp), are not likely to be equally culpable as the source of acceleration. Also, it is one thing to assert that an experimental *S* may learn to make some somatic response which, incidentally, elicits a GSR or a change in heart rate, but it is quite something else to claim that in the same experiment *S*s learn to ***not make*** some somatic response and thus emit fewer GSRs. This is particularly cogent in those studies showing both directions of change within the same *S*s. Nevertheless, the more obvious potential somatic mediators should undoubtedly be examined, observed, or controlled. In the experiment by Shapiro et al. (1964) on operant GSR conditioning, basal skin potential, cardiotachometer, and respiration data were collected during the conditioning sessions and no significant differences in level or trend between the groups were found. There were slightly more GSRs associated with irregular breathing in the instrumental group, but even when these responses were omitted from consideration, the differences between the groups' rates of responding were still significant.

Two recent studies on the possible role of muscular mediation in reward training of the GSR have shed additional light on the problem. Rice (1966) adopted the procedure of presenting visual reinforcement to his *S*s *only if a GSR was emitted in the absence of an observed response in the EMG record taken from S's forearm.* Two types of *S*s were run in this study, a group with a high initial rate of emitted responses and a group with a low initial rate. A significant difference between experimental and yoked control groups was found only in the high-rate *S*s, not in the low-rate *S*s. Two related possibilities exist as explanations of the failure to obtain a significant effect in the low-rate *S*s. First, it may be that instrumental conditionability is correlated with initial response rate and that *S*s with low rates are "poor" conditioners. Second, it is possible that the very low frequency of responding in the low-rate *S*s resulted in a reinforcement "density" (over time) that was below some threshold for instrumental conditioning. It is well to consider that the typical *S* in these studies emits perhaps 3–10 responses per minute and very low-rate *S*s would seldom receive the reinforcement, especially when only GSRs in the absence of EMGs were required.

Van Twyver and Kimmel (1966) measured respiration and forearm EMG in their study of instrumental GSR conditioning. In that study, only responses equal to or greater than 1% of the basal resistance were reinforced. While conditioning effects similar to those previously observed were found in the GSR data, no systematic differences between groups in EMG frequency, respiration rate, or rate of respiration irregularities were noted. Figure 7 presents Van Twyver and Kimmel's data on frequency of emitted GSRs in experimental and control *S*s, with all responses that followed (within 5 seconds) or accompanied EMGs or respiration irregularities omitted. As Figure 7 indicates, the difference between experimentals and controls was quite pronounced

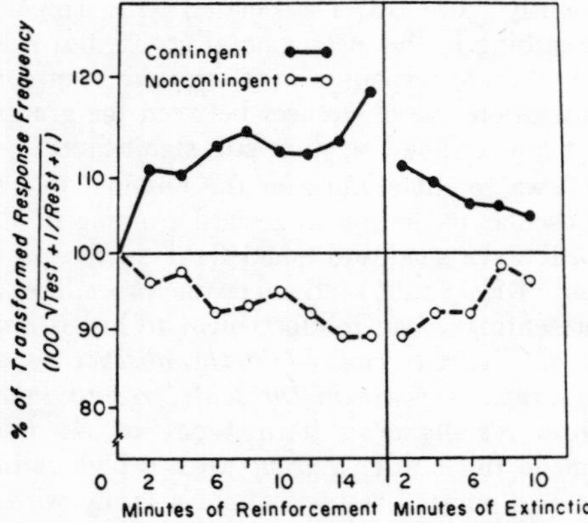

Fig. 7. Relative frequency (percent) of spontaneous GSRs during acquisition and extinction under conditions of operant reward training (omitting all responses preceded or accompanied by EMG or respiration irregularities).

both during acquisition and extinction (and significant in both).

Perhaps the best evidence against the somatic mediation argument has been found in the recent work by Miller and his colleagues (Miller, 1967). In these studies animals under deep curarization (artificially respirated) have shown instrumental conditionability of both heart rate and intestinal contractions. While it is probably impossible to eliminate all skeletal muscular activity without killing the animal, the reduction achieved by deep curarization greatly weakens the strength of the movement-artifact hypothesis. It is of interest to note, since the Miller studies involved rats, that Birk, Crider, Shapiro, and Tursky (1966) have recently reported successful operant GSR conditioning in a human *S* under partial curarization. As those authors have pointed out, the level of curarization employed was, of necessity, not sufficient to completely block the neuromuscular junction, but it certainly ruled out overt movements as contributors to the observed increase in spontaneous GSR frequency.

Summary

At the present time it would appear that Skinner's assumption that autonomically mediated responses cannot be modified instrumentally was both premature and probably incorrect. Apparent instrumental conditioning effects have been reported in recent years using several different autonomically mediated responses and a variety of instrumental conditioning methods. The picture is most clear in the case of reward training, where the major methodological questions of design confounding and possible somatic mediation have been dealt with most adequately. In the case of avoidance conditioning, the possibility of confounding leading to spuriously positive results has not been adequately eliminated. Since several different investigators are currently directing attention to these problems, one may hope that the next few years will provide the theoretical structure that the growing body of positive results now requires.

REFERENCES

Birk, L., Crider, A., Shapiro, D., & Tursky, B. Operant electrodermal conditioning under partial curarization. *Journal of Comparative and Physiological Psychology,* 1966, **62**, 165–166.

CHURCH, R. M. Systematic effect of random error in the yoked control design. *Psychological Bulletin,* 1964, **62**, 122–131.

FOWLER, R. L., & KIMMEL, H. D. Operant conditioning of the GSR. *Journal of Experimental Psychology,* 1962, **63**, 563–567.

GREENE, W. A. Operant conditioning of the GSR using partial reinforcement. *Psychological Reports,* 1966, **19**, 571–578.

GRINGS, W. W. Verbal-perceptual factors in the conditioning of autonomic responses. In W. F. Prokasy (Ed.), *Classical conditioning: A symposium.* New York: Appleton-Century-Crofts, 1965. Pp. 71–89.

HARWOOD, C. W. *Operant autonomic conditioning. Research in progress.* Bellingham: Western Washington College of Education, 1959. Pp. 31–38.

HARWOOD, C. W. Operant heart rate conditioning. *Psychological Record,* 1962, **12**, 279–284.

KIMMEL, E., & KIMMEL, H. D. Replication of operant conditioning of the GSR. *Journal of Experimental Psychology,* 1963, **65**, 212–213.

KIMMEL, H. D., & BAXTER, R. Avoidance conditioning of the GSR. *Journal of Experimental Psychology,* 1964, **68**, 482–485.

KIMMEL, H. D., & HILL, F. A. Operant conditioning of the GSR. *Psychological Reports,* 1960, **7**, 555–562.

KIMMEL, H. D., & STERNTHAL, H. S. Replication of GSR avoidance conditioning with concomitant EMG measurement and *S*s matched in responsivity and conditionability. *Journal of Experimental Psychology,* 1967, **74**, 144–146.

KIMMEL, H. D., STERNTHAL, H. S., & STRUB, H. Two replications of avoidance conditioning of the GSR. *Journal of Experimental Psychology,* 1966, **72**, 151–152.

MANDLER, G., PREVEN, D. W., & KUHLMAN, C. K. Effects of operant reinforcement on the GSR. *Journal of the Experimental Analysis of Behavior,* 1962, **5**, 317–321.

MILLER, N. E. Experiments relevant to learning theory and psychopathology. *Proceedings of the XVIII International Congress of Psychology* (Moscow), 1967, in press.

MILLER, N. E., & CARMONA, A. Modification of a visceral response, salivation in thirsty dogs, by instrumental training with water reward. *Journal of Comparative and Physiological Psychology,* 1967, **63**, 1–6.

MILLER, N. E., & DICARA, L. Instrumental learning of heart-rate changes in curarized rats: Shaping and specificity to discriminative stimulus. *Journal of Comparative and Physiological Psychology,* 1967, **63**, 12–19.

MOWRER, O. H. Preparatory set (expectancy)—A determinant in motivation and learning. *Psychological Review,* 1938, **45**, 62–91.

RAZRAN, G. The observable unconscious and the inferable conscious in current Soviet psychophysiology: Interoceptive conditioning, semantic conditioning, and the orienting reflex. *Psychological Review,* 1961, **68**, 81–147.

RICE, D. G. Operant GSR conditioning and associated electromyogram responses. *Journal of Experimental Psychology,* 1966, **71**, 908–912.

SCHLOSBERG, H. The relationship between success and the laws of conditioning. *Psychological Review,* 1937, **44**, 379–394.

SENTER, R. J., & HUMMEL, W. F., JR. Suppression of an autonomic response through operant conditioning. *Psychological Record,* 1965, **15**, 1–5.

SHAPIRO, D., CRIDER, A. B., & TURSKY, B. Differentiation of an autonomic response through operant reinforcement. *Psychonomic Science,* 1964, **1**, 147–148.

SHEARN, D. W. Operant conditioning of the heart rate. Unpublished doctoral dissertation, University of Indiana, 1960.

SKINNER, B. F. Two types of conditioned reflex and a pseudo type. *Journal of General Psychology,* 1935, **12**, 66–77.

SKINNER, B. F. Two types of conditioned reflex: A reply to Konorski and Miller. *Journal of General Psychology,* 1937, **16**, 272–279.

SKINNER, B. F. *The behavior of organisms: An experimental analysis.* New York: Appleton-Century-Crofts, 1938.

SKINNER, B. F. *Science and human behavior.* New York: Macmillan, 1953.

TROWILL, J. A. Instrumental conditioning of the heart rate in the curarized rat. *Journal of Comparative and Physiological Psychology,* 1967, **63**, 7–11.

VAN TWYVER, H. B., & KIMMEL, H. D. Operant conditioning of the GSR with concomitant measurement of two somatic variables. *Journal of Experimental Psychology,* 1966, **72**, 841–846.

(Received June 23, 1966)

50

Instrumental Conditioning of Autonomically Mediated Behavior: Theoretical and Methodological Issues

Edward S. Katkin and E. Neal Murray

The question of whether or not responses mediated by the autonomic nervous system (ANS) can be conditioned instrumentally has been a subject of continuing controversy because of its theoretical and practical implications. Miller (1961, 1963) pointed out that this issue is of central importance in understanding the neurophysiology of learning, in resolving the knotty problem of whether there is one basic learning process or two, and in providing a parsimonious explanation of psychosomatic symptoms.

Traditionally, learning theorists have assumed that "for autonomically mediated behavior, the evidence points unequivocally to the conclusion that such responses can be modified by classical, but not instrumental, training methods [Kimble, 1961, p. 100]." Explanations for the apparent inability to condition ANS responses instrumentally usually have been founded upon the observation that because the ANS does not interact directly with the external milieu, it is incapable of functioning instrumentally. It is claimed further that the ANS is solely a motor system and, lacking an afferent function, incapable of learning by reinforcemental principles (Smith, 1954). Yet, despite all this testimony an increasing number of positive reports (see Kimmel, 1967) has led to a reopening of the book on instrumental ANS conditioning 25 years after it had been effectively closed by Skinner (1938). Ironically, in the same year that Gregory Kimble (1961) was reiterating the position that instrumental ANS conditioning had not been demonstrated, Gregory Razran (1961, p. 121) was reporting evidence that Lisina, in the Soviet Union, had conditioned vasodilatation instrumentally. Razran's report of Lisina's work stimulated considerable effort in this country; yet when Lisina's work was published in English (1965) and became available for more careful scrutiny by American investigators, it appeared that she did not claim to have conditioned vasodilatation directly. Rather, Lisina concluded that her subjects were able to gain voluntary control over their blood vessels by "using a number of special devices, mainly the relaxation of the skeletal musculature and changing of the depth of respiration [1965, p. 456]." She suggested further that the observed vasomotor changes were in fact unconditioned responses to skeletal and respiratory activity.

Lisina's interpretation of her experiment demonstrates the difficulty of defining clearly the effects of instrumental conditioning procedures on ANS function and is reminiscent of Skinner's earlier (1938) analysis of the problem. Skinner emphasized the point that humans can and do exert voluntary control over their autonomic functions. Think of the child who has learned to cry "real tears" or the boy who has learned not to cry in public.

[1] Preparation of this manuscript was supported, in part, by Research Grant MH-11989 from the United States Public Health Service, National Institute of Mental Health, awarded to the first author, and by Research Grant 50-8921 from the Committee on the Allocation of Research Funds of the Graduate School of the State University of New York, awarded to the second author. The authors are indebted to Roy Lachman for his thoughtful and helpful review and to Neal E. Miller for his comments on an earlier version of the manuscript.

But is this voluntary control exercised directly or chained to a mediating voluntary operant? The following discussion, adapted from Skinner, suggests that instrumental ANS conditioning may be an epiphenomenon associated with skeletal conditioning and/or previous classical conditioning, and the following four cases describe ways in which such apparent instrumental conditioning may occur.

1. *An autonomic response may be an unconditioned response to an external source of stimulation.* For instance, an individual may stick himself with a pin, thereby causing his skin resistance to change. If an experimenter delivered reinforcement upon detecting a change in resistance, the subject might resort to sticking himself with a pin to obtain reinforcement. Admittedly, this pin-sticking strategy seems rather unlikely, either in nature or in laboratories, but there are a variety of less obvious sources of external stimuli (e.g., light, visual patterns) to which a subject might learn to expose himself.

2. *An autonomic response may be an unconditioned response to an internal source of stimulation.* That is, a subject may engage in rapid muscular activity, thereby eliciting unconditioned GSRs. If an experimenter delivered reinforcement upon detecting a GSR, the subject might show an increase in muscular activity, resulting in apparent conditioning of GSR frequency. Indeed, this case is identical to the sequence of events described by Lisina in her study of vasodilatation.

3. *An autonomic response may be a conditioned response to an external source of stimulation.* That is, an individual may read an exciting book or look at an arousing picture, to which autonomic responses have been conditioned previously. Although it is unlikely that an individual in a laboratory would be able to subject himself voluntarily to this type of external stimulation, the following variation of this case suggests that there *are* forms of stimulation which are more susceptible to self-presentation in a laboratory.

4. *An autonomic response may be a conditioned response to an internal source of stimulation.* In this case an individual may engage in subvocal activity (thinking) and this activity may elicit a previously conditioned ANS response pattern. For instance, a subject might have a sexually arousing thought, thereby eliciting a previously conditioned autonomic response. If an experimenter delivered reinforcement upon detecting the autonomic response, the subject might show an increase in the frequency of occurrence of sexually stimulating thoughts, resulting once again in apparent instrumental ANS conditioning.

To illustrate the confounding effect of previously conditioned autonomic responses Skinner referred to work by Hudgins (1933) in which subjects said the word "contract" at the same time that a bright light was flashed on their eye. Subsequently, the mere vocalization of the word "contract" became sufficient to elicit classically conditioned pupillary constriction. This is an example of Case 3 because the spoken word constituted an external stimulus. Had the subject thought the word "contract," the results would conform to the paradigm of Case 4. Thus the subject may be reinforced for skeletal-muscular, verbal, or cognitive activity, and this activity may serve either as a conditional or unconditional stimulus for ANS responding.

In order for the results of studies on instrumental ANS conditioning to be convincing they should be capable of handling the problems raised by these four alternative explanations. Ideally, animal studies might use curare to eliminate skeletal-muscular activity; for that matter, such controls are not out of the question in human research (Birk, Crider, Shapiro, & Tursky, 1966; Smith, Brown, Toman, & Goodman, 1947). Furthermore, the experiments should be designed to reduce the possibility of self-stimulation eliciting previously conditioned autonomic responses.

The purpose of this paper is to review critically the existing evidence for instrumental autonomic conditioning and to evaluate the strength of this evidence in relation to alternative explanations. The authors shall set what they feel to be minimal criteria for the acceptability of evidence for instrumental conditioning. First, there should be some demonstration that the response being reinforced shows an increase in frequency, or amplitude, or probability of occurrence over the level shown in a free-operant period. Second, the experimental design should allow comparisons between experimental groups and appropriate control groups. Finally, the data should be, within reasonable limits, free of obvious alternative explanations, such as those cases described above. There have been a variety of studies on electrodermal, peripheral vascular, and cardiac conditioning. These studies are reviewed separately by response mode.

Instrumental Conditioning of Electrodermal Responses

Rewarding the Unelicited GSR

The first of the recent attempts to condition an autonomically mediated response instrumentally was published in 1960 by Kimmel and Hill who attempted to modify the emission of unelicited GSRs by the contingent presentation of pleasant or unpleasant odors. The criterion for a reinforceable response was a decrease in skin resistance calculated mentally by the experimenter to be equal to or greater than one-half the average magnitude of five responses elicited by electric shocks administered immediately before the session began. The results of this experiment were essentially negative. No differences between experimental and control groups were obtained during the acquisition period, and no overall differences were obtained during extinction. However, one significant effect was found; immediately after the cessation of positive and negative reinforcement, contingent groups showed an increase and noncontingent groups showed a decrease in response frequency, irrespective of the type of reinforcement. This was described by the authors as "quite different from conventional operant conditioning findings [1960, p. 562]."

The next three attempts to modify the frequency of unelicited GSRs (Fowler & Kimmel, 1962; Kimmel & Kimmel, 1963; Mandler, Preven, & Kuhlman, 1962) employed lights as reinforcers. Fowler and Kimmel, reinforcing a GSR which was equal to or greater than the average magnitude of all responses emitted during a 2 minute rest period, reported a significant difference between contingent and noncontingent reinforcement groups during training. However, the significant difference between the contingent and noncontingent groups resulted not from acquisition, but rather from retardation of habituation in the experimental group. At no time in the course of training did the experimental group emit more than 80% of the rate attained in an initial operant period, and no differences were obtained between groups during the extinction period.

Kimmel and Kimmel (1963) replicated the Fowler and Kimmel study with a number of modifications, the most important being an increase in the length of the initial rest period and the reinforcement of *any* GSR that was visually detectable by the experimenter. As in the previous studies the experimental group received reinforcement contingent on the emission of a response, and the control group received an equal number of reinforcers at times when no response was observed. The results of the Kimmel and Kimmel study were far more dramatic than those of the experiment they set out to replicate. The experimental group showed clear acquisition up to 120% of operant level, and the noncontingent control group declined to below 80% of operant level. This pattern of acquisition, which was not obtained in the original Fowler and Kimmel study, may be attributed to either or both of the following factors. First, the longer initial rest period used by Kimmel and Kimmel overcame the difficulty of the general decline in response frequency during the early stages of conditioning found by Fowler and Kimmel. Second, the technique of reinforcing every response detectable used in the second study might have been more effective than that of attempting to make the difficult, instantaneous mental calculations required of the experimenter in the first. It should be noted, however, that in both of these experiments the results are open to interpretation on the basis of experimenter bias (Rosenthal, 1964, 1966, 1967), for in both cases the experimenter was charged with the responsibility of discriminating responses and deciding whether or not to reinforce them. As a matter of necessity the experimenter must have known the difference between the contingent and noncontingent groups and could not have been "blinded." Another problem of interpretation arises from the fact that neither study included direct controls for skeletal-muscular mediation. Although no evidence of gross body movement was observed in the GSR records, there were no electromyogram (EMG) recordings to confirm this. The authors themselves recognized that their "conclusion that autonomically medicated responses can be conditioned instrumentally can be challenged on the basis of an almost infinite number of possible somatic mediators [Kimmel & Kimmel, 1963, p. 213]."

Mandler et al. (1962), reinforcing responses equal to or greater than 500 ohms, found no effects of contingent reinforcement on learning. They also presented evidence suggesting that the general effect of contingent reinforcement was to increase general activation level, as reflected in basal conductance. This activation, they posited, might easily have resulted in skeletal-muscular activity which could have mediated any increase in GSR frequency observed by others. Mandler et al. concluded that instrumental conditioning of autonomically mediated responses remained to be demonstrated.

After the appearance of these studies three reports from the Harvard Medical School appeared (Birk et al., 1966; Crider, Shapiro, & Tursky, 1966; Shapiro, Crider, & Tursky, 1964). In the Shapiro et al. and Crider et al. studies, subjects were told that they should think emotional thoughts and that they would hear a tone worth 5 cents each time the experimenter's apparatus detected such thoughts. In the former study respiration was monitored as a control for movement, and in the latter study an ingenious device was utilized to detect even the slightest of body movements. In neither of these two studies was there any evidence of acquisition, but differential rates of habituation over time occurred. The results of both experiments are clearly susceptible to the Case 4 argument; that is, the GSR might have been a conditioned response to internal stimuli, for it is quite obvious that the subjects were instructed to produce subvocal operants already associated with ANS responses.

In the Birk et al. study a single subject was run under conditions of partial curarization. Here we have the only known attempt in this area to control skeletal artifact by curarization of a human subject. When the acquisition curve during curarization was compared with one obtained from the same subject before curarization it was observed that the curare session yielded *lower* frequencies of response throughout, although both curves showed an acquisitionlike trend. Unfortunately, the authors did not obtain a noncontingent session from this subject with which to compare the two conditioning sessions, and the activating effects (Mandler et al., 1962) of the reinforcing stimulus are not clear. Thus the safest conclusion that can be drawn from this demonstration is simply that the administration of curare resulted in reduced muscle activity and reduced frequency of GSR responding.

Rice (1966) attempted to control for muscle artifact without the use of paralyzing drugs. He set out to control for small amplitude EMG changes in the forearm muscle group that mediated flexion of the fingers from which the GSRs were recorded. Subjects were divided so that half received reinforcement for any GSR emitted and half received reinforcement only for GSRs emitted in the absence of EMG responses. Reinforcement was a light presented in an otherwise dark room. Noncontingent control groups were employed, and in addition data were evaluated as a function of operant level of responding. Rice's findings are complex and must be considered in detail. Figure 1 presents his data for the subjects who were reinforced irrespective of EMG. Consistent with previous reports, there were differential rates of habituation for the contingent and noncontingent groups, with the contingent group demonstrating greater resistance to habituation regardless of operant level. Figure 2 presents the data obtained from subjects who received reinforcement only when no EMG was associated with the response. These data "yielded no significant between groups main effect [Rice, 1966, p. 911]." What they did yield was an interaction between operant level and contingency, resulting from the fact that for one-half of the subjects (low operant) the noncontingent control group actually showed a ***higher*** level of response than the experimental group, whereas for the other half (high operant) the reverse was found. For the high-operant subjects, for whom the trend was in the expected direction, differential habituation but no acquisition was observed. In fact, the only group that approximated acquisition was the low-operant *non*contingent group. Rice observed that the subjects represented in Figure 2 received many fewer reinforcements than those represented in Figure 1 because they received no reinforcement for responses associated with EMG changes and the total time in the experiment was held constant for all subjects. Therefore he ran 18 additional pairs of subjects for longer time periods, reinforcing only those responses which were independent of muscle artifact and delivering to these subjects the same number of reinforcements given to the group which received 100% reinforcement. Under these conditions even the differential habituation trends disappeared, and Rice concluded "the evidence is somewhat equivocal as to whether operant GSR conditioning is possible when only those GSRs given in the absence of preceding muscle tension changes are reinforced [p. 912]."

The next attempt to increase the rate of unelicited GSR activity was made by Van Twyver and Kimmel (1966). The major

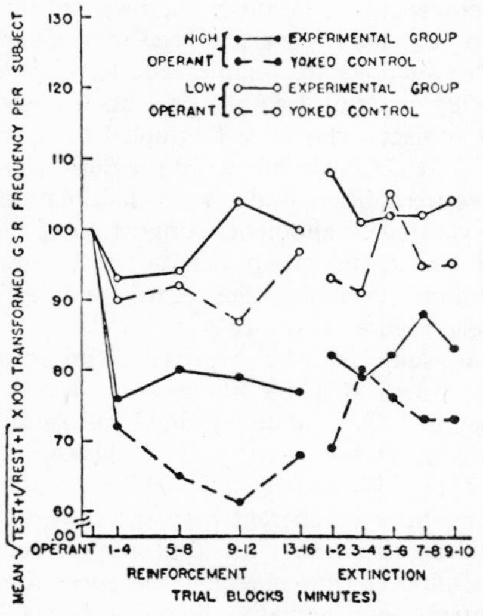

FIG. 1. Mean transformed GSR frequency for subjects in the contingent and noncontingent groups, irrespective of EMG, divided by operant level, during reinforcement and extinction (from Rice, 1966).

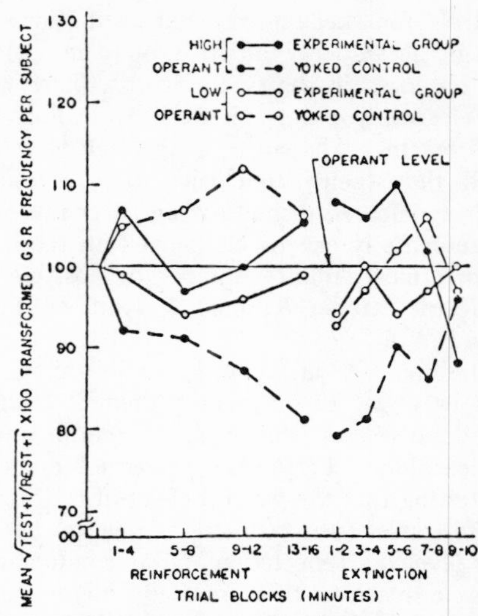

FIG. 2. Mean transformed GSR frequency for subjects in the contingent and noncontingent groups, reinforced only in the absence of EMG, divided by operant level, during reinforcement and extinction (from Rice, 1966).

purpose of their study, too, was to investigate the possibility of somatic mediation in GSR conditioning. Van Twyver and Kimmel controlled carefully for respiration rate, respiration irregularity, and muscle potentials and used a light to reinforce any response equal to or greater than 1% of the basal skin resistance. The EMG and respiration records were evaluated, and it was concluded that there were no differences between groups on any of these possible mediators. Furthermore, all GSRs that were accompanied by forearm EMG activity or breathing irregularity were discarded in a subsequent analysis. The data obtained in this manner were clear—the contingent reinforcement group showed a smooth acquisition curve, and the noncontingent group showed a decay curve. Acquisition was more gradual in this experiment than in the one reported previously by Kimmel and Kimmel (1963), who used no control for muscular mediation. Thus, the Van Twyver and Kimmel study is the only one reviewed which has reported clear evidence for acquisition independent of skeletal-muscular or cognitive mediation.[2]

[2] In addition to the nine studies reviewed thus far, three others have appeared (Gavalas, 1967; Greene,

Two recent reports (Stern, 1967; Stern, Boles, & Dionis, 1966) have raised questions about the adequacy of attempts to reward the unelicited GSR. Stern and his associates have reported that they consistently obtained negative results when attempting to demonstrate operant conditioning of spontaneous GSRs in a manner similar to Kimmel's. They have also suggested a possible explanation for the positive findings of other investigators which incorporates both skeletal-muscular and cognitive mediation effects. Stern has presented empirical evidence obtained from questionnaires given to his subjects suggesting that when control subjects are reinforced only while not responding they tend to develop hypotheses that they are being reinforced for being "relaxed" or "drowsy"; the experimental subjects, on the other hand, develop hypotheses that they are being reinforced for responding, (e.g., "muscular reactions," "slight movements," "thinking about exciting things"). Presumably these hypotheses lead to action, and the action leads to ANS response patterns that may be incorrectly attributed to the reinforcement paradigm employed.

Although Stern's comments provide a basis for reinterpreting data obtained from experiments in which the control subjects are given equal numbers of reinforcements only at times when they are not responding, they are not exhaustive; there are other problems associated with this procedure, as well as with the use of "true" yoked control designs.

The "true" yoked control design, as well as the tactic of using pairs of subjects equated for number of stimulus presentations, establishes an inherent bias against the control group. The systematic manner by which this bias operates has been discussed in general terms by Church (1964), who points out that individual differences in reaction to an event, or stimulus, or reinforcer, are enough to tip the scales in favor of the experimental group over its yoked controls. Church's arguments may be extended to the present context because the events or stimuli in question in these studies could have acted not only as reinforcers, but also as contributors to overall activation level, and it has been demonstrated that increases in activation are positively related to the emission of unelicited GSRs (Greiner & Burch, 1955; Katkin, 1965; Silverman, Cohen, & Shmavonian, 1959). This creates a problem in interpreting the results of these studies inasmuch as the dependent variable itself is frequency of emission of unelicited GSRs.

Consider the situation where the presentation of a light or tone leads to individual differences in activation. To simplify the matter, assume that the stimulus is either "effective" or "ineffective" in increasing a subject's activation level. Table 1 presents the four possible ways in which an experimental subject may be paired with a control subject in terms of these differences in the effectiveness of a stimulus. In Cases 1 and 4 the stimulus induces activation equally for experimental and control subjects, and there is no biasing effect. In Case 2 the light is effective for the experimental subject but not for the control subject; consequently one would expect the experimental subject to emit more GSRs than his paired control. In Case 3 the light is ineffective for the experimental subject but effective for the control subject. Here one might expect the control subject to emit more GSRs than his experimental partner, but there is an additional complicating factor to consider; the experimental subject will make few responses in Case 3 because of the ineffectiveness of the stimulus in raising his activation level. Therefore, the control subject in Case 3 will be *prevented* from being activated by exposure to the stimulus, and his total number of responses will remain low.

This argument refers only to individual

TABLE 1

INDIVIDUAL DIFFERENCES IN THE EFFECTIVENESS OF LIGHT OR TONE AS AN ACTIVATOR

Case	Experimental *S*	Control *S*
1	Effective	Effective
2	Effective	Ineffective
3	Ineffective	Effective
4	Ineffective	Ineffective

Note.—Adapted from Church (1964).

1966; Greene & Nielsen, 1966). None of these demonstrated acquisition independent of potential somatic or cognitive artifact.

differences in the effectiveness of the stimulus as an activator, but the practice of presenting paired subjects with equal numbers of reinforcements can be criticized wherever relevant individual differences among subjects exist. If, for instance, the definition of a criterion response itself were subject to individual variation, the same bias in favor of an experimental group could operate. Such variability in response definition has not been uncommon; in the Van Twyver and Kimmel (1966) study, for example, an unelicited GSR was defined as a 1% change in a subject's basal skin resistance. Clearly by chance some control subjects emitted criterion GSRs more frequently than their matched experimental conterparts but did not have them reinforced. If the effect of presenting the reinforcer were uniformly to increase the probability of emitting criterion responses, and if only the responding of the experimental subject determined the presentation of stimulation, then once again the experimental group profited more than the control group in this situation.

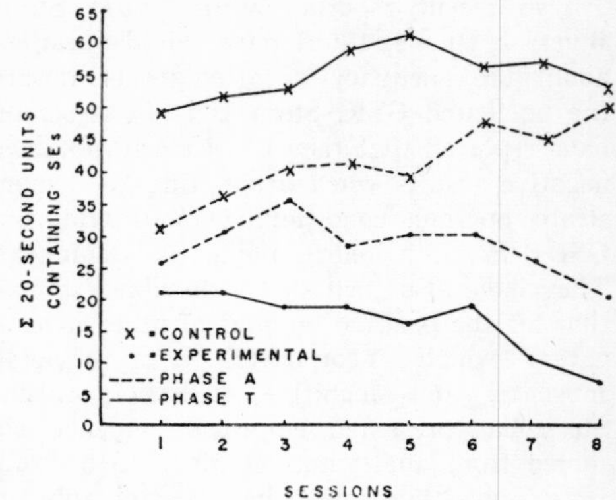

Fig. 3. Changes in frequency of GSR spontaneous emissions (SEs) during Phase A and Phase T (from Senter & Hummel, 1965).

Punishing the Unelicited GSR

Although there have been many attempts to increase the frequency of unelicited electrodermal responses, there have been only two reports of attempts to decrease the frequency of such responses (Johnson & Schwartz, 1967; Senter & Hummel, 1965). In the earlier of these studies, Senter and Hummel administered electric shocks to a subject's forefinger contingent upon the emission of a spontaneous GSR between 600 and 1300 ohms amplitude. Figure 3 contains the data presented by Senter and Hummel; Phase A represents the initial resting period prior to punishment, and Phase T represents the testing period after 15 minutes of training. It is clear from the graph that for the experimental group there are fewer GSRs in Phase T than in Phase A, while for the control group there are more GSRs in Phase T than in Phase A. However, the conclusion that these data represent the results of instrumental suppression of GSR responding is not convincing. In addition to the fact that no EMG or respiration records were taken to control for skeletal-muscular mediation, there is the confounding fact that the experimental and control groups showed marked divergence during the initial adaptation period (Phase A). The data points during the first time period of Phase T are almost indentical with those during the last time period of Phase A for both groups; thus the results of Phase T may well be an extension of the already clearly diverging trend. There is, then, no evidence that the data in Phase T represent the effects of punishment procedures.

Johnson and Schwartz used a loud tone as an aversive stimulus, giving half the subjects contingent and the other half noncontingent punishment. In addition, half the subjects in each group were told that their behavior was related to the tones, and half were told nothing. Johnson and Schwartz obtained marked suppression of GSR frequency for subjects in the experimental group and no change for those in the control group, but no differences were obtained as a function of instructional set. Although Johnson and Schwartz utilized a paired-subject paradigm, which might have biased their results in favor of the experimental group, it must be noted that the magnitude of the difference between their groups was so large that one is tempted to suspect that it would have resulted irrespective of artifactual contributions. They also were prepared for the argument that their results might have been the by-product of skeletal or "cognitive" conditioning, and to control for this they took concomitant EMG readings and interviewed their subjects in depth to investigate possible relationships between "hypotheses" (Dulany, 1962; Spielberger & DeNike, 1966) and the instrumental response patterns obtained. Only one third of their subjects thought that they were able to control the reinforcement, and these subjects were distributed approximately equally among the experimental and control groups. Johnson and Schwartz did not present data for the "aware" and "unaware" subjects separately, nor did they report what percentage of the "aware" subjects had correct hypotheses. Obviously, all of the control subjects had to be incorrect since there was no way in which they could control the reinforcement.

Avoidance Conditioning of the Elicited GSR

In addition to the research described above, four studies dealing with the elicited GSR have appeared. In each of these studies the purpose was to elicit an unconditioned response to a neutral stimulus and to have that response become instrumental in the avoidance of a noxious stimulus.

Kimmel and Baxter (1964) reported the first of these studies. Subjects received a 1000 cps tone followed 4 seconds after offset by a one-second shock. After a single tone-shock pairing the shock could be avoided if a subject emitted a GSR to the tone equal to or greater than the smallest GSR made in a preliminary series of trials used to habituate the subject to the tone. A group of control subjects was given a trial by trial pattern of shocks identical to that of their experimental counterparts. The results of this experiment indicated that both groups showed marked increases in response amplitude during the first three trials, and thereafter the experimental group maintained a higher response amplitude than the control group. Although this experiment was interpreted as demonstrating "that an autonomically mediated response such as the GSR *can* be conditioned with an instrumental avoidance procedure [Kimmel & Baxter, 1964, p. 484]," the Kimmel and Baxter study included no simultaneous measure of muscle activity and employed yoked controls. Recognizing these problems, and especially the methodological one, Kimmel and Sternthal (1967)[3] set out to replicate the Kimmel and Baxter experiment using concomitant EMG measurement and attempting to overcome the individual differences artifact of the yoked control design by matching yoked subjects for responsivity and conditionability. Using essentially the same procedure as Kimmel and Baxter, these investigators obtained nonsignificant results and concluded that the

> results of this study tend to support Church's (1964) contention that the yoked control design may lead to spurious or exaggerated evidence of avoidance conditioning. When only pairs of *S*s who are very closely matched in both GSR responsivity and classical conditionability were used, differences favoring an avoidance group over its yoked controls tended to shrink [1967, p. 45].

One final study has appeared on avoidance conditioning of GSR (Grings & Carlin, 1966).[4] No increase in the magnitude of response elicited by the conditional stimulus was obtained, nor was the trend for the avoidance paradigm to increase the frequency of occurrence of emitted responses to the signal significant. Grings and Carlin's failure to find trends for increased response amplitude is in direct contrast with the findings of Kimmel and Baxter (1964), Kimmel, Sternthal, and Strub (1966), and Kimmel and Sternthal (1967).

Instrumental Control of Peripheral Vascular Responses

Lisina's (1965) report of successful conditioning of vasodilatation has been cited frequently and has been a major impetus for the present interest in instrumental ANS conditioning. Yet, as was mentioned earlier, there was no clear evidence of such conditioning in her report. Lisina set out to condition

[3] Kimmel, Sternthal, and Strub (1966) also attempted two replications of the Kimmel and Baxter paper. They did not include the methodological improvements developed subsequently by Kimmel and Sternthal, and neither of their attempts achieved statistical significance.

[4] In the same study these investigators also attempted punishment of the elicited GSR but obtained equivocal results.

"paradoxical" dilatation by presenting subjects with an electric shock that evoked an unconditioned response of initial constriction followed by vasodilatation. Shocks were promptly terminated upon dilatation but never for constriction. When and only when subjects were presented with exteroceptive feedback concerning the function of the plethysmogram were they able to respond with earlier vasodilatation. Lisina suggested that after they received the exteroceptive feedback, subjects learned to act voluntarily to cause the vascular change.

Since Lisina's initial investigation, only two reports have appeared on the subject, both from the laboratories at Kansas State University (Snyder & Noble, 1965, 1966). Because these papers are quite similar, with almost identical results, only the most recent is discussed. In this study Snyder & Noble reinforced spontaneous vasoconstrictions with a light; the criterion for reinforcement was set separately for each subject so that only responses equal to or greater than those that occurred not more than twice during a rest period were reinforced. Utilizing a yoked control group as well as a base-line control group which received no reinforcement at all, and controlling for respiration and muscular artifact, the investigators obtained clear and highly significant results. The experimental group showed definite increases in frequency of response, the base-line control group showed no change, and the yoked control group, which received reinforcement contingent on not responding, showed a modest *decrease* in frequency of responding. Following the 25 minute acquisition session, half of the experimental subjects received standard extinction procedures and the other half received counterconditioning in which they received reinforcement contingent on not responding; the only extinction trends obtained were for the counterconditioning group. Because the yoked control group showed a decrease from base line as a function of being reinforced for not responding, and the baseline control group showed no change, these data provide some evidence for possible instrumental modification of peripheral vascular

Instrumental Control of Cardiac Rate

Human Subjects

The first report of instrumental modification of human heart rate was published by Shearn (1962), who made delay of shock contingent upon heart-rate acceleration. In this experiment subjects were able to hear their own heartbeats through a loudspeaker, a presentation of exteroceptive feedback similar in some respects to the visual feedback that Lisina had to use to condition vasodilatation. Although Shearn did not run a no-feedback condition with which to compare his results, he interpreted them as positive evidence for successful operant conditioning. Difficulties with this study include the following: (*a*) Shearn reported significant differences in respiratory patterns between the experimental and yoked control groups and suggested that these might have caused reflexive cardiac changes. (*b*) Although he reported evidence for successful acceleration, Shearn added that

> unlike the typical results of studies where the Sidman-type schedule is used with skeletal response, my study showed no reduction in the number of shocks. The mean number of shocks (13.3) in the last session was virtually the same (13.9) as in the first session [1962, p. 531].

(*c*) The yoked control design may have accounted for the slight differences found between the groups.

Harwood (1962) subsequently reported two attempts to condition heart-rate deceleration instrumentally. According to Harwood, both of these attempts were complete failures, and the main thrust of his presentation was to forewarn future investigators of the complexity of attempting such work under "normal" conditions. Harwood suggested, among other things, that a serious assault on the problem should employ multiresponse measurement to facilitate evaluation of cardiovascular activity independent of respiratory or skeletal mediation, or preferably that curarized subjects (animals) be used as a general precaution.

Harwood's negative findings have been criticized by Engel and his associates (Engel & Chism, 1967; Engel & Hansen, 1966) on

reinforcement on a beat-to-beat basis and was thereby severely hampered in delivering contingent reinforcement. To compensate for this, Engel employed sophisticated electronic apparatus to deliver reinforcement on a beat-to-beat basis. Engel and Hansen reported that with this apparatus they successfully slowed the heart rate instrumentally, and Engel and Chism using the same apparatus reported successful heart-rate speeding. In the first of these studies Engel and Hansen used 10 experimental subjects with yoked controls for five of them. Their results indicated that 60% of the experimental subjects showed evidence of heart-rate slowing at statistically significant levels, whereas none of the control subjects did. However, 30% of the experimental subjects showed heart-rate *speeding* at statistically significant levels, and 80% of the control subjects showed significant speeding. In Engel and Chism's paper identical procedures were employed in an attempt to increase rather than decrease heart rate. In this study all experimental subjects ($N = 5$) showed significant increases in rate, but three of five yoked controls also showed similar increases. Interestingly, the increase obtained for the experimental subjects in the speeding study was almost identical with the increase obtained for the control group in the slowing study. Thus it appeared that a heart-rate change which in the first study was obtained under noncontingent reinforcement conditions was in the second study interpreted to be a systematic learning phenomenon resulting from contingent reinforcement. Was there a single factor common to the two studies that might have accounted for these findings? Fortunately, in both studies the investigators interviewed their subjects to determine if they developed any techniques for keeping the light on.

In a detailed analysis of these interviews Murray and Katkin (1968, in press) demonstrated that the subjects' reports of their voluntary actions were highly correlated with the heart-rate changes obtained in both studies, irrespective of reinforcement contingencies. Statistical analyses were presented to support the contention that these two papers failed to demonstrate successful instrumental conditioning of heart rate.

Augmented sensory feedback. In addition to the studies already described, four reports have appeared which explicitly utilized exteroceptive feedback to facilitate the development of voluntary control of heart rate (Brener & Hothersall, 1966, 1967; Frazier, 1966; Hnatiow & Lang, 1965). In the first of these studies Hnatiow and Lang attempted to bring heart-rate *variability* under control. They presented subjects with a visual display of a pointer whose movements were synchronized with the output of a cardiotachometer. The pointer moved on a white field marked by a red stripe in the center. The apparatus was adjusted for each subject so that the red stripe fell at the average heart rate, and subjects were instructed to try to keep the pointer in the center. Reinforcement was defined as the immediate feedback of success or failure. Control subjects were shown a visual display derived from a different subject's heart rate, having no relation to their own autonomic activity. The results indicated clearly that the experimental subjects were able to reduce significantly the variability of their heart rates, while the control subjects were not.

In a study concerned specifically with the effects of augmented sensory feedback on heart *rate,* Brener and Hothersall (1966) also reported positive results. This experiment employed five subjects who were placed in a dim room and presented alternately with green and red lights. During periods when the lights were on subjects received high-pitched tones upon emission of fast heart beats and low-pitched tones upon emission of slow heart beats. During green light periods subjects were instructed to try to produce only high tones and to inhibit low tones; during red light periods they were instructed to try to produce only low tones and to inhibit high ones. Brener and Hothersall found that following training heart rates increased to the green light and decreased to the red light. They concluded that these "results present clear evidence that under conditions of augmented sensory feedback, *S*s rapidly learn to control their heart rates [1966, p. 27]." Recognizing that their results may have been an artifact of learned respiratory changes, Brener and Hothersall (1967) replicated their experiment, using Wood and Obrist's (1964) technique to teach subjects to breathe at a fixed rate and amplitude. They demonstrated that the positive results obtained in the first experiment were replicable, and that they were independent of changes in respiratory behavior. However, they reported that "the possibility that the observed cardiac control was mediated by learned changes in muscle tension remains a problem worthy of empirical investigation [1967, p. 6]." They also concluded that their results were consistent with Hnatiow and Lang's and, taken together, the findings suggest that the extent to which subjects can gain voluntary control over their behavior is a function of the amount of feedback they receive. It seems possible that the feedback in question served to facilitate the learning of skeletal-muscular responses which in turn elicited unconditioned autonomic responses. Neither the Hnatiow and Lang paper nor the two Brener and Hothersall papers investigated this possibility. Brener and Hothersall's work is subject to an additional criticism, for they failed to counterbalance their experiments to control for the possible differential effects of pitch on heart rate. For instance, if high-pitched tones elicited unconditioned heart-rate acceleration, and low-pitched tones elicited unconditioned heart-rate deceleration, then the red and green lights in their experiment may have become conditional stimuli in a classical conditioning paradigm.

A fourth study (Frazier, 1966) that used augmented sensory feedback to facilitate instrumental control of heart rate appears even more clearly to have employed a classical conditioning paradigm. In this experiment Frazier presented alternately to four subjects periods in which a visual stimulus was present and base-line periods in which it was not. During the visual stimulus periods an electric shock was delivered to the subject's left leg after each minute in which the number of heartbeats decreased below the previous minute's total. If heart rate failed to decrease, punishment was omitted. Frazier reported that after training the mere presentation of the visual stimulus reliably evoked heart-rate acceleration; he interpreted this as evidence for instrumental avoidance conditioning. In addition to the fact that no EMG records were obtained to control for somatic mediation, this experiment may also be interpreted as reflecting classical conditioning. Electric shock has beed cited frequently as an elicitor of unconditioned heart-rate acceleration (Graham & Clifton, 1966), and the shock in this experiment appeared only in the presence of the visual stimulus. Thus the visual stimulus may easily have become a conditional stimulus for heart-rate acceleration. As it stands, Frazier's experiment offers no conclusive evidence for avoidance conditioning.

Animal Subjects

Shortly after the publication of the studies on humans described above, four experiments on instrumental conditioning of cardiac rate in curarized rats were reported from Neal Miller's laboratory (DiCara & Miller, 1968; Miller & Banuazizi, 1968; Miller & DiCara, 1967; Trowill, 1967). The first of these studies (Trowill) was undertaken because

> none of the studies yielding positive results has conclusively ruled out the possibility that *S*s learned skeletal responses which had an unlearned tendency to elicit the visceral responses recorded [1967, p. 7].

Trowill's procedure was unique in two ways: animals were curarized deeply, requiring artificial respiration, and electrical stimulation of the medial forebrain bundle in the posterior portion of the lateral hypothalamus was used as the reinforcer. Trowill pretested all subjects on a bar pressing task to ensure that the intracranial stimulation was effective. The criterion for reinforcement was set so that an animal received stimulation when his heart rate was above (or below, depending on group) his predetermined average at a level expected to occur once every 3–5 seconds. Half the subjects were rewarded for rates above average, and half for rates below average. In addition, yoked control subjects were employed for some of the experimental subjects in each group; however, the major criticism of the use of such yoked controls is blunted by the fact that heart rates were conditioned both to increase and decrease. Thus the argument that differential activation could account for heart-rate change is weakened since one cannot logically predict that activation could cause change in both directions. The use of deep curarization was apparently an adequate control for possible somatic mediation, and Trowill's use of electronic apparatus to discriminate criterion responses and to reinforce them avoided the problem of experimenter bias.

Trowill's results indicated that 15 of 19 subjects rewarded for fast heart rates increased their rates (mean increase for all 19 = 18.18 bpm) and that 15 of 17 subjects rewarded for slow heart rates decreased their rates (mean of all 17 = 19.26 bpm). Both of these mean changes were statistically significant; however, for the data obtained from the 11 pairs of yoked subjects in the fast condition and the 9 pairs of yoked subjects in the slow condition the conclusions are less clear. The difference between experimentals and controls in the fast condition was smaller than the difference between groups in the slow condition. Furthermore, when Trowill analyzed for the frequency of correct responses (i.e., number of increases or decreases over baseline, respectively) over training time, he found no significant effect for subjects in the fast condition, but a highly significant effect for those in the slow condition. Thus, it appears that Trowill's results are quite strong for heart-rate slowing, but not at all conclusive for speeding.

Miller and DiCara felt that the magnitude of change obtained by Trowill was so small (5%) as to be "not completely convincing [1967, p. 12]." Therefore they endeavored

> to see (a) whether larger changes in the heart rates of curarized rats can be achieved by "shaping" the responses, i.e., progressively shifting rats to a more difficult criterion after they have learned to meet an easier one, and (b) whether a visceral discrimination can be learned so that the response will be more likely to occur in the stimulus situation in which it is rewarded than in the one in which it is not [1967, p. 12].

To accomplish these aims Miller and DiCara replicated the Trowill experiment with several minor modifications and one major one. The major modification consisted of shifting an animal's criterion for reinforcement by 2% if it improved enough so that it met criterion approximately 10 seconds after a time-in signal was presented. This shift was repeated successively each time the animal met the next criterion; if it failed to meet a new criterion with some consistency, it was shifted down to the prior one. Figure 4 presents the results obtained from 12 rats trained to increase and 11 rats trained to decrease their heart rates over a 90 minute training period. These dramatic findings are clear; both groups showed changes of approximately 100 bpm, and the obtained differences were statistically significant. Furthermore, the differences were not artifacts of the number of brain stimulations received, for this number was almost identical for both groups.

Miller and DiCara observed that the rats tended to increase or decrease their rates not only when the time-in signal (a pattern of tone and light) was on, but also when it was off. This lack of discrimination differed from the more traditional findings on instrumental conditioning of skeletal responses. To determine if the heart-rate response was capable of discrimination training, the investigators subjected 16 of their animals to an additional 45 minutes of training at approximately the same reinforcement criterion attained at the end of the first 90 minutes. At the beginning of this additional training period all 16 animals (eight from the fast

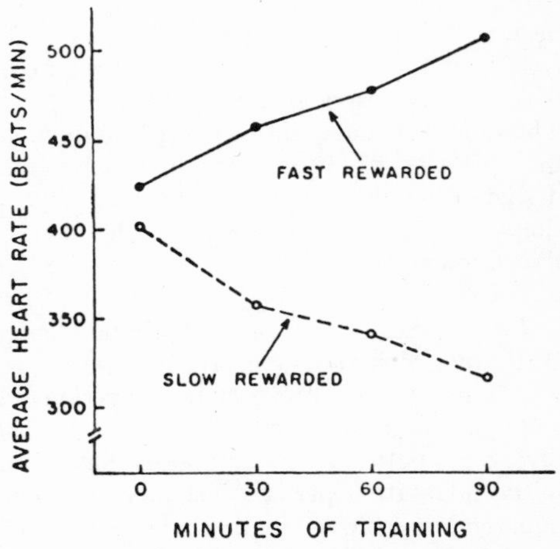

FIG. 4. Change in heart rate for groups rewarded for fast or for slow rates. Each point represents average of bpm during five minutes (from Miller & DiCara, 1967).

somatic mediation they recorded EMG responses. Following essentially the same procedures used in the previous studies conducted in Miller's laboratory, DiCara and Miller obtained evidence that heart-rate changes can be effectively reinforced by escape or avoidance of electric shock. Thus they concluded that "the instrumental learning of a visceral response is not limited to any unique property of the direct electrical stimulation of the brain as a reward [DiCara & Miller, 1968, p. 11]." The EMG records also indicated that there was no evidence of somatic mediation. Taken together with the results of the Miller and Banuazizi experiment, and the earlier Miller and DiCara experiment, these findings lend little support to Black's (1966) suggestion that motor cortex impulses are really the responses being conditioned. Thus Miller and his associates appear to have come as close as is possible to a truly definitive series of experiments establishing the phenomenon of instrumental conditioning of autonomically mediated responses.

Summary and Discussion

The studies reviewed covered two general autonomic functions, electrodermal and cardiovascular activity. With respect to the effects of reward on the frequency of unelicited GSRs, both positive and negative results have been reported; however, much of the evidence contained in these reports was found to be equivocal at best and unacceptable at worst. Those experiments not rendered ambiguous by the use of inappropriate experimental design either showed no evidence for acquisition or were vulnerable to one of the alternative arguments presented in the first section of this paper.

Two studies on punishment of the elicited GSR were reviewed. One of these was shown to be misleading because the experimental and control groups had already established divergent patterns of response during an initial adaptation period. The other obtained apparently positive findings, although the possibility of somatic mediation was not eliminated.

Four studies on avoidance conditioning of the elicited GSR carried out in Kimmel's laboratory were reviewed. Although the first of these was positive, three subsequent replication attempts in the same laboratory failed to achieve significance and suggested that the original positive findings might have resulted in part from inappropriate experimental design.

The reports of instrumental conditioning of cardiovascular function were generally more definitive than those on electrodermal responses; however, many of these experiments also were shown to suffer from inadequate experimental design and/or analysis. Although Lisina's (1965) landmark Russian report of instrumental vascular conditioning was found to be no such report at all, two papers by Snyder and Noble (1965, 1966) provided some evidence for the phenomenon.

A number of studies on cardiac conditioning of humans and animals were reviewed. The few positive reports of instrumental control of cardiac rate in human subjects were subject to reinterpretation on two counts, respiration artifact and skeletal mediation. One of these experiments was clearly confounded by respiration changes. Two others, from Engel's laboratory, were found to be inconclusive, and the evidence presented in them indicated that the obtained effects were probably a function of differential voluntary activity on the part of the subjects. Four recent animal experiments from Miller's laboratory have provided more convincing evidence that cardiac rate can be conditioned instrumentally. These experiments, the clearest to date, were performed on deeply curarized rats.

Although there is some evidence that instrumental ANS conditioning has been demonstrated, the only truly impressive results to date were those obtained in experiments utilizing curarized animals. These same studies also used reinforcers known to be highly *effective*—intracranial reward and aversive electric shock. It is clear that any unequivocal demonstration must be obtained from curarized subjects, for otherwise the fundamental argument of somatic mediation could be invoked. Indeed, if one adheres to Black's (1966) position, the mediation argument can be levied even against studies using curare. The question of effective reinforcement has not received as much attention as that of somatic mediation with the exception of Stern's (1967) expression of surprise at the reports of positive results obtained by experimenters who used lights, presumably weak reinforcers. Clearly, future research is needed to determine just which reinforcers are most effective in producing instrumental ANS conditioning. DiCara and Miller, for instance, utilized electric shock in their experiment to determine if positive results could be obtained with reinforcers other than intracranial stimulation.

Although it seems safe to conclude that instrumental ANS conditioning has been demonstrated in curarized animals, it is not safe to conclude that it has been demonstrated definitively in humans. Theoretically, this distinction may be superfluous; it is likely that conclusive demonstrations on animals are sufficient to resolve the issue at a theoretical level. However, at a practical level a distinction must be drawn between *conditioning* the ANS and *controlling* it (Black, 1966). For those researchers whose primary goal is to gain control over ANS function, and for whom theoretical problems concerning possible mediators and underlying phenomena are less important, it may be unnecessary to demonstrate the pure phenomenon of instrumental conditioning. In fact, it is probably fruitless to pursue further any attempts at providing such demonstrations in humans, because they would require unconscious subjects to eliminate cognitive mediation and complete curarization to eliminate somatic mediation. Indeed, the desired control of autonomic activity might be more efficiently produced by proper reinforcement of both the somatic and the cognitive mediators. That is, for those who want to control autonomic activity, an alternative procedure would be first to determine accurately the relationship between certain voluntary skeletal actions and their associated epiphenomenal autonomic response patterns, and then to reinforce the voluntary responses. What we are suggesting here is simply a program to develop techniques for exploiting what has always been known to psychologists as well as laymen—that one can learn to control his "involuntary" behavior.

REFERENCES

Birk, L., Crider, A., Shapiro, D., & Tursky, B. Operant electrodermal conditioning under partial curarization. *Journal of Comparative and Physiological Psychology,* 1966, **62**, 165–166.

Black, A. H. The operant conditioning of heart rate in curarized dogs: Some problems of interpretation. Paper presented at the meeting of the Psychonomic Society, St. Louis, October 1966.

Brener, J., & Hothersall, D. Heart rate control under conditions of augmented sensory feedback. *Psychophysiology,* 1966, **3**, 23–28.

Brener, J., & Hothersall, D. Paced respiration and heart rate control. *Psychophysiology,* 1967, **4**, 1–6.

Church, R. M. Systematic effect of random error in the yoked control design. *Psychological Bulletin,* 1964, **62**, 122–131.

Crider, A., Shapiro, D., & Tursky, B. Reinforcement of spontaneous electrodermal activity. *Journal of Comparative and Physiological Psychology,* 1966, **61**, 20–27.

DiCara, L. V., & Miller, N. E. Changes in heart rate instrumentally learned by curarized rats as avoidance responses. *Journal of Comparative and Physiological Psychology,* 1968, **65**, 8–12.

Dulany, D. E., Jr. The place of hypotheses and intentions: An analysis of verbal control in verbal conditioning. In C. W. Eriksen (Ed.), *Behavior and Awareness.* Durham: Duke University Press, 1962.

Engel, B. T., & Chism, R. A. Operant conditioning of heart rate speeding. *Psychophysiology,* 1967, 3, 418–426.

Engel, B. T., & Hansen, S. P. Operant conditioning of heart rate slowing. *Psychophysiology,* 1966, 3, 176–187.

Fowler, R. L., & Kimmel, H. D. Operant conditioning of the GSR. *Journal of Experimental Psychology,* 1962, **63**, 563–567.

Frazier, T. W. Avoidance conditioning of heart rate in humans. *Psychophysiology,* 1966, **3**, 188–202.

Gavalas, R. J. Operant reinforcement of an autonomic response: Two studies. *Journal of the Experimental Analysis of Behavior,* 1967, **10**, 119–130.

Graham, F. K., & Clifton, R. K. Heart rate change as a component of the orienting response. *Psychological Bulletin,* 1966, **65**, 305–320.

Greene, W. A. Operant conditioning of the GSR using partial reinforcement. *Psychological Reports,* 1966, **19**, 571–578.

Greene, W. A., & Nielsen, T. C. Operant GSR conditioning of high and low autonomic perceivers. *Psychonomic Science,* 1966, **6**, 359–360.

Greiner, T. H., & Burch, N. R. Response of human GSR to drugs that influence the reticular formation of brain stem. *Federation Proceedings American Societies for Experimental Biology,* 1955, **14**, 346. (Abstract)

Grings, W. W., & Carlin, S. Instrumental modification of autonomic behavior. *The Psychological Record,* 1966, **16**, 153–159.

Harwood, C. W. Operant heart rate conditioning. *The Psychological Record,* 1962, **12**, 279–284.

Hnatiow, M., & Lang, P. J. Learned stabilization of cardiac rate. *Psychophysiology,* 1965, **1**, 330–336.

Hudgins, C. V. Conditioning and the voluntary control of the pupillary light reflex. *Journal of General Psychology,* 1933, **8**, 3–51.

Johnson, H. J., & Schwartz, G. E. Suppression of GSR activity through operant reinforcement. *Journal of Experimental Psychology,* 1967, **75**, 307–312.

Katkin, E. S. Relationship between manifest anxiety and two indices of autonomic response to stress. *Journal of Personality and Social Psychology,* 1965, **2**, 324–333.

Kimble, G. A. *Hilgard and Marquis' conditioning and learning.* (2nd ed.) New York: Appleton-Century, 1961.

Kimmel, E., & Kimmel, H. D. A replication of operant conditioning of the GSR. *Journal of Experimental Psychology,* 1963, **65**, 212–213.

Kimmel, H. D. Instrumental conditioning of autonomically mediated behavior. *Psychological Bulletin,* 1967, **67**, 337–345.

Kimmel, H. D., & Baxter, R. Avoidance conditioning of the GSR. *Journal of Experimental Psychology,* 1964, **68**, 482–485.

Kimmel, H. D., & Hill, F. A. Operant conditioning of the GSR. *Psychological Reports,* 1960, **7**, 555–562.

Kimmel, H. D., & Sternthal, H. S. Replication of GSR avoïdance conditioning with concomitant EMG measurement and subjects matched in responsivity and conditionability. *Journal of Experimental Psychology,* 1967, **74**, 144–146.

Kimmel, H. D., Sternthal, H. S., & Strub, H. Two replications of avoidance conditioning of the GSR. *Journal of Experimental Psychology,* 1966, **72**, 151–152.

Lisina, M. I. The role of orientation in the transformation of involuntary reactions into voluntary ones. In L. G. Voronin, A. N. Leontiev, A. R. Luria, E. N. Sokolov, & O. S. Vinogradova (Eds.), *Orienting reflex and exploratory behavior.* Washington: American Institute of Biological Sciences, 1965.

Mandler, G., Preven, D. W., & Kuhlman, C. K. Effects of operant reinforcement on the GSR. *Journal of the Experimental Analysis of Behavior,* 1962, **5**, 317–321.

Miller, N. E. Integration of neurophysiological and behavioral research. *Annals of the New York Academy of Sciences,* 1962, **92**, 830–839.

Miller, N. E. Animal experiments on emotionally-induced ulcers. In, *Proceedings of the World Conference of Psychiatry, June 4–10, 1962, Montreal,* 1963, **3**, 213–219.

Miller, N. E., & Banuazizi, A. Instrumental learning by curarized rats of a specific visceral response, intestinal or cardiac. *Journal of Comparative and Physiological Psychology,* 1968, **65**, 1–7.

Miller, N. E., & DiCara, L. Instrumental learning of heart rate changes in curarized rats: Shaping, and specificity to discriminative stimulus. *Journal of Comparative and Physiological Psychology,* 1967, **63**, 12–19.

Murray, E. N., & Katkin, E. S. Comment on two recent reports of operant heart rate conditioning. *Psychophysiology,* 1968, in press.

Razran, G. The observable unconscious and the inferable conscious in current Soviet psychophysiology: Interoceptive conditioning, semantic conditioning, and the orienting reflex. *Psychological Review,* 1961, **68**, 81–147.

Rice, D. G. Operant conditioning and associated electromyogram responses. *Journal of Experimental Psychology,* 1966, **71**, 908–912.

Rosenthal, R. The effect of the experimenter on the results of psychological research. In B. A. Maher (Ed.), *Progress in experimental personality research.* Vol. 1. New York: Academic Press, 1964.

Rosenthal, R. *Experimenter effects in behavioral research.* New York: Appleton-Century, 1966.

Rosenthal, R. Covert communication in the psychological experiment. *Psychological Bulletin,* 1967, **67**, 356–367.

Senter, R. J., & Hummel, W. F., Jr. Suppression of an autonomic response through operant conditioning. *The Psychological Record,* 1965, **15**, 1–5.

Shapiro, D., Crider, A. B., & Tursky, B. Differentiation of an autonomic response through operant reinforcement. *Psychonomic Science,* 1964, **1**, 147–148.

Shearn, D. W. Operant conditioning of heart rate. *Science,* 1962, **137**, 530–531.

Silverman, A. J., Cohen, S. I., & Shmavonian, B. M. Investigation of psychophysiologic relationships with skin resistance measures. *Journal of Psychosomatic Research,* 1959, **4**, 65–87.

Skinner, B. F. *The behavior or organisms: An experimental analysis.* New York: Appleton-Century, 1938.

Smith, K. Conditioning as an artifact. *Psychological Review,* 1954, **61**, 217–225.

Smith, S. M., Brown, H. O., Toman, J. E. P., & Goodman, L. S. The lack of cerebral effects of d-tubocurarine. *Anesthesiology,* 1947, **8**, 1–14.

Snyder, C., & Noble, M. Operant conditioning of vasoconstriction. Paper presented at the meeting of the Midwestern Psychological Association, Chicago, April 1965.

Snyder, C., & Noble, M. E. Operant conditioning of vasoconstriction. Paper presented at the meeting of the Psychonomic Society, St. Louis, October 1966.

Spielberger, C. D., & DeNike, L. D. Descriptive behaviorism versus cognitive theory in verbal operant conditioning. *Psychological Review,* 1966, **73**, 306–326.

Stern, R. M. Operant conditioning of spontaneous GSRs: Negative results. *Journal of Experimental Psychology,* 1967, **75**, 128–130.

Stern, R. M., Boles, J., & Dionis, J. Operant conditioning of spontaneous GSRs: Two unsuccessful attempts. Technical Report No. 13, 1966, Indiana

University, Contract Nonr 908-15, Office of Naval Research.

Trowill, J. A. Instrumental conditioning of the heart rate in the curarized rat. *Journal of Comparative and Physiological Psychology,* 1967, **63,** 7-11.

Van Twyver, H. B., & Kimmel, H. D. Operant conditioning of the GSR with concomitant measurement of two somatic variables. *Journal of Experimental Psychology,* 1966, **72,** 841-846.

Wood, D. M., & Obrist, P. A. Effects of controlled and uncontrolled respiration on the conditioned heart rate response in humans. *Journal of Experimental Psychology,* 1964, **68,** 221-229.

(Received August 17, 1967)

51

On the Criteria for Instrumental Autonomic Conditioning: A Reply to Katkin and Murray

Andrew B. Crider, Gary E. Schwartz and Susan Shnidman

Katkin and Murray (1968) have provided the most comprehensive review to date of a large body of studies purporting to demonstrate instrumental conditioning of autonomically mediated behavior. Their apparent conclusion (p. 66) is that studies with curarized animal subjects, especially those emanating from the Miller group, have convincingly demonstrated the phenomenon, contrary to the prevailing views of most learning theorists. The rather larger body of studies with human subjects, however, is almost totally rejected as probably artifactual on one or more grounds.

Although we would not want to claim immunity from criticism for the existing studies with human subjects, we do feel that the problem can be considerably clarified by examining Katkin and Murray's assumptions concerning the conditions under which instrumental autonomic conditioning can be said to occur. Their explicitly stated "minimal criteria" for accepting such demonstrations include (*a*) freedom from alternative explanations, such as peripheral or cognitive mediation; (*b*) an increase in the frequency, amplitude, or probability of the reinforced response over the level shown in a free-operant period; and (*c*) comparisons between experimental groups and appropriate control groups (pp. 53–54). While these criteria have a good deal of face validity, they are used by Katkin and Murray in a rather uncritical manner in their treatment of existing studies on the human level. The following discussion centers on an analysis of these points.

The Mediation Issue

Katkin and Murray indicated that either peripheral or cognitive mediation hypotheses can serve as likely alternative explanations for the effects observed in studies with human subjects. There are, however, a number of logical and empirical difficulties associated with invoking these hypotheses as alternative explanations to the direct operant conditioning of autonomic responses.

Peripheral mediation. As originally outlined by Skinner (1938) and recapitulated by Katkin and Murray, the peripheral mediation model posits the following relationship between a skeletal muscle operant (R^{sm}), an autonomic response (R^{ans}), and a reinforcing event (S^r):

$$R^{sm} \rightarrow S \rightarrow R^{ans} \rightarrow S^r$$

(with an arrow returning from S^r to R^{sm})

[1] The present paper was supported by National Institute of Mental Health Grants MH-08853, MH-04172, MH-07084, and MH-36213. The authors thank David Shapiro and H. J. Johnson for valuable comments on an earlier draft.

[2] Requests for reprints should be addressed to Dr. Andrew Crider, Department of Psychology, Williams College, Williamstown, Massachusetts 01267.

In this paradigm, a skeletal operant produces a source of stimulation which elicits an autonomic response, this in turn being followed by a reinforcing event. The reinforcer is presumed to act to increase the probability of occurrence of the skeletal operant rather than the autonomic reaction. The function of the operant is simply to bring appropriate receptors in contact with sources of stimulation, whether external (e.g., a visual stimulus), or internal (e.g., activation of stretch receptors by tensing). The middle terms of the paradigm ($S \rightarrow R^{ans}$) may represent either an unconditional or conditional reflex, as in the unconditional elicitation of an autonomic reaction with a pin prick or the conditional reaction to a verbal or subvocal stimulus.

Although much invoked, this model is probably inadequate to account for the results of operant autonomic conditioning studies. If the middle terms ($S \rightarrow R^{ans}$) represent an unconditional reflex, the available evidence leads one to expect habituation of the response over time. This may not be true of stimuli of high intensity, as in the defense reflex described by Sokolov (1963), nor of certain homeostatic adjustments such as those involved in strenuous exercise. It would most likely be true, however, of the stimuli produced by minor skeletal movements and the resulting electrodermal and cardiovascular responses involved in the studies in question. Habituation of the GSR to repeated applications of a stimulus is a highly reliable phenomenon (Davis, Buchwald, & Frankmann, 1955), and similar results have been reported for heart rate and peripheral vasomotor reactions (McDonald, Johnson, & Hord, 1964). If, on the other hand, the middle terms are taken to represent a conditional reflex, the response should extinguish due to nonreinforcement. Thus the model would predict a decline in the rate of operantly reinforced autonomic responses over time.

Exactly these results were in fact found in a recent study by Gavalas (1968), which represents the only direct test to date of this mediation model. When subjects were reinforced for GSRs elicited by a deep inspiration, the elicited GSR, as well as other nonspecific GSRs, habituated over training sessions, even though the deep-inspiration operant increased in frequency over the same period. At least two of the operant GSR studies which used 16-minute conditioning periods did report a slight decline in the rate of response after an initial increase in the first 2 minutes of reinforcement (Kimmel & Kimmel, 1963; Schwartz & Johnson, 1969). This is not universal, however (Van Twyver & Kimmel, 1966), and other GSR studies incorporating reinforcement periods of up to 30 minutes indicate that an early decline in response rate ultimately stabilizes around the original operant level (Crider, Shapiro, & Tursky, 1966).

With an additional assumption, this model can be made somewhat more viable. This involves attributing to the reinforcing stimulus (S^r) the capacity to unconditionally elicit the autonomic response being conditioned:

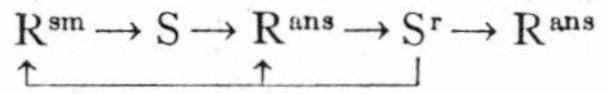

Thus the S^r would serve both as the stimulus limb of an unconditional reflex, impeding the habituation or extinction of the autonomic response, and as an instrumental reinforcer for the prior skeletal operant. The major problem here is that it is unlikely that the reinforcers used in most of the human operant work form powerful enough reflexes to classically reinforce the autonomic change. In classical GSR conditioning, for example, it is notoriously difficult to prevent a decrement over time in the conditional GSR even with the use of painful shock as the reinforcing stimulus (Kimmel, 1964). Certainly no one has yet demonstrated that a dim light such as that used in the operant conditioning studies of the Kimmel group (Fowler & Kimmel, 1962; Kimmel & Kimmel, 1963) or a tone signaling monetary reward such as that used in the Harvard studies (Crider et al., 1966; Shapiro, Crider, & Tursky, 1964) have the ability to classically reinforce the GSR.

A telling difficulty is created for this second model by findings that the same S^r can be used either to increase or to decrease the frequency of a given autonomic response, whether in independent groups of subjects (Engel & Chism, 1967; Engel & Hansen, 1966) or within the same subject (Levene,

Engel, & Pearson, 1968; Shapiro & Crider, 1967). Since the directionality of the conditional response in the classical paradigm is fixed by the nature of the reinforcing stimulus, a mediation model is incapable of accounting for such results.

Aside from these logical problems, there is a dearth of evidence in support of mediation notions. As Katkin and Murray pointed out, Van Twyver and Kimmel (1966) were unable to find differences between groups contingently and noncontingently reinforced for GSR emission in respiration rate, respiration irregularity, or forearm EMG activity, although the contingent group showed an increasing rate of GSRs over time. Similarly, negative results were reported for these somatic variables when vasoconstriction was operantly conditioned (Snyder & Noble, 1968). In addition, Crider et al. (1966) demonstrated clear GSR acquisition effects for several subjects without any apparent difference in gross body movements, measured with a sensitive movement transducer, over a noncontingently reinforced control group. Studies of instrumental heart-rate conditioning have almost universally reported a failure to find skeletal mediators, especially breathing changes, that could account for the obtained results (Engel & Chism, 1967; Engel & Hansen, 1966; Hnatiow & Lang, 1966; Levene et al., 1968).

Cognitive mediation. Given the difficulties with the peripheral mediation issue, the mediation hypothesis has more recently taken the form of a search for central or cognitive factors which could account for the results obtained with human subjects. What seems to be at issue here is the interpretation of postexperimental verbal reports of human subjects. Granting that subjects often report engaging in mental activity during the conditioning experience, the question arises as to the proper evaluation of these data. Katkin and Murray take it as evidence that certain forms of ideation act as conditional stimuli for the elicitation of autonomic responses. They cited, for example, Stern's (1967) finding that subjects reinforced for GSR emission developed hypotheses that they were being reinforced for sundry arousing cognitive responses, while noncontingently reinforced controls developed hypotheses that they were being reinforced for being relaxed or drowsy. This use of Stern's study is particularly unfortunate. Since Stern failed to find any GSR conditioning differences between his groups, the substantive interpretation that Katkin and Murray placed on the correlated verbal reports seems somewhat gratuitous. Stern also pointed out that he included the verbal reports of subjects whose GSR data were discarded from the analysis.

Not mentioned by Katkin and Murray were the observations of Shapiro et al. (1964) that they were unable to find differences between contingently and noncontingently reinforced subjects in either the perceived association between cognitive events and the occurrence of a reinforcer or in the maintenance of ideational activity over time. Gavalas (1967) reported that "none of the subjects reported any knowledge of the purpose of the experiment or believed that they could predict or control the occurrence of the reinforcer [p. 128]," while Schwartz and Johnson (1969) concluded that "postexperimental questioning was not effective in determining what differences, if any, were occurring at a cognitive level as a function of contingent vs. noncontingent reinforcement [p. 31]." In a previous GSR punishment study, Johnson and Schwartz (1967) actually manipulated cognitive set by informing one group of subjects that their behavior would have something to do with the number of reinforcers received during the experiment, while a second group was given neutral instructions. Although the effects for contingent versus noncontingent reinforcement were marked in both groups, no differences were found as a function of instructional set.

Difficulties are also created for this position by the observation that operant autonomic conditioning seems to be specific to the reinforced response. Thus, Black's (1967b) suggestion that the reported autonomic changes could be due to the activation of "central movement processes" which have facilitating effects on autonomic centers was met by the demonstration in curarized animals that intestinal activity could be increased or decreased without correlated heart-rate changes while heart rate could be increased or decreased without correlated intestinal changes (Miller & Banuazizi, 1968). On the human level, Shapiro et al. (1964) and Schwartz and Johnson (1969) found no correlated changes in heart rate when subjects were reinforced for GSR emission; Rice (1966) found no spread of the GSR conditioning effect to finger blood volume; Snyder and Noble (1968) reported an independence between conditioned vasoconstriction and concomitantly measured heart rate, and Gavalas (1967) found no effects on heart rate or finger blood volume when the GSR was operantly conditioned. To account for such results, a cognitive mediation hypothesis would have to posit a different cognitive event (or class of events) for each visceral response pattern resulting from operant conditioning. Needless to say, this possibility far outstrips any known psychophysiological data.

In addition, it is a currently debatable question that cognitive activity per se produces any marked autonomic effects at all. Although there are any number of demonstrations that problem solving, for example, is accompanied by heightened autonomic activity (Kahneman, Tursky, Shapiro, & Crider, 1969), the effect seems to occur only in the context of a requirement to produce a subsequent verbal or motor response. This was first discussed by Weiner (1962), who pointed out that a necessary condition for autonomic arousal in a subject undergoing projective testing was the requirement that he interact in some manner with the experimenter. Subsequent research has shown that when various forms of cognitive activity, such as mental arithmetic or the focused imagination of specific scenes, are carried out without the necessity for a concomitant or later verbal report, a marked attenuation of the autonomic change specific to the task is observed (Johnson & Campos, 1967; Simpson & Paivio, 1966). Such findings are obviously embarrassing for a cognitive mediation position.

Assessing Acquisition Effects

Katkin and Murray specified that "there should be some demonstration that the response being reinforced shows an increase in frequency, or amplitude, or probability of occurrence over the level shown in a free-operant period [p. 53]." This sort of change is one which is usually expected with contingent positive reinforcement and appears straightforward. However, an increase in the frequency of response relative to the level shown in a free-operant period assumes that this base rate is more or less constant at any point in time. Thus, Skinner (1953) pointed out that the behavior in question should reach a "*fairly stable level* against which the frequency of a selected response may be investigated [p. 63, italics added]." If, on the other hand, the free-operant level changes as a function of factors other than the experimental treatment, it is inappropriate to evaluate contingency effects from this initial level.

This problem typifies the case of spontaneous GSR activity. When stimuli are programmed randomly with respect to GSR emission, their frequency declines over time (Johnson, 1963). The problem then becomes one of estimating the change in the probability of response with contingent reinforcement vis-à-vis a declining operant level. The simplest solution is to employ a control group which receives the same instructions and number of stimuli as the experimental group but with the stimuli presented in some noncontingent manner. If it happens that the contingent group shows a heightened rate of response relative to the noncontingent group, the conclusion is drawn that some form of contingency effect is operating.

In rejecting studies which fail to show an increase in response frequency over time, Katkin and Murray implied that if an increase in the absolute rate of response does not occur, an increase in the *probability* of response has not occurred. By taking into account the changing base line, however, this conclusion is generally reversed. To help clarify this notion, the data from an operant GSR study by Fowler and Kimmel (1962) are illustrated in Figure 1A. The contingent group shows a marked decrease in rate during the first 6 minutes of conditioning relative to the initial operant level, followed by a return to this base line with continued reinforcement. The control group shows a slightly larger decrease in the first 6 minutes, with the decline continuing for the remainder of the session. In terms of the criterion used by Katkin and Murray, the data do not show

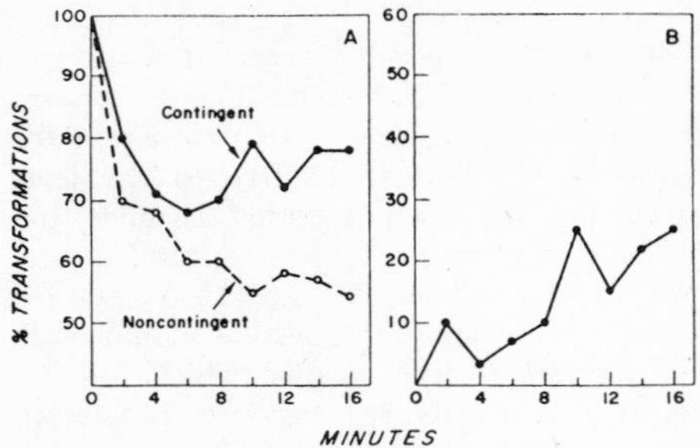

Fig. 1. A. Percentage of transformed GSR frequencies for contingent and noncontingent groups during 16 minutes of stimulation.

B. Difference in percentage response frequencies between contingent and noncontingent groups (after Fowler & Kimmel, 1962).

evidence for conditioning. Figure 1B is a plot of the difference in response frequency between the two groups. This curve corrects for the declining base rate, estimated from the noncontingent group, and demonstrates the relative increase in response probability over time of the contingent group.

Contrary to the conclusion drawn by Katkin and Murray, those studies which have employed contingent and noncontingent reinforcement of spontaneous GSRs (except for Stern, 1967) have all shown evidence for acquisition in terms of an increase in the probability of responding with contingent reinforcement, even though they may not have shown an increase in the actual frequency of responding relative to the initial operant level. Katkin and Murray also failed to cite data that actually meet their criterion for an increase in response frequency. Crider et al. (1966) split a group into high- and low-operant-level subjects and showed that low-operant, contingent subjects increased in absolute frequency over time, while low-operant, noncontingent subjects showed a slight decline. They also ran a group which received 20 minutes of noncontingent reinforcement followed by 20 minutes of contingent reinforcement, and found a decrease in rate with noncontingent reinforcement followed by an increase with contingent reinforcement.

Yoked-Control Designs

The yoked-control design has frequently been used in operant autonomic conditioning studies to evaluate reinforcement effects. The control subject receives the same number and same temporal distribution of reinforcing stimuli as his experimental pairmate. Katkin and Murray, following Church (1964), pointed out that this design may confound random error with treatment effects when the reinforcer is capable of producing the desired behavior change simply as a function of noncontingent stimulation. The potential errors, however, do not occur equally in all operant paradigms.

The essence of the argument is that when the unconditioned effect of the reinforcing stimulus affects the paired subjects unequally, there will be a bias in favor of the experimental member. If the reinforcer is an effective activator for the experimental subject, he will respond frequently and receive many reinforcements. Since the reinforcer is not an effective activator for the control subject, his behavior changes little. If, however, the reinforcing stimulus does not activate the experimental subject, he will respond infrequently and receive few reinforcements. Since the control also receives few reinforcements as a result, the control is prevented from being activated although the stimulus is effective for him. The assumption is, therefore, that a better experimental subject will improve at the expense of a poorer control, but that the better control will not improve when he is yoked with a poorer experimental. These assumptions apply particularly to the reward paradigm.

Katkin and Murray inappropriately applied this analysis to punishment and avoidance paradigms, however. While Church's assumption that a better experimental subject will improve at the expense of a poorer control holds for these paradigms, it is also the case that a better control subject gains when yoked with a poorer experimental, thus equalizing biasing effects. If the nonspecific effect of punishment is to suppress responding, a poorer experimental subject will not suppress and will produce many stimuli for his better control, who will then suppress. In the case of avoidance, if a poorer experimental subject is not activated to avoid, the better control will receive many negative reinforcers and will respond more (Black, 1967a).

Mention should also be made of the case

of punishment where the nonspecific effect of the noxious stimulus is to *increase* responding, as is true with most autonomic variables. The analysis here is equivalent to that of the reward paradigm: A better experimental subject will increase at the expense of a poorer control, but a better control subject will not increase at the expense of a poorer experimental. Since the experimental effect is hypothesized to be a decrease in response rate, however, the bias is in favor of the control group and acceptance of the null hypothesis.

Under conditions where the number of trials is fixed, therefore, the yoked-control design is quite appropriate for punishment and avoidance paradigms. Katkin and Murray failed to appreciate this in their criticisms of the punishment study of Johnson and Schwartz (1967) and the avoidance studies of Kimmel and Baxter (1964) and Kimmel and Sternthal (1967). In reward paradigms, furthermore, potential sources of bias in yoked-control designs can be attenuated by controlling for individual differences in the effectiveness of the reinforcer. Since GSR responsivity to noncontingent stimulation is correlated with resting-level spontaneous response rate (Johnson, 1963), yoked pairs may be matched on this initial level. This was done in a number of operant studies, which nonetheless reported significant contingency effects (Crider et al., 1966; Schwartz & Johnson, 1969).

Conclusion

We have attempted to demonstrate that the bases on which Katkin and Murray are led to reject most of the human operant autonomic conditioning work have less substance than is immediately apparent. What is most striking about the mediation hypothesis, whether phrased in peripheral or centralist terms, is the virtual lack of evidence that presumed mediators can produce the autonomic effects observed under contingent reinforcement regimes. When this is coupled with the logical difficulties that emerge when these hypotheses are more clearly spelled out, it appears that such notions are poor alternatives to the more parsimonious hypothesis of direct instrumental conditioning of autonomic responses.

The problem of assessing those studies that do not show an increase in response rate with contingent reinforcement over an initial operant level is clarified by distinguishing between changes in absolute rate and changes in response probability. In working with autonomic systems which do not readily assume the characteristic of a stable operant level, an assessment of contingency effects can only be made relative to time-dependent trends in this base rate. We have also pointed out that yoked-control designs do not automatically invalidate studies in which they are employed. Church's (1964) critique of this procedure is not fatal to either punishment or avoidance paradigms, and matching for individual differences in the generalized effectiveness of a positively reinforcing event probably attenuates the potential error in reward paradigms.

While the incontrovertible study of operant autonomic conditioning on the human level has yet to be carried out, there exists nonetheless a sizable literature whose import is that human autonomic functions can be effectively controlled with operant procedures. Given the problematic status of the alternative explanations offered by Katkin and Murray, we would further conclude that this control has been effected through the direct instrumental strengthening of the autonomic response itself.

References

Black, A. H. A comment on yoked control designs. Technical Report No. 11, September 1967, McMaster University, Department of Psychology. (a)

Black, A. H. Operant conditioning of heart rate under curare. Technical Report No. 12, October 1967, McMaster University, Department of Psychology. (b)

Church, R. M. Systematic effect of random error in the yoked control design. *Psychological Bulletin,* 1964, **62**, 122–131.

Crider, A., Shapiro, D., & Tursky, B. Reinforcement of spontaneous electrodermal activity. *Journal of Comparative and Physiological Psychology,* 1966, **61**, 20–27.

Davis, R. C., Buchwald, A. M., & Frankmann, R. W. Autonomic and muscular responses and their relation to simple stimuli. *Psychological Monographs,* 1955, **69**(20, Whole No. 405).

Engel, B. T., & Chism, R. A. Operant conditioning of heart rate speeding. *Psychophysiology,* 1967, **3**, 418–426.

Engel, B. T., & Hansen, S. P. Operant conditioning of heart rate slowing. *Psychophysiology,* 1966, **3**, 176–187.

Fowler, R. L., & Kimmel, H. D. Operant conditioning of the GSR. *Journal of Experimental Psychology,* 1962, **63**, 563–567.

Gavalas, R. J. Operant reinforcement of an autonomic response: Two studies. *Journal of the Experimental Analysis of Behavior,* 1967, **10**, 119–130.

Gavalas, R. J. Operant reinforcement of a skeletally mediated autonomic response: Uncoupling of the two responses. *Psychonomic Science,* 1968, **11**, 195–196.

Hnatiow, M., & Lang, P. J. Learned stabilization of cardiac rate. *Psychophysiology,* 1966, **1**, 330–336.

Johnson, H. J., & Campos, J. J. The effect of cognitive tasks and verbalization instructions on heart rate and skin conductance. *Psychophysiology,* 1967, **4**, 143–150.

Johnson, H. J., & Schwartz, G. E. Suppression of GSR activity through operant reinforcement. *Journal of Experimental Psychology,* 1967, **75**, 307–312.

Johnson, L. C. Some attributes of spontaneous autonomic activity. *Journal of Comparative and Physiological Psychology,* 1963, **56**, 415–422.

Kahneman, D., Tursky, B., Shapiro, D., & Crider, A. Pupillary, heart rate, and skin resistance changes during a mental task. *Journal of Experimental Psychology,* 1969, **79**, 164–167.

Katkin, E. S., & Murray, E. N. Instrumental conditioning of autonomically mediated behavior: Theoretical and methodological issues. *Psychological Bulletin,* 1968, **70**, 52–68.

Kimmel, E., & Kimmel, H. D. A replication of operant conditioning of the GSR. *Journal of Experimental Psychology,* 1963, **65**, 212–213.

Kimmel, H. D. Further analysis of GSR conditioning: A reply to Stewart, Stern, Winokur, and Fredman. *Psychological Review,* 1964, **71**, 160–166.

Kimmel, H. D., & Baxter, R. Avoidance conditioning of the GSR. *Journal of Experimental Psychology,* 1964, **68**, 482–485.

Kimmel, H. D., & Sternthal, H. S. Replication of GSR avoidance conditioning with concomitant EMG measurement and subjects matched in responsivity and conditionability. *Journal of Experimental Psychology,* 1967, **74**, 144–146.

Levene, H. I., Engel, B., & Pearson, J. A. Differential operant conditioning of heart rate. *Psychosomatic Medicine,* 1968, **30**, 837–845.

McDonald, D. G., Johnson, L. C., & Hord, D. J. Habituation of the orienting response in alert and drowsy subjects. *Psychophysiology,* 1964, **1**, 163–173.

Miller, N. E., & Banuazizi, A. Instrumental learning by curarized rats of a specific visceral response, intestinal or cardiac. *Journal of Comparative and Physiological Psychology,* 1968, **65**, 1–7.

Rice, D. G. Operant conditioning and associated electromyogram responses. *Journal of Experimental Psychology,* 1966, **71**, 908–912.

Schwartz, G. E., & Johnson, H. J. Affective visual stimuli as operant reinforcers of the GSR. *Journal of Experimental Psychology,* 1969, **80**, 28–32.

Shapiro, D., & Crider, A. Operant electrodermal conditioning under multiple schedules of reinforcement. *Psychophysiology,* 1967, **4**, 168–175.

Shapiro, D., Crider, A., & Tursky, B. Differentiation of an autonomic response through operant reinforcement. *Psychonomic Science,* 1964, **1**, 147–148.

Simpson, H. M., & Paivio, A. Changes in pupil size during an imaginary task without motor response involvement. *Psychonomic Science,* 1966, **5**, 405–406.

Skinner, B. F. *The behavior of organisms: An experimental analysis.* New York: Appleton-Century, 1938.

Skinner, B. F. *Science and human behavior.* New York, Macmillan, 1953.

Snyder, C., & Noble, M. Operant conditioning of vasoconstriction. *Journal of Experimental Psychology,* 1968, **77**, 263–268.

Sokolov, Y. N. *Perception and the conditionea reflex.* Oxford: Pergamon Press, 1963.

Stern, R. M. Operant conditioning of spontaneous GSRs: Negative results. *Journal of Experimental Psychology,* 1967, **75**, 128–130.

Van Twyver, H. B., & Kimmel, H. D. Operant conditioning of the GSR with concomitant measurement of two somatic variables. *Journal of Experimental Psychology,* 1966, **72**, 841–846.

Weiner, H. Some psychological factors related to cardiovascular responses: A logical and empirical analysis. In R. Roessler & N. S. Greenfield (Eds.), *Physiological correlates of psychological disorders.* Madison: University of Wisconsin Press, 1962.

(Received August 6, 1968)

52 Concerning Instrumental Autonomic Conditioning: A Rejoinder

Edward S. Katkin, E. Neil Murray and Roy Lachman

Crider, Schwartz, and Shnidman (1969) interpreted Katkin and Murray's (1968) review of instrumental autonomic conditioning as suggesting that the evidence for the existence of the phenomenon in human subjects has been "almost totally rejected [p. 455]." Quite to the contrary, Katkin and Murray concluded that some work (e.g., Johnson & Schwartz, 1967; Snyder & Noble, 1968) demonstrated such conditioning in a manner difficult to criticize. However, Katkin and Murray's effort was directed at the majority of studies with both animals and humans where alternative explanations could be invoked. These alternative explanations appear reasonable to researchers who emphasize mediation hypotheses and unreasonable to those who do not. Thus it appears that we are in the midst of a controversy based more in epistemology than in data. Although scientific controversies can never be resolved completely within the domain of data, they may be clarified by discussion of their nonempirical bases.

The crux of the present dispute seems to be the failure of Crider et al. to recognize the fiduciary character of their own conceptual framework as well as Katkin and Murray's. Ultimately, the paradigm adopted by any researcher—in accord with Kuhn's (1962) analysis—determines which observations will be accepted as fact and which rejected. As Polanyi (1964) has also concluded, in much of scientific controversy the contending sides "do not accept the same 'facts' as 'facts,' and still less the same 'evidence' as 'evidence' [p. 167]." It cannot be denied that Katkin and Murray minimized the value of descriptive behavioral analyses of instrumental autonomic conditioning in human subjects and that they suggested a broader based interpretation of many of the findings. In response Crider et al. have raised questions concerning four issues discussed in Katkin and Murray's paper: (*a*) the definition of response acquisition, (*b*) the adequacy of the treatment of the yoked-control design, (*c*) somatic mediation, and (*d*) cognitive mediation. Their comments are a welcome addition to the field for they have clarified several complex points. We shall therefore attempt to present certain "facts" and "evidence" which they have failed to perceive as such and which are in danger of being overlooked.

Response Acquisition

Crider et al. have questioned Katkin and Murray's assumption concerning the failure to demonstrate increases in response probability. Clearly the spontaneous GSR rate normally

[1] Preparation of this manuscript was supported, in part, by Research Grant MH-11989 from the National Institute of Mental Health, United States Public Health Service, awarded to the first author. The authors are indebted to David R. Beach for his helpful comments.

[2] Requests for reprints should be addressed to Dr. Edward S. Katkin, Townsend Hall, 4 Administration Road, State University of New York at Buffalo, Buffalo, New York 14214.

decreases over time, and evaluations of response probability should very likely be made against the background of a changing base rate. While the argument proposed by Crider et al. on this issue is sound, the example that they have chosen to illustrate it is far from ideal. The data presented in their Figures 1A and 1B show the most significant divergence between Minutes 8 and 10. Yet the authors of the paper from which these data were obtained (Fowler & Kimmel, 1962) stated clearly that the data points representing Minute 10 to Minute 16 are based on a different subject sample than the data representing the first 8 minutes.

There are obvious alternative research strategies which may be used to cope with this problem, such as Kimmel and Kimmel's (1963) decision to use a rest period sufficiently long to allow for a stable operant level to be achieved. Under these circumstances, response probability can be safely measured against initial operant levels.

Even though Crider et al.'s approach to the evaluation of response probability is eminently acceptable, one cannot necessarily accept their conclusion that most of the GSR studies reviewed have shown evidence for increase in response probability *as a result of instrumental conditioning procedures*. The majority of these studies still remains subject to dispute because of inappropriate experimental design and artifacts of somatic and cognitive mediation.

Yoked-Control Design

Crider et al. have noted that Church's (1964) description of the effects of yoking in avoidance conditioning paradigms is in error. Therefore, Katkin and Murray's description of the inadequacy of the yoked-control design may not be entirely appropriate. Grings and Carlin (1966) also noted this problem, and it has been discussed in detail by Black (1967). The thrust of Crider et al.'s argument concerns Church's third case, in which an experimental subject who is not affected by reinforcement is yoked to a control subject who is affected. Church claims that in this case "the experimental *S* will rarely make a response (by assumption) and the control *S* will not make many responses either because it only occasionally receives an event [1964, p. 124]." Actually, it appears that in the case of autonomic avoidance conditioning, Church is *incorrect,* for the unaffected experimental subject will receive *many* events as a result of not responding. Thus, the affected yoked control will also receive many events resulting in increased responding (because of conditioned stimulus-unconditioned stimulus [CS-UCS] pairing) for the control subject. The inconsistency in Church's argument, as Black (1967) pointed out, can be reconciled if we assume that Church meant to designate the *absence of the UCS* as the event; however, this is also inconsistent, for in the other three cases Church clearly described the *presence* of the UCS as the event.

There is one other way in which the inconsistency can be reconciled. If Church's model were applied to instrumental conditioning of voluntary skeletal responses rather than to involuntary autonomic responses there would be no problem. For instance, in avoidance conditioning if an unaffected experimental subject received many events because he failed to press a bar, then his affected yoked control would also receive many events whether or not he pressed the bar. The CS-UCS pairings for this control subject should not increase bar pressing; rather, given the noxious UCS used in avoidance conditioning, the yoked control would soon *stop* pressing the bar, resulting in no advantage for the control group. Thus, Church's argument against yoked-control designs applies well in the area of skeletal or voluntary response conditioning.

Despite the fact that Crider et al. have presented a clear exposition of an error in Church's original presentation, it is still not clear whether Church's criticism applies to autonomic conditioning. Their conclusion that the yoked-control design is appropriate in studies of avoidance conditioning and suppression of autonomic responses cannot be accepted uncritically.

Church's criticism of yoked designs is valid only when certain assumptions about how response probabilities change are employed. Unfortunately one does not usually know which assumptions are the correct ones in any particular situation. Therefore, he will not know whether yoked designs will be appropriate. In general, if yoked designs are inappropriate under at least one of a given set of plausible assumptions about how conditioning takes place, and if one does not know which set of assumptions is correct, then he cannot employ yoked designs [Black, 1967, p. 8].

Support for Black's general statement may be inferred from the experience of Kimmel and Sternthal (1967), who attempted to replicate earlier findings (Kimmel & Baxter, 1964) of avoidance conditioning of the GSR, controlling for the biasing effects of the yoked-control design used in the initial study. Kimmel and Sternthal interpreted Church along the same lines that Katkin and Murray had, an interpretation which in the light of Grings and Carlin's (1966), Black's (1967), and Crider et al.'s (1969) analysis seems incorrect. Nevertheless, Kimmel and Sternthal (1967) reported that:

the results of this study tend to support Church's (1964) contention that the yoked control design may lead to spurious or exaggerated evidence of avoidance conditioning. When only pairs of *S*s who are very closely matched in both GSR responsivity and classical conditionability were used, differences favoring an avoidance group over its yoked controls tended to shrink [p. 145].

How then are we to interpret Kimmel and Sternthal's report? Perhaps Black's admonition is correct—the biasing effects of the yoked-control design are intricate; when in doubt, avoid it! An alternative explanation may be that the yoking or lack of it was irrelevant, and the phenomenon simply was not replicable. Either way, Kimmel and Sternthal's failure to replicate, in addition to Kimmel, Sternthal, and Strub's (1966) earlier failures, are small comfort for those who would like to believe that they have seen convincing evidence for the avoidance conditioning of autonomic responses in humans.

Somatic Mediation

Crider et al. have stated explicitly the model under which Skinner (1938) first proposed the peripheral mediation hypothesis. They then suggest that the model is not really viable because the R^{ans} term should habituate. On the face of it, their argument seems reasonable and appears to receive confirmation from a recent experiment by Gavalas (1968) concerning respiration and the GSR. Yet other evidence seems to indicate that the assumption of R^{ans} habituation is probably unwarranted for other somatoautonomic relationships. For instance, Lisina (1965) concluded that control of peripheral blood vessels was gained by her subjects by "using a number of special devices, namely the relaxation of the skeletal musculature and changing of the depth of respiration [p. 456]." In addition, Rice (1966) discovered that his findings of operant GSR conditioning were artifacts of associated EMG changes even after extensive trials, indicating that the GSRs (R^{ans}) associated with the small muscle changes did *not*, in fact, habituate. Since Crider et al.'s assumptions about the expected habituation of R^{ans} are reasonable, it is difficult to explain these findings. One possibility is that we cannot be sure in most experimental situations whether we are dealing with orienting behavior, defensive behavior, or homeostatic behavior. Lisina's findings, for instance, may be a reflection of the well-known homeostatic relationship between respiration and cardiovascular activity.

Crider et al. proposed a second model to supplement the original one, but apparently this model was proposed as a straw man, for its proponents proceeded immediately to attack it. For them the most telling difficulty with this new model concerns the inability of the S^r term to elicit an unconditioned response in two directions. Since they are of the opinion that Engel and Hansen (1966) and Engel and Chism (1967) have shown that the heart rate can be conditioned to increase or decrease, they claim that the second model loses its viability. The two papers by Engel and his associates have been evaluated recently (Murray & Katkin, 1968), and it is apparent that they contain no evidence of bidirectional operant conditioning. The only effect demonstrated for experimental groups was an increase in heart rate, a phenomenon easily dealt with by the second model. (Of course, even the increase reported may be questioned on other grounds.) Crider et al. claimed that the data of the Shapiro and Crider (1967) experiment are also inconsistent with their second model. While this may be true, the Shapiro and Crider findings may be handled by the original model if we substitute a cognitive response (R^{cog}) for the R^{sm} term, for the authors readily admit that cognitive activity was present:

> In post experimental interviews, the subjects of the present studies reported a good deal of problem solving activity. Although none claimed to be able to predict the exact occurrence of a feedback tone or a reinforcer, each worked out an idiosyncratic mental strategy to meet the demands of the various schedules [Shapiro & Crider, 1967, p. 174].

Cognitive Mediation

Crider et al. argued that Katkin and Murray exaggerated the importance and relevance of cognitive mediation as a factor in instrumental autonomic nervous system conditioning. The controversy emerging here is reminiscent of the recent intense dialogue concerning the importance of mediation in the instrumental reinforcement of verbal behavior (Katkin, Risk, & Spielberger, 1966; Maltzman, 1966; Spielberger & DeNike, 1966; Verplanck, 1962). The crucial issue in that controversy was not the existence of cognitive mediation, but the extent to which experimenters were able to describe relationships between verbal reports of mediation and the dependent variables under investigation. The same issue seems to be central to the present controversy. For instance, Crider et al. cited two papers by Engel (Engel & Chism, 1967; Engel & Hansen, 1966) as evidence for automatic reinforcing effects independent of mediation. Yet, Murray and Katkin (1968) demonstrated that in those studies clear-cut relationships between subjects' verbal reports and their heart-rate changes were obtained but inadequately described.

The other studies cited by Crider et al. (Gavalas, 1967; Shapiro, Crider, & Tursky, 1964) do not seem to have made any effort to examine systematically the verbal reports of their subjects.[3] Furthermore, in the Shapiro et al. experiment, subjects were instructed specifically to think emotional thoughts. It is inconceivable (from some points of view) that in view of those instructions one could deny the relevance of such emotional thoughts in the analysis of subsequent autonomic responses.

Crider et al. maintained that Katkin and Murray "take it as evidence that certain forms of ideation act as conditional stimuli for the elicitation of autonomic responses [1969, p. 457]" and they suggest that it is currently debatable that cognitive activity *per se* produces any autonomic effects at all. They proceed to cite recent work by Simpson and Paivio (1966) as an example of research which they feel should be embarrassing for a cognitive mediation position. It is unclear to us who should feel embarrassed; Simpson and Paivio concluded quite clearly that although the elimination of the motor response in their paradigm attenuated the pupillary response under investigation, significant autonomic effects were obtained even in the absence of motor responses.

The crucial question here, however, is not the interpretation of any one specific experiment, but the broader issue of the interpretability of any data when viewed through the eyes of descriptive behaviorists. We might ask what Crider et al. would *ever* accept as prima facie evidence for ideation acting as conditional stimuli. It is clear that for most experimenters operating from their epistemological position any evidence offered would be seen as unacceptable. Similarly there are many other scientists (Wann, 1964) for whom any description of human behavior omitting cognitive activity would also be considered unacceptable.

[3] Johnson and Schwartz (1967) were also cited by Crider et al., but they are exempt from our critical comments on this issue. As noted earlier in this paper and in Katkin and Murray's paper, Johnson and Schwartz demonstrated a sophisticated approach to the problem of mediation and have presented some of the best evidence for instrumental autonomic conditioning.

REFERENCES

Black, A. H. A comment on yoked control designs. Technical Report No. 11, September 1967, McMaster University, Department of Psychology.

Church, R. M. Systematic effect of random error in the yoked control design. *Psychological Bulletin,* 1964, **62**, 122–131.

Crider, A., Schwartz, G., & Shnidman, S. On the criteria for instrumental autonomic conditioning: A reply to Katkin and Murray. *Psychological Bulletin,* 1969, 71, 455–461.

Engel, B. T., & Chism, R. A. Operant conditioning of heart rate speeding. *Psychophysiology,* 1967, **3**, 418–426.

Engel, B. T., & Hansen, S. P. Operant conditioning of heart rate slowing. *Psychophysiology,* 1966, **3**, 176–187.

Fowler, R. L., & Kimmel, H. D. Operant conditioning of the GSR. *Journal of Experimental Psychology,* 1962, **63**, 563–567.

Gavalas, R. J. Operant reinforcement of an autonomic response. Two studies. *Journal of the Experimental Analysis of Behavior,* 1967, **10**, 119–130.

Gavalas, R. J. Operant reinforcement of a skeletally mediated autonomic response: Uncoupling of the two responses. *Psychonomic Science,* 1968, **11**, 195–196.

Grings, W. W., & Carlin, S. Instrumental modification of autonomic behavior. *Psychological Record,* 1966, **16**, 153–159.

Johnson, H. J., & Schwartz, G. E. Suppression of GSR activity through operant reinforcement. *Journal of Experimental Psychology,* 1967, **75**, 307–312.

Katkin, E. S., & Murray, E. N. Instrumental conditioning of autonomically mediated behavior: Theoretical and methodological issues. *Psychological Bulletin,* 1968, **70**, 52–68.

Katkin, E. S., Risk, R. T., & Spielberger, C. D. The effects of experimenter status and subject awareness on verbal conditioning. *Journal of Experimental Research in Personality,* 1966, **1**, 153–160.

Kimmel, E., & Kimmel, H. D. A replication of operant conditioning of the GSR. *Journal of Experimental Psychology,* 1963, **65**, 212–213.

Kimmel, H. D., & Baxter, R. Avoidance conditioning of the GSR. *Journal of Experimental Psychology,* 1964, **68**, 482–485.

Kimmel, H. D., & Sternthal, H. S. Replication of GSR avoidance conditioning with concomitant EMG measurement and subjects matched in responsivity and conditionability. *Journal of Experimental Psychology,* 1967, **74**, 144–146.

Kimmel, H. D., Sternthal, H. S., & Strub, H. Two replications of avoidance conditioning of the GSR. *Journal of Experimental Psychology,* 1966, **72**, 151–152.

Kuhn, T. S. *The structure of scientific revolutions.* Chicago: University of Chicago Press, 1962.

Lisina, M. I. The role of orientation in the transformation of involuntary reactions into voluntary ones. In L. G. Voronin, A. N. Leontiev, A. R. Luria, E. N. Sokolov, & O. S. Vinogradova (Eds.), *Orienting reflex and exploratory behavior.* Washington: American Institute of Biological Sciences, 1965.

Maltzman, I. Awareness: Cognitive psychology versus behaviorism. *Journal of Experimental Research in Personality,* 1966, **1**, 161–165.

Murray, E. N., & Katkin, E. S. Comment on two recent reports of operant heart rate conditioning. *Psychophysiology,* 1968, **5**, 192–195.

Polanyi, M. *Personal knowledge: Towards a post-critical philosophy.* New York: Harper & Row, 1964.

Rice, D. G. Operant conditioning and associated electromyogram responses. *Journal of Experimental Psychology,* 1966, **71**, 908–912.

Shapiro, D., & Crider, A. Operant electrodermal conditioning under multiple schedules of reinforcement. *Psychophysiology,* 1967, **4**, 168–175.

Shapiro, D., Crider, A. B., & Tursky, B. Differentiation of an autonomic response through operant reinforcement. *Psychonomic Science,* 1964, **1**, 147–148.

Simpson, H. M., & Paivio, A. Changes in pupil size during an imaginary task without motor response involvement. *Psychonomic Science,* 1966, **5**, 405–406.

Skinner, B. F. *The behavior of organisms: An experimental analysis.* New York: Appleton-Century, 1938.

Snyder, C., & Noble, M. Operant conditioning of vasoconstriction. *Journal of Experimental Psychology,* 1968, **77**, 263–268.

Spielberger, C. D., & DeNike, L. D. Descriptive behaviorism versus cognitive theory in verbal operant conditioning. *Psychological Review,* 1966, **73**, 306–326.

Verplanck, W. S. Unaware of where's awareness: Some verbal operants—Notates, monents, and notants. In C. W. Eriksen (Ed.), *Behavior and awareness.* Durham: Duke University Press, 1962.

Wann, T. W. (Ed.) *Behaviorism and phenomenology: Contrasting bases for modern psychology.* Chicago: University of Chicago Press, 1964.

(Received November 8, 1968)

IX

CLASSICAL CONDITIONING

53
Conditioning and the Voluntary Control of the Pupilary Light Reflex

Clarence V. Hudgins

ABSTRACT

The problem investigated was that of conditioning the pupil of the human eye to verbal stimuli and to other stimuli under the subject's control. It was thought that positive results from such an experiment would lead to a better understanding of so-called voluntary activity.

Procedure. The general plan of the experiment was so organized that a conditioned response could be established in the pupil to a complex stimulus situation and then this complex stimulus could be progressively reduced to a simpler form. The ultimate goal was to substitute for the complex stimulus a verbal process which would be effective even when repeated subvocally by the subject himself. The method by which this plan was carried out was as follows: (1) The pupil was first conditioned to a bell with the light as the unconditioned stimulus. (2) A hand contraction on the part of the subject was next used to close the light and bell circuits. The relaxation of the hand broke the same circuits. Verbal stimuli from the experimenter were used to evoke the hand responses. (3) The bell and the hand responses were finally eliminated leaving only the verbal stimuli from the experimenter as conditioned stimuli. (4) The subjects produced the verbal stimuli vocally, then by whispering, and finally subvocally; and the pupillary responses were measured.

The initial effects of the various stimuli upon the pupil prior to training were determined first. With this done, the training for conditioning was carried out over a period of six days, with an experimental period of 1½ to 2 hours each day. This training was divided into three general steps, each step occupying two experimental periods. In Step I the pupil was conditioned to the bell stimulus. In Step II the pupil was conditioned to a combination of hand responses, the verbal stimuli from the experimenter, and the sound of the bell. In Step III the pupil was conditioned solely to the verbal stimuli from the experimenter. Tests for the nature and amount of conditioning were given at the end of each experimental period. These tests were made exactly as were those for the initial effects of the stimuli prior to training.

The results from 14 subjects indicate that verbal stimuli can become conditioned stimuli for the pupillary response. Conditioned pupillary responses (both dilation and constriction) can be evoked by verbal stimuli from the experimenter or by verbal stimuli self-presented by the subject either vocally, whispered, or subvocally.

Of the 14 subjects 10 were naive and did not know that their pupillary reactions were being conditioned. It is therefore impossible that their verbally controlled reactions should have been influenced by purposes or intentions. Furthermore, inasmuch as there are no afferent nerve endings in the iris, the conditioned pupillary response could not have been controlled by nervous impulses from the responding organ. In spite of these two facts, the verbally controlled conditioned pupillary responses had that appearance of spontaneity and control by the organism which are so characteristic of the behavior called voluntary.

Control experiments indicate: (1) that the pupillary responses described could not be due to changes of accommodation and convergence; (2) that nonsense words and syllables could be substituted for the familiar "contract" and "relax" without affecting the final result; (3) that the conditioned dilation and constriction responses could be established regardless of the conditioned stimulus (contract or relax) which was paired with the unconditioned stimulus; and (4) that the hand responses on the part of the subject were essential in establishing the verbal stimuli as conditioning stimuli.

The conditioned pupillary responses to the verbal stimuli were relatively permanent. No experimental extinction occurred during any of the tests with the verbal stimuli and evidence is given which indicates that after an interval of 15 days the responses were still present. The conditioned responses to the bell and to the hand responses were subject to the usual experimental extinction.

The temporal aspects of the conditioned pupillary response differ from those of the unconditioned light reflex in the following manner:

a. The latency of the conditioned response is from 5 to 10 times larger than the latency of the unconditioned light reflex.
b. The duration of the conditioned response is from 3 to 5 times larger than that of the light reflex.

"An adequate account of what the psychologist has called voluntary action cannot as yet be given because the necessary experimental data have not yet been secured: but the hypothesis is offered that so-called voluntary behavior is essentially a conditioned response having a characteristic latency and temporal course and under the control of self-excited receptor processes."

54 Conditioning the Occipital Alpha Rhythm in Man

Herbert Jasper and Charles Shagass

Several investigators have observed that the occipital alpha rhythm of the human electroencephalogram (E.E.G.) can be depressed or blocked by sound, a previously indifferent stimulus, if it has been repeatedly paired with light, the usual unconditioned stimulus (Durup and Fessard, **8**), (Loomis, Harvey and Hobart, **18**), (Jasper and Cruikshank, **17**), (Cruikshank, **7**). This conditioning of the electrical activity of the occipital cortex to a sound stimulus was observed during the course of other experiments. Cruikshank found, while investigating the latency of alpha block to various light intensities, that a preparatory sound signal, given repeatedly at equal intervals before a light, established in a short while what appeared to be a conditioned response (CR) to the constant time interval, sound-light. Walter (**29**) has spoken of this as 'time conditioning.'

The alpha block has been studied as a simple CR by Loomis, Harvey and Hobart (**18**) with a low tone as the conditioned stimulus. The simple CR to a tone has also been quantitatively studied by Travis and Egan (**28**), who reached the conclusion that the alpha block can be converted into an unstable CR.

Responses which can be conditioned in the classical Pavlovian fashion have been divided into 3 groups by Hilgard and Marquis (**13**). (1), Glandular, smooth muscle and blood responses: such as salivation, galvanic skin response, immunity reaction, nausea and diuresis; (2) relatively involuntary responses in striate muscle: such as flexor reflex, eye movements and change in respiration; (3) semi-voluntary and voluntary responses, such as withdrawal movements to shock, locomotion, and instructed responses. There is some suggestion that a fourth class, 'sensorial responses,' might be added, according to the recent work of Bogoslovski (**4**).

It would be necessary to provide a fifth group, for the response here studied, inasmuch as we are dealing with the electrical activity of the cortical centres themselves. Insofar as we know, the occipital alpha rhythm is independent of peripheral effectors. Although related fundamentally to the visual afferent system the depression of the occipital alpha rhythm is a centrally determined response since it can be induced by hypnosis, attention, etc. (Loomis, Harvey and Hobart, **19**), (Jasper and Cruikshank, **17**).

The purpose of this study has been two-fold. First, to investigate extensively the depression (blocking) of the alpha rhythm in man as a conditioned response; second, to appraise the value of the electroencephalograph as an instrument for the study of conditioning problems, especially in man. The alpha block has here been studied as a response with relation to simple, cyclic, delayed, trace, differential, differential delayed, and backward conditioning.

Technique

Apparatus

A standard amplifier and ink-writing oscillograph were used for brain potential recording. Additional pens were employed for indicating various signals (stimuli and manual responses) on the E.E.G. record. Sound frequencies between 400 and 700 d.v. per second were generated by a beat frequency oscillator. The voltage output was kept constant in all experiments. The light used was a 6 volt flashlight bulb 12 inches in front of the subject's eyes. For manual response, a push button connected to an electromagnetic marker was employed. All stimuli were controlled from outside the sound-proof room used for the subject.

Subjects

There were 34 subjects in all, 24 of whom were male college students. Those with the most regular and continuous alpha rhythm were used repeatedly for different types of experiments. All were unaware of the nature of the experiment to begin with, although some few interpreted it as conditioning.

Procedure

Two electrodes were applied by means of collodion, one to the right occipital area and the other to the right ear. This provided monopolar recording of the occipital brain potentials. The subject was placed on a bed in a completely dark, sound-proof, electrically shielded room. Several sound stimuli were presented in order to avoid the 'startle' effect of sound on the alpha. Then 5 or 6 control trials with sound alone were made. In the backward CR experiments which required some manual response to sound, the subject was told to press the button at every sound he heard.

In all other experiments, following control trials with sound alone, the subject was told to press the button as soon as he could whenever he saw a light, but to remain as relaxed as possible at all times. The manual response alone usually had no effect on the alpha rhythm, as previously shown by Adrian and Matthews (1) and Jasper and Cruikshank (17).[1]

A slow speed of 1 cm per sec. was used in recording because this made for easy visibility of the alpha block. The durations of stimuli, due to lack of an automatic timing device, varied in all experiments $\pm$ 0.3 sec. Test trials were given in random order. The interval between all trials was varied, except of course in cyclic CR, so as to avoid cyclic conditioning.

---

[1] The manual response was introduced because, in preliminary experiments, it was found extremely difficult to obtain a clearly visible CR when the subject remained passive. The effect obtained was in some respects similar to that of Miller and Cole (21) where a more rapid and stable CR was obtained when it was integrated into an action context.

Conditioning was considered established when at least 2 consecutive responses to the conditioned stimulus occurred in which there was a sufficiently long depression of the alpha rhythm to be clearly a response to the stimulus rather than a 'spontaneous' variation.

Results

Simple Conditioned Responses

The first simple conditioned response (CR) was established in about ten paired presentations of sound and light stimuli. Sound alone produced no appreciable effect on the alpha rhythm in control trials before the conditioning was started after initial startle effects were overcome. In the conditioning trials a 500 cycle sound stimulus preceded by 0.7 sec. the light stimulus which lasted 5.3 sec. The sound overlapped the light.

The latent period of the unconditioned alpha block to light was about 0.5 sec. while the latent period of the conditioned response to sound was about the same or slightly longer. (Accurate studies of latency were not possible from records taken at the slow rate of speed used.) The alpha rhythm dropped out just as completely with the conditioned sound stimulus as with the light and the block was of comparable duration (Fig. 1). This CR was not very stable but five consecutive responses were obtained in one subject. In some experiments extinction occurred after as few as three conditioned responses without reinforcement.

Cyclic CR

The periodic repetition of a 4.7 sec. light stimulus with equal intervals of 9.2 seconds between each stimulus resulted in the establishment of a cyclic CR in 10 to 20 trials.[2] In this case the alpha rhythm was blocked following the conditioned time interval without an external stimulus. Actually the cyclic conditioned response usually occurred earlier than the conditioned interval since it was anticipatory (Fig. 2).

The CR was as complete but usually longer than the unconditioned response to light. In one subject after conditioning a CR occurred lasting 90 sec. during which time 7 periodic light stimuli were introduced before the reappearance of the alpha waves. The prolongation of the CR may have been related to a change in emotional attitude of the subject. Cyclic repetition of a stimulus may produce an attitude of expectancy so that a long lasting disturbance is caused by omission of a stimulus in the regular series. Prolonged depression of the alpha rhythm due to emotional disturbances were encountered

[2] The number of trials includes test trials.

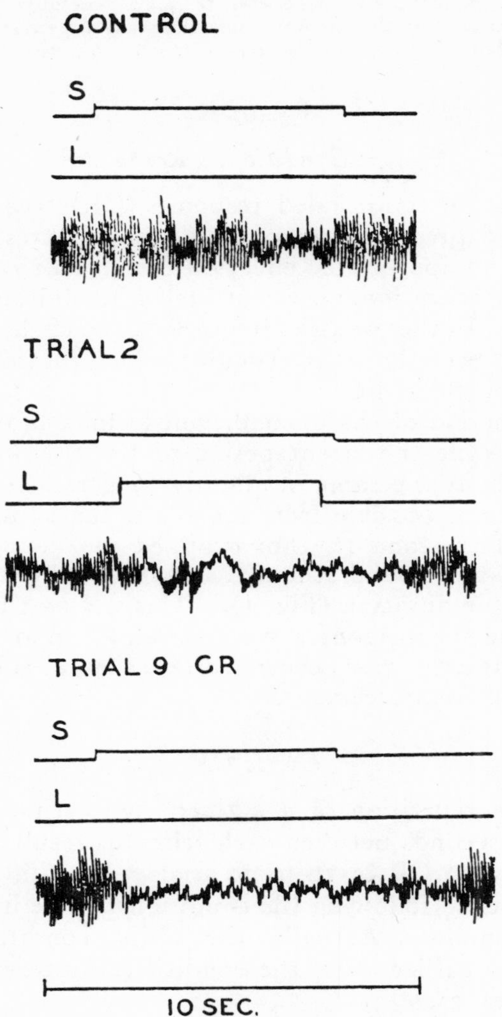

FIG. 1. *Simple CR.* CONTROL shows sound (*S*) without light (*L*) had no effect on alpha waves. (The short depression of the alpha in this sample was similar to 'spontaneous' changes throughout the record in this subject.) TRIAL 2 was second conditioning trial. TRIAL 9 shows blocking of alpha rhythm as CR to sound.

in other experiments but were usually removed after a few reinforced trials.

Extinction of a cyclic CR can be observed only by allowing a long interval to elapse without stimulation, until no further CRs are observed, since 'time' is the conditioned stimulus. In our experiments, after at least 40 to 50 conditioning trials, 2 to 4 successive

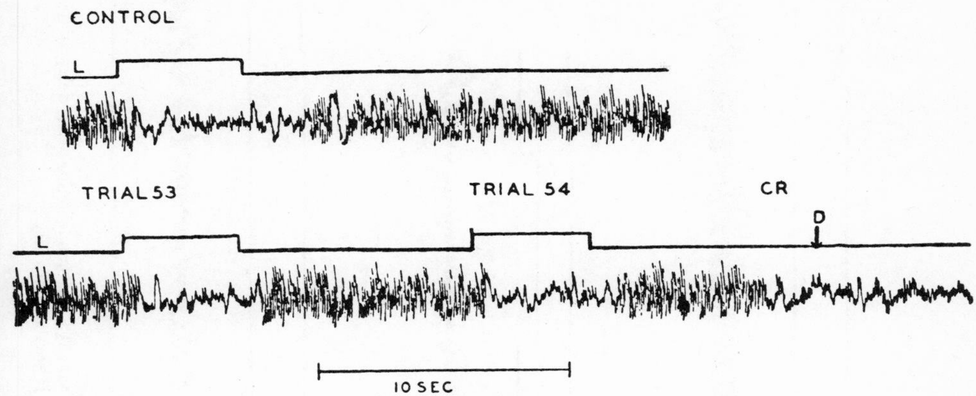

FIG. 2. *Cyclic CR.* CONTROL with single light (*L*) shows no tendency for cyclic repetition of alpha response after light was turned off before conditioning. After 54 repetitions of light at a constant interval (9.4 sec.) one stimulus was omitted at *D* following anticipatory blocking of alpha rhythm which did not return for 90 sec.

cyclic conditioned responses were obtained before extinction, when no reinforcing external stimulus was presented.

Delayed CR

Continuous presentation of sound for some time before light and overlapping it in repeated conditioning trials resulted in a delayed conditioned response to sound (Fig. 3). With delays of the order of 10 sec. a delayed CR was established in about 20 trials. The duration of the sound before the light varied in different experiments between 9.4 and 44.8 sec. The rate of establishment of this form of CR varied with the length of delay; the longer the delay the greater number of trials required.

Delayed conditioned responses were always anticipatory, that is, the response occurred after a delay not quite equal to that used for conditioning. For example in the experiment from which the sample of Fig. 3 was taken conditioning trials were given with a delay of 29.8 sec. The CR occurred after a delay of 27 sec. and lasted until the sound was turned off. The accuracy of these conditioned delay periods has suggested further experiments on the nature of this process with particular reference to the estimation of time, to be reported later.

Extinction was not systematically studied but as many as 4 consecutive delayed CRs were obtained in some experiments with intervals as long as 30 seconds. It was found unnecessary to use Pavlov's method of conditioning first with short delays before extending them.

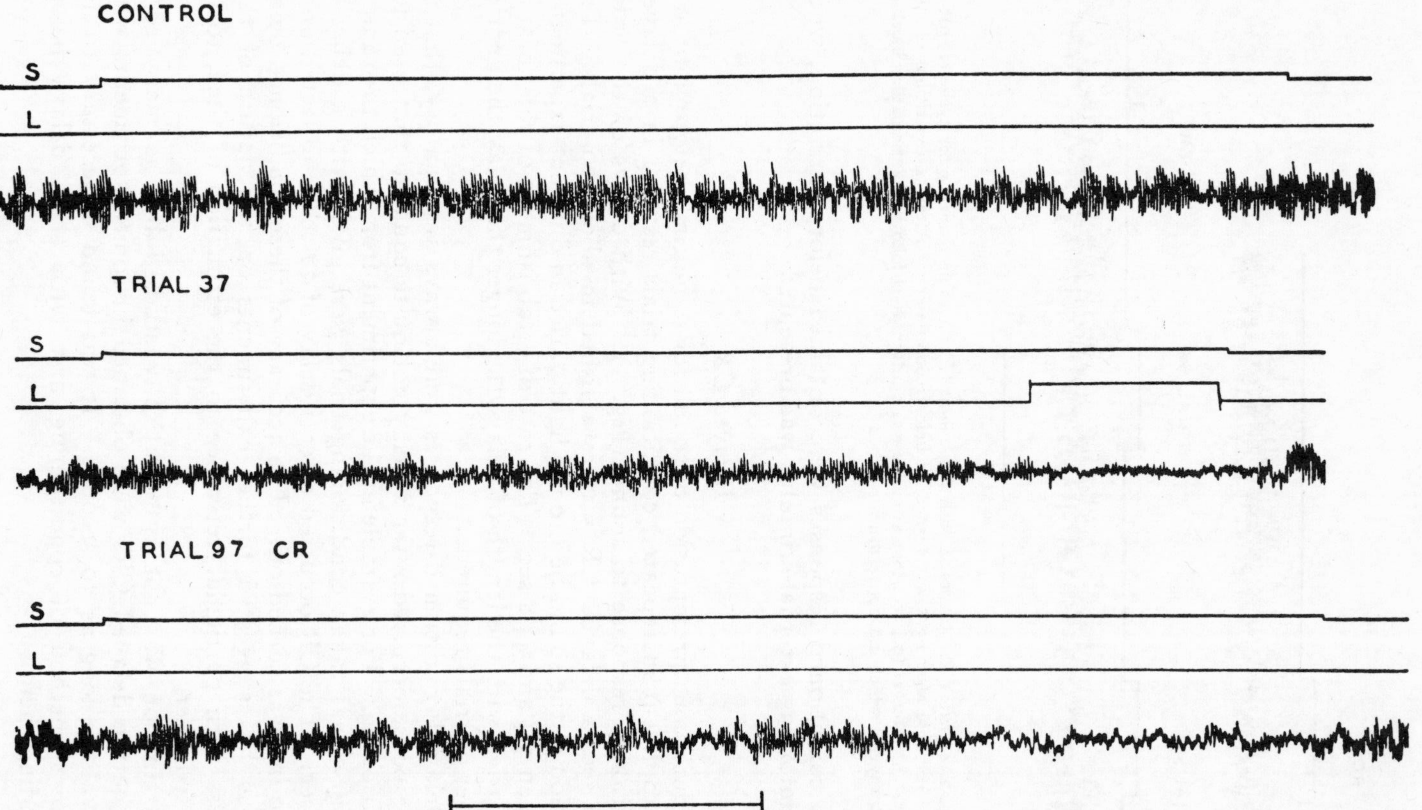

Fig. 3. *Delayed CR.* Control shows no effect of long continuous sound stimulus before conditioning. Trial 37 was conditioning with delay of 30 sec. Trial 97 was test trial showing delayed conditioned blocking of alpha rhythm. It was anticipatory.

Trace CR

A trace CR was established in about 20 trials by stimulating first with a 5 sec. sound (700 cycles) then allowing a 9 to 10 sec. interval to elapse before the unconditioned light stimulus of 5 sec. duration. Controls showed no blocking of the alpha rhythm at any definite interval following a sound stimulus before conditioning. After conditioning, however, there was no immediate response to the sound but after it was turned off a clear block of the alpha rhythm occurred following a delay slightly less than that used for conditioning. The CR was anticipatory as for all such responses involving time. Accuracy in time was less for trace responses than for delayed responses, this being due to the greater prevalence of early responses (as short as one half the trace interval).

Sleep

In both the delayed and trace CR experiments an effect similar to Pavlov's induction of sleep was sometimes observed after 2 or 3 unreinforced trials. The alpha rhythm dropped out, delta waves greatly increased in amount, and the manual response could not be elicited. It was not possible, however, to find the subject asleep in the room on entering. In the 3 cases where this occurred 2 admitted the possibility of having dozed off and one denied it. Whether or not this phenomenon was simply drowsiness brought on by long confinement in a resting position, or due to 'irradiation of inhibition' cannot at this point be ascertained.

Differential CR

The conditioned blocking of the alpha rhythm to sound of a specific frequency (500 cycles) or a differential CR, was produced by pairing the light stimulus only with this frequency and not with others (400 to 475). A simple CR was first established as described above using only the 500 cycle sound. This sound was always reinforced by light while a 400 cycle sound was not. After about 70 trials from the beginning of conditioning an alpha block occurred to the 500 cycle sound alone but not to the 400 alone (Fig. 5). This differentiation is not due to extinction since, in testing, the 400 cycle stimulus was presented first, with no response, followed by the conditioned response to 500 cycles.

After establishing a relatively gross differentiation (100 cycles) the frequency of the non-reinforced sound was approximated more closely to 500 cycles (the reinforced stimulus). Differential CR was obtained for frequencies as close as 475 and 500 cycles per second.

Differential Delayed CR

A delayed CR to a specific tone (700 cycles) was established in about 100 trials. At first a delayed CR to a 700 cycle tone was established as described above. The delay interval was 9.4 sec. Then a 500 cycle sound stimulus was introduced but never followed

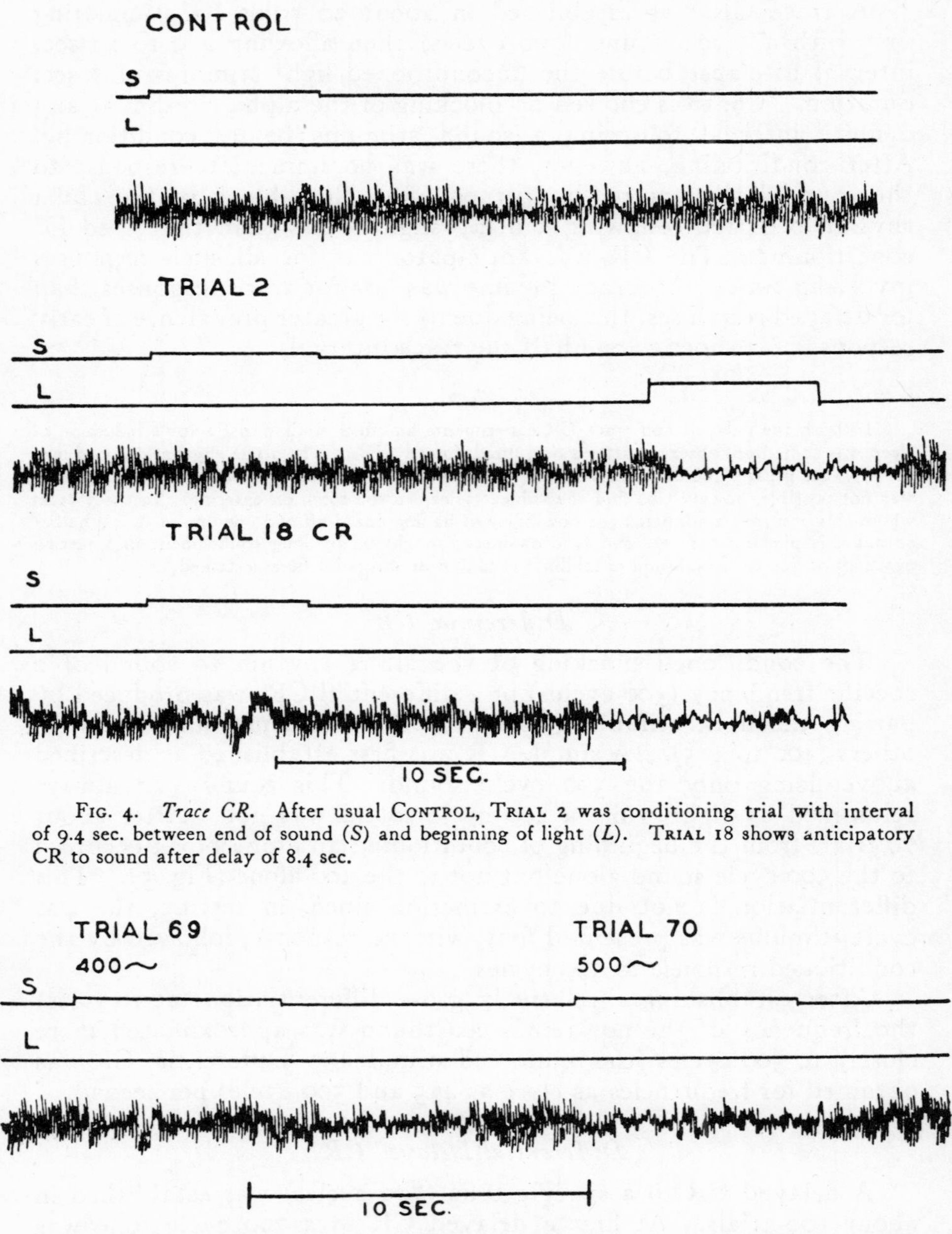

FIG. 4. *Trace CR.* After usual CONTROL, TRIAL 2 was conditioning trial with interval of 9.4 sec. between end of sound (*S*) and beginning of light (*L*). TRIAL 18 shows anticipatory CR to sound after delay of 8.4 sec.

FIG. 5. *Differential CR.* Simple and differential CR to 500 cycle tone has been established (see text). TRIAL 69 shows absence of response to 400 cycles followed by response to 500 cycles in TRIAL 70.

by light while the 700 cycle stimulus was always followed by light after a definite interval. This type of conditioning was hard to establish, as shown by the number of trials necessary, but a clear differential response to the 700 cycle tone after a delay of 9.4 sec. was obtained (Fig. 6).

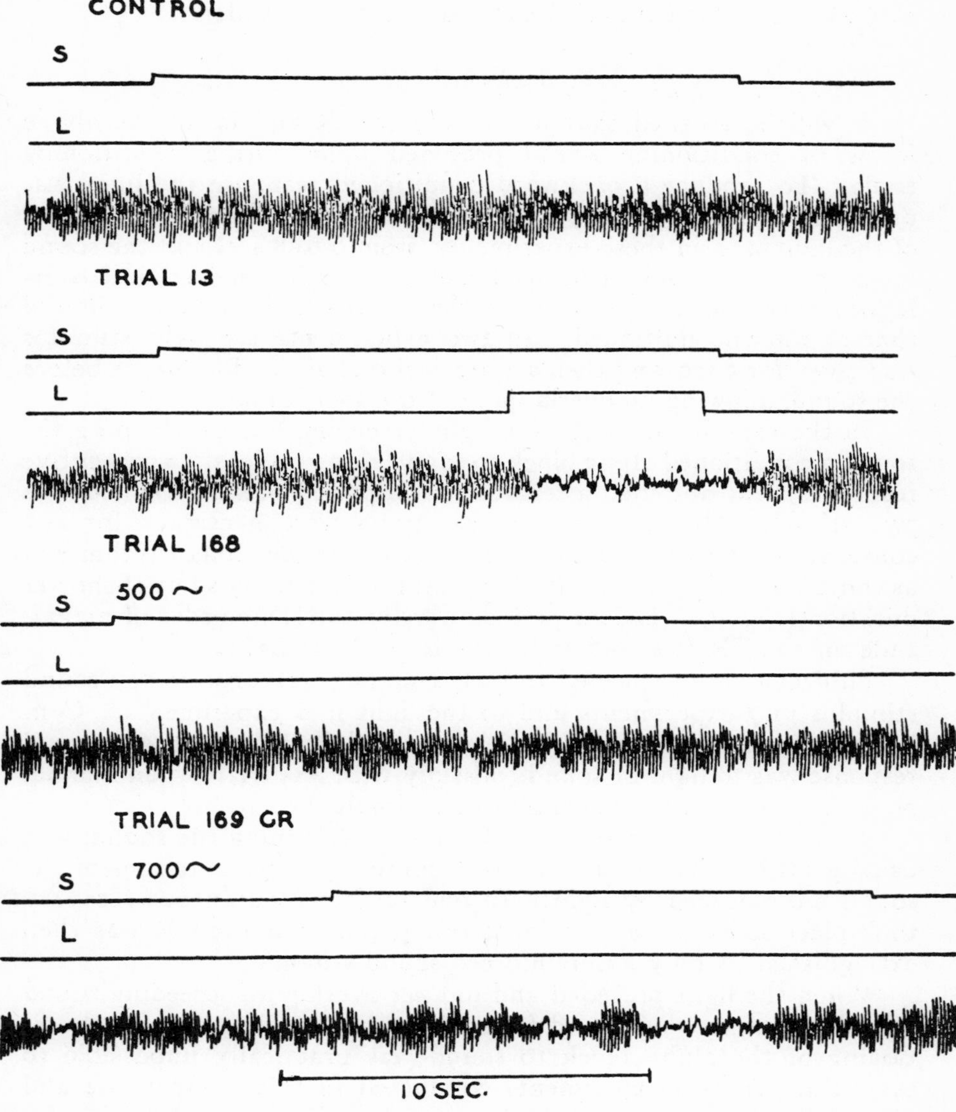

FIG. 6. *Differential Delayed CR.* CONTROL and delayed conditioning trial (TRIAL 13) are similar to those above except that 700 is then reinforced by light and 500 never paired with light. Test TRIAL 168 with 500 then caused no change in alpha waves (it did before differential conditioning) while 700 cycles in TRIAL 169 caused delayed blocking of alpha waves.

Before differential conditioning the 500 cycle tone produced immediately the same response as the conditioned 700 cycle tone. This was simple delayed conditioning to *sound.* There was no response to the 500 cycle stimulus after differentiation was completely established. The response to 500 was always tested before 700 to avoid the possible extinction effect of the reverse order.

Backward CR

It will be recalled that in the simple CR and in all the above forms of conditioning sound preceded light during conditioning trials. To produce a backward conditioned response the light was presented from 0.7 to 2.9 sec. before the sound. Since the duration of the light was, in these experiments, from 6 to 8 sec. and the sound about 5 sec., the unconditioned and the conditioned stimuli overlapped even though the onset of the conditioned stimulus followed that of the unconditioned. In two experiments the light stimulus was given for 5 sec. and then a 1 sec. interval allowed to elapse before the sound, allowing no overlapping of the two stimuli.

In the experiments with the light preceding but overlapping the sound a conditioned alpha block to sound alone (at least 2 consecutive responses) was obtained after about 100 paired presentations of the two stimuli. One hundred to 200 trials were necessary for 4–5 consecutive unreinforced CRs. Depression of the alpha rhythm was as complete in the CR as to the unconditioned stimulus but of shorter duration (Fig. 7). Latent periods were of the same order of magnitude for the CR to sound as for the response to light.

Subjects were requested to give a manual response to the sound stimulus in 7 experiments and to the light in 2 experiments. Conditioning seemed to occur with about equal facility whether manual response was to light or sound (possibly with less trials when manual response is to sound, the conditioned stimulus).

This type of backward CR, the light overlapping the sound, was usually extinguished in about 5 to 6 consecutive presentations of the sound unreinforced by light. In one subject spontaneous recovery took place following a 15 minute rest period, but the CR was then extinguished by only 2 unreinforced sound stimuli.

When the light preceded and did *not* overlap a succeeding sound stimulus, but was separated from it by about 1 sec., definite conditioning of the alpha block to sound was practically impossible to establish. In two experiments conducted in this manner 262 and 387 trials respectively failed to establish sufficient conditioning for 2 consecutive responses to sound alone. It should be noted that, in these experiments, the conditioned stimuli appeared during the

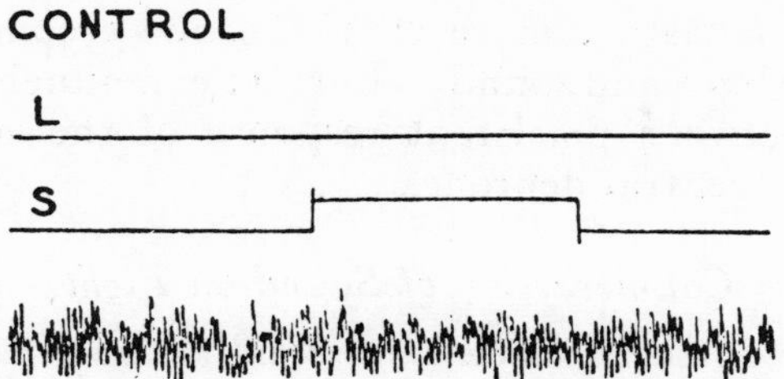

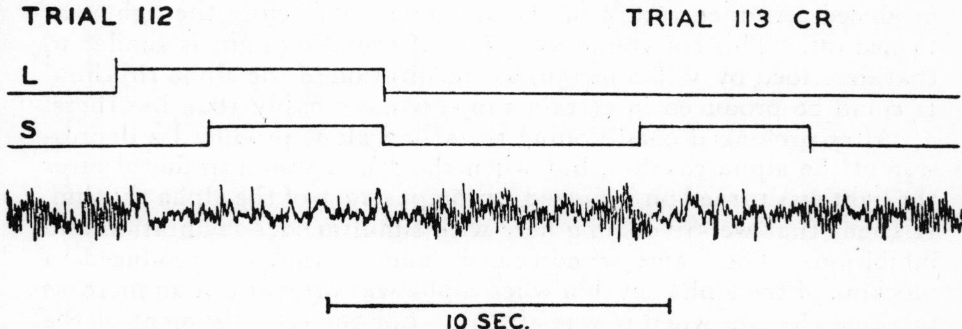

FIG. 7. *Backward CR.* Note inverted order of stimuli, light (*L*) preceeding sound (*S*) by 3.7 sec. in conditioning TRIAL 112. Test TRIAL 113 shows conditioned blocking of alpha rhythm to sound after backward conditioning.

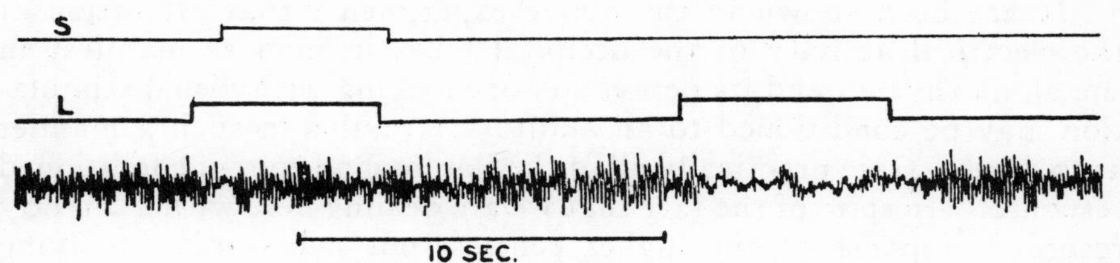

FIG. 8. *Counteraction Effect.* Sound (*S*) presented during a light stimulus (*L*) after backward conditioning caused a prompt return of the alpha rhythm. The succeeding light presented alone caused the usual blocking response.

presence of the alpha rhythm in the electroencephalogram. The sound was consequently associated in time with the *presence* rather than with the *absence* of alpha rhythm so that one might expect the conditioned sound to cause no depression or possibly an increase in alpha waves. That this may occur under certain conditions is shown in the following section.

Under the conditions of these experiments there were a few single depressions of the alpha rhythm to sound stimuli conditioned in this backward manner which may indicate certain excitatory processes of visual stimulation lasting more than the one second interval used here between the light and sound. Further experiments are necessary to distinguish between persistent response of the sense organ and actual 'traces' in central neurones.

Counteraction of Sound on Light

During the course of backward conditioning with overlapping stimuli it was observed that, after at least five conditioning trials, the introduction of a sound stimulus during an alpha block to light produced a sudden return of the alpha rhythm before the light was turned off. This counteraction effect of sound on light is similar to that described by Williams (**30**) as 'facilitation of the alpha rhythm.' It could be produced in certain subjects more easily than in others.

After prolonged conditioning the sound alone produced a depression of the alpha rhythm, but when the sound was introduced after the light was turned on it caused a prompt *return* of the alpha rhythm. It seems that we are dealing here with simultaneous facilitation and inhibition. The same conditioned sound stimulus produced a blocking of the alpha rhythm when alpha was present but an increase in alpha rhythm when it was absent. For the establishment of the counteraction effect, several trials are necessary as for conditioning.

Discussion

It has been shown in the above experiments that alteration of the electrical activity of the occipital lobes in man as manifest in, the alpha rhythm and its depression or blocking with visual stimulation may be conditioned to an auditory stimulus in such a manner as to demonstrate practically all of the Pavlovian types of conditioned response. In spite of the fact that we are dealing here with a directly recorded response of the higher centres, not necessarily involving peripheral effector systems, the alpha block conditioned to sound or to time shows the characteristics of a Skinner (**26**) Type II conditioned response. It is, however, very unstable: appearing with relatively little practice and showing rapid extinction. Some of this instability may be due to the fact that there was practically no motivation and the entire conditioning took place in one sitting. Further experiments with more prolonged training periods and stronger motivation would be necessary to examine the possibility of establishing a more stable response.

The success here obtained with backward conditioning would seem to indicate the value of the electroencephalograph as a sensitive instrument for the study of theoretical problems in conditioning. Pavlov's early view was that conditioning is impossible unless the conditioned stimulus precedes the unconditioned one, but in the Croonian Lecture (23) he stated:

> if the stimulus is introduced after the beginning of the activity, then, although as our present experiments seem to show, a conditioned reflex may also develop, it is insignificant and evanescent; on continuing the procedure, the stimulus, which in this connection we term the neutral agent, becomes inhibitory.

Switzer (27) obtained backward CR of the eyewink in very few trials, but Porter (24) was unable to produce it in 65 trials. Cason (6) using a completely passive method, was also unsuccessful. Bernstein (3) found very few backward CRs for hand withdrawal, as did also Wolfle (31). Woodworth (32) states:

> It is too early to assume a categorical conclusion on backward conditioning, but the concept is certainly not a safe one to assume in explaining the other phenomena of learning.

In reviewing the work on backward association Guthrie (11) expresses the view that backward conditioning is possible if the unconditioned response and the conditioned stimulus overlap. Since these conditions are usually difficult to produce experimentally, backward conditioning is ordinarily relatively unsuccessful. With the alpha rhythm block as a response it is possible to verify Guthrie's theory and to show that backward conditioning can be produced in about 10 times the number of trials necessary for simple CR if the stimuli overlap, but not with separated stimuli.

The counteraction of sound on light in these backward conditioning experiments may be analogous to Pavlov's 'inhibitory stimulus.' In the latter case salivation is arrested by an extraneous 'distracting' stimulus. In our experiments the arrest of the alpha rhythm during visual stimulation is released by an extraneous auditory stimulus but only *after conditioning.*

The counteraction effect apparently develops simultaneously with the conditioned blocking effect even though the one causes a return of the alpha rhythm and the other its disappearance. In our experiments the counteraction effect appeared earlier in the conditioning series than did the conditioned blocking reaction when the light preceded but overlapped the sound, but the two processes seemed to be developing together. If the blocking of the alpha rhythm by visual stimuli be considered due to desynchronization of the 'spontaneous' rhythm of the neurones involved by excitation from incoming afferent stimuli (Jasper, 15) (Bremer, 5) the return of the alpha rhythm by sound might be considered a form of 'inhibition.' Therefore, facilitation (conditioning) and inhibition seem to develop together in conditioning, a concept suggested by Hull (14) and recently given experimental verification and exact mathematical treatment by Graham and Gagné (10).

There is an increasing body of evidence to show that positive conditioning is primarily a facilitation process. Gerard, Marshall and Saul (9) have shown that the electrical response to sound, which is recorded from an auditory nucleus in the mid-brain, is increased by a simultaneous visual stimulus. Martino and Fulchignoni (20) trained a dog to give a conditioned blink response of the left eyelid to red light and of the right eyelid to violet light. On strychninization of the right occipital lobe red light produced clonic movements of the left eyelid followed by generalized epileptoid seizures. Violet light had no such effect. Strychninization of the left occipital lobe enabled the violet light to produce the same result in the left eyelid while red light was ineffective. Hilden (12) in an electromyographic study showed that conditioned hand withdrawal was gradually built up. First only a few motor units were seen to respond before sufficient conditioning for a visible movement.

Some form of facilitation process must also be operating when an auditory stimulus can be conditioned to produce a change in the electrical activity of the occipital lobes indistinguishable from that caused by the unconditioned visual stimulus. Repeated paired presentations of sound and light gradually lower the threshold of certain preexisting synaptic pathways, probably mid-brain, so that auditory stimuli temporarily become effective in desynchronization of the occipital alpha rhythm.

Conditioning studies of the electrical activity of the higher centres are not necessarily identified with those based upon effector responses. Many changes occur in the cortex which are not reflected in peripheral neurones as shown by Adrian and Moruzzi (2), Moruzzi (22), and Jasper and Andrews (16). A most striking demonstration is provided by epileptic discharges of the cortex, which when of moderate amplitude, produce no detectable motor response. However, further increase in cortical discharge is sufficient to cause adequate summation for excitation of motor neurones. Thus cortical activity as directly recorded is related to the 'stimulus-response' system, and principles of conditioning derived from brain potential studies may be applied to general problems of conditioning.

Advantages of electroencephalographic studies of conditioning are principally the facility and rapidity with which various types of CR may be demonstrated in man and its relative independence of voluntary influence. The chief disadvantage is the instability of the CR. Also unless a subject has an almost continuous alpha rhythm which is free from prolonged 'spontaneous' variations, and which responds clearly to visual stimuli, unequivocal conditioned responses cannot be obtained.

SUMMARY

1. The disappearance of the occipital alpha rhythm in man was conditioned to auditory stimuli in the classical Pavlovian manner. The following types of conditioned response were established: simple, cyclic, delayed, trace, differential, differential delayed, and backward. In some cases extinction and spontaneous recovery were shown.

2. After several trials with the backward conditioning technique a sound stimulus reestablished the alpha rhythm if introduced during its depression by light (counteraction effect). With further conditioning, the sound alone produced blocking of the alpha rhythm. This suggests simultaneous 'facilitation' and 'inhibition' processes during conditioning.

3. A state resembling sleep was produced during conditioning of long delay and trace intervals, that is, slow waves appeared in the electroencephalogram and the subjects failed to give the usual manual response.

4. The relation of electroencephalographic studies to general problems of conditioning was discussed.

(Manuscript received October 19, 1940)

REFERENCES

1. ADRIAN, E. D., AND MATTHEWS, B. H. C., The Berger Rhythm; potential changes from the occipital lobes in man, *Brain*, 1934, **57**, 355–385.
2. ADRIAN, E. D., AND MORUZZI, G., Impulses in the pyramidal tract, *J. Physiol.*, 1939, **97**, 153–199.
3. BERNSTEIN, A. L., Temporal factors in the formation of conditioned eyelid reactions in human subjects, *J. genet. Psychol.*, 1934, **10**, 173–197.
4. BOGOSLOVSKI, A. I., An attempt at creating sensory conditioned reflexes in humans, *J. exp. Psychol.*, 1937, **21**, 403–422.
5. BREMER, F., Parenté des diverses ondes électriques de l'écorce cérébrale, *C. R. de la Soc. de biol.*, 1938, **128**, 544–550.
6. CASON, H., Backward conditioned eyelid reactions, *J. exp. Psychol.*, 1935, **18**, 599–611.
7. CRUIKSHANK, R. M., Human occipital brain potentials as affected by intensity-duration variables of visual stimulation, *J. exp. Psychol.*, 1937, **21**, 625–641.
8. DURUP, G., AND FESSARD, A., L'électrencéphalogramme de L'homme, *L'année Psychol.*, 1935, **36**, 1–32.
9. GERARD, R. W., MARSHALL, W. H., AND SAUL, L., Electrical activity of the cat's brain, *Arch. Neurol. Psychiat.*, 1936, **36**, 675–738.
10. GRAHAM, C. H., AND GAGNÉ, R. M., The acquisition, extinction and spontaneous recovery of a conditioned operant response, *J. exp. Psychol.*, 1940, **26**, 251–280.
11. GUTHRIE, E. R., Association as a function of time interval, *Psychol. Rev.*, 1933, **40**, 355–367.
12. HILDEN, A. H., An action current study of the conditioned hand withdrawal, *Psychol. Mon.*, 1937, **49**, 172–204.
13. HILGARD, E. R., AND MARQUIS, D. G., *Conditioning and learning*, N. Y.: D. Appleton-Century Co., 1940.
14. HULL, C., Mind, mechanism, and adaptive behavior, *Psychol. Rev.*, 1937, **44**, 1–32.
15. JASPER, H. H., Cortical excitatory state and synchronism in the control of bioelectric autonomous rhythms, *Cold Spring Harbour Symposium, Quant. Biol.*, 1936, **4**, 320–338.
16. JASPER, H. H., AND ANDREWS, H. L., Brain potentials and voluntary muscle activity in man, *J. Neurophysiol.*, 1938, **1**, 87–100.

17. Jasper, H. H., and Cruikshank, R. M., Electroencephalography II: Visual stimulation and the after-image as affecting the occipital alpha rhythm, *J. gen. Psychol.*, 1937, **17**, 29–48.
18. Loomis, A. L., Harvey, E. N., and Hobart, G. A., Electrical potentials of the human brain, *J. exp. Psychol.*, 1936, **19**, 249–279.
19. Loomis, A. L., Harvey, E. N., and Hobart, G. A., Brain potentials during hypnosis, *Science*, 1936, **83**, 239–241.
20. Martino, G., and Fulchignoni, E., Uber die Bedeutung bedingter Reize (fur den Bahnungsprozess) bei der durch stychninisierung der occipitalen Rinde reflektorisch erzeugten Epilepsie, *Pflüg. Arch. ges. Physiol.*, 1938, **240**, 212–220.
21. Miller, J., and Cole, E., The influence of a voluntary reaction upon the development and the extinction of the conditioned eyelid reaction, *J. genet. Psychol.*, 1936, **48**, 405–440.
22. Moruzzi, G., Contribution a l'électrophysiologie du cortex moteur: facilitation, afterdischarge et épilepsie corticales, *Arch. internat. de Physiol.*, 1939, **49**, 33–100.
23. Pavlov, I. P., *Lectures on conditioned reflexes*, Tr. by W. H. Gantt, N. Y. Int. Publishers, 1928.
24. Porter, J. M., Jr., Backward conditioning of the eye-lid response, *J. exp. Psychol.*, 1938, **23**, 40–410.
25. Shurrager, P. S., and Culler, E. A., Conditioning in the spinal dog, *J. exp. Psychol.*, 1940, **26**, 133–159.
26. Skinner, B. F., Two types of conditioned reflex and a pseudo type, *J. gen. Psychol.*, 1935, **12**, 66–77.
27. Switzer, C. A., Backward conditioning of the lid reflex, *J. exp. Psychol.*, 1930, **13**, 76–97.
28. Travis, L. E., and Egan, J. P., Conditioning of the electrical response of the cortex, *J. exp. Psychol.*, 1938, **22**, 524–531.
29. Walter, W. G., The technique and application of electroencephalography, *J. Neurol. and Psychiat.*, 1938, 359–385.
30. Williams, A. C., Some psychological correlates of the electroencephalogram, *Arch. Psychol.*, 1939, pp. 48.
31. Wolfle, H. M., Conditioning as a function of the interval between the conditioned and the original stimulus, *J. gen. Psychol.*, 1932, **7**, 80–103.
32. Woodworth, R. S., *Experimental Psychology*, N. Y.: Henry Holt & Co., pp. 92–123, 1938.

55 Conditioned Responses of Hippocampal and other Neurons

J. Olds and T. Hirano

INTRODUCTION

Previous studies of neuron changes during conditioning have involved either surgical and paralytic restraints (Yoshii and Ogura 1960; Kamikawa *et al.* 1964) or free behavior in chronic animals (Buchwald *et al.* 1966; Travis and Sparks 1967; Ellison *et al.*, in press). The search for neural traces of critical changes in learning has thus been clouded by the surgical and paralytic procedures or confounded by the correlates of the new conditioned responses. There has also been confusion resulting from the sensitizing effects of the unconditioned stimulus (US) and from changes in attention level which occur as the conditioned stimulus (CS) becomes a sign of a more or less significant unconditioned stimulus. New techniques have permitted us to study conditioned neuron responses that were not correlated with new skeletal movements or with changing states of sensitization or arousal in animals that were neither anesthetized nor paralyzed.

Our object was to obtain evidence which might implicate structures of the hypothalamo-hippocampal system in temporary information storage and in anticipatory or other internal responses divorced from immediate skeletal and arousal patterns. Interest in the hippocampus and its related subcortical structures stemmed from several kinds of experiments which related it in one way or another to temporary information storage. There were first the observations indicating a failure of some information-holding functions in humans after hippocampal damage (Milner 1958). Second, there were difficulties (such as poor extinction, insufficient response variability, and excessive responding to the wrong stimulus) exhibited by animals with hippocampal lesions (Orbach *et al.* 1960; Isaacson and Wickelgren 1962; Niki 1967). Third, there were observations of hippocampal "theta" rhythms during certain phases of learning but not in the early period when the CS was entirely novel (Pickenhain and Klingberg 1967) and not in the later period if at that time the response to the stimulus was fully determined. Finally, there were stimulation studies implicating hippocampus and its related structures in the control of drives (MacLean *et al.* 1962), consummatory responses (Kawakami *et al.* 1967), and positive and negative reinforcements (Ursin *et al.* 1966) which play a major role in marking information for storage or erasure. It seemed that if temporary information storage did occur in hippocampus, then hippocampal neurons might display temporary responses in the course of conditioning; and if anticipatory responses did occur in drive or reward "centers" then these might be observed during the period between the application of the conditioned stimulus and the application of the unconditioned stimulus.

METHODS

The subjects were fifteen rats. Each was prepared with 6–8 indwelling micro-electrodes and a reference lead as described by Olds *et al.* (1969).

Recordings were made simultaneously from 4 probes during 2 sec intervals which included a 1 sec pre-stimulus interval and a 1 sec interval of auditory stimulation.

The CS was a 600 c/sec tone applied for a full

[1] These studies were supported by research grants from the U.S. Public Health Service.

second. The US was a 45 mg food pellet supplied by a noisy food magazine with which the animal was familiar.

During a 500 trial, 3 days pre-conditioning experiment, habituation and pseudo-conditioning procedures were simultaneously employed. That is, the tone was presented repeatedly on a random basis with an average inter-trial interval of 4 min; recordings were made prior to and during the stimulus and movement during a CS application caused its termination. The US was presented repeatedly on a different random schedule with the same average inter-trial interval; there was no correlation between the CS and the US.

During the following 300 trial, 2 days conditioning experiment, the conditions were the same in all respects except that the US was presented not on a schedule of its own but after the completion of each 1 sec CS application. During this period, movement during the CS period or the 1 sec pre-CS period caused a cancellation of the US.

During the pre-conditioning and the conditioning experiments, the animals received all their food from the magazine. During conditioning, the animal had to learn to stand still in order to be fed. The conditioning experiment was continuous with the pre-conditioning experiment that preceded it; that is, there was no pause in the random 4 min schedule at the time of changing from the pre-conditioning to the conditioning program.

Neuronal activity was discriminated and counted automatically by a solid state device which has been described previosuly (Olds 1965, 1967; see also Olds *et al.* 1969).

The data presented are for the last 50 trials of the 500 trial pre-conditioning experiment, and for the last 50 trials of the 300 trial conditioning experiment; a comparison of these data taken before conditioning and after conditioning permitted detection of the changes caused by the conditioning procedure.

The data are presented in terms of the changes in spike firing rate caused by the auditory stimulus; these were quantified in terms of an early "response" (during the first 300 msec of stimulation) and a late "response" (during the last 500 msec of stimulation). In both cases the firing rates during early or late intervals were divided by firing rates during the last 500 msec period before stimulation and thus converted to a percentage of the corresponding base rate.

After the experiments, brains were sectioned and stained so that recording points could be localized.

RESULTS

Evoked responses

There were changes in spike firing frequencies in response to the auditory stimulus before conditioning (see Table I). One-third of the units showed 20% increments in rate during the first 300 msec. This response was short lasting; during the last half of the stimulus only 3% of the neurons (out of a total of 71) showed such increments. There were decrements during both the early and late periods in hippocampus and preoptic area.

The largest responses before conditioning were in dorsal reticular formation, where 43% of the neurons exhibited rate increments amounting to 40% during the first 300 msec. In the ventral reticular formation, hypothalamus, preoptic area and hippocampus the proportion of cases with such increments was much lower (from 5 to 10%).

Conditioned responses

Conditioning caused many neurons which in pre-conditioning tests showed small increments (less than 20%) or decrements during the first 300 msec to exhibit increments above 20% or even above 40% after conditioning. This was particularly marked in the hippocampal and thalamic groups where conditioning caused at least a 5-fold increase in the number of units with firing rates augmented by 40% or more. It doubled the number of such large increments in hypothalamus and preoptic area and reduced the number in reticular formation where these large increments were most numerous in the first place (see Table I).

Conditioning also caused a large increase in the number of neurons whose rates were 20 or 40% above base rate during the last 500 msec of auditory stimulation (see Fig. 1). This was apparent in all groups except in those from the ventral reticular formation. The number of cases with "20 or 40%" responses during the late period increased 10-fold: before conditioning there were

TABLE I

The number of units showing increments and decrements of different sizes caused by auditory stimulation before and after conditioning

Brain area	*N* (units)	Rate increments of 40% or more		Rate increments of 20% to 40%		Rate changes of −5% to +20%		Rate decrements of 6% or more		Number of rats
		Before cond.	After cond.	Before cond.	After cond.	Before cond.	After cond.	Before cond.	After cond.	
First 300 milliseconds										
Dorsal reticular	7	3	2	2	1	2	4	0	0	5
Ventral reticular	10	1	1	3	6	6	3	0	0	8
Thalamus	8	0	3	4	4	4	1	0	0	6
Hypothalamus	10	1	2	2	2	6	5	1	1	6
Preoptic	17	1	2	1	2	11	10	4	3	13
Hippocampus	19	1	5	4	2	10	9	4	3	8
Total	71	7	15	16	17	39	32	9	7	15
Percentage	100	9.9	21.5	22.5	23.9	54.9	45.1	12.7	9.9	
Last 500 milliseconds										
Dorsal reticular	7	0	1	0	2	7	4	0	0	5
Ventral reticular	10	0	0	0	0	9	1	10	0	8
Thalamus	8	0	1	0	4	8	3	0	0	6
Hypothalamus	10	0	1	1	3	7	5	2	1	6
Preoptic	17	0	1	1	2	13	9	3	5	13
Hippocampus	19	0	2	0	4	9	9	10	4	8
Total	71	0	6	2	15	53	40	16	10	15
Percentage	100	0	8.5	2.8	21.1	74.6	56.3	22.5	14.1	

2 such units among the 71 tested, after conditioning there were 21. The relative incidence of units yielding firing rates of 20% or more above baseline during the last 500 msec was increased by conditioning from 0 to 62% in the thalamic group, 0 to 43% in the dorsal reticular group, 10 to 40% in hypothalamus, 0 to 30% in hippocampus, and from 6 to 18% in preoptic area (see Table I).

Even though hippocampus was not at the top of the list of areas yielding large responses after conditioning, the largest *changes* in response caused by conditioning appeared in hippocampus. This is because there were many decremental responses in hippocampus prior to conditioning, and many of these were changed to incremental responses by conditioning. The changes were quantified by subtracting each score after conditioning from the corresponding score after pseudoconditioning control tests (see Table II). There were average increments in the early response in all areas except the dorsal reticular formation where an average decrement of 30% appeared (see Fig. 2). The largest early-response

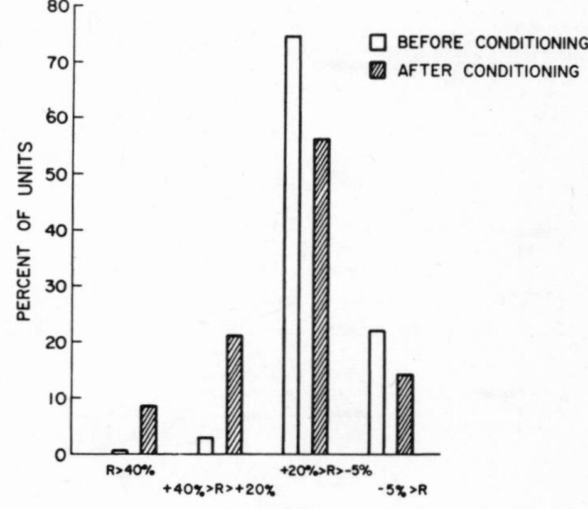

Fig. 1

Changes in firing rates in last 500 msec of stimulation. Incidence of units with supra-normal firing rates of different percentages above baseline before and after conditioning. Symbols are as follows: R > 40% = unit firing rates exceeded control rates by 40% or more; +40% > R > +20% = rates were 20–40% above controls; +20% > R > −5% = rates ranged from 5% below to 20% above controls; −5% > R = rates were 5% or more below control rates.

TABLE II

The mean increments or decrements in percentages caused by auditory stimulation before and after conditioning, with changes due to conditioning

Brain area	N (units)	Early response (first 300 msec) Before cond.	After cond.	Change	Late response (last 500 msec) Before cond.	After cond.	Change	Number of rats
Dorsal reticular	7	51 ± 44	21 ± 19	−30[h] ± 43	10 ± 6	16 ± 20	6 ± 20	5
Ventral reticular	10	20 ± 15	26 ± 13	6 ± 14	4 ± 8	7 ± 7	3[h] ± 3	8
Thalamus	8	21 ± 11	32[a] ± 14	11 ± 9	6 ± 6	18 ± 16	13 ± 16	6
Hypothalamus	10	13 ± 14	22 ± 31	9 ± 36	4 ± 9	11 ± 26	8 ± 32	6
Preoptic	17	7 ± 27	10 ± 22	3 ± 27	1 ± 12	5 ± 19	4[h] ± 23	13
Hippocampus	19	8 ± 21	20 ± 25	13 ± 30	−9 ± 13	12[b] ± 24	20 ± 25	8

[a] Significantly different from pre-conditioning score, 0.05 level, Wilcoxon paired Matched Test.
[b] Significantly different from pre-conditioning score, 0.01 level, Wilcoxon paired Matched Test.
[h] Significantly different from hippocampal group, 0.05 level, Mann-Whitney U Test (Siegel 1956).

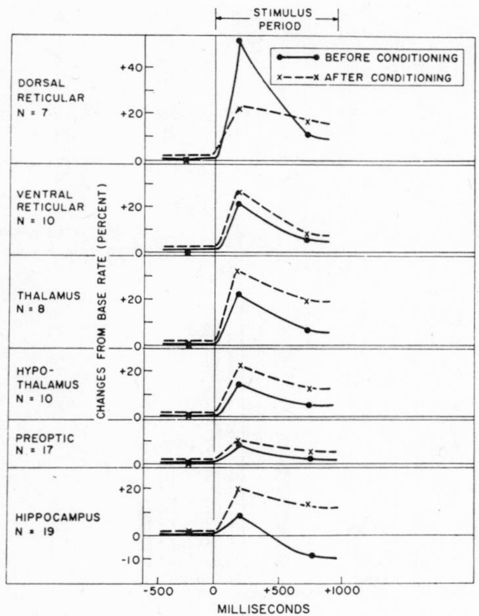

Fig. 2

Averaged incremental or decremental responses before and after conditioning. For each brain area, the rate from baseline for the first 300 and the last 500 msec was computed as a percentage of the baseline rate for the 500 msec prior to stimulation. The average of these scores for the last 50 trials of pseudo-conditioning (before conditioning) and the last 50 trials of conditioning (after conditioning) were used. These scores were averaged for all units in each brain area.

increments averaging 13% appeared in hippocampus; these were reliably different from those of dorsal reticular formation on this score. In second place was an 11% increment in the thalamic group.

All areas exhibited an average increment in the late response (see Fig. 2); however, this increment was small in reticular formation, hypothalamus, and preoptic area where it ranged from 3 to 8%. It was larger in thalamus and hippocampus, amouting to 13 and 20% respectively. The hippocampal change in the late response was highly significant by statistical tests and it was significantly larger than the changes in ventral reticular formation and the preoptic group.

Histological material showed that the hippocampal units were all derived from the ridge of hippocampal pyramids, usually near the boundary of the CA-1 and the CA-3 fields; the dorsal reticular units were about 1 mm lateral to the central grey at its widest point and almost on the boundary line between diencephalon and mesencephalon (see Fig. 3).

DISCUSSION

During conditioning, unit responses consisting of an increase in firing rate were increased in all areas but one, and decremental responses were attenuated or reversed. This was true for early components in all areas but the dorsal reticular formation, and for late components in all parts of the brain tested. Neuronal firing

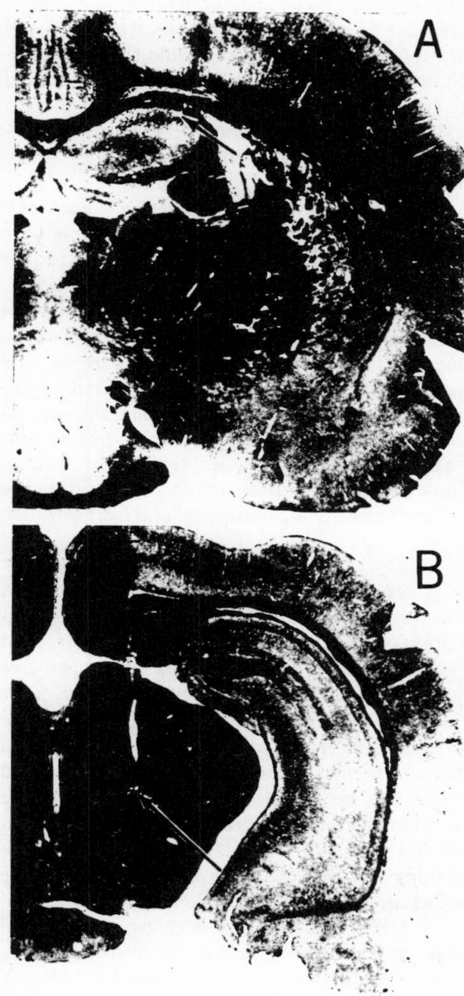

Fig. 3
Histological preparations showing location of probes (*A*: hippocampal, *B*: reticular) whose data are shown in Fig. 2.

rates in hippocampus and thalamus presented, however, the greatest change; those in reticular formation were least affected. As to changes during the course of the 1 sec waiting period, that is during the CS–US interval, there was no special tendency for neurons to become more active toward the point in time where the US would be presented. On the contrary, there was even at the end of conditioning a slight residuum of the initial trend in the opposite direction, that is for neurons to be more active at the onset of the CS than during its last half second. The changes induced by conditioning were, however, most marked in the late period. Many below baseline firing rates which characterized this period after habituation and pseudo-conditioning disappeared and often the early excitatory response lasted into the late period.

Because neuron responses were changed toward an increase in firing rate, suggestive of an excitatory trend, whereas the behavior called for was inhibitory, it would appear that the neuronal activity was not correlated with the amount of overt behavior or with newly learned skeletal movements. Because the stimulus was presented on a random time schedule, catching the animal unaware, it is unlikely that the neuronal activity changes were correlated with orientation responses or postures related to the conditioning. The question arises whether these changes might have been correlated with a tonic muscular state which mediated the behavioral inhibition; this is unlikely as the adoption of a tensed pose, when it was observed after the onset of a trial, regularly tripped the movement detector and caused cancellation of the trial and the data. Because data were taken only for 2 sec periods during which no movement was exhibited, it might be supposed that the data were taken mainly during sleep periods. That the animals were not likely asleep during the conditioning trials was attested by the vigorous capture-response which occurred when the food magazine discharged. In any event, the changes observed were largely in the direction of increased responsiveness to peripheral stimulation, and changes from waking to sleep are regularly in the opposite direction (Evarts 1960). Thus there seems to be little evidence to suggest that any of the new responses observed are correlates of movements, postures, inhibitory processes, or states of sleep. One could, therefore, conclude tentatively that these changes in neuronal firing are correlates of some internal vegetative or preparatory responses related to food, or are correlates of learned motivational or mnemonic traces evoked by the CS. Because the hippocampal responses were suggestively similar to some of those observed during food waiting in a previous experiment (Olds *et al.* 1969), it seems that they might be interpreted in analogy to those data. From these, two characteristics of the hippocampal neuronal activity are relevant: (a) in view of the often large differences between food-waiting and water-waiting response the neurons appeared to be involved in specific anticipatory representations rather than in general activation, inhibition, or arousal; (b) a neuron might participate in an anticipatory response specifically related to food one day and to one specifically related to water several days later. This suggests that if these neuronal activity states are specific representations, they might also be temporary representations. It seems possible, that is, that the neurons involved, instead of bearing some innate relation to food preparatory responses, might be implicated as relatively neutral temporary memory registers which could hold different stimulus–response–reward configurations on different days.

In the quoted set of experiments, interesting accelerating response patterns were observed in the reticular formation, characterized by a crescendo during the last second prior to US application. The largest accelerating response in reticular formation in the present study was very small (see Fig. 4) and one should therefore conclude that such accelerations were more related to the active pedal behavior or the absence of an explicit CS, which differentiated the two experimental studies, than to the "expectancies" which were common to the two experiments.

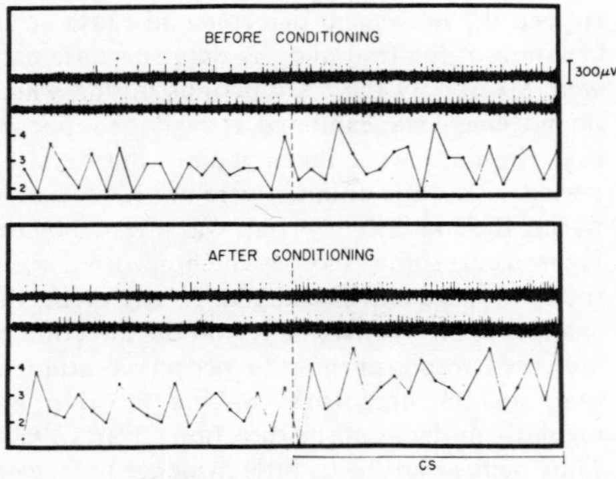

Fig. 4
Dorsal reticular unit activity before and after conditioning. Changes are exemplified by two sample traces (above) and by the average rates for successive 50 msec intervals for each group of 50 trials (below). In the averaged response curve, each point stands for the averaged response rate for a given 50 msec interval stated in arbitrary units; *i.e.*, the digits along the Y axis are some multiple or submultiple of the observed average rates.

If temporary representations are "localized" in hippocampus, what relation might these bear to the organization of behavior? One possibility is that mnemonic traces of a relatively emotional, automatic and temporary character dwell in hippocampal regions whereas another and newer memory system occupies some part of the neocortex. Emotional, automatic and temporary memories might be supposed to characterize the whole memory store among more primitive animals; and in higher species, they might have an important role in constructing and in gating the more unemotional, disengaged and lasting traces.

SUMMARY

Changes in the firing rates of neurons in various subcortical structures induced by application of a 1 sec auditory stimulus were recorded in unrestrained animals during movement-free intervals. Pseudo-conditioning trials (500) were followed by conditioning trials (300). During the latter, a food pellet was presented to the animal by a dispenser after every stimulus, provided there was no detectable movement during the stimulus period. The difference between the response rates before conditioning and after conditioning was considered to be a measure of the effect induced by the conditioning procedure.

1. Except for the early response in the dorsal reticular formation, responses consisting of incremental firing rate were increased in size and number and decremental responses were attenuated or reversed. Even in the dorsal reticular formation the neuronal activity rate during the last half of the stimulus was augmented.

2. Before conditioning an increment in the firing rate of neurons appeared in most areas during the first 300 msec of stimulation after which the firing rate returned almost to control levels or even systematically below control levels in some areas. In most cases, conditioning procedures caused the early increment to be augmented

and the neuronal rate to stay elevated above control levels during the remainder of the 1 sec stimulus (thus reversing the decremental tendency where it had appeared).

3. Incremental changes were largest in hippocampus. These changes were, on the basis of deductive arguments, considered to be independent of skeletal and attentional components of the conditioned response on the one hand, and unrelated to sleep and waking or inhibitory processes on the other. It seemed possible therefore that the changes might be related to a short term or temporary trace of a motivationally significant event.

RÉSUMÉ

RÉPONSES CONDITIONNÉES DES NEURONES HIPPOCAMPIQUES ET AUTRES

Pendant des périodes sans mouvements chez des animaux non maintenus, sont enregistrées au niveau des diverses structures sous-corticales, des modifications du taux de décharge neuronique, induites par un stimulus auditif à 1 sec. Des essais pseudo-conditionnels (500) sont suivis d'essais conditionnels (300). Au cours de ces derniers, une tablette de nourriture est présentée à l'animal par un distributeur après chaque stimulus, à la condition qu'il n'y ait aucun mouvement décelable pendant la période de stimulation. La différence entre les taux de réponse avant et après conditionnement est prise comme mesure de l'effet induit par le processus de conditionnement.

1. A part la réponse initiale du niveau de la formation réticulaire dorsale, les réponses consistant en accroissement de la fréquence de décharge augmentent en taille et en nombre, alors que celles qui s'accompagnent d'une diminution de la fréquence sont atténuées ou inversées. Même dans la formation réticulaire dorsale, le taux d'activité neuronique est augmenté pendant la dernière moitié du stimulus.

2. Avant le conditionnement, un accroissement de la fréquence de décharge des neurones s'observe dans la plupart des aires au cours des premières 300 msec de stimulation, après lesquelles le taux de dècharge revient à peu près aux niveaux de contrôle ou même systématiquement au-dessous de ces niveaux dans certaines aires. Dans la plupart des cas, les processus de conditionnement déterminent l'augmentation de l'accroissement initial, et le fait que le taux de décharge neuronique reste au-dessus des niveaux de contrôle pendant le reste du stimulus à 1 sec (inversant ainsi la tendance décroissante là où elle était apparue).

3. Les variations de l'accroissement sont les plus grandes dans l'hippocampe. Ces changements sont, sur la base d'arguments déductifs, considérés comme indépendants des composantes physiques et d'attention de la réponse conditionnée d'une part et d'autre part, sans relation avec les processus de sommeil et de veille, ou d'inhibition. Il semble ainsi possible que ces changements puissent être liés à une trace à court terme ou temporaire d'un événement significatif de motivation.

The authors are grateful for technical assistance to G. Baldrighi, W. E. Wetzel, H. J. Frey and W. J. Allan, and for computer programming to P. J. Best.

REFERENCES

BUCHWALD, J. S., HALAS, E. S. and SCHRAMM, S. Changes in cortical and subcortical unit activity during behavioral conditioning. *Physiol. and Behavior*, **1966**, *1*: 11–22.

ELLISON, G. D., HUMPHREY, C. L. and FEENEY, D. Some electro-physiological correlates of two behavioral states. *J. comp. physiol. Psychol.*, (in press).

EVARTS, E. V. Effects of sleep and waking on spontaneous and evoked discharges of single units in visual cortex. *Fed. Proc.*, **1960**, *19*: 828–837.

ISAACSON, R. L. and WICKELGREN, W. O. Hippocampal ablation and passive avoidance. *Science*, **1962**, *138*: 1104–1106.

KAMIKAWA, K., MCILWAIN, J. T. and ADEY, W. R. Response patterns of thalamic neurons during classical conditioning. *Electroenceph. clin. Neurophysiol.*, **1964**, *17*: 485–496.

KAWAKAMI, M., SETO, K., TERASAWA, E. and YOSHIDA, K. Mechanisms in the limbic system controlling reproductive functions of the ovary with special reference to the positive feedback of progestin to the hippocampus. In W. R. ADEY and T. TOKIZANE (Eds.), *Structure and function of the limbic system. Progress in brain research, Vol. 27.* Elsevier, Amsterdam, **1967**: 69–102.

MACLEAN, P. D., DENNISTON, R. H., DUA, S. and PLOOG, D. W. Hippocampal changes with brain stimulation eliciting penile erection. In *Physiologie de l'Hippocampe. Colloques Internationaux du Centre National de la Recherche Scientifique, No. 107.* CNRS, Paris, **1962**: 492–510.

MILNER, B. Psychological defects produced by temporal lobe excision. *Res. Pub. Ass. nerv. ment. Dis.*, **1958,** *36*: 244–257.

NIKI, H. Effects of hippocampal ablation on learning in the rat. In W. R. ADEY and T. TOKIZANE (Eds.), *Structure and function of the limbic system. Progress in brain research, Vol. 27.* Elsevier, Amsterdam, **1967**: 305–317.

OLDS, J. Operant conditioning of single unit responses. *Excerpta med., int. Congr. Series No. 87,* **1965**: 372–380.

OLDS, J. The limbic system and behavioral reinforcement. In W. R. ADEY and T. TOKIZANE (Eds.), *Structure and function of the limbic system. Progress in brain research, Vol. 27.* Elsevier, Amsterdam, **1967**: 144–167.

OLDS, J., MINK, W. D. and BEST, P. J. Single unit patterns during anticipatory behavior. *Electroenceph. clin. Neurophysiol.*, **1969,** *26*: 144–158.

ORBACH, J., MILNER, B. and RASMUSSEN, T. Learning and retention in monkeys after amygdala-hippocampus resection. *Arch. Neurol. (Chic.)*, **1960,** *3*: 230–251.

PICKENHAIN, L. and KLINGBERG, F. Hippocampal slow wave activity as a correlate of basic behavioral mechanisms in the rat. In W. R. ADEY and T. TOKIZANE (Eds.), *Structure and function of the limbic system. Progress in brain research, Vol. 27.* Elsevier, Amsterdam, **1967**: 218–227.

SIEGEL, S. *Nonparametric statistics for the behavioral sciences.* McGraw-Hill, New York, **1956,** 312 p.

TRAVIS, R. P. and SPARKS, D. L. Changes in unit activity during stimuli associated with food and shock reinforcement. *Physiol. and Behavior*, **1967,** *2*: 171–177.

URSIN, R., URSIN, H. and OLDS, J. Self-stimulation of hippocampus in rats. *J. comp. physiol. Psychol.*, **1966,** *61*: 353–359.

YOSHII, N. and OGURA, N. Studies on the unit discharge of brainstem reticular formation in the cat. *Med. J. Osaka Univ.*, **1960,** *11*: 1–17.

56

The Observable Unconscious and the Inferable Conscious in Current Soviet Psychophysiology

Gregory Razran

Of the approximately 5,000 reports of experiments on classical conditioning and closely related phenomena that have come out of Russian laboratories since the beginning of the century, approximately 1,500 of which have appeared since 1950, roughly 300—almost all of the since-1950 group—stand out with a particular salience. They stand out, first, because, with only very few exceptions, they are not being duplicated outside the Soviet Union and indeed are hardly known outside it. But more importantly, they stand out because of their own intrinsic worth, in offering highly significant methods and data toward the solution of one of the most vexing problems of contemporary research and thought: the problem of unconscious-conscious interactions and influences in all their applied and systematic ramifications. Somewhat summarily, I would say that these 300 recent Russian experiments contribute greatly toward making the Unconscious observable instead of inferable and the Conscious inferable whether it is or is not observable.

The experiments that I have in mind fall into three separate groups: interoceptive conditioning, semantic conditioning, and studies in the orienting reflex—each group a true supplement to the two others in unraveling and fixing the common problem in its own area. Studies of interoceptive conditioning were begun in the Soviet Union in 1928 but had not gained much empirical momentum until the late '40s; [2] studies of semantic conditioning and of the orienting reflex in its present form are almost wholly a product of Soviet research and thought since 1950.

Interoceptive Conditioning

Definition and Classification

Interoceptive conditioning may best be defined as classical conditioning in which either the conditioned stimulus (CS) or the unconditioned stimulus

[1] Research done under Grant M-2196 of the National Institute of Mental Health, National Institutes of Health, Public Health Service, United States Department of Health, Education, and Welfare.

[2] Bykov's book *The Cerebral Cortex and the Internal Organs,* translated by W. H. Gantt into English in 1957 (Chemical Publishing Co., New York), first appeared in Russian in 1943 and was never revised; the book thus contains reports of only the early experiments, about 10% of the present total.

(US) or both are delivered directly to the mucosa of some specific viscus. When both stimuli are interoceptive, the conditioning might be called intero-interoceptive. When only the conditioned stimulus is interoceptive but the unconditioned one is exteroceptive, the appropriate name might be interoexteroceptive conditioning, while the designation of exterointeroceptive conditioning might be reserved for situations in which only the unconditioned stimulus is interoceptive. Three varieties thus result, of which the second has been most studied, and is to a large extent the most instructive with respect to both theory and practice.

Moreover, the first two varieties, those of interointeroceptive and intero-exteroceptive conditioning, differ very importantly from the third variety of exterointeroceptive conditioning in that only the first two are true cases of interoceptive signalization. That is to say, only they represent a kind of conditioning in which the viscera, the so-called *milieu interieur,* become the signalizers, or initiators and conveyors, of the acquired conditioned information. The third variety, that of extero-interoceptive conditioning, is different. For in it, though the signaled event is interoceptive, the signalizing is exteroceptive: the viscera or the *milieu interieur* become only the direct receivers of the conditioned information, the signalization or transmission remaining embedded in the external stimuli, the *milieu exterieur.* And to complete the roster, one should add that conventional classical conditioning is wholly exteroceptive or exteroexteroceptive, and that it is so even when it involves stimulation of the mucosa of the oral and nasal cavities.[3]

A more detailed classification of interoceptive conditioning (for that matter, also of exteroceptive conditioning) must take further account of not only the type of stimulus but also the type of reaction conditioned, whether it be visceral, skeletal, or, in the case of human subjects, also sensory-verbal; and it should consider the special cases of exteroceptive verbal stimuli in exterointero- and interoexteroceptive conditioning. A total of 15 varieties of interoceptive conditioning is thus possible, and almost all have been demonstrated experimentally in Russian laboratories. Sample illustrative experiments will be presented in a later section.

General Methodology

With animal subjects, the methods of direct delivery of interoceptive stimulations have involved, as might be suspected, the formation of fistulas in the experimental viscus *in situ* and, when needed, also the surgical exteriorization of the viscus. Space forbids discussion of the Russians' special surgical methods and techniques in the area, of which numerous reports have appeared in recent years. Speranskaya's (1953) *Operative Methods and the Conduct of Chronic Physiological Experiments in Dogs,* the Methods sections in the *Fiziologischesky Zhurnal SSSR im. I. M. Sechenova,* and the *Byulleten Eksperimental'noy Biologii i Meditsiny* (available in English since 1956) are recommended. However, the nature and the manner of delivering the interoceptive stimuli need of course be considered. The most typical stimuli are distentions of the lumen of the viscus through inserted fine-walled

[3] When the conditioned or the unconditioned stimulus is a drug that produces only visceral changes, the conditioning should still be classed as exteroceptive because of the exteroception involved in the administration of the drug. However, when this administration factor is ruled out through repeated injections of a physiological solution, the conditioning is interoceptive.

rubber balloons inflated with air or water. The resulting pressures are almost always controlled and recorded on manometers, and the distentions are calibrated, and often further varied with respect to rhythm of delivery, rate of onset, and sometimes also extent of area stimulated (through the use of balloons of different sizes).

Balloons filled with water at different temperatures are frequently further utilized as thermal stimuli, although, when the lumen of the viscus is small, direct luminal irrigation through inflow-outflow tubes seems to be clearly preferred. Irrigation is also the common means of administering chemical stimuli, a large variety of which are being used. Other forms of stimulation are: scratching rhythmically the mucosa with a Pavlov-type stimulator, jets of air or water as tactile stimuli, and electrical stimulations. A standard procedure of interoceptive conditioning has been the differentiation of its conditioned stimuli from related stimuli: that is, continually reinforcing the conditioned stimuli while presenting the related stimuli without reinforcement until the latter cease to produce any conditioning (or produce it to only an insignificant degree).

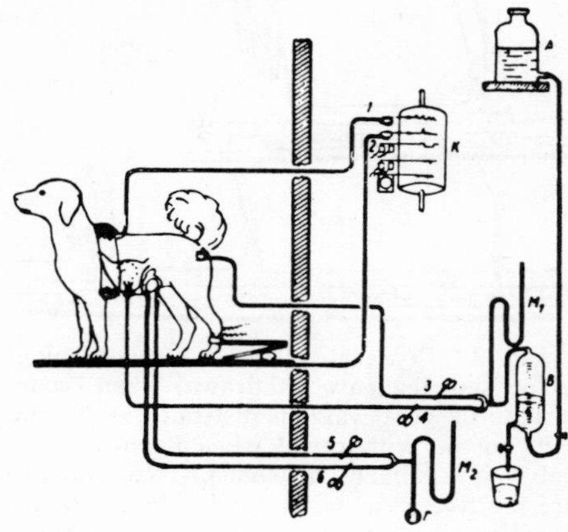

FIG. 1. Conditioning of dogs in Ayrapetyants' laboratory. (Salivary fistula and general arrangement for exteroceptive food conditioning not shown. A—water reservoir; B—graduated vessel; M_1—manometer to measure intragastric pressure; M_2—manometer to measure pressure in ileocecal region; Γ—bulb to inflate the ileocecum; K—kymograph; 1—tube connecting respiratory cuff with Marey tambour; 2—tube connecting apparatus registering paw movement with Marey tambour; 3—tube for inflating rectum; 4—tube for inflating the balloon in the dog's stomach; 5, 6—tubes connecting manometer M_2 with the balloons in the ileocecal region, one balloon on the ileal side of the ileocecal valve and one on the cecal side.)

Figure 1, taken from a report of an experiment in Ayrapetyants' laboratory (Moiseyeva, 1952, p. 95), offers a representative diagram of current Russian practice in conditioned "education" of dogs. As compared with his predecessor of a generation ago, the dog in the figure is equipped not with just one fistula for salivation but with five fistulas: one gastric; one rectal; two ileocecal, one on the ileal and one on the cecal side of the ileocecal valve; and a salivary fistula mentioned in the text but not shown in the figure. (The figure omits not only the salivary fistula but also the entire arrangement of the exteroceptive food conditioning; its intention is presumably to illustrate only the additions to the traditional Pavlov diagrams of dog conditioning.) Moreover, as seen in the figure, provision is made also for recording the dog's respiration and paw movements in response to electric shock and appropriate conditioned stimuli; indeed it is not uncommon for present-day Russian experimenters to register, during conditioning, in addition, the animals' electroencephalographic, electrocardiographic, plethysmographic, psychogalvanic (PGR), oculomotor, and even electroretinographic and pupillographic changes.

The ileocecal fistula in the figure is of special significance in this particular

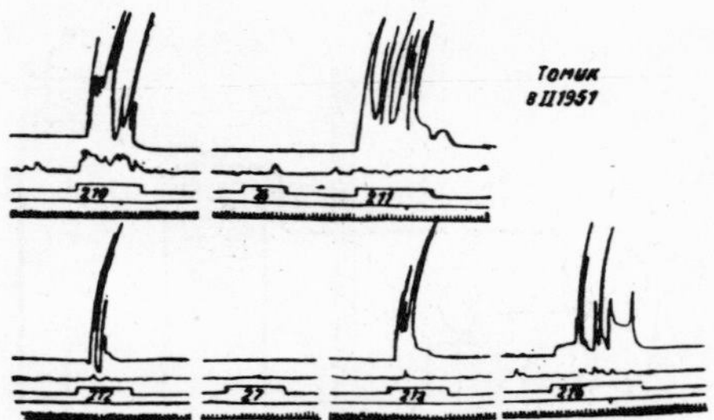

FIG. 2. Ileocecal differential conditioning. (Dog showing paw-withdrawal when cecum next to ileocecal valve is distended at 30-mm. Hg, but no withdrawal when ileum next to valve is similarly distended. No. 210–216 are positive cecal trials, while No. 26–27 are negative ileal trials. Lines from above: paw-withdrawal, respiration, conditioned stimuli, unconditioned stimuli, time in seconds.)

experiment. It has been used by Moiseyeva to train her four dogs to differentiate between ileal and cecal distentions at the same 30-mm. Hg pressure: that is, the animals came to elicit salivation or paw-withdrawal only when one region was distended and not when the other was similarly stimulated. Figure 2 is a kymographic record of such a complete differentiation in dog Tomik. Cecal distentions, positive conditioned Stimuli 210–216, produced paw movement of wide amplitude, whereas ileal distentions, negative Stimuli 26 and 27, failed completely.

With human subjects, on the other hand, two types of techniques in two types of subjects have been used: (*a*) stimulations through fistulas in patients with pre-existing fistulas; and (*b*) swallowing rubber balloons, to which silver electrodes positioned by means of fluoroscopes are often attached, in healthy adult subjects. Figure 3 illustrates the first type of technique. A patient with a urinary bladder fistula volunteered for an experiment in which various unconditioned and conditioned sensory, visceral, and skeletal reactions resulting from, and based upon, distentions of his bladder were thorouoghly investigated by means of an inserted balloon (Ayrapetyants, Lobanova, & Cherkasova, 1952, p. 7). Figure 4 demonstrates the swallowing technique in which, as may be seen, an even wider array of reactions is measured (Makarov, 1959, p. 131). Figure 5 includes two photoplethysmographic records: one of vasoconstriction of the arms when 100 cc. of water at 6°C. was poured into a balloon swallowed by a subject, and one of vasodilatation when the temperature of the water was 43°C. (Pshonik, 1952, pp. 245–246). The significance of using human subjects as checks on, and supplements to, animal experiments needs of course no emphasis.

Sample Experiments with Animals

Ia. Endometrial fistulas were formed in surgically exteriorized horns of three bitches. Glass funnels were fitted into the tops of the openings of the horns to keep out any possible secondary exteroceptive stimulation. When now 10

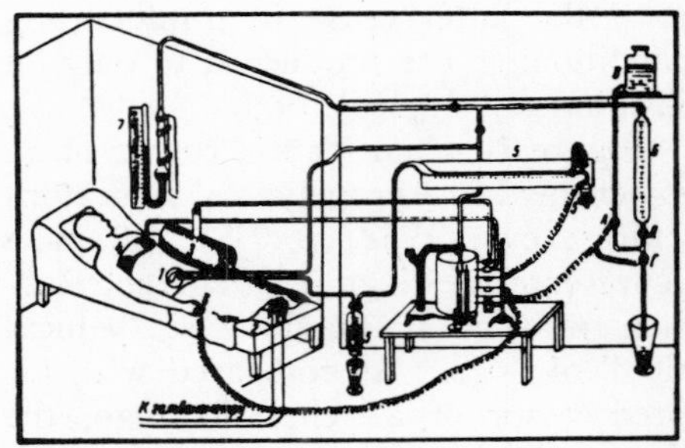

FIG. 3. Apparatus for interoceptive conditioning of human subjects with urinary bladder fistulas. (1—urinary bladder fistula; 2—PGR electrodes; 3—plethysmograph; 4—respiratory cuff; 5—registration of urination; 6—subject's report of sensations; 7—manometer watched by subject; Б—graduated cylinder; B—water vessel regulating pressure; A, Г, Д—valves. Russian words at lower left mean "leading to galvanometer.")

seconds prior to feeding the animals, their uterine interior was scratched rhythmically twice a second, or was irrigated by water at 8–12°C., clear-cut conditioned food reflexes with copious salivation and appropriate motor action came to be established. The conditioned reflexes (CRs) first appeared after six to nine trials and became fully stabilized, yielding 40–80 scale divisions of saliva in 20 seconds, after 25–100 trials. Moreover, after CR stabilization, the CSs of the CRs were well differentiated by the usual method of contrasts: the 2-per-second distentions from 1-per-second ones, and the irrigation with water at 8–12°C. from irrigations with water at 44–48°C. When in one experimental session, the horn of one of the animals was anaesthetized with a 1% solution of tetracaine, the CR almost disappeared for the session, and when a fistula was formed in the animal's second horn and the CS of the first horn was applied to it, considerable CR transfer resulted (Fel'berbaum, 1952).

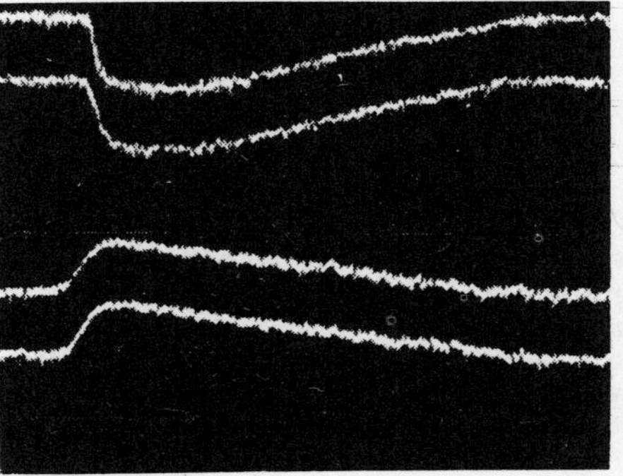

FIG. 5. Changes in volume of arm blood vessels of a human subject when 100 cc. of water at 6°C. or at 43°C. are introduced into his stomach through a rubber tube.

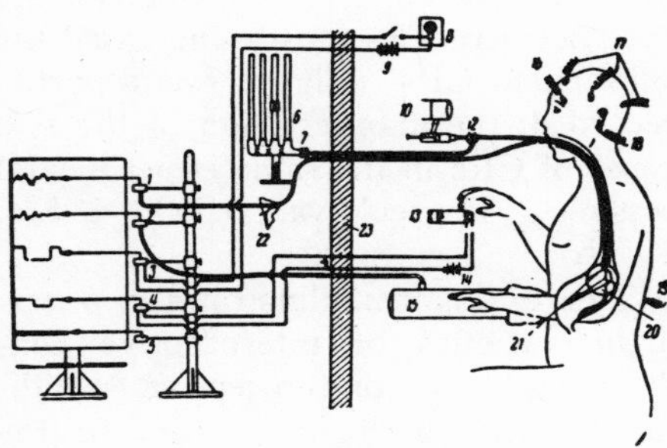

FIG. 4. Swallowing technique for the study of conditioned and unconditioned interoception and their effects on EEG, volume of blood vessels, blinking, and visual sensitivity in human subjects. (1—record of gastric movements; 2—record of arm plethysmogram; 3—marker of light stimulus; 4—subject's report of interoceptive and/or visual sensations; 5—time; 6—burettes to introduce solutions into stomach through opening of Stopcocks 7 and 12; 8—electric bulb; 9 and 14—accumulators; 10—disc of Makarov optic chronaximeter, measuring visual sensitivity; 11—syringe to draw out liquid contents of stomach; 13—key to signal gastric or visual sensations; 15—plethysmograph; 16—electrical registration of blinks; 17—EEG electrodes; 18—indifferent EEG electrode; 19—indifferent electrode for electrical stimulation of the stomach; 20—gastric balloon; 21—electrodes and leads of balloon; 22—terminal of balloon lead; 23—partition between experimenter and subject. Swallowing technique has also been used to study interaction between interoception and respiration, cardiac action, and galvanic skin reflex.)

Ib. The fistulated uterine interior of bitch Zita was stimulated by a jet of air and, after 10 seconds of such stimulation, an electric shock was administered to the animal's right-hind paw. Conditioned paw-withdrawal to the mere air stimulation first appeared after 5 trials and became stabilized after 10–11 trials. Figure 6 shows a kymographic record of such conditioning at Trials 40–42. As may be seen from the figure, the CR is quite unmistakable and surely very much marked; indeed, the extent of the conditioned paw-withdrawal appears to have been no less and even greater than the unconditioned one (Lotis, 1949).

The two experiments thus demonstrate rather clearly the formation and differentiation of interoexteroceptive uterine conditioning of both the vis-

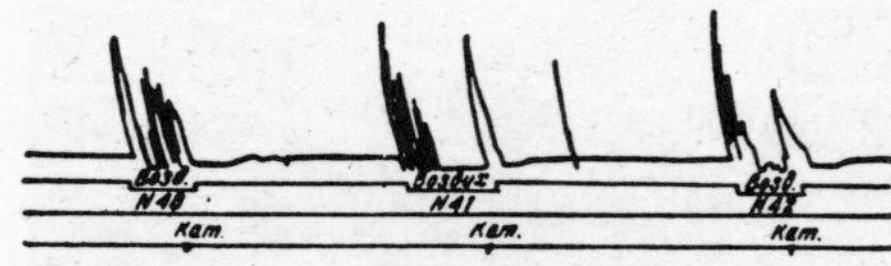

FIG. 6. Conditioned withdrawal of a female dog's paw upon stimulation of her uterus. (Lines from above: paw-withdrawal; conditioned stimulus, stimulation of uterus with a jet of air; ordinal number of application of conditioned stimulus; unconditioned stimulus, electric shock; time in seconds.)

ceral and the skeletal variety—their data corresponding fully to what one finds in reports of conventional extero-exteroceptive conditioning à la Pavlov and à la Bekhterev with the use of bells, metronomes, buzzers, lights, odors, epidermal taction, thermostimulation, and other exteroceptive CSs.

II. Thiry-Vella intestinal loops were formed in five dogs who, in addition, were tracheotomized to permit, by means of inserted cannulae attached to gas chambers, direct (bypassing the oral cavity) regulation of the amount and gaseous composition of inhaled air. After the animals had become fully adapted to the respiratory apparatus, showing no observable special reactions to its use in breathing normal air, the composition of the air was changed to include 10% of carbon dioxide, which, as might be expected, produced a wide variety of hypercapnic respiratory changes and defensive reactions (including the contraction of the spleen). Now, when the changes and the reactions were repeatedly preceded by the rhythmic distentions of the animals' intestinal loops (90–100 distentions in one minute at 60–80 mm. Hg pressure), hypercapnic conditioning was quickly observed.

The first hypercapnic CRs appeared after 3–6 trials and they became quite stable after only 5–16 trials; moreover, the animals readily and successfully differentiated the conditioned distentions from distentions at a rate of 15 per minute at 55–60 mm. Hg pressure, when the latter were presented without reinforcement for 6–15 trials. The same animals conditioned even more readily their hypercapnic reactions to a Tone 1a and differentiated them from stimulation with Tone re (CS generalization was ruled out), and both the exteroceptive and the interoceptive conditionings were in all cases highly resistant to extinction. When the two types of CSs were presented in very close succession, the interoceptive CSs mostly decreased the CRs to the subsequent exteroceptive CSs, whereas the exteroceptive CSs mostly increased the subsequent CRs. Figure 7 is a partial record of the manifestation of the two types of CRs in the same experimental session (Pogrebkova, 1950, 1952a, 1952b).

This experiment illustrates two new main varieties of interoceptive conditioning: interointeroceptive—conditioning the mechanoreceptors of the intestine to stimulations of the arteriochemoreceptors of the lungs; extero-interoceptive—conditioning auditory

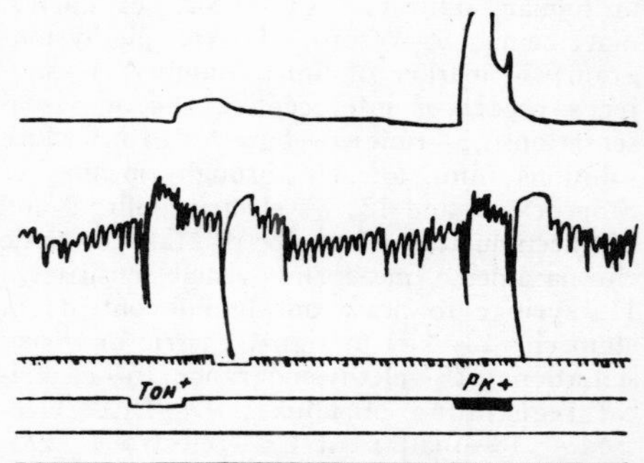

FIG. 7. Hypercapnic extero- and interoceptive conditioning in dogs. (Lines from above: general movement of animal; respiration; lung ventilation; conditioned stimuli, first, auditory, second, distention of intestine; unconditioned stimuli; time in 2-second units.)

stimuli to such stimulations. Moreover, each main variety involved the conditioning of both skeletal and visceral reactions, thus raising the total number of new varieties to four, which, combined with the two varieties of Experiments Ia and Ib, completes the experimental demonstration of all varieties of interoceptive conditioning possible with animal subjects. Again, we might note that the hypercapnic conditioning, both the interoceptive and the exteroceptive, was very fast and stable, a result that should probably be attributed to the unusual arousal nature of the large excess of inhaled carbon dioxide, threatening the very existence of the organism's vital functions and mobilizing fully its adjustive vigilance. And we might add that decremental effects of juxtaposing two kinds of CSs of the same CR, as observed in the results of the juxtaposition of the exteroceptive and the interoceptive CSs in this experiment, are unusual when the two kinds of CSs are both exteroceptive, let us say, visual and auditory. Or, in other words, there is indication here that conditioning to interoceptive stimuli conflicts with conditioning to exteroceptive stimuli in a way that conditioning to exteroceptive stimuli of different modalities does not: that is, interoceptive conditioning is in an important way a class of conditioning by itself, a matter for which more evidence and discussion are adduced in later sections. Finally, we might point to another striking result of the cited study, the finding that while, after prolonged experimentation, the dogs became adapted to breathing the unusual air, the CSs continued to elicit, in a number of cases, widespread aversive organismic reactions.

IIIa. The duodenum of dog Laska was inflated three to four times in 5 seconds with 30–40 cc. of air. On the fifth second a single 4-volt electric

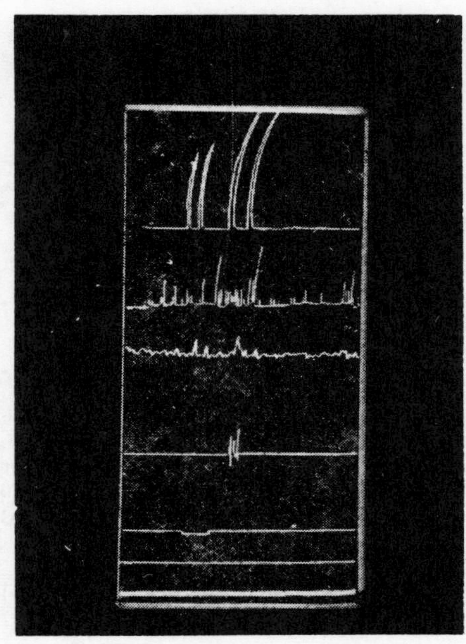

FIG. 8. Second-order interoceptive paw-withdrawal conditioning in dogs. (CS_1 interoceptive, CS_2 exteroceptive. Lines from above: movement of conditioned left-hind paw; movement of nonconditioned right-hind paw; respiratory changes; first-order conditioned stimulus, distention of intestine with 30–40 cc. of air; second-order conditioned stimulus, sound of buzzer; unconditioned stimuli; time in seconds.)

shock was administered to the animal's left-hind paw with electrodes attached to the *articulatio talo cruralis.* Conditioned paw-withdrawal in response to the inflation first appeared on Trial 5. After reinforced Trial 129, the duodenal inflation, now no longer accompanied by shock, was joined by the sound of a buzzer, the sound lasting for 15 seconds and preceding the inflation by 10 seconds. After 18 such trials, the sound of the buzzer came to elicit clear-cut withdrawal of the animal's left paw with some movements of the right paw and respiratory changes, even though the sound was never associated with the administration of the electric shock. Figure 8 unmistakably

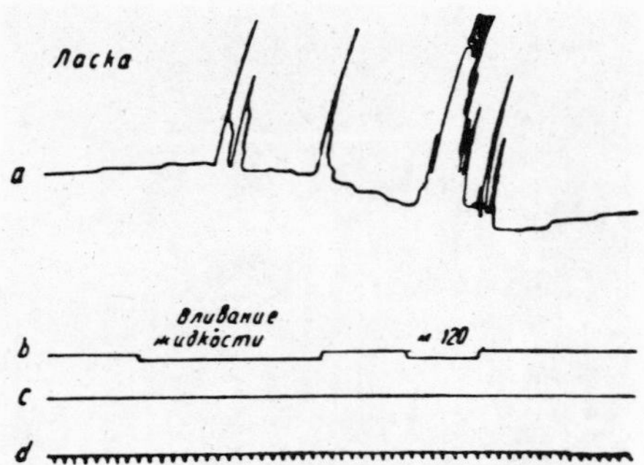

FIG. 9. Second-order interoceptive paw-withdrawal conditioning in dogs. (CS_1 exteroceptive, CS_2 interoceptive. Lines from above: paw-withdrawal; conditioned stimuli, first, irrigation of intestine with water for 15–30 seconds, second, metronome of 120 beats per minute; unconditioned stimuli; time in seconds.)

demonstrates this conditioning based upon a previously conditioned interoceptive CS (Vasilevskaya, 1948, 1950).

IIIa$_1$. Stable conditioned paw-withdrawals, based on the administration of electric shock, to the sound of a metronome of 120 beats per minute were formed in two dogs, who had intestinal loop fistulas, one dog by the method of Thiry and the other by the method of Thiry-Vella. When now prior to the sounding of the metronome, the intestinal loops of the animals were repeatedly irrigated with water at room temperature for 15–30 seconds, clear-cut paw-withdrawals in response to the mere irrigations were noted in one dog after 21 trials and in the other after 46 trials. Figure 9 is a record of the new and of the original conditioned reflex in one dog. The data indicate that the new CR was by no means as stable as the original one; yet, as seen in the figure, it was quite substantial and unmistakable in magnitude. This experiment differs from the one preceding in that in it the second-order, the CS-based, and not the first-order, the US-based, CR was interoceptive (Pauperova, 1952).

IIIb. Distentions of the small intestines of six dogs at 40–100-mm. Hg pressure were accompanied for 7–20 trials by exteroceptive stimulations (sound of a whistle, the flash of a 200-watt lamp, or the odors of ammonia), after which either the distentions or the exteroceptive stimulations were repeatedly accompanied by the administration of an electric shock to one of the animal's paws. When a stable conditioned paw-withdrawal was formed to the shock accompanied stimuli, it was found that the shock unaccompanied stimuli also elicited paw-withdrawal in most cases. Some of these US-unaccompanied CRs were highly resistant to extinction, requiring 3 to 5 days of nonreinforced applications to disappear completely, and no essential differences were noted between cases in which the distentions were the original CSs and the exteroceptive stimuli were the unaccompanied stimuli and cases in which the experimental situation was reversed. Moreover, in three other dogs, clear-cut conditioned paw-movements were formed to "natural" gastric contractions when the contractions were accompanied by the sound of a whistle or the flash of a 200-watt light and the sound or flash were accompanied by a shock to the

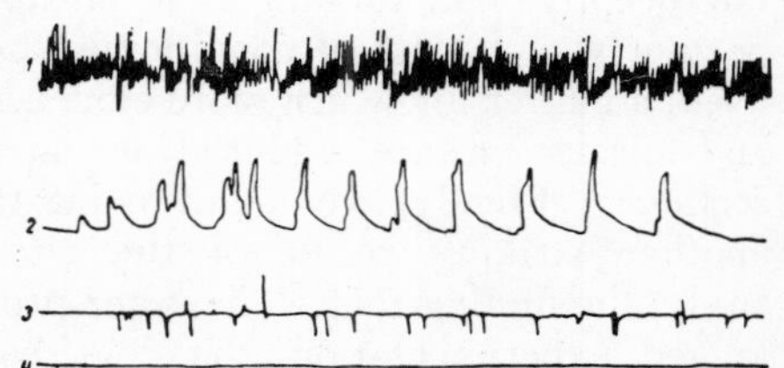

FIG. 10. Conditioned paw movements of a dog to "natural" gastric contractions when the contractions were accompanied by a flash of a 200-watt lamp and the flash was in turn accompanied by an electric shock to the animal's hind paw when its stomach was at rest. (Lines from above: pneumogram; gastric contractions; paw movements; flashing of light, not administered.)

animal's hind paw only when the stomach was at rest. Figure 10 shows the data of such conditioning in one of the three dogs (Goncharova, 1955, p. 165).

The three experiments demonstrate two special types of conditioning: (*a*) second-order conditioning, when conditioning is effected through associating a stimulus with a previously conditioned stimulus (CS_1 is associated with a US and then a CR is formed to CS_2 through associating it with CS_1); and (*b*) associated conditioning (Russian term) or sensory preconditioning (American term), when conditioning is effected by associating a stimulus with a to-be-conditioned stimulus, that is, prior to the latter's CR training (CS_x is associated with CS_y and then a CR is formed to CS_x or CS_y by virtue of the preconditioning association). Both types of conditioning are special in the sense that they are not very readily demonstrable and it is thus important to note that they occur also when either one of the two CSs is interoceptive in nature. The preconditioning or associated conditioning of the paw movements to the gastric contractions is furthermore a special case in that the contractions were, as indicated, natural and not laboratory produced, and are of course motor and not sensory reactions.

A few years ago (Razran, 1955a), the author argued in this journal that second-order conditioning is an infrequent phenomenon and is likely to involve cognitive intervention whenever it occurs. The manifestation of this type of conditioning at the interoceptive level as well as a reanalysis of the evidence for its existence in conventional exteroceptive conditioning, plus an analysis of new evidence at the exteroceptive level, forces the author to retract hereby his former view. The infrequency of the phenomenon appears to be due almost wholly to experimenters' failures to observe certain stimuli sequences that are conducive to its laboratory production.

Fig. 11. Conditioned contraction of the rumen and the reticulum in sheep, with electrical stimulation of the mucosa of the rumen as the US and the sound of a bell as the CS. (Lines from above: contraction of the reticulum, contraction of the lower portion of the rumen, contraction of the upper portion, contioned stimuli, unconditioned stimuli, time in 30-second units.)

Fig. 12. Conditioned contraction of the rumen and the reticulum in sheep, with distention of the rectum as the US and the sound of an electric bell as the CS. (Lines from above: contraction of the reticulum, contraction of the rumen, conditioned stimuli, unconditioñed stimuli, time in 30-second units.)

IV. Fistulas were formed in the rumen, reticulum, and rectum of six karakul sheep. It was found that electrical stimulation of either the animals' epidermis or the mucosae of their rumen, reticulum, or rectum, as well as the distention of their rectum, produced considerable reductions in the periodic contractions of the rumen and reticulum and increases in the rumen-reticulum ratio of contraction. Conditioning was now attempted by ringing a bell a few seconds prior to each of the five aforementioned unconditioned stimulations, and it was successful in all cases. The number of trials needed for the first appearance and for the stabilization of the CR were, respectively: 7–13, 16–24 when the epidermis was stimulated electrically; 6–11, 13–18 when the stimulation was applied

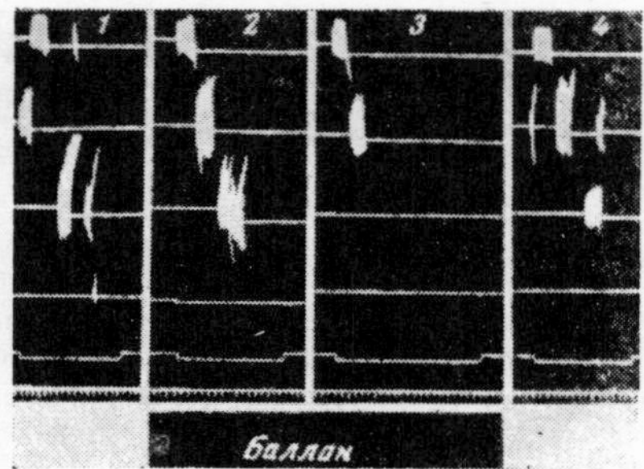

FIG. 13. Dogs' conditioned choices of dishes of 15 cc. of milk of varying salt contents. (CS—gastric inflations with 300–500 cc. of water in inserted rubber balloon. US—administration of 300–500 cc. of 3–5% of a salt solution through a gastric fistula. Strip below main figure shows insertion and removal of balloon, CS. Lines from above: milk with 1%, 2.5%, 3.5%, and 5% of salt contents; response to water dish; and time record.)

to the mucosa of the rumen or reticulum; 17–21 when it was administered to the rectum; and 20–24, 39–44 when the stimulation was the distention of the rectum. Figure 11 is a kymographic record of a CR with electrical stimulation of the rumen as the US, and Figure 12 is one when the US was the distention of the animals' rectum (Dedashev, 1959).

This experiment is of special interest in that it involves special organs as loci of stimulation, the reduction and cessation rather than the production of reactions as CRs, and the use of animals other than dogs as subjects. Moreover, as will be noted, the experiment contains a certain amount of evidence to indicate that the speed of formation of exterointeroceptive conditioning, unlike that of the interoexteroceptive variety (evidence not brought out in presented experiments), may well not be slower than the speed of formation of exteroexteroceptive conditioning.

V. In a pilot experiment with dogs (number not given), the experimenter observed that the salt appetite of the animals was markedly decreased when 300–500 cc. of a 3–5% of an aqueous salt solution had been introduced into their gastric fistulas. The salt appetite was determined through permitting the dogs to choose any of several dishes of 15 cc. of milk containing varying amounts of dissolved salt. When now the administrations of the aqueous salt solutions were preceded by distentions of the animals' stomachs by means of inserted rubber balloons into which 300–500 cc. of water was poured, it was found that the mere distentions became conditioned reducers of the salt appetite. Figure 13 presents the data of such conditioning in one dog. As may be seen from the figure, the animal came to refuse milk of 5% salt contents after the balloon had been in his stomach for awhile but took it when the balloon was removed (Kassil', 1959).

The experiment is significant in demonstrating in preliminary fashion interoceptive conditioning of a comparatively complex choice reaction. A recent American experiment by Cook, Davidson, Davis, and Kelleher (1960) demonstrates 100% avoidance conditioning to jejunal pressure in rats. The experiment is, as far as the reviewer is

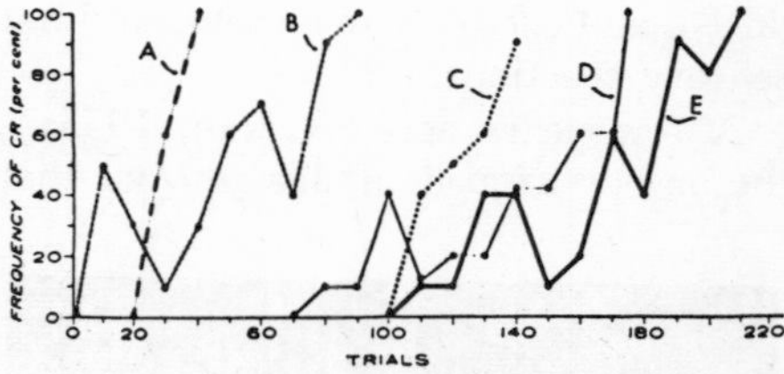

FIG. 14. Representative avoidance acquisition curves for each type of conditioned stimulus. (Each curve represents the results obtained with a single animal. Conditioned stimuli: A—tone, B—acetylcholine, C—*l*-norepinephrine, D—jejunal pressure, E—*l*-epinephrine. Different dogs were used in each experiment. Each point represents the percentage of conditioned leg flexions occurring in blocks of 10 successive trials.) (Reproduced from *Science* by permission)

aware, the first one in true interoceptive conditioning performed in this country, and it would seem worthwhile to present its main data. Figure 14, taken from the report of Cook et al., is self-explanatory.

VI. The urinary bladders of two dogs, a male and a female, were removed, their ureters externalized, and the ureteral orifices sutured to their skin. Normal functioning of the kidney continued. But, as might be expected, there was a loss of the familiar micturition reflex—leg lifting, etc. in the male, and squatting, etc. in the female. Yet, when these dogs saw other dogs urinating, the micturition pattern (getting near a wall, smelling, and leg lifting or squatting) was almost fully restored (Ayrapetyants, 1952, p. 25).

The observation is interesting because it discloses "natural" exterointeroceptive conditioning: that is, conditioning an interoceptively aroused reflex to exteroceptive stimuli—the sights and smells of other dogs—that naturally accompany it.

Sample Experiments with Human Subjects

Ia. Three patients with urinary bladder fistulas volunteered for a series of experiments in which distentions of their bladders, through calibrated inflows of air or of physiological solutions, were thoroughly investigated. Careful recordings were made of: (*a*) the rate and amount of the inflow; (*b*) bladder compression resulting from the contraction of *M. detrusor urinae* and transmitted by means of inserted rubber balloons attached to mercury or spring manometers; (*c*) the rate and amount of urine secretion; and also (*d*) the accompanying respiratory, vascular, and psychogalvanic changes—while, in addition, the subjects (*e*) signaled the time and intensity of their urination "urges." The manometers, equipped with conspicuous dials, were always watched by the patients, so that their scale readings—and column heights or angle displacements—could be used as conditioned stimuli when the manometers were detached (without the patients' knowledge) and the readings were varied independently by the experimenter.

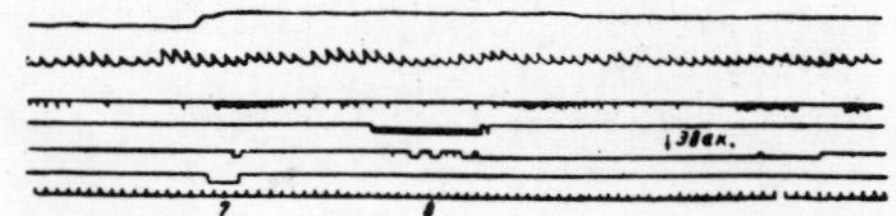

FIG. 15. Interoceptive urinary conditioning in man. (Lines from above: intrabladder pressure; pneumogram; urination; conditioned stimulus, manometer sham reading of 40-mm. Hg; subject's report of urinary urge; unconditioned stimulus, introduction of air into bladder; time in 3-second units.)

Conditioning to the sham readings was effected readily in all cases. The patients began reporting intense urination urges, accompanied by all or most of the aforementioned objective reaction changes, when the manometer readings were high (normal or above normal of what they were while effecting the urge in the training session) even though the inflows were minimal or totally omitted. On the other hand, low or zero readings failed to produce the urge and its accompaniments, even if the inflow was considerably above that which normally produced it; much increased, sometimes double, inflows were now needed to be fully effective (in some cases, the failure was evident only with respect to the urge and the urination, but in others, it meant also absence of any measurable muscular tension—Ayrapetyants et al., 1952). Figures 15 and 16 are records of this conditioning.

Corresponding results were obtained later in patients with intestinal fistulas and with instructions or words as CSs (Ayrapetyants, 1956), while in a pre-

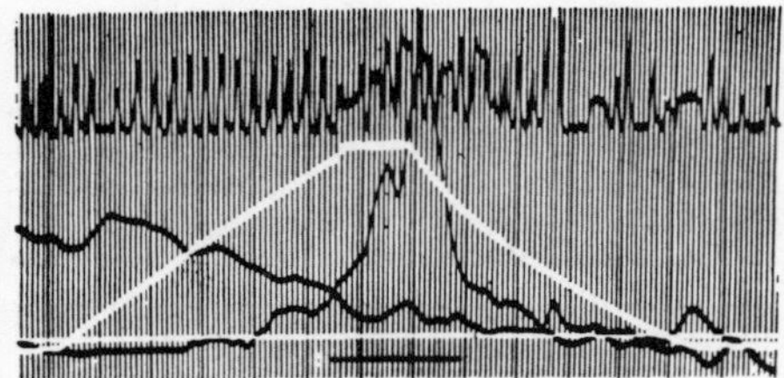

FIG. 16. Interoceptive urinary conditioning in man. (Black lines from above: pneumogram, PGR, intrabladder pressure, subject's report of urinary urge. Upper white line is the conditioned stimulus—the experimenter's moving upward the readings of the manometer. The lower white line is for the unconditioned stimulus—filling the bladder with air; not used in the test, bladder empty.)

liminary pilot study it was shown that a patient with an intestinal fistula, who discriminated accurately an inflow of cold water at 6–8°C. from that of warm water at 40°C., began to report sensations of cold in response to the flash of a red lamp that had been combined eight times with the cold inflow (Ayrapetyants, 1949, p. 37).

Ib. Experimenter noted that when his subjects were told to inhale, they manifested clear-cut plethysmographic arm vasoconstriction in addition to changes in their pneumograms, as may be seen in the left portion of Figure 17. The particular vasoconstriction is apparently a viscerovisceral vasomotor reflex evoked by the respiration. Now, the experimenter told his subjects not to inhale when the word "inhale" was given. But, as may be seen in the right portion of the figure, the vasoconstriction persisted despite the entirely regular pneumogram—it was, presumably, "naturally" conditioned to the word inhale and was not deconditioned—or inhibited—by the instructions "not to inhale when inhale is uttered" (Okhnyanskaya, 1953).

Experiment Ia discloses six varieties of exterointeroceptive conditioning: sensory (report of sensations), motor (respiratory changes), and visceral (vascular and electrodermal reactions), in response to both sensory and verbal CSs. A highly significant special finding of the experiment, and one that no doubt points to the great power of its methodological design, may be noted in Figure 16. As the CSs increased in magnitude—as the experimenter moved the manometer dials upward—changes occurred first in the PGR, then in the intrabladder pressure, and then in respiration and urination urges; when the CSs decreased, the sequence was reversed, except that respiratory fluctuations continued for nearly as long as those of the PGR.

Experiment Ib, which was already indicated as an example of "natural" exterointeroceptive conditioning, seems by all tokens to shed light on two intricate CR mechanisms: (*a*) the chaining of classical conditioning and the CR efficacy of a three-link exteroceptive-interoceptive-interoceptive CR chain when the middle link is omitted, and (*b*) the deconditioning or reversing power of instructions upon previously conditioned verbal stimuli. The experiment is obviously one that could be readily duplicated—and expanded—in any American laboratory.

II. Fine walled, inflatable rubber bal-

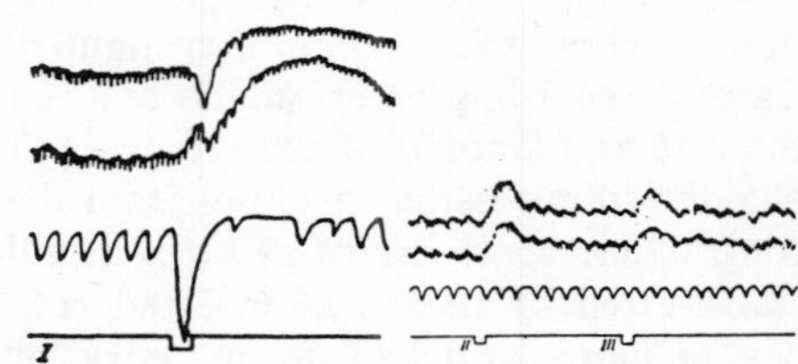

FIG. 17. Conditioned interoceptive vascular changes in human subjects. (Plethysmograms and pneumograms to the sound of the word "inhale" when (*a*) the subjects were told to inhale—left portion of figure—and when (*b*) they were told *not* to inhale when they heard the word "inhale"—right portion of figure.)

loons folded into tubes and equipped with three-pronged silver electrodes positioned by fluoroscopes were swallowed by four healthy human subjects, including the experimenter, P. O. Makarov of Leningrad University. The technique permitted recording a vast array of visceral, skeletal, and verbal-sensory reactions when the stomach was distended, or stimulated electrically. As the stimuli increased in intensity, an extensive period of no-sensation or presensation, in which only visceral and skeletal changes occurred, was noted. With further increases in intensity, came a period of dull and heavy poorly localized sensations of "something"; and with still further increases, a period of fairly clear and fairly localized acute pain. When in the sensation period the active electrodes were positioned 8 cm. apart and each point was stimulated separately, stimulations at intervals of less than 15 milliseconds were reported as single sensations localized between the two stimulated points. When the intervals were 55–105 milliseconds, the reports were of two successive sensations in the same locus; while intervals greater than 105 milliseconds resulted in reports of two successive sensations in two discriminable loci, the correctness of the spatial discrimination being further a regular function of the lengthening of the intervals (Makarov, 1950a, p. 395). Figure 18 is an illustration of the results.

The value of an exact experimental analysis of direct visceral stimulation and sensation in man can hardly be overemphasized, both because of our present almost total ignorance of the field and because of the unique way in which it permits the separation of conscious from unconscious reactivity, a slow-motion way of unfolding a changing panorama that, surely, studies of epidermal and other exteroceptive stim-

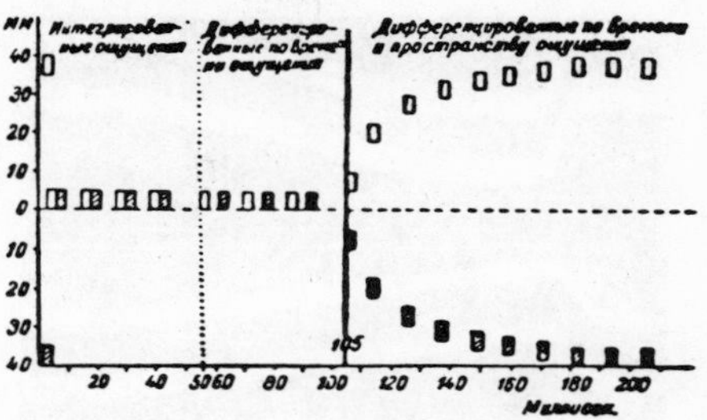

Fig. 18. Two-point threshold in electrical stimulations of the gastric mucosa of human subjects. (From left to right, Russian words mean: "integrated [fused] sensation," "temporal discrimination," "spatial discrimination." Units on abscissa are interstimulation intervals in milliseconds; units on ordinate are distances between stimulation points in millimeters.)

ulations with their low and quick conscious thresholds can by no means emulate.

III. Inflatable gastric tubes were swallowed by three healthy human subjects, 26–32 years of age, to study the interaction of thermal interoceptive and exteroceptive stimulations by means of a modified Novitsky arm photoplethysmograph.[4] After the subjects had become fully adapted to the apparatus and consistently produced only symmetrical records, four types of interoceptive-exteroceptive interactions were investigated: cooling the inside of the stomach by 100 cc. of cold water at 6°C., 10–15 seconds prior to warming the epigastric region with a special 43°C. "thermode"; warming the inside of the stomach with 100 cc. of water at 43°C., 10–15 seconds prior to cooling the outside with a 6°C. thermode; cooling the outside of the stomach with the 6°C. stimulus prior to warming it in-

[4] The Novitsky photoplethysmograph appears to be essentially the same as that described recently by Shmavonian (1959). The Russians have been using this technique extensively for about 10 years in recording digital and forehead changes in blood volumes.

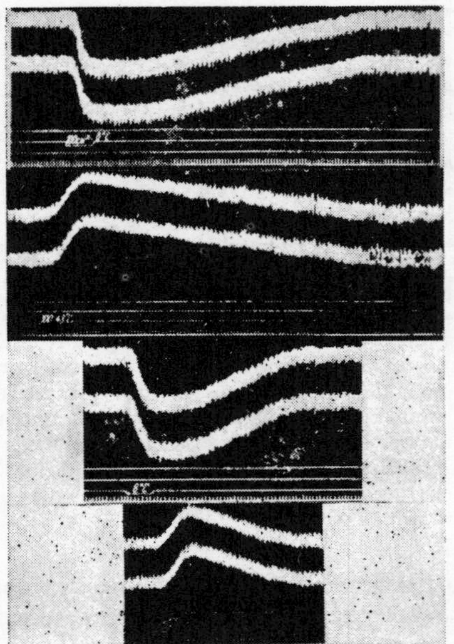

FIG. 19. Arm plethysmograms of human subjects when 100 cc. of water at 6°C. is poured into gastric balloon which the subjects swallowed (upper figure); 100 cc. of water at 43°C. is poured into the balloon (upper middle figure); a 6°C. stimulus is applied to the subjects' epigastric region (lower middle figure); a 43°C. stimulus is applied to their epigastric region (lower figure).

side with the 43°C. water; and warming the outside of the stomach with the 43°C. stimulus before cooling it inside with the 6°C. water.

At first, as might be expected, vasoconstrictions followed by vasodilatations were noted in the first and third series of the experiment and vasodilatations followed by vasoconstrictions in the second and fourth. However, as the experiment continued and interaction set in, the modifications in the first two series of the experiment in which the interoceptive stimuli and reactions preceded the exteroceptive ones contrasted strikingly with those in the last two series in which the sequence was reversed. In both the first and the second series, the preceding reactions—vasoconstriction in the first series and vasodilatation in the second—suppressed the opposing succeeding reactions and indeed became themselves somewhat strengthened, presumably at the expense of the succeeding reactions; whereas in the third and fourth series, the preceding reactions, on the contrary, became clearly weakened and the succeeding reactions strengthened as the unconditioned stimuli of the former were converted into conditioned stimuli for the latter. Or, in other words, in either case the exteroceptive stimuli lost their original functions

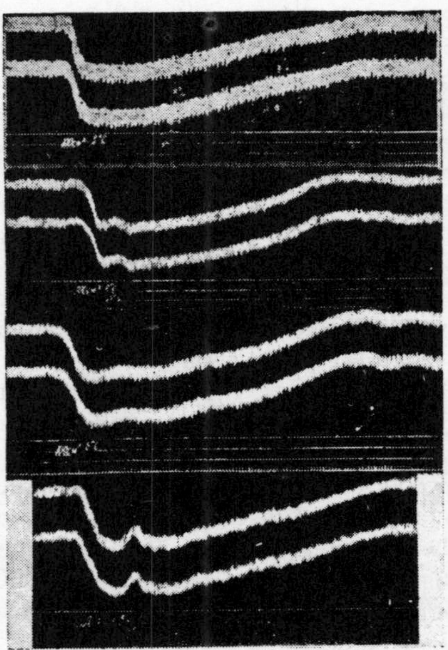

FIG. 20. Arm plethysmograms of human subjects when cooling their gastric mucosa at 6°C. precedes by a few seconds warming their epigastric region at 43°C. (Upper figure—the plethysmograms in response to the cooling prior to its combination with the warming; upper middle figure—the cooling and the warming combined; lower middle figure—cooling after a number of cooling-warming combinations; lower figure—experimenter saying "I am cooling the stomach on the inside" a few seconds before warming the epigastric region at 43°C.)

and the interoceptive reactions became dominant.

Figure 19 is a record of the vasomotor reactions prior to the interoceptive-exteroceptive and the exteroceptive-interoceptive interactions; it shows that the reactions to the two types of stimuli were about equal in magnitude and pattern. Figure 20 demonstrates the dominance of interoceptive reactions when interoceptive cooling and resulting vasoconstriction preceded exteroceptive warming and resulting vasodilatation (first series of experiment). Figure 21 illustrates interoceptive dominance when interoceptive warming and resulting vasodilatation succeeded exteroceptive cooling and resulting vasoconstriction (third series). The results of sequences of interoceptive warming-exteroceptive cool-

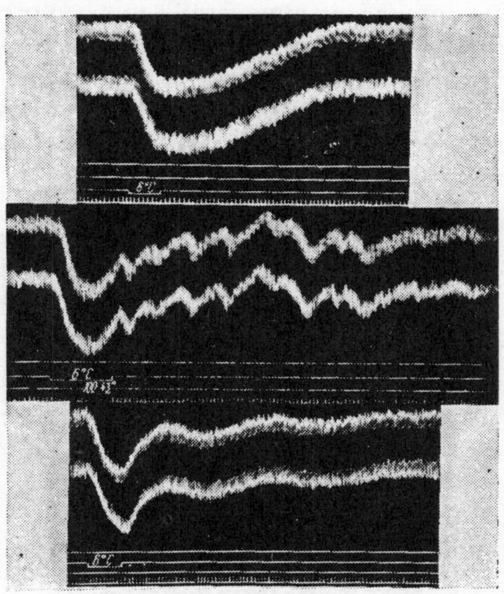

Fig. 21. Arm plethysmogram of human subjects when warming their gastric mucosa at 43°C. succeeds by a few seconds cooling their epigastric region at 6°C. (Upper figure—the plethysmogram in response to the cooling prior to its combination with the warming; middle figure—the cooling and the warming combined, note fluctuation; lower figure—cooling after a number of cooling-warming combinations.)

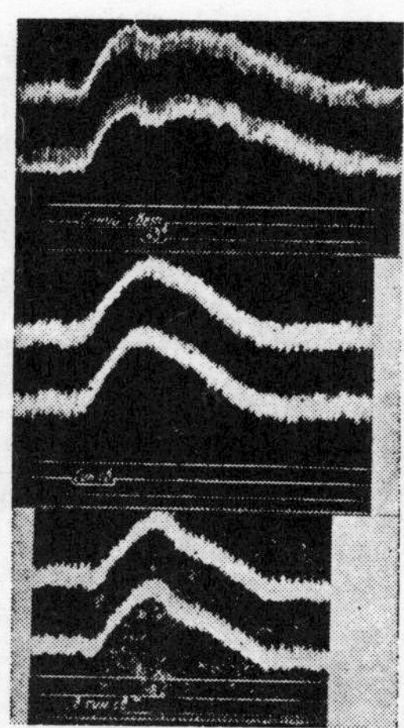

Fig. 22. Conditioned vasodilatation of the blood vessels of the arms of human subjects to the flash of a blue light after the light had been combined a number of times with warming the subjects' epigastric region with a 43°C. stimulus on odd trials of a CR stereotype. (Upper figure—record when the flash of the light was, after a few seconds, reinforced by the warming; middle figure—the flash of light applied alone; lower figure—experimenter saying "I am flashing on a blue light.")

ing (second series) and exteroceptive warming-interoceptive cooling (fourth series) were quite similar. And it might further be noted that: (*a*) as seen in Figure 20, the interoceptive reactions dominated the exteroceptive ones even when the former were produced by verbal CSs as "I am cooling the stomach on the inside" and the latter by their USs, warming at 43°C.; and that (*b*) the modifications in the exteroceptive-interoceptive sequences were of a classical CR type whereas those in the interoceptive-exteroceptive sequences were of an operant CR type. The last statement points to two far-reaching theoretical implications: that the possibility that operant conditioning is somehow related to interoceptively (and proprioceptively) produced (and interoceptively and proprioceptively conditioned) behavior; that the relative

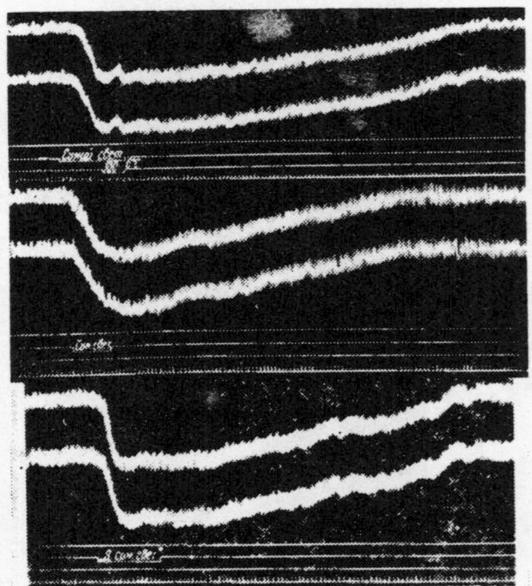

Fig. 23. Conditioned vasoconstriction of the blood vessels of the arms of human subjects (subjects of Figure 22) to the flash of a blue light after the light had been combined a number of times with cooling the subjects' gastric mucosa with water at 6°C. on even trials of the same CR stereotype. (Upper figure—record when the flash was, after a few seconds, reinforced by the cooling; middle figure—flash applied alone; lower figure—experimenter saying "I am flashing on a blue light.")

dominance of a reaction determines in some way the type of conditioning—classical or operant—that the reaction undergoes. The matter will be considered fully in a later theoretical section.

In another part of the experiment, a "vascular neurosis" was formed in three subjects through the disruption of a CR stereotype (pattern) in which the flash of a blue light preceded the warming of the epigastric region at 43°C. on odd trials and the cooling of the gastric mucosa at 6°C. on even trials. Figures 22 and 23 show the clear-cut normal functioning of the CR stereotype. Vasodilatation was produced when the blue light was flashed or the words "I am flashing on a blue light" were pronounced on odd trials (Figure 22), whereas on even trials the same stimuli yielded vasoconstriction (Figure 23). Moreover, similar results were obtained with a CR stereotype in which the flash of a red light preceded the cooling of the epigastric region at 6°C. on odd trials and the warming of the gastric mucosa at 43°C. on even trials (Pshonik, 1952, pp. 240–295). Yet note in Figure 24 what happens when the trial sequence is disregarded and the blue or red light is flashed: wholly irregular and conflicting vasomotion, which, the experimenter adds, was accompanied by vomiting, sensory distortions, and complaints of headache.

The striking and instructive results of the experiment hardly need comment, except to add that while, as was already indicated, interoceptive conditioning—that is, intero-exteroceptive conditioning—is slower in formation than is exteroceptive conditioning, once conditioned, it is generally much more resistant to extinction than is the conventional classical variety, a characteristic that no doubt augments further its relative role in the organism's CR repertory.

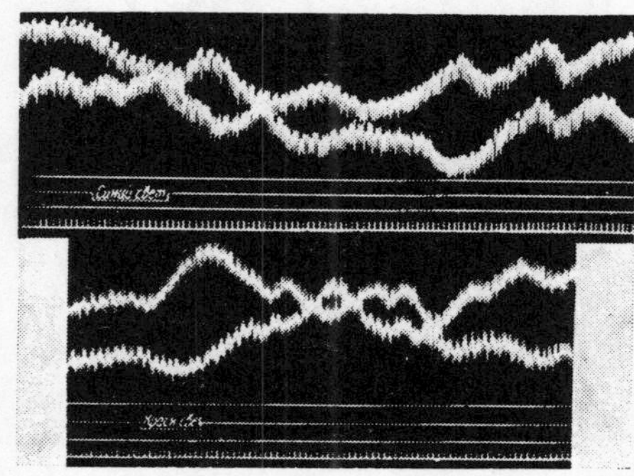

Fig. 24. Marked vascular fluctuation ("vascular neurosis") in response to the flash of the blue light when the cooling and the warming are no longer regularly alternated and the CR-stereotype is thus disrupted (upper figure). Lower figure shows similar results of an experiment in which the flash of a red light was the conditioned stimulus, the epigastric region was cooled, and the gastric mucosa was warmed.

IV. A human patient, whose rectum was removed through surgery and a colostomy installed, continually reported strong defecatory urges and sensations in the area of the nonexistent rectum whenever the attendant removed the feces from his system (Ayrapetyants, 1952, p. 26). Another patient, whose ureters were externalized and sutured to the large intestine, learned to distinguish between urinary and defecatory urges, the distinction being connected in some way, according to the patient's report, with differences between internal sensations following drinking and eating (Bykov, 1943, p. 255).

These two clinical examples by all tokens show "natural" interoceptive visceromotor and sensory conditioning to exteroceptive and interoceptive stimuli (sight of attendant, reported internal stimulation; cf. the observations on the micturition reflexes of the dog and the bitch with removed urinary bladders in the preceding section).

Summary and Special Significance: The Observable Unconscious

The selected 14 experiments on interoceptive conditioning, both animal and human, the experiment on unconditioned interoception in healthy human subjects, and the three clinical observations, in the main, epitomize well, it is believed, the present state of knowledge of the field as reflected in approximately 100 reports of recent Russian experiments. In recapitulation, this knowledge might be compressed into six principal propositions:

1. Unlike the continuum of exteroceptive stimulation which is the body-material of all our conscious experience, the continuum of interoceptive stimulation leads largely to unconscious reactions.

2. Interoceptive conditioning, whether involving conditioned or unconditioned interoceptive stimuli, is readily obtainable and is by its very nature largely unconscious in character.

3. While interoceptive conditioning is by its nature more limited than is exteroceptive conditioning with respect to total kinds and variety of stimulations, the interoceptive kinds of stimulations are, on the other hand, by their very nature much more recurrent, periodic, and organism-bound, making interoceptive conditioning an almost built-in function that is constantly generated and regenerated in the very process of living and acting.

4. Interoceptive conditioning is somewhat slower in formation than is exteroceptive, but, once conditioned, it is more fixed and irreversible (less readily extinguished).

5. When equal but opposing interoceptively produced and exteroceptively produced reactions are juxtaposed, the interoceptive reactions dominate the exteroceptive ones, with the final result that preceding exteroceptive stimuli become conditioned stimuli for succeeding interoceptive reactions whereas preceding interoceptive reactions become strengthened by exteroceptive reactions succeeding them.

6. The juxtaposition of conditioned interoceptive and exteroceptive stimuli of the same conditioned reaction, unlike similar juxtapositions of exteroceptive stimuli of different modalities, produces a certain amount of conflict and decrementation of conditioning.

Of the six propositions, the last three, the more specific ones, might still be in need of further empirical evidence to settle fully their status and parameters. However, there should be absolutely no doubt about the finality of the first three general propositions—the ubiquity, largely unconscious character, and ready all-round conditionability of interoceptive stimulations. And this alone, the author contends, suffices to add a wholly novel dimension to our thoughts about conscious and unconscious reactions, and the Conscious and the Unconscious as such, in experimental-theoretical psychology, Freudian theory, and applied psychosomatics.

A good number of psychologists hold, for instance, to the view that classical conditioning is either wholly a function of cognitive-perceptual expectancies and anticipations, inferred or observed; merely a mechanical subsidiary of integrated goal directed means-end adjustive behavior; or altogether a pseudophenomenon without its own ontological or heuristic status. The clear-cut experiments in interoceptive conditioning in all its attributive forms cited here—and they are, as stated, only a sample—can scarcely be thus interpreted. Indeed, without denying or minimizing the role of perceptual and adjustive factors in higher and more complex levels of learning, one is almost forced to think of a reverse position—notably, one that accords the many empirical facts of simple unconscious interoceptive learning the status of a prime model of all learning and lets the complex varieties seek subsidiary principles when and if the model is inadequate. The aforementioned psychologists, it might well be argued, have not yet, despite many efforts, really given up their traditional tendencies of preferring observations *von oben herab* [from above down] to *von unten herauf* [from below up], and of ignoring, or insufficiently advancing, physiological and surgical means that extend the horizon and expand the dimensions of directly observable behavior. Instead, the argument might run further, these psychologists continue to be rooted in data and concepts that are mainly derived from either analyses of their own private experience (phenomenology) or observations of gross and global behavior without the benefit of the empirics of underlying finer mechanisms. (The last statement pertains not to a neural analysis, but to a finer behavioral one: the almost ever-present parameters of interoception and interoceptive conditioning are behavioral!)

The relevance of interoceptive conditioning to Freudian theory may be even more evident. For, obviously, no matter how far removed Freudian dynamics is from psychophysiological dynamics at present, its eventual material validation—and it is assumed that such validation is at least a future desideratum of the system—will rest largely upon expanded psychophysiological knowledge of the contents of visceral action and of the mechanisms of shifts or associations of such action and contents. Studies of interoception, particularly of human interoception, and of animal and human interoceptive conditioning are clearly concerned with just this. And while the sexual contents of visceral action, at least in human subjects, is not readily experimentable and our general knowledge of contents of visceral action is still quite meager, the experimental information on shiftings of visceral actions is already, as may have been noted, too substantial and too instructive to be passed over by any follower of Freud, whatever his attitude toward physiological validation may be. Consider not only the ready two-way conditionability or shiftability (or displacement or transference) of basic unconscious visceral action, but also the relative irreversibility of the shifts, the evidence of conflicts resulting from the juxtaposition of interoceptive and exteroceptive stimulations, and the dominance of interoceptive reactions over exteroceptive ones of the same measurable magnitudes.

Put somewhat differently, we might say that the many and varied attempts made in the past to synthesize the teachings of Freud with known facts of conditioning (and learning) have suffered from at least three obvious shortcomings: (*a*) They were based on evidence that, with only very few exceptions, came from experiments on conditioning that is conscious in content (that is, of which human subjects are conscious and animals—dogs and rats—are assumed to be conscious) and thus is outside the main operational material of Freud's doctrines. (*b*) The conditioning units of the evidence, at least as far as classical conditioning is concerned, were in the main organismically inconsequential; even the classical unit of salivation is obviously not within the center of the organism's needs and action, not to mention blinking, finger- or paw-withdrawal, knee jerk, galvanic skin reflex, and others. (*c*) The attempts lacked evidence on unconditioned and conditioned organism generated stimulations and the resultant organism generated–environment generated antinomy—an antinomy that, it must be remembered, is only related to, but by no means identical with, the more known exteroceptive somatic-visceral and conscious-unconscious antinomies, and that by all tokens should be of vital significance to the very core of Freudian psychodynamics. The sample experiments cited here purport to obviate the aforementioned shortcomings, although, of course, compared with the magnitude of the Freudian problem, the experiments may offer only pilot formulations and their contributions may so far be largely methodological.

Space forbids special discussion of the relation of applied psychosomatics to interoceptive conditioning. The fact that interoceptive conditioning involves the modifications of functions of vital organs and that it multiplies threefold the possibilities of such modifications should suffice. For, in accordance with the experimental evidence of interoceptive conditioning, not only, let us say, may feelings of anxiety, or its unconscious visceromotor accompaniments, become conditioned stimuli for the production of, let us say, constipation, but also unconscious rectal distentions, or sensations of constipation, may equally be conditioned to bring about conscious and unconscious anxiety or bring about abnormal bile secretion or gout, asthma, dyspnea, hypertension, angina pectoris, or any other functional disturbance. It should, however, be pointed out, with respect to both Freudian psychodynamics and psychosomatics, that, in general, interoceptive conditioning may perhaps best be regarded as primarily a correlate of Freud's so-called primary processes. Freudian symbolism and secondary processes pertain, no doubt, predominantly to the realm of conscious and unconscious verbal and semantic conditioning, to which the next section of the article is devoted.

Semantic Conditioning

Problem and History

Semantic conditioning may be defined as the conditioning of a reflex to a word or sentence irrespective of the particular constituent letters or sounds of the word or the particular constituent words of the sentence: that is, conditioning to meaning (more specifically, to sense or meaning since some semanticists accord the status of meaning only to sentences or propositions). Its experimental analysis is an outgrowth of the object-word (or sensory stimulus-word) and word-object (or word-sensory stimulus) CR transfer noted in the Russian laboratories of Krasnogorsky and Ivanov-Smolensky in the late '20s and early '30s, when, for instance, a conditioned reflex to the sound of a metronome carried over to

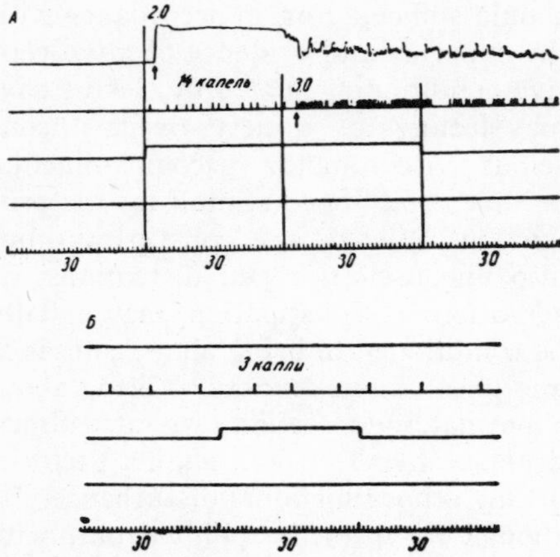

FIG. 25. Positive salivary conditioning to the word *khorosho* [good, well] and negative salivary conditioning to the word *plokho* [poorly, badly, bad] in a 13-year-old boy. (Upper figure data are for *khorosho*, lower for *plokho*. Lines from above in each figure are: motor action of jaw, drops of saliva, conditioned stimulus, unconditioned stimulus, time in seconds; arrow in second line of upper figure refers to beginning of eating. Russian words in the second line of the upper figure mean "14 drops"; in lower figure, "3 drops.")

the sound of the word metronome or vice versa. However, studies of phonetographically unrelated word transfer of conditioning, and indeed of any word-word CR transfer, originated in this country, as did also the term semantic conditioning itself. The author used the term first in *Science* (Razran, 1939b) when he had found that three college students, conditioned to secrete a mean of 249 mg. of saliva per minute to the sight of the words style, urn, freeze, and surf, had transferred their conditioning more to words fashion, vase, chill, and wave than to stile, earn, frieze, and serf.

The author's results were almost wholly duplicated by Riess (1940) who used the galvanic skin reflex and who, in addition, uncovered a positive correlation between semantic conditioning and age (Riess, 1946). In later experiments, the author reported: a crude semantic gradient to traditional association categories (subordinate, contrasts, part-whole, whole-part, coordinate, and supraordinate—Razran, 1949a, 1949b, 1949d, 1949e); more salivary CR transfer from the sentence "Poverty is degrading" to "Wealth is uplifting" than to "Poverty is not degrading" (Razran, 1949c); and a sentential distribution of conditioning in accordance with parts of speech (Razran, 1952). Several other American experimenters engaged in the problem, notably Lacey and his collaborators (Lacey & Smith, 1954; Lacey, Smith, & Green, 1955) with highly instructive results and implications. But, in general, there has been little concerted effort in the area in this country, while the Russians have, in recent years (1952 on), advanced it with special speed and planning.

Sample Russian Experiments

1. Yuri K., 13 years old, was conditioned to secrete a considerable amount of saliva when the word *khorosho* [well, good] was pronounced and to differentiate from it the sound of the word *plokho* [poorly, badly, bad]. Figure 25 shows that when the two words were put in sentences *Khorosho uchenik otvechayet* and *Plokho vorobey poyot,* Yuri secreted, respectively, 14 and 3 drops of saliva in 30 seconds. (The literal translation of the two sentences is: "Well the student answers" and "Poorly the sparrow sings"—Russian word-order is wholly flexible and each of the two adverbs could have been in any position without appearing at all unusual.) Table 1 includes data on the subject's 30-second conditioned salivation to the words *khorosho* and *plokho* and to 16 sentences. Nine of the 16 sentences must have evoked in the subject attitudes of approval (*khorosho*); six, attitudes of disap-

TABLE 1

SALIVATION OF 13-YEAR-OLD BOY TO VARIOUS WORDS AND PHRASES AFTER HE HAD BEEN CONDITIONED POSITIVELY TO THE WORD *Khorosho* [Well, Good] AND NEGATIVELY TO THE WORD *Plokho* [Poorly, Badly, Bad]

Trial No.	Time of experimentation		Words or phrases tested	Test No.	Saliv. dps. in 30 sec.
	Date	Exact time			
1	6/26/52	11:20'00"	*khorosho*	47	9
2		11:25'15"	*Uchenik prekrasno zanimayet-sya* [The pupil studies excellently.]	1	14
3		11:29'15"	*Deti igrayut khorosho* [The children are playing well.]	1	19
4		11:31'15"	*plokho*	11	2
5		11:32'45"	*khorosho*	48	15
6		11:37'00"	*Sovet-skaya Armiya pobedila* [The Soviet Army was victorious.]	1	23
7		11:42'00"	*Uchenik nagrubil ychitel'nitse* [The pupil was fresh to the teacher.]	1	0
8		11:45'15"	*khorosho*	49	18
9		11:49'45"	*Pioner pomogayet tovarischu* [The pioneer helps his comrade.]	1	23
10	7/31/52	10:10'00"	*khorosho*	50	18
11		10:14'00"	*plokho*	12	1
12		10:17'00"	*khorosho*	51	16
13		10:21'00"	*Leningrad—zamechatel'ny gorod* [Leningrad is a wonderful city.]	1	15
14		10:24'30"	*Shkol'nik ne sdal ekzamen* [The pupil failed to take the examination.]	1	2
15		10:26'00"	*khorosho*	52	15
16		10:29'30"	*Brat obizhayet sestru* [Brother is insulting sister.]	1	1
17	8/1/52	11:20'00"	*khorosho*	54	12
18		11:25'30"	*Rybaky poymali mnogo ryby* [The fisherman caught many fish.]	1	18
19		11:31'30"	*Sovet-skaya konstitutsiya—samaya demokraticheskaya* [The Soviet Constitution is the most democratic (of all).]	1	17
20		11:36'30"	*Fashisty razrushili mnogo gorodov* [The Fascists destroyed many cities.]	1	2
21	8/1/52	11:40'30"	*Uchenik razbil steklo* [The pupil broke the glass.]	1	2
22		11:41'30"	*Sovet-sky narod lyubit svoyu Rodinu* [The Soviet people love their Motherland.]	1	17
23		11:45'30"	*Moy drug tyazhelo zabolel* [My friend is seriously ill.]	1	2
24		11:47'30"	*Vrazheskaya armiya byla razbita i unichtozhena* [The enemy army was defeated and annihilated.]	1	24
25		11:51'30"	*Uchenik sdal ekzamen na posredstvenno* [The pupil passed the examination with a mediocre grade.]	1	10

proval or condemnation (*plokho*); and one (the last sentence) was most likely of an intermediate or indeterminate attitude category.

As seen from the table, the conditioned salivation during the action of the "approval" sentences ranged from 14 to 24 drops in 30 seconds. During

TABLE 2

SALIVATION OF 13-YEAR-OLD BOY TO DIFFERENT ARITHMETICAL OPERATIONS AFTER HE HAD BEEN CONDITIONED POSITIVELY TO "10" AND NEGATIVELY TO "8"

Trial No.	Time of experimentation		Arithmetical operations tested	Test No.	Saliv. dps. in 30 sec.
	Date	Exact time			
1	8/12/52	11:10′00″	83 − 73	1	15
2		11:14′00″	5 + 5	1	16
3		11:18′30″	20 − 12	1	2
4		11:20′30″	1000 ÷ 100	1	18
5		11:24′30″	5 × 2	1	19
6		11:28′00″	56 ÷ 7	1	2
7		11:32′00″	24 − 14	1	19
8	8/14/52	13:15′00″	19 − 9	1	7
9		13:19′00″	8 + 2	1	19
10		13:22′30″	48 ÷ 6	1	3
11		13:23′30″	4 × 2	1	2
12		13:24′00″	80 ÷ 8	1	17
13		13:27′30″	112 − 102	1	11
14		13:29′30″	4 + 4	1	3
15		13:31′30″	470 ÷ 47	1	11
16	8/14/52	13:33′00″	99 − 91	1	3
17		13:35′00″	80 ÷ 8	2	21
18		13:38′00″	88 ÷ 11	1	3
19		13:40′30″	35 − 25	1	25

the action of the "condemnation" sentences, the range was from 0 to 2 drops, and during the action of the "intermediate" sentences the salivation amounted to 10 drops for the 30 seconds. No secretion was evident when the experimental sentence was "The pupil was fresh to the teacher," but 24 drops were produced when the sentence was "The enemy army was defeated and annihilated," and 23 drops when it was "The pioneer helps his comrade."

Later, in a second experiment, the same subject formed a positive salivary CR to the word *desyat'* [ten] from which he differentiated the word *vosem'* [eight], and was then confronted with 19 arithmetical problems, to 11 of which the answer was "ten" and to 8 of which the answer was "eight." The results are presented in Table 2, from which it may be seen that the "answer ten" problems produced from 7 to 25 drops of saliva in 30 seconds, with a median of 17 drops, whereas none of the "answer eight" problems yielded more than 3 drops in testing periods of the same duration. The "answer ten" problems were effective even when "eight" was a part of them (Problems 9, 12, and 17).

Finally, in a third experiment, the boy was conditioned positively to the word *vosemnadtsat'* [eighteen] from which he differentiated the word *chetyrnadtsat'* [fourteen], and then presented with arithmetical problems: 9 + 9, 90 ÷ 5, 72 ÷ 4, and 2,232 ÷ 124. His results are given in Figure 26. As may be noted in the figure, 9 + 9 yielded 18 drops of saliva in 30 seconds with a latency of 2 seconds; 90 ÷ 5 produced 13 drops with a 6-second latency; 72 ÷ 4 evoked 9 drops at an 11-second latency; and 2,232 ÷ 124, 2 drops and a latency of 7.8 seconds (Volkova, 1953).

The extremely fine reflection of the subject's thoughts, and no doubt subvocal speech, in the conditioned salivation is indeed very striking.

II. Conditioned vasoconstriction was formed in nine adult human subjects to the words *dom* [house] and *doctor* [doctor] by combining the words with the application of a 10°C. round lead disc to the dorsal side of the subjects'

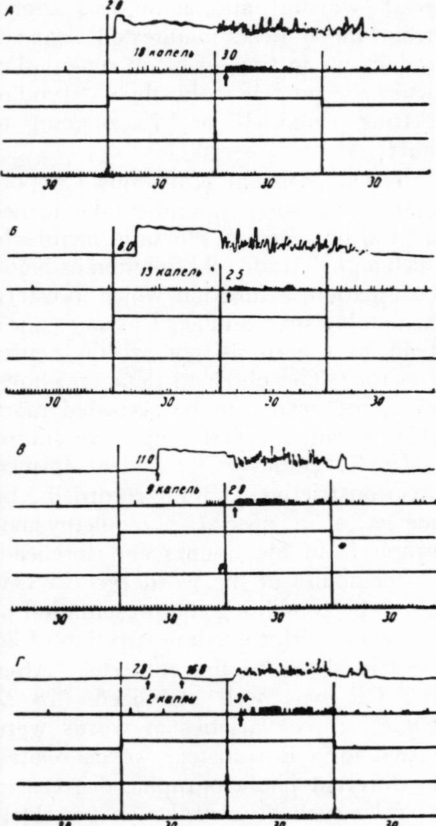

FIG. 26. Semantic salivary transfer in a 13-year-old boy conditioned to the sound of the word *vosemnadtsat'* [eighteen]. (Figures from above down show results when the transfer stimuli were, respectively: 9 +9, 90 ÷ 5, 72 ÷ 4, 2,232 ÷ 124. Lines in each figure from above are: motor action of jaw, drops of saliva, verbal stimulus, unconditioned stimulus, time in seconds; arrows refer to beginning of eating. Russian words in the figures from above down mean: "18 drops," "13 drops," "9 drops," and "2 drops.")

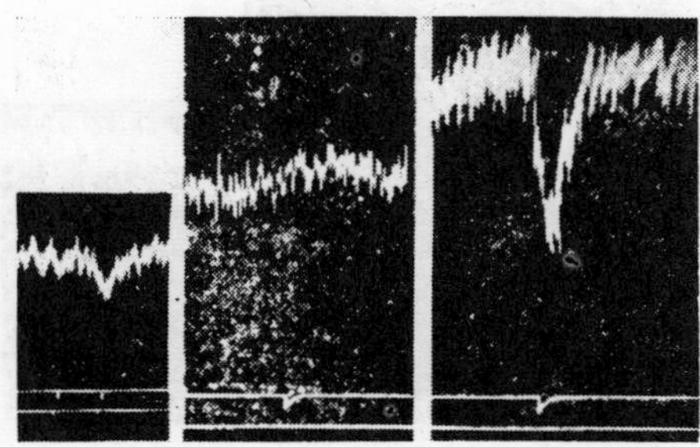

FIG. 27. CR transfer of vasoconstriction from conditioned word *dom* [house] to phonetographically related word *dym* [smoke]. (Left figure—transfer after 8 conditioned trials; middle figure—loss of transfer after 25 conditioned trials; right figure—recovery of transfer 30 minutes after administration of chloral hydrate.)

left arms and recording the conditioned vasoconstriction in both arms photoplethysmographically. CR transfers to phonetographically related words *dym* [smoke] and *diktor* [announcer] and to the semantically related and phonetographically unrelated English word house (subjects knew the language) and Russian word *vrach* [physician] were tested at early and late stages of CR training, and after the administration of 1 gm. of chloral hydrate. Figure 27 shows that the phonetographic CR transfer to the word *dym* was evident only in the early stage of training (8 trials); that it disappeared when the CR was well established (after 25 trials); and that, strikingly, the transfer reappeared 30 minutes after the administration of chloral hydrate to the subjects. Figure 28 brings out the finding that the semantic CR transfer from the Russian word *dym* to the English word house disappeared 30 minutes after administration of the same dose of the drug. The results with the words *dictor* and *vrach* are said to have been quite similar (Shvarts, 1954, 1960).

The experiment thus demonstrates, first, that semantic conditioning is a

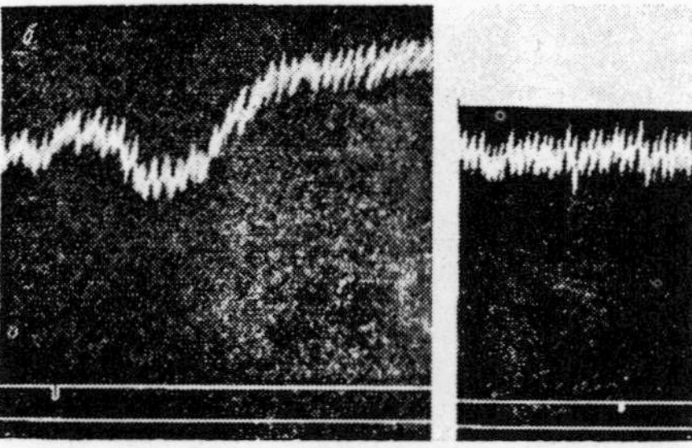

Fig. 28. The effect of the administration of chloral hydrate upon semantic CR transfer of vasoconstriction from conditioned word *dom* to English equivalent house (subjects knew English). (Left figure—transfer before administration of the chloral hydrate; right figure—loss of transfer after the administration of the depressant.)

manifestation of a higher level of our learning potentialities and, second and moreover, that the lower level, the phonetographic manifestation, is not nonexistent in us but is held in abeyance and reasserts itself in periods of lower organismic functioning. As such, the results are seemingly in line with those of Riess (1940, 1946) on the positive correlation between semantic conditioning and age, with Luria and Vinogradova's (1959) comparisons of semantic conditioning in normal and feebleminded children, with some of my own findings, and with what might be expected on general principles.

III. Markosyan (1953) has demonstrated that the reduction in time of blood coagulation produced by the administration of an electric shock could be conditioned in nine rabbits to the sound of a whistle or of a metronome of three beats per second, and that the CR to the metronome could be differentiated so that it was not evoked by beats of one per second. In a later experiment (1958), Markosyan reported that human subjects (he does not state how many) formed coagulation CRs to the sound of a metronome and that these CRs transferred from the sound of the metronome to the sound of the word metronome and to phonetographically related words *metrostroy, metropol', mikrotom,* and *mikroskop*. When now he had conditioned the blood coagulation to the flash of an electric lamp, he found CR transfer to the words *lampochka* [lamp], *fonar'* [lantern], *svet* [light] but not to the word *svist* [whistle], and, again, that after a while the subjects manifested clear-cut reductions in time of blood coagulation when he merely told them "[you're getting a shock]" or "[it is going to hurt]."

The experiment is obviously important in disclosing experimental evidence on the presence of semantic factors in such a phyletically old function as blood coagulation, a function which Western hematologists, unlike Russian ones, tend to regard as not within neural control. The obtained data are, however, too scanty to be accorded more than a tentative and suggestive status.

IV. Seven university students formed vasoconstrictive CRs, recorded by means of a special photoplethysmograph from the fingers and forehead, to the sound of the word *skripka* [violin], by combining the presentation of the word with the administration of an electric shock to the subjects. After the CR was well stabilized (18–27 trials), three varieties of words were tested for CR transfer: words related in different phonetographic degrees to *skripka* such as *skrepka* [paper-clip], *strizhka* [hair cutting, shearing], and *skrytnost'* [reticence, secrecy]; words related in different semantic degrees such as *smichok* [violin bow], *gitara* [guitar], *struna* [string], *mandolina* [mandolin], *arfa* [harp], *baraban* [drum], *orkestr* [orchestra], *sonata,* and several others; and wholly unrelated words such as *stakan* [glass], *lenta* [ribbon], *voda* [water], etc. The main results show transfer to all the semantically related words, phonetographic transfer to only the word *skrepka,* and, of course, no transfer to the unrelated words. The semantic transfer was of two kinds: one in which the transfer words duplicated the vasomotor changes of the conditioned word, viz., vasoconstriction of both the finger and the forehead blood vessels, and one in which the transfer words elicited vasoconstriction of the vessels of the fingers but vasodilatation of the vessels of the forehead. The phonetographic transfer to *skrepka* was only of the second, vasoconstriction plus vasodilatation, kind—a kind which the Russians identify with the arousal of the orienting reflex. By and large, with respect to the semantically related words, the duplicative kind of transfer was manifest when the transfer words were closely related to the conditioned word, and the orienting kind when the relationship was less close.

Other results of the study might be summed up as follows. (*a*) The amount of the CR transfer—that is, the magnitude of the vasomotor change—was largely a positive function of the degree of semantic relationship of the transfer words to the conditioned word. (*b*) In the course of CR training, the transfer became restricted in scope, duplicative transfer changing into the orienting variety and orienting transfer disappearing altogether. (*c*) The scope of the transfer readily widened when a semantically related transfer word was reinforced with the electric shock. (*d*) However, when the reinforced word was unrelated to the conditioned word such as the word *korova* [cow], a new family of semantic-transfer words such as *telyonok* [calf], *loshad'* [horse], *byk* [bull], and *stado* [herd] came into being. (*e*) When a word had a double meaning, one related to the conditioned word and one unrelated, the word manifested transfer when it was applied after related words and no transfer when it was applied after the unrelated words. (*f*) Subjects' knowledge of the purpose of the experiment and experimenter's instruction modified considerably the course of the conditioning and the transfer but did not wholly nullify or reverse it. (*g*) Postexperimentally, subjects sometimes reported that transfer words had been accompanied by shock stimulations, and there was in general little correlation between postexperimental recall of a word and CR transfer to it during the experiment; indeed, some subjects reported words that had not been used by the experimenter.

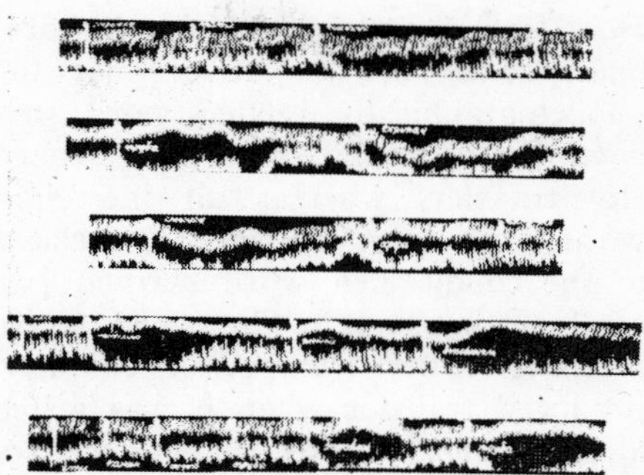

FIG. 29. Verbal transfer of vasomotor CR formed to word *skripka* [violin], with electric shock as the US. (Two subjects—upper four strips, one subject; lowest strip, the other. Lower record—finger plethysmogram; upper record—forehead plethysmogram. Words from upper left: *strizhka* [hair cutting, shearing], *plat'ye* [clothes], *pianino* [piano], *pugovitsa* [button], *truba* [trumpet, chimney], *skripach* [violinist], *smychok* [violin bow], *arfa* [harp], *vilka* [fork], *mandolina* [mandolin], *fleyta* [flute], *violonchel'* [violoncello], *stul* [chair], *kamin* [fireplace], *pechka* [stove], *truba* [trumpet, chimney], *baraban* [drum], *truba* [trumpet, chimney].)

Figure 29 presents data for two subjects (upper four strips for one, lowest strip for the other). As may be seen from the figure, none of the seven unrelated words—*plat'ye* [clothes], *pugovitsa* [button], *kastryula* [saucepan], *vilka* [fork], *stul* [chair], *kamin* [fireplace] and *pechka* [stove]—nor the phonetographically related word *strizhka* [hair cutting, shearing] evoked any transfer, whereas all the eight words that were semantically related to the conditioned word *skripka* [violin] produced it. The word *truba,* meaning either trumpet or chimney, produced transfer when it was tested after *baraban* [drum] but was ineffective when tested after *pechka* [stove] preceded by tests for *kamin* [fireplace] and *stul* [chair]. Again, it may be noted in the figure that in the early stages of the CR training the words for piano, violinist, and violin bow elicited duplicative transfer while the words for trumpet and harp evoked the orienting kind (upper three strips of figure); and that in later stages (fourth strip) the magnitude of the transfer for the words meaning mandolin and violoncello was larger than that for the word meaning flute (Vinogradova & Eysler, 1959).

The cited experiment no doubt contains a comprehensive array of significant findings on semantic conditioning. Unfortunately, however, because of a total absence of statistical treatment of data, only the main results of the existence of semantic CR transfer, the near-absence of phonetographic transfer, the existence of the semantic transfer in both a duplicative and an orienting form, and perhaps also the positive effects of context on semantic CR transfer could be considered demonstrated with any degree of significance.

V. Ten normal school children, 11–15 years of age, and 15 feebleminded children of different degrees of feeblemindedness, 13–17 years of age, were told to press a button when they heard the word *koshka* [cat], which resulted in consistent constriction of the blood vessels of the children's fingers and dilatation of the vessels of their forehead, in response to the mere sound of the word. A large number of words related semantically or phonetographically to *koshka,* and some neutral unrelated words, were now presented to each child to test for resulting transfers of the vascular reactions. Clear-cut differences are stated to have been obtained between the normal and the feebleminded children and among the children of different degrees of feeblemindedness. The normal children transferred their vascular reactions only to semantically related words. The children of moderate degrees of feeblemindedness manifested both semantic and phonetographic transfers, while the transfer of the children of extreme feeblemindedness was only phonetographic. (There are no IQs in the USSR.) Again, it is stated that in the moderately feebleminded children repeated reinforced presentations of the instructional word *koshka* increased the amount of semantic transfer; that a presentation of the word *okoshko* [window] in its own semantic context—after the words *derevo* [tree], *dom* [house], *dver* [door], and *stena* [wall] —decreased phonetographic transfer; and that fatigue in general decreased semantic transfer and increased the phonetographic kind.

The experimenters (Luria & Vinogradova, 1959) present data for 6 of their 25 subjects in six figures. Figure 30 is a composite of four of these figures. The upper strip of the figure shows that a normal child, 11–12 years of age, manifested semantic but not phonetographic transfer—to the words *sobaka* [dog] and *zhivotnoye* [animal] but not to the words *okoshko* [window] and *kroshka* [crumb]. The upper middle strip of the figure shows that the transfer of a 16-year-old "mild oligophrenic" or "debile" child was both semantic and phonetographic—to both *sobaka* [dog] and *kroshka* [crumb];

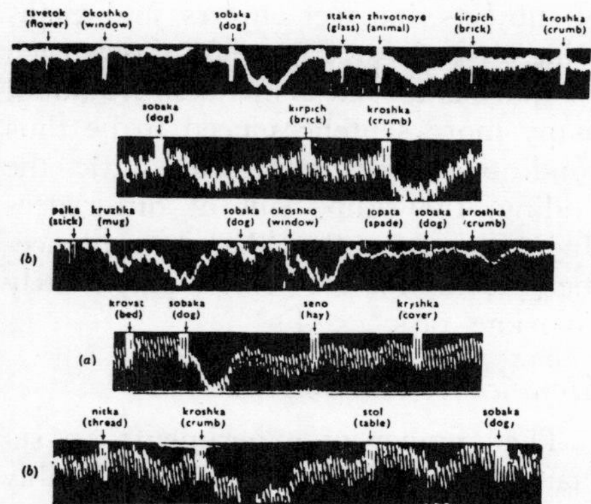

FIG. 30. Verbal transfer of a vascular orienting reaction to word *koshka* [cat] in four school children. (Upper strip—normal 11–12-year-old child; upper middle strip—16-year-old "mild oligophrenic" or "debile"; middle strip—16-year-old "imbecile," greater degree of feeblemindedness; lower middle strip—17-year-old "debile" at beginning of school day; lowest strip—same subject after 6 hours of schoolwork, effects of fatigue.) (Reproduced from the *British Journal of Psychology* by permission)

while the lower middle strip shows that the transfer of a 16-year-old "imbecile" (greater degree of feeblemindedness) was only phonetographic—to *kryshka* [cover] and *kroshka* [crumb] but not to *sobaka* [dog]. The lowest strip of the figure demonstrates that a "17-year-old debile," who transferred his vascular reaction to *sobaka* [dog] but not to *kryshka* [cover] before his school day began, reversed the transfer—to *kryshka* but not to *sobaka*—after 6 hours of schoolwork. (Experimenters' figures not represented in the composite figure pertain to effects of context and of repetition of reinforcement.)

As in the preceding experiment, the data are not reported to have been treated statistically and, as indicated, records of only one subject of each of the three groups varying in intelligence —normal, "debile," and "imbecile"— are presented. Yet, it might be pointed out that the results on the relation of semantic conditioning to intelligence are quite in line with those of Riess (1946) on the relation of this conditioning to age, with some of my own studies, and with common sense; and that the results on the effects of fatigue are quite similar to those of the administration of chloral hydrate in the study by Shvarts (1954) discussed earlier. Moreover, the experiment is of special methodological significance in offering a simple way of studying semantic transfer without recourse to laboratory conditioning. A full report of this experiment and a comprehensive summary of the one preceding appeared in the May 1959 issue of the *British Journal of Psychology*.

VI. Thirty university students, 20–29 years of age, were conditioned to withdraw their fingers when they heard any of three short Russian sentences by combining each of the sentences with the administration of an electric shock to the subjects' fingers. The three sentences were: *Vklyuchayu tok* [I am switching on / the shock], *Rukopis' prochitana* [The manuscript / was read], and *Student vyderzhal ekzamen* [The student / passed / the examination]. (The diagonal bars in the parentheses mark off the equivalents of each Russian word.) After the withdrawal CRs to the sentences were stabilized, CR transfer to the separate words of each sentence was tested. It was found that in the first sentence each of the two words acquired 100% transfer on the first trial; that in the second sentence neither word showed any immediate transfer; while in the third sentence, the word *vyderzhal* [passed] manifested transfer in 87% of the cases, the word *ekzamen* in 50%, and the word *student* in only 10% of the tests. The experimenter attributes the results to differences in total sentential "meaning load" carried by the separate words of the respective sen-

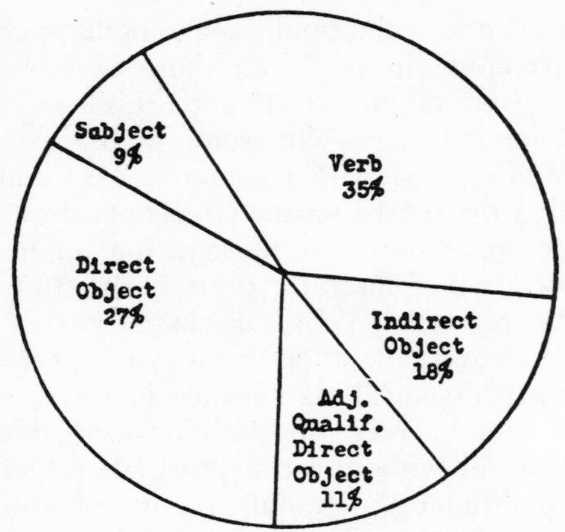

FIG. 31. CR semantic analysis of "I gave him a new ball" and *"Ya dal yemu novy myach."* (Relative CR weights of individual syntactical units.)

tences: that is to say, in the first sentence the total sentential meaning was implied in each of the two words, in the second it was implied in neither, while in the third the sentential meaning was differentially distributed among the three words (Elkin, 1955).

In a previously reported experiment the writer studied the semantic role of parts of speech in salivary conditioning (Razran, 1952). He used three five-word Russian sentences, the meaning of which the subjects learned pre-experimentally, and eight subjects, four of whom were conditioned to the sentences and tested with separate words, and four for whom the sequence was reversed. The sentences were: *Ona khochet kupit' serry koshelyok* [She / wants / to buy / a gray / purse], *Belokuraya devochka nashla zolotoye pero* [The blond / girl / found / a gold / pen], and *Ya dal yemu novy myach* [I / gave / him / a new / ball]. As might be expected, the results of the three sentences did not fully correspond. Yet, they did indicate the preponderant semantic role of verbs and direct objects, the considerable role of infinitive complements, and the rather small role of subjects, indirect objects, and qualifying adjectives. Figure 31 shows such an analysis for one sentence. No doubt many more sentences need to be thus conditioned and tested to settle the finding, the ruling out of differences due to lexical rather than to sentence-membership factors being a particularly laborious task.

Significance and Implications

The number of experiments on semantic conditioning is considerably smaller than that on interoceptive conditioning. Yet in a way these experiments uncover an area of even wider significance, a vast realm of experimentally separable meaning-units continually entering in complex functional relationships with each other and with nonmeaning units, existing both in and out of consciousness, and normally, whether conscious or unconscious, forming controlling rather than controlled systems. Small wonder that the significance of semantic conditioning extends to such varied fields as: the psychology of thinking, ranging from the *unbewusste Bewusstseinslagen* of the Würtzburg school to modern views of cognition as a high level, controlling mechanism of which we may not be aware; Freud's symbolisms, secondary processes, and conscious-unconscious ego relations; cultural anthropology; clinical observations on aphasia and agnosia; practical problems in psychotherapy; general linguistics; general semantics; and such applications as education techniques and the psychology of advertising and of propaganda.

Unfortunately, space permits only a very brief and basic discussion of the topic. Moreover, it seems desirable to postpone the discussion until the topic of the orienting reflex will have been considered. Still, we might mention here that current Russian psychophysiology, quoting Pavlov, accords a *sui generis* status to verbal conditioning in general, regarding it as not merely a case of simple second-order conditioning but as a special higher level "signal-of-signals" CR basis of abstraction of high level generalization. That a word is by its very nature an abstraction, a statement found in Pavlov's writings, is taken very seriously in both its genetic and its ontological aspects as a problem fit for both experimental analysis and clinical research.

The Orienting Reflex

Historical Background

The conditioned stimuli—the sounds of bells, buzzers, whistles, metronomes, etc.; the flashes of lights and the sights of rotating whirligigs and other objects, etc.; the rhythmical scratchings and touchings, etc.; the different thermal stimulations; and the various odors—which Pavlov used to condition the food reactions, and later also the defense reactions, of his dogs were what we call sensory stimuli: that is, stimuli that are accompanied in human subjects by sensory experience, indeed, that *are* the body-material of almost all our ordinary experience. Pavlov, however, was no more concerned with the stimuli's preconditioning sensory experiences than with their postconditioning image experiences—both sensations and images were left in limbo in 1902 when he decided to abandon the term psychic (notably, psychic secretion used by him since 1886) for that of conditioned. Still, the stimuli evoked in the dogs some overt, though often not very pronounced, reactions, and these reactions Pavlov named the "orienting" or the "what-is-it" or the "investigatory" or the "attitudinal" (in Russian, *ustanovychny*) reflex, using the particular names nondifferentially. The stimuli themselves Pavlov called "indifferent stimuli," added evidence that he did not accord their original preconditioned reactions any particular per se significance in his system.

Later on, in the laboratory, Pavlov noted that conditioning proceeded best when the orienting reflex (for convenience, it will be abbreviated as OR) is neither too large nor minimal or absent, and that in the course of CR training the OR tends to disappear—characteristics that might have been correlated with conscious experience by saying, for instance, that the conditioned stimuli must arouse adequate but not overwhelming consciousness and that consciousness tends to disappear as habit develops. But, again, such formulations became, after 1902, wholly alien to Pavlov. Dedicating his entire program of experimentation to, we might say, an isomorphic study of two Cs, the Cortex and Conditioning, he had no use for the isomorphism of the third C, that of Cognition or Consciousness. And, although less radical with respect to the general status of consciousness than early American behaviorists and, unlike them, in no way involved in any attempt to "explain it away," he, nonetheless, was very militant about its invalid methodology. Years before the rise of American behaviorism, in 1904, Pavlov stated that "one must renounce completely psychological formulations and remain standing upon a purely objective ground" (Pavlov, 1928, p. 75; 1949a, Vol. 3, p. 51). And in 1910 he wrote:

> Why had we formerly, like cowards, embraced the subjective method? . . . The subjective method is a method of causeless thinking, psychological reasoning is indeterminate reasoning, recognizing phenomena but not knowing whether they come from here or there. To be content with saying "the dog thought," "the dog wishes" is a fiction, not a disclosure of the cause of the phenomenon. Our [former] psychological explanations were fictitious and groundless. Our [present] objective explanation is truly sci-

entific, always turning to causes, always seeking causes (Pavlov, 1928, p. 165; 1949a, Vol. 3, p. 138).

This total Pavlovian rejection of psychology and of special psychological cognitive-perceptual problems continued for a number of years after Pavlov's death (1936). In time, however, changes occurred in the wake of special problems, new techniques, and modified general points of view. The special problems which contributed to the extension of Pavlovianism to the cognitive-perceptual dimension fall, interestingly, into two contrasting categories: those dealing with predominantly conscious reactions and emphasizing the dimension directly, and those dealing with unconscious ones and emphasizing it by contrast. The extensive and very successful investigations of the Lazarev-Kravkov school of sensory interactions (Razran, 1958, p. 1187), the highly original Makarov experiments (1950a, 1950b, 1952) on interoceptive sensibility, and the aforementioned studies of verbal-semantic conditioning are the outstanding examples of the first category of special problems. Gershuni's experiments (1947, 1955, 1957a, 1957b; Gershuni, Klass, Lukomskaya, Linyuchev, & Segal, 1959) on subliminal auditory conditioning and Pshonik's (1949a, 1949b, 1949c, 1952) on subliminal thermal conditioning, as well as the entire area of interoceptive conditioning, are cases of the second category of problems.

Again, post-Pavlovian CR and UR techniques and methodologies began to move rapidly from what might be called a "monoeffector" to what the Russians call a "polyeffector" stage. That is to say, as may be judged from what has been said so far, the old Pavlov and Bekhterev way of measuring changes in reactivity and conditionability in only one effector such as salivation or paw-withdrawal was replaced by simultaneous recordings of changes in a wide, almost an all-possible, variety of effectors, a state of affairs which almost automatically inducted series of correlations among differing indices of organismic change, and thus also of correlations with the cognitive index. Indeed, in very recent years Russian psychophysiologists have even resorted to such traditional devices as correlating subjects' rankings of stimulations and stimulation changes with objective pointer readings. And spectacular Russian engineering advances multiplied of course the number of possible pointer readings of psychophysiological processes.

Finally, with respect to modified points of view and cognition, one should mention that Pavlov's immediate successor, the late L. A. Orbeli, was a student of the German psychophysicist Hering; that J. S. Beritov (Beritashvili) became much interested in what he called "psychonervous behavior" in his later experiments; and that K. M. Bykov, P. K. Anokhin, and E. A. Asratyan have in general been friendly to psychological approaches and have tended to introduce some innovations in the Pavlovian system (the last two more than the first). And last, but by no means least, there is the fact of the rapprochements between psychophysiology and psychology since 1950—rather, the since-1950 Pavlovianization of psychology—and the deeper penetration of Marxism-Leninism and its emphasis on the emergent efficacy of consciousness, which Pavlov resisted but which his successors could not or would not resist in the same way.

Just how the Russian psychophysiologists' experimental interest in problems of cognition and perception has become, and is continuing to become, closely connected with experimental problems of the orienting reflex is the subject of the last section of the present article. Suffice it to say here only that a recent Russian book, *Perception and the Conditioned Reflex* (Sokolov, 1958c), might as well, and indeed more appropriately, have been named "The Orienting Reflex and the Conditioned Reflex" or "The Orienting Reflex, Perception, and the Conditioned Reflex."

Characteristics

Two collections of experiments wholly devoted to the study of the OR appeared in the Soviet Union recently, one in 1958 and one in 1959. The 1958 collection, entitled *The Orienting Reflex and Orienting-Investigatory Activity* (Voronin, Leont'yev, Luria, Sokolov, & Vinogradova, 1958), consists of 48 reports of experiments divided into six parts: General Problems; the Orienting Reflex and Problems of Reception (Perception); the Phylogeny of the Orienting Reflex; Its Ontogeny, Pathology, and Special Problems in Man. The 1959 collection, named *The Orienting Reflex and Problems of Higher Nervous Activity* (Sokolov, 1959a), comprises only 14 experiments but is nonetheless quite comprehensive; it contains, for instance, a total of 116 figures and a bibliography of 385 titles. And to these 62 OR experiments, should be added at least as many that have been reported in the periodical literature, mostly in the last 5–6 years. There is thus no doubt that the OR is a prime focusing center in current Soviet psychophysiological and psychological research and thought, particularly if one remembers that its experiments purport to provide not only new information but a new look at old information: that is, the OR is put forward as a new concept, although, as will be seen later, in a number of aspects it is really only a renovated one and in some respects it is one that under different names is not at all new to American and Western students of behavior.

The total material of the OR, unlike that of interoceptive and semantic conditioning, does not lend itself readily to summing up and discussing through sample experiments. A better way of presenting it would seem to be an unfolding enumeration and discussion of experimentally established and indicated OR characteristics. Six sets of such characteristics will be considered:

Reactional primacy and holistic specificity. The OR is the organism's normal first reaction to any adequate normal stimulus or, in terms of the stimulus, it is the organism's normal reaction to a normal *novel* stimulus. Novelty does not mean no previous presentations of the stimulus but merely that the stimulus possesses the OR evoking novelty for only a limited number of presentations—the exact number depending upon a variety of parameters to be discussed later—after which the stimulus either evokes no reaction at

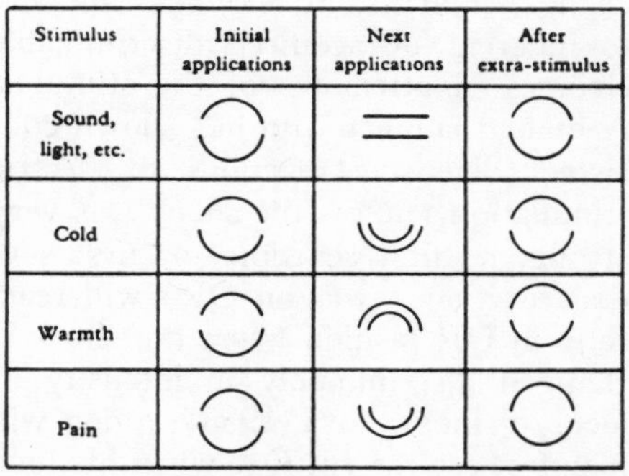

Stimulus	Initial applications	Next applications	After extra-stimulus
Sound, light, etc.			
Cold			
Warmth			
Pain			

FIG. 32. Vasomotor pattern of the orienting reflex in human subjects. (Concave arc above convex arc—constriction of digital blood vessels plus dilatation of forehead blood vessels; double line—no observable vascular change in either digital or forehead blood vessels; two concave arcs—constriction of both digital and forehead blood vessels; two convex arcs—dilatation of both digital and forehead blood vessels.) (Reproduced from the *British Journal of Psychology* by permission)

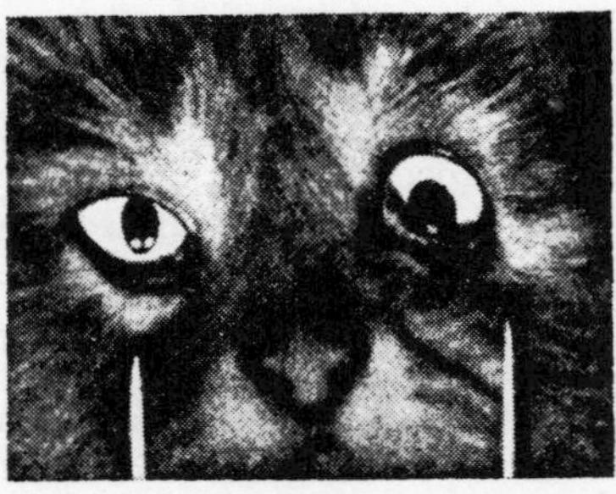

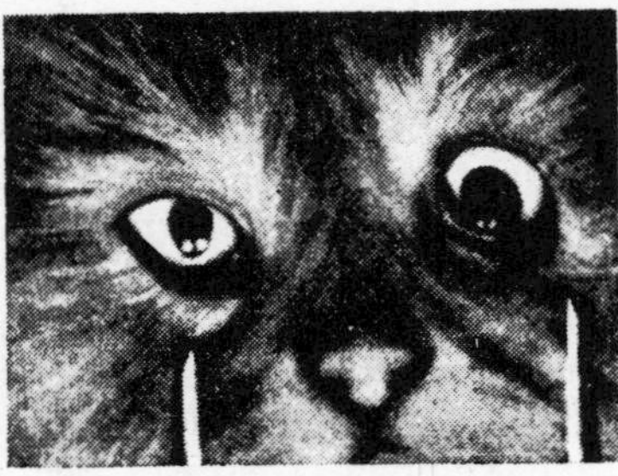

Fig. 33. Pupillary dilatation to sound in cats as part of the orienting reflex. (Left eye is sympathectomized.)

all or a reaction that is different from an OR. And novelty includes also change, that is, it extends liberally to inherent or acquired discriminable changes in attributes of the stimulus. A human subject who has shown consistent defensive reactions to electric stimulation (unless the shock was very strong, its first reaction—or first several reactions—was an OR) will react fully in OR fashion when the shock is changed discriminably in intensity or locus (Vinogradova, 1958), a dog will manifest a clear-cut OR when his food is changed in kind (Anokhin, 1958, p. 17), and a rabbit fully accustomed to a hammock as a vivarium will resume all old OR action upon shock administration (Anokhin, 1958, p. 15).

The OR is in no sense a single reflex but is in all respects a centrally organized, holistic system of a variety of specifically distinguishable visceral, somatic, cognitive, neural, and neuromotor reactions (Sokolov, 1958c). The specificity of the OR reactions pertains to kind, amplitude, latency, and general pattern, and the Russians have been working at full speed to detect and ascertain each distinctness experimentally. At least in human subjects, they seem to have clearly established, as was already indicated earlier, that the dilatation of the blood vessels of the forehead accompanied by the constriction of the digital blood vessels is an OR reaction and only an OR one (Vinogradova, 1958, 1959a, 1959b; Vinogradova & Eysler, 1959; Vinogradova & Sokolov, 1955). Figure 32 (Luria & Vinogradova, 1959), based upon data of a large number of experiments and subjects, illustrates fully this vascular distinctness of the OR in response to stimuli of all modalities, and shows as well the reactional primacy of the OR, its modifiability through repetition, and its reappearance in novel settings. The Russians have also experimented extensively with pupillary dilatation as an OR component. Their data show dilatation to be almost invariably the first reaction to nonvisual stimuli and to occur in about 20% of first reactions to visual stimulations; the 80% of pupil constriction to visual stimuli are interpreted as cases of inhibition of the dilatative OR reaction by the constrictive local-adaptive reflex (Liberman, 1958; Shakhnovich, 1958).

Likewise, the galvanic skin reflex, electroencephalographic desynchronizations, a variety of respiratory changes, and special oculomotor, digital, manual, pedal, and general bodily motor and proprioceptive reactions are being studied carefully and delimited as OR components. Figure 33 (Shakhnovich, 1958, p. 197) shows an OR dilatation of the pupils of a cat's eyes in response to sound, with the muscles of the left iris sympathectomized, and Figure 34 is an oculomotor chart of a subject

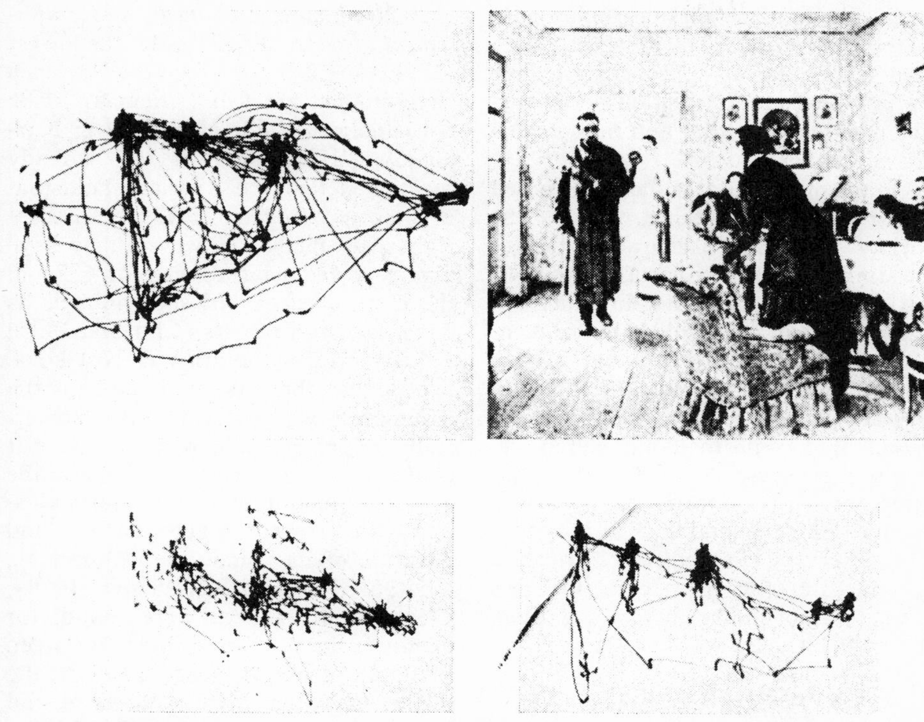

FIG. 34. Eye movement chart of a subject watching painting "'*Nye Zhdali*" [They Did Not Expect Him] by Repin for 10 minutes under three conditions. (Left—uninstructed; lower left—told to think what family was doing when the visitor arrived; lower right—told to estimate ages of members of family.)

looking at a Repin painting "*Nye Zhdali*" [They Did Not Expect Him] under three conditions: without instructions, thinking of what the family was doing when the visitor arrived, and estimating ages of members of the family. Figure 35 (Bronshteyn, Itina, Kamenetskaya, & Sitova, 1958, p. 239) is a record of cessation of respiration in a 72-hour-old human neonate in response to stimulation with a light, which, too, is held to be a specific OR component (non-OR respiratory changes are allegedly characterized not by cessation but by alterations of amplitude, frequency, and pattern). Characteristic stimuli produced cessations of reactions must in general, it is argued, be placed alongside evocations of reactions in determining OR patterns—e.g., the "listening to" of birds and mammals and the "frozen-in-thought" states of apes and children during the solving of problems (Dolin, Zborovskaya, & Zamakhover, 1958). And of course there is in the fully developed OR the specific molar reaction of "turning towards the source of stimulation," a reaction which the author

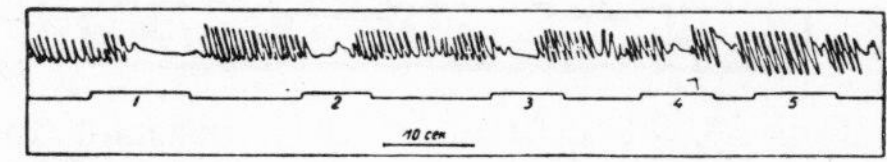

FIG. 35. Respiratory change in human neonate 72 hours after birth, upon stimulation with lights, as part of the orienting reflex. (Note extinction of response after four trials.)

would like to name versive, with a note that this reaction is outside the much used dichotomy of approaching-avoiding or adient-abient reactions—the versive reaction is neither and may indeed precede either.

On the other hand, it should be noted that the OR is specific only to the novelty and change characteristics of the stimuli that evoke it but not to their intrinsic nature, and that in general it might be said that, unlike the alimentary or defensive or sexual reaction patterns of the organism, the OR pattern does not really "manage" the stimuli that come to it but merely reacts to their presence, its reactions thus being more preparatory than consummatory and preadaptive than adaptive. Yet these very characteristics impart a general controlling role to the OR and a function not unlike that of cognition, about which more will be said later.

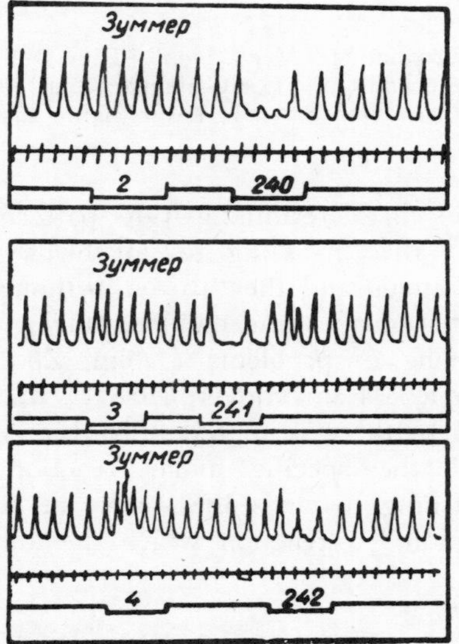

FIG. 36. Persistent (highly resistant to extinction) respiratory reaction of a hare to a rustling sound, and the conditioning of this reaction to the sound of a buzzer. (Lines from above: respiration, time in seconds, application of buzzer and of rustling sound. Numbers denote ordinal trials.)

Evolutionary, ecological, and pathological aspects. Phyletically, the lowest animals the ORs of which the Russians have studied are fish (Chumak, 1958; Gusel'nokov, 1958; Vedyayev & Karmanova, 1958), and it was already mentioned that the Russians hold that in human ontogeny ORs are present right after birth. However, they have also reported near-absence of ORs in the early ontogeny of such animals as dogs, cats, and rabbits (Chumak, 1955; Nikitina, 1954; Nikitina & Novikova, 1958), and there is no doubt that the assumption is that the OR is a phyletically recent reaction pattern. And it is of course a pattern that manifests the greatest amount of evolutionary plasticity with respect to type, variety, and duration of its reactions (Dolin et al., 1958; Vedyayev & Karmanova, 1958). Clear-cut differences were found, for instance, in one experiment between the ORs of monkeys and of dogs, the former showing invariably more and longer lasting reactions and in addition a much greater ratio of somatic to visceral components (Nikitina & Novikova, 1958). And similar results were obtained in comparing the ORs in the first months of normal and of prematurely born children (Polikanina & Probatova, 1955, 1958; Voronin et al., 1958), and in age-specific comparisons of infants (Karlova, 1959) and of baby rats, rabbits, dogs, and monkeys (Volokhov, Nikitina, & Novikova, 1959). Yet, with respect to effectiveness of stimuli that evoke ORs, differences between animals appear to be often more a matter of ecological than of purely phyletic factors. Rustling sounds which are very weak OR stimuli in dogs are very powerful ones with hares. Figure 36 (Klimova, 1958, p. 79) shows that 240 applications of a rustling sound failed to extinguish completely a hare's respiratory OR changes to it, and that furthermore the sound of a buzzer which originally was ineffective became conditioned to the rustle reaction in three trials. Other specific and particularly effective OR evoking stimuli were found to be: the sight of a cat for owls, the odor of rosemary and hunters' decoy sounds for ducks, the displacements of objects in the field of vision and sucking sounds for hares, the sound of wood splintering for beavers, and the sound of waves splashing for fish (Klimova, 1958, p. 77). While it is possible that some of the effectiveness of the particular stimuli was due to "natural" pre-experimental reinforcements, the consistency of the results and the fact that young animals were used in some cases point to congenital ecological differences.

The relation of the OR to human pathology is evident in the difficulty of arousing it in schizophrenic and head injured patients and in feebleminded children. Figure 37 (Gamburg, 1958, pp. 273, 277) demonstrates that a concealed sound (experimenter does not state its nature) which evoked consistent OR somatic and visceral reaction patterns in 29 of 34 healthy subjects, aroused the pattern in only 4 of 42 schizophrenic and in 2 of 28 head injured patients. (As may be seen in the figure, the 5 remaining healthy subjects and the 26 remaining head injured patients responded to the stimulus with defensive reactions, while 15 of the schizophrenics reacted defensively and 23 did not react at all.) Figure 38 (Dolin et al., 1958, p. 57) shows that the sound of a buzzer produced a vascular change in a schizophrenic patient only on the eighth presentation. The results with feebleminded children, compared with normal ones, run in the same direction. In one experiment, all 20 normal children manifested vascular OR reactions to the flash of a light, the sound of a metronome, and the sound of spoken words, but of 40 feebleminded children, only 20, 31, and 35 reacted so to the three respective stimuli (Vinogradova, 1959b, p. 172). Similarly, in another experiment, with 12 feebleminded children and the use of pure tones as stimuli, it was found that one child produced a psychogalvanic OR reaction only on the fifth presenta-

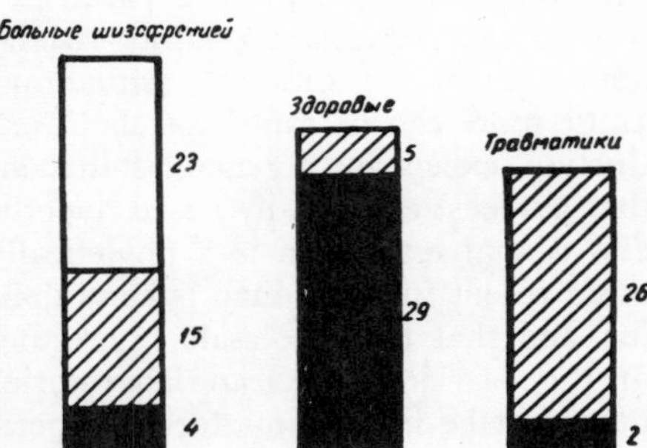

Fig. 37. First reactions of 42 schizophrenics, 28 head injured patients, and 34 normal subjects to equal auditory stimulation. (Blackened area, orienting reflex pattern; diagonal-line area, defensive pattern of reactions; clear area, no reactions. Left bar, schizophrenics; middle bar, normal; right bar, head injured. Numbers of subjects showing each reaction pattern are indicated to the right of the bars.)

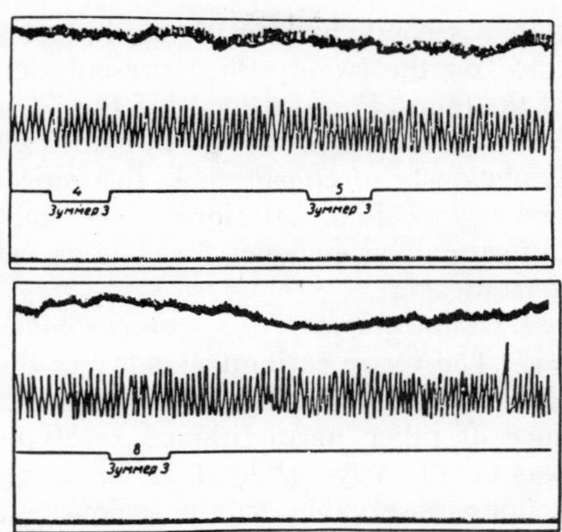

Fig. 38. Absence of vascular and respiratory reactions of a schizophrenic patient to the sound of a buzzer and the appearance of a vascular change only on the eighth stimulation. (No data are given on the first appearance of the respiratory reaction.)

tion, a second child on the ninth, and a third on the twenty-third presentation of the tone (Paramanova, 1959, p. 78). This disruption of the OR in pathology is obviously of considerable theoretical and practical-clinical significance; but, unfortunately, evidence for it is as yet very meager.

Extinguishability and transformability. The ready extinguishability of the OR in dogs in contrast to the persistence of other unconditioned reactions was noted early—at least as far as reactions observable to the naked eye are concerned—in Pavlov's laboratory. Pavlov himself stated that the extinction of the OR "is analogous in details to the extinction of conditioned reflexes" (Pavlov, 1927, p. 256; 1949a, Vol. 4, p. 269) and that "this reflex has points of applications [representations] in both the cortex and the lower-lying regions of the brain" (*ibid.*), and later Orbeli (1949) declared: *"There exist reflexes* [in reference to the OR] *that are innate in origin yet function fully as acquired reactions"* (p. 736). And while present-day data suggest the need of qualifying considerably the generalization of the ready extinguishability of the OR with respect to some stimuli, reactions, and subjects, the same qualification applies equally to the extinguishability of the CR which, too, may be very difficult with largely similar types of stimuli, reactions, and subjects. In other words, Pavlov's original statement about the extinction similarity of the OR and the CR—and we might add also, as will be seen later, his view of the combined cortical-subcortical neural basis of the OR—is in the main still quite valid, although one might well argue with Orbeli and say that extinction similarity of reactions does not mean total functional similarity.

The main indicated parameters of OR extinguishability are: (*a*) direct relationships with the phyletic position of the subjects (Obraztsova, Pomazanskaya, Stel'mach, & Troshikhin, 1958, p. 251; Vedyayev & Karmanova, 1958, p. 204) and the phyletic recency of the specific OR component (Sokolov, 1958b, p. 113)—animals higher in the phyletic scale extinguish their ORs more readily, and the PGR component of the OR extinguishes much more quickly than the respiratory component; (*b*) inverse relationships with intensity of the OR stimulus (Petelina, 1958), amount of brain extirpation in animals (Karamyan, 1958), and organic and functional psychopathology in human subjects (Lichko, 1952; Sokolov, 1959b, p. 48; Usov, 1952)—lesser extinguishability in each of the three cases. And to these should be added: (*c*) the inverse relationship with the ecological significance of the OR stimulus; and (*d*) the unsettled—sometimes direct, sometimes inverse—relationship with intelligence of children as subjects (Polikanina & Probatova, 1958). The first two sets of parameters as well as the fourth parameter operate equally in CR extinguishability (Razran, 1956); while the parameter of ecological significance is not readily applicable to conditioning situations. Little more can be said here about extinction, except for a general statement that our best empirically based hypothesis is that extinction is a phyletically more recent function than is conditioning and that the wholesale bracketing of the two in American laboratories may well be more a matter of experimental convenience than of an empirical and logical cogency.

OR transformability refers to the fact that while, as seen in Figure 32 (the first figure in this section), repeated presentations of some stimuli (and some intensities of stimuli) extinguish the OR, repeated presentations of other stimuli (and other intensities of stimuli) transform the OR into another type of reaction (or, we might say, the OR gives way to another type of reaction). So far, quantitative relations have been worked out only with respect to the transformation of the OR into a defensive reaction, when 20 repeated applications of each of 20 discriminable shock stimuli were administered to adult human subjects (Sokolov, 1958c, p. 64). The criterion of the appearance of the OR was a *disparate* vascular reaction, that is, dilatation of the blood vessels of the forehead plus the constriction of the digital blood vessels; the criterion of the defense reaction was a *conjoint* vascular reaction, that is, constriction of both the digital and the forehead blood vessels. The experiment was performed by E. N. Sokolov, by all tokens the most active student of the topic, and the data are all presented in Figure 39. As may be seen from the figure, the OR first appeared at 2 units of shock but disappeared after the first trial; at 4 units the disappearance took 6 trials, at 6 it took 13 trials, but at 7 units a defense reaction emerged after 17 trials, while at 13 units the defense reaction came after 6 trials, and at 17 units of shock the first reaction was defensive. Quantitative findings on the transformation of the OR into other types of reaction have to date not been reported, but are of course very much needed.

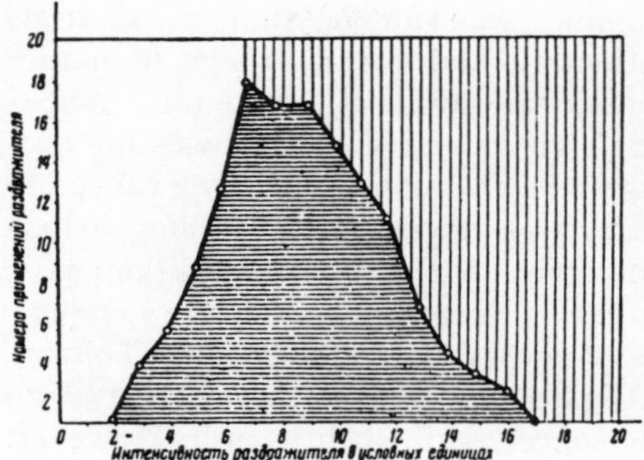

FIG. 39. The transformation of OR vascular patterns upon electrical stimulation into defensive ones as a function of the intensity of stimulus and the number of repetitions. (Clear area, no reaction; horizontal-line area, OR pattern; vertical-line area, defensive pattern. Abscissa, intensity of stimulus; ordinate, number of repetitions.)

Conditionability and reinforceability. That ORs which are difficult to extinguish could act as URs and be readily conditioned to other stimuli was demonstrated, it will be remembered, by the case of the hare whose respiratory OR to a rustling sound was conditioned to a previously inadequate sound of a buzzer (Figure 36). However, even more normal and quickly extinguished ORs could be thus effective. Children's turning to the source of a flash of light could be conditioned to a preceding sound provided the latter is not too loud (Karlova, 1959; Lebedinskaya, 1958; Kasatkin, Mirzoyants, & Khokhitiva, 1953; Mirozyants, 1954), and conversely dogs' or apes' turning to a sound of a buzzer has been known to become conditioned to a preceding change in illumination (Rokotova, 1952). Indeed, it is the contention of Soviet experimenters that all sensory preconditioning or what the Russians call associative conditioning is mediated through OR conditioning (Anokhin, 1958, p. 18; Voronin, 1957, pp. 99–103), while according to Anokhin (1958) and to Polezhayev (1958, 1960) the OR operates in all conventional (food or shock) classical conditioning to complex stimuli as an afferent and feedback central integrator of the stimuli before their actual conditioning to the peripheral food or shock reaction. Moreover, there is the study by Vinogradova (1959a) in which the administration of unavoidable electric shocks to human subjects was preceded by sounds of a metronome or of a pure 50-db. 1,024-cps tone and the resultant vascular conditioning was for a number of trials orienting and not defensive in nature (disparate and not conjoint). Anokhin's views and Vinogradova's findings are clearly of special relevance to American theories of learning which maintain that classical conditioning is a central, sensory, anticipatory, or expectancy event (Woodworth, Tolman, Hilgard), insofar as the OR may well be conceived in such terms. However, the Anokhin-Vinogradova claims extend only to the OR as an intervening preparatory-guiding aid in conditioning, and not as its final idiopathic mechanism.

The reinforceability of the OR refers to the feasibility of strengthening it through some subsequent event in the manner of operant conditioning (though the OR could not readily be conceived of as an operant). Two kinds of experiments are available as evidence. In one, fox cubs were brought to the laboratory to listen to squeaks of mice. The squeaks evoked in the young animals clear ORs which, however, extinguished rather quickly upon repetition. Yet when the cubs were permitted to eat the mice, one meal sufficed to make the OR to the squeaks almost unextinguishable (Biryukov, 1958, p. 24). In the other kind of studies, weakened ORs of human subjects to auditory sensory and verbal stimuli became very much strengthened when the subjects were told that they had to do something such as push a button, count the beats of the metronomes or the letters of the words, etc. when they heard the stimuli—that is, when the stimuli were "functionalized" or "signalized" (Russian term) in a simple "voluntary" reaction time design (Sokolov, 1958c).

Moreover, as has already been indicated in the preceding paragraph, reinforcement of CS evoked ORs continues for some time even in typical classical CR designs. Indeed, it might even be argued that the commonly noted final reduction of the OR in classical conditioning is not unrelated to the loss in rate of CR elicitation in operant conditioning when reinforcement is continuous (100%)—a matter which raises the whole problem of the relation of the two types of CR modifications and which, together with other more general problems of theory, will be discussed in the final section of this article.

Cortical-subcortical basis—vigilance versus investigation. Pavlov's view that the OR is rooted in both the cortex and the subcortex is being supported by a number of—although as yet still few—experiments on the effects of brain extirpation and specific drugs on OR manifestations. The fact is of course that the very conception of the OR as involving the psychogalvanic reflex (see Schwartz' experiment, 1936, 1937, on the relation of Cortex Area 6 to its elicitation), the desynchronization of alpha rhythm, and molar-versive reactions, on the one hand, and pupillary, respiratory, oculomotor, vascular reactions, on the other, means a cortex-subcortex neural expanse—and Pavlov's perspicacity in the matter can only be applauded. In addition, there have been recent attempts by Anokhin as well as by Sokolov to relate the OR *also* to the action of the reticular formation (Anokhin, 1958, p. 15; Sokolov, 1958b, p. 112). Figure 40 (Lagutina, 1958, p. 84) is a localization map of the OR in the brains of 28 cats and 4 apes, obtained by Lagutina with the use of implanted electrodes. The animals were moving freely in the experimental room, and the electrode stimulations of the points marked on the map produced typical ORs of head turning, pupillary dilatation, respiratory and vascular changes, as well as exploratory and manipulative reactions.

As may be seen from the figure,

Lagutina divides the evoked ORs into "general vigilance" and "investigatory" types. Other experimenters have sometimes distinguished between "what-is-it" and "what-to-do" ORs (Kvasov, 1956) or between generalized and specified ones (Sokolov, 1958a, p. 171) or assumed separate orienting and investigatory reflexes or preadaptive and adaptive ones (Bykov, 1958). As yet, however, these divisions and distinctions are far from experimental validation or even clear delineation and are fraught with the difficulty that adaptive and specified reactions are normally integral parts of nutritive, defensive, sexual-reproductive, and other non-OR reaction patterns. Hence, at least for the time being, the reviewer would favor subsuming under the OR only reactions that, it might be said, "hold" but do not "manage" stimuli—thus including investigatory-exploratory reactions but not adaptive-specified ones.

Cognitivity. Discussion of the cognitive nature of innate behavior patterns in animals—for that matter, also of acquired patterns—must obviously relate to indirect arguments and assumptions, and be more than a mere analysis of experimental data. Yet, there is little doubt that if any such pattern is accorded cognitive status, the OR pattern is surely the most likely candidate. Consider the indications of its relatively recent phyletic emergence and its dependence upon the relatively newer regions of the encephalon and of encephalization, its high evolutionary plasticity in both phylogeny and ontogeny, its ready functional lability and modifiability, its disruption in human organic and functional psychopathology, its central organized unity and cohesiveness coupled with its peripheral variety of reactions and, notably, its general function as a preparatory and controlling mechanism of what is to come. Moreover, in human subjects there is also considerable evidence to show that the elicitation of the OR is paralleled by a significant decrease in thresholds of conscious sensitivities (Sokolov, 1958a) and that, in general, positive correlations may be found between the scope and duration of elicited ORs and the scope and duration of conscious experiences (Asafov, 1958; Mikhalevskaya, 1958; Sokolov, 1958a, pp. 170–182; 1958c, pp. 7–38). To be sure, there is also evidence that ORs may operate in human subjects without awareness, particularly in interoceptive stimulations (Mysyashchikova, 1952). However, on the one hand, it is by no means claimed that cognition is a *sine qua non* of OR action but only that fully developed ORs are cognitive, or that cognition "emerges" from the OR as "the material basis." And, on the other hand, as was already indicated, Russian theoreticians are quite friendly to the view of *unbewusste Bewusstseinslagen* and have always been avowed adherents of the concept of levels. Hence, there may well be: reactions without ORs, ORs without cognition, cognition without awareness, and, for that matter, levels of "awarenesses."

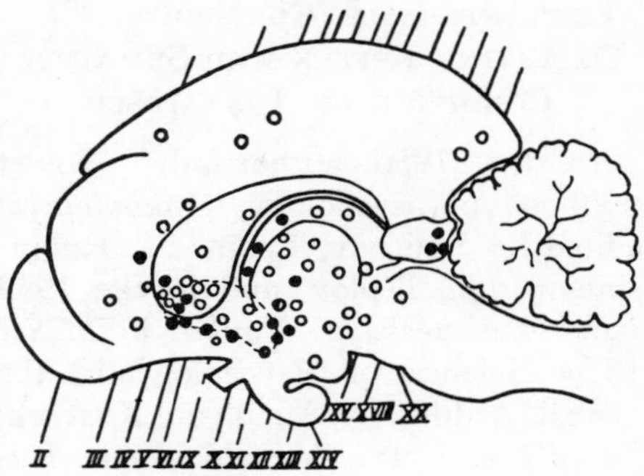

FIG. 40. Localization map of ORs in the brains of 28 cats and 4 apes, with the use of implanted electrodes. (Dots indicate stimulation points which evoked ORs of the "vigilance" type; circles, ORs of the "investigatory" type.)

The Inferable Conscious: The Orienting Reflex and Semantic Conditioning Juxtaposed

In the 1956 authoritative Soviet textbook of psychology, *Psikhologiya,* edited by Smirnov, Leont'yev, Rubinshteyn, and Teplov, and in the 1959 *Psikhologicheskaya Nauka v SSSR* [The Science of Psychology in the USSR], edited by Anan'yev, Kostyuk, Leont'yev, Luria, Menchinskaya, Rubinshteyn, Smirnov, Teplov, and Shemyakin, the now familiar distinction is drawn between the psyche (Russian *psikhika* and *psikhichesky* are on the whole better translated literally even if in a number of contexts "mind" and "mental" sound more English) and consciousness, the former including both conscious and unconscious psychic processes and the latter comprising only the conscious or aware ones. And another, in a way specially Soviet, emphasis is laid on the vast qualitative difference in level and complexity between the human and the animal psyche, the former alone possessing the enormous advantage of the second signal (language) system developed in the course of work and tool using as a means of changing, and not merely adjusting to, the environment. The experiments on both the orienting reflex and semantic conditioning may obviously be interpreted—and indeed are interpreted—as illustrations of these theoretical statements, the OR as an example of the early phylogenetic adjustive *animal* psyche and semantic conditioning as a high level manifestation of the ontogenetically fully developed *human* psyche (it is not assumed that the animal psyche wholly disappears within the human one but only that it becomes transformed, or merely that the two coexist). And both types of experiments, it might further be said, stress the psyche as a controlling mechanism and study it mostly indirectly and physicalistically (or behaviorally). Or, to use the expression of the beginning of this article, both types of experiments deal with the Inferable Conscious "whether it is or is not observable."

A behavioral or physicalistic study of the psyche (mind) and the concept of the Inferable Conscious are of course by no means new or special to Soviet psychophysiology and psychology. American and Western behavioral scientists have surely used the approach for some time, the system of Tolman being perhaps the purest example of it. However, what is significant here is that Soviet scientists appear to be adopting this approach as a sort of synthesis between a Pavlovian methodology and a Marxist-Leninist ontology, and what is even more important is the concentrated cumulation of so many pertinent and ingenious experiments under its egis. Moreover, there are really a number of cardinal differences between, let us say, Tolman's and the Russians' views.

For one thing, as was already indicated, the Russians stress strongly the qualitative distinctness of the human psyche in its special development in the qualitatively distinct second signal system, whereas, according to Tolman, the behavior—we might add, the cognitive behavior—of a rat at choice points in a maze epitomizes almost all basic aspects of psychology.

For another, the Russians are much more involved in studies of minute or molecular behavior and of neurophysiology and in general are much concerned with expanding the realms of observable behavior at all levels rather than with conceptualization at one level and its testing and application at others.

A third difference stems from the consideration that, while at the conscious level present-day Russian stu-

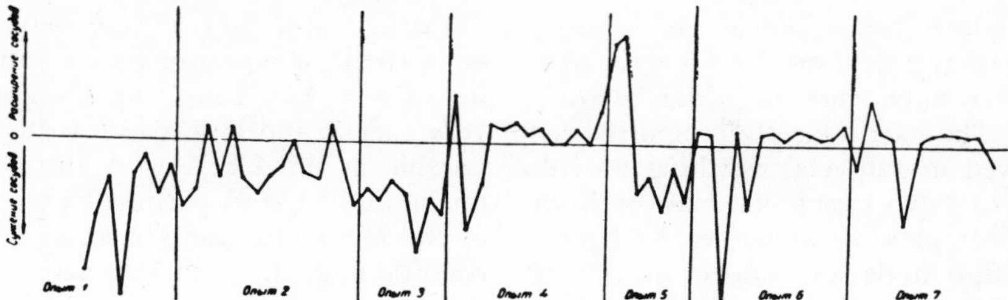

FIG. 41. Transformation of digital vasoconstriction into vasodilatation to electrical stimulation when human subjects were permitted to watch their plethysmographic records and note that dilatation reactions were followed by shock termination.

dents of behavior eschew no psychological category, are much interested in psychic (cognitive) controls, and even resort to direct subjective observations of experience, at the unconscious level they assume nonpsychic material neurobehavioral forces and mechanisms to be largely in full control. In Soviet theory, the psyche is an emergent and is not coextensive with all bodily action (there is quite a touch of Cartesianism in Soviet philosophy).

Fourthly, there is the fact of the Russians' wholesale rejection of the views of gestalt psychology and of Freud, the concept-models which Tolman uses so profusely (sign gestalt, need cathexis, etc.). Present-day Soviet holism is mainly associationistic, related traditionally to Wundt's "creative synthesis," Ehrenfels' *"Gestaltqualität,"* and Pavlov's statement that "associations generate [and govern] *Gestalten* and not *Gestalten* . . . associations" (Pavlov, 1936; 1949b, Vol. 3, p. 46). And, while Soviet psychophysiologists and psychologists grant some status and force to unconscious psychic mechanisms, they are by no means willing to accord these mechanisms Freudian ubiquity and potency—and contents. Finally, there are also significant differences between typical American-Western and Soviet formulations with respect to social aspects of the psyche and behavior, the Russians postulating as a rule the existence of a special social or collective consciousness. Space forbids further discussion of the topic.

One striking Russian experiment on psychic or cognitive control will, however, be cited in conclusion. Five adult human subjects were subjected to rather prolonged and moderately painful electrical stimulations while photoplethysmographic records were taken of changes in the volume of the blood vessels of their arms. As known, such stimulations produce vasoconstriction. However, occasionally and infrequently instances of vasodilatation also occurred in the experiment and when they did the experiment was rigged to terminate the shock, producing, in terms of B. F. Skinner, negative reinforcement. Yet no changes were evident in the subjects' records despite 80 reinforcements. The experimenter then let the subjects watch their records while being stimulated, from which they quickly learned the connection between the vasodilatation and shock termination. And here, as may be seen in Figure 41, a vasomotor transformation from constriction to dilatation was effected in the course of several experimental sessions (Lisina, 1958, p. 341).

In terms of current American psychology, the experiment offers two key findings: that, contrary to assertions, autonomic reactions can be modified by subsequent reinforcement in operant fashion; and that such reinforcement is effective only when cognition is present. The fact that the cognitive act involved in subjects' watching records of their own reactions may well be quite complex, accompanied by behavior differentials, and subject to varying interpretations does not, in the reviewer's opinion, detract from the obvious systematic significance of the results. Nor should the significance of the results be derogated by the consideration that they are no doubt quite specific—are not likely to hold (indeed do not hold) in operant modifications of skeletal reactions, and may well not extend in full to operant modifications of other autonomic reactions. Experimental pioneering is an inductive enterprise.

Overview

Related Areas

Interoceptive conditioning, semantic conditioning, and the orienting reflex are no doubt core areas of current Soviet psychophysiological research, and what has been said thus far has highlighted, it is hoped, considerably the main contours of their empirics and special theory. However, to subsume the three areas under one integrated framework and make it more representative of the topography of the entire field it will be necessary to consider briefly the essence of Soviet basic research in several other related areas such as: (*a*) proprioceptive and what we call (but the Russians do not) operant or instrumental conditioning; (*b*) compound, pattern, systematized, stereotype, and what the reviewer has called configural conditioning; (*c*) verbal conditioning in general, that is, unanalyzed into semantic and phonetographic aspects, the number of studies of which is, as might be expected, many times larger than that of semantic conditioning; and (*d*) phylogenetic and ontogenetic comparisons of conditioning which, too, constitute an area of very special and intensive investigative pursuit in the USSR. A fifth basic, much studied, and pertinent area, that of the neurology and pharmacology of conditioning, has, unfortunately, not yet been sufficiently digested by the reviewer, to permit here even a brief consideration of its essence.

Proprioceptive and operant conditioning. In a pioneer and classic experiment, Krasnogorsky demonstrated in 1911 that passive flexion of the talocrural or of the metatarsal joint of a dog's left-hind paw could become a conditioned stimulus for salivation and food seeking movements. The animal's leg above the paw was placed in a cast while secondary cutaneous stimulation was ruled out through differential conditioning as well as by the fact that the flexion CR was retained after ablation of the animal's ectosylvian and coronary gyri, an operation which, Russian experimenters attest, eliminates cutaneous CRs from extremities. The flexion CR appeared on the twenty-third CS-US combination and at first was generalized to the flexion of the same joint of the right-hind paw. Later on, however, the experimenter succeeded in extinguishing the CR to the flexion of the talocrural joint while forming a CR to the flexion of the metatarsal joint of the same paw (after 74 respective nonreinforcements and 42 reinforcements); and this new CR was differentiated from the flexion of the same joint of the right-hind paw. Another dog, whose right sigmoid gyrus was ablated, was able to form a food CR to the flexion of his right paw, to cutaneous stimulation of his left paw, but not to the flexion of his left paw—thus indicating that this gyrus is involved in the reception of dogs' motor but not cutaneous stimulation, a func-

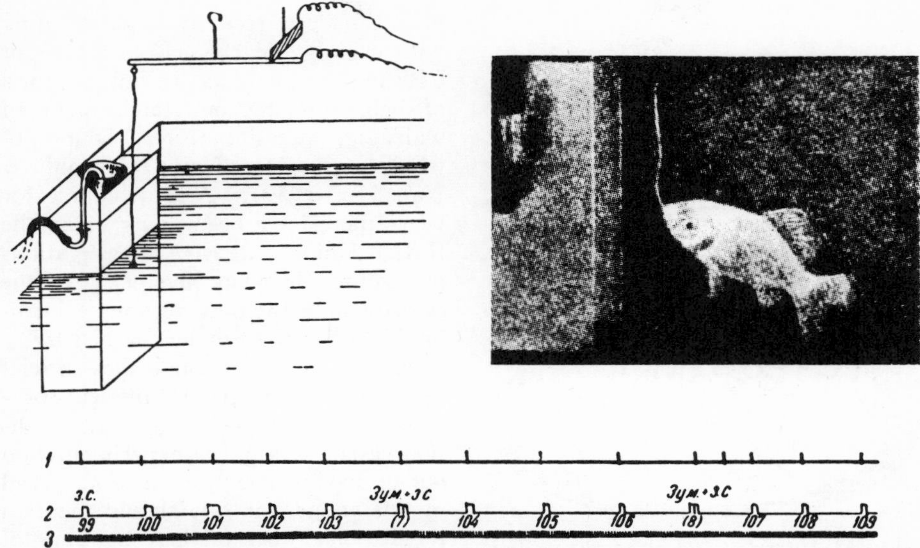

FIG. 42. Discriminated operant food conditioning in fish with a pulling-at-a-suspended-bead technique. (Food is delivered through funnel when bead is pulled, fish pulls when a green light is flashed. 1—kymographic record of pull; 2—application of CS (discriminated operant); 3—time in seconds; зс—green light; зум + зс—buzzer plus green light, attempt to establish conditioned inhibition.)

tion reverse from that of the aforementioned ectosylvian and coronary gyri (Krasnogorsky, 1911).

Krasnogorsky's results were weighed heavily by Pavlov as a discovery of the existence—and locus—of a special motor analyzer, and together with the later studies by Miller and Konorski (1928a, 1928b) and by Konorski and Miller (1930, 1932, 1933, 1934a, 1934b, 1936) formed the basis of his article, "The Physiological Mechanism of So-called Voluntary Movements" (Pavlov, 1936). To be sure, Miller and Konorski, unlike Krasnogorsky, were concerned, as is known, with paw flexion not merely as another kind of classical CS but as another type of CR, the animal's own active production of the flexion as a function of type of subsequent reinforcement and as conditionable to antecedent stimulation—that is, with operant conditioning in Skinner's terms and Type II conditioning in Miller and Konorski's own terminology. However, Soviet experimenters follow Pavlov and, unlike American and even Polish students of the field, link together the Krasnogorsky and the Miller and Konorski types of research (Ayrapetyants & Lichkus, 1957; Petrova, 1941; Skipin, 1940, 1941, 1957, 1959).

Moreover, there is surely no scarcity of experimentation of the runway type of operant conditioning in the Soviet Union. Zelyony, as early as 1913, trained rats to run to a feeding box at the sound of F_2 on a tuning pipe, using smoked paper as a runway to obtain objective records of the pattern of the running in conditioning and extinction. Most of the numerous and highly instructive experiments of Beritashvili (Beritov) and of his many Georgian followers are of the running-to-be-fed, or what the Russians call "free movement," kind of conditioning, as is also a considerable portion of animal studies

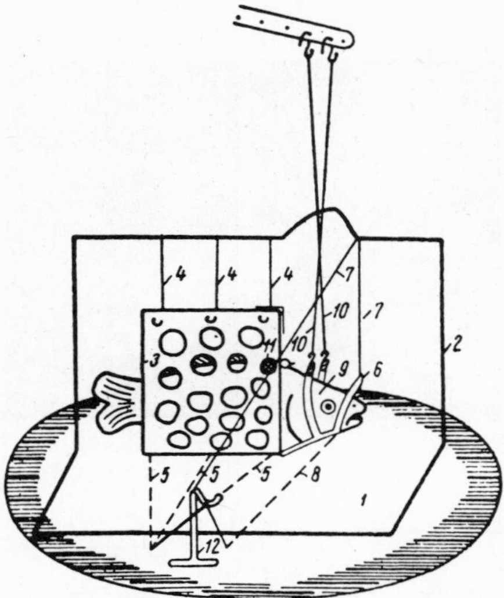

Fig. 43. Apparatus for classical conditioning of fish. (1—base of stand; 2—framework of stand; 3—rubber harness; 4—threads supporting harness; 5—threads moving harness; 6—rubber muzzle; 7—thread supporting muzzle; 8—thread moving muzzle; 9—respiratory cuff; 10—threads connecting cuff with myograph; 11—clamps holding fish in place; 12—anchoring pin for threads.)

of other Soviet CR experimenters, and, incidentally, Pavlov's ill-fated try at CR inheritance. Again, a very extensive series of wholly Skinner-type operant conditioning has been carried on in Voronin's comparative—from fish to chimpanzee—laboratory with the use of a variety of variegated techniques: pulling-at-a-suspended-bead (fish), snapping-at-a-ring (turtles), pecking, lever pressing, drawer opening (birds and mammals), and the like. Figure 42 illustrates the rather interesting pulling-at-a-suspended-bead operant technique used with fish in this laboratory (pull yields calibrated amount of food—Prazdnikova, 1953); while Figure 43 is a diagram of an equally interesting classical CR arrangement with fish (rubber harness, dissolved CO_2 as the US—Chernova, 1953). Indeed, the Voronin laboratory has been recently experimenting with chaining of classical and operant CRs into as many as six links—sound of bell, jumping on stand, sight of whirligig, pressing lever, sound of whistle, grasping of ring (Voronin & Napalkov, 1959). And of course for years there have been going on in the Soviet Union extensive investigations of acquired behavior by means of the operant bulb pressing technique introduced by Ivanov-Smolensky in 1927.

Plainly, what distinguishes Soviet psychophysiology from American psychology with respect to operant conditioning is a lack not of experimentation but of specific conceptualization. And in this connection one should mention the Russians' careful investigations of the comparative cytoarchitectonics of the motor area of the brain, development and differentiation of Area 4, in an evolutionary series of mammals. Roughly, they found the ascending evolution of the development and the differentiation to be: mole, hedgehog, rabbit, rat, horse, pig, badger, bear, tiger, cat, dog, macaque, baboon, orangutang, chimpanzee, and, of course, man (Kukuyev, 1953).

Compound, pattern, systematized, stereotype, and what the reviewer has called configural conditioning. A definition of terms seems to be in order:

> Compound conditioning merely refers to the methodological event in which the CS consists of a group of isolable or isolated stimuli, a + b + c + d + . . . + *n*, administered either simultaneously or in close succession.
>
> Pattern conditioning denotes the behavioral fact that, first, in simultaneous CS groupings the acquired CR characteristics of a group CS may differ from those of any of its components or combination of components and that, second, in the case of successive CS groupings there may be noted also a CR difference when the same group CS is applied in different temporal sequences. Configural conditioning adds the contention that the component-pattern and pattern-pattern

CR differences are not a matter of algebraic summation but of some configural supra-summative process.

The Russian term "systematized conditioning" or "systematism (*sistemnost'*) in conditioning" means no more than the tendency of CRs to become organized into systems and is thus the equivalent of pattern or of configural conditioning.

On the other hand, their term "stereotype" or "dynamic stereotype" conditioning or simply "dynamic stereotypy" refers to a special temporal system or pattern of CRs, consisting of both positive and negative (differential) CSs applied at equal intervals, and always in the same sequence, which becomes so fixed or systematized that the sequential position and not the specific nature of each CS becomes the main determinant of the animal's specific CS reaction, and which, moreover, may produce an experimental neurosis when the system's fixity is radically disrupted.

Now, even a cursory examination of Russian experimental literature reveals a wealth of experiments on compound conditioning—approximately 600 in number—and the fact that such experimentation was begun almost at the very beginning of the CR era. As early as 1906, data in Pavlov's laboratory showed that some stimuli fail to acquire conditioned properties when they are reinforced only in a compound combination even though they may be readily conditioned through separate reinforcements, and that the failure is an inverse function of the relative intensity of the component CSs (Palladin, 1906; Pereltsvayg, 1907; Zelyony, 1907—cf. Humphrey, 1928). Later, in 1910, results in the same laboratory disclosed: that a compound CS may come to elicit a conditioned reaction even though each of its components has no overt CR value when applied alone, that the generalization of compound CRs is relatively much wider than that of each of its components, and that dogs are able to differentiate among temporal sequences of the same four tones (Babkin, 1910; Zelyony, 1910). Again, in 1912, findings in Bekhterev's laboratory were quite clear in indicating that the conditioning of a compound per se is effected at the expense of a gradual elimination of component conditioning, the weaker components being eliminated first, the stronger last (Platonov, 1912). Finally, in Beritov's (Beritashvili) laboratory a dog was conditioned to each component of a four-link simultaneous compound CS while the combined application of all the components—the compound itself—remained inactive, and four dogs were trained to go to one feeding box when a compound of three or four CSs was applied and to another box when one or more components of the compound were administered (Beritov & Bregadze, 1929, 1930; Bregadze, 1937; Bregadze & Tarugov, 1937).

As indicated earlier, the cited experiments are only chronological paradigms of a very much larger number of studies. Their primary results, since duplicated many times, could not be readily questioned. The results, it needs to be added, support in the main quite solidly Ehrenfels' view of *Gestaltqualität,* Pavlov's already quoted expression that "associations generate [and govern] *Gestalten* and not *Gestalten* . . . associations," and Hebb's concept of *t* in the development of superordination. And there is of course no doubt that methodologically compound conditioning is a virtually unique, most objective, and most genetic means of studying this ever-recurring association-versus-gestalt crux of almost everything basic in psychology. Yet, by all tokens, all these experiments and the empirical wealth of their findings have affected but little American research and thought. Hebb, for instance, does not seem to have been aware of them, and Grings and his collaborators (Grings, 1957; Grings & O'Donnell, 1956; Grings & Shmelev, 1959) appear to be the only ones to have attempted a duplication of some results. The obvious reason for this neglect, that of the language barrier, needs to be qualified: since some 22 years ago the writer reviewed a few dozen of these Russian experiments (Razran, 1939c) and even added several of his own (Razran, 1938, 1939d, 1939e, 1939f, 1939g, 1940). Perhaps, it is more a matter of a thought barrier, a systemocentricity of one affluent group ignoring the wealth of the other.

Verbal conditioning in general. Aside from interest in isolable semantic conditioning, current Soviet treatment of verbal conditioning in general is, as was already indicated, quite specific and instructive: the Pavlov concept of a "second signal system" and a "signal of signal*s*," with emphasis on the plural form of the last word. That is to say, verbal conditioning in general (language as such) is not just an array of second-order CRs or signals, of which animals are also capable, but is a special higher level human system characterized primarily by being not a signal of one signal, a simple second-order CR, but a signal of *many* signals, a second-order CR to a compound CS—a system that, put somewhat differently, by its very nature, does not merely link and analyze but synthesizes (abstracts or generalizes) its acquired information. The formulation presumably pertains to the complex R and not the simple S aspect of words: that is, to the emergence of the production and use of language and not to its basic character as a stimulus. And it should be noted that, while Pavlov's view of patternization or *Gestaltung* as a CR resultant was based on a long series of experiments—his article on Dynamic Stereotypy appeared in 1932—his conception of verbal conditioning was merely a general statement made in the last years of his life which for a number of years after his death lay almost fallow with hardly any theoretical and experimental analysis. Since 1950, however (The July Joint Session of the USSR Academy of Sciences and the USSR Academy of Medical Sciences Dedicated to the Problems of the Physiological Teachings of Academician I. P. Pavlov), scores of treatises interpreting all aspects of language, its rise and mechanisms and related genetic, experimental, and clinical studies, "in the light of Pavlov's teachings on the second-signal system," have appeared and a good number of special experiments to validate (or test?) what Pavlov said about verbal conditioning have been devised. A recent series of experiments, all by Kol'tsova (1958), might serve as an example of this type of validative evidence:

1. Ten 20-month-old children were shown a doll 1,500 times in the course of several months. With five of the children, the showing was accompanied by the experimenter's saying "Here is a doll," "Take the doll," and "Give me the doll" and the children's corresponding reactions. With the five remaining children, 30 different sayings and corresponding reactions such as "Look for the doll," "Rock the doll," "Feed the doll," "Seat the doll," and the like were used. In all cases each saying was used an equal number of times. After the training, the experimental doll was placed before each child together with different types of dolls and with a number of other toys, and the child was asked to "Pick a doll." The children of the first group selected the experimental doll as well as a number of other toys. Those in the second group selected the experimental doll as well as the other dolls but no other toys. Moreover, the latent period of the "Pick a doll" reaction was almost four times faster for the children in the second group. Presumably, the highly

variegated stimulations-reactions (the much compounded CR) in the children of the second group produced a full-fledged generalization of "doll" and a differentiation from "nondoll," which the much less varied procedure with the children of the first group failed to do.

2. A book was presented 20 times to nine 19-month-old children, divided into three equal groups. In the first group, only one particular book and one particular saying-reaction were used. In the second group, one particular book but 20 different saying-reactions were employed, while the procedure in the third group involved 20 different books combined with only one type of saying-reaction. As in the preceding experiment, the posttraining test consisted of placing before the children the experimental book together with a number of other types of books and with other objects and asking them to "Pick a book." Generalization-differentiation (abstraction) was best in the second group, considerably poorer in the third, and practically nonexistent in the first, thus indicating both the value of compound conditioning and the superiority of motor-kinesthetic over sensory-visual reactions.

3. Twenty-five small objects were presented individually to 10 26-month-old children, divided into two equal groups. In the first group, the presentation of each of 15 objects was accompanied by the experimenter's saying *"veshch'"* [thing, something, object] and delivering a puff of air to the child's eye producing a blink (see Brackhill in the March 1960 issue of the *American Psychologist* for a diagram of Kol'tsova's eyelid conditioning apparatus), while the presentation of each of the remaining 10 objects was unaccompanied by any extra stimulation. In the second group, in which a different set of 25 objects was used, the showing of all objects was accompanied by the air puff, but only 15 of them by the experimenter's saying *"veshch'."* Posttraining tests were made of both blink conditioning and of generalization-differentiation when all the objects were placed before the children and the experimenter said *"Vozmi veshch'"* [Pick an object. Take something.]. In both tests, the performance of the first group was clearly superior, the second group failing to develop even simple conditioning. The experiment is obviously a significant illustration of the interaction of the first and second signal systems: reciprocal facilitation when the two systems are correlated (first group) and reciprocal interference when there is a discrepancy between them (second group).

4. The presentation of five different and differently colored geometrical figures—red sphere, green pyramid, white cube, blue ring, yellow rectangle—to six 24-month-old children was accompanied by the experimenter's saying *"veshch'"* and delivering an air puff to the children's eyes. The presentation of four other objects—electric switch-key, drumstick, paper chain, iron disc—was accompanied by the experimenter's saying *"pustyak"* [nothing, trifle, trash] and omitting the administration of the air puff. The performance of the children of this group with respect to both conditioning and generalization-differentiation was even superior to that of the children of the first group in the preceding experiment, when the verbal differentiation of the puff unaccompanied objects consisted not of a special differential word but of the omission of the positive word. Yet, when 10 additional geometrical figures were accompanied by the air puff and the saying of *"veshch',"* both the conditioning and the generalization-differentiation were impaired and were restored only when the pre-

sentation of some objects was coupled with the experimenter's saying *"pustyak"* and delivering no air puff. This experiment obviously corroborates more specifically the one preceding, pointing up, in addition, the particular value of balanced differential conditioning and clear verbal differentiation in the development of integrated conditioned-verbal reactions (in Russian terms, the value of a balance between excitation and inhibition in a balanced interaction between the first and second signal systems).

5. Five geometrical figures and four other objects—identical to those of the preceding experiment—were presented to six 24-month-old children. The presentation of the figures was accompanied by the experimenter's saying *"veshch'"* and delivering an air puff to the children's eyes, and of the other objects by saying *"pustyak"* and giving the children candy. With this procedure, in which different reactions were attached to each of the two groups of stimuli, the children did even better both in the conditioning and in the generalization-differentiation of their reactions to the experimenter's saying *"Vozmi veshch'."* And it took 20 additional geometrical figures, rather than 10 as in the preceding experiment, to impair the children's performance.

6. A toy fishing game consisting of a cardboard aquarium, a fishing pole with a magnet, a cardboard pail, and several metal fish was presented a number of times to 14 17-month-old children, divided into two equal groups. The first group was given the opportunity to handle and play with the objects of the game while the second group was only shown them. Neither group had ever seen the objects before and in both groups the experiment comprised two parts, each lasting 10 days. In the first part, the experimenter accompanied the showing or handling of all the objects with saying *"igrushki"* [toys]; in the second part, begun after 13 of the 14 children had mastered the first part (that is, imitated the word and pointed to the game when it was pronounced preceded by "Where is"), each object was shown or handled separately to the accompaniment of the words *"stenki"* [little walls], *"udochka"* [little fishing pole], *"rybka"* [little fish], and *"vedro"* [pail]. The data are quite clear in disclosing differences between the two groups. Four children of the first, the handling-the-objects, group mastered the first part of the experiment in 2 days and the remaining three in 3 days. On the other hand, the task took four children of the second, the shown-the-objects, group 6 days, and two, 9 days; and one did not learn it at all in the 10 days of experimentation. Moreover, the second part of the experiment, requiring imitation of each of the four words and pointing to each of the four objects when the appropriate word was pronounced with the addition of "Where is," was mastered, in the course of 10 days, by all the children of the first group but by none of the second. The implications of this simple experiment hardly need explanation: the role of motor reactions and kinesthesis in the emergence of language. The reviewer wishes, however, that the experimenter had reported more specifically about the equating of the visual and the motor-kinesthetic stimulations.

Phylogenetic and ontogenetic CR comparisons. The number of Russian studies in this area is exceptionally large and the area is indeed the major division of the field cutting across all other categories. The first question that comes to mind is obviously the relation of CR efficacy to evolution, on the one hand, and to maturation, on the other. Twenty-five years ago, review-

TABLE 3

SPEED OF SIMPLE DISCRIMINATED OPERANT CONDITIONING IN FISH, CHICKENS, DUCKS, PIGEONS, RABBITS, DOGS, AND MONKEYS

Animal	*N*	Conditioned reaction	Conditioned stimulus	Conditioning trial			
				First appearance		CR stabilization	
				Range	Median & *SE*	Range	Median & *SE*
Fish	35	Pulling a bead	Green light, bubbling of water, blue light, white light, bell, flickering light	2–35	8±1.52	4–89	21±1.19
Chickens	11	Pecking at a pedal	Bell, green light, beat of metronome, white light bubbling of water	3–20	9±1.79	17–50	32±4.44
Ducks	4	Pecking at a pedal	Beat of metronome	8–19	17.5±2.82	24–102	88±19.96
Pigeons	4	Pecking at a pedal	White light	4–8	4.5±1.03	12–28	24±3.98
Rabbits	36	Pulling a ring with the teeth	White light, tone, bell	2–29	6±1.10	13–120	36±5.24
Dogs	21	Pressing a pedal, pulling out a drawer of a feeding-box, pulling a ring with the teeth	Bubbling of water, bell, metronome-120, whistle, tone, tapping sound	3–87	7±6.04	5–190	45+12.29
Monkeys	14	Pressing a lever, pulling out a drawer of a feeding-box	Bubbling of water, buzzer, metronome, bell, metronome-120, red light	3–35	9±3.20	7–120	32.5±9.66

ing one aspect of this CR efficacy, that of speed of simple conditioning, the present writer was rather specific in stating that "the speed of simple conditioning increases in phylogeny and in the first few years of [human] ontogeny" (Razran, 1935, p. 117), although even then he was puzzled by data showing that conditioning is faster in fish than in amphibians, and he surely was in doubt about the exact role of language in the first years of human conditioning. At present, with available evidence manifoldly multiplied, the 1935 statement surely needs qualification, indeed revision and a good deal of retraction. For one thing, there is the radically masking factor of ecology, the species-specificness of stimuli and reactions such as: (*a*) the vast differences between rabbits' reactions to the sounds of rustling leaves and to a buzzer as already discussed earlier with respect to the OR (Figure 36); and (*b*) the simple fact, pointed out by Voronin (1957), that it is so much more difficult to train a dog to lift his hind paw than his fore paw. For another thing, cumulative evidence of 20 experiments in Voronin's laboratory, summarized and treated statistically in Table 3, clearly shows no differences in speed of simple discriminated operant conditioning among fish, chickens, ducks, pigeons, rabbits, dogs, and monkeys (Voronin, 1954, 1957).

On the other hand, reports of experiments from Voronin's laboratory show also that extinction of the above type of conditioning was considerably slower in the fish than in the other vertebrates tested: 28–78 nonreinforcements versus 6–15 for turtles, 11–84 for chickens, 10–30 for rabbits, 7–15 for baboons, and 8–10 for chimpanzees (Voronin, 1957). (Note the apparent absence of definite evolutionary differences among the other animals.) And there is no lack of evidence to reveal that, aside from differences in mere sensory capacity, evolutionary factors operate in the development of complex conditioning: delayed CRs, complex differentiations, transformation of CSs from positive to negative URs and vice versa, and, above all, pattern or configural conditioning.

A few reports on configural conditioning will be detailed here. Yordanis (1959) compared several classes and orders of vertebrates with respect to inhibition-disinhibition differentiation: that is, imparting to a stimulus a double CR value so that it decreases conditioning when compounded with one CS and increases it when combined with another—a mechanism supposedly basic to effective configural conditioning. He found the task impossible for turtles; difficult for pigeons; somewhat easier for rabbits; and progressively easier for dogs, monkeys, and chimpanzees. Baru (1954) noted that rabbits could be conditioned to differentiate between the various sequences of a successive three-link compound CS—bell, light, buzzer—and between the compound and its components. But in the same laboratory Prazdnikova (1955) encountered difficulties in training fish to develop even one successive compound differentiation, between a three-link CS—bell, light, bubbling of water—and its reverse sequence; while in Beritov's laboratory, dogs, but not rabbits, could be conditioned positively to each of the components of a four-link successive CS—340 + 680 cps tones + two bells of different timbre and intensity—and negatively to the total compound (Beritov & Bregadze, 1929).

Moreover, even with respect to the speed of simple conditioning, an evolutionary factor seems to be quite operative in invertebrates, at least in their first 14 phyla. Finally, it should be added that the problem of the evolutionary reversal of conditionability in amphibians seems to have been solved by the extensive cerebellar and cerebral ablation experiments of Karamyan (1956). Karamyan's data are quite unmistakable in revealing that, while in fish the ablation of the cerebellum abolishes both formed CRs and conditionability and ablation of the cerebrum affects neither, the poor conditionability of amphibians is little affected by either operation; or, in other words, the amphibians are a transitional group of animals in whom the cerebellum has lost power and the cerebrum has not yet acquired it (neuromorphological evidence corroborates this account).

Space prevents special discussion of ontogenetic CR aspects, and perhaps not much needs to be said. In general, ontogeny here repeats phylogeny, but account must obviously be taken of maturation stage at birth. Volokhov (Volokhov et al., 1959) reports studies in his laboratory that the first CR (defensive) in baby guinea pigs appeared on the first day; in pigeons after 6–7 days; in rabbits after 7–11 days; in rats, 10–15 days; in dogs, 17–25 days; and in cats after 22–28 days. And Kasatkin (1957), reviewing a large number of Russian studies on conditioning of human infants, concludes that true CRs appear no earlier than 8–9 days after birth. The specific crux of human ontogeny is of course the emergence of language or verbal conditioning, discussed earlier, and the further consideration that language symbolization or phonetic and semantic conditioning, CR compounding-*Gestaltung,* proprioceptive conditioning, interoceptive conditioning, and the orienting reflex are hierarchically related—a topic that will receive its due in the next section on systematic theory.

Systematic Theory

The systematic theory suggested by the empirics of the three core and four related areas of Soviet psychophysiology—the former reviewed in comparative detail, the latter merely epitomized—pertains, as might be expected, to basic learning or conditioning. And in the writer's opinion, what is suggested may best be formulated as a three-level theory: a first level of simple primary exteroceptive and sensory-interoceptive conditioning that is little modified by proprioceptive and motor-interoceptive CS feedback—the classical variety; a second level of exteroceptive and sensory-interoceptive conditioning that is greatly modified by CS feedback and gives rise to operant conditioning, on the one hand, and to configural conditioning, on the other; and a third level of verbal conditioning which, beginning as simple classical-sensory and operant-motor conditioning, quickly evolves into a special kind of configural conditioning in which sets of surrogate compound CRs emerge, and supraordinate semantic and subordinate phonetic (or phonetographic) units develop. The theory might also be described in terms of three (not two) signal systems: a first signal system of primary environment generated exteroceptive and sensory-interoceptive signals; a second signal system of secondary, response produced proprioceptive and motor-interoceptive signals; and third, a signal of signal*s* surrogate system of words and their equivalents. It will be noted that the levels or signal systems are assumed to exist and operate in hierarchical fashion: first, second-plus-first, third-plus-second-plus-first. The higher levels or signal systems subsume but do not supersede the lower ones, indeed are dependent on the co-presence of the lower ones. And it might be added that, despite century-cumulated evidence and intimate relevance to human control, the third level or signal system is in a lower state of systematization than are the first and the second, particularly the first.

The classical-operant differential. For some time the writer (Razran, 1939a) has maintained that in our present state of knowledge all that can be safely said is that there are two types of conditioning, one in which the modified reaction, *d,* assumes some characteristics of the modifying reaction, *G,* becoming *dG,* and one in which the modified reaction becomes strengthened by the modifying one, becoming *dD.* The writer does not of course object to calling *dG* conditioning classical and, in deference to usage, *dD* conditioning operant or even instrumental (reinforcement conditioning would be better); and he finds, moreover, some general value in the terms respondent, operant, and instrumental as designations of particular classes of reactions. What the writer does consider unwarranted is the conventional assertion that *d,* the modified reaction, is necessarily operant or instrumental in *dD,* or operant, conditioning, and necessarily respondent or noninstrumental in *dG,* or classical, conditioning. Paw-lifting may be *dD,* or operantly, modified even if it is respondently evoked; and *dG,* or classically, modified even if it is operantly emitted (Konorski & Miller, and others). Likewise, the frequency of a dog's scraping a rug while it is being fed or an infant's fingering a button on his mother's robe while being nursed may well be *dD,* or operantly, increased or conditioned, even though the reactions are admittedly noninstrumental; and there is plenty of evidence for *dG,* or classical, CR modifications of instrumental reactions.

In other words, in the writer's view, the main problem of the Pavlov-Thorndike, more modernly, Pavlov-Skinner dichotomy remains unsolved: when does learning or association produce a *dG* and when a *dD* reaction? Or we might say that the dichotomy has thus far been discriminated only semeiologically, in terms of manifestations and functional relationships (Hilgard & Marquis, 1940; Mowrer, 1947; Razran, 1939a, 1955a; Schlosberg, 1937; Skinner, 1935, 1938, 1957, 1959; Solomon & Wynne, 1953) but not etiologically. And it is in the realm of etiology that an analysis of the empirics of current Soviet psychophysiology, with, to be sure, American psychology as a backdrop, is of particular heuristic value.

Four different categories of Soviet CR experiments bear upon the classical-operant problem:

1. Studies of operant conditioning such as paw-lifting and runway reactions in which Russian and Polish, unlike American, experimenters typically take account also of concomitant classical components such as salivation and eating-accompanying motor reactions and in which, moreover, the Russian experimental design is almost always a double and not just a single link: that is, the modified reaction is studied not only as a function of subsequent reinforcement but also as conditionable to antecedent stimulation (the discriminated operant, in Skinner's terms, which, incidentally, the writer finds is neither in essence nor in functional manifestations in any way distinguishable from what Pavlov had studied)

2. Current classical CR experiments in which concomitant CS produced orienting reflexes are carefully measured so that the reaction change to the CS is investigated not only in terms of the acquired CR but also as a change in the CS's own premodified reaction

3. Experiments in which the original reaction to the CS is incompatible with that to the US, as when, for instance, in food conditioning, the CS is an electric shock or some other aversive stimulus

4. Cases in which the effects of the acquired CR on the UR are studied, incremental effects being seemingly related to classical *dG* modifications and decremental effects to the operant *dD* variety. For convenience, control, and theoretical uniformity, only experiments with food as the US (or reinforcement) are being considered in any of the four categories.

The results and implications of the experiments in the first two categories are the very basis of the suggested *dG* classical-*dD* operant differential in the introductory paragraph of this section. For they clearly indicate that while (*a*) CR modifications are eventually of either the *dG* or the *dD* type, (*b*) some reactions are capable only of *dG*-modification, and (*c*) some quickly acquire *dD* characteristics, it is also true that (*d*) *dD* modifications contain *dG* ones as subsistents, (*e*) *dG* modifications become transformed into the *dD* variety when CSs are increased in intensity and thus evoke marked ORs and consequent proprioceptive CS feedback, and (*f*) *dG* modifications manifest themselves in general (except for a special case to be discussed in the next paragraph) only when the ORs of the to-be-modified reactions are either small or readily extinguishable and thus devoid of feedback. Or, more formally and succinctly: *Operant* dD *modification comes into being only when to-be-modified reactions involve a large amount of motor or proprioceptive CS feedback*—which in rough traditional terms may further be simplified as: *Classical conditioning is primarily a modification of sensory reactions and operant conditioning a modification of motor reactions,* with the obvious addition that at a certain level of reactivity the sensory-motor dichotomy is a matter of quantity rather than of quality of stimulation and the needed qualification of the already mentioned special case.

The special case under discussion refers to experiments in the third category in which classical *dG* modification occurs despite undoubted extensive CS feedback involvement. However, a moment's thought reveals that the case is a very special exception hardly refuting the general principle: a modification resulting from the confrontation of two basic and, except for stimulation differences, biologically equally potent antagonistic URs resembling reflex interaction at a spinal level, and thus is quite different from typical CR learning, in which only one reaction, the modifying one, is basic and potent, the other, the modified one, being ancillary, weak in potency, and readily and unantagonistically integrable with the modifying reaction (the question of second-order conditioning and verbal association is a separate matter and need not detain us here). Indeed, it may well be argued that the experiments of the third category illustrate a most primitive, pre-true-conditioning, form of modifiability, characteristic particularly of simple organisms and simple systems whose reaction repertories are both limited and not yet adequately organized into a dominant-subordinate reaction hierarchy. And, again, it may surely be said that in view of its greater universality (occurrence in lowest phyla of the animal kingdom, in few-days-old animals and infants, in interoceptive systems, and lesser dependence on type of CS and UR), lesser disruptability (ablations, drugs), and some other considerations, classical conditioning is a phyletically older form of learning than is operant conditioning.

Finally, the studies in the fourth category point to a rather general yet adaptively very significant classical-operant differential which, as far as the writer is aware, has not previously been noted in either American or Russian CR literature. What may be said to be the quantum of UR reactivity extends in classical conditioning to both the ways of bringing it about and its own magnitude or intensity, whereas in operant conditioning the UR's or reinforcement's own reactivity may actually diminish in the wake of its conditioning. An organism's food or sex reactivity—or sensitivity—will thus increase as a result of the many signals or classical CSs that become attached to it, but may decrease as a result of multiplication and variegation of operant CRs, the *means* of obtaining the food or sex. Some energy seems to be transferred from ends to means. Or, more specifically, in operant conditioning the energy transfer is the reverse from that of what it is in classical conditioning: from ends to means rather than from signals to ends. On the other hand, it must be kept in mind that signals and means are not ultimate terms of CR analysis and, as was already indicated, means involve also some signal CR modifications. Or, in other words, all we may conclude systematically from the fourth category of experiments is that *dG* modifications tend to increase URs and *dD* modifications to decrease them, with the likelihood that in normal life situations the two types of effects neutralize each other and the URs are not radically changed.

Feedback and configural conditioning. The statement in the introductory paragraph of this section that CS feedback "gives rise to operant conditioning, on the one hand, and to configural conditioning, on the other" purports to convey three considerations: (*a*) the obvious fact that by their very nature operant and configural conditionings are altogether different evolvements of simple classical conditioning; (*b*) the postulation that although CS produced proprioceptive feedback is at the base of both types of conditioning, its specific mechanism of action is quite different in each case; and (*c*) the almost obvious deduction that configural conditioning is an evolutionary higher level of learning than is operant conditioning. Only Considerations *b* and *c* will be discussed. It should, for instance, be made clear that the postulation involves the premise that in operant conditioning the feedback is of a strengthening-a-particular-reaction kind whereas in configural conditioning it is of an integrating-a-number-of-reactions type—or, specifically, that in operant conditioning the feedback merely raises the status of a to-be-modified reaction to the extent that subsequent or concomitant URs strengthen its reaction rather than replace it by their own, whereas in configural conditioning feedbacks of a number of to-be-modified reactions combine and integrate to generate a CR that is separately UR modified and by all logic separately represented in the cortex of the organism. Again, with respect to Consideration *c,* it will be remembered that efficacy of configural conditioning correlates positively with ascending classes, orders, and families of mammalian evolution but that efficacy of simple discriminated operant conditioning varies little from fish to chimpanzee. And there is of course no doubt that configural conditioning expands immensely the realm of an organism's CR action, permitting it a vast variety of whole-part and permuted-whole differentiations and integrations of the stimulations of its surrounding environment and thereby also, it should be added, a lesser dependence upon the original nature of the stimulation and a closer tie-in with the organism's own "past." A view that configural conditioning is an organism's apogee of preverbal achievement would be hard to refute.[5]

The argument for proprioceptive feedback as a mediator of configural conditioning rests, however, not only on logic and postulations. There is also a body of empirical evidence. For some time now Russian experiments have disclosed a very substantial direct relationship between efficacy of configural conditioning and extent of CS accompanying, orienting reflexes, which obviously means extent of CS produced proprioception. And, then, there is the corroborative evidence of such studies as those of Kol'tsova, cited earlier, on the importance of kinesthesis in the development of infants' abstractions.

[5] Receptor compounding of stimuli must be distinguished from effector chaining of responses. In the latter case, the proprioceptive CR mechanism is by all tokens primarily a matter of summative integration rather than of suprasummative configuration—a mechanism which seems to the writer to involve too little extra systematic significance to be specifically treated here. It should be noted, however, that the Russians sometimes use the term "chained" instead of "compound" even with respect to stimuli.

The logical argument of "feedback and universals" has been popularized by students of cybernetics but is in need of more concrete amplification in the light of present-day knowledge of conditioning. In the writer's view, three related proprioceptive mechanisms are involved in CR configuring: compression, formalization, and *sui generis* cortical representation. Compression stems from the consideration that secondary proprioceptive stimulation is surely not so specific as the primary exteroceptive and sensory-interoceptive stimulation from which it is derived; or, in Russian terms, that CS accompanying ORs are by no means as specific as the CSs' reflections in respective analyzers. Formalization means that proprioceptive stimulation compares, so to speak, primary sensory stimulations which it receives in relation to itself and hence imparts to these stimulations a special set of formal—we might say extra- or suprasensory—properties such as: sequence, number, intensity and duration ratios, rhythm, proximity, similarity, continuity, and the like (one even thinks of *Einstellungen*). *Sui generis* cortical representation hardly needs enlargement, stating as it merely does that proprioceptive stimulations must have their own cortical representation, what the Russians call the motor analyzer. In fine, all that has been said here is that CS produced proprioception (also motor-interoception) provides the organism with a second recoded and separately stored dimension of acquired conditioned information, and that both logical considerations and empirical evidence point to this dimension as the mediating mechanism of configural conditioning.

Pavlov (1936) considered proprioceptive conditioning the basis of voluntary action, while cognition in general is in current Soviet theory related to either conditioning in general (Bykov & Pshonik, 1949) or to "neutral" orienting reactions (Leont'yev, 1959). In the tradition of earlier American motor theories of consciousness, the writer's thoughts have, however, run in the direction of regarding proprioception and proprioceptive conditioning as the phyletic and neural loci and foei of cognition's emergence as such. (Cognitive control is another matter that may perhaps need linguistic tools to be effective—Razran, 1935, pp. 120–124.) The thoughts are of course in the nature of a speculation, not at present testable in the sense that the relation of proprioception to configural behavior may be tested, and are thus really not integral to the present systematization. Still, it might be mentioned that the speculation is seemingly in line with the role of perception as an outgoing, organism generated activity; with the epistemological views of such philosophers as Perry and R. W. Sellars; is one which the writer has always found privately comfortable; and, incidentally, is also one that is accommodable to Lenin's gnosiology of praxis approximation of truth and activistic consciousness.

The specifics of verbal conditioning. Specific systematization of the experimental evidence of verbal conditioning is a difficult task. For even at the young age when language first begins to appear, full control of pre-experimental and extra-experimental variables is hardly possible, almost total separation of the baby-subject from his human-verbal environment being obviously the prerequisite. And in adult subjects there is the additional uncontrolled and barely controllable factor of the long ingrained fixity of verbal behavior which only very special techniques can pierce to gather glimpses of its developmental dynamics and essence. Plainly, CR experimentalists may well have missed, by a few hundred thousand years, the opportunity of a thorough investigation of language at its base. Still, the fact remains that the CR method is the most basic, most genetic, and most objective in the field, and the aforementioned semantic CR experiments plus studies such as those of Kol'tsova, taken together with related natural and clinical observations, offer some systematic clues.

To begin with, there is no denying that a portion of verbal conditioning is not particularly specific and presents no particular problems. The basic mechanism in conditioning an infant to react appropriately to "Show me your nose" may well be little different from that of conditioning a dog to lift his paw upon command. Likewise, changes in elements of an infant's vocalizations, weakening and elimination of some and strengthening and fixing of others, may well proceed according to typical operant reinforcements and extinctions. Indeed, there is every reason to believe that words and their combinations not infrequently behave as no more than simple CR signals and simple operants even in adult life. Still, these simple mechanisms are doubtless only the birth and fringe characteristics of language dynamics, not its determining core. As stated earlier, even in modifications of purely somatic and visceral reactions of animals, simple CR mechanisms give way, through the development of proprioception, to complex configural ones, and only the efficacy of the latter correlates positively with evolution from fish to chimpanzee. And certainly one could not argue that evolution stops with the advent of man and verbal conditioning or language, eschew the glaring fact that language's main role is exactly that of configuring, or ignore the adduced evidence that proprioception is radically involved in the very rise of true language in early human ontogeny. If proprioception alone yields a second dimension of coding and processing conditioned information and of *sui generis* cortical representation of received stimulation, then language must by all tokens generate or create a third.

The specific higher level evolutionary characteristic of verbal conditioning in true language (that is, apart from and above its function as a simple CR signal or simple operant) is, however, not just that of compounding and configuring but that which might be called surrogation or perhaps better configural surrogation. Words and sentences are not merely configural CRs to different sequences of the same elements of stimulation and compound CRs to compressed classes of stimulations, but are also admittedly duplicates or surrogates of somatic—occasionally also visceral—reactions priorly possessed by the individual. That is to say, full-fledged verbal conditioning, unlike preverbal proprioceptive conditioning, gives rise not just to doubly coded information and to the integration of reactions to different stimulations, but also to double sets of effector reactions to the same stimulation, with consequent conflict, cooperation, and control of one set by the other (not to mention the consideration that verbal reactions generate their own extra proprioception). Moreover, because of inadequate CR capacity, inadequate conditioning, or extinctive effects, the reactive aspects of an individual's words and sentences fall behind their full stimulational values, creating the problem of semantic versus phonetographic (more correctly, phonemographemic) conditioning, that is, of fully compressed and configured verbal CRs versus those in which one or more CR steps have not been developed or had been lost. And again, there is experimental reason to believe that configured verbal CRs become further organized, perhaps through the mediation of their own proprioception, into hierarchies and propositional arrangements (see earlier section on Semantic Conditioning). A whole new realm of what might be called sememes, fully configured semantic CR units anchored to a variety of words or phonemes, thus comes into being and interacts with words or phonemes that have been little or even not at all attached to semantic equivalents. Or, in other words, behavioral psychology must face not only the problem of the verbalization of S-R units but also that of their "semanticization."

In short, verbal conditioning appears by all tokens to recapitulate all forms of CR modifiability—it is not just a higher but the highest level of learning, and not just a third level but a third-plus-second-plus-first level, and, moreover, a third-plus-second-plus-first level interacting continually with a mere second-plus-first and a mere first level. An important phase of the interaction of verbal with nonverbal conditioning is involved in the mechanism of the genesis of verbal control of somatic and visceral reactions. The problem has been recently studied by several Soviet psychologists, notably by Luria, experimenting with the genesis and course of simple voluntary reactions in young children (Luria, 1956). But both space and the complexity of the problem forbid discussion here. However, the writer would like to be permitted to quote a statement he had made 26 years ago (Razran, 1935). He then spoke of "symbolic reactions as a term for verbal-conscious-volitional processes" and said that the individual's

> choosing and rejecting system [is] . . . the sum total of his symbolic "past" that has gradually become dominant over his somatic and visceral "present." Its action and organization are held to proceed ultimately according to the same principles of interdependence, conditioning, and dominance as the somatic and visceral unsymbolic—or little symbolic—system (p. 120).

In the main, the writer's view has changed little since, except perhaps that he would now be less tentative about the conditioning of verbal reactions and would point with pride to our increased evidence of somatic and visceral conditioning.

General Summary—Appraisal

Ten short statements are offered here as a general summary-appraisal of current Soviet psychophysiology. They are offered in lieu of a conventional summary, which would be difficult and which in a compressed theoretic sense was attempted in the preceding section. And they are intended to answer such questions as: What are the general distinctive characteristics of current Soviet psychophysiology? In what ways and why does it differ from corresponding American behavioral psychology? Which of the differences are a function of the Soviets' special socioeconomic and philosophical system and which are merely a result of historical differences in the development of the two disciplines in the respective countries? In which areas are the two disciplines likely to get closer to each other (or should get closer to each other) and in which will (or must) they remain apart? What do Soviet psychophysiologists do that we do not do and should do? And what should we not do or can we not do? More specifically, the first 2 of the 10 statements pertain to the general theory of Soviet psychophysiology and the remaining 8 to its general research program. All statements begin with characterizing epitomized phrases.

A single broad, otherwise nondoctrinaire, associationistic system or ap-

proach. This of course follows Pavlov. But, as noted earlier, most present-day research is in areas which Pavlov only preindicated or which were unknown in his day (interoceptive and verbal conditioning and the orienting reflex are examples of the first category, and EEG correlates of conditioning and the isolation of semantic conditioning illustrate the second). And, while in the early '50s adherence to Pavlov's specific views was strict, it has by now radically loosened, although, to be sure, the associationistic system itself is always upheld.[6] There is no doubt that universal acceptance of a single broad system leads to an economy of effort vis-à-vis, for instance, American practice in which associationistic experimentalists and theoreticians spend so much time in defending and proving their associationism. And it is of course also true that the acceptance of a broad system discourages the development of narrower ones, unlike the vogue in this country where theoreticians are not infrequently tempted to expand a study of a particular aspect of association into an all-embracing general system. Yet, it is quite obvious that our social philosophy cannot permit—and we are glad that it cannot—restriction of any school of psychology.

Little conceptualization. Pavlovian physiology combined with a fixed Marxist-Leninist body-mind view, plus the consideration that Soviet psychophysiologists are by training physiologists and not psychologists, is not conducive to the type of psychological conceptualization that is so characteristic of American behavioral psychology in the last few decades. Hence, it is quite evident that while a large portion of American behavioral psychology has often been mainly concerned with the problem of what *is* associated or conditioned, its Soviet homolog has been busy finding out what *can* be associated or conditioned and, as we have seen, has expanded manifoldly the realm of conditionables—no doubt a very notable contribution. Yet, it is also quite true that Soviet psychophysiological theorization is very unsophisticated—a fact noted by its own critics (Editorial, 1959, p. 163)—and the field is literally clogged with unintegrated empirical findings, just as American behavioristics may well be classed as overloaded with theorizations. A fair statement would seem to be one characterizing American behavioristics as highly overconceptualized and Soviet psychophysiology as markedly underconceptualized.

Little demand from social and clinical psychology. Social psychology in the American sense does not exist in the Soviet Union, while mental illness

[6] The following is a quotation from an editorial in the 1959 March–April issue of the *Zhurnal Vysshey Nervnoy Deyatel'nosti im. Pavlova* [The Pavlov Journal of Higher Nervous Activity]: ". . . much attention is given to interpretations of one or another quotation from Pavlov, whether he was an advocate or an opponent of psychology as an independent science, whether he intended physiology to engulf psychology, and the like. Although we are interested in such topics, it is now more important to us to determine how we ourselves regard the problems in the light of present-day science which, naturally, does not and cannot stand still" (p. 162). And again it might be mentioned that earlier practices of Soviet psychophysiologists and particularly psychologists of stocking their theoretical and even experimental articles with lengthy and numerous quotations from Marx, Engels, Lenin, and Stalin (typically in introductions and conclusions) have by now practically disappeared. In only one of the 17 contributions to the 1959 *The Science of Psychology in the USSR* (Anan'yev et al.) may such quotations be found, and there are none whatsoever in Sokolov's 1958 *Perception and the Conditioned Reflex* and in the 1958 and 1959 volumes on the orienting reflex. To be sure, Pavlov continues to be cited voluminously—but then there is no reason why he should not be.

is by all tokens a problem of much less concern than it is in the United States. There is thus little dilution of research programs and idea systems by the pragmatics of immediate applications and, we might add, by the economics of private psychotherapy, a nonexistent profession. Moreover, while Soviet psychophysiology, like American psychology, is applied widely, as might be expected, to problems of education, for reasons that need not detain us here, professional Soviet educators seem to be able to utilize psychological principles without much dilution, change, and popularization.

Much concern with evolution. Darwin is the Soviets' most respected and accepted non-Soviet scientist, his views being in the main regarded as prolegomena to Marxism-Leninism (schools offer special courses in Darwinism). The evolution of higher nervous activity, that is, phyletic and ontogenetic differences in conditioning, has thus always been a dominant portion of Soviet research and thought. But, evolution is of course an integral constituent of contemporary science anywhere and at one time has indeed loomed large even in specific systematic psychology. Its comparative present-day experimental and theoretical neglect in American comparative and genetic psychology has always seemed to the writer to be one of our most glaring shortcomings.

Little interest in maze, problem box, and other "trial-and-error" types of animal learning. The Soviet-American differential is here no doubt a matter of historical differences in the development of the scientific analysis of learning in the two countries and is unrelated to special characteristics of the Soviet system. It needs, however, to be pointed out that with the advent of Skinner, interest in conventional trial-and-error learning is continually fading even in this country and that, on the other hand, complex operant animal training and its scientific analysis are becoming, as was noted earlier, an integral part of CR programs in Soviet laboratories.

Many studies in compound and configural classical conditioning. The studies are evidently a counterpart of American involvement in maze-and-problem-box and other complex learning, offering the advantage of fully controllable follow-ups of the dynamics of the conversion of the simple into the complex. The negligible number of such experiments in this country is obviously not unrelated to our accelerated neglect of classical conditioning in general—another deplorable characteristic of recent American learning psychology.

Much concern with visceral action. In its CR form this concern has for years been almost a Russo-Soviet monopoly. In its UR manifestation it has become so in recent years with respect to interoceptive stimulations, the numerous studies of which demonstate that no S-R analysis is complete or correct when no account is taken of the almost ever-present interoceptive parameters. A Soviet-American professional differentiation should be noted here: namely, the fact that this type of CR research requires special skills which the American students of conditioning do not usually possess while Americans who possess the skills (gastrologists, urologists) are not as a rule interested in conditioning.

Verbal and semantic conditioning. Soviet intensive interest in the area is, as was mentioned earlier, of recent vintage. Its significance needs no emphasis and it is of course an area which is right in line with American psychologists' research and thought.

The orienting reflex. The topic has been presented in considerable detail in the text and needs no further elaboration, except to point it up as a striking example of directed research—the fact that, while the topic is old, 115 of 164 available experiments pertaining to it have been reported in the last 6 years.

The neurology and pharmacology of conditioning. As noted, the reviewer has as yet not fully digested the experimental material in the area. A preliminary perusal discloses, however, that the area—EEG correlates of conditioning, neuromorphological bases, the effect of drugs and of ablations, and the like—is beginning to receive top priority in the Soviet CR research program, and that apparatus and surgery are by all tokens technically adequate. The area is of course one in which non-Russian research is both extensive and readily synthesizable with what the Russians are doing. And there can of course be no question about the area's most basic systematic and pragmatic significance.

Final Word

Only the first 3 of the 10 statements in the summary-appraisal of current Soviet psychophysiology relate to research programs and approaches which are specific to the Soviet system and philosophy and are not transferrable to our views and practices. The remaining 7 statements deal with areas in which there are no basic differences between what Americans and Russians want to do and think, and in 5 of these areas Russian research is clearly a systematic and pragmatic challenge—we are behind and must catch up.

To be sure, the writer has dwelt only on the strengths of current Soviet psychophysiology. There are surely also weaknesses. However, frankly, the writer is not interested in Soviet weaknesses—let the Russians worry about them. The review is directed to a non-Russian audience and does not involve the general problem of "who is ahead."

REFERENCES

Anan'yev, B. G., Kostyuk, G. S., Leont'yev, A. N., Luria, A. R., Menchinskaya, N. A., Rubinshteyn, S. L., Smirnov, A. A., Teplov, B. M., & Shemyakin, F. N. (Eds.) *The science of psychology in the USSR.* Moscow: Akad. Pedag. Nauk RSFSR, 1959.

Anokhin, P. K. The role of orienting-investigatory reactions in the formation of the conditioned reflex. In L. G. Voronin, A. N. Leont'yev, A. R. Luria, E. N. Sokolov, & O. S. Vinogradova (Eds.), *Orientirovochny refleks i orientirovochno-issledovatel'skaya deyatel'nost'.* [The orienting reflex and orienting-investigatory activity.] Moscow: Akad. Pedag. Nauk RSFSR, 1958. Pp. 9–20.

Asafov, B. D. Changes of the dynamics of the vegetative components of the orienting reflex in the course of application of auditory stimuli of increasing intensity. In L. G. Voronin, A. N. Leont'yev, A. R. Luria, E. N. Sokolov, & O. S. Vinogradova (Eds.), *Orientirovochny refleks i orientirovochno-issledovatel'skaya deyatel'nost'.* [The orienting reflex and orienting-investigatory activity.] Moscow: Akad. Pedag. Nauk RSFSR, 1958. Pp. 123–128.

Asratyan, E., Gutmann, E., & Konorski, J. Mechanisms of motor activity of animals. *Zh. vyssh. nervn. Deyatel.,* 1959, **9**, 301–303.

Ayrapetyants, E. Sh. The interoceptive conditioned reflex. *Trud. Voyen.-Morsk. Med. Akad.,* 1949, **17**, 19–62.

Ayrapetyants, E. Sh. *Higher nervous function and the receptors of internal organs.* Moscow: Akad. Nauk SSSR, 1952.

Ayrapetyants, E. Sh. Materials on the physiology of the internal analyzer in man. *Trud. Inst. Fiziol. Pavlova,* 1956, **5**, 396–406.

Ayrapetyants, E. Sh., & Lichkus, K. V. The formation of temporary connections between interoceptive and proprioceptive stimulations. *Trud. Inst. Fiziol. Pavlova,* 1957, **6**, 142–149.

Ayrapetyants, E. Sh., Lobanova, L. V., & Cherkasova, L. S. Materials on the physiology of the internal analyzer in man. *Trud. Inst. Fiziol. Pavlova,* 1952, **1**, 3–20.

Babkin, B. P. Contributions to the study of the acoustic analyzer. *Trud. Dbshch.*

Russ. Vrach., St. Petersburg, 1910, 77, 197–230.

Baru, A. V. Conditioned motor reflexes to chains of stimuli in rabbits. *Zh. vyssh. nervn. Deyatel.,* 1954, **4**, 712–723.

Beritov, J. S., & Bregadze, A. Physiology of animal behavior: Role of experimental setting in the production of reflex reaction of animals to complex sound stimuli. *Med. Biol. Zh.,* 1929, **5**, 83–101, 131–151.

Beritov, J. S., & Bregadze, A. Physiology of animal behavior: In complex situations: Individual reflex reactions of animals to complex sound stimuli. *Med. Biol. Zh.,* 1930, **6**, 104–113.

Biryukov, D. A. The nature of orienting reactions. In L. G. Voronin, A. N. Leont'yev, A. R. Luria, E. N. Sokolov, & O. S. Vinogradova (Eds.), *Orientirovochny refleks i orientirovochno-issledovatel'skaya deyatel'nost'.* [The orienting reflex and orienting-investigatory activity.] Moscow: Akad. Pedag. Nauk RSFSR, 1958. Pp. 20–25.

Bregadze, A. The formation of individual reactions to a complex of musical tones in dogs. *Trud. Inst. Beritashvili,* 1937, **3**, 415–430.

Bregadze, A., & Tarugov, S. Individual reactions of rabbits to complex sound stimuli. *Trud. Inst. Beritashvili,* 1937, **3**, 431–447.

Bronshteyn, A. I., Itina, N. A., Kamenetskaya, A. G., & Sitova, V. A. Orienting reactions of human neonates. In L. G. Voronin, A. N. Leont'yev, A. R. Luria, E. N. Sokolov, & O. S. Vinogradova (Eds.), *Orientirovochny refleks i orientirovochno-issledovatel'skaya deyatel'nost'.* [The orienting reflex and orienting-investigatory activity.] Moscow: Akad. Pedag. Nauk RSFSR, 1958. Pp. 237–241.

Bykov, K. M. *The cerebral cortex and the internal organs.* Moscow: VMMA, 1943; Medgiz, 1944, 1947, 1954. (Translated into Chinese, Czech, English, French, German, Japanese, and Polish)

Bykov, K. M., & Pshonik, A. T. The nature of the conditioned reflex. *Fiziol. Zh. SSSR,* 1949, **35**, 509–524.

Bykov, V. D. The dynamics of orienting-investigatory reactions during the formation of positive and inhibitory conditioned reflexes and their transformation. In L. G. Voronin, A. N. Leont'yev, A. R. Sokolov, & O. S. Vinogradova (Eds.), *Orientirovochny refleks i orientirovochno-issledovatel'skaya deyatel'nost'.* [The orienting reflex and orienting-investigatory activity.] Moscow: Akad. Pedag. Nauk RSFSR, 1958. Pp. 25–33.

Chernova, N. A. Conditioned respiratory reflexes in fish. *Trud. Inst. Fiziol. Pavlova,* 1953, **2**, 364–369.

Chumak, V. I. Extinction of the orienting reflex of cats in the early post-natal period. *Zh. vyssh. nervn. Deyatel.,* 1955, **5**, 863–872.

Chumak, V. I. Orienting reactions during the formation of conditioned reflexes to a simultaneous application of a positive and a differential stimulus. In L. G. Voronin, A. N. Leont'yev, A. R. Luria, E. N. Sokolov, & O. S. Vinogradova (Eds.), *Orientirovochny refleks i orientirovochno-issledovatel'skaya deyatel'nost'.* [The orienting reflex and orienting-investigatory activity.] Moscow: Akad. Pedag. Nauk RSFSR, 1958. Pp. 231–234.

Cook, L., Davidson, A., Davis, D. J., & Kelleher, R. T. Epinephrine, norepinephrine, and acetylcholine as conditioned stimuli for avoidance behavior. *Science,* 1960, **131**, 990–991.

Dedashev, Ya. P. Exteroceptive and interoceptive conditioned-reflex influences on the motor activity of the reticulum and rumen in sheep. *Fiziol. Zh. SSSR,* 1959, **45**, 1259–1262.

Dolin, A. O., Zborovskaya, I. I., & Zamakhover, Sh. M. The role of the orienting-investigatory reflex in conditioned-reflex activity. In L. G. Voronin, A. N. Leont'yev, A. R. Luria, E. N. Sokolov, & O. S. Vinogradova (Eds.), *Orientirovochny refleks i orientirovochno-issledovatel'skaya deyatel'nost'.* [The orienting reflex and orienting-investigatory activity.] Moscow: Akad. Pedag. Nauk RSFSR, 1958. Pp. 47–60.

Editorial. *Zh. vyssh. nervn. Deyatel.,* 1959, **9**, 161–165.

Elkin, D. G. The characteristics of conditioned reflexes to a complex verbal stimulus. *Vop. Psikhol.,* 1955, **1**(4), 79–89.

Fel'berbaum, I. M. Interoceptive conditioned reflexes from the uterus. *Trud. Inst. Fiziol. Pavlova,* 1952, **1**, 85–92.

Gamburg, A. L. Orienting and defensive reactions in simple and paranoid forms of schizophrenia. In L. G. Voronin, A. N. Leont'yev, A. R. Luria, E. N. Sokolov, & O. S. Vinogradova (Eds.), *Orientirovochny refleks i orientirovochno-issledovatel'skaya deyatel'nost'.* [The orienting reflex and orienting-investigatory activity.] Moscow: Akad. Pedag. Nauk RSFSR, 1958. Pp. 270–281.

GERSHUNI, G. V. The study of subsensory reactions in action of sensory organs. *Fiziol. Zh. SSSR,* 1947, **33**, 393–412.

GERSHUNI, G. V. Conditioned galvanic-skin reactions and alpha-rhythm-depression reactions during subliminal and supraliminal auditory stimulations of human subjects. *Zh. vyssh. nervn. Deyatel.,* 1955, **5**, 665–676.

GERSHUNI, G. V. General results of the study of the activity of the auditory analyzer by means of various reactions. *Zh. vyssh. nervn. Deyatel.,* 1957, **7**, 13–23. (a)

GERSHUNI, G. V. Human auditory discrimination of complex stimuli with increasing amount of information. *Fiziol. Zh. SSSR,* 1957, **43**, 1086–1097. (b)

GERSHUNI, G. V., KLASS, YU. A., LUKOMSKAYA, N. YA., LINYUCHEV, M. N., & SEGAL, A. A. Evaluation of human discrimination of auditory stimuli with increased information and its utilization in the study of the effect of certain pharmacological substances. *Biofizika,* 1959, **4**(2), 158–165.

GONCHAROVA, A. F. Extero- and interoceptive connections between indifferent stimuli. In, *Voprosy sravnitel'noy fiziologii i patologii vysshey nervnoy deyatel'nosti.* Moscow: Medgiz, 1955. Pp. 152–167.

GRINGS, W. W. *Stimulus patterning in learning.* Los Angeles: Univer. Southern California, 1957. (Dittoed Manuscript)

GRINGS, W. W., & O'DONNELL, D. E. Magnitude of response to compounds of discriminated stimuli. *J. exp. Psychol.,* 1956, **52**, 354–359.

GRINGS, W. W., & SHMELEV, V. N. Changes in GSR to a single stimulus as a result of training on a compound stimulus. *J. exp. Psychol.,* 1959, **58**, 129–133.

GUSEL'NOKOV, V. E. The reflection of orienting reactions in the fluctuations of bioelectric potentials in the forebrain of fish, turtles, and pigeons. In L. G. Voronin, A. N. Leont'yev, A. R. Luria, E. N. Sokolov, & O. S. Vinogradova (Eds.), *Orientirovochny refleks i orientirovochno-issledovatel'skaya deyatel'nost'.* [The orienting reflex and orienting-investigatory activity.] Moscow: Akad. Pedag. Nauk RSFSR, 1958. Pp. 209–220.

HILGARD, E. R., & MARQUIS, D. G. *Conditioning and learning.* New York: Appleton-Century, 1940.

HUMPHREY, G. The effect of sequences of indifferent stimuli on a reaction of the conditioned response type. *J. abnorm. soc. Psychol.,* 1928, **22**, 194–212.

KARAMYAN, A. I. *The evolution of the functions of the cerebellum and the cerebrum.* Leningrad: Medgiz, 1956.

KARAMYAN, A. I. Orienting reactions in animals during organic pathology of the central nervous system. In L. G. Voronin, A. N. Leont'yev, A. R. Luria, E. N. Sokolov, & O. S. Vinogradova (Eds.), *Orientirovochny refleks i orientirovochno-issledovatel'skaya deyatel'nost'.* [The orienting reflex and orienting-investigatory activity.] Moscow: Akad. Pedag. Nauk RSFSR, 1958. Pp. 69–75.

KARLOVA, A. N. Orienting reflexes in young children. *Zh. vyssh. nervn. Deyatel.,* 1959, **9**, 37–44.

KASATKIN, N. I. *Outline of the development of higher nervous activity in early human ontogeny.* Moscow: Medgiz, 1951.

KASATKIN, N. I. Early ontogeny of reflex activity of infants. *Zh. vyssh. nervn. Deyatel.,* 1957, **7**, 805–818.

KASATKIN, N. I., MIRZOYANTS, N. S., & KHOKHITIVA, A. P. Conditioned orienting reflexes in children under one year. *Zh. vyssh. nervn. Deyatel.,* 1953, **3**, 192–202.

KASSIL', V. G. Conditioned-reflex influences of the receptors of the stomach on salt appetite in higher animals. *Dokl. Akad. Nauk SSSR,* 1959, **129**(2), 464–467.

KLIMOVA, V. I. The characteristics of components of some orienting reactions. In L. G. Voronin, A. N. Leont'yev, A. R. Luria, E. N. Sokolov, & O. S. Vinogradova (Eds.), *Orientirovochny refleks i orientirovochno-issledovatel'skaya deyatel'nost'.* [The orienting reflex and orienting-investigatory activity.] Moscow: Akad. Pedag. Nauk RSFSR, 1958. Pp. 76–80.

KOL'TSOVA, M. M. *The formation of higher nervous activity of the child.* Moscow: Medgiz, 1958.

KONORSKI, J., & MILLER, S. Méthode d'examen de l'analysateur moteur par les réactions salivo-motrices. *CR Soc. Biol.,* 1930, **104**, 911–913.

KONORSKI, J., & MILLER, S. Odruchy warunkowe analizatora ruchowego. [Conditioned reflexes of the motor analyzer.] *Roczn. Psychiat.,* 1932, 18–19, 308–312.

KONORSKI, J., & MILLER, S. Podstawy fiziologichnej teorij ruchow nabytych. [Physiological bases of a theory of acquired movement.] *Med. dosw. spol.,* 1933, **16**, 95–187, 234–238.

KONORSKI, J., & MILLER, S. Nouvelles récherches sur les réflexes conditionnels moteurs. *CR Soc. Biol.,* 1934, **115**, 91–96. (a)

Konorski, J., & Miller, S. Pewne zagadnienia nauki o wyzszych czynnosciach ukladu nerwowego w zastosowaniu do pedagogiki. [Certain scientific problems concerning the higher activities of the nervous system and their pedagogical application.] *Polsk. Arch. Psychol.,* 1934, **35**(7), 212–231. (b)

Konorski, J., & Miller, S. Conditioned reflexes of the motor analyzer. *Trud. Fiziol. Lab. Pavlova,* 1936, **6**(1), 119–288.

Krasnogorsky, N. I. Inhibition and localization of the cutaneous and the motor analyzers in the cerebrum of dogs. Thesis, Military Medical Academy, St. Petersburg, 1911.

Kukuyev, L. A. Evolution of the nucleus of the motor analyzer and subcortical ganglia. *Zh. vyssh. nervn. Deyatel.,* 1953, **3**, 765–773.

Kvasov, D. G. The muscular apparatus of analyzers. *Fiziol. Zh. SSSR,* 1956, **42**, 621–631.

Lacey, J. I., & Smith, R. L. Conditioning and generalization of unconscious anxiety. *Science,* 1954, **120**, 1045–1052.

Lacey, J. I., Smith, R. L., & Green, A. The use of conditioned autonomic responses in the study of anxiety. *Psychosom. Med.,* 1955, **17**, 208–217.

Lagutina, N. I. The structure of orienting reflexes. In L. G. Voronin, A. N. Leont'yev, A. R. Luria, E. N. Sokolov, & O. S. Vinogradova (Eds.), *Orientirovochny refleks i orientirovochno-issledovatel'skaya deyatel'nost'.* [The orienting reflex and orienting-investigatory activity.] Moscow: Akad. Pedag. Nauk RSFSR, 1958. Pp. 80–86.

Lebedinskaya, E. I. The interrelations between conditioned-orienting and conditioned-reflexes during the formation of a temporary connection between two indifferent stimuli. In L. G. Voronin, A. N. Leont'yev, A. R. Luria, E. N. Sokolov, & O. S. Vinogradova (Eds.), *Orientirovochny refleks i orientirovochno-issledovatel'skaya deyatel'nost'.* [The orienting reflex and orienting-investigatory activity.] Moscow: Akad. Pedag. Nauk RSFSR, 1958. Pp. 86–92.

Leont'yev, A. N. *Problems of psychic development.* Moscow: Akad. Pedag. Nauk, 1959.

Liberman, A. E. New data on the pupillary component of the orienting reflex in man. In L. G. Voronin, A. N. Leont'yev, A. R. Luria, E. N. Sokolov, & O. S. Vinogradova (Eds.), *Orientirovochny refleks i orientirovochno-issledovatel'skaya deyatel'nost'.* [The orienting reflex and orienting-investigatory activity.] Moscow: Akad. Pedag. Nauk RSFSR, 1958. Pp. 145–151.

Lichko, A. E. Characteristics of conditioned and unconditioned defensive and orienting reflexes in acute toxic psychoses. *Trud. Inst. Fiziol. Pavlova,* 1952, **1**, 406–412.

Lisina, M. I. The role of orientation in the transformation of involuntary into voluntary reactions. In L. G. Voronin, A. N. Leont'yev, A. R. Luria, E. N. Sokolov, & O. S. Vinogradova (Eds.), *Orientirovochny refleks i orientirovochno-issledovatel'skaya deyatel'nost'.* [The orienting reflex and orienting-investigatory activity.] Moscow: Akad. Pedag. Nauk RSFSR, 1958. Pp. 339–344.

Lotis, V. I. Conditioned interoceptive uterine reflexes. *Akush. Ginekol.,* 1949, No. 6, 15–19.

Luria, A. R. The regulating role of speech in the formation of voluntary movements. *Zh. vyssh. nervn. Deyatel.,* 1956, **6**, 645–662.

Luria, A. R., & Vinogradova, O. S. An objective investigation of the dynamics of semantic systems. *Brit. J. Psychol.,* 1959, **50**, 89–105.

Makarov, P. O. Pre-excitation and presensation. *Uchen. Zap. Leningr. U., Ser. Biol.,* 1950, **22**(123), 369–399. (a)

Makarov, P. O. A study of interoception in human subjects. *Uchen. Zap. Leningr. U., Ser. Biol.,* 1950, **22**(123), 345–368. (b)

Makarov, P. O. The effect of interoceptive gastric signalizations on the electroencephalogram of man. *Fiziol. Zh. SSSR,* 1952, **38**, 281–287.

Makarov, P. O. *The neurodynamics of man.* Leningrad: Medgiz, 1959.

Markosyan, A. A. Conditioned-reflex changes of blood coagulation. *Zh. vyssh. nervn. Deyatel.,* 1953, **3**, 911–918.

Markosyan, A. A. The interaction of signal systems in the process of blood coagulation. *Zh. vyssh. nervn. Deyatel.,* 1958, **8**, 161–167.

Mikhalevskaya, M. B. The interrelations of the orienting and the conditioned motor reactions of man during the determination of thresholds of visual sensitivity. In L. G. Voronin, A. N. Leont'yev, A. R. Luria, E. N. Sokolov, & O. S. Vinogradova (Eds.), *Orientirovochny refleks i orientirovochno-issledovatel'skaya deyatel'nost'.* [The orienting reflex and orienting-investigatory activity.] Moscow: Akad. Pedag. Nauk RSFSR, 1958. Pp. 151–158.

MILLER, S., & KONORSKI, J. Le phenomène de la généralization motrice. *CR Soc. Biol.,* 1928, **99**, 1158. (a)

MILLER, S., & KONORSKI, J. Sur une forme particulière des réflexes conditionnels. *CR Soc. Biol.,* 1928, **99**, 1155. (b)

MIROZYANTS, N. S. The conditioned orienting reflex and its differentiation in children. *Zh. vyssh. nervn. Deyatel.,* 1954, **5**, 616–619.

MOISEYEVA, N. A. Interoceptive conditioned reflexes from the ileocecal region. In K. M. Bykov (Ed.), *Voprosy fiziologii interotseptsii.* Moscow: Akad. Nauk. SSSR, 1952. Pp. 405–410.

MOWRER, O. H. On the dual nature of learning: A reinterpretation of "conditioning" and "problem-solving." *Harv. educ. Rev.,* 1947, **17**, 102–108.

MYSYASHCHIKOVA, S. S. The extinction of vegetative reactions during the stimulation of the peripheral apparatus of various analyzers. In K. M. Bykov (Ed.), *Voprosy fiziologii interotseptsii.* Moscow: Akad. Nauk SSSR, 1952. Pp. 411–427.

NIKITINA, G. M. Interrelations in the development of orienting and conditioned motor reactions in ontogeny. *Zh. vyssh. nervn. Deyatel.,* 1954, **4**, 406–413.

NIKITINA, G. M., & NOVIKOVA, E. G. The characteristics of the manifestation of orienting reactions in animal ontogeny. In L. G. Voronin, A. N. Leont'yev, A. R. Luria, E. N. Sokolov, & O. S. Vinogradova (Eds.), *Orientirovochny refleks i orientirovochno-issledovatel'skaya deyatel'nost'.* [The orienting reflex and orienting-investigatory activity.] Moscow: Akad. Pedag. Nauk RSFSR, 1958. Pp. 242–248.

OBRAZTSOVA, G. A., POMAZANSKAYA, L. F., STEL'MACH, L. N., & TROSHIKHIN, V. A. Orienting reactions to indifferent and to conditioned stimuli in ontogeny of dogs and rabbits. In L. G. Voronin, A. N. Leont'yov, A. R. Luria, E. N. Sokolov, & O. S. Vinogradova (Eds.), *Orientirovochny refleks i orientirovochno-issledovatel'skaya deyatel'nost'.* [The orienting reflex and orienting-investigatory activity.] Moscow: Akad. Pedag. Nauk RSFSR, 1958. Pp. 248–253.

OKHNYANSKAYA, L. G. A study of the conditioned respiratory-vasomotor reflexes: Respiration as the stimulus of vaso-motion. *Fiziol. Zh. SSSR,* 1953, **39**, 610–613.

ORBELI, L. A. *Problems of higher nervous activity.* Moscow: Akad. Nauk SSSR, 1949.

PALLADIN, A. Formation of laboratory conditioned reflexes to sums of stimuli. *Trud. Obshch Russ. Vrach., St. Petersburg,* 1906, **73**, 393–401.

PARAMANOVA, N. P. The characteristics of the participation of the orienting reflex in the formation of conditioned connections in oligophrenia. In E. N. Sokolov (Ed.), *Orientirovochny refleks i voprosy vysshey nervnoy deyatel'nost'.* [The orienting reflex and problems of higher nervous activity.] Moscow: Akad. Pedag. Nauk RSFSR, 1959. Pp. 77–85.

PAUPEROVA, G. F. Formation of a secondary exteroceptive conditioned reflex on the basis of a primary interoceptive one. In K. M. Bykov (Ed.), *Voprosy fiziologii interotseptsii.* Moscow: Akad. Nauk SSSR, 1952. Pp. 437–442.

PAVLOV, I. P. *Conditioned reflexes: An investigation of the physiological activity of the cerebral cortex.* London: Oxford Univer. Press, 1927.

PAVLOV, I. P. *Lectures on conditioned reflexes.* New York: Liveright, 1928.

PAVLOV, I. P. The physiological mechanisms of so-called voluntary movements. *Trud. Fiziol. Lab. Pavlova,* 1936, **6**(1), 115–118.

PAVLOV, I. P. *Complete works.* Moscow-Leningrad: Akad. Nauk SSSR, 1949. 5 vols. (a)

PAVLOV, I. P. *Pavlov's Wednesdays.* Moscow: Akad. Nauk SSSR, 1949. 3 vols. (b)

PERELTSVAYG, I. Materials to the study of conditioned reflexes. Thesis, Military Medical Academy, St. Petersburg, 1907.

PETELINA, V. V. The vegetative component of orienting reactions of the vestibular, visual, and auditory analyzers. In L. G. Voronin, A. N. Leont'yev, A. R. Luria, E. N. Sokolov, & O. S. Vinogradova (Eds.), *Orientirovochny refleks i orientirovochno-issledovatel'skaya deyatel'nost'.* [The orienting reflex and orienting-investigatory activity.] Moscow: Akad. Pedag. Nauk RSFSR, 1958. Pp. 158–164.

PETROVA, N. K. Materials to the understanding of the physiological mechanism of voluntary movements. *Trud. Fiziol. Lab. Pavlova,* 1941, **10**, 41–50.

PLATONOV, K. I. The formation of associated motor reflexes to simultaneous auditory and visual stimuli in human beings. Thesis, Military Medical Academy, St. Petersburg, 1912.

POGREBKOVA, A. V. Conditioned reflexes to hypercapnia. *Dokl. Akad. Nauk SSSR,* 1950, **73**, 225–228.

POGREBKOVA, A. V. Respiratory intero- and exteroceptive conditioned reflexes and their interrelationship. I: Formation and properties of respiratory intero- and exteroceptive conditioned reflexes. In K. M. Bykov (Ed.), *Voprosy fiziologii interotseptsii.* Moscow: Akad. Nauk SSSR, 1952. Pp. 443–454. (a)

POGREBKOVA, A. V. Respiratory intero- and exteroceptive conditioned reflexes and their interrelationship. II: Correlation of respiratory conditioned reflexes. *Trud. Inst. Fiziol. Pavlova,* 1952, **1**, 103–115. (b)

POLEZHAYEV, E. F. The role of the orienting reflex in the coordination of the activity of the cerebral cortex. In L. G. Voronin, A. N. Leont'yev, A. R. Luria, E. N. Sokolov, & O. S. Vinogradova (Eds.), *Orientirovochny refleks i orientirovochno-issledovatel'skaya deyatel'nost'.* [The orienting reflex and orienting-investigatory activity.] Moscow: Akad. Pedag. Nauk RSFSR, 1958. Pp. 97–111.

POLEZHAYEV, E. F. The problems of the physiological conditions of linkage. *Dokl. Akad. Nauk SSSR,* 1960, **130**, 469–472.

POLIKANINA, P. I., & PROBATOVA, L. E. The development of orienting reactions to auditory stimuli in premature infants. *Zh. vyssh. nervn. Deyatel.,* 1955, **5**, 227–236.

POLIKANINA, P. I., & PROBATOVA, L. E. The problem of the development of the orienting reflex in premature infants. In L. G. Voronin, A. N. Leont'yev, A. R. Luria, E. N. Sokolov, & O. S. Vinogradova (Eds.), *Orientirovochny refleks i orientirovochno-issledovatel'skaya deyatel'nost'.* [The orienting reflex and orienting-investigatory activity.] Moscow: Akad. Pedag. Nauk RSFSR, 1958. Pp. 253–261.

PRAZDNIKOVA, V. The method of studying conditioned food-motor reflexes in fish. *Zh. vyssh. nervn. Deyatel.,* 1953, **3**, 464–468.

PRADZNIKOVA, V. Conditioned food-motor reflexes to chains of stimuli in fish. *Zh. vyssh. nervn. Deyatel.,* 1955, **5**, 901–911.

PSHONIK, A. T. Interaction of extero- and interoceptive conditioned vasomotor reflexes. *Dokl. Akad. Nauk SSSR,* 1949, **67**, 1175–1178. (a)

PSHONIK, A. T. Interaction of extero- and interoceptive conditioned vasomotor reflexes. In, *Problemy Sovetskoy fiziologii, biokhimii i farmakologii.* Moscow: Akad. Nauk SSSR, 1949. Pp. 269–274. (b)

PSHONIK, A. T. The problem of the interaction between exteroceptive and interoceptive vasomotor reflexes. In K. M. Bykov (Ed.), *Problemy kortiko-vistseral'noy patologii.* Moscow: Akad. Med. Nauk SSSR, 1949. Pp. 255–269. (c)

PSHONIK, A. T. *The cerebral cortex and the receptor functions of the organism.* Moscow: GIZ, 1952.

RAZRAN, G. Conditioned responses: An experimental study and a theoretical analysis. *Arch. Psychol.,* 1935, **28** (Whole No. 191).

RAZRAN, G. Studies in configural conditioning: VII. Ratios and elements in salivary conditioning to various musical intervals. *Psychol. Rec.,* 1938, **2**, 370–376.

RAZRAN, G. The law of effect or the law of qualitative conditioning. *Psychol. Rev.,* 1939, **46**, 445–463. (a)

RAZRAN, G. A quantitative study of meaning by a conditioned salivary technique (semantic conditioning). *Science,* 1939, **90**, 89–91. (b)

RAZRAN, G. Studies in configural conditioning: I. Historical and preliminary experimentation. *J. gen. Psychol.,* 1939, **21**, 307–330. (c)

RAZRAN, G. Studies in configural conditioning: II. The effect of subjects' attitudes and of task-sets upon configural conditioning. *J. exp. Psychol.,* 1939, **24**, 95–105. (d)

RAZRAN, G. Studies in configural conditioning: III. The factors of similarity, proximity, and continuity in configural conditioning. *J. exp. Psychol.,* 1939, **24**, 202–210. (e)

RAZRAN, G. Studies in configural conditioning: IV. Gestalt organization and configural conditioning. *J. Psychol.,* 1939, **7**, 3–16. (f)

RAZRAN, G. Studies in configural conditioning: VI. Comparative extinction and forgetting of pattern and of single-stimulus conditioning. *J. exp. Psychol.,* 1939, **24**, 432–438. (g)

RAZRAN, G. Studies in configural conditioning: V. Generalization and transposition. *J. genet. Psychol.,* 1940, **56**, 3–11.

RAZRAN, G. Attitudinal determinants of conditioning and of generalization of conditioning. *J. exp. Psychol.,* 1949, **39**, 820–829. (a)

RAZRAN, G. Semantic and phonetographic generalizations of salivary conditioning to verbal stimuli. *J. exp. Psychol.,* 1949, **39**, 642–652. (b)

RAZRAN, G. Sentential and propositional generalizations of salivary conditioning to verbal stimuli. *Science,* 1949, **109**, 447–448. (c)

RAZRAN, G. Some psychological factors in the generalization of salivary conditioning to verbal stimuli. *Amer. J. Psychol.,* 1949, **62**, 247–256. (d)

RAZRAN, G. Stimulus generalization of conditioned responses. *Psychol. Bull.*, 1949, **46**, 337–365. (e)

RAZRAN, G. Experimental semantics. *Trans. NY Acad. Sci.*, 1952, **14**, 171–177.

RAZRAN, G. A note on second-order conditioning—and secondary reinforcement. *Psychol. Rev.*, 1955, **62**, 327–332. (a)

RAZRAN, G. Operant vs. classical conditioning. *Amer. J. Psychol.*, 1955, **68**, 127–129. (b)

RAZRAN, G. Extinction re-examined and reanalyzed: A new theory. *Psychol. Rev.*, 1956, **63**, 39–52.

RAZRAN, G. Soviet psychology and psychophysiology. *Science*, 1958, **128**, 1187–1194.

RIESS, B. F. Semantic conditioning involving the galvanic skin reflex. *J. exp. Psychol.*, 1940, **26**, 238–240.

RIESS, B. F. Genetic changes in semantic conditioning. *J. exp. Psychol.*, 1946, **36**, 143–152.

ROKOTOVA, N. A. The formation of temporary connections in the cerebral cortex of dogs during the action of several indifferent stimuli. *Zh. vyssh. nervn. Deyatel.*, 1952, **2**, 753–759.

SCHLOSBERG, H. The relationship between success and the laws of conditioning. *Psychol. Rev.*, 1937, **44**, 374–394.

SCHWARTZ, H. G. The effect of experimental lesions of the cortex upon the "psychogalvanic reflex." *Anat. Rec., Suppl.*, 1936, **64**, 42.

SCHWARTZ, H. G. Effect of experimental lesions of the cortex on the "psychogalvanic reflex" in the cat. *Arch. Neurol. Psychiat., Chicago*, 1937, **38**, 308–320.

SHAKHNOVICH, A. R. The pupillary component of the orienting reflex during the action of specific and nonspecific visual stimuli. In L. G. Voronin, A. N. Leont'yev, A. R. Luria, E. N. Sokolov, & O. S. Vinogradova (Eds.), *Orientirovochny refleks i orientirovochno-issledovatel'skaya deyatel'nost'*. [The orienting reflex and orienting-investigatory activity.] Moscow: Akad. Pedag. Nauk RSFSR, 1958. Pp. 191–198.

SHMAVONIAN, B. M. A methodological study of vasomotor conditioning in human subjects. *J. comp. physiol. Psychol.*, 1959, **52**, 315–321.

SHVARTS, L. A. The problems of words as conditioned stimuli. *Byull. eksp. Biol. Med.*, 1954, **38**(12), 15–18.

SHVARTS, L. A. Conditioned reflexes to verbal stimuli. *Vop. Psikhol.*, 1960, **6**(1), 86–98.

SKINNER, B. F. Two types of conditioned reflex and a pseudo-type. *J. gen. Psychol.*, 1935, **12**, 66–77.

SKINNER, B. F. *The behavior of organisms: An experimental analysis.* New York: Appleton-Century, 1938.

SKINNER, B. F. *Verbal behavior.* New York: Appleton-Century-Crofts, 1957.

SKINNER, B. F. *Cumulative record.* New York: Appleton-Century-Crofts, 1959.

SKIPIN, G. V. The nature of the inhibitory (extinctive) process in the higher divisions of the central nervous system of dogs. *Trud. Fiziol. Lab. Pavlova*, 1940, **9**, 459–465.

SKIPIN, G. V. An analysis of the higher nervous activity of dogs by means of a complex motor-secretory methodology. *Trud. Fiziol. Lab. Pavlova*, 1941, **10**, 5–24.

SKIPIN, G. V. Physiological mechanisms underlying the formation of conditioned defense reflexes. *Zh. vyssh. nervn. Deyatel.*, 1957, **7**, 877–888.

SKIPIN, G. V. The interrelation of various forms of conditioned motor-defense reflexes in animals. *Zh. vyssh. nervn. Deyatel.*, 1959, **9**, 429–435.

SMIRNOV, A. A., LEONT'YEV, A. N., RUBINSHTEYN, S. L., & TEPLOV, B. M. (Eds.) *Psikhologiya.* Moscow: UchPedGiz, 1956.

SOKOLOV, E. N. Measuring the sensitivity and the reactivity of the conditioned reflex in connection with the interrelations between orienting and defensive reactions. In L. G. Voronin, A. N. Leont'yev, A. R. Luria, E. N. Sokolov, & O. S. Vinogradova (Eds.), *Orientirovochny refleks i orientirovochno-issledovatel'skaya deyatel'skaya deyatel'nost'*. [The orienting reflex and orienting-investigatory activity.] Moscow: Akad. Pedag. Nauk RSFSR, 1958. Pp. 170–182. (a)

SOKOLOV, E. N. The orienting reflex, its structure and mechanism. In L. G. Voronin, A. N. Leont'yev, A. R. Luria, E. N. Sokolov, & O. S. Vinogradova (Eds.), *Orientirovochny refleks i orientirovochno-issledovatel'skaya deyatel'nost'*. [The orienting reflex and orienting-investigatory activity.] Moscow: Akad. Pedag. Nauk RSFSR, 1958. Pp. 111–120. (b)

SOKOLOV, E. N. *Perception and the conditioned reflex.* Moscow: Moscow Univer., 1958. (c)

SOKOLOV, E. N. (Ed.) *Orientirovochny refleks i voprosy vysshey nervnoy deyatel'nosti.* [The orienting reflex and problems of higher nervous activity.] Moscow: Akad. Pedag. Nauk RSFSR, 1959. (a)

Sokolov, E. N. The orienting reflex. In E. N. Sokolov (Ed.), *Orientirovochny refleks i voprosy vysshey nervnoy deyatel'nosti.* [The orienting reflex and problems of higher nervous activity.] Moscow: Akad. Pedag. Nauk RSFSR, 1959. Pp. 5–51. (b)

Solomon, R. L., & Wynne, C. W. Traumatic avoidance learning: Acquisition in normal dogs. *Psychol. Monogr.,* 1953, **67** (4, Whole No. 354).

Speranskaya, E. N. *Operative methods and the conduct of chronic physiological experiments in dogs.* Moscow: Akad. Nauk SSSR, 1953.

Usov, A. G. Cortical regulation of respiration in senility. *Fiziol. Zh. SSSR,* 1952, **38**, 576–583.

Vasilevskaya, N. E. Interoceptive conditioned reflexes of the second order. *Dokl. Akad. Nauk SSSR,* 1948, **61**, 161–164.

Vasilevskaya, N. E. The formation of second-order interoceptive conditioned reflexes with exteroceptive reinforcement. *Nauch. Byull. Leningr. Gosud. U.,* 1950, No. 26, 21–23.

Vedyayev, F. P., & Karmanova, I. G. Comparative physiology of the orienting reflex. In L. G. Voronin, A. N. Leont'yev, A. R. Luria, E. N. Sokolov, & O. S. Vinogradova (Eds.), *Orientirovochny refleks i orientirovochno-issledovatel'skaya deyatel'nost'.* [The orienting reflex and orienting-investigatory activity.] Moscow: Akad. Pedag. Nauk RSFSR, 1958. Pp. 201–204.

Vinogradova, O. S. The dynamics of the orienting reflex in the process of the formation of conditioned connections. In L. G. Voronin, A. N. Leont'yev, A. R. Luria, E. N. Sokolov, & O. S. Vinogradova (Eds.), *Orientirovochny refleks i orientirovochno-issledovatel'skaya deyatel'nost'.* [The orienting reflex and orienting-investigatory activity.] Moscow: Akad. Pedag. Nauk RSFSR, 1958. Pp. 40–47.

Vinogradova, O. S. The role of the orienting reflex in the formation of conditioned connections in man. In E. N. Sokolov (Ed.), *Orientirovochny refleks i voprosy vysshey nervnoy deyatel'nosti.* [The orienting reflex and problems of higher nervous activity.] Moscow: Akad. Pedag. Nauk RSFSR, 1959. Pp. 86–160. (a)

Vinogradova, O. S. A study of the orienting reflex in oligophrenic children by a plethysmographic technique. In E. N. Sokolov (Ed.), *Orientirovochny refleks i voprosy vysshey nervnoy deyatel'nosti.* [The orienting reflex and problems of higher nervous activity.] Moscow: Akad. Pedag. Nauk RSFSR, 1959. Pp. 161–206. (b)

Vinogradova, O. S., & Eysler, N. A. The manifestations of verbal connections in recording vascular reactions. *Vop. Psikhol.,* 1959, **2**, 101–116.

Vinogradova, O. S., & Sokolov, E. N. Extinction of the vascular component of the orienting reflex. *Zh. vyssh. nervn. Deyatel.,* 1955, **5**, 344–350.

Volkova, V. D. On certain characteristics of the formation of conditioned reflexes to speech stimuli in children. *Fiziol. Zh. SSSR,* 1953, **39**, 540–548.

Volokhov, A. A., Nikitina, G. M., & Novikova, E. G. The development of the vegetative components of the orienting and defensive conditioned reflex in the ontogeny of a comparative series of animals. *Zh. vyssh. nervn. Deyatel.,* 1959, **9**, 420–428.

Voronin, L. G. More about the speed of the formation of conditioned reflexes (reply to E. G. Vatsuro). *Zh. vyssh. nervn. Deyatel.,* 1954, **4**, 756–767.

Voronin, L. G. *Comparative physiology of higher nervous activity.* Moscow: Moscow Univer., 1957.

Voronin, L. G., Leont'yev, A. N., Luria, A. R., Sokolov, E. N., & Vinogradova, O. S. (Eds.) *Orientirovochny refleks i orientirovochno-issledovatel'skaya deyatel'nost'.* [The orienting reflex and orienting-investigatory activity.] Moscow: Akad. Pedag. Nauk RSFSR, 1958.

Voronin, L. G., & Napalkov, A. V. Methodological techniques for the formation of complex systems of conditioned motor reflexes in animals. *Zh. vyssh. nervn. Deyatel.,* 1959, **9**, 788–791.

Yordanis, K. A. Comparative-physiological data on conditioned inhibition and conditioned disinhibition. *Zh. vyssh. nervn. Deyatel.,* 1959, **9**, 126–134.

Zelyony, G. P. The reactions of dogs to auditory stimuli. Thesis, Military Medical Academy, St. Petersburg, 1907.

Zelyony, G. P. Contribution à l'analyse des excitants complexes des réflexes conditionnels. *Arch. Sci. biol., St. Petersburg,* 1910, **15**, 437–453.

Zelyony, G. P. Procède technique pour l'étude des réflexes musculaires conditionnels. *CR Soc. Biol.,* 1913, **75**, 659–660.

(Received March 20, 1960)

YOGA, ZEN, AND AUTOGENIC TRAINING

57

A Study of Certain Relations of Accommodation and Convergence to the Judgment of the Third Dimension

Harvey Carr and Jessie B. Allen

ABSTRACT

While one of the writers of this paper was investigating the problem of the third dimension, controlling the depth location of combined visual images by operating a reflex control over the convergent-accommodatory process, the other, who was serving as subject, insisted that she possessed the ability to locate the image at will, irrespective of the reflex control exercised over the convergence by the conditions of the experiment. Furthermore, it was stated that she had possessed this peculiar ability of voluntary control of depth location throughout life and had often amused herself during youth by voluntarily changing the apparent distances of objects in the visual field. For instance, a house could be made to approach or recede at will. After becoming a psychologist her interests had led her along other lines, and the phenomenon had never been studied.

The results of the following tests show that this subject possesses a direct voluntary control of lenticular accomodation independent of the convergence, and that depth is in her case a function of accommodation and is in no respect *effectively* influenced by convergent changes.

58

Apparent Control of the Position of the Visual Field

Harvey Carr

ABSTRACT

One of my students reported that she possessed the ability of moving the entire visual field upwards. This translocation first occurred involuntarily and after noticing the phenomenon the subject found by trial that it could be repeated at will . . . The phenomenon can be produced voluntarily at any time and under any circumstances . . .An object is momentarily fixated and then slowly raised upwards . . . The entire visual field participates in the movement, and all visual objects keep their relative positions to each other. . . . The translocations may be in any direction and may be initiated and sustained by a movement of either the eyes, head, or body. . . All new objects introduced into the field of vision during the displacement are not perceived.

59 Voluntary Control of the Visual Field

Harvey Carr

ABSTRACT

In a previous paper, I reported two cases of voluntary control over the distance location of visual objects. In order to secure new cases of the phenomenon and to determine the relative frequency of its occurrence. I have canvassed a series of classes in psychology.

On the average one person out of every seven is subject to the involuntary illusion, while only five cases of voluntary control were discovered among 350 students. Of those subject to the involuntary illusion, about one in ten also possesses the power of voluntary control. Several new features in the voluntary phenomenon were reported which are of sufficient interest to merit a short description.

60 Mysticism and Mist in India

Basu K. Bagchi

Religio-Philosophical

India is popularly put within the orbit of the so-called mystic East. In this age of critical analysis it should be clarified, if possible, how much of India's mysticism is naturally evolved and gradually changed religio-philosophical beliefs and practices colored by outer circumstances peculiar to that part of the world; how much, if at all, it is slow multiform accretions to a core of definite psychological experiences, spiritual or para-psychological, if you please, of some individuals who seriously influenced such beliefs and practices, or how much it is product of the two. A complete analysis will be too much of a task here. Some descriptions will be presented and some questions raised without attempting to give categorical answers.

In the Mohenjo-daro civilization (circa 4000 B.C.) unearthed in seven layers about 40 years ago in the western part of India, Sindh, now Pakistan, and other places of the country, contemplative postures as seen in mystic practices have been discovered in stone and seal (24). When the Indo-Aryans settled in northwest India (circa 2000 B.C.-1400 B.C.) their basic scriptures (the earliest one, Rig-Veda with 1017 hymns and other accessory treatises) were composed and passed from mouth to ear. Then came three other Vedas and elaborative treatises. Some of them are the earliest documents of the Indo-Aryan race. There were the nature forces (gods) and there was also postulated the one God, the source of them all (12, 19, 24, 29). The Indian man was seen in the Vedic age engaged in confrontation, adoration and propitiation of them using mystic rituals. Sorcery, magic, ghosts goblins are mentioned too in Atharva-Veda treatise. Primarily based upon the Vedas there evolved later over many centuries from "forest universities" what Oxford orientalist Max Muller called philosophical "rhapsodies"-one hundred and eight upanishad treatises, ten being the earliest (29)*. These are not rhapsodies over ritualism but mystic expressions about man's and the universe's non-material nature, about his supreme goal, about his inherent inseparableness from "That" (tat)-the immanent, transcendent, all-pervading Spirit (Brahman) (12, 19, 27, 29). Some texts describe That in concrete, others in abstract language and often in stirring poetical dialogue or dialect. They even describe That as a loving personal God. But other texts state that That can never be described except negatively by such epithets as "not this," "not this," and yet That is not abstract nothingness, they say. With a bold philosophical stroke as it were some upanishad thinkers depersonalize the cosmic father-mother figure and transform it into an abstract Reality and with no sense of sacrilege thoroughly identify It with the soul of man, though they say man does not realize this identity. But how this separateness came to be—the limited man from the unlimited Man (The Absolute Reality)—is a perpetual challenge to the doctrine of monism. Cosmic illusion (maya), Cosmic sport (lila)?

Through yoga** that indescribable Reality from "Whom" man is seemingly, but as far as he is concerned, actually divorced, could be realized or intuited by him when he is able to delete the darkness of ignorance. This will give him relief from all sorrow, insure true happiness, supernal peace and immortality, whatever these mean. While on earth through wisdom man can rule without ruling his divisive smaller self and feel oneness with everyone and everything, the living and the non-living, in more than emotional glow, poetic vision or over-arching empathy. It is and yet it is not the mystic union. It is the discovery of his true being, which intrinsically is one with All. The illusory veil of separateness drops away. He will extend his helping hand to those who have not yet found their true being. He will love his neighbor as in truth he is none

*The word upanished has two or three meanings, one being secret doctrine.

**Yoga has many meanings: related to the word, yoke, union; union of the individual soul with the universal Soul, means or steps toward that goal, the goal itself, stopping disturbances of the mind (chittwa), inner equilibrium, effort power, skill in action, etc. It has many forms: Hatha Yoga, Raja Yoga, Bhakti Yoga (devotion), Laya yoga (merging), Mantra Yoga (repeating sacred syllables), etc. Upanishads refer to yoga as an effort toward finding Reality. Svetasvatara, Katha and other upanishads, Rishi Yajnavalkya as quoted by Sankar mention practical side of yoga. (11 29, 32).

other than he. Re-birth* which repeats itself until the final discovery is made, is ended.

Other philosophical systems with varying conceptional emphasis but with one fundamental theme—that above all man must find true self-knowledge (prajna), wealth or no wealth, possession or no possession, power or no power, rose along with semi-popular or popular theistic, non-theistic and some strange systems and spread over the land influencing or influenced by diverse elements of Indian culture.

Non-conformist Buddha (6th century B.C.) takes a different road. He quietly but firmly rejects Hindu ritualistic formalism and caste orthodoxy, and completely eschews the discussion of soul, God and immortality and union with any God or gods. He strongly stresses ethical idealism, self-control, self-purification, pursuit of the middle path, non-violence, common sense and compassion toward all. No God to help man, his freedom rests on his own thinking and doing. Buddha does not discard the doctrine of re-birth. Nor does he reject the necessity of prolonged and deep daily meditation on facets of inner being, boundless love, noblest thoughts and ideals etc. This is his mysticism without God. Its end is Nirvana (blowing-out), not annihilation as is often mistakenly interpreted but blessedness and awakening of the self (not in a metaphysical sense) by removing all mental bondages including the over-sticky component of desire (thirst, tanha)—the component which is the root cause of so much suffering. These simple but difficult, straight forward, ideal-man-centered teachings of Buddha underwent subsequent changes, often reversed 180 degrees. They came to be followed by millions of people in the East. Buddha would not even recognize them if he came back.

Departing in some respects from and yet incorporating certain concepts of systems like Sankhya philosophy (like that of many

*Later books postulate that according to lower knowledge (aparavidya), until liberation, the individual soul is conceived as having two other bodies than the material body. When the material body dies, these two subtler bodies carry its seeds or impressions. These seeds or potentialities are the cause of the re-birth of the soul into another body. It is only through true knowledge of its identity with the indescribable Reality (Vedanta philosophy) or of the separation of the soul from the bondage of matter—prakriti—(Sankhya-Yoga philosophy) that the cause effect chain is broken, it is believed, and re-birth is ended. Pending that, re-birth goes on and on with all its miseries and griefs and joys for which the individual alone is responsible. At least according to some systems of philosophy (Sankhya-Yoga) there is no conventional Creator to blame for his sufferings.

independent souls) Patanjali (prior to 500 B.C.), writer of the earliest book on practical yoga (there are several commentaries on this book and dozens of later books), proposes its eight steps, (25). He (11, 29, 27) suggests for first and second, moral and psychological preparations* and two types of physical preparations†. The next step (pratyahara) is to learn mentally to withdraw from all sense perceptions and images, while being still. The last three should be taken together and are called samyama. They are most important: concentration (dharana), meditation (dhyana) and absorption (samadhi). After years of successful practice four types of meditational absorption (savikalpa samadhi) may come, one after the other. First, complete absorption for long lengths of time in the thought of the name, meaning and temporal and spatial associations of any gross object (e.g., light) to the *utter* exclusion of everything else. Second, the state of absorption when the name, meaning and associations concerning a subtle object (i.e. prana, energy) drop off and the object and the process of absorption in it become as it were one. Third, the state when absorption is in "I-ness" or ego without any conceptual notions of I or associations or anything else what-so-ever. This rises as blissful savikalpa samadhi (but no emotional stir in it). Fourth, when the absorption progresses to a still finer aspect of "I-ness", it is asmita savikalpa samadhi. Gradually through this, the mind's distractions disappear and the mind is purified and calm and deep knowledge of the self comes (prajna) like the reflection of the moon at night on a placid lake. It is believed that with the use of samyam (see before) a yogi can come to possess various paranormal powers.**

The fifth or the highest state is nirvikalpa samadhi, called seedless samadhi, i.e., the seeds of previous impressions causative of distractions, bondages, re-birth have been "fried"; they can not sprout any more, unlike in the fourth state. This state, according to Patan-

*Non-violence, veracity, non-stealing, non-receiving; cleanliness contentment; practice of hardiness, study habit, taking God's name, practice of discrimination; non-attachment; friendliness, etc.

†Proper postures, and deep breathing; later books enlist about 84 postures (asana) and 12 types of deep breathing (pranayam), 2 or 3 being the chief ones in each category.

**By themselves these powers, one is warned again and again, are obstacles on the road to the highest state and should not be sought for their own sakes. They are incidental and may give him confidence but no true lift toward his goal. Some of these are: knowledge of the past, present and future, of the sounds of all living beings, of other's minds; of the time of one's own death, of terrestrial or extra-terrestrial systems or events, making

jali, has nothing to do with paranormal powers, *nor* is it the state of mystic union with eternal Spirit or God though theistic or monistic books speak of that, and much of Indian culture is permeated by the thought of this possibility of union or close commerce between the empirical, the manifest and the transcendental, the unseen. In the fifth state the individual soul is untrammeled, it is knowledge (prajna) personified or rather impersonified, it views itself in its pristine, immaterial, immortal, ever-present *unnamable equilibrium.* It is not necessarily relieved of its body-mind vehicle but it is not enslaved any more by the world force of prakriti, it is unfettered and free and has passed beyond all perceptual, emotional, conceptual and volitional chains and distractions. The critic will immediately say "What a vacuum to plunge into, without worldly pleasure, without that flavor of intellection, without the use of reason, without everything one holds dear as man." Many misty Hindus and their western followers sit still with eyes closed, blank the mind, spin occult visions and imagine they are grabbing the gossamer of some convenient, contingent yogic summum bonum. There is nothing wrong with being still but occult visions and thin gossamer, perceptual or conceptual, are not evidently Patanjali's superior samadhi on which he put much higher price. It is believed earthly life of a man is lived and transformed and it can be made compatible with the nearest level of this highest samadhi. But no, it is not occult imageries, according to Patanjali.

Bhagavad-Gita (after 500 B.C.), universally revered book of devotion and yoga with its message of attachmentless action consisting of 700 verses recognizes individual differences in growth lines and advises proportionate emphasis on the type, direction and force of each man's leanings, endowments, needs and circumstances (dharma)—different paths leading to the same goal, (5, 22, 26). It

one invisible going through space, mastery over hunger, thirst and bodily functions; over senses, elements etc. etc. The extent of possible knowledge or mastery is not made clear.

Some contend that talking about such fantastic powers is an illustration of a familiar Indian scriptural or parascriptural custom of stutivada—making exaggerated statements in praise of a desirable condition or system per se (e.g., yoga) to make it attractive to people at first sight. On the other hand, the presentation of a catalogue of these paranormal powers in a matter of fact way in a book of philosophy that solemnly advises the follower to let those powers alone if he wishes to attain the highest goal—realization of the soul—could raise doubts in the mind of those doubting the powers, especially when a thousand and one references to such things occur in Indian literature over centuries. One, of course, does not need to accept them without proof.

certainly encourages devotion—a projection of emotional aspect of man on to the concept of a loving god—Krishna. Gita seeks no reconciliation of different Hindu philosophies on the plane of reason but only on that of intuitional insight. Man who is intrinsically more than man becomes really free when the light of the whole encompasses him. This whole or God or whatever imperfect name is attached to It contradicts, as it were, the Aristotelian logical law of contradiction (5) by being two opposites simultaneously — by being at the same time and beyond time a Person and no-Person, a Being and no-Being, Infinite and one in essence and finite and many in manifestation, recalling Kant's antinomy (self-contradiction) of reason when it attempts to explain the noumenon—that which is beyond phenomenon (*The Critique of Pure Reason*). Reason can take man only so far but intuition (vignan, darshan, anubhuti) can take him farther. This is one of the solid postulates (or truths?) of all mysticism, eastern and western. The ultimate Reality can be argued about in an intellectual exercise but It can *only* be "extrasensorially" perceived and "intra-sensorially and uncommonly" posited and lived while on earth by the pure in heart and wise in action. The unsought reward is inexpressible joy that is not emotional exhilaration, not selfish satisfaction, but an all-enfolding peace that passeth understanding and can interpenetrate active living.

Tolerance towards all types of beliefs — not criticizing them on the one hand or imposing them on others on the other with a hammer, sword, carrot stick or fear of hell, makes for a free atmosphere for man's self-fulfillment. When this tolerance degenerates into over-permissiveness, as is seen in many places including India, and when the axe of reason is not used, jungles of beliefs rise and choke society with only some relieving gasps. The Gita advises the seeking of what is rationally considered good and dispelling of evil but it implies that God is neither what we call good or evil but is more. The wise, the mature intuits this and yet follows the good, makes the world his kin, acts accordingly and is silent.

The chief hero of the Gita, upholder of a righteous cause, is seized with reactive depression in a battlefield and shirks from his duty to defend the cause.* He is advised and inspired by his Lord, Master. He sees a cosmic vision of shattering proportion elimina-

*Many including Gandhi interprets this battle only as an inner one—the battle of the moral man within himself, between good and evil tendency (5).

ting earthly details. When he comes down to solid earth, his emotion-laden uncertain ideas disappear; he performs his earthly duty like a wise man from a higher plain of integration, without elation or attachment, without being a moral or religious automaton but as an awakened functioning unit in this living world organism.

These or similar or divergent philosophic concepts, beliefs and descriptions, not to speak of a thousand and one ramifications of *temple* or *popular* religion with sensible, symbolic, or superstitious elements, have had high priority, serving as monumental guideposts to the billions of Indians over four to six thousand years. How do these concepts, beliefs and practices arise? Do they arise as others of a completely different kind arise elsewhere under different geographic, economic, inter-individual, inter-group situations governed by the same basic inner and outer dominant factors? Do Indian ideas and thought constructs arise out of attempts to dodge or cushion earthly blows, to meet the demands of the hungry, anchor and identity-seeking mind, to regulate its basic drives, emotions and group behavior, to feed the will to believe, the urge to understand and get along in the mysterious, malevolent and merciful universe of man and nature—as other thought-constructs of other men like Christians or even atheists use theirs, though entirely different, for the same purpose? Or alternatively, not denying but granting all these parameters, are there dimensions of experiences of a few people everywhere scattered over the ages that are more intensive, inclusive and incisive than or different from common pleasure of passion, dining, wining, domesticity and social togetherness, than the supreme lift by losing in rarest moods of nature, art or music, in the silence of masterful creative moment, in deep intellectual absorption, or gripping inspiration of all-giving love? Are rare experiences of rare people, Hindu, Christian, Jewish or other prophets or mystics, or lesser but similar or simpler experiences of lesser people precursors, initiators and perpetuators in part, *only in part,* with adverse mixtures, of voluminous theology, cultural patterns, creeds, dogmas and organizations? Or are these experiences delusions, make-beliefs, or at best minor subjective extrapolations and projections of common experiences that serve human needs but *are* not substantively real nor objectively relatable to other known datum? Are the experiences amenable to separation from their respective cultural or verbal garb in which they are displayed by different mystics or different faiths or countries?

Science is conceivably silent, shy or suspicious about these things. It may strike out and try to slash them into shreds. On the other hand it may hold that if people find comfort from believing in religious and mystic experiences why spoil it all by spot-lighting them? And yet science knows that in spite of its hydrogen bomb, mars photography, laser, maser, sub-atomic particles, DNA, RNA, and a million discoveries and inventions it does not know nor pretend to explain all facts nor envisage items that have not yet emerged as facts. It can not even explain the common daily experience of transformation by man of experimentally demonstrable but intrinsically invisible energy, into perceptible three dimensional solid masses like a table which is nothing but energy (13). Its hands are full with its own problems and interests. Why would it bother with things for which adequate foundations of preliminary knowledge have not been laid down to build on as is the case with regular scientific disciplines? Why waste time on exploration of these when the over-anxious believers will mistake exploration for explanation and final commitment? And most importantly, fact-collection in this line in which science may be peripherally interested should be strictly kept apart from value-judgment, normative judgment of ethics and religion, from believing and living what is believed. Scientific recognition or non-recognition of some mystic experiences may inevitably influence this value judgment, this believing and living, rightly or wrongly. Nobody would welcome in this century a repetition of open faith—reason warfare of the middle ages though such a chance is slim. There is no question however that science is slowly winning out: yet personal and social ethics and religious beliefs are far from dead, they need or should not be. But the problem here is different. Is there a higher stimulant of ethics that outflows from paraethical, paranormal states over and beyond and yet infuses general ethics—the ethics, which some believe, is a resultant of dynamic inter-relationship between parent-child-social milieu to start with, in a genetically evolving and advanced animal and not a special sudden gift to man from above?

Famed William James, physician-psychologist of Harvard, in his Gifford lectures at Edinburgh in 1901-1902, (17) gathers in one volume some of the most unusual experiences* of some mystics and

*Consciousness of illumination, ecstasy, conversion phenomenon, twice-borness, healing "miracles", personality transformation (e.g. anhedonia to zest and faith in life), sensory automatisms (blinding photisms) etc.

initial non-mystics of the west, India and other places. In one of the first books of this kind he strives hard not to explain but only to relate them to the subliminal, subconscious realm of man—a realm not all dank, dark or devious. James took a page from Meyer and Prince for this, Freud not yet influential then. Rather than pass adverse judgment on these mystic experiences he almost accepts their validity and utility, pointing out that these experiences most profoundly influenced personal and social ethics for good for these people and others. The spark came from these experiences.

On the other hand, many modern humanists and scientists consider these experiences unnatural or at least unnecessary for establishing a decent community or humanity. *There should be instead, they maintain, naturalistic, common sense foundation of ethics; it does not require a revelatory, prophetic or mystic base.* This reasoning has some value but it is partial. It devalues the need of scientific investigation of mystic experiences—religious or otherwise, and it also fails adequately to put in proper perspective the many contributions of mysticism or mysticism-tinged philosophy and religion of the past, present and possibly of the future to the common experience, inspiration and wisdom of man in spite of some of its confusing conundrums and scurrilous kinks.

Experimental

There are mystics or yogis (Hatha Yogis) in India who are not all "God-intoxicated," but who practice methods which have religious and philosophical origin or implication and by which they claim they can establish unusual control over involuntary functions of the body. To test this we (7, 34, 35) left a portable eight channel polygraph with accessories in the care of Dr. B. K. Anand, Yale-seasoned, Fulton-influenced chairman of the Department of Physiology at the All-India Institute of Medical Sciences in Delhi, at the request of the Indian Council of Medical Research and Rockefeller Foundation. Turning for a while from implanted electrode studies of the so-called "feeding center" and "satiety center" of animal hypothalamus he used Ramananda Yogi in 1959 for two experiments whom we tested for another experiment in 1957 (7, 35). His study was published in 1961 (2). In brief, the Yogi was put in a completely air-tight metallic and glass box (6' x 4' x 4' with brain wave (EEG) electrodes, heart (EKG) electrodes, respiration gas tubes, chest strain gauge for respiration rate and other devices. They were connected to outside machines. Every half hour oxygen (O^2) and carbon dioxide (CO^2) of the air in the box were sampled and analyzed with a Haldane gas volumemeter by the technique of Peters and Van Slyke (1932). Benedict-Roth BMR apparatus was used for control on basal conditions. On the first day he was eight hours inside the unopened sealed box. At the end of three hours the inside candle that had been lighted went out. On the second day he was ten hours in the box (no candle). His basal oxygen consumption outside the box was 19.5 liters per hour, normal range for his surface area. Oxygen consumption inside the box in the first experiment however became reduced to 12.2 liters per hour, in the second to 13.3 liters per hour, a reduction of 37.4% and 31.8% respectively. At one time the reduction was 40% - 50%. His usual heart rate was 85 per minute, inside the box it gradually came down to 60-72 per minute, in the seventh hour to 55 per minute when O^2 concentration was 16.5% and CO^2 was 3.6%. The CO^2 output though higher by the hour was comparatively lower than the basal rate outside the box. In spite of lowered O^2 consumption and higher CO^2 output inside the box there was *no panting, heavy breathing* (*hyperapnea*) as would be the case in ordinary individuals, nor increased heart rate (tachycardia) or electrical changes of the heart in this man. His brain waves showed at first waking alpha waves but later most of the time low voltage fast activity of a few (EEG) biparietal humps but no slow delta waves of deep sleep. Two non-yogi subjects in the box (one in for four hours, the other for seven) did *not show reduced oxygen* consumption compared to their basal rate or showed greater value.

These studies demonstrate beyond doubt that the yogi could voluntarily lower his metabolism for hours without really going to deep sleep and much lower than the reduced (10%-15%) basal metabolic rate of sleep of the average man, and without producing hyperapnea or tachycardia. The authors theorize that by prolonged practice it is possible for the neocortex to condition the autonomic nervous system functions (breathing etc.) through the limbic system of the brain. What does this prolonged practice involve? I had long talks with this yogi in 1957. Only some broad details emerged. I will not deal with his personality. For many years, several hours a day, guided by his preceptor, he performed quiet meditation of a certain type (all details not divulged) concentrating his attention between his eyes and with his tongue rolled back into the posterior wall of his naso-pharynx and with his uvula thrust forward (Kshechari mudra of yogi). The position of the tongue in that site was located by us using iodized oil and X-ray in 1957 (7). It was near or against the inferior surface of the sphenoid bone and sinus about an inch across from the hypothalamus and pituitary region.

On several occasions he claimed he was underground in a pit for several days covered over with loose dirt, maximum twenty-eight days. Everybody mistakenly thought that this man could live *without* oxygen. These initial reports and many other similar reports prompted the above experiment which showed how facts and fiction can be mixed up.

Several years ago an American doctor, Dr. Clarke, not a yogi, reported before a small Otological seminar on the sixth level of outpatient building of the University of Michigan, presided over by the Dean of the Medical School, that as a young man, by accident, he discovered he could do this Kshechari mudra maneuver. On questioning he said that when with this maneuver he kept still for a while he had a feeling or sensation of pleasant "withdrawal" or "quiet well-being". No further elaboration was obtained from him. He showed the maneuver to the group. For years in the past I personally had known this description to be true.

We found a yogi who spent parts of two winters in a Himalayan cave alone, unclad but wrapped in animal skins, spending much of the time in quiet yogic meditation seated on an animal skin. The extreme cold distracted him. He was advised by his teacher to concentrate on warmth and to visualize himself in extremely high temperature situation. He reported gradual success after six months of practice. This man was tested in the Delhi laboratory in an electrically shielded semi-sound proof room with a window on the wall for observation (7, 34). He had EEG, EKG, EMG, GSR electrodes, chest belt with transducer, thermocouples on forehead and elsewhere, plethysmograph and blood-pressure checking devices in appropriate places. On three tests without *any visible maneuver or movement* whatsoever, sitting quietly relaxed and cross-legged on a blanket he voluntarily produced perspiration on the forehead within 1½-10 minutes, accompanied by marked increase in systolic blood pressure of which he was unaware, by slight increases, then decreases in forehead skin temperature (presumably evaporation effects); by marked drops in forehead but only slight drops in palmar skin resistance; and by slight increases in arm skin temperature. This is an interesting autonomic nervous system conditioned response, if we can call that. What are the individual roles of attention and practice, cortex, hypothalamus and related systems?

It was often reported that some yogis could stop the heart. Everybody including physicians thought that it was so. We discovered the truth was in the middle (35). Four Hatha Yogis were tested using the well-known Val-Salva maneuver (holding breath in inspiration or expiration, with chin-lock-jalandhar bandha-, closed glottis). It was interpreted that the three yogis increased the intrathoracic pressure for ten to fifteen seconds necessitating slow return of the venous blood to the heart. The result was that with little blood to pump the heart sounds were diminished, and could not be perceived by stethoscope on the chest, nor pulse felt at the wrists or elsewhere. But the electrical beat of the heart (EKG) went merrily on with QRS complex increased in amplitude in Lead III and decreased in amplitude in Lead I, and also, with right axis deviation of the heart by X-ray (fluoroscope) during inspiration.

The last yogi, a healthy man who did not claim to stop the heart, did, by a slightly different maneuver in addition to chin lock slow the heart rate to 24-25 per minute from 63 per minute for ten seconds with obliteration of the P wave. A well-known cardiologist of the University of Michigan who saw the last man's records told me that he did not know of heart slowing like this by any maneuver as in this case* Carotid sinus massage at the neck or other maneuvers did not precipitate this. Were baro-receptors in the heart involved? If the vagus nerve inhibited the heart rate, what was the mechanism? What part, if any, prolonged (over 5 years) meditational exercise played in the last subject was not clarified.

Walter Miles of Yale University and later chairman of the National Academy of Sciences of Washington arranged the Sterling Fellowship for an Indian student (Behanan) to go to India and learn yoga for a year. Later the findings were presented before the Academy by Miles and worked up into a book by Behanan (8). In 72 days of tests at Yale it was shown that a certain type of deep breathing with or without subsequent concentration exercises for twenty minutes increased oxygen consumption in him 24% over normal.

*An exception is mentioned in a patient (35).

They also produced, according to him, mental calm, relaxation and physical well-being and some emotional control in his daily life. Whether such results could have been achieved by other methods was not proved nor disproved. Some deficiency in conceptual and motor tests was seen immediately following the practices. This large book gives understandable historical, descriptive and philosophical material but only a very small nucleus of scientifically validated fact of one aspect of yoga.

The well-known Hatha yoga purificatory maneuver of naulimudra was photographed (7). It consists in voluntarily churning from side to side, without touching, the abdominal muscles—abdominis rectus and abdominis oblique. It is claimed that by this, acceleration of peristalsis and stimulation of glands and gastro-intestinal tract are effected. (See *Hatha yoga* by S. S. Goswami, Publ. R. N. Fowler, London, 1959). I believe it is useful but control experiments need to be performed.

Many asanas (postures) of Hatha yoga and contortion movements were also photographed (7). Reports of high physical and spiritual gains from all these are probably overdone. Good control experiments are lacking. Recordable effects of a few pranayams (breathing exercises) have been mentioned elsewhere (34). Many pranayams were not studied. We saw the act of voluntarily producing by naulimudra negative pressure in the bladder to suck up half a glass of water through the urethra for cleaning it. The value of this practice is questionable and infection risk high.

It was difficult to secure Raja yogis to sit for experiments, in contrast to Hatha yogis. The former emphasize philosophy, spirituality, religion and meditation and not so much the physical techniques as the latter. However fourteen (7), some better than beginners and two older yogis, co-operated in twenty-five sessions. These were held in electrically shielded rooms at Calcutta University physiological laboratory, at two Delhi laboratories, in a Madras College, at Lonavla laboratory of Swami Kuvalayananda, near Bombay, two private houses and in a high mountain cave. Each sat in the lotus posture, that is, with each leg folded at the knee and the foot placed on top of the opposite thigh, and meditated in *his own way* (this was not divulged) for 14 to 148 minutes, the longest four sessions being over 100 minutes—median duration 49 minutes. With eyes closed they sat entirely motionless in that posture for that length of time—a feat in itself. The respiration rate in two subjects (one was Swami Shantananda) dropped from ten - sixteen per minute to 4 - per minute once and stayed that way for 10 - 16 minutes without pressure on neck—another unusual performance*. On other occasions he pressed his neck with fingers. In other yogis respiration rate dropped but not that much. Palmar electrical skin resistance rise, (or lowering of palm—palm log conductance, which is essentially the same thing, looked at from another angle) a sign of relaxation of the autonomic nervous system, was of course *invariably* present. The median rise was 56% over their control non-meditation periods. Six sessions of four subjects showed the highest rise in electrical resistance, 70%-106% over their control non-meditative periods—*a strong sign* of physiological relaxation, not found in other yogis or non-yogis. Decrease of heart rate and blood pressure was related to skin resistance increase, though not always in the same proportion. No sign of drowsiness or sleep by brain wave (EEG) measurement during these long sessions in the subjects except in one case in whom bi-central theta waves of initial sleep were noticed off and on. Brain waves which were unremarkable, only waves of waking, were well-modulated and regularized, no fast or tensional pattern. On questioning they all claimed to be relaxed and refreshed after each session. Deep probing was not done. Whether or not this habit of deep physiological and mental relaxation without sleep can be utilized as an accessory to emotional and attitude re-education is another subject (3).

An elderly Swami (Vishuddhananda) after leading an active life of doing public good retired to the Himalayan mountains (Rishikesh). He was a bearded, orange-robed, non-organizational, well-respected individual. He co-operated with us after we appealed to him on two occasions though he frankly placed the highest value on inner growth and spirituality rather than on scientific experimentation. In spring we brought the machine to his cave (7) near the flowing Ganges river flanked by four to five thousand feet high mountains and nature's silent grandeur. Two experiments were done on two days. During his meditation for about twenty-five minutes in the cave with no electrical shielding there was the normal well-regulated waxing and waning pattern of waking brain waves,

*It is considered more meaningful to mention pertinent values of physiological parameters in some individuals (7) than their means or averages on only 4 yogis (35). Differences in the selection of values of different periods may also change average values.

no drowsiness or sleep waves. The skin resistance steadily rose 70% over control period. The tone of extremities seemed markedly reduced, and to all appearances he went into a state of complete withdrawal even when EKG and palmar electrodes were being put on during the first experiment. Low intensity sounds within three to five feet like the shuffling of feet were not perceived, according to later subjective report; except for slight shift of the base line on two occasions no blocking of the brain waves or decrease of palmar electrical resistance (sign of change toward alertness) was seen during slight nearby noise. Non-perception of stimuli could not, of course, be objectively checked. How much the voluntary withdrawal from sense perception and thought distraction is possible and how much of its helpful physiological and psychological residue, if any, can be transferred in active life should be the subject of a serious study. No yogi would say that he reached Patanjali's highest samadhi state (see before). And even if they did it would be too much to expect that present electrical machines though they record millionths of a volt of electricity could detect it. Newer approaches may be necessary. Our preliminary results showed that there was room for further investigation.

Anecdotal

Except for one personal encounter, the following reports have been obtained from presumably reliable sources but not independently verified. They are given here for whatever they are worth. In the late afternoon of early hot summer of 1916 an acquaintance of mine came running to tell me of an unusual experience he had just had and asked me to go with him to check it for myself. I walked with him fast for about five blocks to Amherst street of Calcutta and saw a short oldish bare-bodied fakir with grayish hair slowly walking on the sidewalk trailed by about a dozen adults and children. He wore a loin cloth and no shoes, holding a stick over his shoulder from which was hanging a small cloth bundle. Some of the people began to say to him in Hindustani, "Show us that again, show us that again." He kept on walking without talking or apparently paying attention to them. When he came upon a water hydrant on the edge of the sidewalk from which the poor people all around usually drew water for drinking, cooking and bathing he stopped and nonchalantly motioned everyone to push the button of the hydrant and take a palmful of water. We stood around, a few feet from him, each with a palmful of water, fingers curved in. We were alert and curious. For a second or two he was quiet, he did not move his bundle. Then he motioned to us to open our palms and smell what was there. Everybody, including myself, did that and looked at each other in bewilderment. The water smelled like actual rose water. Each of us put his nose to the water many times, smelled the perfume and finally tossed it to the ground.

Was this something like the well-known "rope-trick" of India done through mass hypnotism? Were we hypnotized? After having gone through Scottish Churches College and a University, I had learned to be somewhat sceptical of these things. At least my companion and I did not think or feel that we were hypnotized. The man did not ask or expect any reward or thanks. He kept on walking without looking at us following him. He turned to the right on Harrison Road, a busy thoroughfare and continued on until he came to a street corner where a small building was going up, bamboo scaffolding against it and piles of sand and lime on the ground by it. He suddenly squatted on a sand pile and placed the bundle on it a foot or so to the side. We also squatted almost encircling him a few feet from him. He sat there silent for ten or fifteen minutes looking indifferent, and we looking at him. Then someone spoke, "Show us something again." He said nothing nor made any move. After a while he said in Hindustani, "Pick up some sand and close your fingers around it." We did. He looked blank for a second and then asked us to eat it. On opening our hands each of us found a piece of sandesh where the sand had been and we ate it. Sandesh is a sweetmeat made of milkcurd and syrup and it is soft and somewhat sticky. I wiped my hand on my cotton dhoti cloth and I vaguely remember that the stickiness persisted until I got home. Gradually other people dispersed, I alone following him for many blocks through the highway and side roads toward Howrah bridge over the Ganges. He gave brief answers on the way to a few questions of mine on religion. Finally, unceremoniously, he walked away.

By appointment Dr. Wenger, professor of psychology, University of California, Los Angeles, an attorney, some friends and I visited the large home of a Supreme Court Justice of Bengal in 1957, in the southern section of Calcutta during our yogic research trip in India. In the course of our conversation he told us that some time ago an unknown Sadhu (holy man) visited him. He and his wife and others were sitting in the room where we were sitting. The man questioned the judge about his disbelief in mystical things and wanted to remove his doubts. The judge told us that this man produced in that room a deafening noise of "thunder" for sometime without moving or sounding anything. This sound was not heard by others in the other part of the house or by neighbors. The judge said that he and others did not think they were hypnotized. He also said that he still had in his attic yards and yards of paper with pencil writing on it. At the command of the Sadhu, on the same occasion, a pencil on the table stood up and without being held by any hand wrote before their eyes on the paper answers to many questions put by the judge and others present. Some of the answers were foolish, some noncommittal, some prophetic which proved true later and some exactly correct in relation to matters which the judge alone knew.

In the same year I visited a Sanskrit and English scholar and author, Jagadish Chandra Chatterji in Calcutta. In about 1921 when I first met him he was the cultural minister to the Maharaja of Kashmir state. Years afterwards in the United States he was the head of a small Vedic Institute established with the co-operation of oriental scholars like Lanman of Harvard and others. I had a long talk with him. He assured me repeatedly that he saw with his own eyes as many others did not once but on many occasions, a holy man called Gandha Baba (fragrance creating father) making appear from no-where suddenly without visible means in a roomful of people, fruits of all kinds—bunches of grapes, bananas, mangoes, etc. A beautiful odor would fill the room without anybody bringing in anything fragrant. I have heard about his doing this sort of thing from many sources. This was all incidental to the religious discourses he would be holding with the people in the room.

I heard many times from Shastri Mahasaya, a Sanskrit scholar to whom I am indebted for knowledge of some aspects of yoga, about one particular occasion. This was related by him to others too many times. It was in the home of his well-known married yogi teacher, Shyama Charan Lahiry of Benares. Shastri Mahasaya saw with his own eyes that his teacher's teacher, Babaji, made himself invisible in broad daylight as soon as he stepped across the threshold of the room in which they were all sitting and went out into a wide open yard with a wall around it. The door in the wall that opened into the street was approximately 60 feet from the door of the room from which Babaji came out. All the people were following him closely as he was stepping out. Much later I saw the yard as described. Another learned disciple of Shyama Charan Lahiry also told me about the spiritual realization of Babaji and the latter's paranormal powers like making himself suddenly visible or invisible.

The following is not an anecdote but a common experience known to many students of Lahiry Mahasaya or to his students' students. After a certain type of deep breathing (sahaja pranyam) done for a few minutes with the concurrent up-down-up mental imaging of some points inside the entire spinal column and with the repetition of some symbols along with khechari mudra (see page 11) there arises after months of practice an indescribable "sensation" all around it, top to bottom. It can stay for minutes or hours on end, even when attending to other matters or can be brought on at will or may emerge automatically under favorable circumstances without the breathing or khechari mudra. A quality of withdrawal from usual sense modalities, not actual withdrawal, and experience of in-drawing quiet and restfulness set in. Eye-blinking is reduced. One wants to continue in that state of still freshness, mind refusing to wander much. Some claim that when this is coupled with meditational and other preparations for how long nobody knows this may eventually lead to some type of samadhi. Anatomico-psychophysiological basis of this definite initial experience is unknown, but its practical effect is known, at least its existence. It is not a hear-say.

A knighted Englishman, former Chief Justice of Calcutta High Court, summarized the extensive literature of Tantrik yoga of the 15th century and earlier with its 19th and 20th century forms in a book (36). His pseudonym is Arthur Avalon. Other yogic books also give details similar to those found in this book. I have known yogis who draw a little from modern neuro-anatomy. The Englishman and some modern yogis attempt to combine symbolic old yogic ideas with anatomy and give an esoteric description of the cerebrospinal axis and related structures and their supposed functions for the yogis. It had been postulated that cosmic energy is resident and static at the base of the spine in the form of individuated kundalini force (the symbol being a serpent coiled three and a half times with the tail in its mouth). This force is the immobile support of all ordinary bodily and mental functions. The business of the yogi is to raise this kundalini force through breathing, meditative exercise etc. and make it dynamic and travel up through six energy loops (chakra, different petalled lotuses of different colors and vibrations) inside (?) the spine (shusumna nari-energy conduit) at different levels until it reaches the seventh chakra or sahashrar, or the thousand petalled lotus (6), which is supposed to be the energy counterpart of the cerebrum. The two petalled lotus, called the ajna chakra, the center of command, is next below. It is referred to the point between the eyebrows, but others interpret it as being associated with medulla region. Two other naris (energy conduits), ira and pingala, parallel to shusumna, running down on either side of the spinal column, are considered to mediate vegetative functions, their modern homologue being sympathetic chain ganglia. Others interpret the latter two as referring to right and left nostril through which breathing is effected for the breathing practice. It is believed that when the dynamic force of kundalini rises to the thousand petalled lotus (brain) the yogi attains after prolonged dedication union with cosmic energy, God-consciousness, bliss and mastery over himself. Also different paranormal powers, which are per se obstructions, can come from concentration on different chakras. Are these mixed up anatomy, physiology and esoterics rolled into one? Or are these symbolic of man's march from lower levels to higher levels, from sex to serenity, from sense to reason? Or do they mean something more at least in some fragments?

Ramakrishna Paramhansa who lived near Calcutta in the 19th century and who is revered in India and many circles all over the world as a modern mystic of highest spiritual attainment, used to go into bhava samadhi, a form of ecstasy or trance, times without number sitting or standing in public when he was completely oblivious of the outside world and absorbed in inner experiences for many minutes which he said words could never adequately describe. He preached to his people the fundamental unity of all religions though personally he was a devotee of shakti, the goddess Kali, symbolizing the energy aspect of universe. His life is most sympathetically and interestingly written in a book by Romain Rolland, French Nobel Laureate (30) and by Max Muller (23). He is reported to have paranormal powers too: he suddenly touched with his foot one who later became his foremost disciple, Vivekananda, and made him go into a terrifying vision of nameless void, at another time by a little touch on his chest made him unconscious and later gave him some sort of spiritual awakening (30). Vivekananda was a highly learned over-critical man. Once he told his teacher to his face that the latter's realizations were creations of his sick brain—mere hallucinations. Later he became his ardent follower and a dedicated spiritual and social reformer. He created a real stir in this country when he presented Indian philosophy and culture before the International Parliament of Religions in 1893 in Chicago and in many other places amidst discerning intellectuals and ready-believing masses. William James (17) mentions both of them. Ramakrishna-Vivekananda Society is now a well-established religio-service organization. Many years back I talked with M (Master Mahasaya), superintendent of a Calcutta High School, one of the original disciples and first recorder of Ramakrishna's ethical and religious teachings often given in parables—praised by millions of Indians, literate and illiterate, non-superstitious and superstitious. He verified Ramakrishna's trances. Some westerners while praising his ethical and broad spiritual outlook think of the trances as some sort of neurotic spells like those reported about some Christian mystics—an interpretation which would be offensive to his followers including M and which probably will be incorrect. Unstintedly he gave of himself night and day to whomsoever came for inspiration and instruction. He died of throat cancer. Thus he let the seed of his past life's karma (action) work out in this life without his wanting to use his yogic or spiritual power to curb it though often told to do so: he is liberated forever, so he lets his earthly life finish its course without his trying to interfere with God's will. The latter was his and the usual Indian explanation or were these just statements of final resignation? If these are explanations they will be not understandable to those who do not accept the notion of a past life, and want to know the why of cancer in a liberated man with supposed perfection or to those who do not understand this concept of his surrender to God when there is a possible remedy in his own hands. Will not remedying be God's will under the circumstances? Or is it plain foolishness to talk about self-remedy in cancer no matter who has it—saint or sinner? Yogi may have powers but he can not be all powerful. He can not escape some inexorable laws.

Furnished with a letter from a Municipal commissioner of Madras and personally helped by a retired civil engineer (Narayan Murthi of Rajah Mundri of Andhra Pradesh state in the south) I visited in 1957, a small village of Mummudhivaram, fifty miles from that city. There is a small two-story brick builidng with a backyard enclosed by a high wall with one padlocked rear door. Several rooms of the building are also kept locked. In one room next to the backyard a thirty-five year old ex-cowherd, son of an untouchable, named Bala Yogi, stays secluded for 364 days out of the year doing his meditation. Since 1950, only on the day of Shivaratri celebration, the door opens and he comes out to the porch where he sits motionless for twenty hours and is seen by hundreds of thousands of people of all levels of intelligence and belief. A committee and several high Government officials with the help of public donations had erected this building at the request of Bala Yogi so as to be spared from the crowds. For sixteen years he has performed penances and pursued his religious interest as a seeker without any desire to preach. He has requested that no one bother him and that he be left alone to his spiritual contemplation. Aside from his once-a-year appearance he has given several interviews to his committee, but such communication is made only with gestures. What he actually does inside is not known.

Many unusual happenings are mentioned in a booklet given to me about him and this place by the committee. For one thing it is written, and I heard verbal reports of this too from some members, that he slowly cut down on his consumption of food over a period of years and that since 1950 he stopped eating and speaking. The key to the rear door of the enclosed structure is with his father who lives several hundred yards from there. He told me his son did not eat. I could not find out whether or not there was any hint given by Bala Yogi himself to others by gesture or otherwise that he ate a little or did not eat or void at all. To be sure of this for science I requested permission of the committee to have an investigation made by the Government of India by placing a twenty-four hour guard around the building for a month. I assured them I would see to it that Bala Yogi would not be bothered and no one's spiritual sensibilities would be hurt. For some general reason my request was not granted. Bala Yogi is universally known as a spiritual phenomenon in that part of India. I visited three other Bala yogis who did not talk, were not fasting but were reported much of the time during day and night sitting with eyes closed in meditational posture. Many may pass this up as a mixture of spiritual and schizoid behavior (provided it is not worse) with its dark and bright hues sanctioned by Indian culture.

Sankar Deo, a Sanskrit scholar and member of the Indian Parliament wrote to us on April 20, 1957, about a yogini, Mani Kamma, near Serum, Mysore State, who had been attracting large crowds. He wanted experiments done on this lady. He wrote, "I referred this case for investigation to the Government of India, Ministry of Health, Hyderabad Government and Police Department . . . After all investigations I am totally convinced that she is leading her life without food and water since (sic) 10 years". Because of conflict of dates we could not reach him or her.

Mr. Kuppaswami, professor of psychology, Mysore University wrote on March 8, 1957: "There is a lady yogin in Bose Bungalow, Tiruvannamalai in the Raman Maharshi Ashram (hermitage). She spends most of the time in a trance condition. I saw her in last December. I could not speak to her. You may write to her father and mention my name. If he and the lady agree you can go to test her in the trance condition . . . I may also join you if you do this work in May." We could not go.

Captain D. Majumdar, a respected man in his community, retired member of Indian Medical Service during British regime who did not take much stock in mysticism and thought that most of the religious organizations only fulfilled a popular craving, described the following episode to me and others at Madhupur, Behar, on February 5, 1957. He was the family physician of a neighbor to whose house a yogi, Bharati Baba, used to come once in a while. One day while in that city the healthy yogi announced personally to his friends and disciples that on a certain date at 4 p.m. he would voluntarily leave his body and asked them to come to the house if they could. Sometime before that hour arrived on the specified day the yogi was seen sitting erect in a lotus posture on a woollen rug on the floor, the small of his back slightly supported with a cushion touching the back wall of the room. The ladies and servants observed that for some time there was no movement of the chest in inspiration and expiration. Being alarmed they sent for Dr. Majumdar who hastened to the house. On examining the yogi he found no breathing, no pulse, no heart beat as checked by stethoscope, no corneal reflex. Pupils were examined and they were dilated. There was no response to the noise of the talk nor shuffling of feet by the disciples who had come. Skin color was not unusual, no bleeding from anywhere, his head was straight, neck probably slightly stiff. For four hours he sat like that, statue-fashion, after which the head and body suddenly went limp and fell back toward the wall. No incontinence of urine or feces. Dr. Majumdar could not decide whether to issue a death certificate, which he did after twelve hours. No autopsy.

The phenomenon of sitting absolutely still without the head and body being limp for four hours when there was no sign of life calls for a medical explanation. If it is not a drug-induced death, the mechanism of voluntary vagus inhibition also remains unexplained. In Indian culture a holy man who thinks his work is done in the world and who can make an exit in this way—voluntary demise, not suicide—is praised rather than censured.

Sir M. V. Subhas Rao, retired Madras High Court Justice said to me at his house on April 6, 1957: "I saw Ram Devi of Mangalore go into a trance, while talking, for 15 minutes. The trance may last a week. She often goes without food for a long time. She has a husband. She would not consider being tested by an instrument. Several thousand people come to see her when she visits us." (Reconstructed).

In 1900 a reputable Indologist, Richard Garbe, professor of Sanskrit at Tubingen, wrote a thoughtful and critical article 'On Voluntary trance of Indian Fakirs'. (15). In it he denounced many Hindu fabrications and false claims and later quoted from the writing of a Scottish surgeon, James Braid (9), the western discoverer of hypnotism. Braid made public a detailed letter that he got from Sir Claude Wade, British resident in the court of Maharaja Runjit Singh of Lahore, India, about his and others' personal experiences there. Briefly this is the account. In 1837 a Hatha yogi, Haridas, was placed sitting in a trance condition in a wooden box, 4' x 3' with the Maharaja's seal on its padlock. The box was lowered in the cellar of a small square building three doors of which were bricked over and one door padlocked and stamped with the seal. The key and the seal were kept in the Maharaja's or his minister's possession. Four personal guards of the Maharaja watched the building and the surrounding garden in two hour shifts for twenty-four hours over a period of forty days. The condition of the building was reported every morning and afternoon to Maharaja or his minister. No effort was spared to rule out collusion or fraud. On the day of exhumation after forty days, Sir Claude, a medical doctor, the Maharaja and others saw the seals broken and padlocks opened. Then Haridas' shrivelled cold and stiff body was lifted out of the box. Examined medically there was no pulse felt at the arm or temples and no heart sound. There was a little warmth on the top of his head. For an hour and a half his disciples massaged his body with clarified butter and hot water and normalized the tongue position from Kshechari mudra (see before). Gradually pulse and other vital signs returned. Haridas opened his eyes and in a scarcely audible tone addressed the Maharaja: "Do you believe me now"? This account was also independently given by a German physician, Honigberger, corroborating essentially all the details (16). The distinction between hibernation and this type of trance and relevant matters was discussed by Garbe, physiologists Van Verworn and Preyer (28). Haridas repeated his performance many times before others and the Maharaja. Others have done it too. But above was the most trustworthy report. By Indian standard Haridas was not a spiritual man as his commercial attitude and personal ethics were reported not above reproach.

Speculative

It is more convenient than correct to lump all previously described experiences or phenomena, even if they are genuine, under trance or ecstasy or conditions produced by either or under mysticism which has in reality many shades of meaning and includes many types of practices, hideous and sensible, and belief structures and uncommon realizations. According to Indian thinking there is a wide spectrum in this regard with physical or physiological emphasis at the one end and psychological and spiritual emphasis at the other with shuttling or overlapping at many spots barring all frank deceit and magic. Manifest objective, underlying motive, degrees of interest, levels of attainment and means adapted for the purpose will determine to what part of the spectrum a practitioner or a supposed expert is to belong. Some hints are given but no elaborate guide-lines or measuring sticks to evaluate inner experiences or even samadhi or infinite steps leading to them. Outside of demanded moral preparation for all, paths and by-paths to the goal and substations short of the goal are supposed to be many. For instance, it is said, postures, breathing exercises, eye convergence, meditation on a god or prophet or a symbol, are not at all necessary for every seeker. All this leads to supreme proliferation of individual yogic ideas and to overtangentality without consolidated reason in a field where reason is subordinate.

Next, in yoga or mysticism the role of self-hypnosis is strongly implicated by many western writers. Doubtless this role can not be ruled out. However to state that hypnosis concerns or explains all phenomena subsumed under either is to becloud the picture. Suggestion may produce psychological or behavioral changes but how the latter in special situations (see page 10) can produce physiological homeostasis or extremis or reduced metabolism etc. is certainly not clear. Nor can self-hypnosis in religious or faith framework explain other reported occurrences. The central nervous system substrate of hypnosis or its results has never been given in physiological or anatomical terms but only in facile psychological terms like suggestion. Further, it is recognized that direction of attention is important in yoga, attention is important in all learning situations and initially in habit formation, and attention is also the first step in hypnosis. But for that reason no one should or would elect to put learning and hypnosis in the same basket.

What some yogis are *initially* attempting, in addition to their moral, psychological and physical preparations, is a sort of voluntary de-afferentation and de-conceptualization in order to obtain relaxed (a) attentuated awareness, (b) unwavering concentration on a single or no item, or (c) both at different times or a mixture of the two at the same time. After that lies the veritable uncharted realm they speak of—multifarious valid or supposed experiences, physiologically, mentally or spiritually unique states. An over-simplified hypothetical sketch is given here of a facet of some possible mechanisms involved in a *few* yogic practices. Some of the cited yogis used the tongue maneuver into the naso-pharynx, the tongue probably touching the inferior surface of the sphenoid bone, which is about an inch ventral to the hypothalamic region. Some used pressure on the carotid sinus at the neck manually or by chin-locking or stiffening the neck. All used quiet undirectional attention without sleep for many hours on end daily during their practice for many years in almost immobile posture. Many or probably all used eye-convergence toward the nasion, tip of the nose or a point in front for hours. When taken together all this means that during the practice there is a constant efferent flow of nerve impulses from the entire cortex or at least segments of the cortex including frontal eye-field particularly over cortico-mesencephalic, cortico-bulbar, and aberrant pyramidal tracts (10) in different degrees to the somatic motor component of the cranial nerve nuclei of III, IV, V (motor component), VI, VII, IX, X, XI and XII located in the midbrain, pons and medulla. The impulses may travel between the nuclei themselves through intermediary pathways and eventually go to the respective muscles including those of the eyes, tongue and pharynx. General and special visceral motor discharges may be present too. There will also be continuous sensory feed-back from those muscles* into the cortex of nerve impulses over afferent fibres of the sensory ganglia of the appropriate cranial nerve nuclei. Except possibly for nuclei I, II (allowing for minor diffuse light perception under the eyelids) and VIII (also allowing for minor internal ear noise perception) all other cranial nerve nuclei including V will be afferently active.

So during the practice there is a constant cortico-brain stem-cortical cyclic flow of impulses over motor-sensory pathways with other systems (reticulo-cortical, cortico-reticular) being probably involved in the interplay. Also efferent outflow from the cortex over pyramidal pathways to the peripheral fixed muscles of the yogi body posture and proprioceptive sensory inflow back from them into the cortex will be present over medial lemniscus and thalamus and via spino-cerebellar circuits. But this cortico-peripheral-cortical electrical activity is likely to play a subordinate, subthreshold role to the previously mentioned cortico-brain stem cyclic activity due to strong *attentional* factor during the practice involving the latter system. The role of centrifugal control (from forebrain, brain stem etc.) on afferent inputs from some of the receptor sources including cranial which is now believed to determine to a certain extent what sensory information can be admitted to the central nervous system (? consciousness) can not be ignored. Afferent impulses from the pressed carotid sinuses at the neck will travel over glossopharyngeal fibres (IX) to the tract and nucleus solitarius which will influence the dorso-motor nucleus of the vagus nerve (X) and the diffuse vaso-motor center located in the reticular formation of the medulla causing different grades of inhibition of many vegetative functions (slowing of the heart, respiration etc.). The hypothalamus with its four important tracts (dorsal longitudinal fasciculus with its output and input

*Proprioceptive fibres from extrinsic muscles of the eyes, tongue, pharynx; face are suggested by Pearson, Bowden (14, p. 141).

connections, dorsal hypothalamo-tegmental tract, anterior and posterior hypothalamo-tegmental tracts) can tie into the previously mentioned cortico-bulbar circuit and cranial nuclei and under direct or indirect influence of the frontal or precentral cortex and from below (reticular area) may further regulate different functions including (10, 18) parasympathetic functions.

It has been postulated by many modern investigators now (see 20, 21 etc.) on the basis of numerous stimulation and ablation experiments that barrages of ascending nerve impulses over the non-specific medial reticular formation of the brain stem* influenced by those over regular lateral sensory pathways impinge on the entire cortex through non-specific diffuse thalamic and extra-thalamic projection systems and keep man and animal awake with resulting waking brain waves. Whenever there is inhibition of these barrages of impulses caused by any system sleep intervenes and slow delta waves of sleep appear. So now-a-days it is the custom to talk about "waking center" rather than "sleep center". From this one does not need to assume that the medial reticular brain stem reticular formation** is the center of consciousness; at least it is a very important controling point. It has been compared to a master switch (S. Cobb), that can turn it on or off. Clinical evidence of over-somnolence in tumors of periaqueductal gray, nearby areas or in hypothalamus connected with this area is known.

Taking advantage of the above-mentioned facts and scientific postulates the speculation may not be entirely unjustified that yogi's steady, attenuated, unidirectional awareness*** and attention without or little perceptual or conceptual specificities for prolonged periods on the one hand and without sleep on the other may be mediated by a conditioned dynamic equilibrium, not always stable, as a result of confrontation and organization of descending and ascending discharges at a subcortical pool or pools (?brain stem), with secondary inhibitory influences on the loower vegetative centers and the

*According to some anatomists the arousal function of this system is considered a good physiological concept and except in a general way it does not have an adequate anatomical basis (10, p. 501).

**In the middle ages some investigators thought that region to be the seat of the soul. (see 21, p. 18.). Modern physiological investigations have established its influence not only on the cortex, but on hypothalamus, pituitary mechanisms; thalamus; limbric system; postural and other peripheral conditions (21, p. 20).

***This can not be considered a homologue of elementary awareness of animal thalamus.

periphery. Numerous loose ends and assumptions are in this physiological concept; mechanisms of maintaining over-all neural balance in a complex system with many silent, co-operative and competitive circuits and areas can not be named. I realize that this concept gives only an incomplete picture, it is not an explanation of all yoga. It does not explain the results obtained by other yogic methods than the ones mentioned earlier. Patanjali (11, 32, 37) mentions that there are people who do not have to take routinized yogic steps and still can get the same result. If this is true one wonders about their special constitutional make-up to make this possible.

The present concept has nothing to offer explaining the mechanism of the reported paranormal powers. It has nothing to say how or whether the Vedantic Hindu assumption of the identity of man and absolute Reality is a matter of realization, nor how it can be related to spiritual or psychological ways of finding sanity and balance in this complex and confused world which both yogis and non-yogic desire.

Soma Weiss of Harvard (33) in 1931 was the first to describe thoroughly three types of carotid sinus reflex which is now common medical knowledge. Maybe some day yogic practices and attainments will be evaluated critically by the yogis themselves or probably by scientists* at least to a certain feasible point, and much chaff separated from the wheat, and many beliefs, however necessary, from experiences. To call all yoga and mysticism antiquated nonsense or useless superstition would not be right. "Unlimited scepticism is equally the child of imbecility as implicit credulity" (Braid)**.

*Cf. J.B. Rhine's para-psychological experiments.

**. . . "The divorce between scientific facts and religious facts may not necessarily be eternal as it at first sight seems . . The rigorously impersonal view of science might one day appear as having been a temporarily useful eccentricity rather than the definitely triumphant position which the sectarian scientist at present so confidently announces it to be."—William James (17), page 501.

BIBLIOGRAPHY

[1]Anand, B. K., Structure and function of the limbic system (visceral brain), a review. **Indian J. Physiol. and Pharmacology.** 1957, 1:149-184.

[2]Anand, B. K., Chhinna, G. S. and Singh, Baldev. Studies on Shri Ramanand Yogi during his stay in air-tight box. **Indian J. Medical Research.** 1961, 49:82-89.

[3]Bagchi, B. K. Mental Hygiene and Hindu doctrine of relaxation. **Mental Hygiene.** 1936, 20:424-440.

[4]Bagchi, B. K. Some points of contact between Hindu Religio-philosophical systems and modern concepts of mental hygiene and psychiatry. Presented before Michigan Academy of Sciences, Arts and Letters, April 12, 1952 and Department of Psychiatry, University of Michigan, Nov. 3, 1949. Unpublished.

[5]Bagchi, B. K. A re-evaluation of the philosophy of the Bhagavad-Gita in the setting of modern India and the world. **Mich. Academy of Science, Arts and Letters.** 1950, 34:327-343.

[6]Bagchi, B. K. Consciousness, its aberrations and the electrical rhythm of the brain. **Main Currents in Modern Thought.** 1956, March. 246 East 46 Street, New York City.

[7]Bagchi, B. K. and Wenger, M. A. Electro-physiological correlates of some yogic exercises. **In Proceedings of the First International Congress of Neurological Sciences,** Brussels, 1957. Vol. III, **EEG, Clinical Neurophysiology and Epilepsy.** 707. Ed. Van Bogaert, L. and Radermecker, J., Pergamon Press, London, 1959: 132-149. (The work reported was supported by the Rockefeller Foundation, Rackham Foundation of the Univ. of Michigan, George Washington Univ. (through U.S.A. Air Force contract) and Indian Council of Medical Research of the Govt. of India). See also references 34 and 35.

[8]Behanan, K. T. Yoga, **A Scientific Evaluation.** Introduction by Walter R. Miles. The Macmillan Company, 1937. Paper-back edition, Dover Publications, New York, 270.

[9]Braid, James. Observations on human trance or human hibernation. London and Edinburgh, 1850. There is also a German edition of his selected works, translated by W. Preyer and published by Gebruder Pack, Berlin. (Quoted by Richard Garbe.)

[10]Crosby, Elizabeth C., Humphrey, T. and Lauer, E. W. **Correlative Anatomy of the Nervous System.** The Macmillan Company, New York, 1962, 731.

[11]Dasgupta, S. N. **The Study of Patanjali.** University of Calcutta, 1920, 207.

[12]Dutta, Romesh C., **The Civilization of India.** The Temple Series. J. M. Dent Co., 1900, 361.

[13]Eddington, A. S. **The Nature of the Physical World.** The Macmillan Co., New York, 1929, 361.

[14]Elliot, H. Chandler. **Textbook of Neuroanatomy.** J. B. Lippincott Company, Philadelphia, 1963, 542.

[15]Garbe, Richard. On the voluntary trance of Indian fakirs. **The Monist.** Chicago, U.S.A. 1900, 10:481-500. Translated from German by W. H. Carruth of the University of Kansas.

[16]Honigberger, Johann Martin. **Thirty five years in the East.** English Translation, pub. in London, 1852. (Quoted by Richard Garbe.)

[17]James, William. **The Varieties of Religious Experience.** A study in human nature. Longmans, Green and Co., London, 1929, 534.

[18]Kahn, Edgar A., Bassett, Robert C., Schneider, Richard C. and Crosby, Elizabeth C. **Correlative Neurosurgery.** Charles C Thomas, Springfield, Illinois, U.S.A., 1955, 413. (chapter III on electroencephalography by Bagchi, B. K.)

[19]Macdonnell, A. A. **India's Past. A Survey of Her Literatures, Religions and Antiquities.** The Clarendon Press, Oxford, 1927, 293.

[20]Magoun, H. W. **The Ascending Reticular System.** In **Brain Mechanisms and Consciousness,** Ed. Adrian, E. D., Bremer, F. and Jasper, H. H. Charles C Thomas, Springfield, Ill., U.S.A., 1954, 556.

[21]Magoun, H. W. **The Waking Brain.** Charles C Thomas. Springfield, Illinois, U.S.A., 2nd Edition, 1963, 188.

[22]Max Muller, F. Ed. **Sacred Books of the East.** 50 Vols. Bhagavad-Gita (Vol. 8), Trans. K. T. Telang, Charles Scribner's Sons, New York, 1900, 446.

[23]Max Muller, F. **Ramakrishna, His Life and Sayings.** Longmans, Green and Co. First Edition, London, 1898, New Edition, 1923.

[24]Mookerji, Radha Kumud. **Hindu Civilization.** (From the earliest times to 325 B.C.). Longmans, Green and Co. London, 1936, 351. (Data based on archaelogical and other original sources).

[25]Narain, Raj. Director, **Proceedings of the Seminar on Yoga and Parapsychology.** Mimeographed. Dept. of Psychology-Philosophy. Lucknow University, India, 1962, 109.

[26]Prabhavananda, Swami and Isherwood, Christopher. **Bhagavad-Gita** (English translation). Introduction by Aldous Huxley. Paper back edition. Mentor Book, M103, Pub. The New American Library, 1954, 143.

[27]Prabhavananda, Swami and Manchester, Frederick. **The Upanishads** (English translation of 12 Upanishads). Paper back edition. Mentor Religious Classic, M194, Pub. The New American Library, 1957, 128.

[28]Preyer, W. Uber die Erforschung des Lebens. Jena, 1873. (Quoted by Richard Garbe.)

[29]Radhakrishnan, S. **Indian Philosophy, Library of Philosophy under the General Editorship of J. J. Muirhead.** The Macmillan Co., London and New York, 1923 and 1927. Vols. 1 and 2, 1491.

[30]Rolland, Romain. **New Prophets of India.** Trans. from the French by E. F. Malcolm-Smith. Albert and Charles Boni, New York, 1930, 683.

[31]Smith, Huston. **The Religions of Man.** Paper back edition. Mentor Book, MT 350, Pub. The New American Library, fifth printing, 1962, 336.

[32]Vivekananda, Swami. **Patanjali's Yoga Aphorisms.** (97-228) in **Raja Yoga,** new edition, Brentano's, New York, 1929, 381.

[33]Weiss, Soma and Baker, James B. The carotid sinus reflex in health and disease. Its role in the causation of fainting and convulsions. **Medicine.** 1933, 12:297-354.

[34]Wenger, M. A. and Bagchi, B. K. Studies of autonomic functions in the practitioners of yoga. **Behavorial Science.** 1961, 6:312-324. Pub. M.H.R.I., University of Michigan, Ann Arbor, Michigan.

[35]Wenger, M. A., Bagchi, B. K. and Anand, B. K. Experiments in India on "Voluntary control" of the heart and pulse. **Circulation.** 1961, 24:1319-1325.

[36]Woodroffe, Sir John (Pseudonym Arthur Avalon). **The Serpent Power** (two works on Tantrik yoga translated from Sanskrit with introduction and commentary). Ganesh and Co., Madras, India, 2nd edition, 1924, 320, 183, 154.

[37]Woods, J. H., **The Yoga System of Patanjali,** with two commentaries. Transl. from the Sanskrit. Seventeenth vol. of forty seven vol. Harvard Oriental Series, Ed. by Lanman, E. R. The Harvard University Press, Cambridge, Massachusetts, U.S.A., second edition, 1927, 381. The English translation of sanskrit words and concepts is often quite abstruse.

ADDENDUM

See a recent article by R. K. Wallace of U.C.L.A. entitled, "Transcendental Meditation", published in **Science,** March 27, 1970, describing significant physiological effects. Extensive bibliography.

61

Experiments in India on "Voluntary" Control of the Heart and Pulse

Marion A. Wenger, Basu K. Bagchi, and B. K. Anand

PROMINENT among the many claims of unusual bodily control that emanate from practitioners of Yoga is the ability to stop the heart and radial pulse. Such claims often have been authenticated by physicians, and one "experiment" employed a loud-speaker system so that a large crowd could hear the heart sounds before and after their disappearance. To our knowledge, however, only one investigator had published electrocardiographic results before the work now reported.

In 1935 a French cardiologist, Dr. Thérèse Brosse, took portable apparatus to India and obtained measurements from at least one person who claimed the ability to stop the heart. A published excerpt from her data[4] involving one electrocardiographic lead, a pneumogram, and a pulse wave recording from the radial artery, shows the heart potentials and pulse wave decreasing in magnitude approximately to zero, where they stayed for several seconds before they returned to their normal magnitude. The data were held to support the claim that the heart was voluntarily controlled to a point of approximate cessation of contraction.

During our investigations in India we searched for persons who claimed to stop the heart or pulse, and were cordially assisted by many individuals including the Indian press. We found four. Another claimed only to slow the heart. Of the four, only three consented to serve as subjects, and one of these claimed he was too old to demonstrate heart stopping without a month or so of preparatory practice. Since he was the subject studied by Dr. Brosse in 1935 we were particularly anxious to gain his cooperation and, after considerable persuasion, he consented to demonstrate for us the method he had employed in "stopping the heart" for Dr. Brosse.

Apparatus and Procedures

Our apparatus has been described elsewhere.[1-3] Briefly, it consisted of an 8-channel Offner type-T portable electroencephalograph with appropriate detectors and bridges for DC recording of respiration, skin temperature, electrical skin conductance, and finger blood volume changes. Procedures varied according to the cooperativeness of the subject and other circumstances. For that reason the results are reported for individual subjects.

Results

The first two subjects claimed they could stop the heart. No. 1. Shri Sal Gram, at Yogashram, New Delhi, made four attempts at one session. Only one electrocardiographic lead (III) and respiration were recorded, in what was planned as a preliminary experiment. The subject stood in a semi-crouched posture with the left side pressed against a table. Under retained deep inspiration considerable muscular tension was apparent in neck, chest, abdomen, and arms. Stethoscopically detected heart sounds either disappeared briefly or were obscured by sounds from muscle action. Palpable right radial pulse weakened or disappeared briefly at each attempt to stop the heart. Electrocardiographic records were replete with muscular artifacts but were readable for rate and QRS potential changes. Little change occurred in magnitude of potentials; changes in heart rate were small. There was no indication of heart arrest. The subject refused further cooperation.

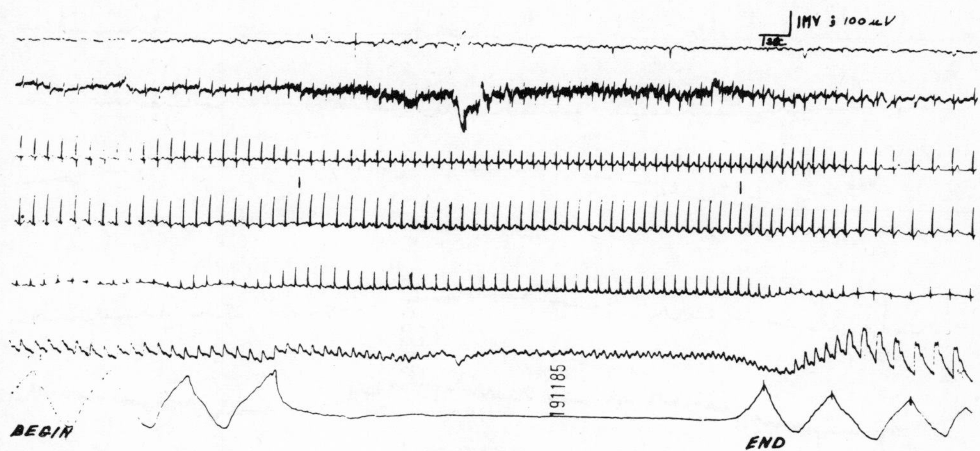

Figure 1

One "heart stopping" attempt by Shri Ramananda Yogi. The channels and variables are (1) muscle potentials, right biceps; (2) muscle potentials, right abdominus rectus; (3) electrocardiographic lead I; (4) electrocardiographic lead II; (5) electrocardiographic lead III; (6) plethysmograph, left index finger; (7) respiration. Calibration was 100 μv/cm. for the electromyogram and 1 mv/cm. for the electrocardiogram. Chart speed was 1.25 cm./sec.

No. 2. Shri Ramananda Yogi, of Andhra, age 33, at All India Institute of Medical Sciences, New Delhi, made seven attempts on 2 days, and additional experiments on a third day during fluoroscopy and x-ray photography,* all in a supine position. This work was initiated in the laboratory of the third author who became interested in the results and was asked to collaborate. His account of the second and third days of experimentation follows:

I examined Shri Ramananda Yogi on March 7, 1957, during two experiments in which he claimed he could stop the heart and pulse. He was investigated during these experiments electroencephalographically and electrocardiographically by Drs. Wenger and Bagchi, who also recorded his finger blood volume, respiration, blood pressure, and muscular activity. In the first experiment, Shri Ramananda Yogi stopped his respiration after taking four deep breaths. The breath was held in inspiration with closed glottis and the chest and abdominal muscles were strongly contracted. By this maneuver the pressure in the thorax was raised. During this period I could feel a very feeble pulse which had a normal rate in the beginning but became quick in the later part of the experiment. The heart sounds could not be heard but one could hear faint murmurish sounds due to the contraction of the thoracic muscles. The neck veins became distended. The breath was held for 15 seconds. This was immediately followed by quickening of the respiration, quick and deep pulse, and loud heart sounds. After a few seconds the heart rate returned to normal. The resting blood pressure had been 130/96 mm. Hg. Immediately following the breath holding it was raised to 210/100.

During the breath-holding period when the pulse was almost imperceptible and no heart sounds could be heard, the electrocardiograph continued to show contractions of the heart. The electrocardiographic pattern showed a slight right axis deviation which disappeared when respiration started again.

In the next experiment, he repeated the same procedure but held the breath in expiration. All other maneuvers were the same. The pulse, although very feeble, could still be felt. No heart sounds could be heard. Venous congestion in the neck took place. The electrocardiograph showed that the heart contractions continued but this time

*The radiologic investigations were conducted at Irwin Hospital with the assistance of Dr. N. G. Gadekar. The authors are grateful to him for his collaboration.

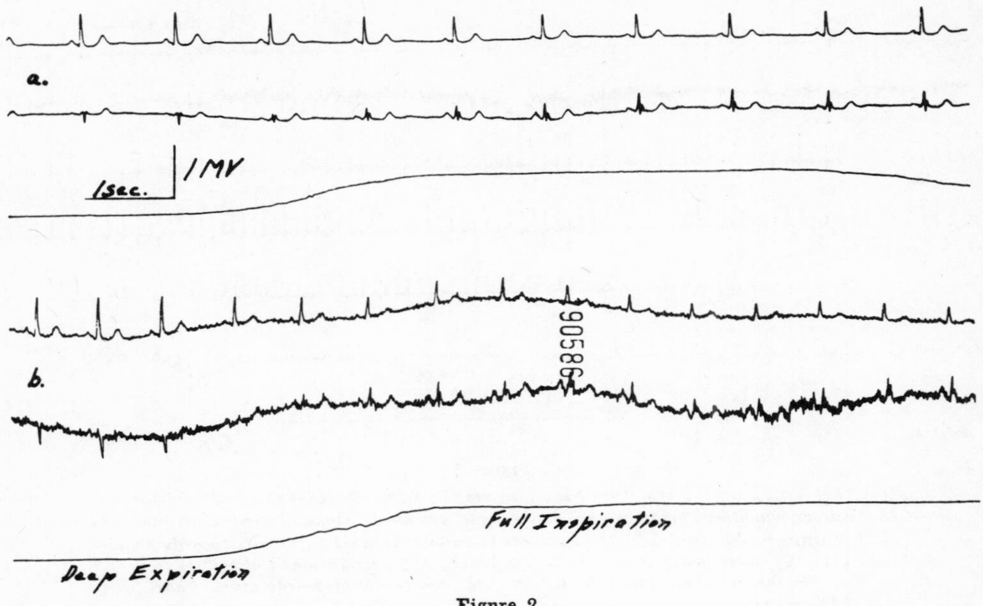

Figure 2

From demonstration by Shri Krishnamacharya of method used in "heart stopping." Electrocardiographic leads I and III, and respiration, are shown in that order (a) at rest, and (b) during demonstration. Calibration was 1 mv./1.5 cm. Chart speed was 2.5 cm./sec.

there was a slight left axis deviation. All other responses were similar to those in the first experiment.

Next day Shri Ramananda Yogi had skiagrams of the chest taken before experimentation and during two experiments, again attempting to stop the heart and pulse, one holding the breath in inspiration, and the other holding the breath in expiration. In both experiments one finds that the maximum transverse measurement of the heart decreases. Normally the maximum transverse measurement was 12 cm. During the first experiment (inspiration) it was decreased to 11 cm. During the second experiment (expiration) it was decreased to 11.5 cm.

Figure 1 shows the first attempt of March 7. During maintained inspiration the QRS potential is seen to decrease in lead I and to increase in lead III. (In the second attempt, during maintained expiration, the potentials increased in lead I and decreased in lead III.) The first two channels show muscle action potentials from the right biceps and the right abdominus rectus at the level of the umbilicus. Marked increases are seen to have occurred in the abdominal recording only. The plethysmographic recording shows that the finger pulse was always detectable, and that the pulse volume was greatly increased immediately after termination of the attempt. No unusual changes were detected in the electroencephalographic records.*

No. 3. Shri T. Krishnamacharya, of Madras, age 67, at Vivekananda College, Madras. This gentleman was the one who had "stopped his heart" for Dr. Brosse in 1935 but would not repeat the attempt for us. He finally consented to demonstrate the method he had employed, but with minimum apparatus attached: a blood pressure cuff, electrocardiographic leads I, II, and III, and a respiration

*This subject had engaged in a number of pit-burial demonstrations but refused us cooperation in that respect. Recently, however, he has cooperated with the third author and Dr. Gulzar Singh in such work. The results are to be reported soon.

Table 1

Variability in Heart Period in Last Phase of Heart Slowing Experiments of Subject no. 4 (HP in seconds)

Experiment no.	1	2	3*	1	2	1	2
Date		4/25		5/23		5/24	
Initial range		0.6-0.9		0.7-0.8		0.7-0.9	
Last 10	1.2	1.2	0.7	1.8	1.3	1.9	1.7
heart periods	1.1	1.6	0.8	1.8	2.1	1.8	2.0
	1.4	1.4	0.9	1.8	*2.5*	1.9	1.9
	2.6	2.0	0.9	*2.8*	2.0	1.9	1.9
	2.5	1.6	1.2	1.9	1.8	1.9	1.8
	1.9	*2.9*	1.6	2.2	2.4	1.8	1.8
	2.3	2.2	1.8	2.1	2.0	1.8	1.8
	2.1	2.3	*2.0*	2.0	2.2	1.6	1.8
	2.0	2.1	1.7	2.0	2.1	1.6	1.8
	1.7	1.6	0.9	1.8	1.4	1.5	1.3
Just after maneuver	0.8	0.9	0.6	1.2	0.9	1.2	0.7
Longest HP	2.6	2.9	2.0	2.8	2.5	2.8	2.9

**Uddiyana* alone

belt; none of which he would tolerate fastened tightly. He said his radial pulse might stop, but his heart wouldn't. The method proved to be similar to that employed by Shri Ramananda Yogi during maintained inspiration. The muscular effort expended was less, but the periods of maintained inspiration were considerably longer. Again, the blood pressure increased, the maximum change being from 128/80 to 140/105. There were no definite "attempt" periods for this subject. He merely permitted us to record data while he reclined and engaged in *pranayama* (breath control) as he pleased. On three attempts to measure blood pressure no sounds could be heard from the brachial artery. On another day with the subject seated, a physician was permitted to palpate both radial arteries and listen to heart sounds stethoscopically.† He reported no absence of heart sounds but at one time the radial pulse was not detectable in either wrist.

We believe the most significant data from these experiments are the changes in QRS potentials in leads I and III (fig. 2). As we previously had found with Shri Ramananda Yogi, during maintained inspiration with glottis closed and with increased tension in abdominal muscles, the heart potentials decreased in lead I but increased in lead III. Again, right axis deviation of the heart is indicated.

No. 4. Shri N. R. Upadhyaya, age 37, at Kaivalyadhama, Lonavla. This gentleman did not claim to stop the heart. He claimed only to slow it. He was a student of Yoga with more than 5 years of training, and had discovered accidentally that he could slow his heart. The maneuver occurred in the reclining position and during maintained inspiration. Just before the attempt a rolled towel was inserted under the lumbar area of the spine which gave a support 4 or 5 inches in height. After inspiration the subject engaged in the Yogic posture known as *uddiyana,* which involves a raising of the diaphragm and an inward distention of the abdomen, and accompanied it with another posture known as *jalandabar bandha,* in which the chin is depressed and extended toward the chest.

We tested him on 3 days, and on the first 2 days applied electrocardiographic chest leads V_2 and V_5 in addition to standard leads I, II, and III. There were only small changes in the magnitude of the QRS potential in any

†The authors are indebted to Dr. S. T. Narasimhan of Madras for his assistance in this experiment and for his genial aid in our work.

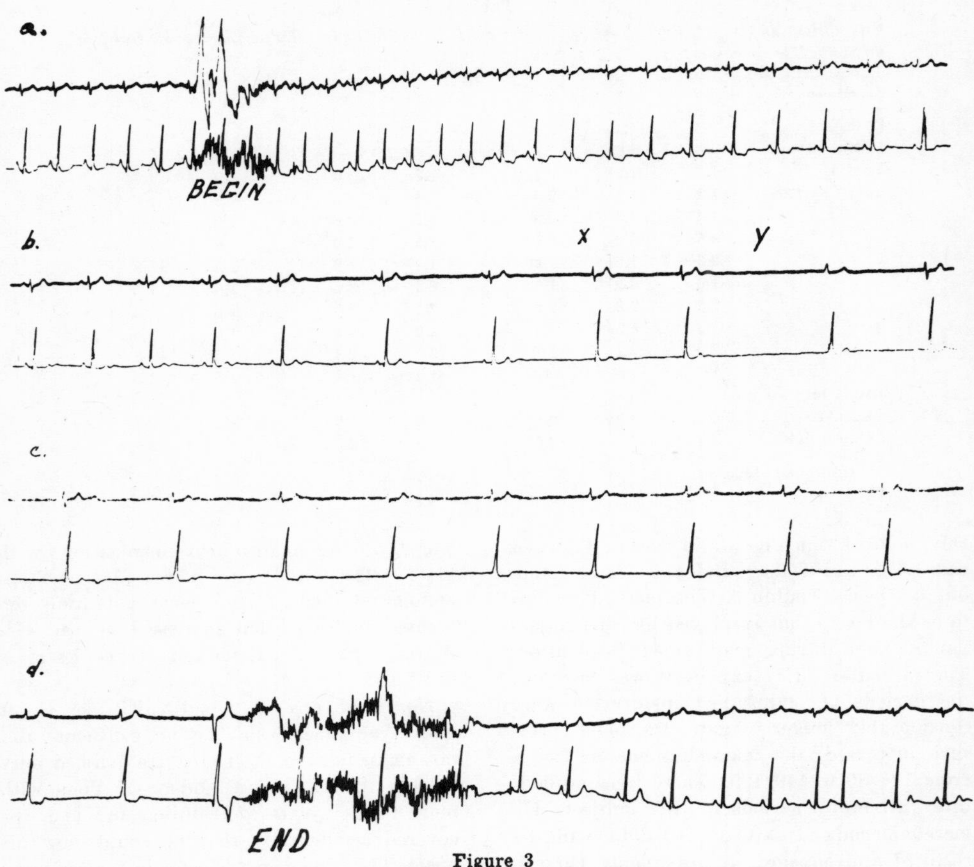

Figure 3

One attempt by Shri Upadhyaya to slow the heart. Sequential portions of the recording of electrocardiographic leads I and III are shown. Calibration was 1 mv/cm. Chart speed was 1.25 cm./sec. The P wave disappeared (x) and was absent for 16 heart cycles. The longest cycle length (Y) was almost 3 seconds.

lead. There was, however, a marked slowing of the heart in each test. In addition to bradycardia we found an increase in the P-R interval and finally a marked decrease or disappearance of the P wave. Thus a nodal rhythm appeared for a few beats before the subject terminated the maneuver. Figure 3 demonstrates the longest P-wave depression we obtained. The longest cycle length was approximately 3 seconds.

Our attempts to discover the mechanisms contributing to these results were not fruitful. Pressure on the carotid sinuses produced an increase rather than a decrease in heart rate. The chin lock alone (*jalandabar bandha*) produced little or no change. *Uddiyana* alone produced some bradycardia and one cycle of almost 2 seconds' duration, as may be seen in the third column of table 1. The other columns of the table show (a) the durations of the last 10 cycle lengths in the six main experiments, (b) the premaneuver range, (c) the first postmaneuver heart period, and (c) the longest recorded heart period. The period-to-

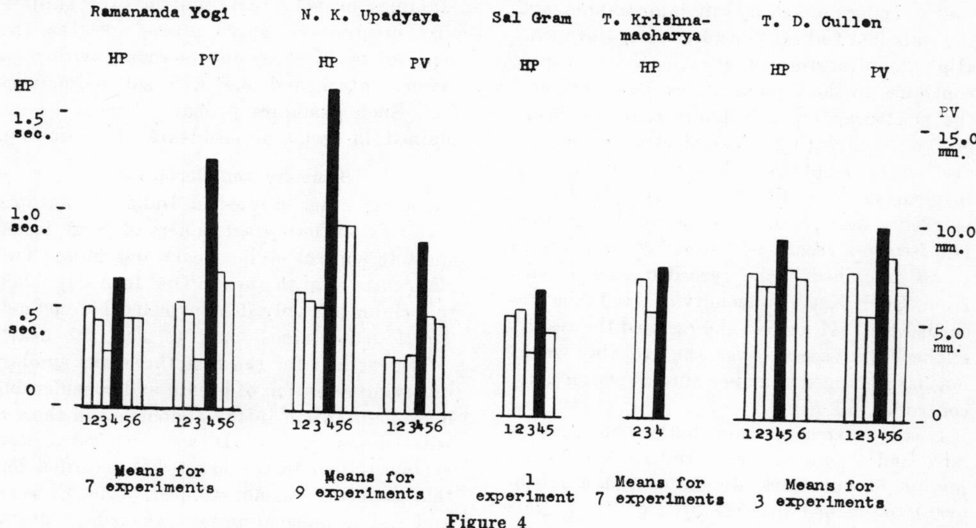

Figure 4

Summary of experiments on control of the heart. Each entry for heart period (HP) and finger pulse volume (PV) represents pooled means for samples of 10 heart cycles from each phase of each experiment. Different phases of the experiments are indicated as follows: 1. Initial rest. 2. After instructions. 3. Fastest HR and lowest PV during maneuver. 4. slowest HR and highest PV during and just after maneuver. 5. 50 HP after sample no. 4. 6. 100 HP after sample no. 4.

period variability is apparent. Of additional interest is the observation that only on the last day of experimentation was the maneuver maintained for 10 or more heart cycles beyond the greatest bradycardia. Apparently experience or amount of attached apparatus, or both, influence the effects of this maneuver. This subject is still available for research. Subsequent work by Indian investigators supports the above observation in that a heart period of 5.6 seconds has now been recorded.[5] Perhaps they will discover the mechanisms underlying the effect.

Discussion

It is obvious that the subjects we tested do not voluntarily control the heart muscle directly. In each instance some striated muscle action intervenes. Through muscular and respiratory control certain changes do occur in circulatory variables. Figure 4 shows changes in heart period and (for some subjects) finger pulse volume for the Indian subjects and for one American (Dr. Cullen)* who attempted to repeat the Valsalva experiment. Decreased heart rate was greatest in the subject who claimed only to slow the heart. The greatest changes in finger pulse volume occurred in Shri Ramananda, who exerted utmost effort to demonstrate "heart stopping." Dr. Cullen employed less effort, although his pattern for heart period is not greatly different from that of Shri Ramananda.

For the first three subjects we assume that by increased tension in the muscles of the abdomen and thorax, and with closure of the glottis, there is developed an increased intrathoracic pressure that interferes *with the venous return to the heart.* With little blood to pump the heart, sounds are diminished, as

*The authors are indebted to Dr. Thomas Cullen, University of California, Los Angeles, for his participation in this experiment and for his assistance in analyzing the data.

well as being masked by muscular sounds, and the palpable radial pulse seems to disappear. High amplification finger plethysmography continues to show pulse waves, however; and the electrocardiograph shows that the heart goes on contracting. The electrocardiograph also shows right axis deviation under deep inspiration. The QRS potentials in lead I are markedly decreased. It seems, therefore, that Dr. Brosse's record of "heart control" is to be so explained. She recorded lead I only from Shri Krishnamacharya. Had she recorded lead III, or had she recalled the results of the Valsalva maneuver, she probably would not have claimed that her subject voluntarily controlled his heart.

It is of interest to know that our conclusions have had a forerunner in India. After completion of our work we discovered a monograph published in 1927 by Dr. V. G. Rele of Bombay.[6] Although he published no data, he writes of electrocardiographic records and x-rays on one subject, and reached conclusions similar to ours, although he extended them more than we care to do.

Our fourth subject demonstrated a different phenomenon that further research may explain. It could be said that he "stopped the heart" for a few seconds. We prefer to assume that by some striated muscular mechanism he stimulated the vagus output to the sinoatrial node, interrupted it, and thus interrupted regular cardiac cycles and a nodal rhythm was briefly established.

In this connection a recent publication in California is of interest.[7] A patient complained of heart slowing when he relaxed. Published electrocardiograms show a period of standstill of approximately 5 seconds followed by a QRS potential with no P component. The report, supplemented by private correspondence, indicates that there were no changes in QRS magnitude but that the P wave at the end of the periods of bradycardia was reduced or absent. Contrary to our Indian subject, he apparently employed no intervening muscular or respiratory mechanism. He merely relaxed. His data call to mind other examples of voluntary control over supposedly involuntary musculature, such as one reported by Lindsley and Sassman[8] with pilomotor control, and one[2] with sudomotor control. Such examples probably are to be explained in terms of accidental conditioning.

Summary and Conclusions

Among other studies in India the authors investigated four practitioners of Yoga in respect to control of the heart and pulse. Two claimed to stop the heart. One formerly made this claim but only demonstrated his method. The fourth claimed only to slow the heart.

The method for the first three was similar, involving retention of breath and considerable muscular tension in the abdomen and thorax, with closed glottis. It was concluded that venous return to the heart was retarded but that the heart was not stopped, although heart and radial pulse sounds weakened or disappeared.

The fourth subject, with different intervening mechanisms also presumably under striated muscle control, did markedly slow his heart. The data indicate strong increase in vagal tone of unknown origin.

References

1. Bagchi, B. K., and Wenger, M. A.: Electrophysiological correlates of some yogi exercises. *In* EEG, Clinical Neurophysiology and Epilepsy. London, Permagon Press, 1959, p. 132.
2. Wenger, M. A., and Bagchi, B. K.: Studies of autonomic functions in practitioners of Yoga in India. Behavioral Sc. 6: 312, 1961.
3. Wenger, M. A., and Bagchi, B. K.: A report on psychophysiological investigations in India. Mimeographed MS for limited circulation.
4. Brosse, T.: A psycho-physiological study. Main Currents in Modern Thought 4: 77, 1946.
5. Kuvalayananda, S.: Kaivalyadhama, Lonavla (Bombay State), India. Unpublished research.
6. Rele, V. G.: The Mysterious Kundalini. Bombay, D. B. Taraporevala Sons & Co., Ltd., 1927.
7. McClure, C. M.: Cardiac arrest through volition. Calif. Med. 90: 440, 1959.
8. Lindsley, D. B., and Sassman, W. H.: Autonomic activity and brain potentials associated with voluntary control of the pilomotors. (MM. arrectores pilorum). J. Neurophysiol. 1: 342, 1938.

62
Electro-Physiological Correlates of Some Yogi Exercises

Basu K. Bagchi and Marion A. Wenger

This is a preliminary report of a pilot investigation on some aspects of Yoga carried on in India for over five months during our leave. Except for the exploratory work of Brosse (13), and of Das and Gastaut (14) there has been, to our knowledge, no systematic electrophysiological investigation of Yoga. Some biochemical and roentgenological studies on certain aspects of Yoga are, however, available (12, 20, 21).

This type of investigation often has peculiar problems of its own, namely: (1) To hunt up informants for securing information on habits and habitats of Yogis, (2) To screen the information, (3) To attempt to locate Yogis of reported competence and experimental amenability, (4) To separate them from learned or non-literate charity-dependent peripatetic or settled sadhus, following mainly a life of religious formalism with or without organizational emphasis, from magicians interested only in monetary reward, and from charlatans, (5) To make the initial contact and try to establish rapport with recommended Yogis after locating them, (6) To obtain their consent, and (7) To evaluate areal accessibility and work conditions. Dozens of informants and Yogic subjects were personally contacted in 17 places in different parts of India, in addition to carrying on a fairly large volume of correspondence from America and in India. A large number of leads on Yogis proved unproductive. Many Yogis were not interested, some would not cooperate. Many could not be reached. It would appear that more time, effort, travel, and expense were involved than would be considered warranted in an adventure like this. That impression would not be exactly true.

Meditative Yogis, even located, usually shy away at the thought of doing their meditative exercises under experimental conditions with electrodes and other trappings attached to their bodies. Further, philosophical and religious aspects of Yoga weigh far more with most of the meditative Yogis than any scientific consideration. They may often be in inaccessible places. The nature of liaison and first contacts with them may mean whether or not they will cooperate. On the other hand, it is less difficult to secure and get the cooperation of Hatha Yogis who emphasize different kinds of postural, breathing and other general or purificatory manoeuvers directly or indirectly tied up with religion. Meditation may be practised and utilized but not necessarily emphasized by them . Some times competent Hatha Yogis are in scattered or relatively out-of-the-way places.

A word about Yoga. Yoga has been an important part of Indian culture since ancient times. It was first described in the early Upanishad literature (1000 B.C - 300 B.C.). It has a rich tradition and numerous variants (15, 18). The word has many meanings. Yoga means union and is related to Latin, yugen, and the word yoke. Different Yogic exercises and methods have stemmed from different philosophical or theoretical principles or premises emphasized by their followers although there is certain underlying similarity in approach in all of them. According to the philosphy of The Bhagavad-Gita (22) Yoga is the union of the individual soul with the universal Soul. According to Patanjali's System, it means stopping the fluctuations and disturbances of the "mind" (Chittva) in order to let the individual soul shine in its own pristine state (3, 25). It does not speak of union with the universal Soul. The word may mean attainment of inner equilibrium (Samatvam) under all circumstances (22). Yoga may refer to the means or steps like physical, moral, psychological, spiritual and meditative disciplines or efforts, or it may refer only to the ends or results just mentioned. Apart from spiritual attainment many unusual powers and practical benefits have been claimed in India for 2 to 3 thousand years to result directly or indirectly from Yogic practices. A scientific inquiry into these is pertinent in this modern age. We did not investigate the psychological, philosophical and spiritual background implications or assumptions of Yoga nor its reported therapeutic effects. Available time compelled us to limit the scope of our inquiry to the electrophysiological concomitants of a few Yogic exercises.

TECHNIQUE AND SUBJECTS

An Offner eight-channel portable (78 lbs) A.D. and D.C. machine of high sensitivity using transistors (not vacuum tubes) and flash light batteries was imported along with accessories such as transducers, bridges, etc. Fig. 1. The machine amplifies electrical potentials about a million times. Only the paper-pulling motor needed outside power source (110 volts, 60 cycles) which could be obtained through an inverter from a 6-volt automobile battery or from a power line, when available. Electroencephalogram (EEG, brain waves), electrocardiogram (EKG, heart waves), electromyogram (EMG, muscle waves), plethysmogram (finger blood volume tracings), respiration, electrical skin resistance, skin temperature, and blood pressure were recorded during different Yogic exercises. This machine with its accessories was taken around the country over four thousand miles and set up in eight different places: two physiological laboratories of two Universities, physics laboratory of a College, two hermitages, one of them having a shielded, semi-sound proof, air-conditioned room, and a Himalayan mountain cave which was relatively difficult of access. It was not possible to explore many supposedly important territories and many leads within the limited time at our disposal.

Comprehensive EEG technique has been described elsewhere (10, 11). But such a technique could not be used because of the necessity of recording the other variables on the 8-channel machine. The following EEG derivations were used, for instance, on the first three channels: left motor-left occipital, right motor-right occipital, and left temporal-right temporal (to determine temporalis muscle tension). EKG lead I was on channel 4, respiration (through a respiration belt-transducer-bridge set-up) on channel 5, continuous skin resistance from two practically non-polarizable zinc-sulfate-paste-zinc-plate palmar leads

Fig. 1

The portable 8-channel Transistor Machine with the bridges and other accessories to obtain simultaneous tracings of 8 electrophysiological phenomena - EEG, EKG, etc. In the center is seated a meditative yogi with his friends and admirers including a high Government official at a New Delhi residence. This picture was taken after the first session of recording. The Yogi consented to be a subject after a great deal of hesitation and discussion regarding the purpose of the research. Later he was most cooperative and allowed three more sessions. This yogi is highly thought of in certain sections of the country.

(through a bridge with 40 microamperes impressed into the subject) on channel 6, plethysmographic cup in one of the fingers (through a transducer and bridge) recording on channel 7, and continuous skin temperature from right index finger or other places (through a thermistor and bridge) on channel 8. The last four determinations were made with D.C. time constant. Proper balancing devices were provided for the last four channels. For details of transducer-bridge assemblies and non-polarizable leads in connection with D.C. recording of the variables of autonomic nervous system see reference (24).

Many variations of the above-mentioned technique and shifting of channels for a particular recording were employed depending upon the kind of experiment done. For instance, during heart experiments sometimes as many as five EKG leads, including three standard leads and two precordial leads (monopolar) were used and only one scalp-to-scalp recording. Some trial measurements of temperature changes of the expired and inspired air through the mouth or through each of the nostrils were made by thermocouples inserted into a regular mouthpiece or hard rubber tube of small lumen, the output being fed into the transistor machine. The pressure of the air was not measured, nor was the calibration systematically graded. Blood pressure measurement could be made from the control room or from a distance by extended non-flexible plastic tubes, about 10 feet long, without going near the subject or being in the same room with him, specially during meditation experiments. The cuff and the stethoscope diaphragm were left attached during the entire experiment to the upper arm and the connecting tubes brought out to the control room, one tube connected to the manometer second tube to the inflating bulb, and the third tube from the diaphragm to the stethoscope. In meditation one blood pressure (B.P.) reading was taken before and one after the practice. In initial meditation experiments several B.P. determinations were made. Inflating the cuff though done from the adjoining room was found disturbing to many subjects, so this practice was discontinued, even though it had served as an excellent comparable stimulus. Usually a control run* before the practice of meditation was made for more or less five minutes with the subject sitting in padmasana or lotus posture with each leg folded at the knee and placed superio-medially on the opposite thigh. Other Yogic postures were sometimes used by the subjects. There was no change of posture during meditation. Altogether 98 sessions on 45 subjects were obtained, more sessions on non-meditative endeavors.

Fourteen subjects sat in meditation doing different techniques of meditation in the above-mentioned posture almost entirely fixed or "immobile" for from 14 to 148 minutes in 25 sessions. Very few movements were noticed during this time. Median duration of meditation was 49 minutes, the longest four being over 100 minutes. There were two highly recommended Yogis (S.P. and S.S.) who had devoted more than 30 years of meditational Yoga and spiritual life. Others were meditative practitioners or primarily Hatha Yogis for from 2 to 30 years from Kaivalyadham (Lonavala P.O.) hermitage, Yogashram of New Delhi and from Calcutta, New Delhi, and Bombay. Subjective account of a sort was taken from each subject about the type and depth of meditation but this will not be elaborated here.

The target of a meditative Yogi is to reach the deepest state of absorptive meditation (samadhi) after undergoing prolonged ethical, self-analytic and purificatory disciplines. According to ancient Patanjali's Yogic book deepest meditation or samadhi has two types, several sub-types and many levels (3, 18, 25). To put in a reconstructed form, it is claimed that a particular sub-type of samadhi is attained when the state of one-pointed, quiet and yet attenuated awareness of "I-ness" becomes so deep or the absorption of "I-ness" in the meaning of an idea, in a God-concept, in a syllable (like the sacred syllabel Aum), in some body zone (like the middle of eyebrow), in a color, form, sound (nad) or

---

*Sometimes the very first few minutes of recording during meditation practice had to be taken as the control run.

experience or in the internal representation thereof becomes so drastically complete that irrelevant and fluctuant specificities of perception or thinking disappear, only the identified "I-ness" with the object of thought remains, and stimuli from the somatic system or external world do not reach the consciousness of the meditator for a short or long time.

A psycho-physiologico-neuroanatomical figure of speech may be employed here. These people are trying as it were to use the attention-phenomenon of the cortex to merge into and stabilize at a different level the "indeterminate" awareness of the deep subcortical mechanism, the highest plane of neural integration ---the centrencephalon (Penfield) (17). Through accessory disciplines buttressed by this prolonged, quiet one-pointed meditation and consequent strengthening of the cortico-sub-cortical dynamism they believe they can attain true self-control, knowledge, and happiness. It is stated that they thus can free the self from false beliefs and attitudes and live attachment-free mature liberated lives, and paranormal powers may become theirs. It is assumed that by the manipulation of the hidden forces of the pran (energy, free translation) of the human microcosm which holds the key to the macrocosm this state can be reached. Old pauranic texts, medieval tantrik or other treatises and many modern books (4, 19) contain elaborate and divergent symbolic accounts and esoteric diagrams and descriptions of the nervous system. These include shatachakra (six lotuses, energy loops) five of which are supposedly in the spinal axis, sahasrar (thousand-petalled lotus, the cerebrum), three main nerve trunks, ira, pingala (? sympathetic nerve chains), sushumna (spine), also brahmi, chitra nerves in addition to the Kundalini (the serpent power, the latent "cosmic energy" (seated at the base of the spine), which can be aroused through different means and made to pierce the six lotuses until it reaches the thousand petalled lotus and brings the highest state to the Yogi.

We were interested in the electro-physiological manifestation of this "awakened" or samadhi state, if it could be found. None of the subjects claimed they reached this state during the experiments. A few stated that during meditational sessions they may have reached at times its precincts.

RESULTS

(a) Meditation

Although the sampling of subjects is small and not of a relatively uniform level, as desired, the following physiological observation may be made.

In effective meditation in an almost "immobile" or fixed posture for a long time (which "immobility" is a feat in itself), the effectiveness being judged from the subjects' report which is far from being comparable or scientifically discriminative, two separate trends are seen:

(1) A trend toward stabilization, with or without minimal fluctuation, or toward lowered rate of the physiological phenomena under consideration. For instance, there were no consistent heart rate changes during meditation compared to control periods. There may be no or 6% to 9% fluctuations in individual cases. (See later for one exception). But that

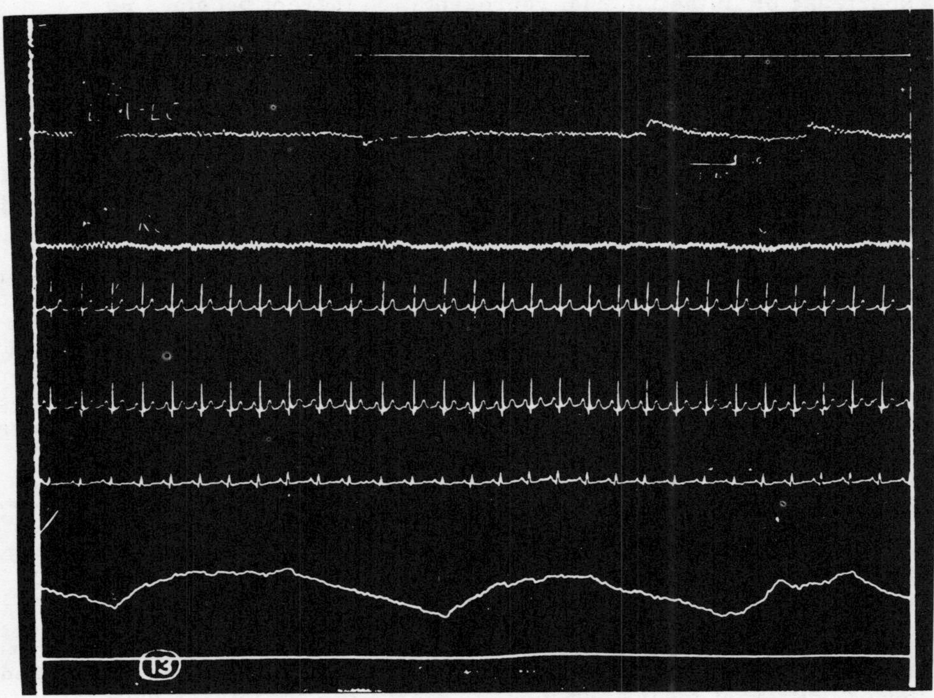

Figure 2.

Control reading on yogic subject, S.S., before he was asked to meditate. This was 12 minutes after electrodes were attached and he was left alone in the room, and 2 minutes following our instructions for him to proceed with meditation. Line 1: blank, line 2: EEG from left motor and occipital, low to medium voltage 10 per second normal waking brain waves (alphas) with some artifacts. Line 3: EEG from right motor-occipital, medium voltage 10 per second waking alphas, Line 4: Electrocardiogram (EKG) Lead I. Line 5: EKG Lead II. Line 6: EKG Lead III. EKG rate is 87 per minute. Line 7: Respiration - upstroke inspiration, downstroke expiration; respiration rate (2 samples) 9 per minute. Respiration rate for 10 minutes previous to this varied between 9 and 12 per minute. The average respiration rate on the previous day during control period was, by contrast, 13 per minute. Line 8; Electrical palmar resistance. No determination of it was made at this time. Three minutes before this it was 199,000 ohms and 10 minutes before this (that is, 7 minutes after electrodes were attached) is was 130,000 ohms (control period). So it is realized that before the subject was actually asked to proceed with meditation his skin resistance had risen by 69,000 ohms (199,000 - 130,000), about 53%. It is entirely possible that the subject had begun relaxation and meditation when he was left alone and before he was actually told to do so. Time between two heavy vertical lines which is one sheet of paper is 22.4 sec. Time calibration of a second and voltage calibration of 50 one-millionths of a volt are on upper right hand corner of this and other figures. In this research the physiological data of every eighth sheet (that is, every 3 minutes) were analyzed besides noting any special changes or other entries on skin temperature, etc. on that sheet or intervening sheets. From each eighth sheet a 10 second sample of EKG, a 2 second sample of EEG and 1 to 3 samples of respiration were counted and then converted into rate per minute or per second.

---

is not true of the respiration rate. In 9 out of 16 sessions of meditation of 12 subjects the respiration rate never went higher than the rate of the control period; it always went down. In the other 7 the rate either went up minimally and transiently and later came down and stayed down or remained about the same as the control period. Also in the 16 sessions average respiration rate during control period was 16 per minute (two subjects' was 10 per minute) whereas the lowest average respiration rate (transient or maintained) during meditation was 12 per minute (23% decrease).* Figs. 2 and 3.

(2) Variability and independence of most of the phenomena as to the extent, rate, instance and direction of change. In meditation respiration in a few instances became for a short or long time (17 to 27 minutes) slow, dropping as low as 50% - 60% of the control period, to 4 - 6 per minute. Or respiration became so shallow during the time that the rate was uncountable. Fig.4. Visual observation several times corroborated this. At that time the heart rate may be minimally slow compared to the persons' control period rate but not to a degree comparable to respiration. The heart rate may not be significantly slow at all in such or other cases. During meditation it never went below 62 per minute. The heart rate increased consistently in only one case as high as 13% as an exception. There is in a large percentage of sessions a tendency toward a drop in respiration rate, as mentioned, but a few hardly showed any change from their control period, while their electrical resistance increased unfailingly.

Blood pressure rose minimally in a few cases, dropped 20 points systolic in two instances at the end and showed no change in others. No abnormal B.P. was seen at any time. Electrical resistance of the palm, as stated, always increased during meditation, for some subjects more, for others less compared with their control period. The median increase was 56%. Six sessions of four subjects showed the highest increase--70% to 106%. The two supposed experts showed the increase as did the two who would not be considered as experts by Indians. The former exhibited questionably less fluctuating character than the latter. As expected,

---

*In 9 sessions (25 - 16) technical difficulty precluded the determination of the respiration rate. One has to watch out that there is no transient Kumbhak or holding of the breath or voluntary slowing of the breath. Unless the respiration rate is consistently slow for quite some time the determinations are rejected.

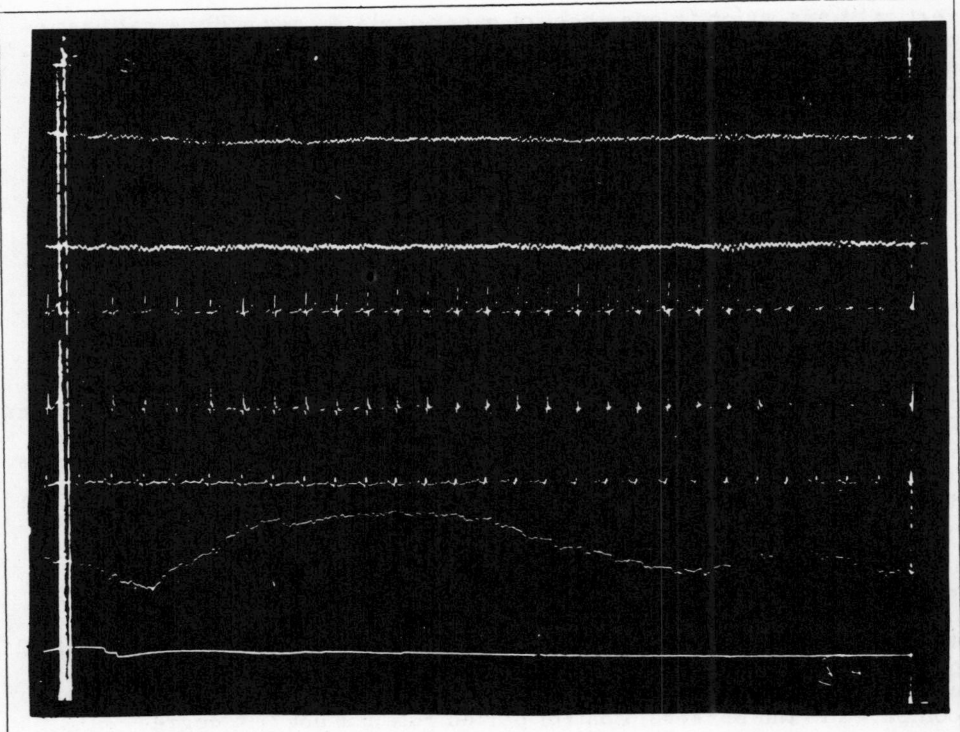

Figure 3.

Same subject as in Fig. 2. Same meditation session without any interruption; 16 minutes after Fig. 2. Subject with eyes closed is quiet and "immobile". Line 1: blank, Lines 2 and 3 show regularized 10 per second alpha waves from the brain. No waves of drowsiness or sleep or tension (fast waves). Lines 4, 5 and 6 are EKG leads as in Fig. 2. EKG rate is 78 per minute. Line 7: respiration (one sample): rate is 5 per minute which is a 44% drop from control period (9 minus 5) in Fig. 2. The next sample showed faster rate. Between Fig. 2 and Fig. 3, (not shown here) 8 respiration rate determinations of 1 to 3 samples each were made -- once 12 per minute (shallow), twice 10 per minute (once shallow), twice 9 per minute, once 5 per minute, and twice very shallow and irregular (uncountable). The respiration line had been continuing all the time between Fig. 2 and Fig. 3. The previous samples give a general idea of the decreased rate of respiration in this intervening period. Line 8: Palmar electrical skin resistance now is 242,000 ohms, an increase of 85% over control period (130,000 ohms).

between subject to subject there was a wide range of skin resistance values.

A few showed a steady increase of electrical resistance to close to 30 minutes after which the physiological variables including palmar resistance began to show up and down changes, indicating probably the

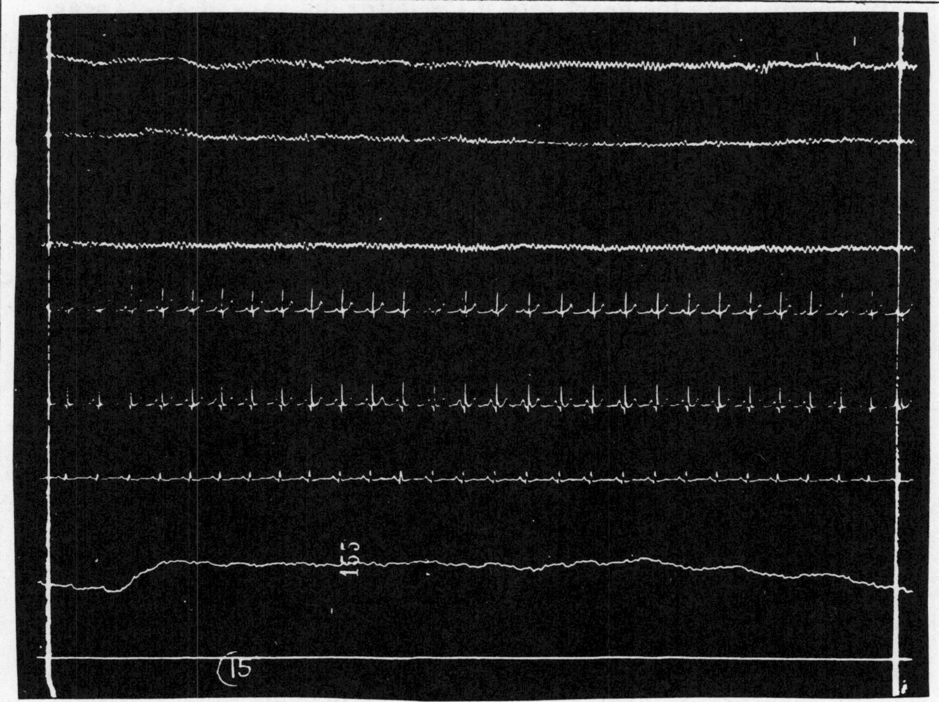

Figure 4.

Same subject as in Figs. 2 and 3. Same meditation session without any interruption 24 minutes after Fig. 3. Subject is still quiet in meditation and "immobile" and has been so in the intervening periods between Fig. 3 and Fig. 4. Line 1: EEG from left and right temporal electrodes, regular 10 per second alphas, no temporalis muscle tension. Lines 2 and 3: same electrodes as in Lines 2 and 3 of Fig. 3. EEG: regular alphas, no waves of drowsiness or sleep nor fast waves. Lines 4, 5 and 6: EKG leads as in Fig. 2 and 3. Heart rate is 81 per minute. Line 7: Shallow respiration line, slightly irregular, rate per minute uncountable. Except for two instances when 6 to 9 per minute respiration rate was seen, this type of shallow, flat and irregular or slow 4 to 5 per second unusual respiration was seen for the entire 24 minutes. Line 8: Skin resistance not taken at this time, but 3 minutes previous to this it was 254,000 ohms, an increase of 95% over control period (130,000 ohms); and about 5% increase over that seen in Fig. 3. Finger temperature before meditation, 28.5° C, 10 minutes after the termination of meditation, 27.5° C, without much intervening change.

limits of the subjects' optimum relaxation or meditation time. If and when the respiratory rate decrease and skin resistance increase went hand in hand the rate of this inverse relation was not necessarily in the same proportion. Also the skin resistance varied more widely from control period than the heart rate if the latter did vary at all. These differential occurrences may in part be due to the following factors: the depth of meditation, the nature of meditation, the physiological limits and fluctuation possibilities of the organs or processes involved, the constitution and the competence of the practitioner and the immediate internal and external circumstances. To what extent practice can bring uniformity or consistency in the profile of the measures investigated may probably be determined in an effective way by a longitudinal study over a long time of the same individual practitioners or by a cross sectional study of some more meditatives who are recognized as experts.

The few instances of slow (4 to 6 per minute) breathing and/or extremely shallow breathing for 17 to 27 minutes in meditation in the face of other stable and regular physiological signs and increase in skin resistance is difficult to explain. The slowing of metabolism may be a factor but except for skin resistance increase concomitant decrease in other functions is not seen. Gas and pressure analysis of the expired air at the time was not done.

Episodes of spontaneous oscillation of the electrical skin resistance line or sudden Galvanic skin reflex of many thousand ohms without any outside stimuli was seen, more or less, in most cases initially, even in the presence of high level of skin resistance and without any corresponding basic change in EEG. Variation of central excitability without any counterpart in consciousness, sub-threshold to EEG, or sudden intermittence of attention may be a cause. But this phenomenon needs further investigation. So does the relationship, if any, between skin resistance changes and varied types of meditational practice at initial stages.

So far meditation has not shown the electrical rhythm of the brain (EEG) that is commonly seen in drowsiness, light sleep, dream (non-specific changes), dreamless deep sleep, or in coma whatever its origin. That is, low voltage 6 - 7 per second waves, central 5 - 7 per second bursts, 13 - 14 per second spindles or very high voltage (150 - 300 one-millionths of a volt) 1 - 3 per second delta waves were not seen as are usually found in drowsiness, light sleep and deep sleep. Nor is the brain rhythm in meditation a homologue of what is found in hibernating animals (flattening of waves). Contrary to a previous report (14), it is the normal waking alpha pattern that is seen in meditation, that is, 8.5 to 11 beats per second, 20 to 100 one-millionths of a volt, sometimes $\frac{1}{2}$ or 1 beat slow, sometimes with good amplitude modulation and regularization but without high amplitude fast waves. Whether or not that finding will stand up with more intensive study of those who have prompt and pronounced command on the voluntary induction of their meditative state in spite of all the trappings on their bodies during the experiment, can not be stated.

The rise of electrical resistance of the skin (which we always found to occur in meditation) has been known for many years to the physiologists as a sign of relaxation. It is also known that this rise does not occur during strong activity of the cerebrum or of the autonomic nervous system. Putting all these findings together and pending a study of a large number of Yogic experts it may be suggested that Yogic meditation is deep relaxation of a certain aspect of the autonomic nervous system without drowsiness or sleep and at the same time a type of cerebral activity without highly accelerated electrophysiological manifestation. More than that can not be said now.

There is, however, another finding in this connection which must be mentioned. If we do not doubt the subjective report of a few meditative subjects it appears that low intensity stimuli like tap or sound or noise within 3 to 5 feet from them, which can be perceived ordinarily and which may or may not cause skin resistance change during meditation often did not reach their consciousness. This non-perception covered in two cases probably a period of more than 25 minutes. Of course there was no objective criterion for this. One of the subjects in the cave (S.P.) went, to all appearances, into the state of complete withdrawal while sitting in meditation almost immediately in his first session even when EKG and palmar electrodes were being put on and his hands were moved passively. The lack of tone in them was marked according to two observers. This state of non-perception which apparently continued for half an hour was admitted on later questioning: sounds and noises of low intensity such as shuffling of feet and moving around to fix the machine within 3 to 5 feet were not perceived by the subject. During this session only skin resistance changes were recorded which went as high as 70% over control period. In the second session other variables were recorded when again low intensity sounds were not perceived according to later subjective report. During such stimuli of low intensity no EEG blocking could be detected. Twice there was some shift of the base line. There were no EEG waves of drowsiness or sleep during this time. All these observations need to be further experimentally checked.

If this non-perception of stimuli is a fact then meditation may represent a twilight state in which stimulus-response reflex probably functions, if it functions at all, in the form of palmar resistance change or EEG blocking, below the cortical level -- below the level of determinate awareness. Or, the prolonged non-perception of such stimuli may be primarily a cortical affair -- the over-all "quiet turbulence" or willed meditation state making ineffective the small ripples caused by extraneous stimuli, with the preservation of subcortical reflex action. If that is so, there must be presistent EEG adaptation phenomenon at the time as commonly seen in EEG of ordinary persons during continuous endogenous or exogenous stimuli (8). Or, this non-perception may be an indirect result of a cortico-subcortical tug-of-war with one-pointedness of conscious will finally merging itself into and stabilizing at a different level the highest integrating subcortical neural mechanism associated with "indeterminate" awareness. These speculations can not be proved nor disproved. However, the neurophysiological significance of the entire problem of meditation in the light of modern studies of the older brain (centrencephalon and limbic system) for the understanding of consciousness would become obvious (1, 2, 6, 16, 17).

(b) Other Experiments:

Only short statements will be made here about each of these. We have recognized the role of pranayama breathing exercises of Yoga (four types of pranayama were studied), uddiyana (raising of the diaphragm), and jalandhar bandh (chin-lock) in the "heart-sound-stopping" experiment. Intra-thoracic pressure raised by holding the breath under this circumstance had its specific effect in that situation causing stoppage of the venous return to the heart. Right or left axis deviation in EKG was noticed depending upon whether the breath was held after deep inspiration or expiration. Afterwards blood pressure increased, systolic 30 to 40

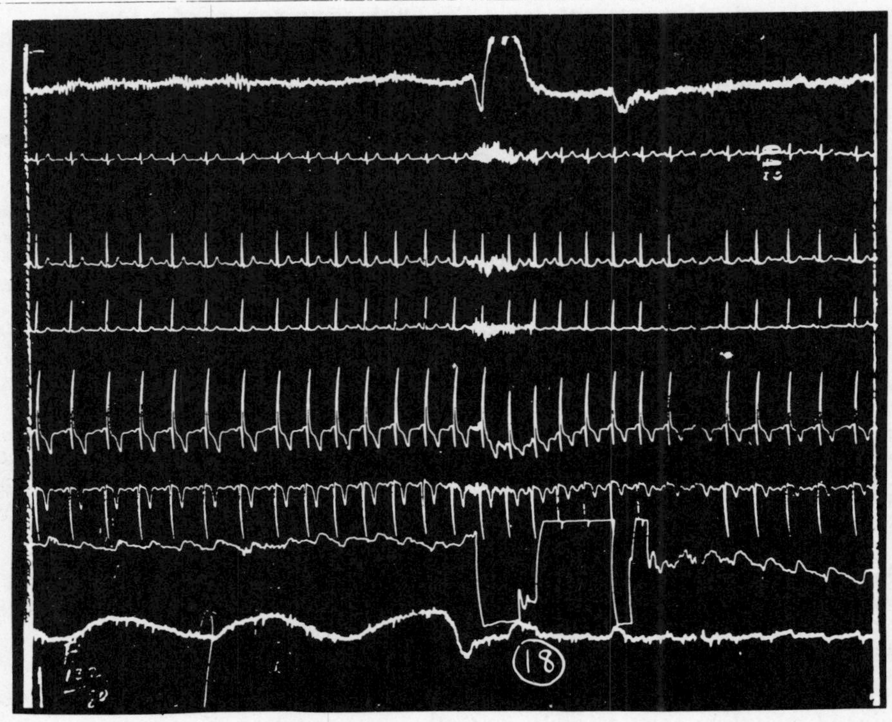

Figure 5.

Heart showing manoeuver, subject N.R.U. Lying on back with a small round pillow placed underneath the lower portion of the chest, the subject holds the breath in inspiration and does chin lock (Jalandhar Bandh), forcibly contracts the abdominal muscles and raises the abdominal pressure. Line 1: EEG, 10 per second and fast activity and artifacts the moment the breath is held. Line 2: abdominal leads for EMG, showing EKG and initial EMG the moment the breath is held. Lines 3, 4 and 5 and 6: EKG leads, last two being precordial leads. Marked right axis deviation is not seen. Heart rate 75 to 63 per minute to the middle of the record before he holds breath. Heart rate begins to slow down from 63 per minute to about 50 per minute toward the end of the picture when the breath is held. Line 7: Plethysmogram (finger blood volume). Slight increase in volume when the breath is held. Line 8: Respiration. Note that about the middle of the record the breath is held as indicated by the flatness of the respiration line. There are some artifacts in Lines 1 and 7.

points, also the finger volume. Heart sound and pulse at wrists as checked by two internist physiologists, were diminished or definitely stopped for a few seconds but not the EKG. The fluroscopic and x-ray evidence during the experiments demonstrated the expected change in the position of the heart - "tubular" heart during holding of breath following inspiration. This is the wellknown Val Salva manoeuver of cardiologists. The spiritual halo that usually surrounds the so-called "heart-stopping"

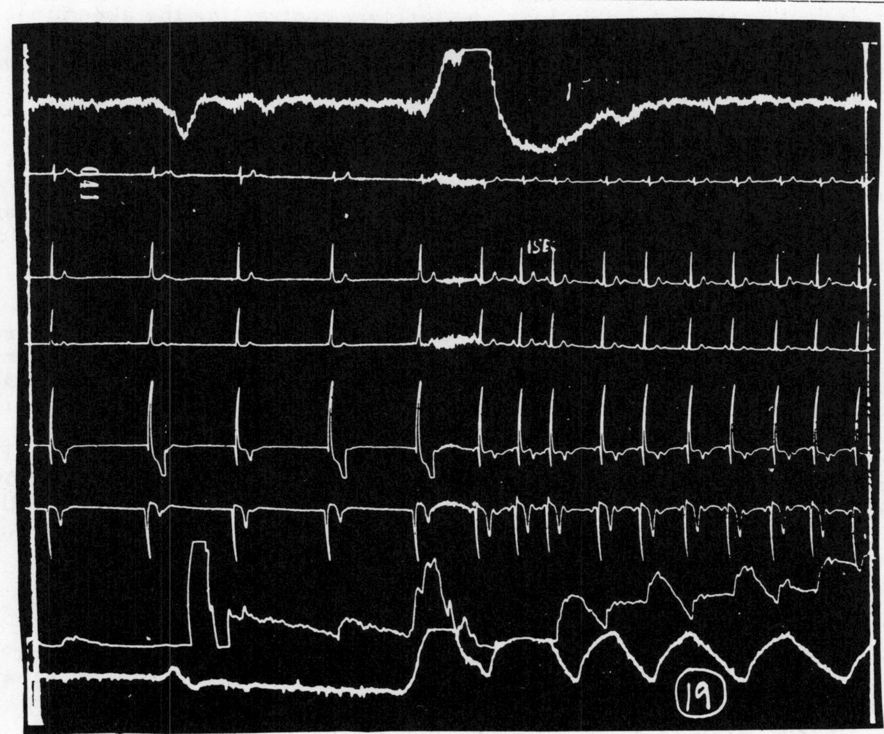

Figure 6.

(cont'd) from Fig. 5. Line 1 & 2: No EEG change, only artifacts. No abdominal muscle tension. Lines 3,4,5,6: slowest heart rate in this figure is 24 per minute, after which it very gradually creeps up toward the initial rate (not shown here) as in early part of Fig. 5. Line 7: Plethysmogram: artifacts during heart slowing and breath holding, also decrease in finger volume. When respiration begins the finger volume line is higher than seen in the control period (initial section of line 7, Fig 5). Line 8: Respiration. First part is flat when the breath is held. When respiration is resumed there is higher excursion of the line compared to initial section of the respiration line in Fig. 5. Blood pressure before heart slowing experiment 132/80, immediately after, 175/70. No subjective symptoms. Sheet numbers were running in reverse order.

experiments in the popular mind in India stems from ideological associations and subjects assertions but the manoeuver itself is simply a physiological one perfected by practice. We have not yet come across any instance of the direct control of the heart without muscular or breathing manoeuver.

In several sessions another man, with some years' practice behind him, was able to slow the beat of the heart to 24 - 25 per minute for over 10 seconds by holding the breath (Kumbhak), lowering the diaphragm, doing the chin lock and forcibly contracting the abdominal muscles, thus raising the intra-abdominal pressure. There was at the end an increase in blood pressure of 43 points systolic and increase of pulse volume but without any strong evidence of right or left axis deviation in EKG as seen in the previous case. (Figs. 5 and 6.) EEG was essentially unaffected. In this vagal effect the role of aortic baroreceptors should be considered. There may be other mechanisms. A control experiment on this subject by way of only manual carotid sinus massage bilaterally for about 30 seconds did not produce heart slowing or B.P. changes. This rules out the possibility of hyperactive carotid sinus reflex which is found in some people. Holding breath or doing chin-lock independently was not particularly effective.

Kapalvati pranayam for five minutes (shallow inspiration and expiration using mostly abdominal muscles, 120 per minute with short apnoeic period between each round of respiration) did not cause abnormal EEG. Nor did deep slow inspiration and expiration of equal duration 3 to 4 per minute lasting for 41 minutes. The latter would not be the case in deep breathing at the rate of 30 - 35 inspirations and expirations per minute. Such breathing for only five or six minutes is known in all EEG laboratories to be sufficient to cause abnormal EEG with clinical symptoms. Bio-chemical analysis of expired air under these two conditions or other pranayam breathing exercises has not been done. Ten to thirty minutes of shirsana (topsy-turvy - standing on head posture) by four subjects caused 16 to 30 point increase in systolic blood pressure and slight speeding of the heart, flushing and sweating of the face. Voluntary sweating of forehead brought about by mental concentration on the forehead with the eyes closed and without apparent exertion or muscular contraction but with increase in blood pressure (20-30 points), fluctuation of breathing rate and increase (not decrease) in the electrical resistance of the palm was seen. Voluntary rolling movement of abdominal muscles (nauli) and relative isolation of abdominal recti muscles and oblique muscles were accomplished with some electromyographic observations. A person could roll his tongue back up into the posterior wall of the naso-pharynx (Khecharimudra). In that position a catheter was inserted through his nose which touched his tongue. To determine the exact location of the latter by x-ray a quantity of iodized oil was injected through the catheter to show the tip of the tongue. Reported significance of Khecharimudra will not be given here.

Sympathetic or parasympathic dominance in reference to different variables mentioned before was studied on two groups of Hath Yoga practitioners in two cities. One group showed more activity of the sympathetic nervous system than the other.

SUMMARY AND CONCLUSIONS

A pilot electrophysiological investigation was carried out in India on 45 Indian Yogic subjects in 98 sessions using a portable 8-channel transistor, battery operated, A.C.-D.C. polygraph of high sensitivity with accessories. Central nervous system and autonomic nervous system changes were recorded as manifested in electroencephalogram, electrocardiogram, electromyogram, skin resistance, skin temperature, blood pressure, breathing rate, finger blood volume during meditation in "immobile" posture for a long time, during four types of pránayam breathing exercises, pulse and heart-sound stopping, heart-slowing, asan (postures), sweat-controlling, relaxation, etc.

This investigation seeks to indicate the general direction in which a particular type of scientific reasearch on Yoga should be carried out rather than present results on the basis of exhaustive studies. Many aspects of Yoga were not touched upon. However, some new findings may be named, for instance, an extreme slowing of respiration rate 4 to 6 per minute or shallow respiration for 17 - 27 minutes during a few meditative sessions, with more than 70% increase of palmar electrical resistance but without change in basic waking EEG and EKG pattern; heart slowing (24 per minute) through particular manoeuvers, etc. (For other results see the body of the article). Mechanisms responsible for these and other results are not altogether clear. Pending further research, it can be said that physiologically Yogic meditation represents deep relaxation of the autonomic nervous system without drowsiness or sleep and a type of cerebral activity without highly accelerated electrophysiological manifestation but probably with more or less insensibility to some outside stimuli for a short or long time.

We are convinced that this type of research should be the business of a team of workers of different specialties and carried on with greater vigor and more organizational emphasis and possibly closer international cooperation than has hitherto been the case. Apart from introducing an objective approach and realistic appraisal in the study of an endogenous discipline of India and apart from exerting a possible influence on people's reactions to some facets of the value system of Indian culture, this kind of research, in spite of its many attendant difficulties specially in securing competent cooperative subjects, needs to be continued as it may throw light on higherto unknown or little emphasized physiological and psychological mechanisms and may help to reemphasize some of the essentials of physical and mental well-being.

ACKNOWLEDGEMENTS

Grateful acknowledgments are made to the following organizations for support or cooperation: The Rockefeller Foundation, Rackham Foundation of the University of Michigan, and George Washington University (through a contract AF 18 (600)-1180 with the U.S. Air Force) and the Indian Council of Medical Research of the Government of India. We are grateful to Dr. B.K. Anand, chairman of the Department of Physiology, All-India Institute of Medical Sciences, New Delhi, India, for his cooperation and interpretation of cardiac findings, to Dr. Gudekar of the same Institute and Irwin Hospital for x-ray studies, to Mr. Gulzar Singh, M.Sc. for technical help, and to other colleagues for their courtesy and consideration. We are also grateful to Dr. N.N. Das of the Physiology Department of the University of Calcutta, to Pandit Prokash Dev of Madir Lane, Delhi, Swami Kuvalayananda, Director of Kaivalya-dham, Lonavla and his colleagues, Principal of Vivekananda College, Madras, and Swami Shantanandji, Swami Purushottamanandji, and other Swamis and Yogis who were gracious enough to act as subjects and also the informants for their cooperation and help. We express our appreciation to Dr. C.G. Pandit, Director of Indian Council of Medical Research and Dr. B.B. Dikshit, Director af All-India Institue of Medical Sciences.

REFERENCES

1. Abramson, H.A. Ed. Conference on Problems of Consciousness; Josiah Macy Jr. Foundation, 1951.
2. Adrian, E.D., Bremer, F. and Jasper, H.H. Consult. Editors. Brain Mechanisms and Consciousness XV + 556. Charles C. Thomas, Springfield, Ill., 1954.
3. Aranya, Hariharananda. Patanjali Yoga Darshan. Transl. and exposition. 4th Edition, 800 p. Calcutta University, Calcutta.
4. Avalon, A. (Sir John Woodroffe). The Serpent Power, Kundalini Shakti (Shat-Chakra-Nirupana and Paduka-Panchaka, Sanskrit Tantrik Texts, Translation, Introduction and Commentary; Sanskrit Text of Shata Chakra Bibriti), Second Revised Edition, X + 320 + 184 + 154. Pub.Ganesh & Co., Madras, India, 1924, First Edition, Luzac & Co., London, 1922.
5. Bagchi, B.K. The Progress of Electroencephalography. Transactions of the Bose Research Institute, 1955, 20: 55-65. Bose Research Institute, Calcutta.
6. Bagchi, B.K. Consciousness, Its Aberrations, and The Electrical Rhythm of the Brain. Main Currents in Modern Thought, March, 1956. Foundation for Integrated Education, Inc., 246 East 46th St., New York 17, N.Y. U.S.A.
7. Bagchi, B.K. A Reevaluation of the philosophy of Bhagavad-Gita in the setting of Modern India and the world. Mich. Acad. Sc., Arts and Letters, 1948. 34: 327-343. (Pub. 1950).
8. Bagchi, B.K. The adaptation and variability of response of the human brain rhythm. J. of Psychology, 1936. 3: 463:485.
9. Bagchi, B.K. Mental hygiene and the Hindu doctrine of relaxation. Mental Hygiene, New York, 1936. 20: 424-440.
10. Bagchi, B.K. Preoperative EEG Localization of Brain Tumors. Chap. XVIII (pp 331-351) In Symposium on Electrochemistry in Biology and Medicine. Pub. for Electrochemical Soc., John Wiley and Sons, Inc. New York, 1955.
11. Bagchi, B.K. EEG Localization of Intracranial Tumors. Chap. III (pp 18-53) In Correlative Neurosurgery by Kahn, E.A., Schneider, R.C., Crosby, E.C. Charles C.Thomas, Springfield, Ill., 1955.

12. Behanan, K.T. Yoga, A Scientific Evaluation. XXIII + 270. Macmillan Co., New York, 1937.
13. Brosse, Thérése. Altruism and Creativity as Biological Factors of Human Evolution. pp 118-145 In Explorations In Altruistic Love and Behavior. Ed. by Sorokin, Pitirim A., pp 349, Beacon Press, Boston, 1950.
14. Das, N.N. and Gastaut, H. Variations de l'activité électrique du cerveau, du coeur et des muscles squélettiques au cours de la médiation et de l'extasé yoguique. In Conditionnement et Reactivité en Electroencéphalographie, pp 211-219. Fifth Colloque at Marseille, 1955. Int. Fed. Soc. EEG and Clin. Neurophysiol. Journal Supplement 6.
15. Macdonnel, A.A. India's Past. A Survey of her Literatures, Religions and Antiquities. Oxford University Press, London, 1927.
16. Moruzzi, G. and Magoun, H.W. Brain stem reticular formation and l activation of the EEG. EEG Clin. Neurophysiol, 1949. 1: 455-473.
17. Penfield, W. and Jasper, H.H. Epilepsy and the Functional Anatomy of the Human Brain. XV + 865. Little, Brown & Co., Boston, 1954.
18. Radhakrishnan, S. Indian Philosophy. Library of Philosophy under the General Editorship of J.J. Muirhead, Vol. I (1923), 684 p.; Vol. II (1927) 807 p., Macmillan and Co., London.
19. Swami S. Kundalini Yoga. Forest University, Rishikesh, P.O., U.P. India.
20. Swami, K. and Karambelkar, P.V. Hydrogen ion concentration and titrable acidity of urinary excretion during Prayanamic Yogic exercises. Yoga Mimamsa, 1956, 6: 9-18, Lonavla, Bombay, India.
21. Swami, K. Ed: Yoga-Mimansa. Vol. I to VI, Embodying three sections in each volume, (1) Various biochemical, roentgenological and pressure studies, during Yogic practices, (2) Translations and philosophical elaborations, (3) Popular accounts. Lonavla, P.O., Bombay, India.
22. Swami, S. Bhagavad-Gita, The. Sanskrit Text, English Translation and Commentary, Fourth Edition, xiii + 418. Almora, India: Advaita Ashram, Mayavati, 1926.
23. Wenger, M.A., Jones, F.N., and Jones, M.H. Physiological Psychology, pp 472, Henry Holt, New York, 1956.
24. Wenger, M.A., Engel, B.T. and Clemens, T.L. Studies of autonomic response patterns: rationale and methods. Behavioral Science, 1957, 2: 216-221.
25. Woods, J.H. The Yoga System of Patanjali, Translation of Sutras and Commentaries. XLI + 381. Vol. 17 of Harvard Oriental Series. The Harvard University Press, Cambridge, Mass. 1927.

63
Some Aspects of Electroencephalographic Studies in Yogis

B. K. Anand, G. S. Chhina, and Baldev Singh

INTRODUCTION

Yogis practising *Raj Yoga* claim that during *samadhi* (meditation) they are oblivious to 'external' and 'internal' environmental stimuli although their higher nervous activity remains in a state of 'ecstasy' *(mahanand)*. Physiological and experimental studies have demonstrated that the basis of the conscious state of the brain is the activation of the reticular activating system through peripheral afferents (Magoun 1958), without which

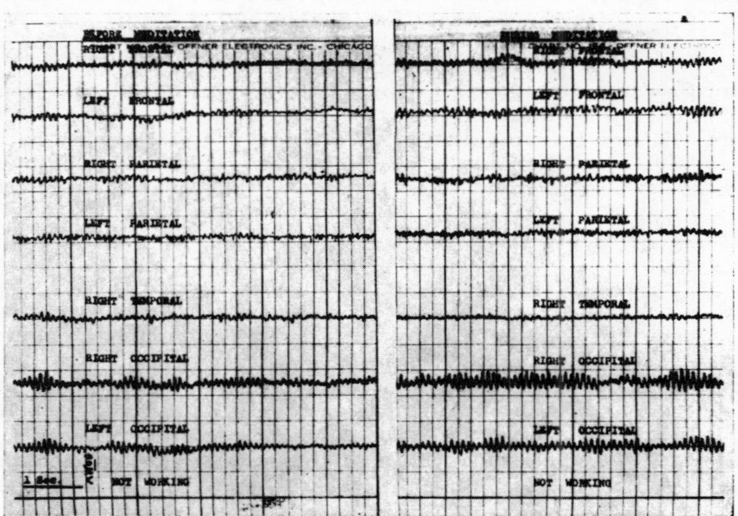

Fig. 1
Monopolar EEG scalp recordings of Shri Ramanand Yogi before meditation and during meditation. Reference electrodes on both ear lobes were joined together. During meditation there is a well marked increased amplitude modulation of alpha activity, especially in the occipital leads, where the amplitude increases from average to a maximum of 50–100 μV. The frequency was between $11\frac{1}{2}$ c/sec and 12 c/sec both before and during meditation.

higher nervous activity passes into the 'sleep' state. Studies were, therefore, undertaken to investigate electroencephalographically the activity of the brain during *samadhi*. Some reports have already appeared on th subject (Das and Gastaut 1955; Bagchi and Wenger 1957) but the physiological mechanism of the state of *samadh* still needs further elucidation.

EXPERIMENTAL SUBJECTS

Scalp EEG was used to study the brain activity o Yogis during *samadhi*. Two types of yoga practitioners who volunteered for this study, were investigated.

1. *Yogis practising meditation*

Four Yogis practising *samadhi* had their EEG re cordings taken before as well as during meditation. Tw of them were exposed to 'external' stimuli which wer photic (strong light), auditory (loud banging noise) thermal (touching with hot glass tube) and vibratio (tuning fork). The effect of these on the EEG activity wa studied both before as well as during meditation. On

---

[1] Aided by a grant from the Indian Council of Medical Research.

[2] Present address: *Tirath Ram Shah Charitable Hospital, Delhi (India)*.

Yogi during *samadhi* also practised "pin-pointing of consciousness" (which means concentrating attention on different points of the vault of the skull).

2. *Yogis with raised pain threshold*

Two Yogis, who had developed increased pain threshold to cold water, were also investigated. They were able to keep their hand in water at 4°C for 45–55 min respectively without experiencing any discomfort. Their EEG records were obtained before and during the period when they kept their hand in cold water.

RESULTS

The changes observed in the EEG records of the Yogis investigated are presented below.

1. *Yogis practising meditation*

All these Yogis showed prominent alpha activity in their normal resting records. During the stage of *samadhi* all of them had persistent alpha activity with well marked increased amplitude modulation (Fig. 1). In addition one of them showed occasional hump activity in the parietal zones (Fig. 2), although he professed to remain awake throughout this period. In both the Yogis who were exposed to 'external' stimulation, all the stimuli blocked the alpha rhythm and changed it to a low voltage fast activity when the Yogis were not meditating. This blocking reaction did not show any adaptation to repetition of the same stimuli. On the other hand, none of these stimuli produced any blockage of alpha rhythm, when the Yogis were in meditation *(samadhi)* (Fig. 3). In the Yogi who concentrated attention on different points of the vertex ("pin-pointing of consciousness"), these attempts were accompanied by well marked 'blinking' responses recorded from the frontal electrodes (Fig. 4).

2. *Yogis with raised pain threshold*

The EEG records of these two Yogis also showed persistent alpha activity, both before and during the period in which the hand was immersed in cold water (Fig. 5). No change in the electrical activity in the parietal leads was observed even when sensory afferents from the hand were expected to be projecting there.

3. EEG studies were also made on a number of beginners in the various yoga practices. It was observed that those who had a well marked alpha activity in their normal resting records showed greater aptitude and zeal for maintaining the practice of yoga.

DISCUSSION

In these experimental observations on Yogis, a persistent and well modulated alpha activity, more marked during *samadhi*, was observed. This alpha activity could not be blocked by various sensory stimuli when the Yogi was in *samadhi*, although it could easily be blocked when he was not meditating. Even during deep meditation when the Yogis appeared quite relaxed and in a sleep-like condition, the EEG record showed only prominent alpha activity. Only in one Yogi was occasional hump activity observed, the alpha rhythm persisting in the rest of the period. Bagchi and Wenger (1957) also found the normal alpha pattern, sometimes with good amplitude modulation, in the EEG records of some Yogis during meditation. Okuma *et al.* (1958) observed in *Zen* practitioners that the alpha waves of these subjects increased remarkably with the progress of their performance, even if their eyes were kept open. Das and Gastaut's (1955) observations on high amplitude fast waves in the EEG records of Yogis during meditation have not been confirmed.

The significance of prominent alpha activity observed

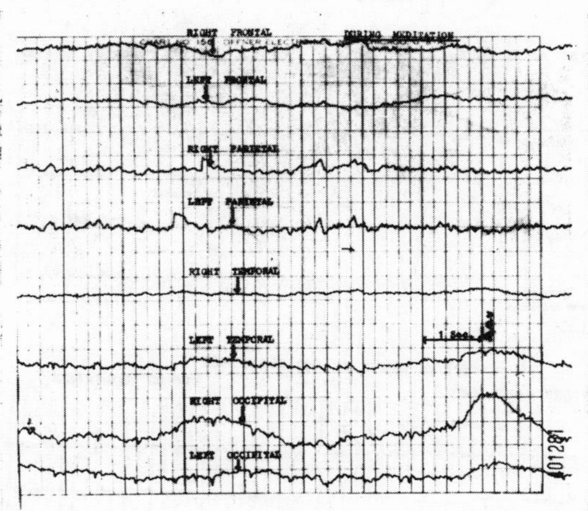

Fig. 2

Monopolar EEG scalp recording of one Yogi during meditation showing occasional hump activity in the parietal zones. For most of the remaining period of meditation he showed prominent alpha activity of 11–12 c/sec.

during meditation is not yet clear. Yogis generally claim that during *samadhi* they are oblivious to their external and internal environments, and in the present experiments their alpha rhythm could not be blocked by external beta rhythms are probably under the control of a subcortical pacemaker, or system of pacemakers. It is, therefore, suggested that the brain activity of Yogis during the stage of *samadhi* has for its basis a type of consciousness

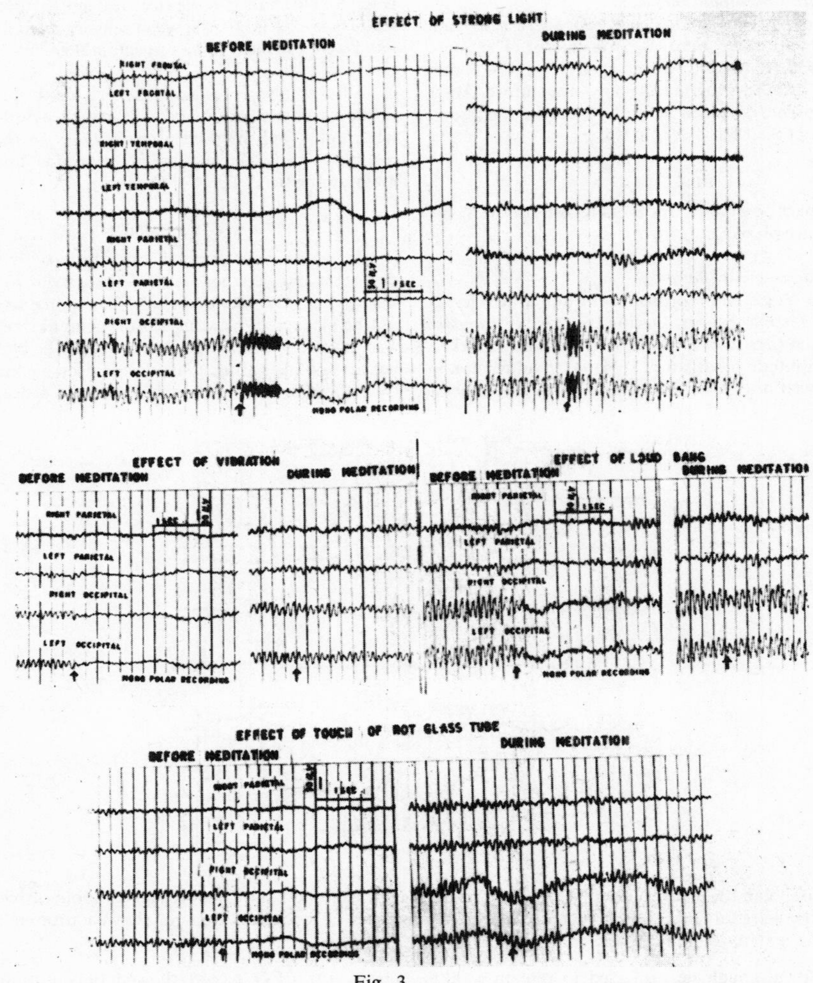

Fig. 3
Monopolar EEG scalp recordings of Shri Ramanand before meditation and during meditation. Photic, vibration, auditory, and thermal stimuli block the alpha rhythm when he is not in meditation. No blockage of the alpha rhythm occurs when he is in meditation.

stimuli. They also did not pass into delta activity. Although the reticular activating system (RAS) is activated by peripheral sensory inputs, it has also been reported that this system is capable probably of some spontaneous or autonomous discharge (Bremer 1954; Dell 1952). The alpha activity may be due to this discharge. Garoutte *et al.* (1958) have also observed that bilateral alpha and which probably depends upon mutual influences between the cephalic RAS and the cortex, and which does not depend upon the activation of RAS through external and internal afferents.

The two Yogis, who could keep their hand without discomfort in ice cold water for long periods, showed alpha activity during this practice. This suggests again ıat these individuals were able to block the afferents from :tivating the RAS and thus remain in alpha activity.

Lastly the observation that those beginners who had well marked alpha activity in their resting records, ıowed greater aptitude for maintaining the practice of ɔga, is quite important and may have some bearing on ıe problem under discussion.

SUMMARY

Four Yogis who practised *samadhi* were investigated ectroencephalographically. It was observed that their :sting records showed persistent alpha activity with creased amplitude modulation during *samadhi*. The pha activity could not be blocked by various sensory imuli during meditation.

Two Yogis, who could keep their hand immersed in e cold water for 45–55 min, also showed persistent alpha :tivity both before and during this practice.

The possible mechanism of these observations has een discussed.

The authors gratefully acknowledge the willing co-peration of the following Yogis for undergoing the arious investigations reported in this paper: Shri .amanand Yogi, Pandit Parkash Dev, Shri Rishi Ram, hri N.S. Paintal, Shri Babu Ram Gupta and Shri Narain 'hander Singh Ray.

REFERENCES

ıAGCHI, B. K. and WENGER, M. A. Electrophysiological correlates of some Yogi exercises. *Electroenceph. clin. Neurophysiol.*, **1957**, *Suppl. 7*: 132–149.

ıREMER, F. The neurophysiological problem of sleep. In *Symposium on brain mechanisms and consciousness.* Blackwell Scientific Publications, Oxford, **1954** pp., 137–162.

DAS, N. N. et GASTAUT, H. Variations de l'activité

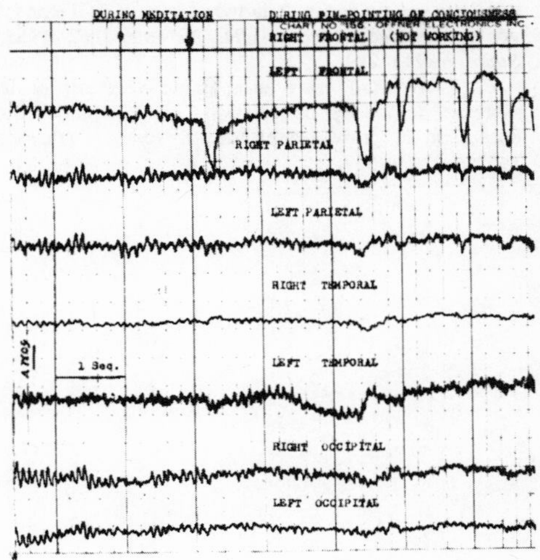

Fig. 4
Monopolar EEG scalp recording of Shri N.S. Paintal during meditation when he started concentrating his attention on different points of the vertex ("pin pointing of consciousness"). It shows a well marked blinking response from the frontal leads during this act.

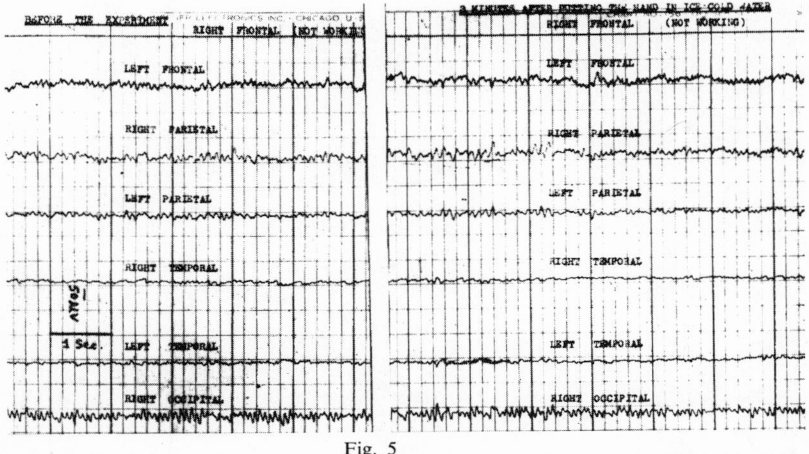

Fig. 5
Monopolar EEG scalp recordings of Shri Babu Ram Gupta who kept his right hand immersed in cold water at 4°C for 55 min. It shows records taken before the start of the experiment and 3 min after placing the hand in water. The alpha activity of 11 c/sec persisted throughout the whole period the hand was kept in water. No discomfort or pain was felt by him throughout this period.

électrique du cerveau, du coeur et des muscles squélettiques au cours de la méditation et de 'l'extase' yoguique. *Electroenceph. clin. Neurophysiol.*, **1955,** *Suppl. 6*: 211–219.

Dell, P. Corrélation entre le système végétatif et le système de la vie de relation, mésencéphale, diencéphale et cortex cérébrale. *J. Physiol. (Paris)*, **1952,** *44*: 471–557.

Garoutte, B. and Aird, R. B. Studies on the cortic pacemakers. Synchrony and asynchrony of bilateral recorded alpha and beta activity. *Electroenceph. cli Neurophysiol.*, **1958,** *10*: 259–268.

Magoun, H. W. *The waking brain.* C. C. Thoma Springfield, Ill., **1958,** 138 pp.

Okuma, T., Kogu, E., Ikeda, K. and Sugiyama, H. Th EEG of Yoga and Zen practitioners. *Electroencep clin. Neurophysiol.*, **1957,** *Suppl. 9*: 51

64

An Electroencephalographic Study of the Zen Meditation (Zazen)

Akira Kasamatsu and Tomio Hirai

INTRODUCTION

It is our common knowledge that EEG undergoes strikes changes in the transition from wakefulness to sleep, and has become one of the reliable ways to assess the state of wakefulness or sleep. In clinical practice, EEG often becomes a good neurophysiological method to find out the disturbance of consciousness. And many studies, both clinical and experimental, on the consciousness have been published during the past 30 years. In recent years electroencephalographic and neurophysiological studies on the consciousness are focused on an understanding of the relationship between the brain mechanisms and consciousness in general (Gastaut, in Adrian et al., Eds., 1954). These studies give rise to an attempt to relate the various electrographic findings with the psychological states and their behavioral correlates (Lindsley, 1952).

The authors have carried out the study on EEG changes during anoxia, epileptic seizures, the exogeneous disorders of the brain and other allied states from neurophysiological and psychological points of view (Kasamatsu & Shimazono, 1957). In the course of our study, it was revealed that a series of EEG changes was observed in the state of attentive awareness during Zen-sitting (Zazen. And what Zazen is like will be explained later, Hirai, 1960; Kasamatsu & Shimazono, 1957). These findings deserve further investigation because of understanding EEG pattern to corresponding psychological state and of interpreting the neurophysiological basis of consciousness. The subject of the present paper is to describe the results of our experiments in detail and to discuss some of the electrographic characteristics in which the mental state in Zen-sitting will be reflected.

Zazen—Zen meditation means the sitting meditation which is a kind of religious exercise in Zen-Buddhism. In Japan there are two Zen sects named Soto and Rinzai. Both sects regard Zazen as the most important training method of their disciples to enlighten their minds. Zen sitting is performed in two basic meditation forms: A full cross-legged sitting and a half cross-legged sitting. During the Zen sitting, the disciple's eyes must be open and look downward about one meter ahead, and his hands generally join. In a quiet room the disciple sits on a round cushion and practices the meditation for about 30 minutes. Sometimes the intensive Zen training is performed 8 to 10 times a day for about one week. This is called *Sesshin* in Zen Buddhism. The disciples do not engage in daily activities but live the religious life following a strict schedule (Hirai, 1960).

By practising Zen meditation it is said that man can become emancipated from the dualistic bondage of subjectivity and objectivity, of mind and body and of birth and death. And he can be free from lust and self-consciousness, and be awakened to his pure, serene and true-self.

This mental state (Satori or enlightenment) will often be misunderstood as trance or hypnosis. It is said that Satori is not an abnormal mental state but one's everyday mind in the Zen sense. Dr. Erich Fromm describes it, "If we would try to express enlightenment in psychological terms, I would say that it is a state in which the person is completely tuned to the reality outside and inside of him, a state in which he is fully aware of it and fully grasps it. *He* is aware of it—that is, not his brain, nor any other part of his organism, but *he*, the whole man. He is aware of *it*; not as of an object over there which he grasps with his thought, but *it*, the flower, the dog, the man, in its or his full reality. He who awakes is open and responsive to the world, and he can be open and responsive because he has given up holding on to himself as a thing, and thus has become empty and ready to receive. To be enlightened means 'the full awakening of the total personality to reality'" (Fromm, Suzuki & de Martino, 1960).

If one asks what this state of mind is concerned with in psychotherapy, it may be said that Zen meditation is the method through which we can communicate with the unconscious. In this context, however, the unconscious does not mean Freud's "unconsciousness." Rather the "unconscious" in Zen is closely related to the unconscious which is stated by Jung, C. G. (Suzuki, 1960) or Fromm, E. (Fromm, Suzuki & de Martino, 1960). In regard to this problem Dr. Daisetsu Suzuki states the meaning of it as "the Cosmic unconscious" (Suzuki, 1959).

At any rate the Zen meditation influences not only the mind but also the body as a whole organism. The authors want to investigate Zen meditation as a subject of psychophysiology, especially that of electroencephálography.

SUBJECTS AND METHODS

EEG was recorded continuously through all stages—before, during and after Zazen with opened eyes. All our EEG data were obtained in the eyes opened state.

As recording electrodes the silver-coated disc electrodes with thin (100μ to 200μ) copper wire in vinyl tube were used, and they were applied with collodion on the scalp of the frontal, central, parietal, and occipital regions in the middle line of the head. These electrodes did not disturb Zen meditation and long-lasting recordings were obtained.

Along with EEG, the pulse rate, respiration and GSR were polygraphically recorded on a San-ei 12 channel ink-writing electroencephalograph. The same experiments were performed for one week during Zen meditation's intensive training (Sessin). These results were useful to confirm the EEG changes in the whole course of Zen meditation.

In order to investigate the functional state of the brain, the responses to sensory stimuli with several modalities were examined. And the blocking time of alpha pattern to repeated click stimulations was measured.

Our experiments were made at the usual Zen-training hall with cooperations of the Zen priests and their disciples as our subjects. But the stimulation experiments were performed in the air-conditioned, sound-free shield room of our laboratory.

The cooperative 48 subjects were selected among the priests and disciples in both Soto and Rinzai sects. Their ages ranged from 24 to 72 years old. According to their experience in Zen training, these subjects could be classified into the following 3 groups:

Group I: 1 to 5 years experience (20 disciples).
Group II: 5 to 20 years experience (12 disciples).
Group III: over 20 years experience (16 priests).

As control subjects, we selected 18 research fellows (23 to 33 years of age) and 4 elderly men (54 to 60 years of age). They have had no experience in Zen meditation, and their EEGs were recorded under the same condition with opened eyes, as the Zen disciples.

RESULTS

1. EEG Changes of Zen-Masters during Zen Meditation

First we shall consider the typical EEG changes of a certain master in detail.* He is a priest with over 20 years' experience in Zen meditation. Before Zen meditation the activating pattern is predominant because his eyes are open (low voltage, fast activity). After Zen meditation has started, the well organized alpha waves of 40–50μV, 11–12/sec. appear within 50 seconds in all the regions and continue for several minutes in spite of opened eyes. After 8 minutes and 20 seconds, the amplitude of alpha waves reaches to 60–70 μV predominantly in the frontal and the central regions. Initially, these alpha waves alternate with the short runs of the activating pattern, but a fairly stable period of the persistent alpha waves ensues during the progress of Zen meditation. After 27 minutes and 10 seconds, rhythmical waves of 7–8/sec. appear for 1 or 2 seconds. And 20 seconds later rhythmical theta trains (6–7/sec., 70–100uV) begin to appear. However, it does not always occur. After the end of Zen meditation alpha waves are seen continuously and 2 minutes later alpha waves still persist. It seems to be the after-effect of Zen meditation.

In the control subjects, EEG changes are not observed; a control subject shows the long-lasting activating pattern of opened eyes. Another 2 control subjects of 58 and 60 years of age also show beta dominant type of EEG with short runs of small alpha waves. But neither increase of alpha amplitude nor decrease of alpha frequency are observed on their EEG of opened eyes. It is not likely that the aging process of the control subjects influences EEG changes during Zen meditation.

Sometimes the theta waves appear as Zen meditation progresses. These changes are clearly shown in the EEG of another Zen master of 60 years of age, large alpha waves with 70–100μV. in amplitude and 8–9/sec. in frequency appear after 24 minutes of his Zen meditation. And 30 seconds later, the rhythmical theta train (6–7/sec. 60–70μV.) begins to appear. The appearance

*Note: Figures illustrating the various EEG changes discussed in this paper have been deleted due to space limitations.—*Editor.*

of the theta train becomes distinct through the stable periods of large and slow alpha waves. From the above-mentioned results it is pointed out that a series of EEG changes in the course of Zen meditation are observed; the activating pattern (of opened eyes) before Zen meditation—appearance of alpha waves at initial stage—increase of alpha amplitude—decrease of alpha frequency—appearance of rhythmical theta train in later stage of Zen meditation.

This series of changes cannot always be observed in all Zen subjects. Some subjects only show the appearance of alpha waves through all the meditation period and others show the typical series of electrographic changes. But from our findings, the changes of electroencephalogram during Zen meditation are classified in the following four stages:

Stage I: a slight change which is characterized by the appearance of alpha waves in spite of opened eyes.
Stage II: the increase in amplitude of persistent alpha waves.
Stage III: the decrease of alpha frequency.
Stage IV: the appearance of the rhythmical theta train, which is the final change of EEG during Zen meditation but does not always occur.

2. EEG Changes and the Degree of Zen Training

In accordance with the subjects' years spent in Zen training, 23 Zen disciples were classified into 3 groups—within 5 years, 5–20 years and over 20 years. Also, the evaluation of the mental states in the Zen sense of these disciples were used and their Zen master divided them into 3 groups; low (L), middle (M) and high (H). This evaluation was made independently without regard to their EEG changes.

Then the authors attempted to relate these degrees of Zen training in the 4 stages (I, II, III and IV) of the EEG changes. Tables 1 and 2 show the results. In the vertical line the stages of EEG changes are plotted and the horizontal line shows the subjects' training years (Table 1). It is clear that the more years spent in Zen training, the more EEG changes are seen. The correlation of EEG changes with mental state, which was evaluated by the Zen master, is shown in Table 2. It shows the close relationship between the evaluation by the master and the degree of EEG changes. From these findings, it will be concluded that the degrees of EEG changes during Zen meditation are parallel with the disciples' proficiency in Zen training. The 4 stages of EEG changes reflect physiologically the mental state during Zen meditation. This will be discussed later.

TABLE 1

Relationship between Degree of EEG Changes and Years Spent in Zazen Training

	Stage			
Rhythmical theta waves	IV	0	0	3
Decrease in alpha frequency	III	3	2	3
Increased alpha amplitude	II	2	1	0
Alpha with eyes open	I	8	1	0
		0–5	6–20	21–40
		Years of Experience in Zazen		

TABLE 2

Relationship between Degree of EEG Changes and Ratings of Disciples' Proficiency at Meditation by the Zen Master.

	Stage			
Rhythmical theta waves	IV	0	0	3
Decrease in alpha frequency	III	0	7	1
Increased alpha amplitude	II	2	1	0
Alpha with eyes open	I	5	4	0
		Low	Medium	High
		Proficiency Rating		

3. EEG Changes during Zen Meditation and Hypnosis

The mental state in hypnosis is generally considered as "trance." Some may think that the mental state of Zen meditation will be a trance-like state. The authors compared the EEG changes in hypnotic trance with those of Zen meditation. In a hypnotized subject, a university student of 20 years, the catalepsy is manifested. Few alpha waves are seen, but the activating pattern is more prominent than EEG in Zen meditation. The series of EEG changes during Zen meditation is not observed in the course of hypnotic trance.

4. EEG Changes during Zen Meditation and Sleep

In the course of EEG recording during Zen training, the disciples sometimes fall into a drowsy state, which becomes clear on the EEG pattern. At this time the click stimulus is given, then the drowsy pattern turns into the alpha pattern and alpha arousal reaction is observed. This electrographical change is usually accompanied with a floating consciousness from sleep to wakefulness, according to the disciple's introspection. This state is different from the mental state in Zen meditation. The sleepiness, which is called "Konchin," is suppressed in Zen training.

As mentioned before the rhythmical theta train appears in some Zen priests during their Zen meditation. The theta train is also seen in the sleep pattern. But the electrographical difference exists between the theta waves in sleep and the rhythmical theta train in Zen meditation. This difference is evident in the following example: A rhythmical theta train is clearly seen on EEG during a certain Zen master's meditation. At this time, the click stimulus is given. The rhythmical theta train is blocked by the stimulation and reappears spontaneously after several seconds later. The alpha arousal reaction, which is often seen by the stimulation in a drowsy state, is not observed. Therefore the rhythmical theta train in this instance has an "alpha activity" (Brazier, 1960) which is similar to the waking alpha rhythm.

Just before falling into sleep or in the hypnagogic stage, large alpha waves are often seen. These waves are similar to that of Zen meditation. But the large alpha waves seen in stage II or III of Zen meditation persist much longer than the pre-sleep pattern. This difference will be discussed later in detail.

5. EEG Response to Click Stimulation in Zen Meditation

In the preceding sections a series of EEG changes in Zen meditation was described. In this section, the authors will deal with the results of alpha blocking to the repetitive click stimuli with regular intervals.

The click stimulation was performed at the stage of long persisting alpha waves of a certain Zen master. To the first stimulus the alpha blocking occurs for 2 seconds. With the regular intervals of 15 seconds, the click stimuli are repeated 20 times, the alpha blocking is always observed for 2–3 seconds. On the other hand the same stimulation is performed in alpha pattern of control subjects with closed eyes. The more the stimulation repeats, the less the alpha blocking time. The same experiments were performed on 3 Zen masters and 4 control subjects and the alpha blocking time to each stimulation was measured. The measurement of the alpha blocking time leads to the following results: In control subjects, the alpha blocking time decreases rapidly, but in Zen masters, the alpha blocking time is fairly constant, though some random changes are seen.

From the above mentioned results it is concluded that there is almost no adaptation of alpha blocking during Zen meditation.

DISCUSSION

It has become apparent in our study that the electrographic changes of Zen meditation are the appearance of alpha waves without regard to opened eyes. These alpha waves increase amplitude and decrease frequency with the progress of Zen meditation. And sometimes the rhythmical theta train appears in the later stage of the meditation. These findings are also parallel with the degree of Zen disciples' mental states in the Zen sense and their years spent in Zen training.

It is common that the mental activities of concentration, mental calculation and efforts to perceive the objects elevate the level of consciousness accompanied by the activating pattern (Adrian & Matthews, 1934; Bartley, 1940; Callaway, 1962; Glass, 1964; Slater, 1960; Walter & Yeager, 1956; Walter, 1950). Thus the activating pattern indicates the augmentation of level of consciousness (Lindsley, 1952; Mundy-Castle, 1958).

Zen meditation is the concentrated regulation of inner mind. It will be, therefore, expected that Zen meditation will bring about the activating pattern. Nevertheless the lowering of cortical potentials is confirmed by our electrographic findings. This is rather paradoxical but is of prime interest to consider a relationship between the physiology of the brain and the level of consciousness.

According to the instructions of Zen meditation, the regulation of inner mind is strongly emphasized. And by obeying the rules of Zen training the well-achieved meditation has been completed. In the well-achieved meditation, it will be said that "concentration" without tension (that is the true concentration) is going on in the utmost inner world of psychic life.

From the electroencephalographic point of view, our results are coincident with EEG changes of lowered consciousness or vigilance. A. C. Mundy-Castle (1953, 1958) states that the persistent appearance of alpha waves indicates the brain function at the time of lowered vigilance. And many empirical observations of alpha waves point out its being not of action but of hypofunction of the brain (Dynes, 1947; Lindsley, 1952).

In attempting to relate the various stages of the EEG pattern to corresponding psychological states and the behavioral correlates, D. B. Lindsley (1952) states that during more or less continuous relaxed state of wakefulness, amplitude modulated alpha waves are characteristic. The same concept is stated by H. Jasper (1941; Jasper & Shagass, 1941) in his sleep-wakefulness continuum; he introduces the concept of the cortical excitatory states reflected on these EEG patterns. According to Jasper's suggestion, it is said that the amplitude modulated alpha waves reflect the lowered level of the cortical excitatory states.

On the other hand, many agents which affect nerve cell metabolism, are known to alter the EEG (Brazier & Finesinger, 1945; Grunthal & Bonkalo, 1940; Jung, 1953). A. Kasamatsu and Y. Shimazono (1957) report that the large and slow alpha waves are observed in the earlier stage of N_2 gas inhalation, just before the loss of consciousness. In this state, subjects experience relaxed consciousness or slightly elevated mood-changes. In the acute alcoholic intoxication the same effects are seen in both the EEG and consciousness (Kasamatsu & Shimazono, 1957). T. Hirai (1960) points out a decrease of the respiratory rate accompanied with the slowing of EEG pattern during Zen meditation. Y. Sugi et al. (1964) report the results of measurements of the respiratory rate, tidal volume and O_2 consumption during Zen meditation. They find a decrease of energy metabolism which is lower than basic metabolism. According to Sugi's suggestion, it may be due to the decrease of energy metabolism in the brain. It is possible that the decrease of energy metabolism also alters the electrographic pattern in Zen meditation.

From the foregoing surveys and discussions, EEG changes during Zen meditation seem to indicate that the cortical excitatory level will be gradually lowered even by the "concentration" of inner mind.

From a psychological point of view, both Zen meditation and hypnotic trance bring about the changes of consciousness. But the trance is called "Sanran" (confusion) and is strictly suppressed in Zen meditation. Therefore, some discussions will be needed about the difference of EEG changes between Zen meditation and hypnotic trance. The authors discovered that there are no definite changes of subjects' electroencephalograms in hypnotic trance. There are many reports concerning the EEG changes in hypnotic trance but many of these indicate that the pattern does not differ from the waking EEG (Kleitman, 1963; Loomis, Harvey, & Habart, 1936). There are no similarities of the pattern in hypnotic trance to EEG changes during Zen meditation.

Some scholars state that the sleep-like changes of EEG, more or less slight, are observed in hypnotic trance. Goldie et al. (Kleitman, 1963) elicited a paradoxical electrographic effect in a drowsy hypnotized subject— the alpha pattern appearing in the opened-eyes condition D. J. Frank (1950) reports that slow activity seen in deep sleep is recorded during the hypnosis. D. B. Lindsley (1952) points out that in accordance with general relaxation, which occurs during hypnotic episodes, there is sometimes an increase in alpha pattern if slight drowsiness supervenes.

K. Fujisawa (1960) studies EEG in the hypnotic state caused by sleep suggestion ("hypnotic sleep") (Baker & Burgwin, 1949; Dynes, 1947) and reveals the low voltage theta pattern which is similar to the drowsy pattern. He also points out that the drowsy-like pattern continues for a fairly long time as far as the rapport with the hypnotized subject is not lost and true sleep ensues if the rapport is lost. It is noticed that the slow rhythm in hypnotic sleep is more similar to the drowsy pattern than the rhythmical theta activity seen in Zen meditation.

Zen meditation is not a sleep from the disciples' introspections. But during Zen training, sometimes the slight drowsiness also can be supervened in a hypnosis. In the transitory state from wakefulness to drowsiness, large alpha waves are prominently seen and are prone to decrease in frequency just before the subject shows slight drowsiness (Oswald, 1959). It may be said, therefore, that the large and slow alpha pattern during Zen meditation is a foregoing pattern of the drowsiness. Perhaps there is the lower threshold in a sweep or span of consciousness during Zen meditation. But in actual sleep, alpha waves recede, spindles burst and slow waves appear, and consciousness is lost. Such a series of electrographic changes does not occur in Zen meditation and consciousness is not lost, since in Zen meditation there is no lack of awareness of things going on externally and internally.

Even in the later stage of Zen meditation, in which the rhythmical theta train is seen on EEG, the sensibility is not lost and in fact the rhythmical theta train shows the marked blocking to sensory stimulation. From these findings, we will show the difference between both schematically in Fig. 1. A series of EEG changes is common at a limit of alpha activities, but the sleep pattern diverges from this series in a downward curve, and turns to deep sleep.

From the foregoing discussions, it can be said that during Zen meditation the level of the cerebral excitatory state is gradually lowered in a way that is different from sleep.

Next we will discuss the alpha blocking in Zen meditation. As described before,

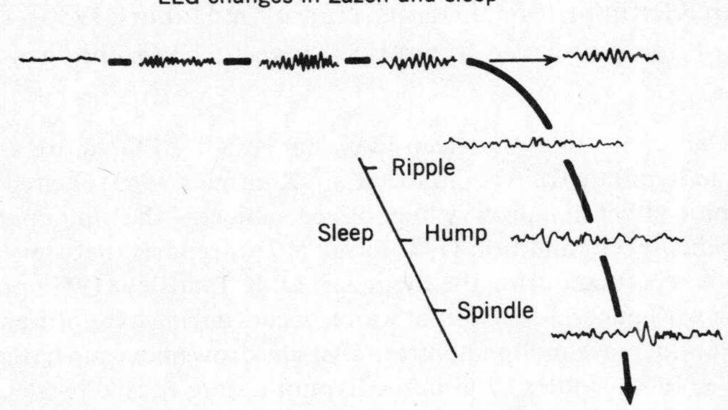

Figure 1 A schematical representation of the difference between EEG changes in Zen meditation and in sleep. A series of EEG changes is common at a limit of alpha activities but the sleep pattern diverges from above horizontal train in a downward curve and turns to deep sleep.

each click stimulus brings about a fairly constant alpha blocking continued for several seconds, even though the stimulation is repeated 20 times at regular intervals. But in control groups with closed eyes, the alpha blocking time is longer at 1st and 2nd stimulations but rapidly decreases and almost diminishes after 3rd or 4th stimulation. So the habituation of alpha blocking is clearly recognized in this ordinary awakening state.

In Zen meditation the alpha blocking is less susceptible to habituation to sensory stimuli than in ordinary waking state. This fact is noteworthy to clarify the arousal state of consciousness in Zen meditation. During Zen meditation "concentration" without tension is maintained in the inner mind of the disciple while keeping the correct sitting form. These mental and physical conditions naturally lead to production of the certain constant experimental circumstances: A kind of concentration subserves the maintenance of a certain level of the consciousness on the one hand, and the sitting meditation form supports the centripetal sensory inflows at a certain level on the other. In these circumstances, it would be supposed that the alpha blocking becomes less susceptible to habituation.

These findings are also supported by the introspection of our subjects in this experiment. The Zen masters reported to us that they had more clearly perceived each stimulus than in their ordinary waking state. In this state of mind one cannot be affected by either external or internal stimulus, nevertheless he is able to respond to it. He perceives the object, responds to it, and yet is never disturbed by it. Each stimulus is accepted as stimulus itself and treated as such. One Zen master described such a state of mind as that of noticing every person one sees on the street but of not looking back with emotional curiosity.

However, it seems to be impossible to consider separately the continuous appearance of alpha waves and the alpha blocking, which is less susceptible to habituation in Zen meditation. The alpha blocking depends upon the cortical excitatory state, conversely the cortical excitatory state closely related to the centrifugal sensory impulses brought about alpha blocking.

Using the arousal reaction of EEG as a criterion response, S. Sharpless and H. Jasper (1956) studied a great variety of characteristics of the habituation. They classified two types of arousal reaction: a longer lasting one more susceptible to habituation and a shorter lasting one less susceptible. This finding, in agreement with other studies (Jung, 1953; Lindsley, 1952) suggests that a longer lasting arousal reaction corresponds to the tonic activation on the cerebral cortex and a shorter lasting one to the phasic activation.

According to Jasper's suggestion, the alpha blocking, which is less susceptible to habituation, seems to be decided by the equilibrium of the tonic and phasic activation on the cerebral cortex. The authors want to postulate that there is an optimal activation mediated by the equilibrium of cortical excitatory state in a broad sweep or span of the waking consciousness. And perhaps its underlying neurophysiological basis may be an interaction between the cerebral cortex and the reticular activation systems of the diencephalic and mesencephalic portions in the brain stem (Moruzzi & Magoun, 1949).

The optimal preparedness for incoming stimuli, which conversely maintains the optimal level of the cortical excitatory state, is well reflected in both the alpha blocking, which is less susceptible to habituation, and in the series of EEG changes, which directs to the slowing of the pattern. These EEG findings persist for a fairly long time and are constant though slight fluctuation is observed. Also these persistant alpha waves can be often seen even after the end of Zen meditation. These findings suggest that in the awakening consciousness, there will be the special state of consciousness, in which the cortical excitatory level becomes lower than in ordinary wakefulness, but is not lowered as in sleep, and yet outer or inner stimulus is precisely perceived with steady responsiveness.

Zen meditation is purely a subjective experience completed by a concentration which holds the inner mind calm, pure and serene. And yet Zen meditation produces a special psychological state based on the changes in the electroencephalogram. Therefore, Zen meditation influences not only the psychic life but also the physiology of the brain. The authors call this state of mind the "relaxed awakening with steady responsiveness."

SUMMARY

Zen meditation (Zazen) is a spiritual exercise held in the Zen sect of Buddhism. Apart from its religious significance, the training of Zen meditation produces changes not only in the mind but also in the body—these influences are of interest to scientific studies, from the standpoint of psychology and physiology.

In the present study the EEG changes accompanied with Zen meditation have been revealed and described in detail. The authors discussed further these electrographic changes in relation to the consciousness with its underlying neurophysiological background, compared with that of the hypnotic trance and sleep.

In our study, 48 priests and disciples of Zen sects of Buddhism were selected as the subjects and their EEGs were continuously recorded before, during, and after Zen meditation. The following results were obtained:

1. The appearance of alpha waves were observed, without regard to opened eyes, within 50 seconds after the beginning of Zen meditation.

These alpha waves continued to appear, and their amplitudes increased. And as Zen meditation progressed, the decrease of the alpha frequency was gradually manifested at the later stage. Further the rhythmical theta train with the amplitude modulated alpha-background was observed in some records of the priests. These EEG changes could be classified into 4 stages: the appearance of alpha waves (Stage I), an increase of alpha amplitude (Stage II), a decrease of alpha frequency (Stage III), and the appearance of rhythmical theta train (Stage IV).

2. These 4 stages of EEG changes were parallel with the disciples' mental states, which were evaluated by a Zen master, and disciples' years spent in Zen training.

3. These electrographic changes were also compared with that of the hypnotic trance and sleep. From the electroencephalographic point of view, the changes of Stages I, II and III could not be clearly differentiated from those seen in hypnagogic state or the hypnotic sleep, though the changes during Zen meditation were more persistent and did not turn into a deeper sleep pattern. The rhythmical theta train is suppressed by click stimulation and turns into a desynchronized pattern, whereas the drowsy pattern turns into alpha waves (the alpha arousal reaction).

4. The alpha blocking to the repeated click stimuli with regular intervals was also examined in Zen meditation with opened eyes and the ordinary conditions of control subjects with closed eyes. The former showed a fairly constant blocking time (3–5 seconds) to every stimuli repeated 20 times and the habituation was not recognized. On the other hand, in control subjects the habituation of alpha waves occurred very quickly. This alpha blocking, which is less susceptible to habituation, is of importance in considering the neurophysiological basis of the mental state during Zen meditation.

These electroencephalographic findings lead to the following conclusions: In Zen meditation, the slowing of the EEG pattern is confirmed on the one hand, and the dehabituation of the alpha blocking on the other. These indicate the specific change of consciousness. The authors further discussed the state of mind during Zen meditation from the psychophysiological point of view.

65

Individual Differences in Response to a Zen Meditation Exercise

Edward W. Maupin

In a previous paper (Maupin, 1962), the literature in English was surveyed in an attempt to extract a coherent picture of Zen Buddhism, its training procedures, and the psychological experiences which result. It was noted that the training seems analogous in some respects to Western insight-oriented psychotherapy. Moreover, the responses to training described in the literature seemed partly interpretable in terms of concepts of ego function which are subject to quantitative measurement. The present study was designed to obtain basic information about the meditation process by correlating differences in the meditation experience with psychological tests of relevant ego functions. Although the nature of the meditation process rather than its therapeutic merits was the subject of investigation, the information developed may be basic to later therapeutic studies.

Zen Buddhism

Zen, a sect of Mahayana Buddhism, originated in China and has played an important role in Japanese culture since its introduction there in the thirteenth century. A variety of training techniques are employed by the Zen master to guide the student to a turning point, *satori,* which appears to be a major shift in the mode of experiencing oneself and the world.

The specific training procedures used are extremely variable, but certain generalizations are possible. The beginning student spends a portion of the day sitting motionless and engaging in concentration. The object of concentration varies and may be changed as the student progresses. The aim of the practice is to suspend the ordinary flow of thoughts without falling into a stupor. Usually there is an initial phase in which concentration is extremely difficult. With greater success, relaxation and a kind of pleasant "self-immerson" follow. At this point internal distractions, often of an anxiety-arousing kind, come to the fore. Herrigel (1953) indicates that the only way to render this disturbance inoperative is to "look at it equably and at last grow weary of looking." Eventually another phase begins, in which concentration is accompanied by a sense of calm stillness, a sense of energy and vitality, and a feeling of invulnerability. This state of mind is traditionally described with the analogy of a mirror, which reflects many things yet is itself unchanged by them. Much of this is reminiscent of the stance taken by the psychoanalytic patient, who experiences his associations as derivatives of his own mental proc-

[1] A PhD dissertation study. The author wishes to thank Edward S. Bordin and the members of his research committee for their help with the study and Eileen Flamm for editorial assistance.

[2] Now at the Neuropsychiatric Institute, Center for Health Sciences, University of California, Los Angeles.

esses regardless of the reality of the objects represented. An observing attitude can be maintained until anxiety or other affects become too intense. One difference has been pointed out by Sato (1958), who notes that the Zen student, unlike the analysand, does not dissect what he is experiencing with ideational operations. The ideas, if they appear, need not be grasped or verbalized. This writer has suggested that Zen meditation procedures, just because they are nonlogical and nonverbal, may enable one to deal with certain problems which are inaccessible to verbal communication—for example, problems hinging on pre-Oedipal experience antedating the development of language and logical structure.

To our knowledge, long-term alterations in personality factors as a result of Zen training have not been systematically observed, although meditation in conjunction with psychoanalysis has facilitated therapeutic progress in some instances (Fromm, 1959; Kondo, 1958). Conversely, individual variations in response to meditation training have not been studied with respect to their dependence on personality factors or ego functions. It is the latter problem with which the present study is concerned. Personality factors are considered the independent variables of the study; response to meditation is the dependent variable.

Methods and Procedures

The strategy of the present study is this: if a specific function, such as attention, is important in the meditation process, then individuals who are relatively efficient in this function will tend to meditate more successfully and hence respond more extensively during the first few weeks of practice. The importance of various functions in determining response to meditation can then be judged by correlating response to meditation with appropriate measures of these functions. This section describes the subjects who participated in the experiment, the meditation exercise and the procedure for scaling response, and finally the procedures for assessing relevant personality functions.

Sample

Male subjects were recruited by means of an advertisement in the campus newspaper offering instruction in a Zen meditation exercise to persons interested enough to spend additional time taking tests. Thirty-nine men responded and were seen for an initial interview. A series of psychological tests was given in later appointments, after which the meditation sessions were begun. Eleven of the 39 subjects dropped out either before the meditation sessions began or before completing enough sessions to permit a good estimate of response. The remaining 28 subjects, all of whom completed at least nine meditation sessions, were included in the sample.

As might be expected, many of the subjects appeared to have a therapy-seeking motivation. They evidently felt enough subjective discomfort to look, sometimes explicitly, for new ways of dealing with personal problems. (Overt psychosis was not encountered, and no subject presented any other symptoms which seemed grounds for exclusion.) In other ways, too, the sample seemed atypical of the general college male population. The rater, who had scored Rorschach protocols for several other studies, commented on the unusually high incidence of primary-process responses to the test in this sample. It is not clear that any of the biases operating in the sample would have a consistent effect on the relationships between experimental measures. The generally high level of motivation to carry out the meditation instructions, while unusual, should facilitate better understanding of the meditation process. With regard to prior knowledge of Zen, some subjects had read the popular works on the subject, but the extent of their reading had no relationship with their response to the exercises.

Meditation Exercise

The meditation exercise used in this study required concentration on the breathing as described by Kondo (1958). The 45-minute practice sessions, followed by short interviews, were conducted each weekday for a 2-week period. Before the first sessions, the subject was shown the correct sitting position—back straight, not resting on the chair back, feet flat on the floor—and given the following instructions:

> While you are sitting let your breathing become relaxed and natural. Let it set its own pace and depth if you can. Then focus your attention on your breathing: the movements of your belly, not your nose or throat. Do not allow extraneous thoughts or stimuli to pull your attention away from the breathing. This may be hard to do at first, but keep directing your attention back to it. Turn everything else aside if it comes up.
>
> You may also find yourself becoming anxious or uncomfortable. This is because sitting still and concentrating like this restricts the ordinary ways of avoiding discomfort. If you have to feel uncomfortable, feel uncomfortable. If you feel pleasant, accept that with the same indifference. Gradually, perhaps not today or even after several days, it will grow easier for you to concentrate. I will come and tell you when the time is up. Do you have any questions?

These instructions were repeated before the second session. At subsequent sessions the subject entered the experimental room, adjusted the chair height with firm cushions, and began concentrating without comment. The practice room was dimly lit and sparsely furnished. Two chairs were arranged so that subjects could not see one another but faced an unadorned wall. The experimenter remained in an adjacent room while one or a pair of subjects practiced the breathing exercise. If someone asked what was supposed to happen as a result of the practice he was told:

> I cannot tell you that, mainly because I don't know that much about you. Whatever happens will come from within you. It is better not to be distracted by preconceived ideas about it. The important thing is to accept whatever happens. Don't move, and keep your attention on your breathing.

After each session the interview was initiated with the question, "How did it go?" Further questions by the experimenter were limited to asking for clarification and for an estimate of the degree of concentration. These interviews were recorded verbatim and serve as the basis for scaling response.

Procedure for Scaling Response to Meditation

Frequently, especially after the early meditation sessions, subjects simply reported difficulty in concentrating, and their remarks indicated that little else had happened. Most subjects, however, proceeded beyond this point, reporting a variety of experiences in later sessions more or less comparable to those described in Zen literature. For purposes of scaling, five responses patterns were distinguished which appeared to recur in the protocols. The description of these patterns, as submitted to the raters, is as follows:

Type A. The identifying characteristic of these sessions is dizziness, some type of "befogged" consciousness, which occurs when a subject begins to concentrate. Typically, subjects report feeling dizzy or having sensations "like going under anesthetic" or "like being hypnotized." The experience is somewhat unpleasant, and it is common to find the subject retreating from the task of concentrating into increased thinking.

Type B. These are sessions in which the subject reports feeling quite calm and relaxed. His concentration need not have been very sustained.

Type C. In this pattern pleasant body sensations occur. They are often reminiscent of hypnotic phenomena and sometimes involve feelings which would be strange in the ordinary states of waking consciousness. They may be rather erotic-sounding as they are described. Sensations like "vibrations" or "waves" are often reported, or the subject feels his body is "suspended" or "light." Typically, concentration is somewhat more sustained than in Type A or B sessions.

TABLE 1

RESPONSE PATTERNS REPORTED BY GROUPS OF HIGH, MODERATE, AND LOW RESPONSIVENESS TO MEDITATION

		Pattern[a]				
Group	n	A	B	C	D	E
High	6	0	2	3	4	6
Moderate	10	2	3	8	5	0
Low	12	6	8	0	0	0
Total	28	8	13	11	9	6

[a] A, refers to "dizziness," B refers to "relaxation", C refers to "pleasant body sensations," D refers to "vivid breathing," E refers to "detachment."

Type D. The distinguishing feature of this pattern is the vivid way in which the breathing is experienced. Often the belly movements seem larger, or the subject feels "filled with air." Concentration seems almost effortless in such sessions.

Type E. This is the experience described by Herrigel (1953, pp. 56 ff.). It appears to be a very lucid state of consciousness which is deeply satisfying. There is a "nonstriving" attitude, and the subject is able to take a calmly detached view of any thoughts and feelings which happen to emerge. Concentration seems to be easy and fairly complete. A frequent accompaniment of this pattern is extensive loss of body feelings.

The procedure for scaling subjects began with identifying the patterns they reported after each session. Rating was done independently by two advanced graduate students in clinical psychology who were unaware of the hypotheses of the study. The two raters agreed that 214 of the 330 session reports failed to show any of the five patterns. The remaining 116 sessions, then, were critical for identifying response. Here the judges agreed in only 54 cases, or 48%. Compared with the expectancy of chance agreement, however, this obtained agreement is significant ($\chi^2 = 137.4$, $df = 3$, $p < .001$).

The next step in scoring response to meditation consisted of grouping subjects according to their "highest" response in any of the first 10 sessions. Three groups were defined: a "high-response" group, consisting of subjects who reported a Type E response (detachment) at least once; a "moderate-response" group, including subjects who failed to report a Type E response, but reported a Type C or a Type D response (pleasant body sensations or vivid breathing) at least once; and a "low-response" group composed of subjects who reported nothing more than relaxation or dizziness (Types A or B). With this grosser grouping of subjects, the raters agreed on the placement of all but 3 (89%) of the 28. Table 1 presents the number of subjects assigned to each group as well as the number of subjects in each group who indicated a given pattern of response at least once. The subjects within each group are to be considered as having tied scores on a 3-point ordinal scale ranging from high to low response to meditation. Some support for the ordinal relationship is adduced from the fact that high-response subjects experienced not only the detached pattern but also reported the body alterations of the moderate-response group fairly frequently. This high group also had fewer uneventful (unscorable) sessions than the others.

Personality Factor Appraisal

Five aspects of personality functioning were considered in the study: receptive attention, concentration, breadth of attention deployment, tolerance for unrealistic experience, and capacity for adaptive regression.

Attention. The forward portion of the Digit Span subtest of the Wechsler-Bellevue Scale was used to measure the availability of receptive attention (Rapaport, Gill, & Schafer, 1945). The administration procedure followed that of Blackburn and Benton (1957). High scores were interpreted as reflecting a free availability of attention.

Concentration. A continuous additions test (Wells & Ruesch, 1945) was used in an attempt to measure concentration, conceived as a more effortful excluding function. Mean and standard deviations of sums completed per 15-second interval during 60 intervals served as the two experimental scores. A high mean or low variability were interpreted as signs of sustained concentration, hence a positive correlation between response to meditation and mean sums and a negative correlation between response to meditation and deviations of sums were predicted.

Scanning Control. Studies reported by Schlesinger (1954) and Gardner, Holzman, Klein, Linton, and Spence (1959) suggest individual differences in characteristic breadth of attention deployment. The criterion measure for this cognitive control principle, the size estimation task used by these authors, was included in the present study. Large-average errors of estimation were interpreted as indicating narrow attentional focusing ("concentration"), whereas low-average errors indicated habitual broad scanning.

Tolerance for Unrealistic Experience. Three measures were used to assess the reaction of subjects in the face of unrealistic experience. The first is the Rorschach measure used by Klein, Gardner, and Schlesinger (1962). Tolerance on this dimension refers to acceptance of experiences which do not agree with what is known to be "true," while intolerance implies resistance to such precepts or cognitive experiences. In the Rorschach ratings, major weight was placed on the subject's attitude toward his responses rather than on formal scores. The tolerant person was characterized as one who naturally and comfortably accepted the blots as an opportunity for projection. The intolerant person, in contrast, was one who maintained a critical and cautious attitude of reality testing toward his responses.

These criteria were applied to the Rorschach records by two judges trained in the use of the system for a previous study (Lichtenstein, 1961). Records which could not be assigned by the raters to either the tolerant or the intolerant categories were placed in a "mixed" category which was interpreted as a midpoint on the scale. The judges agreed in their ratings of 19 of the 28 records. In only two instances was a record rated tolerant by one and intolerant by the other. Computing reliability by means of the rank correlation coefficient, tau, gives .63 ($p < .001$).

The results of Klein et al. (1962) and Lichtenstein (1961) suggested a link between the tolerance dimension and the rate of alteration of reversible figures. The four figures used appeared to vary markedly in the extent to which tolerance influenced response. In the present study the Necker cube, Rubin figure-vase, and windmill and staircase figures were presented, the subject looking into an illuminated box at the figures 28 inches away. Number of reversals for each figure during the 30-second presentation plus the total number of reversals for all four figures were the scores used.

The third estimate of the subject's tolerance for unrealistic experience was the amount of autokinetic movement he reported. The median estimated movement for 20 trials was the experimental score. It was expected that subjects who are less comfortable in such an unstructured situation would give small estimates of movement.

Capacity for Adaptive Regression. During the course of meditation, the student must pass through a phase of internal distractions in order to reach the stillness of meditation. We infer that these distractions will occasionally express the drive-dominated content or organization of primary-process cognition. Furthermore, feelings such as omnipotence erupt, and the thinking function must be abandoned. Clearly, then, lowering of the level of psychic functioning is involved, and the concept of "regression in the service of the ego" may be applicable. Two measures of the ability to regress were therefore included in the present study: Holt's (Holt & Havel, 1961) scoring for primary-process thinking on the Rorschach, and visual imagery during free association.

The Rorschach scoring system is designed to identify not only the emergence of primary-process form and content but also the adequacy with which these manifestations are controlled. Goldberger (1958) related the system to capacity for regression in the service of the ego in a study of reactions to perceptual isolation. He distinguished three general modes of handling primary-process material: (*a*) a mode which admits such derivatives into awareness in a modulated, controlled fashion without disruption; (*b*) a mode which wards off the emergence of such material into awareness through defensive operations; and (*c*) a mode in which the person, although finding primary-process derivatives threatening, is unable to prevent their appearance in consciousness due to inadequate defenses. The Rorschach scores, rated by a clinical psychologist trained for another study, were used to identify these modes, considered to lie along a single dimension. Subjects were identified as showing a "large amount of controlled primary process" if they ranked high on

TABLE 2

Rorschach Tolerance Ratings of Response Groups

Group	n	Number of subjects rated			tau
		Tolerant	Mixed	Intolerant	
High response	6	4	2	0	
Moderate response	10	4	2	4	
Low response	12	2	5	5	.37*

* $p < .01$.

amount of primary-process responses and also high on effectiveness of defense. They were considered to show a "small amount of controlled primary process" if they ranked low on "amount" and high on "effectiveness." They showed a "small amount of poorly controlled primary process" if they ranked low on "amount" and low on "effectiveness," and they showed a "large amount of poorly controlled primary process" if they ranked high on "amount" and low on "effectiveness."

Whether or not the subject tended to report spontaneous visual imagery during a short free-association task served as a second measure of capacity for adaptive regression. It was felt that the emergence of imagery represents a clear shift away from the primarily verbal thought of adults to a genetically more primitive mode of recognition. The free-association task was that used by Gardner et al. (1959). The subject is asked to report, for 3 minutes, everything that came to mind after hearing each of two stimulus words: DRY and HOUSE. Imagery (for example, "I see a picture of mud") was identified in 23 of the 56 work protocols, and each subject received 1 point for each of his 2 protocols in which imagery was identified, thus giving a 3-point scale. Average percentage of agreement between the three raters was 92.8.

Correlations between Measures. Since tolerance for unrealistic experience and capacity for adaptive regression seem to be overlapping concepts, correlations between measures of the two were calculated. The tau correlation between the Rorschach rating of tolerance and the Rorschach measure of capacity for regression was .18, but the actual relationship seemed to be curvilinear. The "amount of primary process" and "effectiveness of control" components of the latter dimension were each correlated with tolerance .28 ($p < .05$) when taken individually. Subjects responding with a small amount of poorly controlled primary process were thus likely to be identified as "tolerant." The visual imagery scores were correlated with tolerance ratings (tau = .34, $p < .01$) better than with the Rorschach dimension (tau = .04). There appears to be overlap between the three measures, yet not so much that they can be considered to be measuring identical phenomena.

Results

Inasmuch as linear relationships were predicted between the ordinal scale for response to meditation and the measures of the independent variables, Kendall's rank-correlation coefficient, tau (Siegel, 1956), was used to express these relationships. A two-tailed probability of .05 was the level of significance required to reject the null hypothesis in all comparisons.

Meditation and Attention Functions

It was expected that response to meditation would be correlated positively with digit span scores (measuring receptive attention) and average sums in continuous additions (measuring active concentration) and negatively with variability scores in continuous additions (measuring fluctuations in active concentration). None of these hypotheses was borne out. No relationship was found, either, between response to meditation and size estimation error (measuring scanning control or breadth of attentional scanning). A relationship may exist between receptive attention and response to meditation but of a more complex type than originally thought. Both high- and low-response subjects received higher Digit Span scores than the moderate-response group. Tested post hoc with Kruskal-Wallace one-way analysis of variance, this curvilinear relationship gives $H = 6.38$ ($p < .05$).

Meditation and Tolerance for Unrealistic Experience

A subject's comfort in the face of unrealistic experience was expected to influence his ability to meditate in the early phases of practice. This led to the prediction that response to meditation would be positively correlated with the Rorschach measure of the tolerance control principle, with the frequency of perceived reversals of reversible figures and with the amount of autokinetic movement reported. As shown in Table 2, the Rorschach measure of tolerance was positively correlated with response to meditation (tau = .37, $p < .01$). Autokinetic movement and figure reversals were not significantly related to response.

Meditation and Capacity for Adaptive Regression

The two measures used to assess capacity for adaptive regression were the Rorschach measure based on amount and degree of primary-process thinking and visual imagery during free association. As shown in Tables 3 and 4, both of these measures were positively correlated with response to meditation. Because of the complex nature of the Rorschach dimension, the relative contributions of its various components was explored in a multivariate chi-square analysis (Lancaster, 1949). It appears that the main contribution to the relationship with response to meditation comes from the "effectiveness of control" component ($\chi^2 = 7.82$, $df = 2$, $p < .02$), but the contributions from the "amount" component and the interaction between "amount" and "control," while not significant, are still noteworthy ($\chi^2 = 4.26$ and 3.84, respectively, with $df = 2$ in each case).

TABLE 3

Rorschach Manifestations of Primary Process in the Response Groups

Group	n	Number of subjects receiving score[a]				tau
		1	2	3	4	
High response	6	5	1	0	0	
Moderate response	10	1	3	4	2	
Low response	12	1	3	3	5	.49**

[a] The scores refer to the following points on the dimension: 1 = Many primary-process responses, well controlled; 2 = few primary-process responses, well controlled; 3 = few primary-process responses, poorly controlled; 4 = many primary-process responses, poorly controlled.
** $p < .001$.

TABLE 4

Visual Imagery Scores of the Response Groups

Group	n	Number of subjects with scores of:			tau
		2	1	0	
High response	6	5	0	1	
Moderate response	10	2	3	5	
Low response	12	1	5	6	.35*

* $p < .01$.

Discussion

Certain inferences about the nature of the meditation process may be drawn from the correlations between response to meditation and measures related to regression in the service of the ego. As suggested elsewhere (Maupin, 1962), *satori* seems to fit into the class of psychologically adaptive regressions described in the psychoanalytic literature. The meditation process may then be conceptualized as a sequence of states of regression, each of which develops functions on which succeeding states must depend. For example, the early response pattern includes a kind of relaxed drowsiness in which primary-process derivatives appear. The ability to deal with them in an accepting, undisruptive fashion enables the student to get through to the next stage, past alterations in body feelings, to "mirrorlike" detachment, in which more "regressed" elements may emerge. The "nonstriving" quality of this state, presumably a lesson of the previous phase, constitutes a safeguard against maladaptive reactions, even though further impulse derivatives are emerging. *Satori* is said to grow out of this stage, and we may guess that it is based on certain internal safeguards developed there.

While the correlations between response to meditation and the Rorschach measures of tolerance for unrealistic experience and capacity for controlled primary process underline the relevance of such experience in the meditation process, it seems unlikely that attention functions are as irrelevant to meditation as the obtained correlations would indicate. More probably, when issues related to capacity for adaptive regression are resolved, a subject can easily learn to use attention properly. One way to study this problem further would be to observe the response to meditation of two groups, both having high capacity for regression but differing in their attention skills.

The response scale used in the study by no means taps all aspects of the response observed. Subjectively felt benefits similar to those resulting from relaxation therapies were reported by several subjects. The amount of such benefit did not appear to be correlated with level of response. Subjects in the high- and moderate-response groups occasionally mentioned the emergence of very specific and vivid affects other than anxiety while they were practicing. These included hallucinoid feelings, muscle tension, sexual excitement, and intense sadness. Although subjects differed in the extent to which they tried to puzzle out the meanings of such experiences, none appeared to reach any deeper understanding of the material which emerged. On the other hand, several reported an increased awareness of previously elusive subjective events when they occurred. Perhaps the period of observation was too short to permit the development of deep insight, or perhaps meditation leads naturally to a different type of outcome—an acceptance of feelings without further inquiry into their significance.

REFERENCES

Blackburn, N. L., & Benton, A. L. Revised administration and scoring of the digit span test. *Journal of Consulting Psychology,* 1957, **21**, 139–143.

Fromm, E. Psychoanalysis and Zen Buddhism. *Psychologia,* 1959, **2**, 79–99.

Gardner, R. W., Holzman, P. S., Klein, G. S., Linton, Harriet P., & Spence, D. P. Cognitive control: A study of individual consistencies in cognitive behavior. *Psychological Issues,* 1959, **1**, 1–186.

Goldberger, L. Individual differences in the effects of perceptual isolation as related to Rorschach manifestations of the primary process. Unpublished doctoral dissertation, New York University, 1958.

Herrigel, E. *Zen in the art of archery*. New York: Pantheon, 1953.

Holt, R. R., & Havel, Jean. A method for assessing primary and secondary process in the Rorschach. In Maria Rickers-Ovsiankina (Ed.), *Rorschach psychology*. New York: Wiley, 1960. Pp. 263–315.

Klein, G. S., Gardner, R. W., & Schlesinger, H. J. Tolerance for unrealistic experience: A study of the generality of a cognitive control. *British Journal of Psychology,* 1962, **53**, 41–55.

Kondo, A. Zen in psychotherapy: The virtue of sitting. *Chicago Review,* 1958, **12**(2), 57–64.

Lancaster, H. O. The derivation and partition of χ^2 in certain discrete distributions. *Biometrika,* 1949, **36**, 117–129.

Lichtenstein, E. The relation of three cognitive controls to some selected perceptual and personality variables. Unpublished doctoral dissertation, University of Michigan, 1961.

Maupin, E. W. Zen Buddhism: A psychological review. *Journal of Consulting Psychology,* 1962, **26**, 362–378.

Rapaport, D., Gill, M., & Schafer, R. *Diagnostic psychological testing*. Vol. 1. Chicago: Year Book Publishers, 1945.

Sato, K. Psychotherapeutic implications of Zen. *Psychologia,* 1958, **1**, 213–218.

Schlesinger, H. J. Cognitive attitudes in relation to susceptibility to interference. *Journal of Personality,* 1954, **22**, 354–374.

Wells, F. L., & Ruesch, J. *Mental examiners' handbook*. New York: Psychological Corporation, 1945.

(Received March 18, 1964)

66

Autogenic Training: Method, Research and Application in Medicine

W. Luthe

Introduction

Autogenic Training is a psychophysiologic form of psychotherapy which the patient carries out himself by using passive concentration upon certain combinations of psychophysiologically adapted stimuli. In contrast to the other methods of psychotherapy, autogenic training approaches and involves mental and bodily functions simultaneously. Passive concentration on Autogenic Standard Formulas can be so tailored that a normalizing influence upon various bodily and mental functions will result. From a neurophysiologic point of view there is clinical and experimental evidence indicating that certain changes of cortico-diencephalic interrelations are the functional core around which autogenic training revolves (1).

About 40 years ago the founder of the method, J. H. Schultz, psychiatrist and neurologist in Berlin, wrote the first publications about clinical and experimental observations of what he called "autogenic organ exercises" (2, 3). In 1932, the first edition of *Autogenic Training* became available (4). Since then, 10 German editions have appeared (5), translations into Spanish (6), Norwegian (7), and French (8), as well as a recent American edition (1). During the last three decades autogenic training has become widely known in Europe and today it is regarded as a valuable standard therapy in various fields of medicine. It has also been integrated into the training programs of many universities (1, 9–12).

The steady increase of interest in autogenic training is reflected by the progressively increasing number of publications of a clinical and experimental nature (13). During each of the last three years more than 100 articles on the subject were published in medical journals and books. It is interesting, however, that only about one per cent of a total of about 1,000 publications were written by English speaking authors (1).

*For further information see *Autogenic Therapy,* Luthe, ed., Grune & Stratton, Inc.

BACKGROUND OF THE METHOD

The beginning of autogenic training stems from research on sleep and hypnosis carried out in the Berlin Institute of the renowned brain physiologist Oskar Vogt during the years 1890 to 1900. Vogt observed that intelligent patients who had undergone a series of hypnotic sessions under his guidance were able to put themselves for a self-determined period of time into a state which appeared to be very similar to a hypnotic state. His patients reported that these "autohypnotic" exercises had a remarkable recuperative effect (1, 4, 5).

At the time he observed that these short-term mental exercises, when practiced a few times during the day, reduced stressor effects like fatigue and tension. Other disturbing manifestations, as for example headaches, could be avoided and the impression was gained that one's over-all efficiency could be enhanced. On the basis of these observations Vogt considered such self-directed mental exercises to be of definite clinical value. He called them "prophylactic rest-autohypnoses" (*Prophylaktische Ruhe-Autohypnosen*).

Stimulated by Vogt's work (14) J. H. Schultz became interested in exploring the potentialities of autosuggestions. His aim was to find a psychotherapeutic approach which would reduce or eliminate the unfavorable implications of contemporary hypnotherapy, such as the passivity of the patient and his dependency on the therapist.

During subsequent years, while investigating the question of hallucinations in normal persons, Schultz collected data which appeared to link up with Vogt's prophylactic mental exercises (4). Many of Schultz's hypnotized subjects reported to have experienced, almost invariably, two types of sensations: a feeling of heaviness in the extremities often involving the whole body and frequently associated with a feeling of agreeable warmth. Schultz concluded that the psychophysiologic phenomena related to the experience of heaviness and warmth were essential factors in bringing about the changes from the normal to a hypnotic state.

The next question was whether a person could induce a psychophysiologic state similar to a hypnotic state by merely thinking of heaviness and warmth in the limbs. The systematic pursuit of this question was the actual beginning of autogenic training. Under certain technical circumstances and by the use of passive concentration on verbal formulas implying heaviness and warmth in the extremities, Schultz's subjects were able to induce such a state, which appeared to be similar to a hypnotic state.

The self-directed nature of the approach had a number of clinical advantages over the conventional techniques of hypnosis, among them, the active role and the responsibility of the patient in applying the treatment and the elimination of dependence on the hypnotist.

METHOD

From Schultz's clinical work a number of useful verbal formulas gradually evolved which, according to their more bodily or mental orientation, formed two basic series of mental exercises: the *Standard Exercises* and the *Meditative Exercises.*

The six *standard exercises* are physiologically oriented. The verbal content of the standard formulas is focused on the neuromuscular system (heaviness) and the vasomotor system (warmth); on the heart, the respiratory mechanism, warmth in the abdominal area, and cooling of the forehead.

The *meditative exercises* are composed of a series of seven exercises which focus primarily on certain mental functions and are reserved for trainees who master the standard exercises.

Later, as more clinical and experimental data became available, a number of complementary exercises specifically designed for normalization of certain patho-functional deviations evolved. These were called *special exercises.*

Psychophysiologically, autogenic training is based on three main principles: (a) reduction of exteroceptive and proprioceptive afferent stimulation; (b) mental repetition of psychophysiologically adapted verbal formulas; and (c) mental activity conceived as "passive concentration."

A reduction of afferent stimuli requires observation of the following points: the exercise should take place in a quiet room with moderate temperature and reduced illumination; restricting clothes should be loosened or removed; the body must be relaxed, and the eyes closed, before the mental exercises are begun. Three distinctive postures have been found adequate: (a) the horizontal posture; (b) the reclined arm-chair posture; and (c) the simple sitting posture. All three training postures require careful consideration of a number of points. When certain details are not observed, disagreeable side-effects or after-effects and ineffective performance of the exercises have been reported.

The first exercise of the *autogenic standard series* aims at muscular relaxation. The functional theme of the verbal formula is heaviness. Right-handed persons should start out with passive concentration on "My right arm is heavy." Left-handed persons should begin with focusing on the left arm.

During the very first exercises about 40 per cent of all trainees will experience a feeling of heaviness predominantly in the forearm. During subsequent periods of regular training, the whole arm becomes heavy and the feeling of heaviness will spread to other extremities. This spreading of a certain sensation (the heaviness, tingling, warmth) to other parts of the body is called the "generalization phenomenon." Along with the development of the generalization phenomenon, passive concentration on heaviness will be extended to the other arm or the homolateral leg. Usually the heaviness training continues until heaviness can be experienced more or less regularly in all extremities. This may be achieved within two to eight weeks. Clinical investigations of larger groups of trainees, however, indicate that about 10 per cent of the patients do not experience a sensation of heaviness. This fact is one of the reasons why patients should be told that the *Heaviness Formula* (and others) functions merely as a technical key to bring about many different functional changes in the brain and bodily system, and that a sensation of heaviness may or may not occur.

Furthermore, it has been found helpful to tell a patient that many changes of bodily functions occur (see section on experimental data) which one cannot feel. It is also important for the patient to know that according to experimental observations, the exercises are effective as long as they are performed correctly, even if one does not feel anything at all. Apart from this it is necessary that the therapist is familiar with the therapeutic problems resulting from different forms of *autogenic discharges* and processes of *autogenic abreaction* which may start while the patient is in an autogenic state (15–18).

Subsequently, passive concentration on warmth is added, starting, for example with "My right arm is warm." This formula aims at peripheral vasodilation. Depending on the generalization of the feeling of warmth in other limbs, the training progresses until all extremities become regularly heavy and warm. This training may take another period of from two to eight weeks.

After having learned to establish the feeling of heaviness and warmth, the trainee continues with passive concentration on cardiac activity by using the formula "Heartbeat calm and regular." Then follows the respiratory mechanism with "It breathes me," and warmth in the abdominal region: "My solar plexus is warm." The final exercise of the physiologically oriented standard exercises concerns the cranial region which should be cooler than the rest of the body. Here, one applies the formula "My forehead is cool."

The time usually needed to establish these exercises effectively varies between four and ten months.

The trainee's attitude, while repeating a formula in his mind, is conceived as "passive concentration." Passive concentration may best be explained in comparison with what is usually called "active concentration." Concentration in the usual sense has been defined as "the fixation of attention," or "high degree of intensity of attention," or "the centering of attention on certain parts of experience." This type of mental activity involves the person's concern, his interest, attention, and goal-directed investment of mental energy and effort during the performance of a task and in respect to the functional result.

In contrast, passive concentration implies a casual attitude during the performance and with regard to the functional result. Any goal-directed effort, active interest, or apprenhensiveness must be avoided. The trainee's casual and passive attitude toward the psychophysiologic effects of a given formula is regarded as one of the most important factors of the autogenic approach. Furthermore, the effectiveness of passive concentration on a given formula depends on two other factors namely (a) the mental contact with the part of the body indicated by the formula (for example, the right arm); and (b) keeping up a steady flow of a filmlike (verbal, acoustic or visual) representation of the autogenic formula in one's mind. Passive concentration on a formula should not last more than 30 to 60 seconds in the beginning. After several weeks the exercises may be extended to three or five minutes; after a few months up to 30 minutes and longer.

The state of passive concentration is terminated by applying a three-step procedure, namely (a) flexing the arms energetically, then (b) breathing deeply, and (c) opening the eyes. Usually three exercises are performed in sequence, with about a one minute interval between each of them.

After the standard exercises have been mastered satisfactorily, one may train to modify the pain threshold in certain parts of the body or train the time sense for waking up at a specific time. The therapy may be continued by applying autogenic principles for approaching specific functional disorders or even certain organic diseases. A number of special formulas and procedures have been worked out for meeting the therapeutic requirements of various functional and organic disorders like bronchial asthma, writer's cramp, hemorrhoids, brain injuries, esophagospasm, pruritus and others.

The meditative exercises should not normally be started until after six to 12 months of standard training, and the trainee should be able to prolong the autogenic state up to 40 minutes without experiencing any disagreeable side-effects or after-effects.

The meditative series begins with passive concentration on phenomena of visual imagination, as for example the spontaneous experience of certain colors. Later, the trainee may focus on seeing all colors at will. When that is achieved, the meditative series continues with visual imagination of objects. This training phase may take several weeks before results are obtained. It is followed by imagining abstract concepts like "happiness" or "justice" in different sensual modalities (musical, chromatic, plastic). Still later, one may meditate on one's own feelings and, in contrast, try to evoke the image of another person. Finally, at the deepest level of meditation, an interogatory attitude may be assumed in expectation of answers from the unconscious.

Autogenic training at the meditative level may be applied as what has been called "Nirvana Therapy" (4, 5) in clinically hopeless cases (for example advanced cancer) or in monotonous and desperate situations as may occur under exceptional circumstances. The meditative exercises have also been found to be of particular value in depth-dimensional psychotherapy. In general, it has been observed that the effects of the more physiologically oriented standard exercises are reinforced by the meditative training. However, the meditative exercises are not introduced to the average patient. The average clinical therapy centers on the standard formulas in combination with special exercises and intentional formulas specifically designed to meet the therapeutic requirements of relevant functional or organic disorders.

EXPERIMENTAL DATA

From experimental data and clinical results we know that passive concentration on the standard formulas induces multidimensional changes of a mental and organismic nature. In principle, two categories of effects may be distinguished: immediate effects, occurring during passive concentration on the different formulas, and effects resulting from practice of autogenic exercises over periods of weeks and months. Information about the immediate effects during the exercises is still incomplete. However, the experimental data available indicate clearly that each of the standard formulas induces physiologic changes of certain autonomic functions which are coordinated by diencephalic mechanisms.

During passive concentration on heaviness, Siebenthal (19, 20), Wittstock (21), and Eiff and Jörgens (22) recorded a significant decrease of *muscle potentials.* Along the same lines Schultz found a significant reduction of the patellar response during passive concentration of heaviness in both legs (4, 5). Determinations of motor chronaxie (musc. extensor digit. comm. dexter) by Schultz, Lewy, and Gaszmann (4, 5) indicated that the intensity of the stimulus has to be increased during the heaviness exercise because the excitatory threshold rises from its resting value.

FIGURE 1

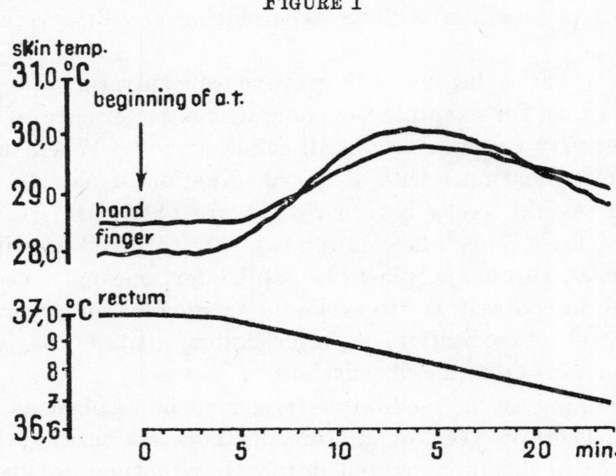

Changes in peripheral circulation during passive concentration on heaviness and warmth have been verified by a number of independent authors (1, 3, 4, 23, 24). The most extensive study was carried out at the University of Würzburg by Polzien (25, 26, 27). Polzien found that the rise of the skin temperature was more pronounced in distal parts of the extremities than in the more proximal areas. Simultaneously variable changes in the rectal temperature were recorded. Depending upon the subject, and the duration of passive concentration, the increase of skin temperature in the fingers varied between 0.2 and 3.5°C. These findings are in accordance with other results reported by Siebenthal (19) and Müller-Hegemann (28). Using special devices, both authors independently recorded an increase of weight in both arms during passive concentration on heaviness. The measured increase of weight has been ascribed partly to the relaxation of regional muscles and partly to an increase of blood flow in the arm (1).

More recently, Marchand (29, 30) demonstrated that the standard exercises and passive concentration on warmth in the liver area induce certain changes in the trainee's blood sugar level. During the first three standard exercises there is a slight increase of blood sugar. The fourth standard exercise (It breathes me) coincides with a slight drop in blood sugar, which is followed by another slight increase during passive concentration on "My solar plexus is warm" (fifth standard exercise). Subsequently passive concentration on warmth in the liver area is associated with a significant rise. The control values obtained after termination of the exercises indicate a sharp drop of blood sugar values, which, however, are slightly higher than the control values determined before starting the standard exercises. *White cell counts* during this investigation (24 subjects) indicated that the first four standard exercises are associated with a slight but progressive decrease in white cell values. This trend was reversed during the fifth standard exercise and during passive concentration on warmth in the liver area which was associated with a marked increase. The highest white cell values were obtained three minutes after termination of the exercises. Subsequent determinations corresponded to values obtained before starting the exercises (29, 30).

Various electroencephalographic studies (1, 15, 18, 31–38) during passive concentration on the standard formulas revealed that the different standard exercises and the autogenic state were associated with certain changes which are similar to but not identical with, patterns occurring during sleep or hypnosis (18, 38).

According to the observations reported by P. Geissmann and C. Noel (36) no true psychogalvanic reactions appeared during the standard exercises in completely relaxed trainees; certain reactions which were observed in a number of subjects seemed to be due to difficulties related to the experimental arrangement.

A systematic study of the respiratory changes occurring during the standard exercises revealed a significant decrease of the respiratory frequency which was associated with a gradual and significant increase of the thoracic and abdominal respiratory amplitude and a corresponding significant augmentation of the inspiration/expiration ratio (1, 13, 38, 39). Furthermore, it was observed that passive concentration on heaviness in the limbs is associated with a significant decrease of the respiratory volume and that the different standard formulas may produce a number of qualitative changes of the trainee's respiratory pattern. In asthmatic patients an almost instantaneous normalization of a disturbed pattern of respiratory innervation has been observed frequently (1).

The close physiologic and topographic relations between respiratory and circulatory mechanisms stimulated further studies of the effect of the standard exercises on cardiac activity (1), blood pressure (1, 13), the electrocardiogram (1, 13, 40) and certain variables more closely related to metabolic processes (1, 25–27, 29, 30, 41–43) In a group of normotensive subjects it was found (1,13) that passive concentration on heaviness produces a slight but significant decrease of the heart rate (5 to 10%) and a tendency toward lowering of the blood pressure. In hypertensive patients regular practice of the two first standard exercises usually produces a significant drop of the systolic (10–25%) and the diastolic (5–10%) blood pressure (1, 13).

FIGURE 2

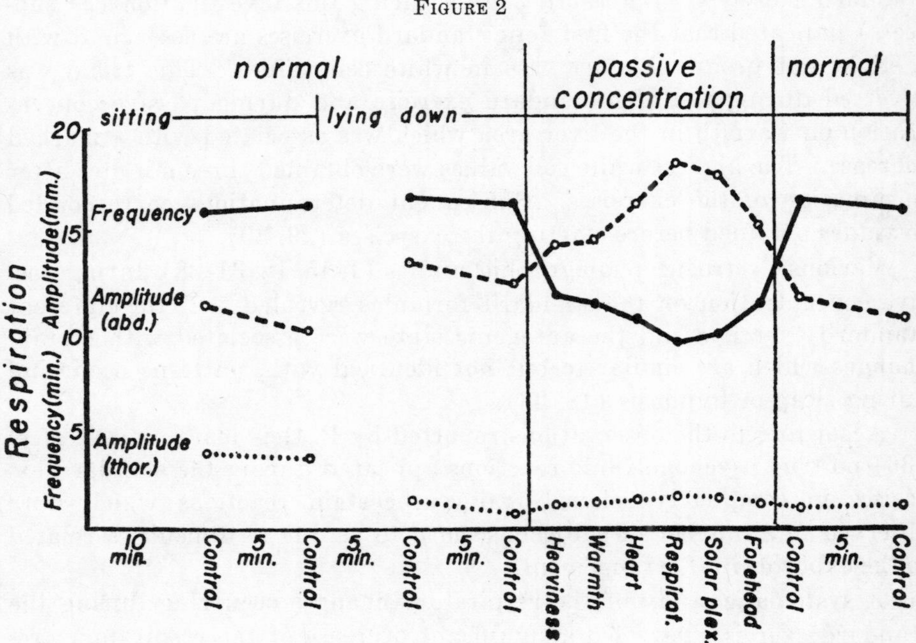

Electrocardiographic changes during autogenic standard therapy were reported by various authors (1, 13, 35, 36, 40, 44). The relevant observations may be summarized as follows: during passive concentration on heaviness (and warmth) the heart rate usually decreases. In relatively few cases an increase of the heart rate has been observed. This paradoxic reaction is regarded as resulting from *autogenic discharges* (16, 17, 18).

During the Third World Congress of Psychiatry in Montreal (1961), Polzien reported that 28 out of a group of 35 patients with confirmed ST-depressions showed an elevation of the ST-curve and an increase of the T-wave by .05 mV or more during the first standard exercise. In five cases the ECG remained unchanged and two patients reacted with further deterioration. In a control group of 20 patients with normal curves, an elevation of the ST-curve or the T-wave by .05 mV or more was observed in 10 trainees. It is of particular interest that a correlation between the heart rate and the ST and T-wave changes did not exist. This finding is in contrast to the physiologic correlation which normally exists between the heart rate and the elevation of the "ST segment-T wave phase." In other words, it is not possible to explain the elevation of the "ST segment-T wave phase" as observed during autogenic training, by the simultaneously occurring changes (decrease, increase) of the trainee's heart rate (40).

More recent investigations carried out at the University of Würzburg have verified the normalizing effect of the standard exercises on certain hyperthyroid conditions (45). Other experimental studies dealing with the effect of autogenic training on bodily work and subsequent recuperation have been carried out at the University of Leipzig (42, 43).

Briefly, the experimental data indicate that passive concentration on physiologically oriented formulas influences autonomic functions which are coordinated by diencephalic mechanisms. Both clinical results and experimental data indicate that autogenic training operates in a highly differentiated field of bodily self-regulation and that with the help of autogenic principles it is possible to use one's brain to influence certain bodily and mental functions effectively. It is evident that this type of psychophysiologic manipulation requires proper training, adequate medical background knowledge, critical application, and systematic control of the effects of the treatment (10). Furthermore, I hope it is quite clear that autogenic training is neither a simple relaxation technique nor a self-persuasive approach as applied by Coué (46).

The long-range effects resulting from regular practice of the standard exercises are manifold and depend largely on the psychophysiologic constellation of the individual and the nature of the patient's condition. Briefly, one could say that a gradual process of multidimensional optimalization develops. This process is reflected in psychodynamic changes which can be verified by physiologic measurements and projective tests.

In line with reports on gradual changes in the patient's behavior (1, 13, 47, 48), I have observed a characteristic pattern of projective changes, for example, in the Drawing-Completion test: Progressive differentiation of the projective responsiveness, increase of output, more shading, elaboration of details, stronger pressure of lines, increase of dynamic features, better integration and composition of the drawings, less rigidity, fewer inhibitions, faster performance, and better adaptation to the different stimuli. Corresponding changes have been observed in the Draw-A-Person test (47).

Our observation that a patient's progressive improvement jumps ahead after four to eight months of regular practice of the standard exercises is reflected objectively by the patient's performance in the control tests which I administer at regular intervals during autogenic standard therapy. With respect to these clinical observations it is of particular interest that the EEG also reveals significant differences between trainees who have practiced autogenic exercises for two to four months and others who have practiced the standard exercises for much longer periods (15, 18). Subjects practicing two to four months show an EEG pattern similar to the EEG pattern seen in states of "predrowsiness," as for example, bursts of anterior theta waves with a tendency to spatial generalization in anterioposterior direction in association with a preserved alpha activity. In contrast, subjects with longer training periods (6 to 36 months) pass very rapidly from the pattern of a normal state to a pattern characterized by (a) a flattening of the baseline pattern with theta oscillations; (b) the alpha main frequency shows an increase of rapidity (1 unit/sec.); and (c) brief paroxysmal bursts of theta waves in temporal-posterior derivations (15, 18, 31–38, 44, 48).

These electroencephalographic differences between short-period and long-period trainees seem to indicate that the regular practice of the standard exercises over longer periods of time bring about certain functional changes in the trainee's brain.

CLINICAL APPLICATION

Autogenic training has been applied to patients suffering from a variety of psychosomatic disturbances, a number of mental and behavior disorders, certain organ diseases and the psychophysiologic effects resulting from mental and bodily stress in general.

Disorders of the Respiratory Tract

Autogenic therapy has yielded good results in the treatment of various disorders of the respiratory tract. The following results (1) of a group of 150 adult patients suffering from bronchial asthma and treated with autogenic training were reported: 99 (66%) had no further complaints during a control period varying between six to 50 months, 38 (25%) patients had significantly improved and slight or no improvement was observed in 13 (9%) cases.

The application of autogenic training to asthmatic children and teenagers is limited by a number of factors. Children below the age of nine usually have difficulty maintaining passive concentration effectively. This is due partly to intruding thoughts and partly to the anxiety developed by some children during the exercises. The best results have been obtained with children aged ten to thirteen. Teenagers tend to regard the exercises as somewhat ridiculous, and usually are unable to collaborate seriously.

FIGURE 3

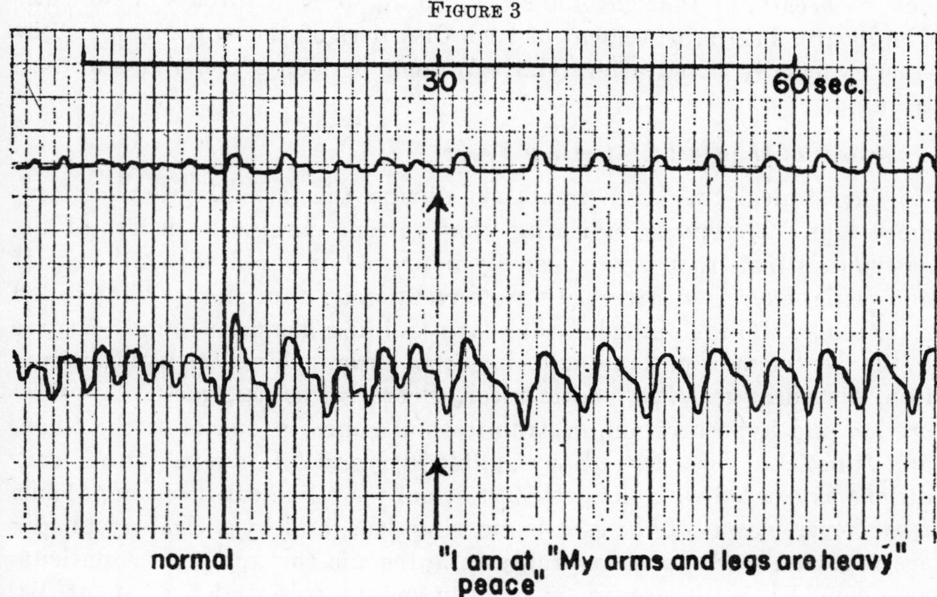

Patients practicing autogenic training gradually become calmer and appear to be more at ease and to feel more relaxed. Interpersonal relations improve and the frequency and volume of emotional outbursts diminish or cease completely. Sleep disturbances are reduced, and the patients learn to deal more effectively with symptoms like insomnia, headaches, and lack of appetite. The standard exercises have been applied successfully and without complications in patients with: pulmonary and bronchial tuberculosis, pneumothorax, temporary blocking of the phrenic nerve, pneumoperitoneum, pneumolysis, pleurisy, and thoracoplasty. Cavitation is not considered a contraindication.

Particularly sensitive patients who previously had developed anxiety and tension before the refilling of their pneumothorax were much less concerned, less tense, and more at ease before and during the refill-procedure. A few detailed case studies (1) indicate that pulmonary hyperemia as indicated by passive concentration on ''My lungs are warm'' appears to be responsible for improvements (for example disappearance of cavities) after conventional therapy had failed. However, so far no statistical data are available to demonstrate reliably that this organ-specific approach has a particularly marked effect on the course of tuberculous processes.

It has been frequently observed that patients undergoing autogenic training become less susceptible to colds, to nervous coughing or to shortness of breath, or that chronic susceptibility to sore throats or sinusitis disappears. The improvement of such disorders of the upper respiratory tract is usually noted after relatively long periods of regular standard training.

Disorders of the Gastrointestinal Tract

The clinical literature published during the last 30 years contains more than 100 reports indicating that various chronic disorders of the gastrointestinal tract were markedly improved or cured with the help of autogenic exercises after other conventional approaches had failed repeatedly. So far autogenic standard training has been found effective in the treatment of patients suffering from chronic disorders like constipation, diarrhea, gastritis, peptic ulcer, ulcerative colitis, gastric hypersecretion and hyposecretion, gastrointestinal neuroses, spasm of the colon, and functional disorders of the gallbladder and bile ducts. Other conditions associated with the gastrointestinal tract like cardiospasm, globus hystericus, spasms of the esophagus, singultus, and hysterical dysphagia are reported to respond favorably to autogenic therapy. Most patients note considerable improvement or even spontaneous and permanent disappearance of functional disturbances during advanced phases of autogenic standard therapy. In cases in which there is little or no satisfactory improvement after the patient has practiced the standard exercises for at least three months, the application of additional organ-specific formulas in combination with patient adapted intentional formulas (for example, ''My throat is wide, swallowing does not matter'') is indicated (1).

The application of autogenic training in the clinical treatment of different forms of chronic nonspecific gastritis (atrophic, superficial, hypertrophic) has been found of particular value. There are no reliable data which prove that autogenic therapy promotes recovery from gastric ulcers. However, in patients suffering from duodenal ulcer, a significant decrease in the number of relapses has been observed in those patients who learned and continued to practice autogenic training after leaving the hospital (49, 50).

Disorders of the Circulatory System

Autogenic therapy of certain functional and organic alterations of the heart and vascular system usually requires certain modifications of the standard approach. As a rule, passive concentration on the heart (third standard exercise) should be placed at the end of the standard series. This modified sequence was adopted after frequent observations that passive concentration on the heart, when practiced too early (for example, after heaviness and warmth training), may intensify the patient's ailments or lead to unpleasant effects during or after the exercises. The onset of palpitation, extrasystoles, feelings of uneasiness, and anxiety have been reported. A number of patients react with increased awareness of their cardiac disturbance and consequently tend to develop resistance to the therapy.

FIGURE 4

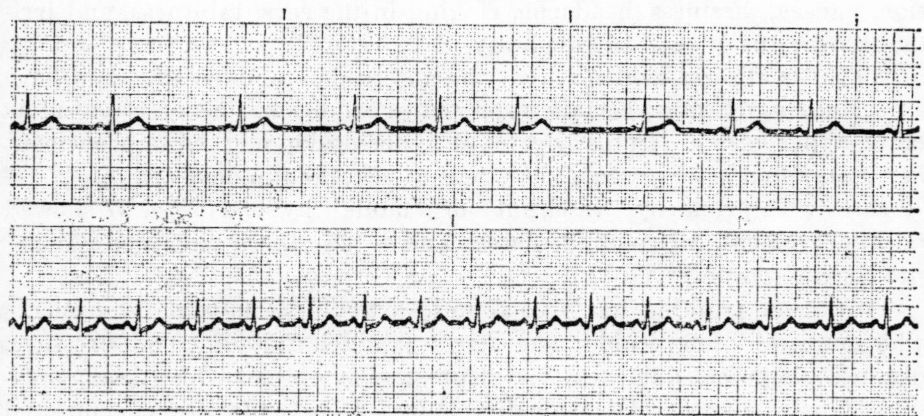

Autogenic therapy of patients suffering from anginal attacks due to different precipitating causes has yielded very rewarding results. Marked relief and a reduction in attacks have been also reported in cases with considerable coronary insufficiency on a sclerotic basis (1). The best results have been obtained when functional factors were predominant. It has been reported that patients with severe coronary spasms had no further complaints after a few months of autogenic training.

Clinically, the prophylactic value of the autogenic approach appears to be of particular importance. In 31 patients practicing autogenic training, Laberke (50) did not observe any myocardial infarctions during a control period of one to four years. In a control group of 30 nontrained patients there were four infarctions.

Sinus arrhythmia, premature contractions (extrasystoles), and disorders resulting from fast discharges of stimuli (tachycardia, racing heart) have been reported to respond favorably to autogenic therapy (1, 13). Palpitations and even accelerated activity of the heart resulting from hyperthyroid conditions have been normalized within a few weeks (1, 13, 45). A variety of functional disorders of the heart (for example, neurocirculatory asthenia, irritable heart, effort syndrome, emotional tachycardia) tend to respond very well to autogenic training (1, 44). After three to five weeks of standard training (heaviness and warmth) most patients note some significant improvement. They report that they feel more relaxed, calmer, and that their sleep is better. Hypochondriac attitudes (for example, frequent observation of heart rate) subside. Palpitations, "jumps" or "skips," attacks of tachycardia, pains in the precordial region, and feelings of oppression in the chest were less frequent and less disturbing. A number of concomitant symptoms characteristic of a neurotic pattern, such as respiratory disorders, meteorism, constipation, nausea, dizziness, headache, ringing in the ears, faintness, and irritability gradually tend to fade away.

There is clinical evidence that the autogenic standard exercises have a normalizing effect on certain hypotensive conditions and on essential or labile hypertension. In many cases of hypertension marked improvement has been noted after six to eight weeks of regular standard training. Of 79 cases with primarily "essential" or "labile" hypertension, 37 showed no improvement, 19 responded well, and in 29 others some definite improvement was observed (1).

Autogenic training has been found particularly helpful for patients who manifest disagreeable side effects (sudden fall of blood pressure, dizziness, impotence, flushes) and depressive symptoms while taking antihypertensive drugs. Occasionally, however, patients are encountered whose diastolic and systolic blood pressure increase during passive concentration on heaviness and warmth. This paradoxic reaction is considered to be due to the release of *autogenic discharges* affecting relevant mechanisms of the vasomotor system. For this reason it is advisable to control the blood pressure of persons undergoing autogenic therapy. Patients who regularly respond with an increase of blood pressure during the standard exercises should discontinue autogenic training.

The normalizing influence of autogenic exercises on autonomic self-regulation, and the possibility of increasing at will the circulation in the extremities, make autogenic training a valuable therapeutic tool in the treatment of various disorders of the peripheral circulation (for example, intermittent claudication, Buerger's disease, ischemic neuritis, Raynaud's phenomenon, scleroderma, frostbite, cold hands and feet). Clinical and experimental observations indicate that different forms of passive concentration enhance peripheral circulation to a variable degree. From a physiologic point of view, experimental evidence supports the assumption

FIGURE 5

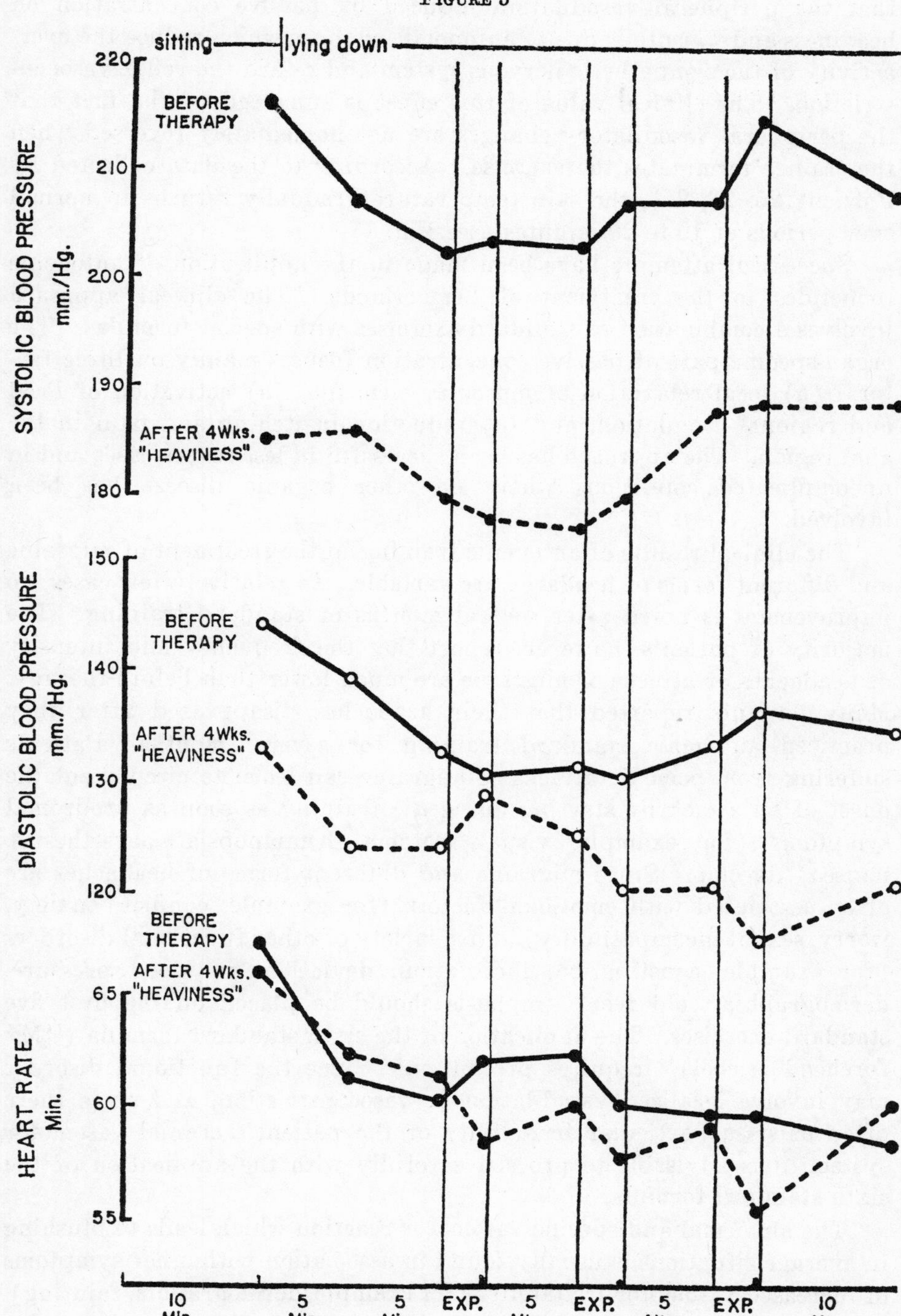
sitting
lying down
SYSTOLIC BLOOD PRESSURE
mm./Hg.
220
210
200
190
180
BEFORE THERAPY
AFTER 4Wks. "HEAVINESS"
DIASTOLIC BLOOD PRESSURE
mm./Hg.
150
140
130
120
BEFORE THERAPY
AFTER 4Wks. "HEAVINESS"
HEART RATE
Min.
65
60
55
BEFORE THERAPY
AFTER 4Wks. "HEAVINESS"
10 Min.
5 Min.
5 Min.
EXP.
5 Min.
EXP.
5 Min.
EXP.
10 Min.

that the peripheral vasodilation induced by passive concentration on heaviness and warmth is due to autonomic changes which reduce the over-activity of the sympathetic nervous system and retard the reflex vasoconstriction. The clinical value of this effect is enhanced by the fact that the peripheral vasodilatory changes are not immediately reversed when the trainee terminates the exercise. According to the data collected by Polzien (25, 26, 27), the skin temperature gradually returns to normal over periods of 10 to 20 minutes (see Fig. 1).

Successful attempts have been made in the application of autogenic principles to the treatment of hemorrhoids. The clinical approach involves a combination of standard exercises with special formulas. The organ-specific part of passive concentration focuses mainly on three factors: (a) local relaxation of muscular elements, (b) activation of local and regional circulation, and (c) reduction in itching and pain in the anal region. The approach has been successful in less severe cases and in uncomplicated conditions where no other organic disease has been involved.

The clinical results of autogenic training in the treatment of migraine and different forms of headache are variable. In relatively few cases, no improvement is noted after several months of standard training. The majority of patients, however, report that the frequency and intensity of headaches or attacks of migraine are much lower than before therapy. Many patients reported that their headaches disappeared after they practiced autogenic standard training for several months. Patients suffering from periodic attacks of migraine can learn to circumvent the onset of an attack by starting autogenic training as soon as prodromal symptoms (for example, visual scotoma, hemianopsia, paresthesias, nausea) develop. Since migraine and different forms of headaches are often associated with emotional factors (for example, conflict, anxiety, worry, sexual incompatibility) and a variety of other functional disorders (for example, constipation, indigestion, deviations in blood pressure, dermographia, cold feet), emphasis should be placed on the first five standard exercises. The application of the sixth standard formula ("My forehead is cool)" requires precaution. Since the functional disorder may involve localized vasodilation or vasoconstriction, and since there often exists a particular irritability of the patient's cranial vasomotor system, it is advisable to proceed carefully with the application of the sixth standard formula.

The abnormal and specific vasomotor reaction which leads to blushing in average situations is usually found in association with other symptoms of increased vasomotor irritability (for example, dermographia, fainting). The disorder has been conceived as a disturbance in the person's psychophysiologic adaptation to certain environmental situations. Both the increased vasomotor irritability and the person's fear that blushing may occur subside under the influence of autogenic standard therapy. After two to three months of regular standard training the patients tend to note that they are less emotionally involved and remain, without effort, much calmer in situations which previously induced blushing. In addition to the standard exercises, two specific formulas have been used: "My neck and shoulders are warm," and "My feet are warm." These two formulas are considered as physiologic extensions of the warmth training and are designed to divert the abnormal circulatory changes to areas of the body which are acceptable to the patient. Under everyday circumstances the patient learns to control blushing by "switching on" "My feet are warm" (or "My neck and shoulders are warm") as soon as a situation arises where blushing may occur. The physiologic effects of this formula may be supported by adding "My forehead is cool."

Endocrine Disorders

The clinical value of autogenic training in the field of endocrine dysfunctions and associated metabolic deviations is associated with its normalizing effect on self-regulatory mechanisms of the autonomic nervous system. The reduction in tension and the gradual smoothing of neurotic symptoms and unfavorable emotional reactive-affective patterns are additional advantages in cases of endocrine or certain other metabolic disorders precipitated, and aggravated, by mental or emotional stress.

So far, the autogenic approach has been applied mainly to patients with dysfunctions of the thyroid gland, diabetes mellitus, conditions related to latent tetany and gynecologic disorders associated with endocrine derangements.

It has been reported repeatedly that the regular practice of autogenic exercises has a normalizing influence on deviations in blood sugar and glycosuria. In diabetic trainees significant decreases of blood sugar have been observed and it was necessary to reduce the habitual dosage of insulin by 10 to 20 units per day. It is for this reason that autogenic training should not be applied to diabetic patients when regular control is not possible.

In hyperthyroidism, autogenic standard therapy has been applied alone or as an adjunct, together with other clinical approaches. The warmth exercises usually require careful adaptation. In many cases a reduced step-by-step approach is necessary because of the patient's increased vasomotor irritability. Too early introduction of the heart exercise may easily cause an increase in palpitation or other cardiac complaints. In difficult cases it is advisable to postpone the heart exercise until the end of the standard series; instead, emphasis should be placed on practicing the fourth standard exercise ("it breathes me"). Beneficial effects are commonly noted by the time an advanced heaviness training is reached. As autogenic training progresses, a number of typical manifestations like sweating, hyperkinesis, tremor, nausea, vomiting, diarrhea, and irritability tend to subside gradually. After six to 12 weeks of regular training, a significant increase in body weight and a progressive normalization of the hyperthyroid condition as reflected by laboratory tests have been reported (1,44,45).

SUMMARY AND CONCLUSIONS

The scope of this paper does not permit discussion of other clinical and nonclinical fields of aplication of autogenic therapy. The experience of almost 40 years of clinical application of autogenic training may be summarized as follows:

1. The successful application of autogenic standard therapy in medical and psychiatric disorders is influenced by the nature of the disorder and also varies with the clinical condition of each individual case.
2. Autogenic training can be used alone or as an adjunct to psychotherapy with about 80 to 90 per cent of adults of all ages. It is known that the applicability of autogenic training is influenced by the age factor, intelligence, and over-all development. The method has been successfully used with some intelligent children at the age of six; very good results have been observed in children from eight to twelve; many difficulties were encountered in teenagers (thirteen to sixteen) and there is some evidence that patients beyond the age of fifty have more difficulties than patients between eighteen and forty-nine (1, 10, 11, 12).
3. In comparison with other psychotherapeutic approaches, relatively little time is required for therapy. After each step of the standard exercises has been introduced to a patient, only periodical control sessions are required for guiding him. Apart from this the patient carries out his own therapy by performing a number of mental exercises for about ten minutes three times a day.
4. Group therapy is possible. Outpatients do best in groups of less than ten, while patients who are under continuous medical supervision may practice autogenic training in larger groups. The application of autogenic exercises to more than 20 persons should be discouraged because adequate medical control is not possible (10).
5. The effectiveness and progress of the therapy can be controlled by physiologic and psychologic tests (1, 13, 44, 47, 48).

6. Autogenic training has been most frequently applied in nonpsychiatric disorders. Clinical results demonstrated that autogenic training has been effective and helpful in the treatment of (a) disorders of the respiratory tract; (b) disorders of the gastrointestinal tract; (c) disorders of the cardiovascular system and vasomotor disturbances; (d) disorders of the endocrine system; (e) disorders of the urogenital system; (f) disorders of pregnancy; (g) skin disorders (for example, allergic conditions, pruritus, verruca vulgaris; (h) ophthalmologic disorders and blindness (glaucoma, scotoma, certain forms of squint; (i) neurologic disorders (certain neuromuscular disorders, brain injury, and epilepsy). A variable 60 to 90 per cent of patients suffering from lond-standing disturbances like insomnia, headaches, bronchial asthma, chronic constipation, and the like were cured, or improved considerably, within periods ranging from two to ten months of standard training (1).

7. It has been observed that autogenic training is very helpful in the treatment of behavior disorders and motor disturbances as for example, stuttering, writer's cramp, enuresis, hysterical dysphagia, singultus, globus hystericus, blushing, certain states of anxiety, and phobia. Patients have reported over periods ranging from a few weeks to several months that their anxiety, insecurity, and neurotic reactions are smoothing out or had gradually lost their significance. Generally, an increase of emotional and physiologic tolerance, with a considerable decrease of the previous need for reactive affective discharges, has been reported. Social contact becomes less inhibited and more natural. Interpersonal relations have been reported to be warmer and more intimate with certain persons and less emotionally involved with others (1).

8. It has been observed that with the help of autogenic standard exercises unconscious material becomes more readily available. Forty-eight per cent of trainees reported spontaneously that they dreamed more frequently. The impression was gained that dream and memory material are more readily produced by trainees than by other patients. Free association seems to be facilitated (1, 16, 17).

9. Many trainees have reported that their intellectual efficiency had increased. Furthermore, it has been noted that increase of bodily resistance develops against all kinds of stress (1).

10. Until further information is available, autogenic training should not be applied in acute psychoses, postpsychotic states, chronic involutional psychoses, initial stages of schizophrenia, in oneiroid conditions, certain forms of epilepsy, "clouded" or "twilight states" (*psychische Dämmerzustände*), constant and multiple paradox reactions, and certain medical conditions (for example, bleeding peptic ulcer, acute myocardial infarction, insufficiently controlled diabetes (10).

11. Autogenic training is not indicated as a treatment for recalcitrant psychopaths, for patients with dementia or oligophrenia, and for patients with a lack of motivation to get better (1, 10, 12).

Clinical and experimental observations gathered over the past 35 years have indicated that the physiologic changes occurring during autogenic exercises are of a highly complex and differentiated nature, involving autonomic functions which are coordinated by diencephalic mechanisms. The physiologic changes which occur during the standard exercises coupled with the fact that the regular practice of autogenic training over longer periods of time has a normalizing influence on a great variety of bodily and mental disorders led to the conclusion that autogenic training exerts a therapeutic action on certain mechanisms which are of pathofunctional relevance for many different types of bodily and mental disorders. In summarizing my experimental and clinical findings I hypothesized (1, 13, 18) that the therapeutic key factor lies in a self-induced (autogenic) modification of cortico-diencephalic interrelations, which enables natural forces to regain their otherwise restricted capacity for self-regulatory normalization. The hypothesis implies that the function of the entire neurohumoral axis (cortex, thalamus, reticular system, hypothalamus, hypophysis, adrenals) is directly involved and that the therapeutic mechanism is not unilateraly restricted to either bodily or mental functions.

REFERENCES

1. Schultz, J. H. and Luthe, W. *Autogenic Training. A Psychophysiologic Approach in Psychotherapy.* Grune & Stratton, New York. 1959.
2. Schultz, J. H. Über Narkolyse und autogene Organübungen, zwei neue psychotherapeutische Methoden. *Med. Klin.*, 22: 952, 1926.
3. ————. Über selbsttätige (autogene) Umstellungen der Wärmestrahlung der menschlichen Haut im autosuggestiven Training. *Dtsch. Med. Wschr.*, 13: 571, 1926.
4. ————. *Das Autogene Training.* Thieme Verlag, Stuttgart, 1932.
5. ————. *Das Autogene Training.* Tenth edition. Thieme Verlag, Stuttgart, 1961.
6. ————. *El Entrenamiento Autogeno.* Ed. Cientifico Medica, Barcelona, Lisboa, Madrid, 1954.
7. ————. *Øvningshefte for Autogen Training.* Olaf Norlis Forlag, Oslo, 1956.
8. ————. *Le Training Autogène.* Presses Universitaires de France, Paris, 1958.
9. Durand de Bousingen, R. Enseignement du Training Autogène en France, In *Proc. Third Internat. Congress of Psychiatry,* Montreal, 1961. University of Toronto Press, Toronto, 1962.
10. Luthe, W. International Coordination of Autogenic Training. In *Proc. Third Internat. Congress of Psychiatry,* Montreal, 1961, See reference 9.
11. Müller-Hegemann, D. and Kohler, C. Eight Years Experience with Autogenic Training. In *Proc. Third Internat. Congress of Psychiatry,* Montreal, 1961. See reference 9.

12. Müller-Hegemann, D. and Kohler-Hoppe, C. Über neuere Erfahrungen mit dem Autogenen Training. *Psych. Neurol. med. Psychol.*, 3: 97, 1962.
13. Luthe, W. Physiological and Psychodynamic Effects of the Autogenic Training. In *Topical Problems of Psychotherapy.* Stokvis, B., Ed. S. Karger, Basel, New York, 3: 174, 1960.
14. Schultz, J. H. Oskar Vogt in der Geschichte der medizinischen Psychologie Deutschlands. *Nervenarzt,* 2: 41, 1951.
15. Geissmann, P., Jus, A. and Luthe, W. Neurophysiologic and Psychophysiologic Aspects of the Autogenic State. In *Proc. Third Internat. Congress of Psychiatry,* Montreal, 1961. See reference 9.
16. Luthe, W. The Clinical Significance of Various Forms of Autogenic Abreaction. In *Proc. Third Internat. Congress of Psychiatry,* Montreal, 1961. See reference 9.
17. ————. Signification clinique de diverses formes d'abrèactions autogènes. *Rev. Méd. Psychosom.*, 4: 3, 1962.
18. Luthe, W., Jus, A. and Geissmann, P. Autogenic State and Autogenic Shift: Psychophysiologic and Neurophysiologic aspects. *Acta Psychother.*, 1962 (in press).
19. Siebenthal, W. Eine vereinfachte Schwereübung des Schultz'schen autogenen Trainings. *Ztschr. Psychother. Med. Psychol.*, 2: 135, 1952.
20. Schultz, J. H. Diskussionsbemerkungen zu den Arbeiten R. Kraemer und v. Siebenthal. *Ztschr. Psychother. Med. Psychol.*, 5: 214, 1952.
21. Wittstock, W. Untersuchungen über cortikale Einflüsse auf körpereigene Aktionsströme. *Psychiatr. Neurol. Med. Psychol.*, 2/3: 85, 1956.
22. Eiff, A. W. and Jörgens, H. Die Spindelerregbarkeit beim Autogenen Training. In *Proc. Third Internat. Congress of Psychiatry,* Montreal, 1961. See reference 9.
23. Binswanger, H. Beobachtungen an entspannten und versenkten Versuchspersonen. Ein Beitrag zu möglichen Mechanismen der Konversionshysterie. *Nervenarzt,* 4: 193, 1929.
24. Stovkis, B., Renes, B. and Landmann, H. Th. Skin Temperature under Experimental Stress and During Autogenic Training. In *Proc. Third Internat. Congress of Psychiatry,* Montreal, 1961. See reference 9.
25. Polzien, P. Die Änderung der Temperaturregulation bei der Gesamtumschaltung durch das Antogene Training. *Ztschr. exp. Med.*, 125: 469, 1955.
26. ————. *Über die Physiologie des hypnotischen Zustands als eine exakte Grundlage für die Neurosenlehre.* Karger, Basel, New York, 1959.
27. ————. The Influence of Autogenic Training on Thermoregulation. In *Proc. Third Internat. Congress of Psychiatry,* Montreal, 1961. See reference 9.
28. Müller-Hegemann, D. Über die cortico-viscerale und die psychosomatische Betrachtungsweise in der Psychotherapie. *Psychiatr. Neurol. Med. Psychol.*, 2/3: 33, 1956.
29. Marchand, H. Die Suggestion der Wärme im Oberbauch und ihr Einfluss auf Blutzucker und Leukozyten. *Psychother.*, 3: 154, 1956.
30. ————. Das Verhalten von Blutzucker und Leukozyten während des Autogenen Trainings. In *Proc. Third Internat. Congress of Psychiatry,* Montreal, 1961. See reference 9.
31. Franek, B. and Thren, R. Hirnelektrische Befunde bei gestuften aktiven Hypnoseübungen. *Arch. Psychiatr. Nervenkr.*, 181: 360, 1948.

32. Heimann, H. and Spoerri, T. Elektroencephalographische Untersuchungen an Hypnotisierten. *Monatsschr. Psychiatr. Neurol.* 4: 261, 1953.

33. Israel, L. and Rohmer, F. Variations électroencéphalographiques au cours de la relaxation autogène et hypnotique. In *La relaxation.* Aboulker, P., L. Chertok and M. Sapir, Eds. Expansion Scientif. Francaise, Paris. 1958, pp. 88–98.

34. Israel L., Geissmann, P. and Noel, C. Modifications des rythmes électroencéphalographiques au cours de la relaxation, observées à l'analyse de fréquence. *Rev. Méd. Psychosom.*, 3: 133, 1960.

35. Jus, A. and Jus, K. Étude polygraphique du Training Autogène. *Rev. Méd. Psychosom.*, 3: 136, 1960.

36. Geissmann, P. and Noel, C. Electrocephalographic Study with Frequency Analysis and Polygraphy of Autogenic Training. In *Proc. Third Internat. Congress of Psychiatry,* Montreal, 1961. See reference 9.

37. Jus, A. and Jus, K. Polygraphic Study of Autogenic Training. In *Proc. Third Internat. Congress of Psychiatry,* Montreal, 1961. See reference 9.

38. Luthe, W. Autogenic Training: Method, Research and Application in Psychiatry. *Dis. Nerv. Syst.*, 23: 383, 1962.

39. ————. Experimentelle Untersuchungen über den Einfluss des Autogenen Trainings auf die Atmung. I. Mitteilung: Frequenz- und Amplitudenänderungen bei normalgesunden Personen. *Ztschr. Psychother. Med. Psychol.*, 3: 89, 1958.

40. Polzien, P. Electrocardiographic Changes during the First Standard Exercise. In *Proc. Third Internat. Congress of Psychiatry,* Montreal, 1961. See reference 9.

41. ————. Respiratory Changes during Passive Concentration on Heaviness. In *Proc. Third Internat. Congress of Psychiatry,* Montreal 1961. See ref. 9.

42. Hiller, J., Müller-Hegemann, D. and Wendt, H. Über Auswirkungen des Autogenen Trainings auf Erholung und physische Leistung. In *Proc. Third Internat. Congress of Psychiatry,* Montreal, 1961. See reference 9.

43. ————. Experimentelle Untersuchungen über den Einfluss des autogenen Trainings (A.T.) auf die Leistung. *Med. Sport,* 1: 5, 1962.

44. Schultz, J. H. and Luthe, W. Autogenic Training. In *Proc. Third Internat. Congress of Psychiatry,* Montreal, 1961. See reference 9.

45. Polzien, P. Therapeutic Possibilities of Autogenic Training in Hyperthyroid Conditions. In *Proc. Third Internat. Congress of Psychiatry,* Montreal, 1961. See reference 9.

46. Coué, E. *Die Selbstbemeisterung durch bewusste Autosuggestion.* Schwabe Verlag, Basel, Stuttgart, 1961.

47. Luthe, W. Zur psychotherapeutischen Verlaufskontrolle durch projektive Tests bei Autogenem Training. In *Aktuelle Psychotherapie.* Speer, E., Ed. J. F. Lehmanns Verlag, Munich, 1958, pp. 159–168.

48. ————. Entrenamiento autógeno. Modificaciones psicofisiológicas e indicaciones clinicas. *Rev. lat. amer. hipn. clin.*, 2: 61, 1962.

49. Laberke, J. A. Studie zur Genese und Therapie des Gastritis-Syndromes. *Therapie d. Gegenwart,* 10/11: 350, 1950.

50. ————. Erfahrungsbericht über eine psychosomatische Kombinationsbehandlung bei sogen. inneren Krankheiten. In *Die Vorträge der 3. Lindauer Psychotherapiewoche 1952.* Thieme Verlag, Stuttgart, 1953, pp. 108–122.

67

"Shavasan": A Yogic Exercise in the Management of Hypertension

K. K. Datey, S. N. Deshmukh, C. P. Dalvi, and S. L. Vinekar

Hypertension increases the morbidity and mortality in man. Extensive actuary data have shown that increase in mortality is directly proportional to the severity of blood pressure. In spite of diagnostic advances in recent years, the etiology of hypertension still remains relatively obscure. The vast majority of patients fall into the category of "essential hypertension" and specific therapy is not available for these cases. Drugs are therefore necessary for these patients.

The antihypertensive drugs available so far are by no means ideal and have many disadvantages. Any new method for reducing blood pressure is, therefore, most welcome. Management of hypertension by measures other than drugs has been envisaged by Bilgutay *et al.*[1] In this communication, we present a new approach, the management of hypertension with "shavasan," a yogic exercise.

MATERIALS AND METHODS

From the Department of Cardiology, K. E. M. Hospital, 47 hypertensive patients are included in this study, 37 men and 10 women. Their ages varied from 22 to 64, with an average of 46 years.

Their "original" systolic blood pressure ranged from 160 to 270 mm Hg and the diastolic from 90 to 145 mm Hg (average 186/115 mm Hg).

The etiology of hypertension was essential in 32, renal in 12 and arteriosclerotic in 3 cases. These patients were divided into three groups (table 1).

Most of the patients on drug therapy were under treatment for an average period of about 2 years at the Hypertension Clinic of the K. E. M. Hospital, Bombay. During this period their drug requirement had been stabilized. Any attempt in reducing the dosage of drugs caused rise in blood pressure. The patients who had not received any antihypertensive drugs were first given placebo tablets for at least a month before teaching them this exercise.

The blood pressure of patients not receiving antihypertensive drugs was recorded in the recumbent position only. However, in those on antihypertensive drugs, it was recorded in three positions: recumbent, sitting and upright.

Paper presented at The Joint Annual Meeting of the American College of Angiology and International College of Angiology, Las Vegas, 1967.

Data are recorded in detail in tables 2, 3, and 4. The mean blood pressure was calculated as diastolic blood pressure plus one-third the pulse pressure. The average mean blood pressure in the recumbent position in group I and the average of the mean blood pressures in all the three positions of groups II and III are shown in table 5.

The symptoms in these patients varied. Giddiness was present in 30, headache in 28, chest pain in 12 (angina 7), palpitation in 12, breathlessness on exertion in 10, exhaustion in 10, insomnia in 8, irritability and nervousness in 8.

All the investigations were carried out in these patients before the commencement of the yogic exercise and at regular intervals thereafter. Electromyogram of the frontalis muscle was recorded before and during the exercise in some patients to confirm the muscular relaxation during the exercise (fig. 1).

These investigations were repeated periodically. Patients were instructed to attend the Cardiac Centre every day to learn the exercise; once they had learned it correctly, they were advised to attend weekly to check their blood pressure and the correct technique of the exercise.

TABLE 1

Three groups of patients

Group	Etiology of Hypertension			Total
	Essential	Renal	Arterio-sclerotic	
I. No drugs	5	2	3	10
II. Blood pressure well controlled with drugs	17	5	0	22
III. Blood pressure inadequately controlled in spite of drugs	10	5	0	15
Total	32	12	3	47

TECHNIQUE OF EXERCISE

The patient was allowed a light breakfast an hour before the exercise. He wore light and loose clothes while performing the exercise, which was performed as follows.

The patient lies in the supine position, lower limbs 30 degrees apart and the upper making an angle of 15 degrees with the trunk, with the forearms in the midprone position and fingers semiflexed. The eyes are closed with eyelids drooping. The patient is taught slow, rhythmic diaphragmatic breathing with a short pause after each inspiration and a longer one at the end of each expiration. After establishing this rhythm, he is asked to attend to the sensation at the nostrils, the coolness of the inspired air and the warmth of the expired air. This procedure helps to keep the patient inwardly alert and to forget his usual thoughts, thus becoming less conscious of the external environment, thereby attaining relaxation. The patient is asked to relax the muscles so that he is able to feel the heaviness of different parts of the body. This is achieved automatically once the patient learns the exercise. The exercise is performed for 30 min. An experienced supervisor checks that there is no movement of any part of the body, except rhythmic abdominal movements. Physical relaxation is checked from time to time by lifting the extremities and letting them go, to observe their flaccidity. Most of the patients learn the exercise correctly in about 3 weeks. The pulse, blood pressure and respiration were recorded before and after the exercise. After patients learn the exercise correctly, the respiratory rate is usually between 4 and 10 per min. (fig. 2).

TABLE 2

Group I: without drugs (10 patients)

Patient No.	B.P.* after Placebo for 1 Month (Initial B.P.)	Mean Initial B.P.	B.P. after "Shavasan"	Mean B.P. after "Shavasan"	Difference in Mean B.P. before and after "Shavasan"
	mm Hg	*mm Hg*	*mm Hg*	*mm Hg*	*mm Hg*
1	150/110	123	130/85	100	23
2	170/110	130	160/80	107	23
3	165/95	118	200/90	127	−9
4	200/110	140	170/90	117	23
5	210/90	130	145/90	108	22
6	200/110	140	150/90	110	30
7	180/120	140	130/80	97	43
8	210/110	143	140/90	107	36
9	180/130	147	140/90	112	35
10	180/105	130	110/75	87	43
Average		134.1 (134)		107.2 (107)	26.9 (27)

* B.P. = blood pressure.

Mean B.P. = Diastolic Blood Pressure + one-third pulse pressure. B.P. in recumbent position (no drugs).

The dosage of drugs in patients of groups II and III was adjusted according to the response. The percentage of reduction of drugs was calculated in each case from the original drug requirement.

RESULTS

Subjective improvement. The majority of the patients showed improvement in their symptoms. Headache, giddiness, nervousness, irritability and insomnia disappeared in almost all the patients. Even the other symptoms became less marked, and in general the patients experienced a sense of well being after this exercise.

Objective improvement. In group I (10 patients) the average mean blood pressure of 134 mm Hg was reduced to 107 mm Hg (an average reduction of 27 mm Hg. This was statistically significant (p 0.05) (table 6)). Figure 3 shows a typical response. In group II, in the 22 patients who were well controlled with drugs, the average mean blood pressure was 102 mm Hg. At this stage, yogic exercise was taught. As the blood pressure was well controlled,

TABLE 3

Group II: 22 patients adequately controlled with drugs, given "shavasan" and attempts made to reduce the drug requirement which was more or less stable before giving "shavasan"

Patient No.	Initial B.P.*			Initial Mean B.P. (Average)	B.P. with Drugs			Mean B.P. with Drugs (Average)	B.P. after "Shavasan"			Mean B.P. after "Shavasan" (Average)	Significant Drug Reduction†
	Upright	Sitting	Recumbent		Upright	Sitting	Recum-bent		Upright	Sitting	Recum-bent		
	mm Hg			*mm Hg*	*mm Hg*			*mm Hg*	*mm Hg*			*mm Hg*	
1‡	155/120	160/120	165/120	133	120/80	120/80	125/85	95	135/80	140/80	140/80	99	No
2‡	185/120	190/120	195/120	143	130/80	130/80	130/80	97	130/80	130/80	135/85	99	No
3‡	160/120	160/120	160/120	133	130/80	135/80	140/80	99	135/80	140/85	140/85	101	No
4	175/95	180/95	180/95	123	135/80	130/80	140/80	106	135/90	135/90	140/90	106	No
5P	180/135	190/135	195/135	154	120/80	130/80	140/85	98	115/80	125/85	130/85	96	Yes
6	160/120	160/120	160/120	133	140/90	140/90	140/90	107	120/70	125/80	125/80	92	Yes
7P‡	155/110	160/110	165/110	127	120/80	140/85	150/85	105	135/80	140/90	140/90	104	No
8	190/125	180/125	180/125	144	135/85	130/90	130/90	103	130/85	135/90	135/90	103	No
9P	160/125	165/115	170/115	132	130/80	140/85	140/90	102	120/70	140/85	140/90	99	No
10	195/130	195/130	190/130	152	130/80	135/80	130/80	99	130/75	130/80	135/80	96	Yes
11‡	175/115	170/115	170/115	134	115/85	120/85	120/90	97	140/90	140/90	140/90	107	No
12P	200/150	200/150	200/150	167	120/85	125/90	130/95	102	120/80	140/90	140/90	102	Yes
13	180/110	180/110	180/110	133	135/85	140/85	145/85	103	130/85	130/90	130/90	102	Yes
14	200/120	200/120	200/120	147	140/85	140/85	145/85	104	135/70	135/70	135/70	94	Yes
15	140/100	145/100	150/100	115	135/85	140/90	145/90	105	135/85	135/85	140/85	102	Yes
16	160/100	160/100	160/100	120	135/85	135/85	135/85	102	135/85	135/85	135/85	102	Yes
17	210/110	215/110	220/110	145	135/85	140/85	140/85	103	130/85	135/85	130/85	101	Yes
18P	190/105	185/105	185/105	133	120/80	130/90	135/90	100	120/70	135/85	135/85	97	Yes
19	210/110	210/110	210/110	133	135/90	135/90	135/90	105	120/80	120/80	120/80	93	Yes
20‡	160/115	160/115	160/115	130	135/80	135/80	135/80	98	135/85	140/85	140/90	104	No
21	200/125	200/130	200/130	152	140/90	140/90	140/90	107	130/75	135/80	135/85	98	Yes
22	180/105	185/105	185/105	131	140/90	140/90	140/90	107	135/85	135/90	135/90	104	Yes
Average				137				102				100.04 (100)	

* B. P. = blood pressure. P = patients taking drugs capable of producing postural hypotension. Mean B.P. = diastolic blood pressure + one-third pulse pressure.

† Significant reduction in drug requirement = Drug requirement reduced by 33 per cent or more of the original requirement. This was seen in 13 patients (these are indicated "yes"). The average reduction was by 68 per cent after "shavasan" in these 13 patients.

‡ = irregular (six patients).

TABLE 4

Group III: 15 patients not adequately controlled by drugs, given "shavasan" and attempts made to reduce blood pressure and drug requirement whenever possible

Patient No.	Initial B.P. before Drug			Initial Mean B.P.	B.P. with Drugs			Mean B.P.* with Drugs	B.P. after "Shavasan"			Mean B. P. after "Shavasan"	Drug Requirement Significantly Reduced†
	Upright	Sitting	Recumbent		Upright	Sitting	Recumbent		Upright	Sitting	Recumbent		
		mm Hg		*mm Hg*		*mm Hg*		*mm Hg*		*mm Hg*		*mm Hg*	
1	225/120	230/120	235/120	157	170/80	170/80	170/80	110	120/70	120/70	120/70	87	Yes
2P	200/120	200/120	200/120	147	140/90	215/105	215/105	130	140/90	210/105	220/105	130	No‡
3P	200/120	200/120	200/120	147	130/85	170/105	180/110	120	120/80	165/100	180/115	115	Yes
4	190/130	190/130	190/130	150	165/120	165/120	165/120	135	130/80	145/80	145/80	100	Yes
5	180/120	180/120	180/120	140	155/100	155/100	155/100	118	130/100	130/100	130/100	110	Yes
6	195/130	190/130	185/130	150	160/100	160/100	160/100	120	130/80	130/80	130/80	97	Yes
7P§	180/120	180/120	180/120	140	140/95	160/115	165/120	125	120/85	140/90	155/95	106	Yes
8	195/120	195/120	195/120	145	150/105	150/105	150/105	120	145/95	140/95	140/95	111	No
9P	170/120	170/120	170/120	137	130/80	170/105	180/110	119	130/90	170/105	180/105	120	No
10P	220/140	220/140	220/135	165	145/95	165/130	170/130	132	140/90	170/105	170/105	120	No
11§	160/100	160/100	160/100	120	155/75	150/75	150/75	100	135/85	140/85	140/85	103	No
12P§	230/140	230/140	230/140	170	145/95	160/115	160/120	125	145/90	165/110	170/115	123	No
13	200/120	200/120	200/120	147	150/105	150/105	150/105	120	125/90	130/85	130/85	101	No
14P	200/120	200/120	200/120	147	140/90	175/95	180/100	118	145/85	175/100	210/100	122	No‡
15P§	180/130	180/130	180/130	147	130/80	145/95	160/95	108	110/75	145/95	150/100	105	No
Average				147.2 (147)				120				110	

* Mean B.P. = diastolic blood pressure + one-third pulse pressure. P = patients taking drugs capable of producing postural hypotension.

† Significant reduction in drug requirement = drug requirement reduced by 33 per cent or more of the original requirement. This was seen in six patients (these are indicated as "yes"). The average reduction was by 72 per cent in these six patients.

‡ Drug requirement slightly increased (2 patients).

§ Irregular or unable to perform the exercise correctly (4 patients, 2 + 2).

TABLE 5

The average mean blood pressure of the three groups

Group	Mean Blood Pressure before Any Therapy (Original Blood Pressure)	Initial Mean Blood Pressure with Placebo or Drugs
	mm Hg	
I	136	Placebo, 134
II	137	Drugs, 102
III	147	Drugs, 120

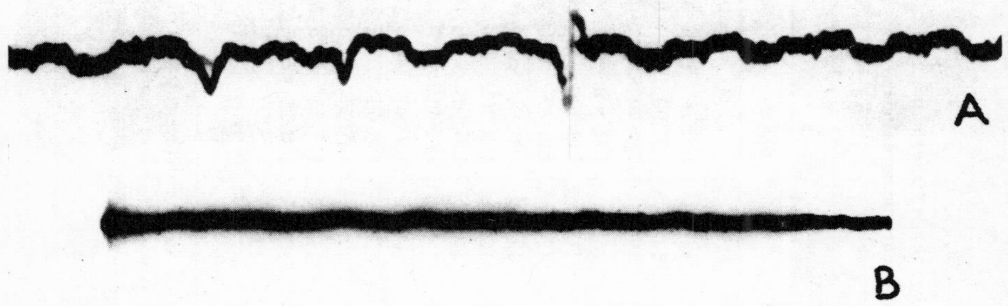

Fig. 1. Electromyogram from frontalis muscle (*A*) before and (*B*) during "shavasan" (yogic exercise) (patient supine, eyes closed).

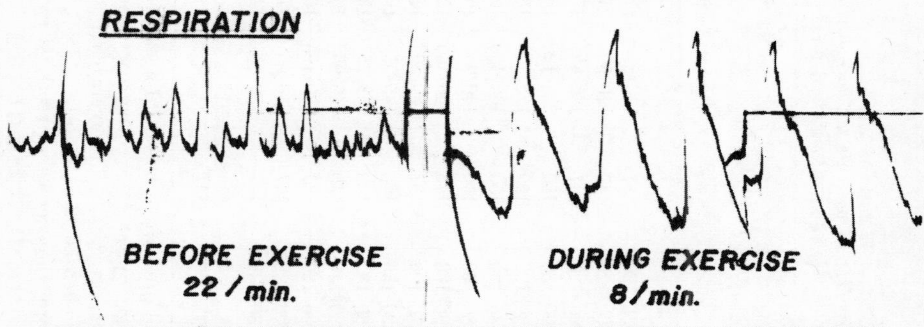

Fig. 2. Physiogram showing a fall in respiratory rate during "shavasan" yogic exercise

no attempt was made to reduce it further; only the drug requirement was gradually reduced, keeping the mean blood pressure constant. It was possible to reduce the average drug requirement to 32 per cent of the original, in 13 patients (59 per cent of the cases). This is statistically significant (p 0.05) (table 7). Of the remaining nine patients in whom the drug requirement

TABLE 6

The objective improvement in 10 patients without drugs

Blood Pressure	Average
	mm Hg
Initial mean	134
After "shavasan"	107*
Reduction	27

* *p* 0.05.

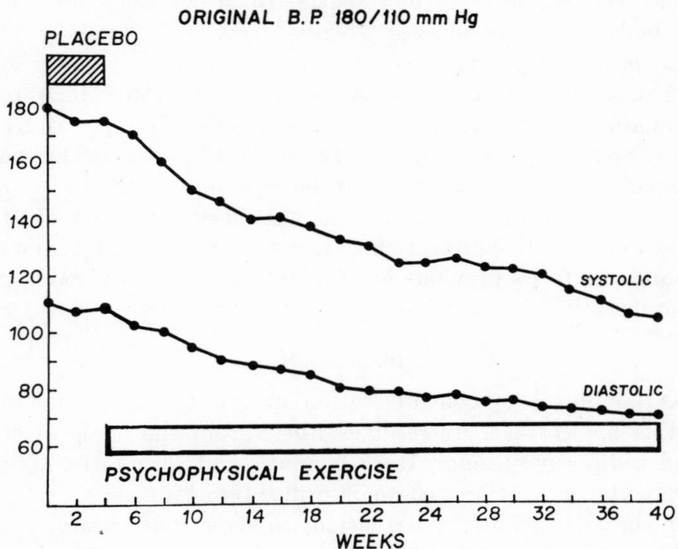

FIG. 3. Typical response of blood pressure with "shavasan" (yogic exercise) (patient from group I).

TABLE 7

Objective improvement in groups II and III (37 patients on drugs)

Group and No. of Patients	Average Initial Mean Blood Pressure	Average Mean Blood Pressure after "Shavasan"	Drug Requirement
	mm Hg		
II (22)	102	100	32% in 13 (59%)*
III (15)	120	110	29% in 6 (40%)

* *p* 0.05.

could not be reduced, six were irregular in performing the exercise but the remaining three were regular. The average mean blood pressure after the yogic exercise in this group of 22 patients was about the same (102 to 100 mm Hg). Out of 15 patients in group III, the drug requirement could be reduced to 29 per cent of the original in 6 patients (40 per cent) (table 7). The dose was unchanged in 7 (of these, 2 were irregular and 2 could not perform the exercise correctly). The dose had to be slightly increased in 2 other patients, who were regular with their exercise. The average mean blood pressure after the yogic exercise in this group of 15 patients had come down from 120 to 110 mm Hg.

Etiology and response relationship. Although there is a slight difference in the favorable response in cases of essential (62.5 per cent) and renal hypertension (42 per cent), it is not statistically significant. However, of the three patients of arteriosclerotic hypertension, none responded.

Of the 22 patients who did not respond to this therapy, 10 were either irregular or unable to perform the exercise correctly. Hence, failure to obtain a significant response was seen in only 12 out of 47 patients. One patient, who was very irregular in the beginning, responded only after he performed the exercise regularly. Another patient, who was very well controlled with this exercise, was not able to perform the exercise regularly for 1 month and the blood pressure rose to its previous level. However, this could be brought down again after reinstating the exercise and doing it regularly.

DISCUSSION

The hypothalamus is a control station of the autonomic nervous system and maintains homeostasis. It receives fibers from the cerebral cortex, both directly and indirectly through the thalamus, and also forms a part of the ascending reticular activating system described by Magoun.[2]

Whatever the etiology of hypertension, there is, *pari passu,* a rise in the level at which the homeostatic mechanisms operate and these react against both rise and fall of blood pressure.[3] Thus, the regulatory mechanism in the hypothalamus is probably set at a higher level in hypertension, and if it can be reset at the normal level, the hypertension may be controlled.

Pompeiano and Swett[4] have shown that a low rate, low intensity, monotonous stimulation of peripheral afferent nerves brings about electroencephalographic synchronization and even sleep, by decreasing the tonic activity of the ascending reticular activating system. Further, Bronk *et al.*[5] have demonstrated rhythmic variations in the potential in the hypothalamus by stimulating the afferent visceral nerves through which sympathetic responses may be elicited. Thus, it is possible to influence the hypothalamus by impulses through afferent somatic and visceral nerves.

While performing this exercise, the person relaxes with slow, rhythmic diaphragmatic breathing, thereby reducing the frequency and intensity of the proprioceptive and enteroceptive impulses. Further, he remains inwardly alert but is less conscious of the external environment. It is therefore postulated that this exercise probably influences the hypothalamus through the continuous feedback of slow, rhythmic proprioceptive and enteroceptive impulses and tends to set it at a lower level, thereby reducing the blood pressure.

The ultimate results of such therapy will depend in individual cases on the forces of factors raising the blood pressure by influencing the hypothalamus as against the therapeutic potentials of this exercise.

There was significant response both in essential as well as renal hypertension. However, there was no statistical difference between them. No significant response was seen in patients with arteriosclerotic hypertension.

The essential requirements for good response are: correct technique, regularity in performing the exercise and a quiet environment. The disadvantages are: (1) the daily duration of the exercise, which is 30 min., is rather long and (2) it is difficult to perform this exercise correctly in the presence of nasal congestion.

SUMMARY

"Shavasan," a yogic exercise, was given to 47 patients with hypertension of various etiologies. A significant response was obtained in about 52 per cent of the patients. There was no response in patients with arteriosclerotic hypertension. The exercise is easy to perform, has no side effects and requires no equipment. There was symptomatic relief and a sense of "well being" in the vast majority of the patients. This therapy opens a new avenue in the management of hypertension.

K. K. Datey, M.D., F.I.C.A., F.R.C.P. (Lond.)
Director, Dept. of Cardiology
K. E. M. Hospital, Bombay-12, India

REFERENCES

1. Bilgutay, A. M., and Lillehai, C. W.: Treatment of hypertension with implantable electronic device. *J. A. M. A., 191:* 649, 1965.
2. Magoun, H. W.: In *The Ascending Reticular Activating System.* Research Publication of the Association for Nervous and Mental Disease, Vol. 30. Williams & Wilkins, Baltimore, 1950, p. 480.
3. Smirk, F. H.: *High Arterial Pressure.* Blackwell, Oxford, 1957, p. 327.
4. Pompeiano, O., and Swett, J. E.: E.E.G. and behavioral manifestations of sleep induced by cutaneous nerve stimulation in normal cats. Arch. Biol., *100:* 311, 1962.
5. Bronk, D. W., Lewy, F. H., and Larrabee, M. G.: The hypothalamic control of sympathetic rhythms. Am. J. Physiol., *116:* 15, 1936.

68 A Physiological Model of Phobic Anxiety and Desensitization

M. H. Lader and A. M. Mathews

INTRODUCTION

OVER the last decade increasing attention has been paid to the use of treatments said to be based on learning theory principles, generally termed 'behaviour therapy'. Implicit in the utilization of learning theory models is the hypothesis that neurotic behaviour results from a conditioning process; in its most simple form this has been assumed to occur when a previously neutral stimulus is repeatedly associated, presumably by chance, with unconditioned stimuli producing excessive emotional reactions, (Eysenck and Rachman, 1965). Treatment is therefore based on methods found effective in reducing fears which have been produced by similar conditioning under experimental conditions; that is, extinction and reciprocal inhibition (Wolpe, 1958).

In clinical practice the most widely accepted behaviour therapy technique is that of systematic desensitization in the treatment of phobic anxiety. This technique has been the subject of a large number of recent studies (Rachman, 1967) and consists of the presentation of graded phobic stimuli (usually in imagination) while the patient is deeply relaxed.

The purpose of the present paper is to propose an alternative model of the genesis of phobic anxiety, based in part on experimental work using psychophysiological measures of arousal, and to derive from this an account of the mechanism of desensitization which is distinct in many respects from that offered by Wolpe. The mechanisms proposed are not intended to be complete in themselves since in the interests of simplicity, the relevance of such factors as symbolic and other cognitive processes, social learning and inter-personal relationships has not been considered.

THE RELATIONSHIP BETWEEN 'AROUSAL' AND PHOBIC ANXIETY

In the context of this paper, the concept of arousal will be employed in a non-theoretical way; we will use the term to refer to one particular aspect of physiological activity e.g. the skin conductance level and the frequency of spontaneous skin conductance fluctuations.

* Supported by the Medical Research Council, U.K. directly to M.H.L. and by grant no. G963/173 to A.M.M. The authors are indebted to Dr. S. Rachman for helpful comments and suggestions.

However, cautious extension of the limited physiological interpretation of arousal to the behavioural field is used in the thesis to be proposed and is warranted by experimental evidence regarding spontaneous skin conductance fluctuations (Lacey and Lacey, 1958; Burch and Greiner, 1960).

The relationship between physiological measures of arousal and anxiety has been reviewed on several occasions (e.g. Malmo, 1957; Martin, 1961; Lader, 1968). However, few successful attempts have been made to utilize such measures in the evaluation of behaviour therapy. This lack may be partly attributed to the confusing and contradictory results obtained by different workers using disparate measures, conditions of measurement, and clinical groups. Recent work (Martin, 1960) suggests that rather more consistent results are obtainable if patients are studied as relatively homogeneous clinical groups; for example 'anxiety states' and 'specific phobics' are more profitably considered as separate groups, rather than being subsumed under the rubric 'anxious patients'.

In a previous publication, Lader and Wing (1966) reported that patients with anxiety states showed many more spontaneous fluctuations in their palmar skin conductance tracings than normal Ss matched for age and sex; anxious patients habituated much more slowly to a series of 20 identical auditory stimuli as assessed by the GSR. A more recent extension of this work (Lader, 1967) demonstrated that agoraphobic and socially phobic patients show only slightly fewer spontaneous fluctuations than do patients with anxiety states, while specific phobics (e.g. of animals or objects) have more fluctuations than normals but are still well below the levels of the other phobic groups. Complete figures for all groups tested are given in Table 1.

TABLE 1.

Clinical group	Number in sample	Spontaneous fluctuations per minute	Rate of habituation*
Anxiety state	16	6.8	29
Agoraphobia	19	6.0	39
Social phobia	18	6.2	39
Specific phobia	19	2.8	68
Normals	75	1.5	72

*The higher the value given, the more rapid the rate of habituation.

Another experimental finding was that the rate of habituation of the GSR was inversely related to the number of spontaneous fluctuations (arousal level). It was postulated that if the level of arousal at any moment is low, the intrusion of a repetitive or continuing stimulus would have only a transitory effect in raising the level of arousal further and the ensuing habituation would be rapid. This is shown diagrammatically in the lower part of Fig. 1. However, if the ongoing level of arousal is high, habituation would be slow, especially at first (upper part of Fig. 1).

With this model it can be seen that a critical level of arousal would be predicted above which a repetitive stimulus would not be accompanied by any habituation; instead the level of arousal would become higher with each successive stimulus producing a "positive

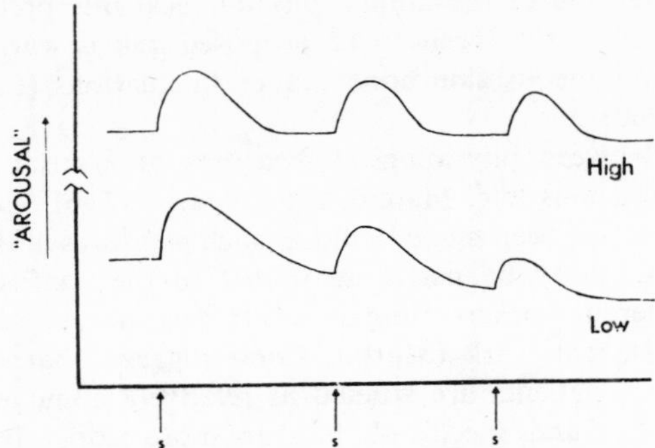

FIG. 1. Diagrammatic representation of the relationships found between level of arousal (in terms of spontaneous skin conductance fluctuations) and habituation to repeated stimuli (s). The upper part shows the responses of a highly aroused individual, and the lower part the responses of a S at a lower level of arousal.

feed-back" mechanism. The result of this is that even a minimal stimulus (when repeated) will produce a series of responses which will maintain the over-arousal. This situation is represented diagrammatically in Fig. 2. The 'level of arousal' of an individual is equated, for the purposes of the diagram, with the number of spontaneous fluctuations of skin conductance occurring at rest. The 'critical level' for this individual is represented by a hatched area above which habituation is slow or non-existent depending on how intense the experimental stimulus is for the individual.

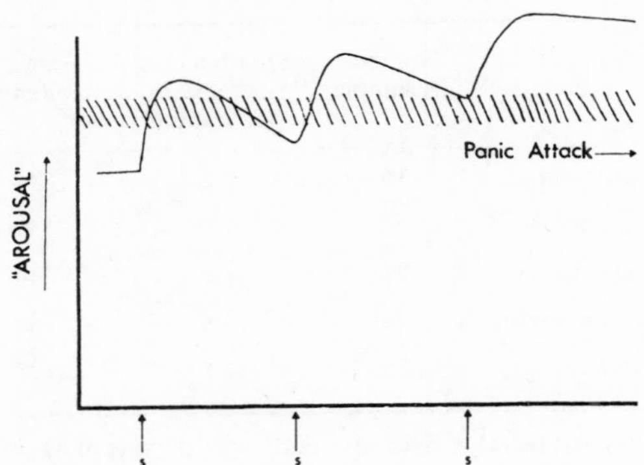

FIG. 2. Diagrammatic representation of the postulated genesis of a panic attack resulting from repeated stimulation of a highly aroused individual.

We conceptualise the reactions of a patient suffering from an anxiety state as essentially similar to this latter example, in that constant stimulation of even a mild nature would be sufficient to maintain, or further elevate, a pathologically high level of arousal, experienced as anxiety.

Although the reaction of pathological anxiety to environmental or even ideational stimuli may seem a remote extrapolation from prolonged habituation to auditory stimuli, it is suggested that the orientation reaction, the defensive reaction (Lynn, 1966), and the panic reaction may be viewed as responses lying along a single continuum in terms of their physiological and behavioural components. Which response occurs may depend not only on the intensity of stimulation but also on the level of arousal prior to stimulation. The panic reaction has previously been described as a disorganised attempt to reduce the level of stimulation (Schultz, 1964).

Clearly, almost any commonly occurring stimuli, under these conditions, may very quickly produce a conditioned response of anxiety; but the model used here would lead us to predict that any improvement following desensitization to these stimuli would be short-lived in the absence of a relatively permanent decrease in arousal.

In the case of agoraphobia (and social phobia), the average number of spontaneous fluctuations of skin conductance, and therefore presumably the level of arousal, is only slightly lower than in anxiety states, which might seem inconsistent in the terms of the present hypothesis, in view of the lack of completely generalized anxiety in agoraphobics. This difficulty may be resolved by postulating that the resting level of arousal in agoraphobic patients is close to, but below, their 'critical level'; and also by taking into account the differences in the *arousing* properties of different stimuli.

It would follow from this that stimuli of low arousing properties, e.g. of low intensity, would not increase arousal above the critical level, but that stimuli of moderate arousing properties would do so. When stimuli of this type occur frequently the possibility of positive feed-back taking place may be assumed to be high, thus producing a panic attack. Under these conditions the agoraphobic may be supposed to behave in the same way as previously described in anxiety states.

Stimuli associated with high arousing properties are generally of high intensity, complexity, novelty etc. (Berlyne, 1960). Stimuli of this type are predominantly found outside the home, in crowds, shops, while travelling, and in any unfamiliar place; or in other words the very situations typically most feared by agoraphobics. It is not suggested that this represents a complete explanation of agoraphobic symptoms, as stimuli of a covert or symbolic nature may also be involved in determining arousal level. We cannot use this model to explain, for example, why agoraphobics can often tolerate many more situations when accompanied by someone they know or when driving instead of walking. However, it is consistent with the characteristic clustering of avoided situations, not normally found, for instance, in patients with animal phobias. Again, repetition of these conditions may be expected to result in the rapid learning and generalization of fears to relatively milder stimuli, and the effects of such learning may be alleviated by behaviour therapy. However, the model used here would again require that prediction that this improvement would readily break down under stressful stimulation leading to frequent relapses.

Patients with specific fears have the lowest resting rate of spontaneous fluctuations of skin conductance, not significantly different from normal Ss, and habituation to auditory stimuli also follows a normal pattern. These patients, then, seem to be characterized by abnormal defensive or phobic responses to only precisely defined types of stimuli, presumably as a consequence of some kind of conditioning process. However, it has been shown (Marks and Gelder, 1966) that these patients differ from other phobics in that their fears are uniformly of early onset (mean age 5 yr), so that specific phobics may be viewed as those whose 'normal' childhood fears have failed to diminish as they get older. One possible hypothesis that could account for this failure is represented in Fig. 3. Here a stressful stimulus (e.g. feared object) presented to one individual produces an increase in arousal sufficient to exceed the 'critical level' so that habituation to that specific stimulus is slow or absent; while in another individual an initial response not exceeding his critical

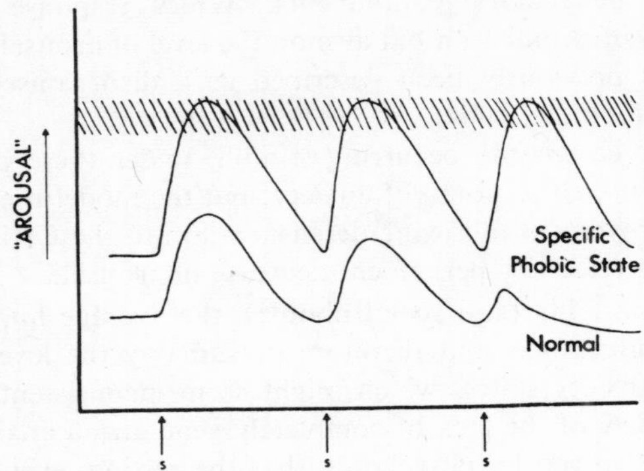

FIG. 3. Diagrammatic representation of a postulated mechanism to account for the persistence of specific phobias.

level does not prevent normal process of habituation. This difference could be brought about by either a relatively large initial response (innate or conditioned) or by individual variation in critical level and resting level of arousal. Unlike the conditions previously discussed, in the case of specific phobics it would be expected that successful behavioural treatment should be permanent, since once the occurrence of phobic stimuli is no longer accompanied by an abnormal increase in arousal, habituation should proceed normally. The continuation of orientation reactions to novel presentations of phobic stimuli following treatment would, of course, be expected, but should not lead to relapse provided that the critical level is not exceeded. This expectation does not seem inconsistent with the few attempts to use skin conductance measures in the assessment of desensitization treatment (e.g. Hoenig and Reed, 1967).

In summary, the form that a particular phobic reaction takes is supposed to result from an interaction between level of arousal and conditioning. In the case of agoraphobic and social phobic patients the primary pathology is seen as a chronic state of overarousal with conditioning of anxiety to environmental stimuli playing a minor part. In the case of monosymptomatic fears of specific objects or situations, conditioning of anxiety, or its failure to extinguish, is thought to be of relatively greater importance than any permanent state of high arousal. This distinction is posited to underlie the different response rate of each phobic group to behavioural treatment.

WOLPE'S THEORY OF THE MECHANISM OF SYSTEMATIC DESENSITIZATION

Wolpe (1958) states that the desensitization procedure produces results by inhibiting unadaptive anxiety. Thus, "if a response antagonistic to anxiety can be made to occur in the presence of anxiety evoking stimuli so that it is accompanied by a complete or partial suppression of the anxiety responses the bond between these stimuli and the anxiety responses will be weakened". He postulates that the presentation of each item in the imaginal hierarchy elicits a small amount of anxiety which is counteracted by relaxation on the part of the S. This antagonism he terms "reciprocal inhibition" using a Sherringtonian analogue. The extension of the use of the term "reciprocal inhibition" is unfortunate. Sherrington (1947) uses the term to refer to inhibition occurring simultaneously with excitation but in different loci, viz. neurone pools in the spinal cord. For example, excitation causing flexion of the leg muscles also causes inhibition of the nerves innervating the extensors of the limb so that the leg is able to flex smoothly. However, Wolpe implies by the use of the term the quenching of one response by a super-added incompatible one. Furthermore he postulates that reciprocal inhibition, that is, the strength of the "negative habit" of not responding in an anxious way to the stimulus would increase with repetition and would result in a diminution in the intensity of the phobic reaction.

Wolpe regards anxiety and relaxation as mutually incompatible states. This would seem a reasonable premise, but the issue becomes clouded by the stress on *muscular* relaxation. Jacobson (1964) has argued that emotional reactions are not merely accompanied by, but are dependent on, increased muscular activity. It may be seen that this is a restatement of the James-Lange hypothesis. Jacobson's corollary is that if relaxation can be established, then emotional responses will be damped down by virute of diminution in proprioceptive impulses. Although the possibility that a James-Lange type of mechanism may contribute to the experience of emotion cannot be discounted (Gellhorn, 1964; Schachter, 1966) the bulk of the evidence would suggest that this is not the only mechanism. Certainly muscular activity has not been conclusively demonstrated as causal to emotional experience. Part of the confusion may be semantic. It is assumed that because the patient complains of feeling tense there is increased tension in at least some muscles. This is by no means so and is tantamount to assuming that a verbal parallelism is a physiological equivalence. Indeed, we have observed on several occasions that forearm electromyographic activity in an anxious patient can be zero while autonomic activity is marked and the patient reports feeling anxious. Conversely, we have obtained records in which forearm EMG has increased following relaxation training, although the patient reports a decrease in subjective anxiety. The relative inefficacy of muscle relaxant drugs such as mephenesin in the treatment of anxiety states underlines this point. Therefore, the induction of *muscular* relaxation need not necessarily produce any diminution in the anxiety as experienced by the S.

Whatever the precise mechanism of 'relaxation', a weight of evidence has accrued demonstrating that *both* relaxation and the graded presentation of stimuli are essential to the technique (Kondas, 1967; Lang and Lazovik, 1963; Lang, Lazovik and Reynolds, 1965; Lomont and Edwards, 1967; Moore, 1965; Paul, 1966; Rachman, 1965, 1966, 1967). In most of these experiments, separate groups of comparable phobic patients were treated with relaxation alone, relaxation plus desensitization and usually one or more control groups which received no treatment or psychotherapy or presentation of irrelevant stimuli. In general, the groups treated with desensitization and relaxation were the only ones to respond significantly. This is in accord with the requirements of the reciprocal inhibition model used by Wolpe.

AN ALTERNATIVE MODEL OF THE MECHANISM OF DESENSITIZATION

Using the assumptions already outlined, i.e. the similarity between the physiological components of the orientation and phobic reactions, and that hyperarousal may be experienced as anxiety, an alternative and more parsimonious model of the mechanism of desensitization can be put forward. This proposes that systematic desensitization may be regarded as habituation occurring when the rate of habituation is maximal; that is, when the level of arousal is as low as possible consistent with clear consciousness. Thus, the need to train the patient to relax is not because of the muscular relaxation induced *per se*, but because the instructions used and conditions inherent in relaxation training subserve the function of engendering a low level of arousal. Preliminary work (unpublished) strongly supports the idea that while successful training in relaxation (in terms of subjective experience) is not inevitably accompanied by a diminution of EMG, it is accompanied by a significant drop in the number of spontaneous fluctuations of skin conductance.

It can be seen that the model proposed is similar to the theory that desensitization is extinction but with the all-important difference that level of arousal must be contemporaneously as low as possible; and it is of interest here that a recent review of the habituation phenomenon found no grounds for a distinction in the mechanisms underlying extinction and habituation (Thompson and Spencer, 1966).

In a recent review of desensitization, Rachman (1967), commenting on Lomont's (1965) arguments in favour of extinction, picked out three studies which "justify the conclusion that reciprocal inhibition produced by relaxation is superior to extinction as a method of reducing fear". These three studies—Lomont and Edwards (1967), Rachman (1965) and Davison (1968)—should therefore be briefly considered in the evaluation of what may be termed the maximal habituation hypothesis. In each of these three experiments, the control group receiving extinction procedures in the absence of reciprocal inhibition were presented with imaginal items from the fear hierarchy alone, (Rachman, 1965; Davison, 1968) or in combination with a control procedure involving maintaining steady muscular tension in place of relaxation (Lomont and Edwards, 1967). It is apparent that all of these control groups were subjected to procedures likely to maintain, or even raise their general level of arousal while experimental Ss were relaxed, presumably leading to reduced arousal, so that a test of the maximal habituation hypothesis is not provided.

An apparently more crucial experiment supporting the greater efficacy of reciprocal inhibition procedures was carried out by Gale, Strumfels and Gale (1966) using three groups of rats matched on the basis of their acquisition of a conditioned fear response to a sound stimulus. The rats were split into three groups: (a) a group who were presented with the sound without reinforcement (extinction): (b) a group who were presented with the sound as they fed (reciprocal inhibition): and (c) a no treatment group. The fear response was abolished in the reciprocal inhibition group more rapidly than in the extinction group. However, all the rats were food deprived which is known to result in increased arousal measured by behavioural or physiological criteria; in the animals given food as part of the reciprocal inhibition procedure, this elevated level of arousal would be lowered. Consequently even this experiment is not crucial to the maximal habituation hypothesis.

It may be objected that both the reciprocal inhibition and maximal habituation model require treatment to take exactly the same form; that is, alternate presentations of graded phobic stimuli, and relaxation. Clearly if precisely the same predictions follow from both models, nothing new has been said, although it could be argued that the maximal habituation model is more parsimonious and avoids the difficulties involved in the Hullian concept of conditioned inhibition. However, if the hypothesis is to be of practical use it must either account for experimental findings more satisfactorily than reciprocal inhibition, or must lead to new predictions.

PREDICTIONS ARISING FROM THE MAXIMAL HABITUATION HYPOTHESIS

(1) As has already been outlined, marked differences in response to behaviour therapy in different phobic groups would be expected on the basis of physiological differences. Marks and Gelder (1965) and Gelder, Marks and Wolff (1967) have offered evidence that this is the case, and that while improvement rates of 80–90 per cent are found in patients with monosymptomatic phobias, only about 60 per cent of agoraphobic patients improve appreciably, despite fairly lengthy treatment. It could be countered that since agoraphobic patients frequently have widespread phobias and suffer from a greater degree of generalised anxiety reciprocal inhibition of phobic responses will naturally be more difficult and take a longer time. However, no explanation along conventional behaviour theory lines would seem to account for this marked disparity in clinical presentation as well as their different response rates.

(2) The hypothesis allows a more direct means of predicting treatment outcome than by clinical group, since the rate of habituation as determined by response to auditory stimuli should correlate with habituation to phobic stimuli in desensitization. Evidence is available that outcome to desensitization is correlated with habituation rate to a significant degree (Lader, Gelder and Marks, 1967). In addition a generally high level of arousal as measured by fluctuations of skin conductance would lead to the prediction of a poor prognosis, unless training in relaxation was successful in producing a sufficient reduction.

(3) Following the successful application of desensitization treatment, it can be predicted from what has already been said, that the probability of relapse will depend crucially on the general level of arousal of the individual patient. Relapse may be expected even in a successfully treated monosymptomatic phobia if phobic stimuli should coincide with a temporary period of high arousal. The phenomenon of dishabituation in dogs has been fully described by Pavlov (1927). In the case of human Ss two interesting anecdotal accounts are given by Glaser (1966) in which experimental Ss who had completely habituated to a painful stimulus showed a sudden spontaneous recovery of physiological response under stressful circumstances, in one case being in front of an audience (P.40) and in another following local riots (P.43).

(4) An alternative method of reducing level of arousal would be through the use of drugs, and the short acting barbiturate methohexitone sodium has been used in recent studies

(Friedman, 1966; Friedman and Silverstone, 1967) with apparent success. Although it is premature to state that drug-assisted desensitization is superior to the usual form, it is claimed to produce more rapid results. Such a finding would certainly be consistent with the maximal habituation hypothesis but may present some difficulty for the account of desensitization given by Wolpe (1958). Since methohexitone sodium is a depressant drug the accumulation of conditioned inhibition should be slowed if more than the absolute minimum dose required to inhibit anxiety is given. Indeed, it is not clear whether the drug reaction should even be considered a response incompatible with anxiety in the sense normally used in reciprocal inhibition.

It may also be added that the methohexitone sodium technique has the unavoidable disadvantage that a delay of 15–20 sec must take place before any drug can reach the brain and have a pharmacodynamic effect. This time lag would have to be taken into account in desensitization technique if the reciprocal inhibition model is used.

(5) Eysenck and Rachman (1965) in describing the mechanism of desensitization state that "with each evocation and subsequent dampening down of the anxiety responses, conditioned inhibition is built up". Evoking actual anxiety responses is not a necessary part of the maximal habituation hypothesis so that it is predicted that desensitization with drugs, or relaxation, could be carried out successfully in the absence of any subjective anxiety at all on the part of the patient. A further difference following from this is that, whereas Wolpe would predict that anxiety inhibiting responses should be concentrated in the interval immediately *following* the presentation of a phobic stimulus, maximal habituation should occur if relaxation or drug injection occurs immediately *before* presentation, in sufficient quantity to prevent any arousal of anxiety at all during presentation.

(6) Lomont (1965) has raised two points in his discussion of extinction and desensitization which apply equally well to the maximal habituation model. Firstly, the phenomenon of spontaneous recovery, demonstrated by Agras (1965) to occur during desensitization, is well established in extinction and habituation studies, but has no clear place in the reciprocal inhibition model.

Secondly, the question of massed vs. spaced practice, touched on by Lomont, has not yet been fully examined despite the difference in predictions, arising from reciprocal inhibition and habituation/extinction. A study by Ramsay, Barends, Breuker and Kruseman (1966) appears to have demonstrated that spaced practice in desensitization is superior in its effects to massed practice, contrary to the effects usually found in habituation and extinction. Some reservations should be made concerning this, however, as in this study inter-item interval was in fact the same in both groups, the only difference being that the spaced practice groups were given four periods of practice with four hierarchy items per session, while the massed group had two periods of 8 items.

The evidence relating to the effect of inter-trial interval on extinction is in fact contradictory. Studies by Teichner (1952) and Cole and Abraham (1962) suggest that this may be in any case a small effect and extinction rate depends more on the interaction with inter-trial interval during acquisition. In the case of habituation rate, the evidence does indicate more rapid habituation within sessions, the shorter the inter-stimulus interval, with the magnitude of response showing exponential decay (Thompson and Spencer, 1966). There does not seem to be a very marked effect due to length of time between sessions on absolute rate of habituation (Montagu, 1963) although the size of initial response in repeated sessions becomes progressively less. In order to predict from the maximal habituation hypothesis the result of the Ramsay *et al.* (1966) experiment, we need to know more about the effect of massed practice in habituation on spontaneous recovery and generalisation, including the interaction between inter-stimulus interval, number of stimuli used per session and amount of time between sessions.

It seems possible that the optimum method of obtaining relatively long-lasting habituation is in fact to give periods of massed practice until habituation appears complete, then allowing a rest interval, followed by further periods of massed practice to deal with spontaneous recovery. This is amenable to direct experimental testing.

Clearly, further work in this area would be useful, not only for the purpose of deciding a theoretical issue, but also because of the possibility of immediate practical gains in desensitization treatment.

CONCLUSION

It is appreciated that many of the hypotheses put forward in this paper must be regarded as highly speculative at the present time. However, it is hoped that at least some of the views are of heuristic value and that the habituation model of desensitization will be of practical therapeutic utility.

REFERENCES

AGRAS, W. S. (1965) An investigation of the decrement of anxiety responses during systematic desensitization therapy. *Behav. Res. & Therapy* **2**, 267–270.

BERLYNE, D. E. (1960) *Conflict, Arousal and Curiosity*. McGraw-Hill, New York.

BURCH, N. R. and GREINER, T. H. (1960) A bioelectric scale of human alertness: concurrent recordings of the E.E.G. and G.S.R. *Psychiat. res. Rep. Am. Psychiat. Ass.* **12**, 183–193.

COLE, M. and ABRAHAM, F. (1962) Extinction and spontaneous recovery as a function of amount of training and extinction inter-trial interval. *J. physiol. Psychol.* **55**, 978–982.

DAVISON, G. (1968) Systematic desensitization as a counter-conditioning process. *J. abnorm. Psychol.* (In press).

EYSENCK, H. J. and RACHMAN, S. (1965) *Causes and Cures of Neurosis.* Routledge and Kegan Paul, London.

FRIEDMAN, D. E. I. (1966) A new technique for the systematic desensitization of phobic symptoms. *Behav. Res. & Therapy* **4,** 139–140.

FRIEDMAN, D. E. I. and SILVERSTONE, J. T. (1967) Treatment of phobic patients by systematic desensitization. *Lancet* **1,** 470–472.

GALE, D. S., STRUMFELS, G. and GALE, E. N. (1966) A comparison of reciprocal inhibition and experimental extinction in the psychotherapeutic process. *Behav. Res. & Therapy* **4,** 149–155.

GELDER, M. G., MARKS, I. M. and WOLFF, H. H. (1967) Desensitization and psychotherapy in the treatment of phobic states: a controlled enquiry. *Br. J. Psychiat.* **113,** 53–73.

GELLHORN, E. (1964) Motion and emotion: the role of proprioception in the physiology and pathology of the emotions. *Psychol. Rev.* **71,** 457–472.

GLASER, E. M. (1966) *The Physiological Basis of Habituation.* Oxford University Press, London.

HOENIG, J. and REED, G. F. (1966) The objective assessment of desensitization. *Br. J. Psychiat.* **112,** 1279–1283.

JACOBSON, E. (1964) *Anxiety and Tension Control. A Physiological Approach.* Lippincott, Philadelphia.

KONDAS, O. (1967) Reduction of examination anxiety and "stage-fright" by group desensitization and relaxation. *Behav. Res. & Therapy* **5,** 273–281.

LACEY, J. I. and LACEY, B. C. (1958) The relationship of resting autonomic activity to motor impulsivity. *Proc. Ass. Res. nerv. ment. Dis.* **36,** 144–209.

LADER, M. H. (1967) Palmar conductance measures in anxiety and phobic states. *J. psychosom. Res.* **11,** 271–281.

LADER, M. H. (1968) Psychophysiological aspects of anxiety. In *Studies of Anxiety: A Symposium,* (Ed. LADER, M. H.) Royal Medico-Psychological Association Special Monograph, London (In press).

LADER, M. H., GELDER, M. G. and MARKS, I. M. (1967) Palmar conductance measures as predictors of response to desensitization. *J. psychosom. Res.* **11,** 283–290.

LADER, M. H. and WING, L. (1966) *Physiological Measures, Sedative Drugs, and Morbid Anxiety.* Oxford University Press, London.

LANG, P. J. and LAZOVIK, A. D. (1963) Experimental desensitization of a phobia. *J. abnorm. soc. Psychol.* **66,** 519–525.

LANG, P. J., LAZOVIK, A. D. and REYNOLDS, D. J. (1965) Desensitization, suggestibility, and pseudotherapy. *J. abnorm. Psychol.* **70,** 395–402.

LOMONT, J. F. (1965) Reciprocal inhibition or extinction? *Behav. Res. & Therapy* **3,** 209–219.

LOMONT, J. F. and EDWARDS, J. E. (1967) The role of relaxation in systematic desensitization. *Behav. Res. & Therapy* **5,** 11–25.

LYNN, R. (1966) *Attention, Arousal and the Orientation Reaction.* Pergamon Press, Oxford.

MALMO, R. B. (1957) Anxiety and behavioral arousal. *Psychol. Rev.* **64,** 276–287.

MARKS, I. M. and GELDER, M. G. (1965) A controlled retrospective study of behaviour therapy in phobic patients. *Br. J. Psychiat.* **111,** 561–573.

MARKS, I. M. and GELDER, M. G. (1966) Different ages of onset in varieties of phobia. *Am. J. Psychiat.* **123,** 218–221.

MARTIN, B. (1961) The assessment of anxiety by physiological behavioral measures. *Psychol. Bull.* **58,** 234–255.

MARTIN, I. (1960) Somatic reactivity. In *Handbook of Abnormal Psychology.* (Ed. EYSENCK, H. J.) Pitman, London.

MONTAGU, J. D. (1963) Habituation of the psycho-galvanic reflex during serial tests. *J. psychosom. Res.* **7,** 199–214.

MOORE, N. (1965) Behaviour therapy in bronchial asthma: a controlled study. *J. psychosom. Res.* **9,** 257–274.

PAUL, G. L. (1966) *Insight versus Desensitization in Psychotherapy.* Stanford University Press, Stanford.

PAVLOV, I. P. (1927) *Conditioned Reflexes, an Investigation of the Physiological Activity of the Cerebral Cortex.* Oxford University Press, London.

RACHMAN, S. (1965) Studies in desensitization—I. The separate effects of relaxation and desensitization. *Behav. Res. & Therapy* **3,** 245–251.
RACHMAN, S. (1966) Studies in desensitization—II. Flooding. *Behav. Res. & Therapy* **4,** 1–6.
RACHMAN, S. (1967) Systematic desensitization. *Psychol. Bull.* **67,** 93–103.
RAMSAY, R. W., BARENDS, J., BREUKER, J. and KRUSEMAN, A. (1966) Massed versus spaced desensitization of fear. *Behav. Res. & Therapy* **4,** 205–207.
SCHACHTER, S. (1966) The interaction of cognitive and physiological determinants of emotional state. In *Anxiety and Behavior* (Ed. SPIELBERGER, C. D.). Academic Press, New York.
SCHULTZ, D. P. (1964) *Panic Behavior.* Random House Press, New York.
SHERRINGTON, C. S. (1947) *The Integrative Action of the Nervous System.* University of Cambridge Press, Cambridge.
TEICHNER, W. H. (1952) Experimental extinction as a function of the intertrial intervals during conditioning and extinction. *J. exp. Psychol.* **44,** 170–178.
THOMPSON, R. F. and SPENCER, W. A. (1966) Habituation: a model phenomenon for the study of neuronal substrates of behavior. *Psychol. Rev.* **73,** 16–43.
WOLPE, J. (1958) *Psychotherapy by Reciprocal Inhibition.* Stanford University Press, Stanford.

XI

VOLUNTARY CONTROL AND HYPNOSIS

69

Physiological Response to the Suggestion of Attitudes Specific for Hives and Hypertension

David T. Graham, J. D. Kabler, and Frances K. Graham

THE STUDY reported in this article is the third in a series designed to test a specificity-of-attitude hypothesis for psychosomatic disease proposed by Grace and Graham in 1952[6]. The method of these experiments was the manipulation of attitude (the independent variable) by direct suggestion under hypnosis while measurements were made of changes in the physiological state (the dependent variable). The advantages of this method are that the independent variable is defined by the operations of the experimeter, and that no inference as to the subjects' feelings or the depth of hypnosis is required. Previous experiments[7,15] yielded results supporting the hypothesis; the design of the experiment to be described provides more information about methodological questions and replicates the previous work.

The hypothesis states that there is associated with each psychosomatic disease a specific attitude toward the life events that precipitate the first appearance or later exacerbations of the disease. "Attitude" is defined as (1) how a person perceives his own position in a situation—what he feels is happening to him, and (2) what action, if any, he wishes to take. The attitudes are different for different diseases, but all persons with a given disease have the same attitude. On the basis of clinical study of 128 patients, Grace and Graham[6] described attitudes associated with 11 diseases.

In the present experiment two of these attitudes were employed, one associated with hives and the other with hypertension, and five dependent variables, skin temperature, diastolic and systolic blood pressure, pulse rate, and respiratory rate were measured. The specificity-of-attitude hypothesis predicted that there would be a greater rise in skin temperature and a smaller rise in diastolic blood pressure during suggestion of the attitude associated with hives than during suggestion of that associated with hypertension. No predictions were made concerning changes in the remaining three physiological variables.

The experimental design, a double two-by-two Latin square, permitted analysis of variance for treatment effect (the hypothesis being tested) and for day (first or second), order of presentation of attitude, intersubject differences, and differences between two experimenters.

From the Departments of Medicine, Pediatrics and Psychiatry, University of Wisconsin School of Medicine, Madison, Wis.

Supported by Grant M-2011 from the National Institute of Mental Health, U. S. Public Health Service.

Presented at the annual meeting of the American Psychosomatic Society, Mar. 25, 1960, Montreal.

Thanks are expressed to Mr. Paul Rasmussen for technical assistance.

Received for publication November 7, 1960.

Subjects, Apparatus, and Procedure

Subjects

Twenty male volunteer University students were paid for serving as subjects. Their ages ranged from 18 to 30, with a mean of 21. A brief preliminary interview on the day of the first experimental session for each subject was used to screen out any with overt emotional disturbance or other handicap.

Apparatus

Experiments were conducted in an air-conditioned laboratory whose temperature was kept 23°C., with a variation of not more than 1.5°C. during any one session. The subject was comfortably seated in an adjustable lounge chair. Skin temperature was measured by a thermistor fixed to the dorsum of the left hand and connected to a galvanometer that permitted readings accurate to within .01°C. Temperature changes were recorded continuously on an ink-writing polygraph. Systolic and diastolic blood pressures in the right arm were recorded at intervals of approximately 3 min. by an automatic sphygmomanometer that responded to sound signals over the brachial artery. This device made it unnecessary to touch the subject during the experiment. Only three blood pressure measurements were made during each 10-min. period in which an attitude was in effect and each 10-min. control period, since it was felt that more frequent measurements might produce vascular congestion in the arm, with consequent interference with the production of the blood pressure sounds. The lead of the electrocardiogram was recorded continuously on another channel of the polygraph. The respiration rate was recorded continuously on the polygraph by a strain gage attached to an elastic strap that was placed around the chest in such a way that subjects reported no subjective interference with breathing.

Procedure

The attitudes used in the present study were those previously found to be associated with urticaria and essential hypertension.[6] Each subject was given both attitude suggestions on the first experimental day. Half of the subjects had the hives attitude first on the first day; the other half had the hypertension attitude first. The orders were reversed on the second day. Subjects were assigned to these presentation orders by means of a table of random numbers, with the restriction that an equal number were assigned to each order for each of the two experimenters. Each experimenter conducted the experimental sessions on both days for half of the subjects. He induced hypnosis and made the attitude suggestions while an assistant monitored the apparatus.

The procedure on each of the experimental days included a prehypnotic period of 20–45 min. during which the physiological variables were recorded and allowed to stabilize. The experimenter instructed the subject and attempted to allay his anxiety during this time. Hypnosis was then induced by the method of ocular fixation, and several methods for testing the depth of the trance state were employed. After 10–25 min. the subject was given relaxation suggestions for a 10-min. period. This was the first control period. The selected attitude suggestion was then given continuously for 10 min. This was the first experimental attitude period. There followed, each for 10 min., a second control period, the second experimental attitude period, in which the second attitude was suggested, and a relaxation period, after which the subject was awakened.

The suggestion given for the hives attitude was as follows: "You feel mistreated, unfairly treated, wrongly treated. There is nothing you can do about it, nothing you even want to do about it. You are thinking only of what happened to you. You just have to take it. You are the helpless, innocent victim of unfair, unjust treatment. Nobody should do a thing like that to you. That is the way you're feeling; that feeling gets stronger and stronger."

Just before this suggestion was begun, the subject was told that he would be burned with a match. A match was immediately struck, and the *unlighted* end touched to the skin of the right hand, the hand to which the thermistor was not attached.

The suggestion for the hypertension attitude was as follows: "We're going to do something to you; you have to be ready, on guard, prepared for whatever it is. It might be a burn, an electric shock, a needle stick. It's going to come; you just have to sit there and wait and try to be ready for whatever it is. You feel that you may be attacked and hurt at any instant; it may be painful, it may be dangerous. You feel that you are in danger. You're threatened every instant, you have to watch out."

For reasons discussed in previous papers,[7, 15] an experimental session was not conducted if the subject showed: (1) a drop in skin temperature of more than 1.5°C. between the time he entered the laboratory and the induction of the first attitude, or (2) a drop of skin temperature to 30.0°C. or less during the same period. If either of these conditions was present on a subject's first experimental day, the experiment was interrupted and the subject dismissed from the study. Five subjects were rejected in this way. If either of these conditions occurred on the second experimental day, the subject was to return on a third day, at which time the procedure planned for the second day would be carried out. This happened once. It should be emphasized that these criteria for exclusion of subjects were formulated in advance of the preceding experiment in this series. One subject showed a steep rise in skin temperature during the induction period. This had not been encountered previously, but the possibility had been anticipated and the decision to dimiss the subject, should this occur,[7] had been made. Furthermore, if a subject at any time during the experimental procedure opened his eyes or said that he did not feel that he was hypnotized, the experiment was to be discontinued at that point, and the results discarded. This occurred once. No statements made by the subjects *after* the conclusion of the experiment were considered in evaluating the results. In short, our definition of hypnosis for the purposes of this study was that the subject pass the initial tests for a trance state and that he then remain quietly in his chair with his eyes closed throughout the procedure.

Treatment of Data

Data obtained from a total of 80 attitude suggestions, given in 40 days of experimenting, were available for analysis. Five parameters were used to describe each of the five physiological variables: the last control reading, the rate of change during the control period, the mean absolute change during the attitude period, the maximal absolute rise during the attitude period, and the rate of change during the attitude period.

Although continuous records of pulse, respiration, and temperature were obtained, only readings taken at 2-min. intervals were used in the data analysis. As in the previous experiments, the last control reading was taken 1 min. before attitude induction, and there were four readings at 2-min. intervals before this point (during the control period) and five readings at 2-min. intervals after this point (during the attitude-induction period). Blood pressure was not recorded continuously; the three readings obtained in each period were used in the data analysis.

Change was measured from the last control reading and was expressed as an absolute gain or loss, in degrees for temperature, in millimeters of mercury for blood pressure, in beats per minute for heart rate, and in number per minute for respirations. Change in physiological variables has sometimes been expressed as a percentage or relative change, but as Lacey[10] pointed out, this form of expression frequently distorts the data. He recommended instead that the experimental level be adjusted for the regression on the control level. Scatter graphs of the data obtained in the experiment being reported indicated that the lines of regression closely approximated a slope of 1.0. With a slope of 1.0, the absolute change equals the adjusted change, and little improvement in accuracy would be gained by correcting for the exact regression. Changes were computed separately for each subject, so that in determining the change from control to attitude period, each subject served as his own control.

The mean absolute change during the attitude period was the mean of the changes at 1, 3, 5, 7, and 9 min. in the case of the continuously recorded variables and the mean of the changes at approximately 2, 5, and 8 min. for systolic and diastolic blood pressure. The maximal absolute rise during the attitude period was the difference between the last control reading and the highest attitude reading. Both mean change and maximal rise were considered, since the latter has frequently been preferred to averages in the analysis of physiological data. [1,11,14,16]

In addition to the average absolute change during the attitude period, the average rate of change per minute (or slope of the best-fitting line) was computed for both the control and attitude periods. The first experiment in this series showed that when temperature was rising or falling during the control period, the trend was continued during the attitude period, although it was attenuated by attitude suggestions. When reliable trends are manifested, slope may be a more sensitive measure of change than average magnitude. In any case, it becomes important to determine whether groups are equated for control-period trends. Glickstein has made this point in a recent discussion of the effect of base line changes.[8]

For these reasons, the slope measure was adopted in this and the two preceding experiments. The control slope was computed by averaging the slope of the lines passing through each of the control points and the last control reading. The attitude slope was similarly the average slope of the lines passing through each attitude point and the last control point. In the case of temperature, only the slope of the last 6 min. of the attitude period was used. This procedure was also used in the previous experiment to eliminate the effects of an initial drop in temperature, interpreted as an undifferentiated startle response to stimulation.

TABLE 1. SIGNIFICANCE OF EFFECTS OF SUGGESTION (HIVES AND HYPERTENSION ATTITUDES), AS EVIDENCED BY *F* VALUES FROM ANALYSIS OF VARIANCE OF 5 PHYSIOLOGICAL VARIABLES, USING 2 CONTROL-PERIOD MEASURES AND 3 ATTITUDE-PERIOD MEASURES

	Control measures		*Attitude measures*		
Variable	*Last reading*	*Rate of change*	*Max. diff.*	*Average diff.*	*Rate of change*
Temp.‡	<1.0	<1.0	8.45†	4.73*	5.45*
DBP‡	<1.0	3.78	6.51*	5.27*	6.44*
SBP‡	<1.0	<1.0	<1.0	<1.0	<1.0
HR‡	<1.0	<1.0	1.10	2.16	1.30
Resp.‡	<1.0	<1.0	<1.0	<1.0	<1.0

*$p<.05$.
†$p<.01$.
‡In this and Tables 2–6, "Temp." stands for "temperature," "DBP" and "SBP," for "diastolic" and "systolic blood pressures," respectively, "HR," for "heart rate," and "Resp.," for "respiration rate."

Results

The analyses of variance for each of the five dependent physiological variables are summarized in Tables 1–6. Tables 1–5

TABLE 2. SIGNIFICANCE OF ORDER EFFECTS, AS SHOWN BY *F* VALUES FROM ANALYSIS OF VARIANCE OF 5 PHYSIOLOGICAL VARIABLES, USING 2 CONTROL-PERIOD MEASURES AND 3 ATTITUDE-PERIOD MEASURES

	Control measures		*Attitude measures*		
Variable	*Last reading*	*Rate of change*	*Max. diff.*	*Average diff.*	*Rate of change*
Temp.*	1.03	<1.0	6.42†	4.88†	1.36
DBP	<1.0	3.04	1.81	1.07	<1.0
SBP	<1.0	2.48	<1.0	<1.0	<1.0
HR	3.79	3.52	<1.0	<1.0	<1.0
Resp.	2.36	<1.0	<1.0	<1.0	<1.0

*See Table 1.
†$p<.05$.

TABLE 3. SIGNIFICANCE OF DAY EFFECTS, AS SHOWN BY *F* VALUES FROM ANALYSIS OF VARIANCE OF 5 PHYSIOLOGICAL VARIABLES, USING 2 CONTROL-PERIOD MEASURES AND 3 ATTITUDE-PERIOD MEASURES

	Control measures		*Attitude measures*		
Variable	*Last reading*	*Rate of change*	*Max. diff.*	*Average diff.*	*Rate of change*
Temp.‡	1.18	15.89†	<1.0	2.34	3.68
DBP	2.93	<1.0	1.34	<1.0	1.19
SBP	4.19*	<1.0	<1.0	1.05	1.39
HR	29.52†	<1.0	<1.0	<1.0	<1.0
Resp.	2.36	2.04	3.48	2.72	2.17

*$p<.05$.
†$p<.01$.
‡See Table 1.

show effects, respectively, of the suggestions, the order in which the suggestions were made, the day, subject differences, and experimeter differences on the two control period measures and the three attitude period measures for each of the five physiological variables. Table 6 records the error terms for each of the 25 analyses.

Effects of Suggestions

The analyses show that control periods preceding induction of the hives attitude did not differ significantly from those preceding induction of the hypertension attitude either in the height of the last control reading or the rate of change of a variable during the control period (Table 1).

In contrast to the lack of difference during the control periods, there were significant differences during the attitude periods in both temperature and diastolic blood pressure. All three measures of change during the attitude periods—the mean absolute change, the maximal rise, and the rate of change per minute (or slope)—showed a significant difference between the hives attitude and the hypertension attitude. As shown in Table 7, these differences were in the direction predicted by the hypothesis. There was a relatively greater rise in skin temperature with the hives attitude and a relatively greater rise in diastolic blood pressure with the hypertension attitude. Figures 1 and 2 show the course of the changes in these variables during the control and attitude periods.

According to the results of a one-tailed t test, the maximal diastolic blood pressure, the average diastolic blood pressure,

TABLE 4. SIGNIFICANCE OF SUBJECT EFFECTS, AS INDICATED BY *F* VALUES FROM ANALYSIS OF VARIANCE OF 5 PHYSIOLOGICAL VARIABLES, USING 2 CONTROL-PERIOD MEASURES AND 3 ATTITUDE-PERIOD MEASURES

	Control measures		*Attitude measures*		
Variable	*Last reading*	*Rate of change*	*Max. diff.*	*Average diff.*	*Rate of change*
Temp.‡	4.20†	2.36†	1.43	1.39	1.15
DBP	13.98†	1.34	2.54†	1.89*	2 02*
SBP	9.44†	1.05	1.13	<1.0	<1.0
HR	5.30†	<1.0	1.32	1.16	1.08
Resp.	13.22†	1.25	2.97†	4.06†	4.83†

*$p<.05$.
†$p<.01$.
‡See Table 1.

TABLE 5. SIGNIFICANCE OF EXPERIMENTER EFFECTS, AS MANIFESTED BY *F* VALUES FROM ANALYSIS OF VARIANCE OF 5 PHYSIOLOGICAL VARIABLES, USING 2 CONTROL-PERIOD MEASURES AND 3 ATTITUDE-PERIOD MEASURES

	Control measures		*Attitude measures*		
Variable	*Last reading*	*Rate of change*	*Max. diff.*	*Average diff.*	*Rate of change*
Temp.‡	4.43*	<1.0	5.11*	4.16	1.75
DBP	<1.0	<1.0	1.56	<1.0	<1.0
SBP	<1.0	<1.0	<1.0	<1.0	<1.0
HR	1.99	1.21	3.40	4.20	5.08*
Resp.	<1.0	3.28	11.51†	8.26*	9.37†

*$p<.05$.
†$p<.01$.
‡See Table 1.

TABLE 6. ERROR TERMS (RESIDUAL VARIANCE) FROM ANALYSIS OF VARIANCE OF 5 PHYSIOLOGICAL VARIABLES, USING 2 CONTROL-PERIOD MEASURES AND 3 ATTITUDE-PERIOD MEASURES

	Control measures		*Attitude measures*		
Variable	*Last reading*	*Rate of change*	*Max. diff.*	*Average diff.*	*Rate of change*
Temp.*	.671	.003	.058	.037	.002
DBP	32.3	2.80	23.2	20.7	1.95
SBP	52.0	5.28	53.5	51.7	4.93
HR	42.8	1.08	44.9	28.0	1.53
Resp.	2.56	.117	27.2	8.74	.358

*See Table 1.

and the slope of diastolic blood pressure during the attitude period all showed a significant rise over control period values ($p<.01$ in each case). The maximal skin temperature during suggestion of the hives attitude was also significantly greater than during the last control reading ($p<.01$). The average skin temperature during the period was not significantly greater than the last control reading but there is an initial drop in temperature at the beginning of the attitude suggestions, and an average of temperature for the whole attitude period includes this artifact. Comparison of the slope of the temperature curve during the hives attitude with that during preceding control periods also did

TABLE 7. Temperture and Diastolic-Blood-Pressure Response During Hives and Hypertension Attitude-Periods

Response	*Hives*	*Hypertension*
Temperature (°C.)		
Maximal change	+ .199	+ .043
Average change	+ .023	− .070
Rate of change (last 6 min.)	+ .022	− .003
Diastolic blood pressure (mm. Hg)		
Maximal change	+4.425	+7.175
Average change	+ .650	+2.984
Rate of change	+ .112	+ .904

not reveal a significnt increase (i.e., acceleration), if all hives attitude periods were considered together. As Fig. 3 and 4 show, however, there was a difference in the behavior of control-period temperatures from Day 1 to Day 2. If only Day 2 was considered, when the control temperature curve was almost flat, the attitude slope was significantly greater than the control slope ($p<.05$). Failure of slope to increase during the hives attitude period if the control slope was rising (as it was on Day 1) was also found in the first experiment of this series[7] and has been observed by Plutchik and Greenblatt.[13] It may represent a manifestation of the "law of initial values."[10] In other words, if the temperature is already rising rapidly, the rate of rise cannot be expected to increase much.

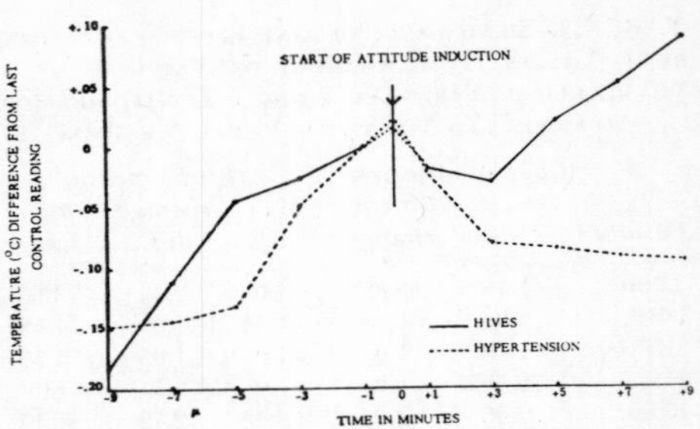

Fig. 1. Mean differences between hand temperature at the last control reading (1 min. before start of attitude suggestion) and at other times indicated.

No predictions were made concerning the differential effects of the two attitude suggestions on systolic blood pressure, pulse, or respiration, and no significant treatment effect was seen. Pulse and respiration rates were more rapid and systolic blood pressure was lower during the attitude period than during the control period. In the preceding experiment,[15] all three of these physiological variables rose when attitude suggestions were made but, in agreement with the present findings, there were no significant differential changes.

Order Effects

Order of presentation was not an important source of variability. It had a significant effect only on temperature and,

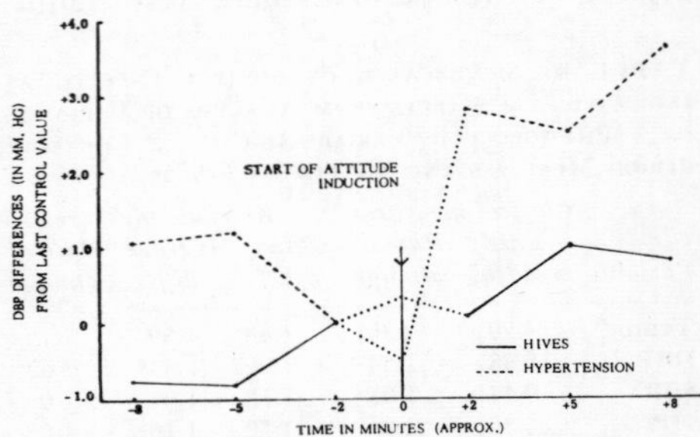

Fig. 2. Mean differences between diastolic blood pressures at last control reading (about 2 min. before start of attitude suggestion) and at other times indicated.

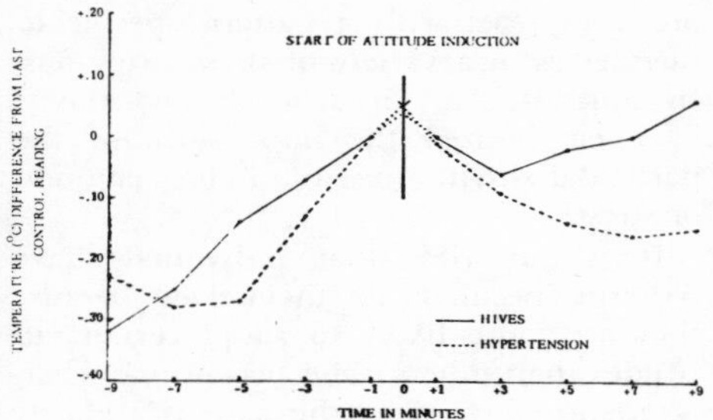

Fig. 3. Hand temperatures on Day 1.

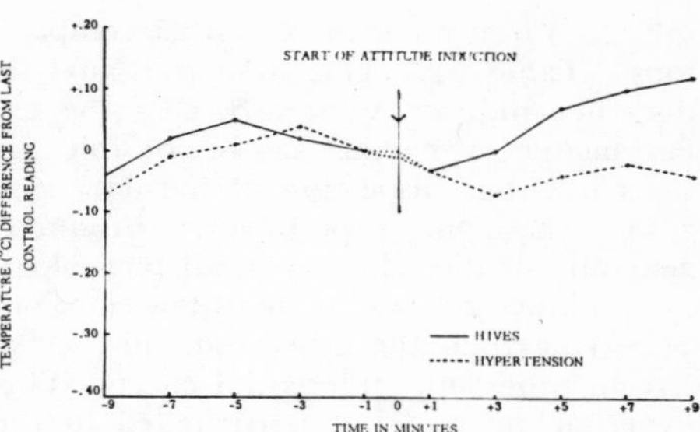

Fig. 4. Hand temperatures on Day 2.

at that, on only two of the five temperature measures (Table 2).

Day Effects

The experimental design permitted a test of differential effect between the first and second days. The only significant effects were on control measures, and these were in the direction that might be anticipated if anxiety was greater on the first day. On the whole, subjects had lower temperature values, higher blood pressure readings, and higher pulse and respiratory rates on the first day. The mean values of the last control readings on each of the 2 days is shown in Table 8. Day differences in the last control readings were significant for pulse rate and systolic blood pressure, and day differences in control slope were significant for temperature (Table 3). Graphs of temperature changes for each day separately illustrate the greater stability during Day 2 control periods as well as a clear differentiation of attitudes on both days (Fig. 3 and 4).

There was no significant difference owing to the day of the experiment in any of the attitude measures. While the subjects may have been more anxious, apprehensive, or otherwise uncomfortable on their first exposure to the experiment, the measured responses to the attitude suggestion were not affected.

TABLE 8. Mean Values of Last Control Readings on Each Experimental Day

Variable	*Day* *1*	*2*
Temperature (°C.)	34.1	34.2
Diastolic blood pressure (mm. Hg)	71.7	69.5
Systolic blood pressure (mm. Hg)	121.8	118.5
Heart rate (beats per min.)	79.4	71.4
Respirations per min.	16.3	15.8

Subject Effects

Intersubject differences were an important source of variability. For all five of the physiological variables, large and significant differences in the control readings existed among the subjects. These differences were less when the various measures of change were considered; they were significant only for diastolic blood pressure and respiration rate (Table 4).

Experimenter Effects

The experimenters could not introduce inaccuracy or bias in reading physiological responses since these were recorded on a polygraph monitored by an assistant. However, in allaying initial anxiety, inducing hypnosis, and giving attitude suggestions, the tone of voice and general manner of the experimenters could differ. Any experimenter effect was confounded with subject effects, since each experimenter conducted both sessions with half of the subjects instead of half of the sessions for all subjects. For this reason, the present design gave a relatively insensitive estimate of experimenter effect because it required testing against within-subject variance rather than error variance. Despite this, the experimenters were shown to introduce significant variation in 6 of the 25 comparisons (Table 5). The statistical analysis does not indicate what aspect of the experimenters' behavior was responsible for the effect, but inspection of the data suggested that one experimenter produced generally smaller changes in all physiological variables and was perhaps less emotion-provoking than the other but more effective in producing differential effects. This experimenter had not participated in the two earlier studies.

Discussion

The general problem of specificity in psychosomatic disease and the relation of the present approach to other proposed solutions was discussed by Grace and Graham.[6] Hamilton has also provided a critical review of the available evidence supporting any specificity approach.[9] It is sufficient in this context to point out that the specificity-of-attitude hypothesis is not incompatible with other proposed solutions to the specificity problem. Because the attitudes can be verbalized more or less readily and because they concern events close in time to the onset or exacerbation of an illness, the hypothesis is perhaps more susceptible to experimental testing than are most in this field.

The problem of specificity in psychosomatic disease is intimately related to the question of specificity of emotions. The prevailing view until recently has been that emotions cannot in general be differentiated on the basis of physiological changes that accompany them. However, in the past decade, the specificity position has been bolstered by a number of studies showing relatively low intercorrelations among physiological variables—They do not rise and fall together as an "emotion" syndrome but are more or less independent.[4]

Two main factors have been proposed by other authors to account for such specific patterns of physiological response. Lacey, in particular, emphasized the role of individual differences,[11] and Malmo's work[12] also suggested that individuals are prone to react with a pattern specific to themselves in a variety of stress situations. In contrast, Ax,[1] Schachter,[14] and Davis[3] have emphasized the importance of the particular situation used to induce emotion or stress.

It is our view that individuals have patterns peculiar to themselves because they are more likely to adopt certain attitudes than others. One person may characteristically perceive himself as unfairly treated and helpless, while another may characteristically perceive himself as needing to be "on guard." Similarly, some situations may tend to elicit similar response patterns in most subjects because they are more likely to be perceived in one way than in another by most persons having a similar cultural background. For instance, Schachter[14] and Ax[1] attempted to devise situations that would elicit different emotions, fear and anger; judging from their subjects' reports, they were quite successful in doing so. Many of the stress situations employed in research with individual differences allow, by contrast, more opportunity for the individual to adopt his own idiosyncratic perceptions. Funkenstein, in fact, relied on this possibility when he used a single mental-work-failure task to elicit the several "emotions" that he wished to study.[5]

The specificity-of-attitude hypothesis, then, is compatible with a demonstration of either individual-specific or situation-specific patterns but suggests that the major effects of both of these factors may be accounted for by specific attitudes. An advantage of the attitude hypothesis is that it is not confined to a statement that different attitudes will produce some significantly different physiological pattern, but rather predicts the kind and direction of the physiological changes that should occur.

Differences in response to the two attitude suggestions might be thought to reflect differences in general "emotionality," or "arousal" such that responses to the hypertension suggestion were "stronger" than those to the hives suggestion. The study reported did not produce definite evidence against this explanation. The two previous studies,[7, 15] however, showed clearly that differences in response to attitude sugestions are not merely quantitative, along the same continuum, but are different in kind. For example, suggesting the "hives" attitude produced a rise in skin temperature, as in the present series of experiments, whereas suggesting the attitude associated with Raynaud's disease produced a significant drop in skin temperature.

The experiments reported tested the attitude hypothesis experimentally rather than by correlational methods. The experimental method requires that the independent variable be under the control of the experimenter and not be a variable inherent in the subject that is measured if it can be identified. For this reason, the method of producing "stress" and then correlating physiological changes with subjects' reports was not utilized. The method used by Schachter[14] and Ax[1], that of devising specific stress situations, was considered, but was not chosen because it would be difficult to structure situations sufficiently to produce in the majority of subjects attitudes as specific as those with which we were concerned. Direct suggestion of an attitude to a hypnotized subject was therefore attempted. This not only permitted manipulation of the independent variable but had the advantage of being applicable to a wide variety of situations. That such suggestions would prove sufficiently strong to elicit any substantial physiological response had been uncertain, but the results of the three experiments in the series indicate that they were. A casual comparison of the magnitude of all physiological responses to attitude suggestions under the conditions employed in our experiments with those of others indicates that the hypnotic suggestions may be slightly more effective than Lacey's anticipation-of-stress situations but produce less response than the actual stress situations used by Lacey[11] and others.

Several points of methodological interest have been answered by the design of this study. Day of testing and, with one exception, order of presentation did not affect significantly the physiological response to the attitudes. There were significant differences between the first and the second day during control periods, but these differences disappeared when attitude suggestions were introduced. Aside from the main attitude effects, the largest source of variance was intersubject differences. Again, however, individual-specific effects were largely restricted to the control measures. The level of all physiological variables and, in the case of temperature, the rate of change were a reliable characteristic of individuals over the four control periods measured. In response to the attitudes, only diastolic blood pressure and respiratory rate showed significant individual differences.

Our experimental procedure aimed to restrict the range of idiosyncratic individual perceptions. The results suggest that when this was done, the effect of intersubject differences was reduced. That all variance associated with individual subjects could be successfully eliminated by this method is not implied. In the first place, the absence of significant intersubject differences may be as easily the result of insensitive experimental methods as of any other cause. It should also be noted that we are speaking only of the magnitude of change of a single physiological variable and not of complex patterning of response.

Since subject differences contribute some variance, even when it is not statistically significant, experimental designs gain in efficiency if this source is isolated. Another source of variability that is rarely treated is that due to the experimenter. Again, a design that removes this effect will prove more efficient than one that does not. The gain in efficiency may be estimated by comparing error terms in the present experiment with those obtained previously.

Standard deviations for the five physiological variables were reported in the second experiment of this series.[15] For diastolic blood pressure, systolic blood pressure, and heart rate, they were within the range of standard deviations that Lacey reported.[11] The standard deviations obtained in the experiment reported in this article were approximately one-half as large; Engel,[4] who has also used a Latin square design, reported figures similar to ours. Since the efficiency of an experiment is reciprocally related to the standard deviation squared,[2] the present design is 400 per cent as efficient as the design of the second experiment in this series.

Summary and Conclusions

A specificity-of-attitude hypothesis for psychosomatic disease was tested experimentally by means of hypnotic suggestion of attitudes. Twenty healthy male subjects were given two attitude suggestions on each of 2 days, with the order of presentation reversed on the second day. The two attitudes employed were those associated with hives (the subject felt that he was being unjustly treated and could think of nothing he wanted to do about it), and hypertension (he had to be on guard against bodily assault). It was predicted that skin temperatures would rise more with the hives suggestion than with that for hypertension and that diastolic blood pressure would rise more with the hypertension than with the hives suggestion. Systolic blood pressure, respiratory rate, and pulse rate were also measured. No predictions of differential effects of the two attitudes on these variables were made.

The predictions were confirmed. Mean change, maximal rise, and rate of change of skin temperature during the hives suggestion were significantly greater than the corresponding changes during the hypertension suggestion. All three measures of change in diastolic blood pressure were significantly greater during the hypertension than during the hives suggestion. There were no differential effects of the two attitude suggestions on systolic blood pressure, heart rate, or respiratory rate.

The experimental design also permitted an analysis of the variance associated with factors other than the two attitude suggestions. There were significant differences during the control periods but not during attitude periods between the first and second day of experimenting. Intersubject differences were also important during control periods but were less so for attitude periods. Some significant variability was associated with the order of presentation of attitudes and with the two experimenters.

1300 University Ave.
Madison 6, Wis.

References

1. Ax, A. F. The physiological differentiation between fear and anger in humans. *Psychosom. Med. 15:*433, 1953.
2. Cochran, W. G., and Cox, G. M. *Experimental Designs,* ed. 2. New York, Wiley, 1957.
3. Davis, R. C. Response patterns. *Trans. New York Acad. Sc. 19* (Series 11):731, 1957.
4. Engel, B. T. Stimulus response and individual response specificity, *A.M.A. Arch. General Psychiat. 2:*305, 1960.
5. Funkenstein, D. T., King, S. H., and Drolette, M. E. *Mastery of Stress.* Cambridge, Harvard, 1957
6. Grace, W., and Graham, D. T. Relationship of specific attitudes and emotions to certain bodily diseases. *Psychosom. Med. 14:*243, 1952.
7. Graham, D. T., Stern, J. A., and Winokur, G. Experimental investigation of the specificity of attitude hypothesis in psychosomatic disease. *Psychosom. Med. 20:*446, 1958.
8. Glickstein, M. Temporal patterns of cardiovascular response. *A.M.A. Arch. Gen. Psychiat.* 2:12, 1960.
9. Hamilton, M. *Psychosomatics.* New York, Wiley, 1955.
10. Lacey, J. I. The evaluation of autonomic responses: Toward a general solution. *Ann. New York Acad. Sc. 67:*123, 1956.
11. Lacey, J. I., and Lacey, B. C. Verification and extension of the principle of autonomic response-stereotypy. *Am. J. Psychol. 71:*50, 1958.
12. Malmo, R. B., Shagass, C., and Davis, F. H. Symptom specificity and bodily reactions during psychiatric interview. *Psychosom. Med. 12:*362, 1950.
13. Plutchik, R., and Greenblatt, M. Tempera-

ture changes of the skin: A function of initial level. *Am. J. Psychol. 69:*403, 1956.

14. SCHACHTER, J. Pain, fear and anger in hypertensives and normotensives. *Psychosom. Med. 19:*17, 1957.
15. STERN, J. A., WINOKUR, G., GRAHAM, D. T., and GRAHAM, F. K. Alterations in physiological measures during experimentally induced attitudes. *J. Psychom. Res. 5:*73, 1961.
16. WENGER, M. A., ENGEL, B. T., ESTESS, F. M., and SONNENSCHEIN, R. R. Autonomic response patterns during intravenous infusion of epinephrine and nor-epinephrine. *Psychosom. Med. 22:*294, 1960.

70

Yoga, "Yogic Feats," and Hypnosis in the Light of Empirical Research

Abdulhusien S. Dalal and Theodore X. Barber

Both yoga and hypnosis have been commonly thought to involve altered states of consciousness and to give rise to unusual phenomena. However, empirical studies reviewed in this paper indicate the following: (a) Since the hypothesized altered states (yogic samadhi and hypnotic trance) have not as yet been adequately denoted, they cannot, at present, be viewed as scientific concepts. (b) The observable phenomena that are popularly associated with the term *yoga* (e.g., fire-walking, burial-alive, lying-on-nails, and voluntary control over autonomic functions) and with the term *hypnosis* (e.g., hallucinations, age-regression, amnesia, and the human-plank feat) are quite different from what they seem to be superficially; are not mysterious at all; and are readily explainable. A methodological critique of studies in these areas is presented and suggestions are made for further research.

Both yoga and hypnosis are said to involve altered states of consciousness or trance states (Behanan, 1937; Das, 1963; Williams, 1948). Also, both yoga and hypnosis have been popularly associated with unusual phenomena. The yogi is thought to be able to control various autonomic functions, to walk across a burning pit, to be buried alive, and to lie on nails. Similarly, the hypnotic subject is said to be able to hold a heavy weight on his abdomen while lying stretched in midair with supports only at his head and ankles. Also, it is commonly believed that suggestions given to a deeply hypnotized subject produce insensitivity to pain, hallucinations, regression to childhood, amnesia, deafness, blindness, colorblindness, and many other unusual phenomena. This paper will critically analyze (a) the altered states of consciousness or trance states that are said to be involved in yoga and in hypnosis and (b) the phenomena that are supposedly manifested by yogis and by hypnotic subjects. The discussion is based on (a) our recent empirical research in hypnotism (Barber, in press; Barber, Dalal and Calverley, 1968; Dalal, 1966), (b) the first author's studies in yoga for nearly two decades, including seven years devoted exclusively to the practice of yoga at Sri Aurobindo Ashram in Pondicherry, India, and (c) a review of all available empirical studies pertaining to the so-called yogic feats that have been published during the past two decades.

The Terms Yoga and Yogi

Before we proceed to the main discussion, it is necessary to clarify the terms

[1] Work on this paper was supported in part by a research grant (MH-11521) to T. X. Barber from the National Institute of Mental Health, U.S. Public Health Service. Reprint requests to Medfield Foundation, Harding, Massachusetts, 02042.

yoga and *yogi*. Two main usages of the term *yoga* may be distinguished:

1. Strictly speaking, yoga refers to a system of beliefs and practices whose goal is to attain a "union" of the individual self with Supreme Reality or the Universal Self. There are several systems of yoga in this sense (Das, 1963; Wood, 1959; Woods, 1927). Each has an ethical-religious discipline at the core and each sees the trance state of samadhi as the final step in attaining "union" with Supreme Reality. However, each system of yoga emphasizes somewhat different processes for the attainment of samadhi and "union." For instance, the eightfold system of Raja Yoga, taught by Patanjali, includes self-control (yama), religious observances (niyama), physical postures (asana), regulation of the breath (pranayama), suppression of the flow of external sense impressions (pratyahara), concentraton or fixed attention on an object (dharana), meditation or contemplation of an object for a long period of time (dhyana) and, finally, a state of absorption or trance in which the person is no longer conscious of his concentration (samadhi).

2. In its popular usage, especially current in the West, the term *yoga* has come to be associated more or less exclusively with the physical postures (asana) and the regulation of breathing (pranayama). This popular usage is not found in Indian philosophy. An individual who merely practices the postures and the breathing exercises (as some do for health or therapeutic purposes), without a concomitant consecration to a spiritual discipline and to the goals of yoga, is simply practicing a few yogic exercises; he is *not* practicing yoga in the strict sense.

The term *yogi* has a strict meaning, a more popular meaning, and also a fallacious meaning. Strictly speaking, a yogi is a person who has ostensibly attained the goal of yoga, namely, samadhi and a "yoking with" or "union" with Supreme Reality. In more popular usage, the term *yogi* refers to a person who practices a system of yoga regardless of whether or not he has attained samadhi or "union." In its fallacious sense, the term *yogi* refers to an individual who labels himself as a yogi but who has *not* seriously practiced any system of yoga. The so-called yogic feats which will be discussed later in this paper are not performed by yogis in the strict sense of the term, although they are performed by some who have not attained samadhi and by self-styled yogis.

Hypnotic and Yogic Trance

Both hypnosis and yoga are said to involve a state of trance. The hypnotic trance state has been traditionally regarded as giving rise to "hypnotic phenomena" such as analgesia, hallucination, and amnesia. However, the yogic trance state of samadhi, contrary to popular belief, has not been regarded in the Indian tradition as necessarily related to so-called yogic phenomena such as voluntary control over autonomic functions, fire-walking, or lying-on-nails. Though the latter phenomena have been popularly subsumed under the term *yoga*, the Indian spiritual tradition has always looked down upon the pursuit of such unusual capacities as worldly and vainglorious. Since yoga aims at freedom from the "world-illusion" though the achievement of samadhi, the attainment of unusual psychophysiological capacities and the attainment of samadhi tend to be mutually exclusive goals. The confusion between these two disparate goals appears to be due to the following:

1. Certain practices such as physical postures (asana) and breathing exercises (pranayama) have been associated with both goals. However, those who seek to develop unusual psychophysiological capacities take to a highly intensive and al-

most exclusive practice of such exercises, whereas those who aspire to attain samadhi follow highly variable systems of yoga which may, but often do not, include such exercises. In fact, at the Ashram where the first author lived and practiced yoga, hardly anyone among the nearly one thousand practitioners of yoga was seriously concerned with the exercises pertaining to physical postures and breath regulation, although a few performed them for purposes of physical health.

2. Another reason for the widespread misconception of yoga as the ability to perform unusual feats lies in the fact that the general public in the West and also in India typically comes in contact only with self-styled yogis who demonstrate such feats, whereas yogis who seek to attain samadhi tend to lead inconspicuous lives. In fact, judging from the first author's experiences, a very large number of those who claim to possess unusual "yogic" capacities have not practiced any form of yoga. Their public demonstrations of "yogic powers" often involve deception and they typically use the term *yoga* as a cloak for mendicancy—e.g., for promoting the sale of quack medicines and talismans.

Problems in Denoting Hypnotic and Yogic Trance

Attempts to denote the hypnotic trance state objectively have not been successful. Over a period of many years researchers have sought a physiological index that would differentiate the hypnotic trance state from the waking state but have failed to find one (Barber, 1961; Crasilneck and Hall, 1959; Gorton, 1949; Levitt and Brady, 1963). In the absence of a physiological index, the hypnotic trance state is usually inferred from hypersuggestibility; that is, from a high level of response to suggestions of analgesia, hallucination, age-regression, amnesia, and the like. However, the latter criterion gives rise to circular reasoning: the hypnotic trance state is inferred from the high level of response to suggestions and, vice versa, the hypnotic trance state is used to account for the high level of response to suggestions from which it is inferred (Barber, 1964a; Chaves, 1968). Some investigators maintain that the hypnotic trance state can be denoted in terms of certain subtle but objective signs such as passivity, disinclination to talk, and fixed facial expression (Erickson, Hershman and Secter, 1961; Gill and Brenman, 1959). However, these signs are also unsatisfactory because they are functionally related to explicit or implicit suggestions to sit quietly for a period of time with eyes closed and to become relaxed, drowsy, sleepy, and passive (Barber, in press; Barber and Calverley, in press). Thus, when the hypnotic trance state is inferred from passivity, disinclination to talk, fixed facial expression, and other signs of this type, the state is judged to be present because the subject is responsive to suggestions and the inferred state is used circularly to account for the subjects responsiveness to suggestions.

There are practical difficulties in denoting the yogic trance state of samadhi. In the first place, the state appears to be extremely rare. Secondly, a yogi who has ostensibly attained samadhi is not likely to condescend to cooperate in a scientific investigation because in the eyes of such a yogi the work of the scientist is trivial and futile, if not presumptuous and sacrilegious.[2]

[2] Although the first author is convinced from his personal experiences that those yogis who believe they have attained samadhi are not likely to agree to serve as subjects in scientific experiments, it is certainly possible to enlist the cooperation of some practitioners of yoga who have not attained samadhi. However, researchers should be careful to ascertain what yogis mean by samadhi when they use this term to describe their experiences, because in several Indian vernaculars the term is employed rather loosely and is practically interchangeable with the term *dhyana* (meditation or contemplation).

Verbal Reports of Trance States Treated as Dependent Variables

Since we do not have any clear-cut physiological indices of the hypnotic trance state, and since criteria based on overt responses to suggestions are entangled in circular reasoning, it appears that the least equivocal criterion for inferring the presence of a hypnotic trance state is the subject's report that he has experienced it. Similar considerations appear to apply to the yogic trance state of samadhi. Consequently, it appears that the appropriate scientific or empirical approach to the hypnotic and yogic trance states consists in specifying the conditions under which *reports* of having experienced them are elicited and in delineating the contents of these states as they are *reported* by subjects.

The chief experiences that are reported to be associated with the hypnotic trance state are feelings of relaxation, feelings of unreality, feelings of automaticity and compulsion, alterations in body image, and certain unusual sensations (Barber and Calverley, in press; Field, 1965; Ludwig and Levine, 1965). Recent studies which aimed to delineate the conditions under which reports of having been in hypnotic trance are elicited indicate that such reports are functionally related in a complex way to many denotable antecedent variables. These effective antecedents include the subject's preconceptions of what the hypnotic trance state is like, his pre-experimental expectations concerning whether and to what degree he will be hypnotized, the degree to which he responds to suggestions and has new experiences during the hypnotic session, and cues transmitted by the experimenter to suggest that the subject was or was not hypnotized (Barber, in press; Barber and Calverley, in press; Barber, Dalal, and Calverley, 1968). Furthermore, subjects' reports pertaining to their having experienced hypnotic trance are also partly dependent on subtle situational variables such as the wording of the questions that are used to elicit their reports (Barber, Dalal, and Calverley, 1968). Consequently, it appears that subjects may differ in their reports pertaining to whether or not they experienced hypnotic trance even though they had very similar experiences during the hypnotic session.

The samadhi experience is couched in complex and abstract terms, though it is said to be more concrete and real than sense experience. The state has been described as a triune experience of supreme Truth-Consciousness-Bliss (sat-chit-ananda) pertaining to an altogether different realm of experience and therefore impossible to describe except by metaphors and paradoxes. However, the most important element of the samadhi experience is thought to lie in the merger of personal consciousness with the Universal Self. The Universal Self, in turn, is conceived paradoxically as both immanent in all forms of existence and as transcending the manifest universe.

Some or all of the previously mentioned variables related to subjects' reports of having been hypnotized (which have their obvious correlates in yoga) may play a role in the reported experience of samadhi. But for reasons already stated, researchers have not as yet found means to test empirically the hypothesis that reports of samadhi are functionally related to such factors as preconceptions, pre-expectations, and suggestions from the spiritual guide (guru). Nor does it appear feasible to investigate empirically the role of numerous variables that are thought to be associated with the attainment of samadhi such as practices for self-control including sexual abstinence, physical postures, regulation of the breath, and various forms of concentration and meditation including long-continued fixation on an object and continual repetition of a sacred incantation.

However, several topics that have been recently approached empirically may shed some light on the reported experience of samadhi. One relevant topic has been labeled by Maslow as *peak-experiences*. According to Maslow, a substantial number of normal individuals in Western civilization report that they have had experiences as follows: "These moments were of pure, positive happiness when all doubts, all fears, all inhibitions, all tensions, all weaknesses, were left behind. Now self-consciousness was lost. All separateness and distance from the world disappeared... Perhaps most important of all, however, was the report in these experiences of the feeling that they had really seen the ultimate truth, the essence of things, the secret of life, as if veils had been pulled aside" (Maslow, 1962). Intensive studies with individuals who report these kinds of experiences can be conducted scientifically. Since there is considerable similarity in the reported contents of peak-experiences and of the samadhi experience, reliable data concerning one of these experiences should contribute to understanding of the other.

Another new research area that seems to be related to samadhi has been labeled by Deikman (1963, 1966) as *experimentally-induced contemplative meditation*. Deikman instructs his subjects along the following lines:

> "This experiment will explore the possibilities of seeing and experiencing when you cease thinking altogether and concentrate your attention on only one thing: the blue vase on the table in front of you. Look at the vase intently, focus all your interest on it, try to perceive the vase as directly as possible but without studying it or analyzing it. Have your entire mind concentrated on the vase, at the same time remain open to the experience—let whatever happens happen. All thinking must come to a stop so that your mind becomes quiet. Do not let yourself be distracted by thoughts, sounds, or body sensations—keep them out so that you can concentrate all your attention on the vase" (Deikman, 1966).

Some experimental subjects who were given these instructions stated that they experienced striking changes in perception of the self and of objects and Deikman hypothesized that analogues of mystical experiences could also be achieved. These exploratory studies lead to the interesting possibility of studying experimentally some aspects of the samadhi experience. Further studies along these lines could also employ other forms of contemplative meditation besides perceptual concentration and could investigate other reported aspects of *experimentally-induced contemplative meditation* besides unusual perceptions (Maupin, 1965).

Finally, we might also point out that intensive studies of the "psychedelic" experience, reported by some individuals who have received LSD, mescaline, psilocybin, and similar drugs, may also further our understanding of samadhi.

Having discussed the trance states that are said to be involved in both hypnosis and yoga, let us now turn to an analysis of the observable phenomena that are supposedly exhibited by hypnotic subjects and by yogis or self-styled yogis.

Hypnotic Phenomena and So-Called Yogic Phenomena

A striking similarity between the phenomena attributed to hypnosis and to yoga, as revealed by empirical studies, is that many of the phenomena are not what they at first appear to be. Recent work has underscored this fact with regard to many of the phenomena traditionally subsumed under the label *hypnosis*. For instance, "hypnotic amnesia" is essentially different from amnesia as it is commonly understood. In "hypnotic amnesia," the "forgotten" material (a) is recognized more quickly than neutral control material, (b) exerts the same objective effects (such as practice effects or retroactive inhibition effects) as material that is clearly remembered, and (c) is readily verbalized by the subject when he is given explicit or implicit permission to do so (Barber, 1962b; Barber and Calverley, 1966; Graham and Patton, 1968; Patten, 1932; Sturrock, 1966; Thorne, 1967; Williamsen, Johnson, and Ericksen, 1965). Similarly, the "hypnotic deaf" subject differs from the deaf person in that he is responsive to auditory stimulation. For instance, "hypnotic deafness" does not prevent the stuttering that is normally produced by delayed auditory feedback (Barber and Calverley, 1964b; Kline, Guze, and Haggerty, 1954; Kramer and Tucker, 1967; Sutcliffe, 1961). To cite yet another instance, it has been popularly believed that in "hypnotic age-regression" the subject goes back psychophysiologically to the suggested age, relives the previous events, and manifests behavior appropriate to the suggested age. However, standard psychological and physiological tests typically show important discrepancies between the behaviors of the "hypnotically-regressed" subject and the behaviors expected at the suggested age (Barber, 1962a; Barber, in press). Similarly, data which we have reviewed in detail elsewhere (Barber, 1961, 1963, 1964b, 1964c, 1965b, in press) show that there are crucial differences between "hypnotically-suggested" and non-suggested or "naturally-occurring" analgesia, dreaming, hallucination, blindness, and color-blindness.

Empirical investigations of so-called yogic phenomena also show them to be different from what they are naively conceived to be. Thus in "voluntary stoppage of the heart" claimed by a few yogis or self-styled yogis, no voluntary control over the heart musculature is directly involved and the heart does *not* stop. At best, a few individuals in India have demonstrated an ability to slow the heart by utilizing methods "involving retention of the breath and considerable muscular tension in the abdomen and thorax, with closed glottis" (Wenger and Bagchi, 1961). At times, these maneuvers retard venous return to the heart, the cardiac sounds cannot be detected with a stethoscope, and the palpable radial pulse seems to disappear. However, electrocardiographic and plethysmographic records show definitely that the heart is in continuous action (Wenger and Bagchi, 1961; Wenger, Bagchi and Anand, 1961).[3]

The feat of fire-walking also takes on a somewhat different appearance in the light of the following observations: "In south India and in other locations it [fire-walking] seems to be a common practice in connection with religious or other celebrations. We are informed, however, that the coals are well covered with ashes and that the 'walk' is more like a 'run.' Moreover, the participants are not yogis; they are the young men of the villages. And, since they have seldom or never worn shoes, the epidermis of the soles of their feet is reported to us as being ⅛″ to ¼″ in thickness. There may be fire-walkers with tender feet, but we have not seen them or heard about them in India [Wenger and Bagchi, 1961]."

Apparently, no case of fire-walking by a yogi has as yet been documented whereas numerous non-yogis have rather easily performed the "feat." Among the latter is a British university student who walked in bare feet with perfect safety across a 12

[3] Wenger and Bagchi (1961) pointed out a possible artifact in an earlier study by Brosse (1946) whose published EKG records of an Indian subject showed a gradual reduction in heart potentials to near zero. They noted that "during breath retention there is a deviation in position of the heart so that potentials in lead I are decreased, but those in lead III are increased. Dr. Brosse's record of 'heart control' involved lead I only. Had she measured any other lead her conclusions concerning voluntary control of the heart probably would have been negative as are ours."

foot trench with a surface temperature of 800° C. (Brown, 1938; Ingalls, 1939). Price (1937) discussed this demonstration as follows:

"There was not the slightest trace of blistering . . . each foot was in contact with the embers for not more than about a third of a second. This time factor (plus *confidence* and a steady, deliberate placing of the feet) is the secret of fire-walking. The low thermal conductivity of the wood ash may be a contributory factor. We received no evidence that 'faith' or a specially induced mental state played any part in the performance."

The fire-walking feat has also been performed before competent observers in the United States by Coe (1957), a chemist living in Florida. With eight steps he crossed bare-footed a 14-foot pit of red-hot coals without suffering pain or forming blisters. After reviewing the literature of physics and chemistry to discover possible principles that might be involved, he proffered the following hypothesis: "First there is natural moisture present on the surface of the skin. Under intense heat the skin sweats, and the moisture, which enters the spheroidal state, is converted to vapor which occupies 22.4 molar volumes of the original moisture. This is why a microscopically thin layer of moisture can be a protection. . . . The hotter the object the longer the spheroidal state is maintained, and the greater the protection afforded by the cushion of vapor [Coe, 1957]." Whether or not Coe's hypothesis accounts for the phenomenon can be determined in further research.

Writers such as Behanan (1937) state that, by intensive practice of certain types of yogic exercises, voluntary control can be acquired over the diaphragm, the anal sphincters, the recti abdominis muscles, and the muscles involved in vomiting. Careful research is needed to validate these claims. At the present time the contentions are based on hearsay evidence; they have not as yet been documented scientifically. Wenger and Bagchi (1961), who have carried out the most extensive scientific investigation in this area, were unable to find anyone in India who could demonstrate unequivocal *voluntary* control over the anal or urethral sphincters, although they did find "those who employ a catheter or other form of tube and, with its help, are able to draw up water into the bladder or lower bowel." Wenger and Bagchi found one person in India who could regurgitate voluntarily. However, this person stated that he did not have any special training; he had discovered accidently that he could vomit at will and used it occasionally as a cleansing device.

Wenger and Bagchi also found a yogi who could produce perspiration on his forehead within 1½ to 10 minutes after he began concentrating. They placed this feat in context as follows: "This man had spent parts of two winters in caves in the Himalayas. During such periods, usually alone and unclad except for an animal skin, much of his time was spent in meditation. . . . The cold distracted him, and his teacher advised him to concentrate on *warmth* and to visualize himself in extremely high temperature situations. . . . He reported gradual success *after about six months of practice*. Later he found that in a moderate climate the same practices produced not only increased sensations of warmth but perspiration."

Wenger and Bagchi also found one subject who was able to defecate at will. They stated that "He employed a particular position and movement of the tongue to aid in this control. Again the control had only been gained after long practice. In all of these forms of control it seems reasonable to assume that conditioning had occurred." Exactly what the hypothesized "conditioning" involved, however, is not clear and requires careful study (Katkin and Murray, in press; Kimble, 1961; Razran, 1961).

A few yogis or self-styled yogis have "buried themselves alive" (Rao, *et al.*, 1958). A well-publicized case of a yogi who buried himself for 62 hours was documented by Vakil (1950), who pointed out that the feat did not demonstrate that the yogi was able to survive with little or no oxygen for the following reasons. The oxygen requirement of a man lying perfectly still is about 2 ½ cubic feet of air per hour. Therefore, the oxygen content of the cubicle in which the yogi was entombed, measuring 5½′ x 4½′ x 8′, was sufficient to support a person substantially more than 62 hours.

A common belief is that yogis are able to lie on a bed of nails and to thrust metal spikes into their flesh. However, the first author, during his seven years' training in yoga, did not hear of any practitioner of yoga who claimed to be able to perform such feats. Rather, they are performed by a small number of fakirs and dervishes (religious mendicants) who are often confused with yogis. Although Wenger and Bagchi (1961) found no individuals in India "who stuck needles into their bodies," they found one person who lay on beds of nails. Wenger and Bagchi noted that "For the latter the task was easy. By practice he had learned to relax the muscles of the back, and the nails were close together and not very sharp." It should also be noted that the apparent "analgesia" or the stoicism may at times be due to the fact that some of these individuals who lie on nails are addicted to hashish or opium (Rawcliffe, 1959). When the nails are not blunt and when no drugs are taken, the performance demonstrates a not too unusual ability to tolerate a certain amount of pain. The French author, Heuzé, who demonstrated the feat of sticking hatpins through his flesh, insisted that such performances can be learned by most individuals who possess average patience and fortitude (Heuzé, 1926).

The "feats" discussed above are not an integral part of training in yoga. There is evidence to indicate, however, that the breathing and postural exercises, which are part of some but not all systems of yoga, may be accompanied by significant autonomic changes and may give rise to enduring changes in autonomic balance with possible beneficial effects on mental and physical health (Wenger and Bagchi, 1961). These possible changes in autonomic system functioning which may be produced by some yogic exercises, and also by related exercises such as those involved in "autogenic training" (Gorton, 1959; Schultz and Luthe, 1959), appear to offer rewarding possibilities for further empirical investigations.

Research Methodology

Critiques of research in hypnosis (Barber, 1962c, 1967, in press) indicate that much of the early experimentation in this area was vitiated by faulty methodology. Several of the criticisms that have been delineated with respect to hypnotic research are equally applicable to research pertaining to yoga or to so-called yogic feats.

Perhaps the most common fallacy in both areas is the assumption that special factors have to be invoked in order to explain the observed phenomena. For instance, it has been commonly assumed that the "hypnotic phenomenon" termed the *human-plank feat* (in which the subject lies stretched in midair with only his head and feet resting on supports) involves a special factor which comes into play during the hypnotic trance. However, empirical evidence demonstrates that this "feat" can be performed by at least 80% of unselected control subjects who are firmly told to make their body rigid and to keep it rigid (Barber, in press; Collins, 1961). Again, it has been commonly supposed that hypnotic hallucinations are rare phenomena that can be elicited only in a deep hypnotic trance. No well-defined criteria of what constitutes a hypnotic hallucination have been specified, so that in practice the phenomenon is said to be elicited when the subject testifies that he sees or hears something which is not actually present. However, experimental studies show that a considerable number (at least one-third) of unselected control subjects also testify that they see objects and hear sounds that are not present when suggestions intended to produce hallucinations are given to them in a non-hypnotic setting (Barber and Calverley, 1964a; Bowers, 1967; Spanos and Barber, 1968). Along related lines, no systematic attempts have been made to ascertain to what extent the so-called yogic feats can be elicited from untrained individuals. Evidence cited previously in connection with fire-walking and burial-alive suggests that some of these "feats" require a certain amount of courage, but no special training.

A related methodological error common to hypnosis and yoga is the failure to employ a control for the assumed independent variable. For instance, "hypnotic blindness" is said to produce alterations in physiological processes related to vision such as elimination of the EEG alpha block response which normally follows visual stimulation. However, experimental studies with controls indicate that the same effect can be demonstrated in non-hypnotic subjects (Barber, 1961). A study demonstrating a similar effect on alpha blocking due apparently to yogic meditation (Anand, Chhina and Singh, 1961) is vitiated by the same failure to employ a control for the assumed independent variable. In fact, it appears that, with a few possible exceptions, studies on yogis lack scientific validity due to failure to employ necessary controls and are useful only for suggesting lines of future inquiry.

Still another methodological pitfall is the use of the same subjects under the experimental and control conditions. Several studies in hypnotism suggest that under these circumstances subjects are apt to respond differently under the experimental (hypnotic) and control treatments because they think they are expected to or should perform differently under each treatment (Barber, 1962c, 1967, in press; Orne, 1959; Pattie, 1935; Wolberg, 1948). The same problem is present in studies on yoga. For instance, using himself as a subject, Behanan (1937) compared the effects obtained when certain yogic breathing exercises were practiced alone and when they were combined with a 20-minute period of concentration. The differences in results under the two conditions could not be unequivocally viewed as treatment effects; they could have been produced by Behanan's divergent expectations concerning the results that would be obtained under the two treatments. In brief, since subjects participating in studies on hypnosis or on yoga typically except divergent results under the experimental and control conditions, it is advisable to assign one group of subjects to the experimental condition and a different group to the control condition. Of course, such an independent-groups experimental design is maximally efficient when the subjects allocated randomly to each group are matched on relevant variables such as on base-level response to a suggestibility scale.

Finally, research in hypnosis as well as in yoga tends to be characterized by inadequately-defined terms. The ambiguity of terms such as *amnesia* and *hallucination* in the field of hypnosis has been pointed out above. Similarly, various writers on yoga assign different connotations to terms such as *meditation*. The contradictory findings pertaining to EEG correlates of meditation reported by Anand et al. (1961), Das and Gastaut (1955), and Wenger and Bagchi (1961) may be due to disparate techniques of meditation involved in the three investigations.

Overview

Yogi samadhi and hypnotic trance have been popularly regarded as states of consciousness in which an individual has access to unique experiences and to unusual capacities. However, empirical studies appear to indicate the following: (a) attempts to denote these states of consciousness independently of the subject's verbal report have not proved successful; (b) a critical scrutiny of the observable phenomena that are popularly associated with the term *hypnosis* (e.g., hallucination, age-regression, amnesia, and the "human-plank feat") and with the term *yoga* (e.g., fire-walking, burial-alive, lying-on-nails, and voluntary control over various autonomic functions) tends to divest them of the mystery with which they have been enveloped; (c) some of the assumptions regarding the factors effective in producing the phenomena have failed to be validated by empirical studies; and (d) several conclusions drawn from past studies of the phenomena are invalidated by methodological errors.

It may be fruitful for present-day researchers to direct greater attention to the experiences that are associated with yogic samadhi and with hypnotic trance. However, since the experiences are inferred from the subjects' verbal reports, the findings of such studies will necessarily apply to the *verbal reports* rather than to the experiences per se. Such caution and scepticism are demanded by the fact that reports of having had certain types of experiences during the hypnotic session depend on numerous factors such as on cues given by the hypnotist concerning what he believes the subject experienced (Barber, in press; Orne, 1959). This important finding in hypnosis was made possible by questioning traditional assumptions regarding the hypnotic state. It is necessary to pose similar questions with regard to experiences subsumed under the term *samadhi* in order to ascertain if factors similar to those affecting the subjects' reports in hypnosis also operate here.

With regard to the explanation of observable phenomena, a beginning has been made in hypnosis in that a series of antecedent variables underlying hypnotic hypersuggestibility has been delineated (Barber, 1965a, in press). In yoga, the few studies carried out so far suggest that some of the yogic exercises may be accompanied by significant changes in the autonomic nervous system and may give rise to enduring changes in autonomic balance. Further studies are needed to isolate the independent variables in yogic exercises which are effective in producing autonomic changes. Investigations of this type may prove to be at least as fruitful as those which isolate the variables involved in performing the so-called yogic feats.

REFERENCES

Anand, B. K., Chhina, G. S., and Singh, B. Some aspects of electroencephalographic studies in yogis. *Electroencephalography and Clinical Neurophysiology,* 1961, 13, 452–456.

Barber, T. X. Physiological effects of "hypnosis." *Psychological Bulletin,* 1961, 58, 390–419.

Barber, T. X. Hypnotic age regression: A critical review. *Psychosomatic Medicine,* 1962, 24, 286–299. (a)

Barber, T. X. Toward a theory of hypnosis: Posthypnotic behavior. *Archives of General Psychiatry,* 1962, 7, 321–342. (b)

Barber, T. X. Experimental controls and the phenomena of "hypnosis": A critique of hypnotic research methodology. *Journal of Nervous and Mental Disease,* 1962, 134, 493–505. (c)

Barber, T. X. The effects of "hypnosis" on pain: A critical review of experimental and clinical findings. *Psychosomatic Medicine,* 1963, 25, 303–333.

Barber, T. X. "Hypnosis" as a causal variable in present-day psychology: A critical analysis. *Psychological Reports,* 1964, 14, 839–842. (a)

BARBER, T. X. Hypnotic "colorblindness," "blindness," and "deafness." *Diseases of the Nervous System,* 1964, 25, 529–537. (b)

BARBER, T. X. Toward a theory of "hypnotic" behavior: Positive visual and auditory hallucinations. *Psychological Record,* 1964, 14, 197–210. (c)

BARBER, T. X. Experimental analysis of "hypnotic" behavior. A review of recent empirical findings. *Journal of Abnormal Psychology,* 1965, 70, 132–154. (a)

BARBER, T. X. Physiological effects of "hypnotic suggestions": A critical review of recent research (1960–64). *Psychological Bulletin,* 1965, 63, 201–222. (b)

BARBER, T. X. "Hypnotic" phenomena: A critique of experimental methods. In J. E. Gordon (Ed.) *Handbook of clinical and experimental hypnosis.* New York: Macmillan, 1967. Pp. 444–480.

BARBER, T. X. *A scientific approach to "hypnosis."* Princeton, N. J.: Van Nostrand, in press.

BARBER, T. X., and CALVERLEY, D. S. An experimental study of "hypnotic" (auditory and visual) hallucinations. *Journal of Abnormal and Social Psychology,* 1964, 68, 13–20. (a)

BARBER, T. X., and CALVERLEY, D. S. Experimental studies in "hypnotic" behavior: Suggested deafness evaluated by delayed auditory feedback. *British Journal of Psychology,* 1964, 55, 439–466. (b)

BARBER, T. X., and CALVERLEY, D. S. Toward a theory of "hypnotic" behavior: Experimental analyses of suggested amnesia. *Journal of Abnormal Psychology,* 1966, 71, 95–107.

BARBER, T. X., and CALVERLEY, D. S. Multidimensional analysis of "hypnotic" behavior. *Journal of Abnormal Psychology,* in press.

BARBER, T. X., DALAL, A. S., and CALVERLEY, D. S. The subjective reports of hypnotic subjects. *American Journal of Clinical Hypnosis,* 1968, 11, 74–88.

BEHANAN, K. T. *Yoga: A scientific evaluation.* New York: Dover, 1937.

BOWERS, K. S. The effect of demands for honesty on reports of visual and auditory hallucinations. *International Journal of Clinical and Experimental Hypnosis,* 1967, 15, 31–36.

BROSSE, T. A psycho-physiological study. *Main Currents in Modern Thought,* 1946, 4, 77–84.

BROWN, B. G. A report on three experimental fire-walks by Ahmed Hussain and others. *University of London Counsel for Psychical Investigation,* 1938, Bull. 4.

CHAVES, J. F. Hypnosis reconceptualized: An overview of Barber's theoretical and empirical work. *Psychological Reports,* 1968, 22, 587–608.

COE, M. R., JR. Fire-walking and related behaviors. *Psychological Record,* 1957, 7, 101–110.

COLLINS, J. K. Muscular endurance in normal and hypnotic states: A study of suggested catalepsy. Honor's thesis, Department of Psychology, University of Sydney, 1961.

CRASILNECK, H. B., and HALL, J. A. Physiological changes associated with hypnosis: A review of the literature since 1948. *International Journal of Clinical and Experimental Hypnosis,* 1959, 7, 9–50.

DALAL, A. S. An empirical approach to hypnosis: An overview of Barber's work. *Archives of General Psychiatry,* 1966, 15, 151–157.

DAS, J. P. Yoga and hypnosis. *International Journal of Clinical and Experimental Hypnosis,* 1963, 11, 31–37.

DAS, N. N., and GASTAUT, H. Variations de l'activité' électrique du cerveau du coeur et des muscles squelettiques au cours de la méditation et de'l'extase' yogique. *Electroencephalography and Clinical Neurophysiology,* Suppl., 1955, 6, 211–219.

DEIKMAN, A. J. Experimental meditation. *Journal of Nervous and Mental Disease,* 1963, 136, 329–343.

DEIKMAN, A. J. Implications of experimentally induced contemplative meditation. *Journal of Nervous and Mental Disease,* 1966, 142, 101–116.

ERICKSON, M. H., HERSHMAN, S., and SECTER, I. I. *The practical applications of medical and dental hypnosis.* New York: Julian Press, 1961.

FIELD, P. B. An inventory scale of hypnotic depth. *International Journal of Clinical and Experimental Hypnosis,* 1965, 13, 238–249.

GILL, M. M. and BRENMAN, M. *Hypnosis and related states.* New York: International Universities Press, 1959. Pp. 38–39.

GORTON, B. E. The physiology of hypnosis. *Psychiatric Quarterly,* 1949, 23, 317–343 and 457–485.

GORTON, B. E. Autogenic training. *American Journal of Clinical Hypnosis,* 1959, 2, 31–41.

GRAHAM, K. R., and PATTON, A. Retroactive inhibition, hypnosis, and hypnotic amnesia. *International Journal of Clinical and Experimental Hypnosis,* 1968, 16, 68–74.

HEUZÉ, P. *Fakirs, fumistes et cie.* Paris, 1926.

INGALLS, A. G. Fire-walking. *Scientific American,* 1939, 163, 135–138.

KATKIN, E. S., and MURRAY, E. N. Instrumental conditioning of autonomically mediated behavior: Theoretical and methodological issues. *Psychological Bulletin,* in press.

KIMBLE, G. A. *Hilgard and Marquis' conditioning and learning.* (2nd ed.) New York: Appleton-Century-Crofts, 1961.

KLINE, M. V., GUZE, H., and HAGGERTY, A. D. An experimental study of the nature of hypnotic deafness: Effects of delayed speech feedback. *Journal of Clinical and Experimental Hypnosis,* 1954, 2, 145–156.

KRAMER, E., and TUCKER, G. R. Hypnotically suggested deafness and delayed auditory feedback. *International Journal of Clinical and Experimental Hypnosis,* 1967, 15, 37–43.

LEVITT, E. E., and BRADY, J. P. Psychophysiology of hypnosis. In J. M. Schneck (Ed.) *Hypnosis in modern medicine.* (3rd ed.) Springfield, Ill.: Charles C Thomas, 1963. Pp. 314–362

LUDWIG, A. M. and LEVINE, J. Alterations in consciousness produced by hypnosis. *Journal of Nervous and Mental Disease,* 1965, 140, 146–153.

MASLOW, A. H. Lessons from the peak experiences. *Journal of Humanistic Psychology,* 1962, 2, 9–18.

MAUPIN, E. W. Individual differences in response to a Zen meditation exercise. *Journal of Consulting Psychology,* 1965, 29, 139–145.

ORNE, M. T. The nature of hypnosis: Artifact and essence. *Journal of Abnormal and Social Psychology,* 1959, 58, 277–299.

PATTEN, E. F. Does post-hypnotic amnesia apply to practice effects? *Journal of General Psychology,* 1932, 7, 196–201.

PATTIE, F. A. A report on attempts to produce uniocular blindness by hypnotic suggestion. *British Journal of Medical Psychology,* 1935, 15, 230–241.

PRICE, H. Kuda Bux. *Spectator,* 1937, 158, 808.

RAO, H. V. G., KRISHNASWAMY, N., NARASIMHAIYA, R. L., HOENIG, J., and GOVINDASWAMY, M. V. Some experiments on a "yogi" in controlled states. *Journal of All-India Institute for Mental Health,* 1958, 1, 99–106.

RAWCLIFFE, D. H. *Illusions and delusions of the supernatural and the occult.* New York: Dover, 1959.

RAZRAN, G. The observable unconscious and the inferable conscious in current Soviet psychophysiology: Interoceptive conditioning, semantic conditioning, and the orienting reflex. *Psychological Review,* 1961, 68, 81–147.

SCHULTZ, J. H., and LUTHE, W. *Autogenic training.* New York: Grune & Stratton, 1959.

SPANOS, N. P., and BARBER, T. X. "Hypnotic" experiences as inferred from subjective reports: Auditory and visual hallucinations. Harding, Mass.: Medfield Foundation, 1968.

STURROCK, J. B. Objective assessment of hypnotic amnesia. Paper presented at the meeting of the Eastern Psychological Association. New York, April 15, 1966.

SUTCLIFFE, J. P. "Credulous" and "skeptical" views of hypnotic phenomena: Experiments in esthesia, hallucination, and delusion. *Journal of Abnormal and Social Psychology,* 1961, 62, 189–200.

THORNE, D. E. Memory as related to hypnotic suggestion, procedure, and susceptibility. Unpublished doctoral dissertation, University of Utah, 1967.

VAKIL, R. J. Remarkable feat of endurance by a yogi priest. *Lancet,* 1950, 2, 871.

WENGER, M. A. and BAGCHI, B. K. Studies of autonomic functions in practitioners of Yoga in India. *Behavioral Science,* 1961, 6, 312–323.

WENGER, M. A., BAGCHI, B. K., and ANAND, B. K. Experiments in India on "voluntary" control of the heart and pulse. *Circulation,* 1961, 24, 1319–1325.

WILLIAMS, G. W. Hypnosis in perspective. In L. M. LeCron (Ed.) *Experimental Hypnosis.* New York: Macmillan, 1948. Pp. 4–21.

WILLIAMSEN, J. A., JOHNSON, H. J., and ERIKSEN, C. W. Some characteristics of posthypnotic amnesia. *Journal of Abnormal Psychology,* 1965, 70, 123–131.

WOLBERG, L. R. *Medical hypnosis.* Vol. I. New York: Grune & Stratton, 1948. P. 49.

WOOD, E. *Yoga.* Baltimore: Penguin Books, 1959.

WOODS, J. H. *The Yoga-system of Patanjali.* Cambridge: Harvard University Press, 1927.

71

Hypnosis Reconceptualized: An Overview of Barber's Theoretical and Empirical Work

John F. Chaves

For the past 100 years there has been a dominant conceptualization or paradigm concerning the topic hypnosis. This paradigm, which might be called hypnotic state theory, includes assumptions, methodological guidelines and implicit criteria for the selection of meaningful research topics. The pivotal assumption of this theory is that an altered state of consciousness, the hypnotic trance state, is instrumental in eliciting a wide range of behaviors which, due to their assumed antecedents, are called hypnotic behaviors. Recently, however, an entirely new paradigm has emerged. Although concerned with these same behaviors, the new paradigm rejects the fundamental assumptions and methodological guidelines which constitute the traditional viewpoint.

The new approach to the topic hypnosis has evolved largely through the work of T. X. Barber and is reflected in his 72 published papers and in a forthcoming book (Barber, in press). One purpose of this paper is to provide a concise summary of this new paradigm and to review some of the pertinent supporting evidence. A second purpose is to clarify some misconceptions which have appeared in discussion of Barber's work. These misconceptions have resulted largely from a failure to recognize that Barber rejects the fundamental assumptions of hypnotic state theory. The clarification of these misunderstandings is related to a more general problem pertaining to the effects of anomalous data and theories on traditional scientific paradigms. A brief examination of this problem may provide a useful framework for viewing the relationship between Barber's paradigm and hypnotic state theory.

Science and Innovation

Kuhn (1962) has convincingly argued that the progress of science is marked by two types of activity. One, which he calls "normal science," consists of re-

[1]Supported by an NASA predoctoral traineeship. I wish to thank Dr. T. X. Barber for the hours he spent with me in discussions of his work. I would also like to thank John D. McPeake for valuable assistance in formulating the major points made in this paper.

search and theorizing which is well articulated with the prevailing scientific paradigms. Normal science, in fact, is concerned with the elaboration and extension of traditional paradigms. In the area of hypnosis, normal science consists of such activities as the search for physiological and behavioral correlates of the hypnotic state and the identification of those factors which make one "susceptible" to hypnosis. These problem areas clearly derive from the underlying assumption of hypnotic state theory, namely, that a hypnotic trance state is instrumental in eliciting hypnotic behaviors.

A second phase of scientific activity emerges with the occurrence of reconceptualizations which do not articulate with the prevailing paradigms. These revolutionary paradigms require not only a reconstruction of traditional theory, but also a reevaluation of earlier empirical findings and a redefinition of admissible problems.

A frequent result of the interaction of new and traditional paradigms is a failure of those working within the traditional framework to recognize that anything new or anomalous has occurred. Traditional formulations undergo *ad hoc* modifications which turn the unexpected into the predicted. For example, the post-Copernican persistence of the Ptolemaic conception of the solar system can be attributed to the ease with which epicycles could be added in an *ad hoc* manner to account for new data.

In psychology, perhaps the most significant paradigmatic revolution has been the development of behaviorism, which, since its inception, has redefined the field of psychology including its admissible problems and its methodology. Barber's new paradigm has similar implications for the more limited area of research on hypnosis. Thus, an understanding of Barber's work requires recognition of the fact that the assumptions which have provided the basis for the traditional analysis of hypnotic behaviors are being opened to question. Furthermore, his rejection of these assumptions has methodological implications as well as implications for the kinds of questions which can be asked about hypnotic behaviors. In summary, it should be realized that Barber's paradigm is *fundamentally* different from the traditional paradigm and cannot be understood in terms of traditional constructs and assumptions.

Barber's Critique of the Hypnotic State Construct

Barber's contribution in hypnosis has antecedents in the work of Dorcus (1937), Pattie (1935, 1941, 1950), White (1941), Sarbin (1950), and Sutcliffe (1960, 1961). Although these investigators criticized some of the extreme claims made for hypnosis, they did not critically analyze the *fundamental* assumptions underlying hypnotic state theory. One exception to the latter generalization is Sarbin, who stressed the importance of role-taking in hypnotic performance. Although Sarbin objected to the hypnotic state construct, largely because it is mentalistic, it was Barber who (a) showed more precisely the difficulties and contradictions encountered in using this construct and (b) demonstrated that so-called hypnotic behaviors are functionally related to denotable antecedent variables (Dalal, 1966; Richman, 1965). It is these two features of Barber's work which provide the basis for the new hypnosis paradigm and which will be examined and compared with the traditional approach. First, let us examine Barber's critique of the hypnotic state construct.

The hypnotic trance state, although the subject of much research for the past century, has proven to be quite elusive. No reliable physiological index of the hypnotic state has been found (Barber, 1961; Crasilneck & Hall, 1959; Gorton, 1949; Levitt & Brady, 1963) and the behavioral criteria which investigators have been forced to rely on are ambiguous and not agreed upon (Barber, in press). Although there is disagreement with respect to the criteria that are used to infer that a hypnotic state is present, in most instances this state is inferred from an elevation in the subject's level of response to test suggestions, e.g., suggestions for body rigidity, hallucination, age regression, anesthesia, and amnesia. A second set of criteria used to infer the presence of the hypnotic state include a hypnotic-like appearance on the part of the subject and subject's testimony that he was hypnotized. Barber (1964a, 1965b, in press) has pointed out that logical and empirical considerations cast doubt on both the use of these criteria to infer the existence of a "special state" (the hypnotic state) and the scientific usefulness of the hypnotic state construct as an explanation for so-called hypnotic behavior.

Barber's Logical Critique

Barber (1964a, in press) has cogently argued that the hypnotic state construct is imbedded in circular reasoning. The presence of the hypnotic state is typically inferred from the fact that the subject shows a high level of response to test suggestions of the type included in standardized scales, e.g., the Davis-Husband Scale, the Friedlander-Sarbin Scale, the Stanford Susceptibility Scales. Circular reasoning is difficult to avoid here because the hypnotic state is used to account for the high levels of response to test suggestions from which it is inferred. Investigators using standardized scales of suggestibility to infer the presence of the hypnotic state find it difficult if not impossible to avoid the tautology that subjects manifest high response to suggestions because they are in a hypnotic state and we know that they are in a hypnotic state because they show high response to suggestions! Inferring the presence of a hypnotic state from the subject's hypnotic-like appearance (e.g., appearance of lethargy or extreme relaxation) or from his testimony that he was hypnotized also leads to the same type of circular reasoning. There is evidence to show that (a) a hypnotic-like appearance is itself primarily a response to suggestions for relaxation, drowsiness, sleep, and passivity and (b) subjects typically infer that they were hypnotized by observing that they were responsive to suggestions for relaxation, drowsiness, and sleep and/or to suggestions for limb rigidity, anesthesia, amnesia, and so on

(Barber & Calverley, 1966b). In brief, the presence of the hypnotic state is inferred from criteria which are inseparable from responses to suggestions, and circularity is difficult to avoid because the hypnotic state construct is used to explain responses to suggestions.

A related logical difficulty pertains to the construct hypnotic depth or trance depth. This construct can be used, within the traditional paradigm, to account for all of the variance in hypnotic experiments. When subjects do not respond to suggestions or respond only to a few suggestions, their failure to respond is frequently attributed to their not being in a sufficiently deep trance. When subjects prove to be highly suggestible, it is inferred that they are in a deep hypnotic trance. As one might suspect, there is no independent criterion for trance depth; it is inferred from the same behaviors it purports to explain. Trance depth is clearly an *ad hoc* formulation with virtually no explanatory capability.

From the above and other logical considerations Barber (in press) has drawn two important conclusions: (a) tautological reasoning can be avoided only if investigators are able to delineate criteria for the hypnotic state which are independent of response to suggestions; (b) until such time, however, appeal to the notion of hypnotic state or trance depth does not constitute a scientifically adequate explanation of so-called hypnotic behaviors.

Barber's Empirical Critique

Barber's empirical critique of the hypnotic state construct rests on two considerations: the lability of the criterial behaviors used to infer this state and the failure to find high correlations among these behaviors. Both of these considerations will be discussed briefly in turn.

Barber has presented evidence that some of the behaviors which have been regarded as principal features of the hypnotic state are not as invariant as previously thought. Orne (1959), for example, who has perhaps been most explicit about the behavioral correlates of the hypnotic state, has argued that, ". . . hypnosis . . . [is] . . . a change in the subjective experience of hypnotized individuals." He specifically stresses three aspects of subjects' testimony concerning their experiences: (a) trance logic (referring to incongruous fusings of perceptions and hallucinations), (b) testimony that the subject feels a compulsion to follow suggestions provided by the experimenter, and (c) testimony that the hypnotic state was experienced as basically different from the normal waking state. Barber, Dalal, and Calverley (1966), however, have demonstrated that the extent to which subjects do, indeed, report or exhibit some of these effects depends on the specific wording of questions submitted to them. Since this experiment by Barber, *et al.* is a pioneering effort in this area and since subjects testimony has become an increasingly important criterion of the hypnotic state (cf. Tart & Hilgard, 1966), this experiment will be considered in some detail.

In the experiment presented by Barber, *et al.*, 53 female nursing students who had previously been rated as good hypnotic subjects (passing at least 5 of the 8 test suggestions on the Barber Suggestibility Scale) were randomly assigned to three experimental groups. Subsequent to a hypnotic induction procedure, subjects in each group were asked slightly different questions concerning their experiences under hypnosis. Subjects in Group A were asked positively worded questions such as ". . . did you feel you *could* resist the suggestions?" and "Did you experience the hypnotic state as basically *similar* to the waking state?" Subjects in Group B were given negatively worded questions such as, ". . . did you feel you *could not* resist the suggestions?" and "Did you experience the hypnotic state as basically *different* from the waking state?" Subjects in the third group were given questions which were both positively and negatively worded.

The subjects' testimony depended to a marked degree on how the questions were worded. When the questions were phrased negatively (Group B), subjects' reports were in agreement with Orne's (1959) findings. That is, most subjects (83%) testified that they could not resist responding and (72%) that they experienced the hypnotic state as basically different from the waking state. However, when questions were positively phrased (Group A), only a small minority (22%) testified that they could not resist responding and (17%) that the hypnotic state was basically different from the waking state.

These data suggest the need for a close scrutiny of post-experimental inquiries designed to elicit testimony concerning subjects' experiences. Furthermore, given the lability of these responses, it appears doubtful whether testimony that the subject felt a compulsion to follow suggestions and experienced the hypnotic state as different from the waking state can be seriously regarded as principal features of the presumed trance state. Similarly, the extent to which subjects exhibit trance logic may depend on the manner in which they are questioned. Consider the example, used by Orne (1959), of subjects given suggestions to hallucinate a man sitting in a chair. Will the subjects respond in the same way when asked, "Do you see the chair through the man?" and when asked, "Does the man block the chair?" The experimental data presented by Barber, Dalal, and Calverley (1966) suggest that rather different testimony will be elicited by such variations in the wording of the question.

Another variable which appears to affect subjects' testimony that they were hypnotized is the extent to which the experimenter leads them to believe that they were hypnotized. In another part of the investigation mentioned above (Barber, *et al.*, 1966), it was found that subjects who were told by the experimenter that they appeared to be deeply hypnotized rated themselves as more deeply hypnotized than an uninstructed group who, in turn, rated themselves as more deeply hypnotized than subjects who were told that they did not appear to be hypnotized. These differences in self-ratings of hypnotic depth were obtained in spite of the fact that subjects in all three groups performed at comparable levels on a standardized scale of suggestibility. These data suggest that inferences about the experimenter's expectations which are made by the subjects can affect testimony concerning their subjective experiences.

In short, it appears that many denotable variables affect subjects' testimony concerning their experiences. Until the effects of these variables can be adequately assessed, there seems to be little justification for the use of subjects' testimony as an index of the hypnotic state.

A second empirical consideration which casts doubt on the hypnotic state construct concerns the relationships between the three classes of behaviors from which the hypnotic state is inferred: (a) the subject's hypnotic-like appearance, (b) his responses to explicit test suggestions, and (c) his testimony that he was hypnotized. Although low to moderate correlations (typically accounting for about 10% to 25% of the variance) have been found between these sets of behaviors, subjects may be responsive to test suggestions without showing a hypnotic-like appearance, may show a hypnotic-like appearance and testify that they were not hypnotized, and may testify that they were hypnotized without manifesting a high level of response to test suggestions (Barber, in press; Barber & Calverley, 1966b). These data are far from what would be expected if these criterial behaviors are, indeed, indicative of some underlying state. Furthermore, even the low correlations which have been obtained can be parsimoniously attributed to the fact that subjects who exhibit one type of "hypnotic" behavior, such as responding to test suggestions, may take this fact as evidence that they were hypnotized.

In summary, many sources of evidence, both logical and empirical, have been presented by Barber to support the contention that the hypnotic state is a superfluous construct which does not add to our understanding of the behaviors to be explained or the conditions which are necessary to elicit them. The hypnotic state construct, as traditionally used, appears to serve functions analogous to those served by the construct of phlogiston in chemistry, ether in physics, and entelechy in biology. It is incumbent upon investigators who wish to use the construct hypnosis to provide an unambiguous, non-tautological denotation and to demonstrate that this construct enters into functional relations with the behaviors of interest. Many denotable antecedent variables, which are discussed next, have been shown to be functionally related to so-called hypnotic behaviors. It has not been demonstrated, however, that the hypnotic state is one of these variables.

Instrumental Antecedent Variables

Since Barber's paradigm does not assume that the state of hypnosis is instrumental in eliciting so-called hypnotic behaviors, he does not ask, as do many other researchers, "What is the nature or the essence of hypnosis?" Instead, he begins his analysis with an entirely different question: "What variables included in or associated with procedures labeled as hypnotic inductions are instrumental in eliciting behaviors traditionally labeled as hypnotic?" To answer this question,

Barber and his associates (Barber, 1965a, in press; Barber & Calverley, 1964g, 1965a, 1965b) have isolated a series of discrete variables that are typically included in hypnotic induction procedures. These variables include the following: (a) the situation is defined to the subject as hypnosis; (b) instructions designed to produce positive motivation in the situation are administered; (c) suggestions of eye-closure, relaxation, drowsiness, and sleep are administered; and (d) it is suggested that the subject will find it easy to respond to further suggestions. A number of studies conducted by Barber and his associates have assessed the effects of each of these antecedent variables on three aspects of hypnotic behavior, namely, (a) responses to standardized test suggestions, (b) hypnotic-like appearance (e.g., fixed facial expression, apparent relaxation, apparent passivity), and (c) testimony of having been hypnotized. Barber (in press) has also hypothesized that antecedent variables external to the hypnotic induction procedure may yet influence the degree to which each of these behaviors is manifested. Included among these antecedent variables are: *a set of instructional-situational variables* (such as the specific wording of the suggestions put to the subject and the tone of voice in which suggestions are administered); *a set of subject variables* (the subject's personality characteristics and his attitudes toward the experiment); and *a set of experimenter variables* (experimenter's personality characteristics and his expectations concerning the subjects' performance). Experiments which have demonstrated functional relations between these antecedent variables and the three aspects of hypnotic behavior will be briefly summarized next.

Antecedent Variables Affecting Responses to Test Suggestions

Simply defining the situation to the subject as "hypnosis" tends to facilitate responses to test suggestions for hallucination, body immobility, selective amnesia, etc. (Barber & Calverley, 1964g, 1965a). That is, *with other variables held constant*, subjects told they have been assigned to the hypnosis group show a higher level of suggestibility than those told they have been assigned to the nonhypnotic (control) group. Also, subjects who are told that they are in a hypnosis situation *and are also told* repeatedly that they are becoming progressively more relaxed, drowsy, and sleepy show a slightly higher level of suggestibility than those told only that the situation is hypnosis (Barber & Calverley, 1965a). A surprising finding has been that exhortative statements to try to respond maximally together with statements that it is easy to experience the suggested effects (these two variables in combination are referred to by Barber as "task-motivational instructions") generally increase suggestibility to the same extent as a treatment which includes all of the variables commonly subsumed under the term "hypnotic induction" (Barber, 1965b, 1965c).

Other variables included in "hypnotic induction" procedures do not appear to significantly influence responses to test suggestions. For example, it does not appear to matter whether the subject closes his eyes, nor does it matter whether the instructions are given by means of a tape recorder or in person by the experimenter (Barber & Calverley, 1964b, 1965a). However, higher levels of suggestibility are obtained when the instructions are administered in a forceful tone rather than lackadaisically (Barber & Calverley, 1964d).

Variables external to the "hypnotic induction" procedure also affect responses to test suggestions. For instance, when the situation is defined as a test of gullibility subjects are far less responsive than when it is defined as a test of imagination (Barber & Calverley, 1964c, 1964e). More work needs to be done to delineate the effects of such complex variables as the experimenter's personality and prestige which could also play important roles.

Many researchers have attempted to assess the effects of enduring personality characteristics on response to suggestions. Conflicting and nonsignificant findings have dominated this area of inquiry (Hilgard, 1965; Hilgard & Bentler, 1963; Barber, 1964c). After a careful review of more than fifty investigations that have been conducted in this area, Barber (1964c) presented evidence which indicates that (a) attitudinal-motivational variables are more important than enduring personality characteristics such as neuroticism, introversion, dominance, or ego-strength in determining response to suggestions and (b) to the extent that personality variables play a role, they probably do so by interacting with attitudinal-motivational and other situational variables.

Antecedent Variables Affecting Hypnotic-like Appearance

Recent experimental data indicate that the extent to which subjects manifest a hypnotic-like appearance depends to an important degree on their pre-experimental attitudes and expectations concerning "hypnosis" (Barber, *et al.*, 1966). Moreover, the relaxation, passivity, and lack of spontaneity which characterize a hypnotic-like appearance can be directly related to suggestions of relaxation, drowsiness, and sleep included explicitly in the hypnotic induction procedure and implicitly in the definition of the situation as hypnosis (Barber, in press).

Antecedent Variables Affecting Testimony of Having Been Hypnotized

Orne (1959) contended that a successful hypnotic induction procedure results in an altered state of awareness which subjects will unhesitatingly describe as basically different from the normal waking state. Similarly, Tart and Hilgard (1966) contend that "*S*'s report that he feels hypnotized to some degree is primary data about the presence or absence of hypnosis, if not a criterion of hypnosis." In contradistinction, Barber treats the subject's testimony that he is or is not hypnotized as a dependent variable which has to be functionally related to denotable antecedent variables. Although the relationships are complex, Barber's recent experiments (Barber, in press; Barber & Calverley, 1966b; Barber, Dalal, & Calverley, 1966) indicate that at least four classes of antecedent variables affect subject's testimony that he is or is not hypnotized. These include:

(a) subjects' preconceptions of what hypnosis is supposed to involve, (b) the extent to which they respond to suggestions, (c) the specific wording of the questions which are submitted to subjects to elicit their testimony, and (d) whether the experimenter states or implies that he believes that the subject was, in fact, hypnotized.

In summary, Barber's approach has stimulated a series of experiments designed to determine which of the many variables present in situations commonly termed "hypnotic" are instrumental in eliciting high response to test suggestions, a hypnotic-like appearance, and testimony of having been hypnotized. The data now available indicate that (a) many antecedent variables play an important role and (b) further studies are needed to delineate the main effects and the interactive effects of *all* important antecedent variables.

Further Analyses of Responses to Test Suggestions

Much of the interest in the topic hypnosis can be attributed to an apparently unique set of responses that are elicited by suggestions, for example, by suggestions of body rigidity, hallucination, age regression, analgesia, deafness, blindness, colorblindness, enhanced strength and endurance, enhanced learning proficiency, amnesia, and so on. To clarify further Barber's paradigm, let us first look closely at the explicit and implicit assumptions which underly the traditional analysis of these responses (which we are subsuming under the shorthand label of "responses to test suggestions").

The assumptions underlying the traditional analysis of the aforementioned responses have typically not been made explicit. Furthermore, different investigators appear to make somewhat different assumptions concerning these responses. Nevertheless, two assumptions appear to have sufficient consensus to be regarded as representative of the traditional approach:

(1) The fundamental assumption, which provides unity to the traditional paradigm, is that the hypnotic state is instrumental in eliciting suggested hallucination, age regression, analgesia, deafness, amnesia, and so forth.

(2) Suggested psychological and physiological alterations are, in some sense, viewed as "genuine." The interpretations of the word "genuine" in this context have varied from statements to the effect that the subject really believes he is experiencing the suggested effect and thus is not merely deceiving the experimenter, to claims that suggested deafness, blindness, and amnesia are indistinguishable from naturally-occurring deafness, blindness, and amnesia.

Each of the above assumptions, which appear explicitly or implicitly to underly traditional hypnotic state theory, will be discussed briefly.

(1) It has already been noted that validation of the assumption that the hypnotic state is instrumental in eliciting responses to test suggestions requires that this state be unambiguously and nontautologically denoted and that it be shown to enter into functional relations with the responses. Available evidence described earlier and reviewed in detail by Barber (1967a, in press) suggests

that, if the hypnotic state does, indeed, stand for anything distinct, it is not an important factor in eliciting responses to suggesions. In order to demonstrate that the hypnotic state plays a role in eliciting these responses, it is first necessary to exclude the possibility that the obtained responses are attributable to a host of other variables, such as motivational, situational, and instructional variables, which are *known* to affect responses to test suggestions. Unfortunately, these relevant variables are frequently not controlled or randomized in experiments designed to support hypnotic state theory. This problem will be discussed in more detail when the topic of experimental design is considered.

Relevant here are a series of studies from Barber's laboratory in which concern was with responses to test suggestions for age-regression, amnesia, analgesia, colorblindness, deafness, and enhanced motor and cognitive proficiency (Barber, 1962b, 1962c, 1963, 1964b, 1965a, 1966; Barber & Calverley, 1966a, 1966d). These studies showed that a high level of response to the aforementioned test suggestions can be elicited either by (a) administering brief "task-motivational instructions" (that is, exhortative statements which urge the subject to cooperate and which assure the subject that if he cooperates he will find it rather easy to experience the suggested effects) and by (b) a "formal hypnotic induction procedure" which includes "task-motivational instructions" *plus* repeated suggestions of relaxation, drowsiness, sleep, and of entering a state of deep hypnosis. Attempts to account for these anomalous findings within the traditional paradigm are discussed later in this paper.

(2) The assumption that suggested psychological and physiological alterations are indistinguishable from their non-suggested counterparts was made by many early investigators and has been implied by some recent workers. In an investigation of suggested deafness Erickson (1938) concluded that "there was produced [in hypnotic subjects] a condition not distinguishable from neurological deafness by any of the ordinarily competent tests employed." Weitzenhoffer (1953) also claimed that ". . . when bilateral auditory anesthesia is suggested to a subject who is in the *deepest* possible trance state, who is *fully participating* in the trance situation and in whom a *specific set* has been deeply and carefully established, one may expect alterations of sensory functions which cannot be distinguished from those resulting from organic causes" (p. 148).

Barber's recent investigations pertaining to suggested analgesia, amnesia, age regression, colorblindness, and deafness strongly indicate, however, that the suggested conditions do not produce many of the objective consequences normally associated with these conditions. For example, subjects reporting total or partial deafness in response to suggestions of deafness (given with and also without a hypnotic induction procedure) showed stuttering and mispronunciations under conditions of delayed auditory feedback in the same way as persons with normal hearing (Barber & Calverley, 1964f). Whenever one or more of the objective consequences normally associated with these conditions is manifested in response to suggestions subsequent to a hypnotic induction procedure, similar consequences have been obtained by administering the same suggestions to motivated control subjects. For instance, suggestions of colorblindness, given with and also without a hypnotic induction procedure, will elicit responses on the Ishihara Test for Colorblindness which resemble some of the responses given by the congenitally colorblind (Barber & Deeley, 1961). Furthermore, these "colorblind" responses will be given by a comparable number of subjects in the hypnotic and the control group. These and other studies which have been reviewed by Barber (1961, 1965c) cast doubt on the "genuineness" of these suggested responses when "genuineness" is interpreted in the strictest sense.

The problem of "genuineness" has, in part, developed because the criteria used to infer that the suggested effects have taken place are not agreed upon. For instance, some investigators use the term analgesia when subjects (or patients) testify that they do not feel pain and do not exhibit gross signs of discomfort when normally noxious stimuli are administered. Other investigators reserve the term analgesia for instances when subjects report no pain *and*, in addition, do not manifest physiological signs, such as increased lability of respiration, heart rate, and systolic blood pressure, which are normally present when pain is experienced and normally absent when analgesic drugs are administered. The importance of distinguishing between these criteria has been emphasized by recent investigations (Barber, 1963, 1965d) which have demonstrated that suggestions, given with or without a hypnotic induction procedure, exert a powerful effect on subjects' verbal reports, but produce only small changes on a limited number of physiological measures. Analogous findings with other suggested alterations indicate the importance of individually assessing the effects of suggestions on each of the criteria—verbal reports, overt behavior, autonomic variables—that are used to infer that these alterations have occurred.

BARBER'S METHODOLOGICAL CRITIQUE

Barber's critique of traditional hypnotic state theory is inseparable from his broad-ranging methodological critique of research conducted in this area. To do justice to the broad scope of Barber's new approach, we will now briefly consider one part of his methodological critique, specifically, that part which pertains to experimental designs. Barber's (1962a, 1967a, in press) major criticisms can be summarized as follows:

(1) Prior to Hull's (1933) work in this area, almost all experimental research utilized a few selected "good" hypnotic subjects and no control group. A substantial number of present-day clinical studies also fail to use a control group. Apparently, these studies *assume* that any unusual behavior exhibited by hypnotic subjects can be attributed to the presumed hypnotic state. This design is clearly not acceptable in terms of the rigorous requirements of modern scientific method.

(2) Subsequent to Hull's work, a substantial number of experimental investigators employed a control group which was composed of unselected subjects or of non-suggestible or "unhypnotizable" subjects. The behaviors of these controls were compared to those of selected "good" hypnotic subjects or suggestible subjects. Unfortunately, this design has several weaknesses. When differences in the performance of experimental (hypnotic) and control subjects are obtained, it is impossible to determine whether the differences are due to the treatment (the hypnotic induction procedure) or to preexisting differences in suggestibility. Furthermore, in many experiments, control subjects have not been asked to perform the criterion behaviors (i.e., to hallucinate or to be amnesic for certain material) under the assumption that "awake" subjects would not respond to these suggestions. The assumption that non-hypnotic subjects do not respond to these kinds of suggestions is invalid. For example, Barber and Calverley (1964a) found that in a group of 78 volunteer female subjects, 33% testified that they saw a suggested (non-existent) object and 54% testified that they heard a suggested (non-existent) sound. These subjects were simply given firm, direct suggestions, without a hypnotic induction procedure or motivating instructions.

(3) Other investigators have used a within-subjects design, that is, the performance of a single group of subjects was compared under hypnotic and waking treatments. Since this design takes into consideration the high correlation between hypnotic and waking suggestibility, it is especially sensitive to small increments in suggestibility that may be associated with hypnotic induction procedures (Hilgard & Tart, 1966). However, this design also has an important weakness. The use of a within-subjects design to assess the effects of hypnotic and waking treatments requires that these treatments be independent. There is evidence to indicate, however, that subjects under a waking treatment who expect that they will subsequently also receive a hypnotic treatment perform differently from those who do not (Sutcliffe, 1960; Barber, 1962a). The typically lower level of performance by waking subjects who expect that they will later be "hypnotized" has been hypothesized to be due to "holding back" or to anxiety concerning the forthcoming experience with hypnosis (Zamansky, 1966). The specific interpretation is not important here. What is important is the fact that expectancy of hypnosis is an important variable and should be taken into careful consideration when subjects are used "as their own controls" in hypnotic research.

(4) A design used primarily by Orne (1959) and his coworkers compares the performance of selected "good" hypnotic subjects with that of selected "unhypnotizable" subjects who are instructed to simulate hypnosis and to act as if they experience the suggested effects. This design has the advantage of partialing out experimental artifacts associated with the "demand characteristics" of the experiment. Thus, behaviors which are attributable, for example, to subjects' expectations about hypnosis should be manifested by both the simulating and the

hypnotic subjects (Orne, 1959). Orne has interpreted the differences in performance between the hypnotic and simulating groups as reflecting the "essence" of hypnosis. However, Barber (1962a, in press) has noted that there are two weaknesses in the design used by Orne which obviate such an interpretation:

(a) The hypnotic and simulating treatments are confounded with preexisting differences in suggestibility. Thus any differences in performance under the hypnotic and simulating treatments can be attributed to differences in suggestibility existing prior to the treatments rather than to the presence or absence of the presumed state of trance.

(b) Instructions given to the simulating group strongly imply that they are to go through the motions of responding to suggestions of analgesia, hallucination, amnesia, etc. but are *not* to experience the suggested effects. Instructions and suggestions given to the hypnotic group state explicitly that the subjects are expected to experience the suggested effects. Consequently, differences in responses to suggestions of analgesia, hallucination, amnesia, etc. can be attributed to differences in instructions received by the two groups rather than to the presence or absence of the presumed state of hypnosis. In order to assess the effects of these differences in instructions (assuming that subjects were *randomly* assigned to the hypnotic and simulating groups) it would be necessary to include a group of hypnotic subjects who are also instructed to act *as if* they experienced the suggested effects.

In brief, until the effects of subject variables (preexisting differences in suggestibility) and instructional variables can be assessed, it would appear difficult to claim that differences between hypnotic subjects and simulators reflect the "essence" of hypnosis.

To overcome the difficulties associated with each of the designs mentioned above, Barber has utilized an independent groups design with subjects *randomly assigned* to each of the experimental groups. Depending on the nature of the problem, the subjects are drawn from parent populations which are either unselected or selected on the basis of suggestibility. In any case, the experimental treatments are not confounded with preexisting differences in suggestibility as was noted above. Furthermore, in Barber's recent studies, many combinations of treatment and instructional variables are assessed by using a *factorial design.* Some of the advantages of using independent groups in a factorial design can be seen from the contrasting conclusions drawn by Williamsen, Johnson, and Eriksen (1965) and by Barber and Calverley (1966d) with respect to suggested amnesia.

Williamsen, *et al.* found that hypnotic subjects who were given suggestions of amnesia showed more forgetting than waking subjects who did *not* receive suggestions of amnesia. Also, the hypnotic subjects showed *less* amnesia than waking subjects instructed to act *as if* they had forgotten. It was concluded that the "hypnotic trance induction" played an important role in producing these differences. Barber and Calverley (1966d), however, noted that many variables were confounded in the Williamsen, *et al.* study. For example, suggestions of amnesia were given to hypnotic subjects but not to waking subjects. Furthermore, suggestions to act *as if* the critical material had been forgotten were given to waking subjects but not to hypnotic subjects. While it is possible that the "hypnotic trance induction" was an important variable, it is also possible that the results were due to differences in instructions and suggestions received by the hypnotic and waking groups. Similarly, whether or not subjects testify that they have forgotten may depend on other variables such as the tone of voice or the permissiveness with which the suggestions for amnesia are given, variables not assessed in the Williamsen, *et al.* study. To assess the effects of these and other variables, Barber and Calverley (1966d) conducted a complex 2 × 2 × 4 factorial experiment which included the three experimental groups used by Williamsen, *et al.* plus 13 additional experimental groups.

The three groups that were included in both the Williamsen, *et al.* and the Barber and Calverley experiments (a hypnotic group given suggestions of amnesia, a waking simulation group, and a waking control group) *performed in the same way in both experiments.* Although Barber and Calverley replicated the results obtained by Williamsen, *et al.* with three experimental groups, the results obtained by Barber and Calverley with their 13 additional experimental groups cast serious doubt on the earlier conclusion that "hypnotic trance induction" (and the presumed trance state) plays an important role in eliciting amnesic performance. In general, when hypnotic and waking subjects were given identical suggestions pertaining to amnesia (e.g., permissive suggestions to try to forget, authoritative suggestions that they would forget, or suggestions to act as if they had forgotten), their performance did not differ significantly. Differences in performance were dependent, not on the presence or absence of hypnotic induction procedures, but on the wording of the suggestions for amnesia; for example, subjects receiving authoritative suggestions differed from those receiving suggestions to simulate regardless of whether or not they had received a hypnotic induction. In brief, the Barber and Calverley experiment indicated that (a) an important variable in producing amnesic performance is simply the presence or absence of suggestions to forget, (b) differences in the wording of suggestions for amnesia produce differences in amnesic performance, and (c) whether or not subjects have received a hypnotic induction procedure is not an especially important variable in eliciting acceptance of suggestions to forget.

Criticisms of Barber's Paradigm

A search of the literature yielded eight papers that presented criticisms of Barber's work (Bowers, 1966; Conn & Conn, 1967; Evans, 1966; Hilgard, 1964, 1965; Hilgard & Tart, 1966; Ludwig & Lyle, 1964; Tart & Hilgard, 1966). These critical papers have attempted to deal with Barber's paradigm in two ways.

First, they have criticized it from the assumptions of hypnotic state theory. Secondly, they have criticized specific experimental findings. These papers are examined with the hope of clarifying the relationship between Barber's approach and hypnotic state theory.

Attempts to Conceptualize Barber's Paradigm within Hypnotic State Theory

In several publications Hilgard has made critical comments with respect to Barber's work. For example, Hilgard (1964) writes as follows: ". . . all practicing hypnotists today know that hypnosis is *not* sleep, and that a mention of sleep need not be used at all in inducing hypnosis. Three other metaphors are used: concentration of attention, relaxation, and imagination. Hence, when any one of these is used there is a kind of hypnotic induction involved . . . Barber commonly asks the subject to imagine that what he tells him is actually the case, and this is a variety of induction, even though it is not so defined" (p. 16). Hilgard appears to be saying that Barber's control condition, in which subjects are asked to imagine, is a type of hypnotic induction and, thus, Barber's results with "control" subjects can be attributed to hypnosis. Evans (1966) has also stated a similar argument. These attempts to subsume Barber's results under hypnotic state theory lead to several difficulties. First of all, if subjects who are instructed to imagine vividly are being exposed to a hypnotic induction procedure, where would we draw the line separating those instructions which presumably constitute a hypnotic induction procedure from those which do not? Secondly, there is little justification for labeling instructions to imagine as a "hypnotic induction" unless we are able to demonstrate that they produce a state of hypnosis. However, Barber's critique of the methods used to infer the presence of the presumed hypnotic state have *not* been criticized by Hilgard or by Evans. These questions thus remain open and provide a serious challenge to hypnotic state theory.

In another recent paper Hilgard's characterization of Barber's work is misleading. Hilgard (1965) claims that in Barber's approach ". . . the role of induction is minimized, and the trance state is also minimized" (p. 20). Two points are relevant here. It is somewhat of an understatement to say that Barber "minimizes" the trance state. Instead, Barber has argued that this construct has been ambiguously and tautologically denoted and has impeded progress in delineating relevant variables. Barber's work simply cannot be interpreted in terms of the trance state construct. It is also misleading to contend that Barber minimizes the role of hypnotic induction. One of the most important aspects of Barber's work has been the systematic delineation of relevant variables, including those which comprise procedures that have been traditionally termed hypnotic inductions. These "hypnotic induction" variables, as stated previously, include, for example, (a) definition of the situation as hypnosis, (b) task-motivational instructions, and (c) suggestions of relaxation, drowsiness, and sleep. Barber has shown that *each* of these variables plays an important role in eliciting the behaviors which are to be explained. Although these variables are instrumental in eliciting a wide range of behaviors, hypnotic state theorists have not succeeded in demonstrating that they are also instrumental in eliciting a special state of consciousness (hypnosis or hypnotic trance).

Ludwig and Lyle (1964), Bowers (1966), and Evans (1966) have argued that just because the same results are obtained under hypnosis and task-motivation conditions does not demonstrate that these conditions are the same or that their effects are mediated by similar mechanisms. This argument, however, does not consider the fact that, while many denotable variables, including task-motivational instructions, are known to play an important role in eliciting responses to suggestions, the construct "hypnosis" has not as yet been denoted or operationalized with sufficient clarity so that it can be shown to be necessary or even helpful in eliciting these behaviors. Moreover, until a non-tautological denotation can be provided for the construct "hypnosis," it is difficult to see how the effects of "hypnosis" can be examined.

Another recent critique by Conn and Conn (1967) does provide some hypotheses which need to be tested. They criticize Barber for failing to consider important variables including "primitive transference motivations," "need gratifications," "human variables," and "the willingness and trust of the subject." Barber (1967b) has agreed that these (and other) variables should be evaluated empirically; however, he has also pointed out that, before these (and other) variables can be examined, they need to be clearly denoted and operationalized.

Each of the above criticisms was aimed at one aspect of Barber's work. However, his contribution includes three major aspects: (a) a delineation and critical examination of the assumptions underlying hypnotic state theory, (b) specification of the nature of the behaviors that are to be explained, and (c) isolation of antecedent variables that are instrumental in eliciting these behaviors. Further critical examinations of Barber's work by hypnotic state theorists are in order; however, such critiques should consider all three aspects of his endeavors.

Criticisms of Barber's Experimental Findings

In recent studies, Hilgard and Tart (1966; Tart & Hilgard, 1966) have reported data which they interpret as casting doubt on some of Barber's results and interpretations. Since Hilgard and Tart's experiments may be important, they are considered in detail.

In one recent paper, Hilgard and Tart (1966) hypothesized that Barber's failure to find differences between the effects of task-motivational instructions and a hypnotic induction procedure was due to failure to use a design sensitive enough to detect the differences. To overcome this presumed difficulty, Hilgard and Tart first assessed subjects' responses to a series of standardized test suggestions under one of three conditions: (a) a waking condition in which subjects were told that they would be tested while awake for responsiveness to suggestions;

(b) an imagination condition in which subjects were told they would be given a test of imagination and were asked to try to imagine; and (c) a hypnotic induction condition. In a second session all three groups were tested for suggestibility under a hypnotic induction condition. Small but statistically significant increases in suggestibility were found in the second session (under the hypnotic induction condition) for subjects who had been tested under the waking and imagination conditions in the first session but not for those who had initially been given the hypnotic induction. These data, together with results from a replication designed to assess the role of expecting hypnosis, led Hilgard (1965) to conclude ". . . there is indeed an effect to be attributed to hypnotic induction *beyond the effect of motivating instructions* . . ." (p. 43). This conclusion is not warranted by Hilgard and Tart's data. The instructions given to subjects under the waking and imagination conditions were *not* the same as those given under Barber's task-motivational condition but, rather, were almost identical to those used by Barber for *unmotivated controls.* Thus, Hilgard and Tart *confirmed* Barber and Calverley's (1962, 1963a, 1963b) findings that, as compared to *an unmotivated control treatment*, hypnotic induction procedures produce a small increment in suggestibility. However, Hilgard and Tart's data are not directly relevant to another finding reported by Barber and Calverley (1962, 1963a, 1963b), namely, that the set of variables labeled as "task-motivation instructions" and the set of variables labeled as "hypnotic induction" generally produce a comparable gain in suggestibility.[2] Furthermore, even if a procedure labeled as a "hypnotic induction" is found at times to be more effective than "task-motivation instructions" in facilitating suggestibility (Edmonston & Robertson, 1967), this does not demonstrate that the "hypnotic state" or "trance" raised response to suggestions because no one has as yet succeeded in denoting the "hypnotic trance state" by indices which are independent of response to suggestions.

The results of another recent experiment by Tart and Hilgard (1966) have led them to criticize Barber for failing to control for subjects who "spontaneously slip into hypnosis." Tart and Hilgard worked with a group of subjects who had previously shown a high level of suggestibility under a condition of waking imagination. The subjects were tested again for suggestibility but were now told not to drift into hypnosis and "were immediately brought back to full wakeful-

[2]The experiment presented by Hilgard and Tart (1966) was recently replicated *and also extended* by Barber and Calverley (1966c). (The experiment was extended in that Barber and Calverley included additional groups that were exposed to task-motivational instructions.) In general, Barber and Calverley found that (a) task-motivational instructions facilitated suggestibility above the level found under an imagination or unmotivated control treatment, (b) a hypnotic induction procedure (which included task-motivational instructions) also facilitated suggestibility above the level found under an imagination or unmotivated control treatment, and (c) although the level of suggestibility obtained with a hypnotic induction procedure (that included task-motivational instructions) tended to be slightly higher than that obtained with task-motivational instructions alone, the two treatments did not differ significantly in their effects on suggestibility.

ness by rousing them" whenever they said that they were drifting into a borderline state between sleep and wakefulness or that they were drifting into a state of light hypnosis. When the subjects were thus maintained in an aroused condition, they typically failed to manifest a high level of suggestibility. Tart and Hilgard concluded from these data that "in order for *S* to respond well to suggestibility tests under waking-imagination conditions, he had to slip into a borderline or hypnotic state." The assumption underlying this conclusion, as stated by Tart and Hilgard, may be stated: "*S*'s report that he feels hypnotized to some degree is primary data about the presence or absence of hypnosis, if not a criterion of hypnosis." However, serious problems are encountered if we are to infer that a subject is in a state of hypnosis because he says so. A series of experiments has demonstrated that the subject's testimony that he is in a state of hypnosis is functionally related to such denotable antecedent variables as (a) his preconceptions concerning what hypnosis involves, (b) the extent to which he has responded to suggestions, (c) the wording and tone of voice of the questions that are submitted to him to elicit his testimony, and (d) the degree to which the experimenter states or implies that he believes the subject is hypnotized (Barber, in press; Barber & Calverley, 1966b; Barber, Dalal, & Calverley, 1966). In brief, subjects' reports that they do or do not feel hypnotized have been shown to be functionally related to a series of denotable variables and, as Barber (1967b) has pointed out, "It now remains to be demonstrated, it cannot be assumed *a priori* without evidence, that *S*s' testimony that they are 'in' or 'out' [of hypnosis] is also functionally related to another factor—namely, the presence or absence of hypnosis."

There seems to be even more difficulty with the notion that subjects spontaneously slip into hypnosis. If subjects do, indeed, slip in and out of hypnosis as readily as is implied by Tart and Hilgard and if the hypnotic state is relevant to behavior, it seems that much research in psychology has been negated since very few if any investigators control for this variable. Carried to its logical conclusion, it would be necessary for all experimenters in psychology—regardless of whether they are studying learning, reaction time, psychophysics, psychophysiology, or whatever—to ensure periodically that their subjects had not slipped into hypnosis. Until further evidence has been provided, Tart and Hilgard's (1966) contention that, "The fact that *E* has not administered an induction procedure should not be taken as ensuring that *S* is wide awake and in a normal state" must remain open to question. More specifically, constructs such as "normal state" and "hypnosis," as used by hypnotic state theorists, must be unambiguously and non-tautologically denoted before questions pertaining to "spontaneous hypnosis" can be seriously entertained.

Conclusion

Barber has developed a new paradigm for hypnosis which challenges the fundamental assumptions of hypnotic state theory. The implications of this behavioral paradigm are broad; they include the formulation of new questions concerning the behaviors of interest and revisions in methodology and in the interpretation of earlier empirical findings.

Unfortunately hypnotic state theorists have misinterpreted this paradigm by viewing it in terms of their own constructs and assumptions. This has led to the development of *ad hoc* constructs such as "spontaneous hypnosis" which are assumed to account for Barber's anomalous data.

Kuhn (1962) has shown that these misunderstandings are to be expected when the fundamental assumptions of the prevailing normal science are challenged. Resolution of these misunderstandings can only occur with recognition of the need for a reevaluation of the assumptions of hypnotic state theory. It is hoped that this paper has contributed to a recognition of this need.

REFERENCES

BARBER, T. X. Physiological effects of "hypnosis." *Psychological Bulletin*, 1961, 58, 390-419.

BARBER, T. X. Experimental controls and the phenomena of "hypnosis:" a critique of hypnotic research methodology. *Journal of Nervous and Mental Disease*, 1962, 134, 493-505. (a)

BARBER, T. X. Hypnotic age regression: a critical review. *Psychosomatic Medicine*, 1962, 24, 286-299. (b)

BARBER, T. X. Toward a theory of hypnosis: posthypnotic behavior. *Archives of General Psychiatry*, 1962, 7, 321-342. (c)

BARBER, T. X. The effects of "hypnosis" on pain: a critical review of experimental and clinical findings. *Psychosomatic Medicine*, 1963, 25, 303-333.

BARBER, T. X. "Hypnosis" as a causal variable in present-day psychology: a critical analysis. *Psychological Reports*, 1964, 14, 839-842. (a)

BARBER, T. X. Hypnotic "colorblindness," "blindness," and "deafness." *Diseases of the Nervous System*, 1964, 25, 529-537. (b)

BARBER, T. X. Hypnotizability, suggestibility, and personality: V. A critical review of research findings. *Psychological Reports*, 1964, 14, 299-320. (c)

BARBER, T. X. The effects of "hypnosis" on learning and recall: a methodological critique. *Journal of Clinical Psychology*, 1965, 21, 19-25. (a)

BARBER, T. X. Experimental analyses of "hypnotic" behavior: a review of recent empirical findings. *Journal of Abnormal Psychology*, 1965, 70, 132-154. (b)

BARBER, T. X. Measuring "hypnotic-like" suggestibility with and without "hypnotic induction;" psychometric properties, norms, and variables influencing response to the Barber Suggestibility Scale (BSS). *Psychological Reports*, 1965, 16, 809-844. (c)

BARBER, T. X. Physiological effects of "hypnotic suggestions:" a critical review of recent research (1960-64). *Psychological Bulletin*, 1965, 63, 201-222. (d)

BARBER, T. X. The effects of hypnosis and suggestions on strength and endurance: a critical review of research studies. *British Journal of Social and Clinical Psychology*, 1966, 5, 42-50.

BARBER, T. X. "Hypnotic" phenomena: a critique of experimental methods. In J. E. Gordon (Ed.), *Handbook of clinical and experimental hypnosis*. New York: Macmillan, 1967. Pp. 444-480. (a)

BARBER, T. X. Reply to Conn and Conn's "Discussion of Barber's 'Hypnosis as a causal variable . . . .'" *International Journal of Clinical and Experimental Hypnosis*, 1967, 15, 111-117. (b)

BARBER, T. X. *A scientific approach to "hypnosis."* Princeton, N. J.: Van Nostrand, in press.

BARBER, T. X., & CALVERLEY, D. S. "Hypnotic" behavior as a function of task motivation. *Journal of Psychology*, 1962, 54, 363-389.

BARBER, T. X., & CALVERLEY, D. S. The relative effectiveness of task-motivating instructions and trance-induction procedure in the production of "hypnotic-like" behaviors. *Journal of Nervous and Mental Disease*, 1963, 137, 107-116. (a)

BARBER, T. X., & CALVERLEY, D. S. Toward a theory of hypnotic behavior: effects on suggestibility of task motivating instructions and attitude toward hypnosis. *Journal of Abnormal and Social Psychology*, 1963, 67, 557-565. (b)

BARBER, T. X., & CALVERLEY, D. S. An experimental study of "hypnotic" (auditory and visual) hallucinations. *Journal of Abnormal and Social Psychology*, 1964, 63, 13-20. (a)

BARBER, T. X., & CALVERLEY, D. S. Comparative effects on "hypnotic-like" suggestibility of recorded and spoken suggestions. *Journal of Consulting Psychology*, 1964, 28, 384. (b)

BARBER, T. X., & CALVERLEY, D. S. The definition of the situation as a variable affecting "hypnotic-like" suggestibility. *Journal of Clinical Psychology*, 1964, 20, 438-440. (c)

BARBER, T. X., & CALVERLEY, D. S. Effect of *E*'s tone of voice on "hypnotic-like" suggestibility. *Psychological Reports*, 1964, 15, 139-144. (d)

BARBER, T. X., & CALVERLEY, D. S. Empirical evidence for a theory of "hypnotic" behavior: effects of pretest instructions on response to primary suggestions. *Psychological Record*, 1964, 14, 457-467. (e)

BARBER, T. X., & CALVERLEY, D. S. Experimental studies in "hypnotic" behavior: suggested deafness evaluated by delayed auditory feedback. *British Journal of Psychology*, 1964, 55, 439-446. (f)

BARBER, T. X., & CALVERLEY, D. S. Toward a theory of hypnotic behavior: effects on suggestibility of defining the situation as hypnosis and defining response to suggestions as easy. *Journal of Abnormal and Social Psychology*, 1964, 68, 585-592. (g)

BARBER, T. X., & CALVERLEY, D. S. Empirical evidence for a theory of "hypnotic" behavior: effects on suggestibility of five variables typically included in hypnotic induction procedure. *Journal of Consulting Psychology*, 1965, 29, 98-107. (a)

BARBER, T. X., & CALVERLEY, D. S. Empirical evidence for a theory of "hypnotic" behavior: the suggestibility-enhancing effects of motivational suggestions, relaxation-sleep suggestions, and suggestions that the subject will be effectively "hypnotized." *Journal of Personality*, 1965, 33, 256-270. (b)

BARBER, T. X., & CALVERLEY, D. S. Effects on recall of hypnotic induction, motivational suggestions, and suggested regression: a methodological and experimental analysis. *Journal of Abnormal Psychology*, 1966, 71, 169-180. (a)

BARBER, T. X., & CALVERLEY, D. S. *Fact, fiction, and the topic hypnosis: multivariate analysis of behavioral dimensions.* Harding, Mass.: Medfield Found., 1966. (b)

BARBER, T. X., & CALVERLEY, D. S. *Hypnotic induction procedure, imagination instructions, task-motivational instructions, and suggestibility: a replication and extension of experiments by Barber and Calverley (1962-63) and Hilgard and Tart (1966).* Harding, Mass.: Medfield Found., 1966. (c)

BARBER, T. X., & CALVERLEY, D. S. Toward a theory of "hypnotic" behavior: experimental analyses of suggested amnesia. *Journal of Abnormal Psychology*, 1966, 71, 95-107. (d)

BARBER, T. X., DALAL, A. S., & CALVERLEY, D. S. *The subjective experiences of hypnotic subjects.* Harding, Mass.: Medfield Found., 1966.

BARBER, T. X., & DEELEY, D. C. Experimental evidence for a theory of hypnotic behavior: I. "Hypnotic" color-blindness without hypnosis. *International Journal of Clinical and Experimental Hypnosis*, 1961, 9, 79-86.

BOWERS, K. Hypnotic behavior: the differentiation of trance and demand characteristic variables. *Journal of Abnormal Psychology*, 1966, 71, 42-51.

CONN, J. H., & CONN, R. N. Discussion of T. X. Barber's " 'Hypnosis' as a causal variable in present day psychology: a critical analysis." *International Journal of Clinical and Experimental Hypnosis*, 1967, 15, 106-110.

CRASILNECK, M. B., & HALL, J. A. Physiological changes associated with hypnosis: a review of the literature since 1948. *International Journal of Clinical and Experimental Hypnosis*, 1959, 7, 9-50.

DALAL, A. S. An empirical approach to hypnosis. *Archives of General Psychiatry*, 1966, 15, 151-157.

DORCUS, R. M. Modification by suggestion of some vestibular and visual responses. *American Journal of Psychology*, 1937, 49, 82-87.

EDMONSTON, W. E., JR., & ROBERTSON, T. G., JR. A comparison of the effects of task motivational and hypnotic induction instructions on responsiveness to hypnotic suggestibility scales. *American Journal of Clinical Hypnosis*, 1967, 9, 184-187.

ERICKSON, M. H. A study of clinical and experimental findings on hypnotic deafness: I. Clinical experimentation and findings. *Journal of General Psychology*, 1938, 19, 127-150.

EVANS, F. J. Current developments in experimental hypnosis. Paper presented at the meeting of the American Psychological Association, New York, September 2, 1966.

GORTON, B. E. The physiology of hypnosis. *Psychiatric Quarterly*, 1949, 23, 317-343, 457-485.

HILGARD, E. R. The motivational relevance of hypnosis. In M. R. Jones (Ed.), *Nebraska Symposium on Motivation, 1964.* Lincoln, Nebr.: Univer. of Nebraska Press, 1964. Pp. 1-40.

HILGARD, E. R. *Hypnotic susceptibility.* New York: Harcourt, Brace, & World, 1965.

HILGARD, E. R., & BENTLER, P. M. Predicting hypnotizability from the Maudsley Personality Inventory. *British Journal of Psychology*, 1963, 54, 63-69.

HILGARD, E. R., & TART, C. T. Responsiveness to suggestions following waking and imagination instructions and following induction of hypnosis. *Journal of Abnormal Psychology*, 1966, 71, 196-208.

HULL, C. L. *Hypnosis and suggestibility.* New York: Appleton, 1933.

KUHN, T. S. *The structure of scientific revolutions.* Chicago: Univer. of Chicago Press, 1962.

LEVITT, E. E., & BRADY, J. P. Psychophysiology of hypnosis. In J. M. Schneck (Ed.), *Hypnosis in modern medicine.* (3rd ed.) Springfield, Ill.: Thomas, 1963. Pp. 314-362.

LUDWIG, A. M., & LYLE, W. H., JR. Tension induction and the hyperalert trance. *Journal of Abnormal and Social Psychology*, 1964, 69, 70-76.

ORNE, M. T. The nature of hypnosis: artifact and essence. *Journal of Abnormal and Social Psychology*, 1959, 58, 277-299.

PATTIE, F. A. A report of attempts to produce uniocular blindness by hypnotic suggestion. *British Journal of Medical Psychology*, 1935, 15, 230-241.

PATTIE, F. A. The production of blisters by hypnotic suggestions: a review. *Journal of Abnormal and Social Psychology*, 1941, 36, 62-72.

PATTIE, F. A. The genuineness of unilateral deafness produced by hypnosis. *American Journal of Psychology*, 1950, 63, 84-86.

RICHMAN, D. N. A critique of two recent theories of hypnosis: the psychoanalytic theory of Gill and Brenman contrasted with the behavioral theory of Barber. *Psychiatric Quarterly*, 1965, 39, 278-292.

SARBIN, T. R. Contributions to role-taking theory: I. Hypnotic behavior. *Psychological Review*, 1950, 57, 255-270.

SUTCLIFFE, J. P. "Credulous" and "skeptical" views of hypnotic phenomena: a review of certain evidence and methodology. *International Journal of Clinical and Experimental Hypnosis*, 1960, 8, 73-101.

SUTCLIFFE, J. P. "Credulous" and "skeptical" views of hypnotic phenomena: experiments in esthesia, hallucination, and delusion. *Journal of Abnormal and Social Psychology*, 1961, 62, 189-200.

TART, C. T., & HILGARD, E. R. Responsiveness to suggestions under "hypnosis" and "waking-imagination" conditions, a methodological observation. *International Journal of Clinical and Experimental Hypnosis*, 1966, 14, 247-256.

WEITZENHOFFER, A. M. *Hypnotism: an objective study in suggestibility.* New York: Wiley, 1953.

WHITE, R. W. A preface to the theory of hypnotism. *Journal of Abnormal and Social Psychology*, 1941, 36, 477-505.

WILLIAMSEN, J. A., JOHNSON, H. J., & ERIKSEN, C. W. Some characteristics of post-hypnotic amnesia. *Journal of Abnormal Psychology*, 1965, 70, 123-131.

ZAMANSKY, H. S. The effect of expectancy for hypnosis on prehypnotic performance. Paper read at Eastern Psychological Assn annual meeting, New York, April 15, 1966.

72

The Effects of "Hypnosis" on Pain

Theodore X. Barber

Experimental and clinical studies concerned with the effects of "hypnotically-suggested analgesia" in surgery, in labor, and in chronic pain are critically evaluated. The review suggests that "hypnotic analgesia" at times produces not a reduction in pain but an unwillingness to state directly to the hypnotist that pain was experienced or a temporary "amnesia" for the pain experienced. In other instances, suggestions of pain relief given under "hypnosis" produce some degree of diminution in anxiety and pain as indicated by reduction in physiological responses to noxious stimuli and by reduction in requests for pain-relieving drugs. The data suggest that "the hypnotic trance state" may be an extraneous variable in ameliorating pain experience in situations described as "hypnosis;" the critical variables appear to include: (a) suggestions of pain relief, which are (b) given in a close interpersonal setting.

A NUMBER OF INVESTIGATORS[34] contend that "hypnotically-suggested analgesia" lessens or entirely prevents pain, while others[82] are of the opinion that hypnotic suggestions produce verbal denial of pain experience without affecting pain and suffering. This paper critically evaluates the effects of "hypnosis" on pain. Relevant clinical and experimental studies are reviewed to answer two questions: (1) Does "hypnotically-suggested analgesia" refer to reduction of pain, to verbal denial of the pain experienced, or to a combination of both of these effects? (2) Of the many independent and intervening variables subsumed under the term "hypnotically-suggested analgesia"—e.g., the suggestions of analgesia, "the hypnotic trance state," the close relationship between patient and physician—which are effective and which are superfluous to producing "pain relief"?

From the Medfield Foundation, Medfield, Mass., and the Division of Psychiatry, Boston University School of Medicine, Boston, Mass.

The writing of this paper was made possible by Research Grant MY4825 from the National Institute of Mental Health, U.S.P.H.S.

Received for publication Oct. 3, 1962.

Denotations of Critical Terms

The Term "Hypnosis"

When investigators report that "analgesia" was produced under "hypnosis" or in a "hypnotized" subject, they appear to be saying in an abbreviated way that pain and suffering were ameliorated in a subject who was selected as meeting criteria of "hypnotizability;" who was placed in "a state of trance;" and who was given suggestions of pain relief by a prestigeful person with whom he had a close interpersonal relationship.[17] This confounding of a number of independent and intervening variables under the single term "hypnosis" leads to serious problems. It may be that one or two of these variables (e.g., suggestions of pain relief given in a close interpersonal setting) are sufficient to reduce pain and that the other variables—"the hypnotic trance state," the selection of subjects as "hypnotizable"—are extraneous. In the following discussion we shall at first use the word "hypnosis" as it is commonly used, to refer to all of these variables in combination. After we have reviewed clinical and experimental investigations concerned with "hypnotically-suggested analgesia," we shall turn again to the term "hypnosis" and will place this concept under critical analysis.

The Term "Pain"

"Pain" is a multidimensional concept. First, "pain" refers to an unpleasant sensation which varies not only in intensity (from "mild" to "excruciating") but also in quality (from the lancinating sensation associated with trigeminal neuralgia, to the burning sensation found in causalgia, to the deep, aching sensation of abdominal cramps). Secondly, the term "pain" subsumes not only these various "sensations of pain" but also a "reaction pattern" which is generally categorized by such terms as "anxiety" or "concern over pain." Although these two components of the "pain experience"— "sensation of pain," and "anxiety" or "reaction to pain" —are normally intimately interrelated, a series of studies, summarized below, suggests that they can be partly dissociated under certain conditions.

Beecher[25, 27] has presented cogent evidence that similar wounds which presumably produce similar "pain sensations" may give rise to strikingly different "reaction patterns." He studied 215 seriously wounded soldiers in a combat zone hospital. Two-thirds of the men did not show signs of suffering, were in an "optimistic, even cheerful, state of mind," and refused pain-relieving drugs. This apparent lack of anxiety and suffering was not due to shock, and it was not due to a total "pain block;" the men were clear mentally and complained in a normal manner to rough handling of their wounds or to inept venipunctures. Beecher compared the "reaction pattern" of the wounded soldiers with the reactions shown by 150 male civilians who had undergone major surgery. Although the postoperative patients were suffering from less tissue trauma, only one-fifth of these patients (as compared to two-thirds of the soldiers) refused medication for relief of pain. The striking difference in reaction to injury in the two groups was apparently due to differences in the significance of the wound. The soldier viewed his wound as a good thing; it enabled him to leave the battlefield with honor. The civilian viewed his surgery as a calamitous event. Beecher[28] notes that "one cannot know whether in the above instances [of the wounded soldiers] the pain sensation or the reaction to pain is blocked; however, since the conscious man badly wounded in warfare often does not suffer at all from his great wound, yet is annoyed by, and suffers apparently normally from, a venipuncture, one can conclude that the nervous system can transmit pain sensations but that somehow the reaction to them is the altered element."

Hill *et al.*[76, 77] and Kornetsky[86] have presented evidence to support the hypothesis that pain relief following morphine administration is closely related to "relief of anxiety" or "reduction in fear of pain." Cattell,[43] Beecher,[28] and Barber[11, 12] have reviewed other studies which suggest that morphine and other opiates at times alleviate suffering by minimizing "anxiety" and "concern over pain" without necessarily elevating the pain threshold or altering "awareness of pain." Data indicating that placebos also at times ameliorate "pain experience" by alleviating "anxiety" or "reaction to pain" have been reviewed by Beecher[20] and by Barber.[11]

Additional evidence that "the reaction component" of the "pain experience" can be at least partly dissociated from "pain sensation" is found in the effects of such surgical procedures as prefrontal leukotomy and topectomy of Brodmann's areas 9 and 10. These and other operations on the frontal areas at times appear to ameliorate intractable pain by alleviating "anxiety," "worry," and "concern over pain."[11, 57, 97] Leukotomized patients characteristically state that their pain is the same, but it does not bother them anymore. Investigators who have studied the effects of frontal operations appear to agree with Ostenasek's[113] conclusion that "when the fear of pain is abolished, the perception of pain is not intolerable."

The above and other data[11, 12, 132] suggest that in attempting to delineate the effects of "hypnotically-suggested anesthesia or analgesia,"* it may be more relevant to focus on the "reaction" component of the "pain experience" rather than on "pain sensation per se." If "hypnotically-suggested analgesia" relieves "anxiety" or "concern over pain" but does not affect pain as a sensation or exerts only an indirect effect or a minor effect on pain sensation, it can be said that it (1) affects a major component of the "pain experience" and (2) it may be exerting as much effect on "pain experience" as powerful analgesics such as morphine.[132]

"Hypnotic Analgesia," "Posthypnotic Amnesia," and Denial of Pain

A series of clinical reports indicates that suggestions of analgesia given under "hypnotic trance" at times result not in a reduction in pain and suffering but in an apparent amnesia for the pain and suffering experienced. In a number of investigations[37, 63, 80, 93, 114, 128] the hypnotized patients cried, moaned, or showed signs of shock during surgery or parturition but maintained afterwards that they had forgotten the experience. For instance, Schultze-Rhonhof[128] reported that obstetric patients who had received extensive antenatal training in entering "deep trance" showed overt behavioral signs of pain during labor—some groaned, others cried, others showed marked agitation—but the patients maintained on awakening that they were not aware of having suffered. This investigator interpreted his findings as indicating that hypnotic suggestions of pain relief rarely if ever produce a complete suppression of pain: "In the majority of cases, the complete analgesia which is claimed on awakening is the result of the amnesia."

Raginsky[124] refers to cases of minor surgery performed under hypnosis in which the patients appeared "amnesic" immediately after surgery; however, when hypnotized at a later date, the patients could ". . . usually recall the site of pain and describe accurately the pain experienced at the time of the operation." Myers,[112] Perchard,[121] and Dorcus and Shaffer[53] have also presented data indicating that "posthypnotic amnesia" for pain experienced during surgery or during parturition is temporary and easily reversible.

Other findings, recently reviewed in detail elsewhere,[19] also indicate that "posthypnotic amnesia" is labile and superficial. These findings include the following:

1. With few if any exceptions, investigators report that "amnesic" hypnotic subjects recall the "forgotten" events if the hypnotist states, "Now you remember."[36] Subjects who have been deeply hypnotized also recall the "forgotten" happenings when given *implicit* permission to remember. Such tacit permission may be

*Since "anesthesia" (insensibility to all stimuli) also includes "analgesia" (insensibility to painful stimuli), studies concerned with relief of pain by hypnosis rarely made a distinction between these terms, and the terms will be used interchangeably in the present review.

given by asking, "Do you remember?," with the intonation that the subject is permitted to remember;[143] by giving a "hint;"[110] by instructing the subject to allow his hand to write automatically;[110, 144] and so forth.

2. Experimental evidence indicates that "amnesic" hypnotic subjects *recognize* the material which they claim not to remember; this recognition is indicated by overt behavior—e.g., avoidance of "amnesic" material but not of similar control material[6, 144]—and by alterations in pulse and respiration when presented with the "forgotten" material but not when presented with comparable control material.[31]

3. Experimental evidence indicates that "somnambulistic" subjects who show "complete amnesia" when interviewed by the hypnotist show very little if any effects of the "amnesia" when tested by indirect methods which do not depend on verbal reports, such as assessment of practice effects or of retroactive inhibition effects.[101, 108, 117]

4. "Amnesic" hypnotic subjects characteristically make such statements as: "I haven't any inclination to go back over it;" "I do remember but I can't say;" "I know it but I can't think about it—I know what it is but I just kind of stop myself before I think of it."[32, 146] These and other remarks made by "amnesic" hypnotic subjects can be interpreted as supporting ". . . not a dissociation theory, but rather a motivational theory, a theory that such amnesia is due to an unwillingness to remember, an attempt to occupy oneself with other things than an effort to recall."[118]

The above and other data[19] suggest that "posthypnotic amnesia for pain" may be more labile and temporary than is at times supposed and may be difficult to differentiate from purposive denial of the pain experienced or from unwillingness to admit to the hypnotist that pain was experienced. It should be noted here that the verbal reports of "good" hypnotic subjects often appear to be closely correlated with what the hypnotist leads the subject to believe he is expected to report.[7, 8, 10, 13, 16–18] If the hypnotist implies when interviewing the subject that he should state that no pain was experienced, the "'good" hypnotic subject may comply on a verbal level even though pain was experienced. On the other hand, if the hypnotic subject is given a means of stating what occurred without at the same time directly contradicting the hypnotist's explicit suggestions and the hypnotists's apparent desires and expectations, he may give a different report. Kaplan[82] has presented an interesting case study which can be interpreted along these lines. A highly trained hypnotic subject was placed in "a very deep trance" and given two suggestions: that his left arm was analgesic and insensitive and that his right hand would continuously perform automatic writing. The "analgesic" left arm was pricked four times with a hypodermic needle; when receiving this stimulation, the subject's right hand wrote, "Ouch, damn it, you're hurting me." After a minute or two, the subject asked the experimenter, "When are you going to begin," apparently having "forgotten" that he had received the painful stimuli. Kaplan interpreted these findings as indicating that hypnotic suggestions of analgesia produce ". . . an artificial repression and/or denial of pain, but that at *some level* pain *is* experienced—moreover, experienced as discomfort at that level."

The motivation for denial of pain is present in the hypnotic situation. The physician has invested time and energy hypnotizing the patient and suggesting that pain will be relieved; expects and desires that his efforts will be successful; and by his words and manner communicates his desires and expectations to the patient. The patient in turn has often formed a close relationship with the physician-hypnotist and would like to please him or at least not to disappoint him. Furthermore, the patient is aware that if he states that he suffered, he is implying that the physician's time and energy were wasted and his efforts futile. The situation is such that even though the patient may have suffered, it may be difficult or disturbing for him to state directly to the physician-hypnotist that he experienced pain and it may be less anxiety provoking to say that he did not suffer.

It should be noted that the motivation to deny pain is not necessarily a function of the patient's having been hypnotized. Similar findings may be obtained in any situation, hypnotic or nonhypnotic, in which the physician invests time and effort attempting to support the patient and to ameliorate the patient's suffering. These conditions making for denial of pain appear to be present, for instance, in situations described as "natural childbirth." Mandy *et al.*[102] have presented data indicating that the "natural childbirth" patient who reports to the physician and to the physician's associates that she was "delighted with natural childbirth" may state, when interviewed by an independent observer, that her delivery was more painful than she had anticipated or believed was necessary "but she couldn't admit it to the house staff for fear of disappointing them."

Carefully controlled studies are needed in which patients who have ostensibly experienced "hypnotic analgesia" are interviewed not only by the hypnotist but also by a person who is not associated with "hypnosis" and to whom the patient is willing to confide. It can be hypothesized from the data presented above that some hypnotic subjects who deny pain or who appear to have amnesia for pain when questioned by the hypnotist will state that they experienced pain and that they suffered when interviewed by a person whom they trust and who is not associated with the "hypnosis."

Overt Behavioral Reactions as Criteria of "Analgesia" or "Pain Relief"

The data cited above suggest caution in using the hypnotic patient's verbal report as given to the hypnotist as an index of pain relief. Caution is also necessary in using the hypnotic patient's lack of overt behavioral reactions to noxious stimulation as indicating that pain and suffering have been abolished; as noted above, the hypnotic subject is often motivated to please the hypnotist or to try not to disappoint the hypnotist and this may at times be sufficient for him to try to inhibit overt signs of pain such as moaning, wincing, or restlessness.

The findings presented by Javert and Hardy[81] with respect to "natural childbirth" may also apply to the patient undergoing labor under "hypnosis." The subjects consisted of 26 untrained labor patients and 5 patients who had been "trained" in "natural childbirth" by others (not by Javert or Hardy). During labor the untrained patients showed evidence of anxiety and pain, while the "natural childbirth" patients appeared relatively "serene." Between uterine contractions both groups were asked to compare the pain of labor with the pain produced by application of radiant heat to the forelimb. (These measurements were made in both groups prior to the administration of analgesic or anesthetic drugs). The "natural childbirth" patients did not differ from the untrained patients in estimates of pain intensity; both groups rated the pain of labor as relatively severe and equal in maximal intensity to blister-producing thermal stimulation. Javert and Hardy interpreted these findings as indicating that the regimen known as "natural childbirth" produces a "satisfactory reaction pattern" but has little if any effect on the intensity of the pain experienced during labor.

An additional consideration should be noted here: Velvovski *et al.*[139] claims that from 7 to 14% of *unselected* patients in the Soviet Union give birth without medications, without showing signs of pain, anxiety, or suffering, and without receiving any training or preparation. Are more than 14% of *unselected* hypnotized patients able to perform this feat? No data are available to answer this question; reports concerned with the effects of hypnosis in parturition are in all cases based on volunteers or *selected* patients.

The proportion of *selected* hypnotic patients able to deliver without medications and without exhibiting signs of suffering may not greatly exceed the 7 to 14% of *unselected* patients which Velvovski claims can deliver in this way without any training at all. Although some investigators[46, 107, 109] report that more than one-third of selected hypnotic patients are able to deliver without anodynes, others[4, 40, 147] find that no more than 14% of patients who volunteer for hypnosis training are able to deliver without medicaments and without showing gross signs of pain.

Similar considerations may apply in surgery: The proportion of *selected* patients who are able to undergo surgery with "hypnoanesthesia" alone may not greatly exceed the proportion of *unselected* patients who were able to undergo surgery in the preanesthetic period without manifesting signs of pain. Data presented by Trent,[138] Leriche,[98] Elliotson,[58] and Chertok[44] indicate that although some surgical patients, prior to the advent of anesthetics, struggled and screamed, a small proportion of patients ". . . bravely made no signs of suffering at all." Although it is often stated that at the present time approximately 10% of the population is able to undergo surgery under "hypnotic trance," Wallace and Coppolino[141] note the following:

> Our percentage of success in the complete substitution of hypnoanesthesia for chemoanesthesia has been less than the previously quoted 10 per cent. There have not been any published series of cases in which a statistical analysis would indicate that approximately 10 per cent of the patients are able to withstand a surgical intervention with hypnoanesthesia as a sole modality. Therefore, it is our conclusion that the 10 per cent estimate is an often-repeated but unsubstantiated quantity and that the true percentage of successful cases is much below that figure.

The above data suggest two conclusions:

1. Caution is necessary in accepting the hypnotic patient's verbal report or lack of overt behavioral reactions as valid indices that the patient did not suffer. The hypnotic situation is often structured in such a manner that the patient is motivated to inhibit overt signs of pain and to deny pain experience.
2. The proportion of *selected* hypnotized patients who are able to undergo labor or surgery without manifesting motoric signs of pain and without receiving anodynes may not greatly exceed the proportion of *unselected* patients who are able to do the same thing without any preparation at all. Careful controls are needed to determine if the effects attributed to "hypnosis" are due to the selection of patients.

Physiological Indices of Anxiety and Pain

The data presented above suggest that an objective index of pain which is difficult or impossible to affect voluntarily is needed in studies concerned with "hypnotically-suggested analgesia." Unfortunately there appears to be no single index and no combination of indices which unequivocally indicate the presence or absence of pain and suffering. However, a series of studies demonstrate that a satisfactory, although not conclusive, objective index of anxiety and pain consists of an alteration in one or more systemic physiological functions which are difficult to alter by voluntary effort.

In normal subjects painful stimulation almost always produces alterations in one or more of the following: blood pressure, heart rate, respiration, digital vasomotor tone, skin resistance, and degree of tension in localized muscles.[24, 28, 42, 48, 68, 69, 129] Although nonpainful stimuli at times also produce alterations in these physiological indices, they rarely produce the same degree or the same pattern of alteration as painful stimuli.[24, 59, 92] There is also evidence to indicate that morphine, meperidine, nitrous oxide, and other analgesics and anesthetics drastically reduce these normally expected responses to noxious stimulation. The galvanic skin response to painful stimulation is apparently markedly reduced by morphine at low dose levels (8 mg.)[3] and is apparently abolished by nitrous oxide anesthesia,[38] by meperidine (100 mg.), and by morphine at higher dose levels (20–100 mg.).[3] It also appears that morphine (8–16 mg.) and codeine (32–64 mg.) reduce the vasoconstriction response to noxious stimulation to near the vanishing point[129] and that the elevation in blood pressure which normally follows painful stimulation is eliminated by anesthetic doses of barbiturates.[68]

Before turning to experimental studies which used physiological variables to assess the effects of "hypnotically-suggested analgesia," two considerations should be emphasized. (1) Subjects differ in their physiological patterns of response to the same noxious stimulus, and the same subject may show different patterns of physiological response to different types of noxious stimuli.[92] When physiological variables are used to assess "hypnotic analgesia," it is necessary to take inter- and intrasubject variability into account. (2) Alterations in physiological variables during painful stimulation appear to be more closely correlated with the "anxiety" or "reaction" component of the pain experience than with "pain sensation per se."[21, 54, 74, 99, 127] This consideration, however, is not a major objection to the use of autonomic indices to assess the effects of "hypnotically-suggested analgesia." "Anxiety" or "concern over pain" appears to be a major component of the total pain experience, and if "hypnotic analgesia" reduces anxiety and concern over pain, it can be said to exert an important effect on pain experience even if it does not significantly affect pain as a sensation.[11, 12, 28, 72, 76, 77, 86, 132]

Experimental Studies of "Hypnotic Analgesia"

Dynes[56] monitored heart rate, respiratory rate, and change in skin resistance in response to pinch and pinprick in 7 "trained somnambules" under control conditions and after suggestions of analgesia were given under trance. The noxious stimulation produced an average increase in respiratory rate of 3 cycles per minute under the control condition and of 1 cycle per minute under the trance condition. Heart rate showed a mean increase of 2½ beats per minute under the control condition and failed to show an increase under the trance condition. All subjects showed galvanic skin responses (GSR) of the same order of magnitude under the control and trance conditions. This study is open to at least one major criticism: The stimuli were always administered first under the control condition and then under trance. As Shor[132] has pointed out, since physiological reactions to painful stimulation generally show a habituation or adaptation effect,[48, 64, 74, 131] tending to decrease during a second and subsequent stimulations, the possibility was not excluded that a similar reduction in heart rate and respiratory rate might have been observed during the second stimulation if the subjects had not been placed in "hypnotic trance" and had not been given suggestions of analgesia. This experiment and the other experiments reviewed below were concerned primarily with the effects on pain experience of explicit suggestions of analgesia given under "hypnotic trance." A number of studies[54, 130, 132, 134, 145] also assessed the effects of "hypnotic trance per se" on physiological responses to noxious stimuli; these studies were unable to demonstrate differences in autonomic reactivity to painful stimulation under a waking condition and a trance condition which did not include explicit suggestions of analgesia.

Sears[130] employed facial flinch, respiratory depth, respiratory variability, pulse amplitude, pulse variability, and GSR as indices of pain. Seven carefully selected "deep-trance" subjects participated. The pain stimulus consisted of a sharp steel point pressed against the calf of the leg for 1 sec. with a pressure of 20 oz. without breaking the skin. This stimulus was first applied in a waking control series to determine which of the physiological variables were reliable indices of pain. In a subsequent hypnosis series the subjects were placed in deep trance, suggestions of anesthesia were given for the left leg, the right leg was employed as a control, and the stimulus was applied alternately to the two legs. In a third series of experiments (voluntary inhibition), the subjects were instructed to try to inhibit reactions to the painful stimulus.

Sears presented the following findings with respect to the critical hypnosis series: When the painful stimulus was applied to the "anesthetic" leg, the hypnotized subjects showed significantly less facial flinch, respiratory depth, respiratory variability, pulse variability, and GSR than when the stimulus was applied to the control leg. The amplitude of the pulse did not differ significantly when the "anesthetic" and control limbs were stimulated.

Sears' findings have been generally in terpreted as a convincing demonstration of the effect of hypnotic analgesia on physiological reactions to painful stimuli.[70, 75, 84, 142] However, Shor[132] has recently reanalyzed Sears' data and found that some of the computations were incorrect. Shor's analysis shows that respiratory depth, pulse variability, and pulse amplitude were not significantly different when the stimulus was applied under trance to the "anesthetic" and control limbs. Facial flinch, respiratory variability, and GSR differed significantly under the "anesthetic" and control conditions. A further problem arose when the probabilities for the waking control series were recomputed: Shor found that in this series respiratory variability was not significantly different before and after painful stimulation and was thus of questionable adequacy as a criterion of physiological response to painful stimulation. In brief, Shor's careful reanalysis of Sears' original data indicates that only 3 measures instead of 6 as reported originally were significantly affected by hypnotically suggested analgesia. However, of these three measures, one (respiratory variability) was of questionable adequacy under the conditions of the experiment as an index of response to painful stimulation and another (facial flinch) is not a physiological variable and is amenable to voluntary control. Sears' major finding, then, was that the GSR to painful stimulation was reduced by 22% under hypnotically suggested analgesia. This mean reduction in GSR was due to 4 of the 7 subjects; the other 3 subjects showed a GSR of the same order of magnitude when the stimulus was applied to the "anesthetic" and control limbs.

As mentioned above, Sears performed an additional series of experiments in which the same subjects were instructed to try to inhibit all responses to the painful stimulus. In this voluntary inhibition series significant physiological reactions were found and the subjects showed facial flinch. Sears interpreted these findings as indicating that "Voluntary inhibition of reaction to pain does not present a picture even remotely resembling the reaction under true hypnotic anesthesia." However, the subjects' failure to inhibit flinching renders this conclusion questionable. In pilot studies Sears had found that the flinch response to the stimulus could be inhibited "by most people with little difficulty." As Hull[80] has pointed out, since the flinching response is normally under voluntary control, it appears possible that the "trained" hypnotic subjects participating in the Sears experiment did not actually try to suppress reactions to pain when instructed to do so. The Sears study thus appears to be open to the same criticism that applies to other studies in "hypnosis" which employed "trained" hypnotic subjects "as their own controls," namely, when a single group of "trained" hypnotic subjects is tested under both the experimental and the control conditions, it is difficult to exclude the possibility that the subjects may purposively give an inferior performance under the control condition in order to comply with what they correctly or incorrectly surmise are the wishes or the expectations of the experimenter.[17, 133]

Doupe *et al.*[54] studied the effect of hypnotically suggested analgesia on the vasoconstriction response to painful stimulation. Eight subjects were used, but data are presented only on 5 subjects. These 5 subjects participated in 11 experiments. After the subject was deeply hypnotized, digital vasodilatation was produced by placing his legs in warm water. Suggestions were then given that one arm was insensitive and analgesic with the understanding that the alternate arm would remain normally sensitive. Pin-prick stimulation (and, at times, ice stimulation) was then applied alternately to the "anesthetic" and normal limbs. From 6 to 40 stimulations were applied to each limb in each experiment. Eight of the 11 experiments failed to show a significant difference between the "anesthetic" and normal limbs in vasoconstriction response to the noxious stimuli. In the remaining 3 experiments, stimulation of the "anesthetic" limb produced less vasoconstriction than stimulation of the control limb, the reductions ranging from 36 to 40%. Doupe *et al.* also recorded respiration and pulse in these experiments but did not present the data obtained on these measures. They state only that "No significant changes in pulse rate were recorded" and "A slight alteration in respiratory rhythm was caused by stimuli applied to either [the "anesthetic" or normal] side, but this tended to be greater when the normal side was stimulated."

Brown and Vogel[38] compared physiological responses to noxious stimulation under hypnotically suggested analgesia, waking-imagined analgesia, local analgesia produced by Novocain,* and general anesthesia produced by nitrous oxide. Three pain stimuli were used (lancet, weighted thumbtack, and water at 49° C.); three physiological measures were monitored (GSR, pulse, and blood pressure); and 3 carefully selected "deep-trance" subjects participated. The authors presented the results in the form of raw data without statistical analysis. From these data they deduced the following general conclusions: (1) Waking-imagined analgesia may be as effective as suggestions of analgesia given under trance in reducing physiological responses to noxious stimulation. (2) Nitrous oxide anesthesia is totally dissimilar to hypnotically suggested analgesia; nitrous oxide anesthesia but not hypnotically suggested analgesia abolishes physiological reactions to noxious stimulation. It is difficult to determine from the raw data presented in the report if these general conclusions are justified. However, a careful analysis of Brown and Vogel's data has recently been performed by Shor,[132] who reports the following: (1) Physiological responses to the noxious stimuli did not differ significantly under hypnotically

*Winthrop Laboratories, New York, N. Y.

suggested, waking-imagined, and Novocain analgesia. Given the small number of subjects and the variability of the data, it was not possible for statistically significant effects to emerge. (2) With respect to the conclusion that nitrous oxide anesthesia is totally dissimilar to hypnotic analgesia, it appears that this is valid for the galvanic skin response, but it is not clear if it also applies to the pulse and blood-pressure responses. The GSR to noxious stimulation dropped out under nitrous oxide but not under hypnotic analgesia. (3) With respect to the conclusion that waking-imagined analgesia may be as effective as hypnotically suggested analgesia in attenuating physiological reactions to noxious stimulation, it appears that what is being said is that since neither waking-imagined nor hypnotically suggested analgesia had any measurable effect, they both by default had about equal effectiveness.

In addition to performing reanalyses of the data of previous experiments, Shor[132] also carried out an experimental study of his own. The experimental group consisted of 8 "somnambulistic" subjects; the control group consisted of 8 subjects who had demonstrated in a series of preliminary sessions that they were not susceptible to hypnosis. Prior to the experiment proper all subjects chose a level of electric shock which they found painful but which they were willing to tolerate with equanimity for an extended series of experiments. Each subject was then presented with his chosen level of electric shock under 5 experimental conditions while skin resistance, respiration, and heart rate were recorded continuously on a polygraph. The experimental conditions (counterbalanced to control for order effects) were as follows: (1) wake control (the effect of the wake state alone); (2) hypnotic control (the effect of hypnosis alone); (3) wake inhibition (voluntary suppression of reactions to pain in the waking state); (4) hypnotic inhibition (voluntary suppression of reactions to pain in the hypnotic state); and (5) hypnotically suggested analgesia. The experimental group ("somnambulistic" hypnotic subjects) was hypnotized under Conditions 2, 4, and 5; the controls (subjects insusceptible to hypnosis) were instructed to pretend as if they were hypnotized under these three experimental conditions. Shor presented the following findings: (1) The experimental group did not show significantly different physiological responses to the noxious stimuli under any of the five experimental conditions. (2) The control group also failed to show significant differences in physiological responses under any of the experimental conditions. (3) There appeared to be a trend (not significant) for over-all reactivity to be less under the waking inhibition condition. Shor concluded that his data offered no support to the hypothesis that hypnotically suggested analgesia has special effects on physiological responses to painful stimuli which are beyond the bounds of waking volitional control.

Since skin-resistance change (GSR) is easily monitored and is markedly responsive to painful and to anxiety-arousing stimulation, it has been employed in an extensive series of studies concerned with the effects of "hypnotic analgesia." Five early studies which used the GSR as the sole criterion of physiological response to pain reported contradictory findings. Peiper,[120] working with 4 subjects, Prideaux,[123] with 4 subjects, and Levine,[98] with 1 subject, reported that noxious stimulation applied to a skin area for which analgesia had been suggested under trance produced a normal GSR. Georgi,[65] working with 3 subjects, and Moravcsik,[111] with 1 subject, reported that hypnotic suggestions of analgesia reduced the GSR to painful stimuli. In these early studies the experimental procedures are not presented in detail, and the data are not analyzed statistically. Two recent studies, summarized below, were carried out more rigorously; here again, contradictory results were obtained.

West *et al.*[145] monitored skin resistance in an extensive series of repeated sessions with 7 subjects. (A total of 45 experimental sessions was held, an average of more than 6 sessions per subject.) Each experimental session included a waking control condition followed by a hypnosis condition. Under the control condition each subject received a series of painful stimuli of increasing intensity produced by radiant heat applied to a forelimb; following these control trials, hypnosis was induced, suggestions were given that the limb was anesthetic, and the painful stimuli were again presented in the same order. The mean GSR to the painful stimuli was significantly reduced under the hypnotic-analgesia condition for all subjects, the reductions ranging from 26 to 67%. West *et al.* note that the GSR was at times reduced, even when ". . . there was no alteration in pain perception, according to subjective reports," and during the control periods a stimulus evoking reports of relatively severe pain at times failed to produce a GSR. The findings thus appear to be consistent with earlier reports[64] that the galvanic skin response to noxious stimulation may be more closely related to the "threat-content" or "anxiety" aroused by a noxious stimulus rather than to "pain perception per se." This study appears to be open to one major criticism: The control trials always preceded the hypnotic trials with 3 subjects and were only "occasionally" reversed with the other 4 subjects; since the GSR to noxious stimulation tends to decrease over a series of trials,[48, 64, 74, 131] the effects of hypnotically suggested analgesia may have been confounded with possible adaptation effects.[132] In a recent experiment which controlled adaptation effects, Sutcliffe[134] presented contradictory findings.

Sutcliffe recorded the galvanic-skin response to noxious stimulation under normal waking conditions, hypnotically-suggested analgesia, and waking stimulation of analgesia. Adaptation of the GSR to the noxious stimuli was controlled by employing different subjects under each experimental condition. In pre-experimental sessions, 24 subjects were given a series of electric shocks, and a level of shock was established which invariably produced pain. The subjects were then randomly assigned to three experimental groups with 8 subjects (4 "somnambulists" and 4 "nonsomnambulists") in each group. Group 1 received 4 electric shocks at intervals of 1 min. under normal waking conditions. Group 2 received the 4 electric shocks after suggestions of analgesia were given under trance. Group 3 received the 4 shocks after receiving instructions under waking conditions to act as if the shocks were nonpainful. The GSR to the shocks was the same under the waking control condition, the hypnotic analgesia condition, and the waking acting condition. The "somnambulists" did not differ from the "nonsomnambulists" under any of the experimental conditions. Sutcliffe's study also included three additional experimental conditions designed to determine if hypnotically-hallucinated shock or waking-acting as if receiving electric shock produce a GSR similar to that found when electric shock is actually received. Hypnotically-hallucinated shock did not produce a GSR comparable to that found during actual shock; waking-acting as if receiving a shock produced a GSR of the same order of magnitude as actual shock.

Most of the experiments described above were limited in scope as follows: (1) The pain-producing stimuli—pinprick, electric shock, or radiant heat applied to a limb for no more than 3 sec.—were of brief or momentary duration. (2) Pain reactivity under hypnotically suggested analgesia was compared with reactivity under an *uninstructed* waking condition. Although four of these experiments[54, 56, 130, 145] found that suggestions of analgesia given under "trance" were more effective than no-instructions in reducing physiological responses to noxious stimulation of short duration, this does not demonstrate that "hypnotic trance" was a necessary condition in producing this effect. As Brown and Vogel[38] hypothesized, it may be possible to produce a similar reduction in physiological reactivity to painful stimuli by instructing a control group to try to imagine that a noxious stimulus is nonpainful. Barber and Hahn[24] recently presented a carefully conducted experiment designed to test this hypothesis. The Barber and Hahn experiment was performed as follows.

Prior to the experiment proper, a standardized suggestibility scale was administered under nonhypnotic conditions to 192 female students. The 48 most "suggestible" subjects (ranking in the upper quartile with respect to scores on the suggestibility scale) were selected to participate in the critical experiment. These selected subjects, who were homogeneous with respect to sex, age, social background, and level of pre-existing suggestibility, were allocated at random to one of four experimental conditions (hypnotically suggested analgesia, uninstructed condition, control condition, and waking-imagined analgesia) with 12 subjects to each condition. Subjects assigned to the hypnosis condition were given a standardized 20-min. trance induction procedure followed by a series of tests to assess suggestibility. All subjects in this group appeared to enter trance (i.e., appeared drowsy and showed psychomotor retardation and lack of spontaneity and initiative) and responded positively to the test suggestions. Suggestions were then given for a period of 1 min. to induce anesthesia of the left hand; following these suggestions the hypnotized subject immersed the "anesthetic" hand in water near the freezing point (2°C.) for 3 min. (Previous investigators[29, 59, 90, 149] had reported that "aching pain" is elicited in normal subjects by water near the freezing point within 10–60 sec.; if the stimulus is not removed, pain continues for 2–4 min. before adaptation sets in; and the intensity of the pain experienced is closely related to increments on such physiological variables as heart rate, systolic and diastolic pressure, and respiratory variability.) Subjects assigned to the uninstructed, control, and waking-imagination conditions were not hypnotized. Under the uninstructed and control conditions the subjects were simply asked to immerse the left hand in water: the uninstructed group immersed the hand in water near the freezing point (2°C.) for 3 min., and the control group immersed the hand in water at room temperature for the same period of time. Subjects allocated to the waking-imagined analgesia condition were instructed and motivated for a 1-min. period to imagine a pleasant situation when the noxious stimulus (water at 2°C.) was applied ("... When your hand is in the water, try to imagine that it is a very hot day, that the water feels pleasantly cool, and that your hand is relaxed and comfortable...").

Soon after stimulation all subjects completed a questionnaire designed to assess subjective experiences. This questionnaire yielded the following findings: (1) the hypnosis and waking-imagination groups did not differ in subjective reports, stating that, on the average, the stimulus was experienced as uncomfortable but not painful. (2) The hypnosis and waking-imagination groups differed significantly from the uninstructed group, which rated the stimulus as painful, and from the control group, which rated the stimulus as not uncomfortable. Physiological variables (heart rate, skin resistance, forehead-muscle tension, and respiration) monitored prior to and during stimulation were analyzed in terms of Lacey's autonomic lability scores[91] to control for differences in base (prestimulus) levels of physiological functioning. This analysis showed the following: (1) The hypnosis and waking-imagination groups did not differ on any physiological response to the noxious stimulus. (2) As compared to the uninstructed condition, both hypnotically suggested analgesia and waking-imagined analgesia were effective in reducing muscle tension and respiratory irregularities during the noxious stimulation. (3) Under hypnotically suggested analgesia and waking-imagined analgesia, muscle tension but not respiratory irregularity was reduced to the low level found under the control condition. (4) Heart rate and skin-resistance level during the period of noxious stimulation did not differ under the hypnotic analgesia, waking-imagined analgesia, and uninstructed condition; under these conditions subjects showed significantly faster heart rate and signfiicantly lower skin resistance than under the control condition. In brief, the Barber and Hahn experiment found that hypnotically suggested analgesia is effective in attenuating pain experience as indicated by subjective reports and by reduction in forehead muscle tension and respiratory irregularities; although pain experience is reduced under hypnotic analgesia, it is not abolished; and the experience of pain appears to be as effectively mitigated by waking-imagined analgesia as by hypnotically suggested analgesia.

The findings reviewed above appear to indicate that hypnotically suggested analgesia at times has some effect on physiological reactions to noxious stimuli, but this effect is by no means as drastic as is implied in previous reviews.[70, 75, 84, 142] Brown and Vogel,[38] Sutcliffe,[134] and Shor[132] failed to reject the null hypothesis of no difference in autonomic responses to painful stimuli under hypnotically suggested analgesia and a waking control condition. Doupe *et al.*[54] found that hypnotically suggested analgesia reduced the vasoconstriction response to pin prick in three experiments but failed to do so in eight experiments. Sears[130] observed a 22% reduction in mean GSR to noxious stimuli under hypnotic analgesia. In Dynes'[56] experiment, hypnotically suggested analgesia reduced the expected increase in heart rate, and in respiratory rate, by 2½ beats per minute, and by 2 cycles per minute, respectively. West *et al.*[145] observed a 26–67% reduction in galvanic-skin response to painful heat under hypnotically suggested analgesia. However, in the Dynes experiment and in the West *et al.* experiment, the hypnotic trials almost always followed the control trials, and it appears possible that some of the observed reduction in autonomic reactivity associated with hypnotic analgesia was produced by adaptation to the stimuli. Further, in the experiments presented by Dynes,[56] Sears,[130] Doupe *et al.*,[54] and West *et al.*,[145] in which physiological reactivity was reduced under hypnotic analgesia, the comparison was made with an *uninstructed* waking condition. Barber and Hahn[24] also found that as compared to an uninstructed condition, hypnotically suggested analgesia reduced some physiological responses to noxious stimulation (muscle tension and irregularities in respiration); however, these investigators also found that instructions given under waking conditions to imagine a pleasant situation when noxious stimulation was applied were as effective as suggestions of analgesia given under hypnotic trance in producing these effects.

A substantial number of "hypnotic-analgesic" subjects participating in the above experiments manifested gross physiological responses to relatively "mild" pain stimuli such as pin prick. This raises a crucial question: Does the "hypnotic analgesic" subject undergoing surgery show autonomic responses indicative of anxiety or pain? In searching the literature, no studies were found which presented data on a series of physiological variables recorded continuously during surgery performed under "trance." A small number of surgical studies were found which presented a few discontinuous pulse or blood-pressure measurements; these studies are reviewed below.

Surgery Under "Hypnotically Suggested Analgesia"

Discussions concerned with the effectiveness of "hypnotically suggested analgesia" in surgery[2, 39, 75, 124, 136] generally follow an outline as follows: It is first stated that the effectiveness of "hypnosis" is beyond dispute since Esdaile performed amputations and many other major operations "painlessly" under "mesmeric trance" in India during the years 1845 to 1851; the authors then present a few subsequent cases of surgery preformed under "trance" and then conclude that "hypnotic analgesia" produces a drastic reduction in pain experience. The argument to support this contention almost always relies heavily on Esdaile's series.

Although Esdaile's cases are generally referred to as "painless" surgery performed under "trance," a close look at Esdaile's original report[61] suggests that his operations may not have been free of anxiety and pain. Esdaile did not claim that all or even a majority of his patients remained quiet during surgery.[103] Some patients showed "disturbed trances" and others awakened from "trance": "She moved and moaned" (p. 200); "He moved, as in an uneasy dream" (p. 204); "About the middle of the operation he gave a cry" (p. 222); "He awoke, and cried out before the operation was finished" (p. 232); "The man moved, and cried out, before I had finished. . . . on being questioned he said that he had felt no pain" (pp. 145-146). Esdaile claimed that many of his operations were successful even though pain may have been experienced because the patients forgot the pain: ". . . the trance is sometimes completely broken by the knife, but it can occasionally be reproduced by continuing the process, and then the sleeper remembers nothing; he has only been disturbed by a night-mare, of which on waking he retains no recollection" (pp.145-146).

In 1846, the governor of Bengal appointed a committee consisting of the inspector-general of hospitals, three physicians, and three judges to investigate Esdaile's claims.[35] Esdaile removed scrotal tumors from 6 carefully selected patients who had been placed in "mesmeric trance" by "passes" made over the body over a period of about 6–8 hr. (Three additional patients who were to undergo surgery before the committee were dismissed when it was found that they could not be mesmerized after repeated attempts extending up to 11 days.) The committee reported that during surgery 3 of the 6 patients showed "convulsive movements of the upper limbs, writhing of the body, distortions of the features, giving the face a hideous expression of suppressed agony; the respiration became heaving, with deep sighs." The other 3 patients did not show gross signs of pain; however, 2 of these 3 showed marked elevations in pulse rate during the surgery on the order of 40 beats per minute.

In brief, it appears that some of Esdaile's surgical cases awakened from "trance" and suffered and some remained in "trance" but showed either "a hideous expression of suppressed agony" or marked tachycardia. However, a certain number of Esdaile's surgical patients did *not* show overt signs of pain and stated on awakening that they had *not* suffered. Although this is indeed remarkable, caution should be exercised in generalizing from these cases. In the first place, the proportion of Esdaile's patients that fell into this category cannot be determined from the data presented in his report. Secondly, if facilities had been available for recording blood pressure, pulse, skin resistance, and other autonomic variables continuously, it appears possible that these patients may also have shown physiological reactions indicative of anxiety and pain. Thirdly, it cannot be assumed that these patients would have moaned or cried during the surgery if they had *not* been in "mesmeric trance;" although many of Esdaile's nonmesmerized surgical patients cried and struggled, his report suggests that a few of his surgical patients who could *not* be placed in "mesmeric trance" did *not* show gross signs of pain (pp. 214-215).

Following Esdaile's report, scattered cases have been published of surgery performed under "hypnosis."[36, 47, 49, 50] Typically these reports state that an operation was performed under hypnotically suggested analgesia, e.g., dental extraction, avulsion of fingernail, incision of infected digit,[126] removal of cervicouterine tumors,[125] and the ". . . cooperation of the patient was perfect, the operation was painless and there was no post-operative pain"[126] or the patient ". . . woke up without pain or any physiological disturbance."[125] The procedures employed and the patient's overt behavior and subjective reports are not presented in detail, and physiological measures monitored during the surgery are not reported.

The few reports that present some physiological data suggest the possibility that the "hypnotic-analgesic" surgical patient may experience some degree of anxiety and pain. Finer and Nylen[62] presented a successful case of excisions and skin grafts performed upon a severely burned patient under "hypnoanesthesia;" although the patient did not show overt motoric signs of pain, blood pressure and pulse showed significant elevations. Kroger and Kroger and DeLee[87, 88] employed "hypnotic analgesia" in the removal of breast tumor, in subtotal thyroidectomy, in excision biopsy for breast tumor, and in cesarean section and hysterectomy; no physiological data are presented with the exception of the cesarean section and hysterectomy; in this case, Kroger and DeLee[88] write that during the surgery the blood pressure varied from 125/85 to 80/60 and pulse varied from 76 to 100. Taugher[135] has presented 3 cases of surgery (tonsillectomy, curettage, and cesarean section) performed under "trance." Although the patients did not complain of pain, blood pressure and pulse showed marked variability; in the cesarean section, for instance, blood pressure varied from 140/90 to 80/20, and pulse rate varied from 86 to 120.

Mason[106] has presented a case of mastoplasy performed under hypnotically suggested analgesia. With the exception of sodium amytal, administered the night before surgery, no other medications were given. During the operation the entranced patient did not show noticeable signs of pain; on awakening, she appeared to be amnesic for any pain that may have been experienced. Mason writes that at some point during the operation—the precise time is unspecified—the patient's "...pulse rate stabilized at 96 and respiratory rate at 24 per minute," with the implication that these measures may have been unstable prior to this period.

In other recent surgical cases the effects of hypnotically suggested analgesia were confounded with the effects of sedative and analgesic drugs. Marmer,[105] for instance, employed hypnosis in an extensive series of surgical cases (bunionectomy, laminectomy, thyroidectomy, hemorrhoidectomy), but substantial quantities of analgesic agents (nitrous oxide, meperidine, caudal block with lidocaine) were always used, no control cases are reported, and it is difficult to separate the effects of "hypnosis" from the effects of the drugs. Tinterow[137] has presented 7 cases of hypnotic surgery (cesarean section, bilateral vein ligation, vaginal hysterectomy, debridements and skin grafts, hemorroidectomy, appendectomy, and open-heart surgery); in most of these cases, secobarbital, atropine sulfate, chlorpromazine, and promethazine were administered singly or in combination. Similarly, Owen-Flood[114] presented a case of appendectomy performed under "hypnoanesthesia" in which the effects of "hypnosis" were confounded with the effects of a regular dose of scopolamine and one-half the routine dose of morphine.

In other surgical cases, hypnotic suggestions of analgesia were sufficient to produce a satisfactory reaction pattern during part of the operation, but chemical agents were required before surgery was completed. Anderson[2] reports that an entranced subject showed little if any overt signs of pain at the commencement of an abdominal exploration; however, before the operation was completed, the patient ". . . practically broke his hypnotic trance," and thiopental was administered. Butler[39] presented similar findings concerning an abdominal exploration: As the fascia was being incised, the hypnotized patient showed signs of pain and was given cyclopropane.

The above data suggest the possibility that surgery performed under "hypnotic trance" may not be as painless and as free from anxiety as has at times been supposed. Although highly selected subjects were used in all of these studies, some subjects showed physiological reactions which appear to be indicative of anxiety and pain, others "broke the trance," and others required the assistance of chemical agents. These findings appear not to contradict Bernheim's[30] contention that "hypnotism only rarely succeeded as an anesthetic, that absolute insensibility is the exception among hypnotizable subjects, and that the hypnotizing itself generally fails in persons disturbed by the expectations of an operation." The findings also do not contradict Moll's[110] contention that "a complete analgesia is extremely rare in hypnosis, although authors, copying from one another, assert that it is common."

Additional studies are needed to delineate more precisely the effects of hypnotically suggested analgesia on surgical pain. Such studies should meet the following minimum requirements: (1) A series of physiological variables—blood pressure, pulse, skin resistance, respiration—should be recorded simultaneously and continuously during surgery performed on two groups of subjects, one group undergoing the surgery under hypnotically suggested analgesia and the other under chemical anesthesia. (2) The two groups should be matched as closely as possible with respect to such background variables as age, sex, and social class and with respect to type of surgery. (3) The data should be analyzed by appropriate statistical techniques[55, 91] to take into account differences in physiological base levels under hypnotic analgesia and chemical analgesia. The findings reviewed above suggest that if these minimal requirements are included in surgical studies, it will be found that "hypnotic analgesic" subjects show significantly greater physiological reactions indicative of anxiety and pain than anesthetized subjects.

Reduction in Anodyne Requirements as an Index of Pain Relief

Some patients in labor, some postoperative patients, and some terminal cancer patients who are given suggestions of pain relief under "hypnotic trance" state to the hypnotist that their pain has been reduced or abolished. Since the statements of the hypnotic subject, as given to the hypnotist, do not always correspond to the true state of affairs,[9, 15, 17] a number of investigators have focused on a reduction in the hypnotic patient's need for anodynes as a somewhat more objective and somewhat more reliable index of pain relief.

August[5] compared drug requirements during labor of 850 trained hypnotic patients who had chosen "hypnosis" as the preferred form of anesthesia and 150 control patients who had refused hypnosis.

The control group received an average of 53.7 mg. of meperidine (Demerol*) and 22.7 mg. of barbiturates (Seconal† or Nembutal‡); the hypnotic group received, on the average, 30.3 mg. of meperidine and 2.2 mg. of barbiturate. Abramson and Heron[1] compared narcotics requirements during labor of 100 hypnotic patients and of 88 controls picked at random from the hospital files. The hypnotic group had participated on the average in 4 prelabor hypnotic training sessions, each session requiring a period of 30 min.; the controls had been delivered previously by other obstetricians and had not received antenatal training. The control group on the average received 123.6 mg. of meperidine; the hypnotic group received an average of 103.5 mg. of meperidine, a reduction of 16%

The studies of August[5] and Abramson and Heron[1] are open to a number of criticisms: (1) The hypnotic group consisted of volunteers who may have represented a selected group of patients who were likely to be more cooperative during labor. (2) The obstetricians gave more time and attention to the hypnotic patients than to the control patients. (3) The hypnotic group was apparently given medication only on demand, while the control group received medicaments more or less routinely. Perchard[121] has carried out a large-scale study which attempted to control some of these variables. A total of 3083 primigravidas were observed, of whom 1703 did not volunteer for antenatal classes. The other 1380 primiparas, who volunteered for classes, were assigned to three experimental treatments as follows. Group 1 (268 patients) received three instructional talks concerning parturition plus a visit to the labor wards. Group 2 (126 patients) was given the three instructional talks, plus a visit to the labor wards, plus three physical relaxation classes conducted by a physiotherapist. Group 3 (986 patients) received the three instructional talks, the visit to the labor wards, plus three training sessions in hypnosis. In the hypnotic training sessions this group was given practice in entering trance; practice in responding to suggestions of anesthesia; suggestions that labor would be painless; and suggestions that amnesia would follow the labor. (Fifty-six per cent, 26%, and 18% of the subjects in Group 3 were rated as "good," "moderately good," and "poor" hypnotic subjects, respectively.) There were no significant differences among the four groups (nonvolunteers, Group 1, Group 2, and Group 3) in: duration of labor; calmness, relaxation, and cooperation during labor; number of patients judged to have had severe pain; incidence of amnesia for labor; and proportion of patients eager to have more children. There was a small difference in the amount of sedation requested during labor: 40% of the hypnosis group and 32, 34, and 35% of Groups 1, 2, and nonvolunteers, respectively, requested less than 100 mg. of meperidine. (The 40% figure for the hypnotic group was increased to 44% in the subgroup rated as "good" hypnotic subjects.) Perchard concluded that "It would appear that no detectable benefits were derived from the simple relaxation exercises and that not more than a limited subjective benefit with slightly reduced need for sedation resulted from the hypnosis."

Papermaster *et al.*,[115] Bonilla *et al.*,[33] and Laux[95] assessed the effects of hypnotically suggested pain relief on narcotics requirements in postoperative cases. Papermaster *et al.*[115] worked with 33 unselected patients undergoing major abdominal surgery. An attempt was made to hypnotize each patient 3 times, twice prior to and once after surgery; during the hypnosis sessions it was suggested that the area of incision would produce no postoperative discomfort. A matched con-

*Breon Laboratories, New York, N. Y.
†Eli Lilly and Company, Indianapolis, Ind.
‡Abbott Laboratories, North Chicago, Ill.

trol group, consisting of 33 patients undergoing similar surgery but not receiving hypnosis training, was selected from the hospital files. The hypnotic group requested and received an average of 4.21 doses of meperidine (50 mg. per dose) postoperatively as compared to 7.57 doses for the control group, a reduction in narcotics requirements of 45 per cent. The authors do not present data for the individual subjects, stating only that the range in doses of meperidine received varied from 0 to 44 and from 0 to 29 in the control and hypnosis groups, respectively.

Bonilla *et al.*[33] worked with 10 male patients undergoing uncomplicated arthrotomy of the knee. Each patient participated in from 1 to 4 30-min. hypnotic sessions prior to surgery and received suggestions that he would experience no postoperative discomfort; in some instances, hypnotic sessions were also conducted in the postoperative period. This group was compared on postoperative narcotics requirements with 40 preceding male patients undergoing uncomplicated arthrotomy for similar knee afflictions. The control group received an average of 360 mg. of meperidine postoperatively as compared to 275 mg. for the hypnosis group, a reduction of 24%.

It appears that in the Papermaster *et al.*[115] and Bonilla *et al.*[33] studies the hypnotic group received medicaments only on demand while the control group was given medication routinely. Laux[95] presented an experimental study which controlled this factor. Forty veterans undergoing urological surgery were assigned either to an experimental hypnosis group (20 subjects) or to a nontreated control group (20 subjects). The two groups were matched with respect to type of surgery, age, sex, and socioeconomic status. The experimental subjects received suggestions intended to relieve postoperative pain in 3 presurgery and 1 postsurgery hypnosis sessions. Criteria for postoperative pain relief included: (1) number of requests for anodynes; (2) amount of drugs given; and (3) the charge nurse's evaluation of the amount of pain suffered. The assessment period extended over 5 days. During the first postoperative day the number of requests for anodynes by the hypnosis group was 34 per cent less than for the control group. There were no significant differences between the two groups on any of the criteria during the remaining 4 days of the assessment period.

Butler,[39] Cangello,[41] and Perese[122] assessed the effect of suggestions given under "trance" on pain associated with terminal cancer. Butler found that after a series of intensive trance sessions with 12 selected hypnotizable cancer patients, 1 patient showed a 50% reduction in narcotics requirements for a few days and another showed a 100% reduction for 3 weeks. (Of the remaining 10 patients, 8 manifested subjective relief of pain during and, at times, for a brief period following the trance sessions.) Cangello[41] reported that after a series of intensive hypnotic sessions 18 of 31 selected cancer patients manifested from 25 to 100% reduction in narcotics for a period extending from 2 days to 12 weeks. Perese[122] reported that "hypnosis" was "useful" in relieving pain in 2 of 16 cancer patients and that with another 4 patients it diminished narcotic requirements "slightly." In these studies the physicians worked intensively with their hypnotic patients, and a control group receiving a similar amount of attention was not used for comparison. It is thus difficult to determine to what extent the reported pain relief was due to the support the patients received from the physician and to what extent it was due to other factors subsumed under the term "hypnosis." This factor—the support and attention received by the patient from the physician—will be discussed again below.

In summary, the studies reviewed

above appear to indicate that hypnotically suggested pain relief produces some degree of reduction in anxiety and pain in some patients undergoing surgery or parturition and in some patients suffering from postoperative pain or cancer pain. However, these studies also suggest that although pain experience is at times ameliorated, it is only in very rare cases abolished. A more precise statement of the effects of hypnotically suggested analgesia in surgery, in labor, and in chronic pain appears to be that when given suggestions of pain relief under "hypnotic trance," some patients are able to endure whatever degree of pain is present, are not overly anxious, and do not seem to suffer to the degree expected when anxiety is present.

The Effects of "Hypnotic Suggestions" on "Functional" or "Conditioned" Pain

Although it appears that hypnotic suggestions rarely if ever abolish pain experience in conditions in which noxious stimulation is continually present—e.g., in surgery, in chronic pain—this does not exclude the possibility that hypnotic suggestions may at times eliminate some types of pain, specifically, those types of pain which appear to be produced by a "conditioning or learning process." Dorcus and Kirkner[52] have presented experimental findings which support this contention. These investigators worked with two groups of selected patients: a group of 5 males suffering from pain associated with spinal-cord injuries and a group of 5 females suffering from chronic dysmenorrhea. (No pathology could be found in the latter group that could account for the chronic painful menstrual condition.) Each of the spinal-cord cases participated in approximately 16 hypnotic sessions; the dysmennorheics participated in from 1 to 5 hypnotic sessions. The method of treatment included: induction of hypnotic trance; suggestions of anesthesia to needle pricks and burns; suggestions to induce "hallucinatory pain;" suggestions to remove the "hallucinatory pain;" and posthypnotic suggestions that whenever pain arose in the waking state, it would disappear immediately. The spinal-cord cases showed a reduction in requests for anodynes and reported less pain, but none were free from pain. The dysmenorrheics, on the other hand, ". . . were relatively free from pain upon discontinuance of therapy and have remained relatively free from pain for at least two years." The authors presented the following interpretation of these findings:

> We believe that dysmenorrhea is a conditioned process brought about in the following manner. Pain above threshold levels has been present at some time during menstruation. When the experience has once occurred, such changes as extra-cellular edema, basal temperature change, muscle tonicity, vascular changes, and breast change which were originally associated with the painful experience reinstate the pain even in the absence of the organic factors that originally brought it about. . . . In the dysmenorrheic, when we break the chain of expectancy and tension, we break down the conditioned process, whereas in the spinal nerve injury cases we are not destroying a conditioned process, but suppressing the primary pain-arousing mechanism. This is held in abeyance only insofar as the factors that tend to focus the individual's attention on the pain is concerned and in that respect the pain may appear abated. It does not remain inhibited because the source is continually present.

Dorcus and Kirkner's findings with respect to dysmenorrhea may be relevant to other "functional" painful conditions such as certain types of headaches or backaches. There is evidence to indicate that some headaches are associated with "emotional tension, anxiety, and conflict" and with prolonged contraction of the muscles of the head and neck, and that alleviation of the "conflicts and anxieties" and/or relief of the muscle hyperfunction at times relieves the headache.[151] There is also evidence to indicate that some backaches are associated with sustained contraction in the muscles of the back; the sustained skeletal-muscle hyperfunction is one component of a more generalized pattern of response to "anxiety, hostility, and conflict;" and the backache may be ameliorated by relieving either the "anxiety" or the muscular contractions.[78] The findings presented by Dorcus and Kirkner,[52] Wolff,[151] and Holmes and Wolff[78] suggest the hypothesis that some types of headaches and backaches can be effectively relieved by suggestions (given with or without "hypnotic trance") intended to eliminate the tension–anxiety–conflict pattern and the sustained muscle contractions in the neck or back. Experiments are needed to test this hypothesis.

Significant Variables in "Hypnotic Analgesia"

The general conclusion indicated by this review is that some degree of reduction in pain experience can at times be produced by suggestions given under "hypnosis." The question may now be raised: Which of the many variables subsumed under the concept of "hypnosis" are effective and which are irrelevant to producing this effect? To answer this question, it is necessary first to specify the referents of the term "hypnosis."

Although formal definitions of "hypnosis" and "hypnotized" differ widely, in practice the terms are used more or less interchangeably and appear to derive meaning from a consensual frame of reference; that is, when it is stated that subjects were "hypnotized" or "placed in hypnosis," it is implied that: (a) one of various types of procedures that have been historically categorized as "trance inductions" was administered and (b) the subjects manifested a number of characteristics which by consensus are presumed to signify the presence of "the hypnotic trance." These two interrelated referents of the term "hypnosis" can be further specified as follows.

1. Investigators agree that a wide variety of procedures can be classified as "trance inductions." At the present time such "induction procedures" generally include verbal suggestions of relaxation, drowsiness, and sleep, and often also include some type of "physical stimulation" such as the sound of a metronome or eye fixation on a "hypnodisk." However, other types of "induction procedures," comprehensively described by Pattie[119] and by Weitzenhoffer,[143] have been used in the past and are at times used now, including hyperventilation, compression of the carotid sinus, stimulation of "hypnogenic zones," and use of "passes" or "hand gestures." Although the administration of one of these "induction procedures" appears to be necessary to induce an inexperienced subject to enter "the hypnotic trance," a consensus exists that after a subject has had experience with or "training" in "hypnosis," he may be induced to enter "trance" by a drastically abbreviated "induction procedure" consisting of a prearranged signal or cue word.

2. Numerous attempts have been made to find physiological indices of "the state of trance" which is said to be produced when the "induction procedure" is "successful." These attempts have failed to yield an acceptable criterion[15] and the presence of "the trance state" is inferred from the subject's observable characteristics and behaviors. These "trance characteristics" according to Erickson *et al.*[60] include a loss in mobility, tonicity throughout the body, rigid facial expression, and literalness in response. Other investigators list similar indices. Pattie[119] refers to ". . . passivity, a disinclination to talk . . . a great degree of literal-mindedness, and a lack of spontaneity and initiative." Weitzenhoffer[143] notes that "There seems to be some agreement that hyp-

notized individuals, even when behaving in a most natural manner, still show a constriction of awareness, a characteristic literal-mindedness, some psychomotor retardation, and possibly a degree of automatism." Gill and Brenman[66] similarly write that entranced subjects who have been instructed to behave as if they are not hypnotized show ". . . momentary lapses into somewhat stiff or frozen postural attitudes . . . an impression of a slight slowing down of the pace of bodily movement . . . [and] a fleeting glazing of the eyes, the 'unseeing look' normally found in reverie or in a 'brown study.' "

It has often been assumed that "hypnotic trance," as inferred from the characteristics and behaviors described above, is crucial to producing "pain relief" by suggestions. A series of recent investigations, summarized below, suggest that this assumption is open to question.

"The Hypnotic Trance" as a Factor in "Hypnotic Analgesia"

The presence of "hypnotic trance" is not sufficient to produce "analgesia" by suggestions. Esdaile[61] presented cases of patients manifesting many if not all of the characteristics of "deep trance" who "shrunk on the first incision" and showed normal responses to painful stimulation. Winkelstein and Levinson,[148] Anderson,[2] Butler,[39] Liébault[100] and others also found that some "deeply entranced" patients did not respond positively to suggestions intended to produce pain relief. The crucial question, however, is not, Is "hypnotic trance" sufficient to produce "analgesia" by suggestions?, but, Is "hypnotic trance" a necessary or an extraneous factor in producing this effect? Contrary to what the early literature on "hypnosis" might lead one to expect, recent studies indicate that subjects who are in "a very light trance" and subjects who are not "in trance" are often as responsive and at time more responsive to suggestions of pain relief than "deep-trance" subjects.

Barber and Hahn[24] found that waking control subjects instructed to imagine a pleasant situation during painful stimulation showed as much reduction in pain experience, as indicated by subjective reports and by reduction in muscle tension and respiratory irregularities, as "entranced" subjects given suggestions of anesthesia. Von Dedenroth[140] has presented a series of cases in which patients who manifested the characteristics of "deep trance" did not respond to suggestions of pain relief, and patients who appeared at best to be "in a light hypnoidal state" and patients who insisted that they were not hypnotized at all, showed dramatic relief of stubborn headache or underwent dentistry without analgesics or anesthetics even though these agents had been demanded consistently for prior dental work. Von Dedenroth interpreted his data as indicating that "each instance of hypnotherapy is dependent upon the patient's inner responsiveness and the character and nature of his motivation rather than upon trance level or depth." Lea *et al.*[96] arrived at a similar conclusion in an investigation concerned with the effects of "hypnosis" on chronic pain: "We assumed that our success would depend upon the depth of hypnosis, but, to our surprise, we found that this was not necessarily the case. As a matter of fact, two of our best patients obtained only light to medium trances, and significant responses were noted in even the very lightest hypnoidal states." Along similar lines, Cangello[41] found in a study of the effects of "hypnosis" on pain associated with cancer that "an individual who entered a deep trance might be unable to obtain relief of pain while another who was at best in a hypnoidal or light state experienced complete pain relief." Laux[95] presented comparable results in an experimental investigation on postoperative pain: "Some of those who appeared to be the most deeply hypnotized had marked pain, and some who showed little response to the hypnosis had little pain and attributed their comfort to the effects of hypnosis."

Comparable findings have been presented in a series of recent studies employing "hypnosis" in obstetrics. Michael[107] found that some patients who at best attained only "a very light hypnotic trance" underwent labor without medications and without manifesting overt signs of pain while others who attained " a deep trance" experienced severe pain and required standard doses of narcotics. Winkelstein[147] observed that "Some women, hypnotized only to the lightest degree managed their delivery successfully, while others, deep in the somnambulistic state were unable to cope with the discomfort of labor." Similarly, Mody[109] noted no relationship in his sample of 20 selected patients between "the depth of hypnosis" and the degree of pain experienced during parturition.

The data cited above suggest that "the hypnotic trance state" may not be a critical factor in producing "pain relief" by suggestions. The data reviewed below suggest that the critical factors in so-called "hypnotic analgesia" may include: (a) suggestions of pain relief; which are (b) given in a close interpersonal setting.

The Interpersonal Relationship

Butler[39] attempted to relieve pain associated with carcinoma in 12 selected patients who were able to attain "a medium or deep trance." Each patient received suggestions of pain relief in a series of trance sessions held daily and at times 2–4 times per day. Ten of the 12 patients stated that their pain was reduced during and, at times, for a brief period following the hypnotic sessions; however, when "hypnosis" and the relationship between patient and physician were terminated, the patients showed a return of the original pain syndrome. The significant finding in these cases was that when "hypnosis" was discontinued, but the physician continued to give the same amount of personal attention to the patient, the patient continued to show pain relief. Marmer[104] has also pointed to the attention and support given to the patient as a significant variable, writing that "The realization that the anesthesiologist is willing to invest time, effort, warmth and understanding in an attempt of hypnosis will give most patients added security and trust in the physician and will result in decreased tension and anxiety." Lea *et al.*[96] reported similar observations in a study on chronic pain: "At times it was hard to decide whether benefit was actually being derived from hypnosis itself or such extraneous factors as the secondary gain a patient would derive from an unusual amount of personal attention from the hypnotherapist."

Recent reports concerned with the effects of "hypnosis" on the pain of parturition also emphasize the significance of interpersonal factors. In a study with 200 obstetrical patients, Winkelstein[147] found that to produce some measure of pain relief by suggestions, it was necessary for the physician to devote a great amount of time and attention to each patient. This investigator de-emphasized the importance of "the trance state" in producing pain relief by suggestions, pointing to the following variables as crucial: (1) the suggestions themselves; (2) the mental attitude of the patient toward pregnancy and delivery; (3) the will to succeed; (4) the confidence of the patient in the procedure as well as in the obstetrician; and (5) the patient–obstetrician rapport. Chlifer[45] had similarly observed that the effectiveness of suggestions of pain relief in labor is not correlated with "the depth of trance;" pain may be ameliorated by suggestions given to nontrance subjects; and "the success of verbally induced analgesia is closely related to the personality of the subject and the relationship established between the doctor and the parturient woman." After wide experience in the use of "hypnosis" for relief of labor pain, Kroger and Freed[89] proffered the hypothesis that if a close relationship exists between patient and obstetrican, about 10–15% of nonmedicated patients will be free of discomfort during labor even though the hypnotic trance state is not induced.

The above studies suggest that the critical factors in so-called "hypnotic analgesia" may include "suggestions of pain relief" given in a close interpersonal setting. The interpersonal variable has been emphasized above; the "suggestions of pain relief" require further comment.

"Suggestions of Pain Relief" as a Critical Factor in "Hypnotic Analgesia"

The effects of "suggestions of pain relief per se" have at times been confounded with the effects of "hypnotic trance." In a number of studies[1, 5, 33, 115] the experimental group was placed "in hypnotic trance" and then given suggestions to relieve pain; the control group was not placed "in trance" and was not given pain-relieving suggestions. These studies failed to exclude the possibility that the effective factor in ameliorating pain in the experimental group was not "the hypnotic trance" but "the suggestions of pain relief per se;" if the control group had been given suggestions of pain relief without trance, it might also have shown a reduction in pain experience. Supporting evidence for this supposition is found in the Barber and Hahn[24] experiment in which a nontrance control group given instructions or "suggestions" intended to ameliorate pain showed a similar reduction in pain experience as entranced subjects given suggestions of anesthesia.

Sampimon and Woodruff[126] have presented data indicating that direct suggestions given without "hypnotic trance" are at times sufficient to alleviate pain. In 1945 these investigators were working under primitive conditions in a prisoner of war hospital near Singapore. Anesthetic agents were not available, and "hypnosis" was employed for surgery. Two patients could not be "hypnotized;" since the surgical procedures (incision for exploration of abscess cavity and extraction of incisor) had to be performed without drugs, Sampimon and Woodruff proceeded to operate after giving "the mere suggestion of anesthesia." To their surprise they found that both patients were able to undergo the normally painful procedures without complaints and without noticeable signs of pain. These investigators write that "As a result of these cases two other patients were anesthetized by suggestions only, without any attempt to induce true hypnosis, and both had teeth removed painlessly." Other workers[14, 20, 22, 23, 67, 85, 150] have presented comparable findings with respect to the effectiveness of "direct suggestions" given without the induction of "hypnotic trance."

Similar findings have been presented in studies concerned with the effects of placebos. Hardy *et al.*[74] found that 2 subjects given an inactive drug with the suggestion that it was a strong analgesic showed elevations in pain threshold over 90% above the control levels; blisters were produced in these subjects without reports of pain. Beecher,[26] Dodson and Bennett,[51] and others[79, 83, 116] have presented evidence indicating that about one-third of postoperative patients receive "satisfactory relief" of pain when inert agents are administered as pain-relieving drugs. Laszlo and Spencer[94] found in a study with 300 cancer patients that "over 50 per cent of patients who had received analgesics for long periods of time could be adequately controlled by placebo medication." Although few if any studies on the "placebo effect" report detailed data concerning the relationship between patient and physician and the suggestions given to the patient, it appears likely that in some if not many of these studies the patients were given suggestions of pain relief in a close interpersonal setting.

Indications for Further Research

To determine the significance of "hypnotic trance" as a factor in relieving pain by suggestions, additional experiments are needed which control three critical variables noted above: (1) the selection of subjects; (2) the interpersonal relationship between subject and experimenter; and (3) the suggestions of pain relief per se. These experiments should be conducted as follows:

1. The effects of "hypnotic trance" should not be confounded with differences between subjects. In a number of studies cited above,[1, 5, 33, 115, 132] subjects meeting criteria of "hypnotizability" were assigned to the "trance" treatment, and unselected subjects or nonhypnotizable subjects were assigned to the control treatment. The criterion used for selecting the experimental group, that the subjects were "hypnotizable," is difficult if not impossible to differentiate from an interrelated implicit criterion, namely, that the subjects were highly responsive to suggestions with or without "hypnotic trance." If suggestible subjects are allocated to the "trance" treatment and less suggestible subjects to the control treatment, it is impossible to determine if greater response to suggestions of pain relief in "entranced" subjects, as compared to control subjects, is due to their being in "trance" or to their being more suggestible to begin with. To control this factor, it is necessary that subjects be randomly assigned to the "trance" and nontrance treatments from an original group of subjects who show a similar level of suggestibility.[17, 133]

2. Subjects allocated to the "trance" and nontrance treatments should be given comparable time and attention by the experimenter and should have a comparable opportunity to form a close relationship with him.

3. Both groups should be given similar suggestions of pain relief, one group to be given the suggestions under "trance" and the other under nontrance conditions.

The data presented in this review suggest that if these critical variables are controlled, it will be difficult to reject the null hypothesis of no difference in response to suggestions of analgesia in nonentranced and "deeply entranced" subjects.

Variables Intervening Between Suggestions of Pain Relief and Reduction in Pain Response

A number of investigators have postulated that suggestions of analgesia are effective in diminishing subjective and physiological responses to pain if and when they lead the subject to stop thinking about or to stop attending to the pain. Liébault[100] hypothesized that the process of suggested analgesia can be described simply as the focusing of attention on ideas other than those concerning pain. Young[152] presented a similar hypothesis: Pain relief produced by "hypnosis" or by suggestive procedures is due to a "taking of an attitude and consequently *refusing* to feel the pain or even to take cognizance of it." August[5] postulated that "Hypnoanesthesia results from directing attention away from pain response towards pleasant ideas." These hypotheses receive some support from a recent experimental study[24] which found that the subjective and physiological responses to painful stimuli which characterize "hypnotic-analgesic" subjects can be elicited from control subjects by instructions to think about and to imagine a pleasant situation when noxious stimulation is applied.

The intervening variables in so-called "hypnotic analgesia" may be similar to those which presumably operate in the placebo situation and in other nontrance situations in which pain experience is abated without medications. These intervening variables have been summarized succinctly by Cattell:[43]

The intensity of the sensation produced by a painful stimulus is determined to a large extent by circumstances which determine the attitude towards its cause. If there is no worry or other distressing implications regarding its source, pain is comparatively well tolerated, and during important occasions injuries ordinarily painful may escape notice. On the other hand, in the absence of distraction, particularly if there is anxiety, the patient becomes preoccupied with his condition, and pain is badly tolerated.

It appears unnecessary to hypothesize additional intervening variables in so-called "hypnotic analgesia." In any situation (hypnotic or nonhypnotic) in which anticipation or fear of pain is dispelled, and "anxiety" is reduced, and the subject does not "attend to" or "think about" the painful stimulus, noxious stimulation is apparently experienced as less painful and less distressing than in situations in which "anxiety" and "concern over pain" are present.[11, 28, 71–73, 76, 77, 86, 132]

Summary

1. In some instances, suggestions of pain relief given under "hypnotic trance" appear to produce some degree of diminution in pain experience as indicated by reduction in physiological responses to noxious stimuli and by reduction in requests for pain-relieving drugs. In other instances, however, "hypnotically suggested analgesia" produces, not a reduction in pain experience, but an unwillingness to state directly to the hypnotist that pain was experienced and/or an apparent "amnesia" for the pain that was experienced.

2. The motivation for denial of pain is present in the hypnotic situation. The physician has invested time and energy hypnotizing the patient and suggesting that pain will be relieved; expects and desires that his efforts will be successful; and communicates his desires to the patient. The patient in turn has often formed a close relationship with the physician-hypnotist and does not want to disappoint him. The situation is such that even though the patient may have suffered, it is at times difficult or disturbing for him to state directly to the physician that pain was experienced and it is less anxiety provoking to state that he did not suffer.

3. A series of experiments that monitored heart rate, skin resistance, respiration, blood pressure, and other physiological responses which are normally associated with painful stimulation found that in some instances "hypnotically suggested analgesia" reduced some physiological responses to noxious stimuli and in other instances physiological responses were not affected. However, experiments which found reduced autonomic responses to noxious stimuli under "hypnotic analgesia" compared reactivity under the hypnotic condition with reactivity under an *uninstructed* waking condition. In a recent carefully controlled experiment in which physiological reactions to painful stimulation were compared under (a) "hypnotically suggested analgesia" and (b) a waking condition in which subjects were instructed to imagine a pleasant situation when noxious stimulation was applied, it was found that both conditions were equally effective in reducing subjective and physiological responses to painful stimulation.

4. Studies concerned with surgery performed under "hypnoanesthesia alone" rarely present any physiological data; the small number of studies that presented a few pulse or blood pressure measurements suggest the possibility that "hypnotic-analgesic" subjects undergoing surgery may show autonomic responses indicative of anxiety and pain. In other studies concerned with surgery performed under "hypnosis" the effect of "hypnotically suggested analgesia" was

confounded with the effects of sedative and analgesic drugs.

5. The data appear to indicate that in surgery, in chronic pain, and in other conditions in which noxious stimulation is continually present, pain experience is at times reduced but is rarely if ever abolished by "hypnotically suggested analgesia." However, the data also indicate that suggestions given under "hypnotic trance" (and possibly without "hypnotic trance") may at times drastically reduce or eliminate some painful conditions, such as dysmenorrhea and certain types of headaches and backaches, which appear to be produced by a "conditioning or learning process."

6. This review suggests that the critical variables in so-called "hypnotic analgesia" include: (a) suggestions of pain relief, which are (b) given in a close interpersonal setting. Additional research is needed to determine if "the hypnotic trance state" is also a relevant variable. Further experiments should control: (a) the preexisting level of suggestibility among subjects assigned to the "trance" and control treatments; (b) the interpersonal relationship between subject and experimenter; and (c) the suggestions of pain relief per se. The data reviewed suggest that if these variables are controlled, it will be found that suggestions of pain relief given either to waking control subjects or to "deep-trance" subjects produce a comparable reduction in pain experience.

Medfield Foundation
Medfield, Mass.

References

1. ABRAMSON, M., and HERON, W. T. An objective evaluation of hypnosis in obstetrics. *Amer. J. Obstet. Gynec. 59:*1069, 1950.
2. ANDERSON, M. N. Hypnosis in anesthesia. *J. Med. Ass. Alabama 27:*121, 1957.
3. ANDREWS, H. L. Skin resistance change and measurements of pain threshold. *J. Clin. Invest. 22:*517, 1943.
4. ASIN, J. The utilization of hypnosis in obstetrics. *J. Amer. Soc. Psychosom. Dent. Med. 8:*63, 1961.
5. AUGUST, R. V. *Hypnosis in Obstetrics.* McGraw-Hill, New York, 1961.
6. BANISTER, H., and ZANGWILL, O. L. Experimentally induced visual paramnesias. *Brit. J. Psychol. 32:*30, 1941.
7. BARBER, T. X. Hypnosis as perceptual-cognitive restructuring: I. Analysis of concepts. *J. Clin. Exp. Hypnosis 5:*147, 1957.
8. BARBER, T. X. Hypnosis as perceptual-cognitive restructuring: II. "Post"-hypnotic behavior. *J. Clin. Exp. Hypnosis 6:*10, 1958.
9. BARBER, T. X. Hypnosis as perceptual-cognitive restructuring: IV. "Negative hallucinations." *J. Psychol. 46:*187, 1958.
10. BARBER, T. X. The concept of "hypnosis." *J. Psychol. 45:*115, 1958.
11. BARBER, T. X. Toward a theory of pain: Relief of chronic pain by prefrontal leucotomy, opiates, placebos, and hypnosis. *Psychol. Bull. 56:*430, 1959.
12. BARBER, T. X. "Hypnosis," analgesia, and the placebo effect. *J.A.M.A. 172:*680, 1960.
13. BARBER, T. X. Antisocial and criminal acts induced by "hypnosis": A review of experimental and clinical findings. *A.M.A. Arch. Gen. Psychiat. 5:*301, 1961.
14. BARBER, T. X. Experimental evidence for a theory of hypnotic behavior: II. Experimental controls in hypnotic age-regression. *Int. J. Clin. Exp. Hypnosis 9:*181, 1961.
15. BARBER, T. X. Physiological effects of "hypnosis." *Psychol. Bull. 58:*390, 1961.
16. BARBER, T. X. Hypnotic age regression: A critical review. *Psychosom. Med. 24:*286, 1962.
17. BARBER, T. X. Experimental controls and the phenomena of "hypnosis": A critique of hypnotic research methodology. *J. Nerv. Ment. Dis. 134:*493, 1962.
18. BARBER, T. X. Toward a theory of "hypnotic" behavior: The "hypnotically-

induced dream." *J. Nerv. Ment. Dis. 135:*206, 1962.

19. BARBER, T. X. Toward a theory of hypnosis: Posthypnotic behavior. *A.M.A. Arch. Gen. Psychiat.* 7:321, 1962.
20. BARBER, T. X., and CALVERLEY, D. S. "Hypnotic behavior" as a function of task motivation. *J. Psychol. 54:*363, 1962.
21. BARBER, T. X., and COULES, J. Electrical skin conductance and galvanic skin response during "hypnosis." *Int. J. Clin. Exp. Hypnosis* 7:79, 1959.
22. BARBER, T. X., and DEELEY, D. C. Experimental evidence for a theory of hypnotic behavior: I. "Hypnotic color-blindness" without "hypnosis." *Int. J. Clin. Exp. Hypnosis* 9:79, 1961.
23. BARBER, T. X., and GLASS, L. B. Significant factors in hypnotic behavior. *J. Abnorm. Soc. Psychol. 64:*222, 1962.
24. BARBER, T. X., and HAHN, K. W., JR. Physiological and subjective responses to pain-producing stimulation under hypnotically-suggested and waking-imagined analgesia. *J. Abnorm. Soc. Psychol.*, in press.
25. BEECHER, H. K. Pain in men wounded in battle. *Ann. Surg. 123:*96, 1946.
26. BEECHER, H. K. The powerful placebo. *J.A.M.A. 159:*1602, 1955.
27. BEECHER, H. K. Relationship of significance of wound to pain experienced. *J.A.M.A. 161:*1609, 1956.
28. BEECHER, H. K. *Measurement of Subjective Responses.* Oxford Univ. Press, New York, 1959.
29. BENJAMIN, F. B. Effect of aspirin on suprathreshold pain in man. *Science 128:*303, 1958.
30. BERNHEIM, H. *Suggestive Therapeutics.* Associated Booksellers, Westport, Conn., 1957, p. 116. (Original date of publication: 1887).
31. BITTERMAN, M. E., and MARCUSE, F. L. Autonomic responses in posthypnotic amnesia. *J. Exp. Psychol. 35:*248, 1945.
32. BLUM, G. S. *A Model of the Mind.* Wiley, New York, 1961, p. 162.
33. BONILLA, K. B., QUIGLEY, W. F., and BOWERS, W. F. Experience with hypnosis on a surgical service. *Milit. Med. 126:*364, 1961.
34. BRAID, J. *Neurypnology.* Churchill, London, 1843.
35. BRAID, J. Facts and observations as to the relative value of mesmeric and hypnotic coma, and ethereal narcotism, for the mitigation or entire prevention of pain during surgical operations. *Med. Times 15:*381, *16:*10, 1847.
36. BRAMWELL, J. M. *Hypotism: Its History, Practice, and Theory.* Julian Press, New York, 1956, p. 106. (Original date of publication: 1903).
37. BROCA, P. Note sur une nouvelle méthode anesthésique. *C. R. Acad. Sci. (Par.) 49:*902, 1859.
38. BROWN, R. R., and VOGEL, V. H. Psychophysiological reactions following painful stimuli under hypnotic analgesia, contrasted with gas anesthesia and Novocain block. *J. Appl. Psychol. 22:*408, 1938.
39. BUTLER, B. The use of hypnosis in the care of the cancer patient. *Cancer 7:*1, 1954.
40. CALLAN, T. D. Can hypnosis be used routinely in obstetrics? *Rocky Mountain Med. J. 58:*28, 1961.
41. CANGELLO, V. W. Hypnosis for the patient with cancer. *Amer. J. Clin. Hypnosis 4:*215, 1962.
42. CANNON, W. B. *Bodily Changes in Pain, Hunger, Fear, and Rage.* Appleton, New York, 1915.
43. CATTELL, M. The action and use of analgesics. *Res. Publ. Ass. Nerv. Ment. Dis. 23:*365, 1943.
44. CHERTOK, L. *Psychosomatic Methods in Painless Childbirth.* Pergamon, New York, 1959, pp. 3-4.
45. CHLIFER, R. I. Verbal analgesia in childbirth. *Psychotherapia (Kharkov)* 307, 1930 (Cited by Chertok[44]).
46. CLARK, R. N. Training method for childbirth utilizing hypnosis. *Amer. J. Obstet. Gynec. 72:*1302, 1956.
47. COCHRAN, J. L. The adaptability of psychosomatic anesthesia for the performance of intermediate surgery on certain types of patients. *Brit. J. Med. Hypnotism 7:*26, 1955.
48. COHEN, L. H., and PATTERSON, M. Effects of pain on heart rate of normal and schizophrenic individuals. *J. Gen. Psychol. 16:*273, 1937.

49. Cooper, S. R., and Powles, W. E. The psychosomatic approach in practice. *McGill Med. J. 14:*415, 1945.
50. Crasilneck, H. B., McCranie, E. J., and Jenkins, M. T. Special indications for hypnosis as a method of anesthesia. *J.A.M.A. 162:*1606, 1956.
51. Dodson, H. C., Jr., and Bennett, H. A. Relief of postoperative pain. *Amer. Surg. 20:*405, 1954.
52. Dorcus, R. M., and Kirkner, F. J. The use of hypnosis in the suppression of intractable pain. *J. Abnorm. Soc. Psychol. 43:*237, 1948.
53. Dorcus, R. M., and Shaffer, G. W. *Textbook of Abnormal Psychology,* ed. 3. Williams & Wilkins, Baltimore, 1945.
54. Doupe, J., Miller, W. R., and Keller, W. K. Vasomotor reactions in the hypnotic state. *J. Neurol. Psychiat. 2:* 97, 1939.
55. Dykman, R. A., Reese, W. G., Galbrecht, C. R., and Thomasson, P. J. Psychophysiological reactions to novel stimuli: Measurement, adaptation, and relationship of psychological and physiological variables in the normal human. *Ann. N.Y. Acad. Sci. 79:*43, 1959.
56. Dynes, J. B. An experimental study of hypnotic anesthesia. *J. Abnorm. Soc. Psychol. 27:*79, 1932.
57. Dynes, J. B., and Poppen, J. L. Lobotomy for intractable pain. *J.A.M.A. 140:* 15, 1949.
58. Elliotson, J. *Numerous Cases of Surgical Operations Without Pain in The Mesmeric State; With Remarks Upon the Opposition of Many Members of the Royal Medical and Chirurgical Society and Others to the Reception of the Inestimable Blessings of Mesmerism.* H. Baillière, London, 1843, pp. 15-17.
59. Engel, B. T. Physiological correlates of pain and hunger. Doctoral dissertation, Univ. of California at Los Angeles, 1956.
60. Erickson, M. H., Hershman, S., and Secter, I. I. *The Practical Application of Medical and Dental Hypnosis.* Julian Press, New York, 1961, pp. 55-58.
61. Esdaile, J. *Hypnosis in Medicine and Surgery.* Julian Press, New York, 1957. (Originally entitled *Mesmerism in India* and published in 1850.)
62. Finer, B. L., and Nylen, B. O. Cardiac arrest in the treatment of burns, and report on hypnosis as a substitute for anesthesia. *Plast. Reconstr. Surg. 27:*49, 1961.
63. Franke, U. Amnesie und Anaesthesie bei der Hypnosegeburt. *Dtsch. Med. Wschr.* 874, 1924. (Cited by Chertok[44])
64. Furer, M., and Hardy, J. D. The reaction to pain as determined by the galvanic skin response. *Res. Publ. Ass. Nerv. Ment. Dis. 29:*72, 1950.
65. Georgi, F. Beiträge zur Kenntnis des psycho-galvanischen Phänomens. *Arch. Psychiat. 62:*571, 1921.
66. Gill, M. M., and Brenman, M. *Hypnosis and Related States.* Internat. Univ. Press, New York, 1959, pp. 38-39.
67. Glass, L. B., and Barber, T. X. A note on hypnotic behavior, the definition of the situation and the placebo effect. *J. Nerv. Ment. Dis. 132:*539, 1961.
68. Goetzl, F. R., Bien, C. W., and Lu, G. Changes in blood pressure in response to presumably painful stimuli. *J. Appl. Physiol. 4:*161, 1951.
69. Gold, H. The effect of extracardiac pain on the heart. *Res. Publ. Ass. Nerv. Ment. Dis. 23:*345, 1943.
70. Gorton, B. E. The physiology of hypnosis. *Psychiat. Quart. 23:*317, 457, 1949.
71. Hall, K. R. L. Studies of cutaneous pain: A survey of research since 1940. *Brit. J. Psychol. 44:*279, 1953.
72. Hall, K. R. L., and Stride, E. The varying response to pain in psychiatric disorders: A study in abnormal psychology. *Brit. J. Med. Psychol. 27:*48, 1954.
73. Hall, K. R. L. Pain and suffering. *South African Med. J. 31:*1227, 1957.
74. Hardy, J. D., Wolff, H. G., and Goodell, H. *Pain Sensations and Reactions.* Williams & Wilkins, Baltimore, 1952.
75. Heron, W. T. Hypnosis as an anesthetic. *Brit. J. Med. Hypnotism 6:*20, 1954-1955.
76. Hill, H. E., Kornetsky, C. H., Flanary, H. G., and Wikler, A. Effects of anxiety and morphine on discrimination of intensities of painful stimuli. *J. Clin. Invest. 31:*473, 1952.

77. Hill, H. E., Kornetsky, C. H., Flanary, H. G., and Wikler, A. Studies on anxiety associated with anticipation of pain. I. Effects of morphine. *A.M.A. Arch. Neurol. Psychiat. 67*:612, 1952.
78. Holmes, T. H., and Wolff, H. G. Life situations, emotions and backache. *Res. Publ. Ass. Nerv. Ment. Dis. 29*: 750, 1950.
79. Houde, R. W., and Wallenstein, S. L. A method for evaluating analgesics in patients with chronic pain. *Drug. Addict. Narcot. Bull.* App. F., 660, 1953.
80. Hull, C. L. *Hypnosis and Suggestibility: An Experimental Approach.* Appleton-Century-Crofts, New York, 1933, p. 252.
81. Javert, C. T., and Hardy, J. D. Influence of analgesics on pain intensity during labor (with a note on "natural childbirth"). *Anesthesiology 12*:189, 1951.
82. Kaplan, E. A. Hypnosis and pain. *A.M.A. Arch. Gen. Psychiat. 2*:567, 1960.
83. Keats, A. S. Postoperative pain: Research and treatment. *J. Chron. Dis. 4*: 72, 1956.
84. Kirkner, F. J. "Control of sensory and perceptive functions by hypnosis," in *Hypnosis and its Therapeutic Applications,* edited by R. M. Dorcus. McGraw-Hill, New York, 1956.
85. Klopp, K. K. Production of local anesthesia using waking suggestion with the child patient. *Int. J. Clin. Exp. Hypnosis 9*:59, 1961.
86. Kornetsky, C. Effects of anxiety and morphine in the anticipation and perception of painful radiant heat stimuli. *J. Comp. Physiol. Psychol. 47*:130, 1954.
87. Kroger, W. S. "Introduction and supplemental reports," in *Hypnosis in Medicine and Surgery,* by J. Esdaile. Julian Press, New York, 1957.
88. Kroger, W. S., and DeLee, S. T. Use of hypnoanesthesia for cesarean section and hysterectomy. *J.A.M.A. 163*:442, 1957.
89. Kroger, W. S., and Freed, S. C. *Psychosomatic Gynecology.* Free Press, Glencoe, Ill., 1956, p. 130.
90. Kunkle, E. C. Phasic pains induced by cold. *J. Appl. Physiol. 1*:811, 1949.
91. Lacey, J. I. The evaluation of autonomic responses: Toward a general solution. *Ann. N.Y. Acad. Sci. 67*:123, 1956.
92. Lacey, J. I. "Psychophysiological approaches to the evaluation of psychotherapeutic process and outcome," in *Research in Psychotherapy,* edited by E. A. Rubinstein and M. B. Parloff. American Psychological Ass., Washington, D. C., 1959.
93. LaFontaine, C. *L'Art de Magnétiser ou le Magnétisme Animal,* ed. 3. Baillière, Paris, 1860.
94. Laszlo, D., and Spencer, H. Medical problems in the management of cancer. *Med. Clin. N. Amer. 37*:869, 1953.
95. Laux, R. An investigation of the analgesic effects of hypnosis on postoperative pain resulting from urological surgery. Doctoral dissertation, Univ. of Southern Calif., 1953.
96. Lea, P. A., Ware, P. D., and Monroe, R. R. The hypnotic control of intractable pain. *Amer. J. Clin. Hypnosis 3*:3, 1960.
97. Le Beau, J. Experience with topectomy for the relief of intractable pain. *J. Neurosurg. 7*:79, 1950.
98. Leriche, R. *The Surgery of Pain.* Williams & Wilkins, Baltimore, 1939, pp. 55-56.
99. Levine, M. Psychogalvanic reaction to painful stimuli in hypnotic and hysterical anesthesia. *Bull. Johns Hopkins Hosp. 46*:331, 1930.
100. Liebault, A. A. Anesthésie par suggestion. *J. Magnetisme* 64, 1885. (Cited by Chertok[44])
101. Life, C. The effects of practice in the trance upon learning in the normal waking state. Bachelor's thesis, 1929 (Cited by Hull[80]).
102. Mandy, A. J., Mandy, T. E., Farkas, R., and Scher, E. Is natural childbirth natural? *Psychosom. Med. 14*:431, 1952.
103. Marcuse, F. L. Hypnosis in dentistry. *Amer. J. Orthodont. Oral. Surg. 33*:796, 1947.
104. Marmer, M. J. Hypnoanalgesia: The use of hypnosis in conjunction with

chemical anesthesia. *Anesth. Analg. (Cleve.) 36:*27, 1957.

105. Marmer, M. J. *Hypnosis in Anesthesiology.* Thomas, Springfield, Ill., 1959.
106. Mason, A. A. Surgery under hypnosis. *Anaesthesia 10:*295, 1955.
107. Michael, A. M. Hypnosis in childbirth. *Brit. Med. J. 1:*734, 1952.
108. Mitchell, M. B. Retroactive inhibition and hypnosis. *J. Gen. Psychol. 7:*343, 1932.
109. Mody, N. V. Report on twenty cases delivered under hypnotism. *J. Obstet. Gynec. India 10:*3, 1960.
110. Moll, A. *The Study of Hypnosis.* Julian Press, New York, 1958, pp. 105, 125, 248. (Original date of publication: 1889.)
111. Moravcsik, E. E. Experimente über das psychogalvanische Reflexphänomen. *J. Psychol. Neurol. 18:*186, 1912.
112. Myers, F. W. H. *Human Personality and its Survival of Bodily Death,* vol. 1, Longmans, Green, New York, 1954, p. 181. (Original date of publication: 1903.)
113. Ostenasek, F. J. Prefrontal lobotomy for the relief of intractable pain. *Johns Hopkins Hosp. Bull. 83:*229, 1948.
114. Owen-Flood, A. "Hypnosis in anaesthesiology," in *Hypnosis in Modern Medicine,* edited by J. M. Schneck, Thomas, Springfield, Ill., 1959.
115. Papermaster, A. A., Doberneck, R. C., Bonello, F. J., Griffen, W. O., Jr., and Wangensteen, O. H. Hypnosis in surgery: II. Pain. *Amer. J. Clin. Hypnosis 2:*220, 1960.
116. Papper, E. M., Brodie, B. B., and Rovenstine, E. A. Postoperative pain; its use in comparative evaluation of analgesics. *Surgery 32:*107, 1952.
117. Patten, E. F. Does post-hypnotic amnesia apply to practice effects? *J. Gen. Psychol. 7:*196, 1932.
118. Pattie, F. A. "Theories of hypnosis," in *Hypnosis and its Therapeutic Applications,* edited by R. M. Dorcus. McGraw-Hill, New York, 1956, p. 8.
119. Pattie, F. A. "Methods of induction, susceptibility of subjects, and criteria of hypnosis," in *Hypnosis and its Therapeutic Applications*, edited by R. M. Dorcus. McGraw-Hill, New York, 1956.
120. Peiper, A. Untersuchungen über den galvanischen Hautreflex (psychogalvanischen Reflex) im Kindesalter. *Jahrb. Kinderheilkunde 107:*139, 1924.
121. Perchard, S. D. Hypnosis in obstetrics. *Proc. Royal Soc. Med. 53:*458, 1960.
122. Perese, D. M. How to manage pain in malignant disease. *J.A.M.A. 175:*75, 1961.
123. Prideaux, E. The psychogalvanic reflex. *Brain 43:*50, 1920.
124. Raginsky, B. B. The use of hypnosis in anesthesiology. *J. Pers. 1:*340, 1951.
125. Rose, A. G. The use of hypnosis as an anaesthetic, analgesic, and amnesic agent in gynaecology. *Brit. J. Med. Hypnotism 5:*17, 1953.
126. Sampimon, R. L. H., and Woodruff, M. F. A. Some observations concerning the use of hypnosis as a substitute for anesthesia. *Med. J. Australia 1:*393, 1946.
127. Sattler, D. G. Absence of local sign in visceral reactions to painful stimulation. *Res. Publ. Ass. Nerv. Ment. Dis. 23:*143, 1943.
128. Schultze-Rhonhof, F. Der hypnotische Geburtsdämmerschlaf. *Zbl. Gynäk.* 247, 1922. (Cited by Chertok[44])
129. Schwartz, A. M., Sata, W. K., and Laszlo, D. Studies on pain. *Science 111:*310, 1950.
130. Sears, R. R. Experimental study of hypnotic anesthesia. *J. Exp. Psychol. 15:*1, 1932.
131. Seward, J. P., and Seward, G. H. The effect of repetition on reaction to electric shock. *Arch. Psychol.* #168, 1934.
132. Shor, R. Explorations in hypnosis: A theoretical and experimental study. Doctoral dissertation, Brandeis Univ., 1959.
133. Sutcliffe, J. P. "Credulous" and "sceptical" views of hypnotic phenomena: A review of certain evidence and methodology. *Int. J. Clin. Exp. Hypnosis 8:*73, 1960.
134. Sutcliffe, J. P. "Credulous" and "skeptical" views of hypnotic phenomena: Experiments on esthesia, halluci-

nation, and delusion. *J. Abnorm. Soc. Psychol. 62:*189, 1961.

135. TAUGHER, V. J. Hypno-anesthesia. *Wisconsin Med. J. 57:*95, 1958.
136. TINTEROW, M. M. The use of hypno-analgesia in the relief of intractable pain. *Amer. Surg. 26:*30, 1960.
137. TINTEROW, M. M. The use of hypnotic anesthesia for major surgical procedures. *Amer. Surg. 26:*732, 1960.
138. TRENT, J. C. Surgical anesthesia, 1846-1946. *J. Hist. Med. 1:*505, 1946.
139. VELVOVSKI, I. Z., PLATONOV, K. I., PLOTITCHER, V. A., and CHOUGOM, E. A. *Psychoprophylactic.* Leningrad, Medguiz, 1954. (Cited by Chertok[44])
140. VON DEDENROTH, T. E. A. Trance depths: An independent variable in therapeutic results. *Amer. J. Clin. Hypnosis 4:*174, 1962.
141. WALLACE, G., and COPPOLINO, C. A. Hypnosis in anesthesiology. *New York J. Med. 60:*3258, 1960.
142. WEITZENHOFFER, A. M. *Hypnotism: An Objective Study in Suggestibility.* John Wiley, New York, 1953.
143. WEITZENHOFFER, A. M. *General Techniques of Hypnotism.* New York, Grune, 1957, p. 347.
144. WELLS, W. R. The extent and duration of post-hypnotic amnesia. *J. Psychol. 2:*137, 1940.
145. WEST, L. J., NIELL, K. C., and HARDY, J. D. Effects of hypnotic suggestion on pain perception and galvanic skin response. *A.M.A. Arch. Neurol. Psychiat. 68:*549, 1952.
146. WHITE, R. W. A preface to the theory of hypnotism. *J. Abnorm. Soc. Psychol. 36:*477, 1941.
147. WINKELSTEIN, L. B. Routine hypnosis for obstetrical delivery: An evaluation of hypnosuggestion in 200 consecutive cases. *Amer. J. Obstet. Gynec. 76:*152, 1958.
148. WINKELSTEIN, L. B., and LEVINSON, J. Fulminating pre-eclampsia with Cesarean section performed under hypnosis. *Amer. J. Obstet. Gynec. 78:*420, 1959.
149. WOLF, S., and HARDY, J. D. Studies on pain: Observations on pain due to local cooling and on factors involved in the "cold pressor" response. *J. Clin. Invest. 20:*521, 1941.
150. WOLFE, L. S. "Hypnosis in anesthesiology," in *Techniques of Hypnotherapy,* edited by L. M. LeCron. Julian Press, New York, 1961.
151. WOLFF, H. G. *Headache and Other Head Pain.* Oxford Univ. Press, New York, 1948.
152. YOUNG, P. C. An experimental study of mental and physical functions in the normal and hypnotic state. *Amer. J. Psychol. 37:*345, 1926.

73 EEG Alpha Rhythms and Susceptibility to Hypnosis

Perry London, Joseph T. Hart, and Morris P. Leibovitz

DESPITE a rich literature of anecdotal and clinical material on the relationship of hypnosis to physiological functions, especially to events in the central nervous system, the research findings are highly equivocal. Most relevant studies have been concerned with shifts in brain wave patterns, as measured by the electroencephalogram (EEG). The studies have attempted to identify the underlying processes which accompany the observed or reported events characteristic of passage between waking and hypnotic states of consciousness. With some exceptions, however, most studies have failed to demonstrate EEG correlations of the hypnotic state[1,2]. Similarly, most studies of physiological functioning have failed to find clear-cut changes in the autonomic nervous system resulting from "hypnosis *per se*, that is, without further verbal instructions"[3].

So far, most studies of hypnosis, including most physiological studies, have concentrated entirely on the identification of state differences, and have failed to take adequate account of subject differences which might be critically important to the experimental phenomena. The most important subject variable in this connexion is susceptibility to hypnosis. A large number of reliable instruments exist for measuring this very stable trait both in children and adults[4]. Few experimenters have, however, troubled to sort subjects systematically according to susceptibility, nor have they used both high and low-susceptible subjects in the experimental manipulations. This means, in effect, that most hypnotic experiments use only subjects who are very highly susceptible to hypnosis.

A considerable number of experiments on performance differences have shown, however, that high and low-hypnotic susceptibles often differ considerably in operant performance measures of strength, endurance, psychomotor coordination, and rote memory. These differences are sometimes further widened or obscured by the application of hypnosis or other experimental instructions to both kinds of subjects. London and McDevitt[5] found, moreover, that high and low-susceptibles differed significantly on a number of base-rate measures of autonomic nervous system functions, as well as differing under experimental training conditions. The present experiment applied the same methodology to the examination of EEG alpha rhythms. It is part of a larger study on training in the autoregulation of brain functions[6] and the behavioural and physiological correlations of imagery (unpublished observations of Leibovitz). The present report concerns only the observation of differences in the duration of operant alpha rhythm between high and low-susceptibles in two test conditions.

Volunteers for an experiment on "brain waves and hypnosis" requiring 3 h over two different sessions were recruited from college employment offices and through press releases to local newspapers. Of the 154 women—ranging in age from 16 to 61—who volunteered for the experiment, 125 completed the full test battery.

The test battery included: group and individual tests of visual and other imagery; the Harvard group scale of hypnotic susceptibility (HGSS)[7], administered by tape recording; and an individually administered EEG. The group tests were always given first, usually one week before the individual session. The basic pieces of equipment used for the EEG recording were: a Grass model 7, 4-channel polygraph, equipped with 7PI preamplifiers; a Kronhite model 330*A* bandpass filter tuned to 8–13 Hz. The filter was connected to the polygraph by a lead from *J*7 of channel 1; a Tektronix model 502*A* oscilloscope was used for display of the output signal; the amplified output of the oscilloscope was connected to an amplitude-sensitive trigger. The triggering level was established by adjusting the height of the oscilloscope signals to activate the trigger for unattenuated alpha signals, but not for the filter attenuated signals outside the alpha band. From the trigger, outputs went to a Gra-lab model 172 timer for tuning the alpha duration and to a signal marker on the polygraph which permitted the polygraph operator to visually check the accuracy of the trigger.

The EEG recording session was divided into ten trials of 2 min in which the subject was told only to relax with eyes closed, followed by a single EEG sample taken over 2 min during a visual imagery task with eyes closed. During the recording session the subject was seated in a comfortable reclining chair inside a large Faraday cage (2·44 m^3) in a quiet, darkened room.

EEGs were recorded from four sites: channel 1, left occipital to left ear; channel 2, right frontal to right ear; channel 3, left occipital to left frontal; and channel 4, left frontal to left ear. Silver flat disk electrodes by Grass were used; the occipital placement was 2 cm up and 2 across from the inion, the frontal 2 cm up and 2 across from the nasion. Only channel 1 was monitored on-line through the filter–trigger–timer to obtain alpha scores. These scores were read directly in s/sample period of 120 s; these quanta were used as the alpha indices.

For purposes of analysis, HGSS scores were grouped in four categories representing the tenth, twenty-fifth, seventy-fifth and ninetieth percentiles, respectively, of the score distribution. Scores of 0–4 represented low susceptibility, 5–7 moderately low, 8–11 moderately high, and 12, which is the scale's ceiling, represented high susceptibility. Mean EEG alpha durations among the groups

Table 1. MEAN EEG ALPHA DURATIONS (S/120S) BY HYPNOTIC SUSCEPTIBILITY SCORES

Hypnotic susceptibility		EEG alpha duration	
Score	*N*	Operant	Visual imagery
0–4	25	44·5	48·0
5–7	25	63·9	65·1
8–11	67	50·2	56·1
12	8	84·6	84·6

(Table 1) were compared by analysis of variance, which showed that the alpha rhythm durations differed significantly both under operant conditions where no special instructions were imposed ($F = 4 \cdot 89$, $P < 0 \cdot 005$, two-tailed test) and in conditions where the subjects were instructed to engage in active visual imagery throughout the period of measurement ($F = 3 \cdot 30$, $P < 0 \cdot 05$, two-tailed test). Simple effects tests, as expected from Table 1, further indicated that the most significant differences are a function of the highest and lowest susceptibility groups: the two middle groups do not differ significantly from each other.

Despite the differences in the numbers of trials and instructions, the relationship of alpha duration to scores of susceptibility to hypnosis is almost identical on both the operant and visual imagery conditions. This finding may be of independent interest in its bearing on the controversy about whether deliberate mental activity interferes with the alpha productivity of untrained subjects.

The principal finding, however, that very susceptible people have high alpha durations, gives an important clue about why previous studies have failed to find alpha differences between waking and hypnotic states. Because most experiments use highly susceptible subjects, and because the highly susceptible may routinely produce high alpha under waking conditions, then no change may be observed under hypnosis merely because a ceiling effect is inherent in the subject selection process. In any case, the finding supports the idea that hypnotic susceptibility is a stable personality trait with measurable physiological substrates. (And it is in accord with recent findings about the importance of individual differences in responsiveness to the conditioning of sleep and wakefulness[8].) Future experiments on differences in state should take these findings into account.

This work was supported by grants from the US National Institutes of Mental Health to J. T. H. and P. L., and by a US Public Health Service research scientist development award to P. L.

PERRY LONDON

University of Southern California.

JOSEPH T. HART

University of California at Irvine.

MORRIS P. LEIBOVITZ

University of Southern California.

[1] Weitzenhoffer, A. M., *Hypnotism: an Objective Study in Suggestibility* (Wiley, New York, 1953).

[2] True, R. M., and Stephenson, C. W., in *Clinical Correlations of Experimental Hypnosis* (edit. by Kline, M. V.) (Thomas, Springfield, Illinois, 1963).

[3] Edmonston, W. E., in *Handbook of Clinical and Experimental Hypnosis* (edit. by Jordon, J. E.) (Macmillan, New York, 1967).

[4] Hilgard, E. R., *Hypnotic Susceptibility* (Harcourt, New York, 1965).

[5] London, P., and McDevitt, R. A., *AMRL-TR*-67-142 (W-P AF Base, Ohio: Aerospace Medical Research Laboratories, 1967).

[6] Hart, J. T., paper presented at the seventh Annual Meeting of the Society for Psychophysiological Research, San Diego, October 20–22 (1967).

[7] Shor, R. E., and Orne, Emily C., *Harvard Group Scale of Hypnotic Susceptibility, Form A* (Consulting Psychologists Press, Palo Alto, California, 1962).

[8] Fenton, G. W., and Scotton, L., *Brit. J. Psychiat.*, **113**, 1283 (1967).

XII

VOLUNTARY CONTROL, CONSCIOUSNESS, AND PHYSIOLOGY

74 Altered States of Awareness

Ernest R. Hilgard

A man knows that he belongs to the human race because of his conscious experiences; the existentialists tell us that what he most dreads is death, which means the end to experience. He values his experiences, but he is frightened by some of them, too; in any case they provide for him the meaning of his life, his joys, his sorrows, his hopes, and his defeats. Why, then, are psychologists so timid in talking about conscious states, about awareness, when they are the initial subject matter?

The answer is, in one sense, simple: psychologists are trying to be scientists like other scientists, and they will do anything to be proper about it, even to the point of denying some of their subject matter. The early scientific psychologists, who did accept consciousness as their subject matter, so sterilized it through their formal methods of introspection that those who wanted an enriched psychology broke with that tradition. It was the behaviorist Watson who put a chapter on personality in his textbook, not the mentalist Titchener. But behaviorism proceeded then to deny consciousness and carried the revolt against sterility too far, so that new strictures arose against some of the topics of psychology. At the risk of some oversimplification, we may recognize three contemporary trends that represent efforts to be scientific about psychology and still to be open to its range of potential topics. One is a continuing battle for purism of the kind that Watson's behaviorism started. This now takes the form of an attack on intervening variables of all kinds, but particularly on what is called "reification," or making "things" out of traits or states or conscious experiences. This position we may call *contemporary behaviorism,* to note that it has changed some since Watson through the efforts of Skinner, Bandura, and others. A second tendency, at the other extreme, is to throw off preoccupation with scientific objectivity in order to celebrate human values and to revel in experience, hence to look for the outreaches of human potentiality, consciousness expansion, and rich encounters. There are many variants of this, but to use the name of a society dedicated to this standpoint, we may call this *humanistic psychology*. The third tendency is the more moderate one of throwing off the "you mustn't say" strictures of behaviorism without departing from objectivity, and without rejecting an enlightened operationism. Miller, Galanter, and Pribram (32) somewhat whimsically called such a position "subjective behaviorism," but I prefer to use an older term and call it *contemporary functionalism.* The older functionalist, typified by James Rowland Angell, was always willing to use consciousness, behavior, physiology, and mathematics, and I see no reason why the moderates of today should not recognize this ancestry. It does not mean that there have been no advances since Angell for this group, just as there have been advances since Watson for the avowed behaviorists.

THE PROBLEM OF STATES AND AWARENESS OF THEM

Skinner has been the spokesman for the empty organism, thus representing those who object to dealing with nonobservables in the form of either introspections or inferences. Some behaviorists permit verbal reports to serve as objective evidence for inner states, but Skinner is careful to have them refer to *stimuli,* not to states. For instance, in his book on verbal behavior: "Consider, for example, the complex tact, *I am hungry.* This is emitted under the control of relevant (usually private) stimuli" (44, p. 151).

Those who carry on the behaviorist tradition in the study of personality and social psychology attack *traits* and *states* as misleading ways in which to describe the continuities in personality (33), preferring always to point to specific learned behavior (2).

> From the viewpoint of social behavior theory the results viewed in this and previous chapters reflect the empirically unjustified assumptions of trait and state theory and not merely the limitations of measurements (33, p. 148).

Others carry the same general line of attack against descriptions of present awareness according to states of consciousness. In an avalanche of papers, Barber has taken a position against hypnosis as a state and prefers to write the word "hypnosis" in quotation marks as though it represented something mythical (3, 5). Sarbin, too, has taken up cudgels against named states of awareness; he finds myths not only in the concept of "hypnosis," but in "hallucination" and "schizophrenia" (38), and also "anxiety" (39).

The arguments of Skinner, Bandura, Mischel, Barber, and Sarbin deserve to be taken quite seriously, for they have all been able to go a long way toward advancing our understanding of important phenomena while embracing their position. I am prepared to recognize this, even though the general position is one that I cannot endorse.

So much for contemporary behaviorism. The other extreme, of variants on existentialism and humanistic psychology, not only is found in the popular movements of the time toward consciousness expansion, psychedelic experiences, and experimentation with drugs or Eastern religions, but is found also in the serious psychological literature. The forms are quite varied (8, 11, 30, 34, 37, 42). While there is something refreshing about this protest against "scientism," just where it will lead in producing a new paradigm for psychology is uncertain. The prior history of various returns to a "naive" or "phenomenalistic" psychology in the past does not augur well for its survival, even though it will doubtless have a leavening influence.

The middle ground—what I am calling contemporary functionalism—has had an interesting upsurge over the last 10 years. A careful historical analysis would probably find a growing distaste for stimulus-response concepts, already criticized by Dewey many years ago (13), but joined in by others from time to time (19, 32, 48). Information theory for a time appeared to provide a new model, at the time that other influences were converging to produce an emphasis upon a cognitive psychology. Mathematical models were precise but "neutral" on many of the classical issues. Computer simulation came in and made respectable new studies of thinking that appeared more promising than the earlier introspective ones. In any case, many voices began to be heard (14, 24, 31). Investigators began to turn to neglected fields of study, to attention, imagery, hallucinations, and other "subjective" phenomena which had earlier made behaviorists uncomfortable. Those who began to look into these neglected topics had themselves been brought up on behaviorism and operationism, and they did not find that they had to strain their scientific integrity or objectivity in redirecting the focus of their investigations. They merely had to have confidence that psychology's tools were now sharp enough, its position as a science firmly enough established, that they could venture into the study of "subjective" phenomena without loss of scientific status.

Thus psychologists do not march forward with a common front, but they veer toward a contemporary behaviorism, at one extreme, toward humanistic psychology at the other extreme, or to the middle ground of contemporary functionalism. I wish now, from the vantage point of the mid-position, to examine waking consciousness and its alterations in sleep and dreams and in hypnosis.

WAKING AS A COMPLEX STATE

By a normal waking state we mean one in which we can report accurately what is happening in the environment about us and can use this information to control our behavior. We do not do these things well when we are asleep, or drugged, or delirious. Under those other conditions most of us are willing to say that someone is in an altered state, even though the borderlines are hard to define.

A little reflection will show us that the waking state itself is far from simple. When we carry on a conversation, we are simultaneously listening and formulating a reply, and even while replying we are thinking of our next maneuver. Thus we are not doing just one thing at a time, and how many things we can do at once, and how well we do them, pose problems for experimental study. This is the kind of thing that sets the functionalist to work; subjective observation has set him some problems that he hopes he can solve by objective methods. His preference for operationism does not make him deny the problems. He performs experiments on listening to more than one conversation at a time, on one message to one ear and another to the other, on reciting aloud what is coming in over earphones, with informative results (10). It takes a little effort to remain alert, even when we are not very tired, and to avoid vacant staring we do a good deal of squirming about, chewing of pencils, lighting a pipe, adjusting clothing, or talking merely to eliminate silence. My point here, as against those who deny "states," is that I see no reason to deny that being awake is a state, even though it is hard to give it a fully satisfactory definition.

SLEEPING AND DREAMING AS ALTERED STATES

With the discovery of the rapid eye movement state of sleep (the REM state) and another state of sleep, defined negatively as non-rapid eye movement (the NREM state), we now understand quite well that there are at least three states where before we thought there were only two. That is, the sleep state is divided (in addition to several levels of depth) into two quite distinct states with different properties. But I should like to point out something here: it did not take electroencephalographic (EEG) studies to tell us that there were two states, such as waking and sleeping, even though it was sometimes hard to tell when a person was asleep (or merely playing possum). If he was trustworthy, he would tell you when he was simulating sleep, although sometimes, when really asleep, he might deny having been asleep. It is important to remember this when we come to the discussion of hypnosis, for the dismissal of hypnosis as a state because we don't know its physiology would be like dismissing sleep as a state *before* we knew its physiology. Lack of perfect correlation and ill defined border conditions do not prevent useful experiments from reaching satisfactory inferences through convergent operations (15).

Fortunately, the study of sleep and dreaming has advanced greatly since the discovery of REM sleep by Aserinsky and Kleitman (1), and the questions of "state" are essentially now empirical ones. Stoyva and Kamiya (47) have made a very useful contribution to the discussion of a strategy for the study of consciousness by citing in some detail the recent history of research on sleeping and dreaming. Extending the notion of convergent operations, they have shown how useful the interchange has been between the physiological indicators (chiefly, but not exclusively, EEG and REM) and dream recall. There would be no way of knowing whether the eye movements were related to dreaming except by asking the subject. At the same time there is no certainty with respect to the subject's reported dream content. Only by exercising ingenuity in relating the two kinds of data (physiological response and verbal report) can the true state of affairs be inferred. Some of the work of Kamiya (cited in 29) on the operant control of the EEG alpha rhythms is referred to as a further illustration of the usefulness of combining objective measures with verbal reports. Under the conditions in which the alpha is "on," the subject commonly reports that he is relaxed and not experiencing visual imagery: he usually finds long periods of time in continuous alpha to be quite pleasant (35). There are always some uncertainties and options in any single set of correspondences from two orders of data, but in this case there are three kinds of data—operant procedures, physiological indicators, and verbal reports—so that more complex convergences are possible, leading to greater confidence about the inferences to a "mental state." Here we have a very good illustration of contemporary functionalism in that the curiosity about the mental state is fully as great as that of any phenomenologist, yet the rules of scientific evidence are followed as faithfully as by the more classical behaviorist who does not wish to make inferences respecting states of consciousness.

Our "inventory" of states now adds up to the three-fold cycle of waking, REM, and NREM, through which everyone goes every day, and now another set (alpha and non-alpha) through which people also go, but which they can learn to control in the laboratory, or in certain kinds of exercises, as under a Zen master.

HYPNOSIS AS AN ALTERED STATE

Hypnosis is commonly considered to be a "state," perhaps resembling the state in which the sleepwalker finds himself, hence the term "somnambulist" as applied to the deeply hypnotized person. William James, in discussing the significance of suggestibility in hypnosis, endorsed the notion of a special state: "The suggestion-theory may therefore be approved as correct, provided we grant the trance-state as its prerequisite" (25, p. 601).

In a very lucid analysis of what happens during hypnotic induction and within the established hypnotic state, Kubie and Margolin (27) agreed with those who, like James, found it congenial to recognize the hypnotic state as an altered state of awareness. They pointed out that the usual hypnotic procedures (including immobilization and monotony) tend to cut off communications with the outside world, except for the voice of the hypnotist. The subject then confuses the hypnotist's words with his own thoughts. Following the steps of induction, in which the ego boundaries have been blurred, the established hypnotic state emerges, including a partial reexpansion of ego boundaries. Gill and Brenman (16), quoting this earlier analysis favorably, also distinguished between the fractionation of the ego in the induction phase and a reintegration within the established hypnotic state, although still at a partially regressed level.

As in the case of other "states" the conception of an established hypnotic state has not gone unchallenged. The two most vocal challengers are Barber and Sarbin. Barber stated, for example:

> Further research aiming to delineate the factors making for high response to suggestions should be careful not to confound the variable labeled as "trance" with such variables as instructions designed to produce positive motivation to perform well on assigned tasks and instructions or suggestions to perform the criterion behaviors. We can predict that when the varia-

bles discussed in this chapter are kept distinct and unconfounded, the variable labeled "hypnotic trance" will prove to be extraneous to producing "analgesia," "hallucinations," "amnesia," and so on (4, p. 472).

Sarbin, extending his earlier role-theoretical analysis in hypnosis in collaboration with Andersen, clearly rejected the concept of trance (or other "state" concepts):

> One feature that differentiates role-theoretical and other social psychological formulations of hypnosis from the more credulous theories is the nonemployment of the trance either as a descriptive device or as a mediational process (40, p. 342).

To get at the problem of the nature of hypnosis, we have conducted a number of experiments in our laboratory on hypnotically suggested analgesia. I wish to use them to illustrate the puzzling nature of these phenomena, and the futility of jumping to conclusions about task motivation, demand characteristics, or role enactment in accounting for the detailed events that occur.

Some of the preliminary accounts appeared in this journal (20), but a number of experiments since then have added quantitative sophistication to the suggestions there offered (17, 18).

My collaborators and I have used two methods to induce pain, and this fact in itself turns out to be important in order to avoid finality in interpreting and generalizing from experiments. The first method is that of the *cold pressor test,* in which a hand and forearm are placed in circulating ice water. When the water is actually at 0°C, an immersion period of 30 to 40 seconds is sufficient to produce intense pain (23, 49). The second method is that of *ischemic pain,* in which a tourniquet is applied above the elbow, the subject exercises for a time by squeezing a dynamometer, and then waits while the pain in the forearm rises (28, 45). The pain often does not begin right away, and there may be little or no pain for a matter of minutes, but then it begins to rise and eventually the forearm becomes excruciatingly painful, so that the tourniquet has to be removed.

Both of these pains lend themselves to lawful verbal reports. The courses of verbal reports on a scale from 0 up, with 10 the point at which they would very much wish to have the hand taken out of the water or the tourniquet removed; both fit a power function with time, in accordance with the psychophysical expectations found by Stevens (46) for so many sensory functions. Both kinds of pain can be reduced under the suggestion of hypnotic analgesia, the amount of reduction in pain correlating with the degree of susceptibility to hypnosis as measured by the standard hypnotic susceptibility scales. The correlations are of the order of $r = .50$ or .60, large enough to be adequately significant with the sizes of the samples that we use. These results suffice to indicate the reality of pain reduction under hypnosis; the correlation with measured hypnotic susceptibility indicates that the individual differences in susceptibility not only are enduring over time, but are predictive of measures not used in the original measuring scales, which had no strictly analgesia items, the most closely related being an item on anosmia to ammonia. There is nothing in what I have reported that would cause disagreement by Barber or Sarbin, except that Barber would wish to know whether induction of hypnosis was really necessary; both he and Sarbin accept the main point that there *are* persistent individual differences which are highly relevant to hypnotic-type responses.[2] Thus the question of "state" is not necessarily involved.

Accepting the logic of converging operations, we were desirous of having a meas-

[2] Barber commonly implies that the individual differences are motivational, but this is an inference on his part (4, p. 451), and others disagree (36).

ure of physiological response that correlated with the verbally reported pain. We used several measures of cardiac response and breathing, but, without going into the details, we found a repeatedly monitored measure of systolic blood pressure in the middle finger of the nonstimulated hand to be the most satisfactory.[3] For example, in experiments on cold pressor pain not involving hypnosis, but with water at four temperatures (15°C, 10°C, 5°C, and 0°C) we were able to demonstrate a correspondence between the rise in blood pressure both with time in the water and with temperature, the blood pressure curves paralleling very closely the verbally reported pain. We found the same relationship of rising blood pressure with reported pain in the ischemia experiment. So far there is a parallelism between the results with cold water and ischemia, but here the parallelism stops.

In replicated experiments in the ice water, using both waking controls and a control for hypnosis *without* analgesia as well as hypnosis *with* suggested analgesia, we found that blood pressure *rose more for the hypnotic analgesia condition* than for the waking condition, in the same subjects. Without further evidence, one might conjecture that the reports of hypnotic analgesia were not genuine, and that perhaps we had a kind of "lie detection" experiment here, with the higher blood pressure associated with the reports of "no pain" under hypnosis. However, the same effect was found for the low hypnotizables who were reporting as much pain as ever, so that this explanation is not satisfactory. Apparently there was some kind of excitement associated with the expectations engendered by the instructions, and this was communicated to those who were to become analgesic as well as to those who did not become analgesic. Also the shock of cold water may have had reflex effects uncontrolled by the hypnotic suggestions. The results show how generalizations from single experiments can be misleading, such as the interpretation that the analgesic effect owes to the relaxation and nonanxiety of the hypnotic subject (43). Our hypnotic subjects without instructions for analgesia were relaxed, too, but suffered pain, and if we use rise in blood pressure as indicating the opposite of relaxation, our most analgesic were the least relaxed and most anxious of all.

We also ran a carefully designed experiment using simulators to see the effect on their blood pressure as they deliberately attempted to "fool" the experimenter into believing that they were analgesic. This is possible in a short exposure to the ice water; a willing subject can be trained to act as though he feels no pain. The experiment (recently completed and as yet unpublished) was extremely carefully controlled to be sure that the experimenter had no knowledge as to whether or not his subject was a "true" hypnotic subject or a subject "simulating" hypnosis, but the subject, after the experiment was over, reported back to the staff member who had given the instructions for simulation in the first place. The ones posing as hypnotic subjects in no case found the pain relieved (though occasionally it was slightly reduced) while the subjects selected on the basis of prior evidence of analgesia felt no pain throughout the experiment. What about the blood pressure? There was a clear difference between the blood pressure under the analgesia conditions for the "true" and "simulating" subjects. Both increased their blood pressure over the waking condition, but the simulators increased theirs by an amount that was significantly greater than the amount by which the "trues" increased theirs. In terms of formal design,

[3] Credit for the blood pressure-measuring device, and for the ischemic pain experiments, goes primarily to John Lenox, who has kindly permitted me to refer to his work before he has had the opportunity to publish it. A number of the ideas in this paper have come from him and other collaborators in these experiments.

therefore, the experiment worked out all right, showing that "simulators" and "trues" differ, but the details remain to be explained. Let me now state some propositions regarding the relationship between blood pressure and pain and hypnosis: *1) When pain is felt in the cold pressor experiment in the normal waking state, there is a tendency for blood pressure to rise in an amount correlated with the amount of experienced pain. 2) In the cold pressor experiment combined with hypnotic instructions blood pressure may rise independent of the amount of felt pain.*

What this amounts to—and I wish to state this rather firmly—is that the cold pressor experiment is not a good place to look for physiological correlates of pain and of pain reduction under hypnosis. We have used far more subjects than earlier investigators and, I believe, more stringent controls. Our results make the findings of an experiment such as that of Barber and Hahn (7) not wrong but indecisive on the issues being studied.

Had we stopped with the cold pressor experiment, we would have made generalizations that were misleading, for the ischemic pain experiments turned out differently. In them, the highly susceptible subjects not only felt no pain under suggested hypnotic analgesia, for durations of ischemia half again as long as the time at which they were writhing in pain in the waking state, but *the blood pressure did not rise.* This is not a result of subject selection, for these subjects came from the subjects who had participated in the cold pressor experiment, where they responded as the others did. An effort to use the simulator design in this experiment failed: we were not able to coax or cajole or train simulators to deny pain long enough under the more severe stress of ischemia to serve the design of a simulation experiment. The end result is, of course, the same: showing the genuineness of the hypnotic analgesia for those subjects who can experience it. It shows also that the "demand characteristics" of the experiment, the motivation to please the experimenter, are not sufficient, *unless* the subject can really rid himself of the pain as a consequence of the hypnotic procedures. This leads to the third assertion about blood pressure and pain: *3) In the ischemic experiment, the failure of the blood pressure to rise, which is correlated with absence of a verbal report of pain, supports the reality of the hypnotic analgesia.*

These are the unexpected results that take the scientist off of the debating platform and send him back to the laboratory. Why should there be this difference between the effects of hypnosis on blood pressure in the cold pressor experiment and in the ischemia experiment? There are many variables, and these have to be explored: the time factors, the effects of temperature confused with the effects of pain in the cold pressor test, the amount of prior experience of the subjects. These are not matters for debate but for study. At the same time, even though these questions are open they do not deny the ultimate finding: highly susceptible hypnotic subjects, following hypnotic induction and with the suggestion of analgesia, felt no pain under circumstances in which they would normally feel excruciating pain, and their physiological responses also attested to the absence of pain.[4]

With these data before us we are ready to turn to the question of whether our subjects were in a "state" of hypnosis. It is very easy to lose the focus of theoretical argument. Very often the argument against "state" becomes an argument against the reality of the phenomena, and our experiments settle (unless they are subsequently challenged) the empirical reality of hypnotic pain reduction. I doubt that this is any longer at issue, because of the abun-

[4] The heart rate in ischemia also confirmed the absence of pain, although it served no better than blood pressure in the cold pressor experiment.

dant clinical evidence in addition to the laboratory evidence. But this is not the basic argument over "state." Our data would have to be supplemented by a demonstration that these same subjects, when not in a "state" of hypnosis, could not yield the phenomena. They do not yield them when they are awake; that is, these are not cases of people insensitive to pain. But would they yield them if exhorted, cajoled, their imaginations stirred up, their attention directed elsewhere? Here we are in difficult territory, because these are all methods of inducing the "state" (if such there be) in highly susceptible subjects. Actually the question is not a very decisive one, for surely people capable of producing hypnotic phenomena can produce them in various ways, through self-hypnosis, at a signal set up through posthypnotic suggestion, and so on. It would not matter very much whether, after demonstration of the ability to achieve hypnotic analgesia in the ischemic pain situation, a subject could do it without special induction procedures. I would expect him to—but he might do it by entering the state of hypnosis, and this is the nub of the question. Just as we know that he experiences pain by asking him, so we ask him about hypnosis. He might be lying in either case, but he usually tells the truth; in any case the experimenter exercises such ingenuity as he can to find out what is so. If he pushes the subject for the truth, as Bowers (9) did, the simulators will confess to falsifications, while the truly susceptible subjects will not. This led Bowers to conclude that many of the experiments designed to prove that task-motivated behavior was the equivalent of hypnotically suggested behavior prejudged the issue by accepting behavioral compliance as the criterion of similarity of response. His own experiments led him to believe that over and above this compliance hypnosis represents "an altered state within which suggestions have a peculiarly potent effect."

Careful interviews with subjects prior to and following hypnosis have begun to throw considerable light on the cognitive and affective features of the hypnotic involvement, and its relationship to experiences outside hypnosis (22). To deny the relevance of such observations because of preconceptions as to the appropriate methods of science is to throw away highly pertinent and illuminating data.

ALTERED STATES AS CATEGORIES *VS.* STATES AS CAUSES

The world is too complex and confusing to study all at once, so that scientists "stake their claims" through some sort of classification of phenomena. This persists in the names of fields of science such as physical sciences, life sciences, social sciences. Such classificatory schemes are in the interests of simplification and order and need not be divisive. Thus the separation of life sciences from physical sciences does not prevent the emergence of the fields of biochemistry or of biophysics, which resist classification in the old rubrics. But classification serves other purposes of more scientific relevance, as in distinguishing phases (ice, water, steam), with rules relating them and their transitions. The same is true in the classification of states of awareness. It is convenient to talk about waking and sleep, about sobriety and drunkenness, about reality orientation and reality distortion, without thereby disturbing objective reporting of what is known about the phenomena under study.

The main reason that these classificatory labels are objected to is that they occasionally run the danger of becoming explanations. The controversy over instincts that has gone on intermittently for a half-century illustrates this. Instinct was (and still is!) a convenient term for species-specific behavior, when used in a classificatory sense, to describe, for example, characteristic nest building of birds according to species. But if a pigeon returns home *be-*

cause of a homing instinct, we get into a little trouble, and there is some circularity, for the way we know it has a homing instinct is that it returns home. There still is no harm in pointing out that some pigeons have this instinctive tendency and others do not. The objection is not to instinct as a classificatory word but to instinct as an explanatory one.

This same argument holds for words like sleep and hypnosis. In drawing some parallels between their treatment, I am not endorsing the interpretation of hypnosis as a partial sleep, although that theory of Pavlov's continues to have supporters in many parts of the world. I am simply using sleep as a familiar illustration to justify ways of talking about hypnosis, which is less familiar.

If one wishes to argue that sleep is not a state, I have no quarrel with that, because then the argument is not over fact but over words. If sleep is not a state, then hypnosis is not a state either. But if sleep *is* a state, which I suppose most of us would accept, then maybe hypnosis is a state, too. Let us see where we are on empirical grounds.

Suppose we apply to sleep the same critical attacks that have been made on hypnosis. First, it is very difficult to define sleep; as Kleitman (26) pointed out, it is usually defined negatively as the opposite of wakefulness. It is hard to distinguish it from coma, for example, except by some sort of operational definition: you can be aroused from sleep, but not from coma. (This departs from a "state" definition and hence can be used to argue against sleep as a state.) Or suppose you build some sort of behavioral test for sleep, such as what the person tells you when you wake him up ("I must have been asleep, because I didn't hear you come in"), whether or not he reports a dream, or some more "objective" measures such as whether or not he snores or how loud a bell it takes to rouse him. These would not work out too badly, but they would be subject to the same criticisms as those leveled against hypnotic scales. For one thing, any of these measures can be "faked." A person can imitate snoring, even though he is not asleep; he can report a dream when he had none. Therefore, we do better to abandon the notion of "sleep," write the word in quotation marks, and stick to the input-output measures that we use in experiments concerned with the phenomena associated with a person who is lying in bed at night with his eyes closed.

Those who might wish to save sleep as a state, but still reject hypnosis, have another line of argument. Because of the EEG we now have a physiological definition of sleep and hence are on much more secure ground when we study, for example, the question of whether or not sleep-learning is possible. This is a good argument, but it does not completely escape the circularity that the EEG measures of depth of sleep can only be validated by behavioral measures. It turns out, in fact, that it is more difficult to be aroused from stage 1 than from "deeper" stages, if REMs are occurring in stage 1. This evidence comes not from physiology but from behavior, although, to the extent that there are lawful correlates, everything remains in order. It would be a great help to the researcher on hypnosis if there were physiological indicators, such as the EEG in sleep, to help define more precisely what condition the subject is in. The search for such indicators, as illustrated by our use of the blood pressure measures in relation to pain, is valid and should be pressed. But the question I am raising is this: would we have denied sleep as a "state" before the EEG measures were available? I think not. Perhaps hypnosis research lags behind sleep research by a decade or two. Hence I do not believe that the absence of clear physiological correlates of hypnosis is crucial in the answer to the question of whether or not it represents a state.

The temptation to use state as a cause

or an explanation is another matter. Even though snoring and dreams typically occur during sleep, it does not help very much to say that sleep causes them, for of course it is possible to sleep without snoring or dreaming. Still it does no harm to point out that their presence is *more likely* during sleep. This is an empirical question, and similar empirical questions arise in relation to hypnosis: is typical hypnotic behavior more likely when the conditions have been favorable for entering a hypothesized hypnotic state? The empirical answer is not as easy as might be supposed. In one of our investigations concerned with this problem we found that it was necessary to use careful designs to demonstrate rather small increments as a result of hypnotic induction (21). As Tart and I pointed out, the experimental design must take into account the following considerations: 1) many nonhypnotizable people are not going to change following attempted induction; 2) some highly susceptible subjects do not require induction to enter hypnosis and yield all the phenomena, and hence they do not change between waking suggestion and hypnosis; 3) the argument for change with induction therefore rests on a minority of subjects who respond only slightly to waking suggestion but are very responsive to suggestion after induction procedures. Under these circumstances, small groups of unselected subjects are quite likely to lead to no statistical difference attributable to induction, especially if extraneous demands toward increasing voluntary compliance are added in the nonhypnotic condition. While our conditions, favorable to detecting small changes, led to the conclusion that the effects of induction could be demonstrated, the matter is still subject to some empirical controversy (6). Although the question of the effects of induction continues to be of interest, it is not crucial to the question whether or not some subjects can enter a state of hypnosis. That has to be determined on more subtle grounds, using converging operations which will doubtless include subjective report, overt behavior, and physiological indicators.

Suppose the results turn out that there are phenomena associated with the hypnotic state that are not as likely when the subject is not in that state. It is possible to accept all of this without making the state of hypnosis the cause of the behavior, any more than sleep causes snoring or dreams.

Note also that the use of a classificatory rubric does not require that there be no differentiation within that rubric. Thus sleep turns out not to be one state but at least two (REM and NREM). This does not negate a differentiation between sleep and waking: both the REM subject and NREM subject are still asleep. I doubt if there will turn out to be just one state of hypnosis; subjects capable of deep relaxation under hypnosis are also capable of alert hyperactivity, and only careful study will reveal what it is that they have in common physiologically. But there is no harm in still defining the area of experimental study as that of hypnosis, just as the student of REM and NREM sleep still studies sleep.

It would be possible to go on to the definition of other kinds of altered states of awareness, such as those engendered by fever or by drugs. These cover a wide range from stupor to ecstasy. All the arguments over states as descriptive and as explanatory apply here as well. The work of Schachter and Singer (41) shows how careful we have to be before we assign emotional states on the basis of the kind of drug ingested. This does not require us to argue that emotion is a myth.

CONCLUSION

The arguments against altered states of awareness follow the pattern of scientific logic historically associated with American behaviorism, continuing today in some forms of contemporary behaviorism and social behavior theory. Those who take these positions tend to dislike traits and states and intervening variables generally, particularly if the names have survived from the subjective vocabulary of an earlier era. Revolt against this earlier behaviorism has taken two forms: a more extreme rejection in the form of humanistic psychology, and a milder corrective in the form of a modern functionalism, emphasizing convergent operations. The last method has worked well in the study of sleeping and dreaming, in the relating of subjective experiences to the control of the alpha rhythm, and there is no reason why it should not work well in the study of hypnosis. The method permits the identification and classification of states of awareness, without assigning to these states any unusual causal properties. The inference of a state is coherent with the kind of ordering of disparate data that advanced sciences ultimately achieve.

REFERENCES

1. Aserinsky, E., and Kleitman, N. Regularly occurring periods of eye motility, and concomitant phenomena, during sleep. Science, *118:* 273–274, 1953.
2. Bandura, A., and Walters, R. *Social Learning and Personality Development.* Holt, Rinehart, and Winston, New York, 1963.
3. Barber, T. X. "Hypnosis" as a causal variable in present-day psychology: A critical analysis. Psychol. Rep., *14:* 839–842, 1964.
4. Barber, T. X. Hypnotic phenomena: A critique of experimental methods. In Gordon, J. E., ed. *Handbook of Clinical and Experimental Hypnosis,* pp. 444–480. Macmillan, New York, 1967.
5. Barber, T. X. Reply to Conn and Conn's "Discussion of Barber's 'Hypnosis as a causal variable...'" Int. J. Clin. Exp. Hypn., *15:* 111–117, 1967.
6. Barber, T. X., and Calverley, D. C. Toward a theory of "hypnotic" behavior: Replication and extension of experiments by Barber and co-workers (1962–65) and Hilgard and Tart (1966). Int. J. Clin. Exp. Hypn., *16:* 179–195, 1968.
7. Barber, T. X., and Hahn, K. W., Jr. Physiological and subjective responses to pain-producing stimulation under hypnotically suggested and waking-imagined "analgesia." J. Abnorm. Soc. Psychol., *65:* 411–418, 1962.
8. Barron, F. *Creativity and Personal Freedom.* Van Nostrand, New York, 1968.
9. Bowers, K. Hypnotic behavior: The differentiation of trance and demand characteristic variables. J. Abnorm. Psychol., *71:* 42–51, 1966.
10. Broadbent, D. E. *Perception and Communication.* Pergamon, New York, 1958.
11. Bugental, J. F. The challenge that is man. J. Humanistic Psychol., *7:* 1–9, 1967.
12. Chaves, J. F. Hypnosis reconceptualized: An overview of Barber's theoretical and empirical work. Psychol. Rep., *22:* 587–608, 1968.
13. Dewey, J. The reflex arc concept in psychology. Psychol. Rev., *3:* 357–370, 1896.
14. Galanter, E., and Gerstenhaber, M. On thought: The extrinsic theory. Psychol. Rev., *63:* 218-227, 1956.
15. Garner, W. R., Hake, H. W., and Eriksen, C. W. Operationism and the concept of perception. Psychol. Rev., *63:* 145–159, 1956.
16. Gill, M. M., and Brenman, M. *Hypnosis and Related States.* International Universities Press, New York, 1959.
17. Hilgard, E. R. Pain as a puzzle for psychology and physiology. Amer. Psychol., *24:* 103–113, 1969.
18. Hilgard, E. R. A quantitative study of pain and its reduction through hypnotic suggestion. Proc. Nat. Acad. Sci. U. S. A., *57:* 1581–1586, 1967.
19. Hilgard, E. R. *Theories of Learning,* 1st ed. Appleton-Century-Crofts, New York, 1948.
20. Hilgard, E. R., Cooper, L. M., Lenox, J., Morgan, A. H., and Voevodsky, J. The use of pain-state reports in the study of hypnotic analgesia to the pain of ice water. J. Nerv. Ment. Dis., *144:* 506–513, 1967.
21. Hilgard, E. R., and Tart, C. T. Responsiveness to suggestion following waking and imagination instructions and following induction of hypnosis. J. Abnorm. Psychol., *71:* 196–208, 1966.
22. Hilgard, J. R. *Personality and Hypnosis: A Study of Imaginative Involvements.* University of Chicago Press, in press.
23. Hines, E. A., and Brown, G. E. A standard stimulus for measuring vasomotor reactions: Its application in the study of hypertension. Proc. Staff Meetings Mayo Clin., *7:* 332, 1932.
24. Holt, R. R. Imagery: The return of the ostracized. Amer. Psychol., *19:* 254–264, 1964.
25. James, W. *Principles of Psychology,* vols. 1 and 2. Holt, New York, 1890.
26. Kleitman, N. *Sleep and Wakefulness,* rev. ed. University of Chicago Press, Chicago, 1963.
27. Kubie, L. S., and Margolin, S. The process of hypnotism and the nature of the hypnotic state. Amer. J. Psychiat., *100:* 611–622, 1944.
28. Lewis, T., Pickering, G. W., and Rothschild, P. Observations upon muscular pain in intermittent claudication. Heart, *15:* 359–383, 1931.
29. Luce, G. G., and Segal, J. *Sleep.* Coward-McCann, New York, 1966.
30. Maslow, A. H. *Toward a Psychology of Being.* Van Nostrand, Princeton, N. J., 1962.

31. McClelland, D. C. The psychology of mental content reconsidered. Psychol. Rev., *62:* 297–302, 1955.
32. Miller, G. A., Galanter, E., and Pribram, K. H. *Plans and the Structure of Behavior.* Holt-Dryden, New York, 1960.
33. Mischel, W. *Personality and Assessment.* Wiley, New York, 1968.
34. Murphy, G. *Human Potentialities.* Basic Books, New York, 1958.
35. Nowlis, D. The control of EEG alpha through auditory feedback and the associated mental activity. Unpublished manuscript, 1968.
36. Orne, M. T. Hypnosis, motivation, and compliance. Amer. J. Psychiat., *122:* 721–726, 1966.
37. Otto, H. A., ed. *Human Potentialities: The Challenge and the Promise.* Warren H. Green, St. Louis, 1967.
38. Sarbin, T. R. The concept of hallucination. J. Personality, *35:* 359–380, 1967.
39. Sarbin, T. R. Ontology recapitulates philology: The mythic nature of anxiety. Amer. Psychol., *23:* 411–418, 1968.
40. Sarbin, T. R., and Andersen, M. L. Role-theoretical analysis of hypnotic behavior. In Gordon, J. E., ed. *Handbook of Clinical and Experimental Hypnosis,* pp. 319–324. Macmillan, New York, 1967.
41. Schachter, S., and Singer, J. E. Cognitive, social, and physiological determinants of emotional state. Psychol. Rev., *69:* 379–399, 1962.
42. Severin, F. T., ed. *Humanistic Viewpoints in Psychology.* McGraw-Hill, New York, 1965.
43. Shor, R. E. Physiological effects of painful stimulation during hypnotic analgesia under conditions designed to minimize anxiety. Int. J. Clin. Exp. Hypn., *10:* 183–202, 1962.
44. Skinner, B. F. *Verbal Behavior.* Appleton-Century-Crofts, New York, 1957.
45. Smith, G. M., Lawrence, D. E., Markowitz, R. A., Mosteller, F., and Beecher, H. K. An experimental pain method sensitive to morphine in man: The submaximum effort tourniquet technique. J. Pharmacol. Exp. Ther., *154:* 324–332, 1966.
46. Stevens, S. S. Matching functions between loudness and ten other continua. Percept. Psychophysics, *1:* 5–8, 1966.
47. Stoyva, J., and Kamiya, J. Electrophysiological studies of dreaming as the prototype of a new strategy in the study of consciousness. Psychol. Rev., *75:* 192–205, 1968.
48. Thurstone, L. L. The stimulus-response fallacy in psychology. Psychol. Rev., *30:* 354–369, 1923.
49. Wolf, S., and Hardy, J. D. Studies on pain: Observations on pain due to local cooling and on factors involved in the "cold pressor" effect. J. Clin. Invest., *20:* 521–533, 1941.

75

Electrophysiological Studies of Dreaming as the Prototype of a New Strategy in the Study of Consciousness

Johann Stoyva and Joe Kamiya

An observation which has long impressed the authors is how the area of dream research suddenly became more acceptable—and interesting—to psychologists following the discovery of the rapid eye movement (REM) indicator by Aserinsky and Kleitman (1953). Why the sudden gain in respectability for this type of introspective evidence, a gain which was really quite striking when one considers the precarious position which reports of private events have long held in experimental psychology, in this country at least? Was the enthusiasm justified, or were the supposed benefits largely illusory?

The authors' position is that there were good reasons for the marked gain in the scientific status of dream studies. Admittedly, not everyone would agree with this position. Hall (1966), for example, feels that we are in danger of physiologizing the dream out of existence. Malcolm (1959) also expresses dissenting views—he considers the electrophysiological studies of dreaming as largely irrelevant to the study of the dream as a psychological experience. Many others, as Fisher (1965) and Rechtschaffen (1964) point out, have concentrated purely on the physiology of the REM state, ignoring the experiential concomitants.

The authors differ from both of these positions, maintaining that a crucial

[1] The preparation of this paper and certain of the research projects described in it were supported partly by a National Institute of Mental Health Postdoctoral Fellowship to the first author in the Interdisciplinary Training Program at the University of California Medical Center, United States Public Health Service Grant Number 5T1MH-7082; by Grant Number MH-05069-03; and by United States Public Health Service National Institutes of Health Grant Number 5 RO 1 0H00 213-02 and Office of Pesticides Program Contract Number 86-65-62.

[2] The authors would like to express their gratitude to Allan Rechtschaffen, Donald Stilson, David Foulkes, Thomas Budzynski, and David Metcalf for their many valuable criticisms and suggestions.

point in the new status of dream studies is that the combined use of verbal report and physiological evidence has given renewed construct validity to the hypothetical internal state of dreaming. Subsequently—given the validity of REMs as an indicator of dreaming and of localizing the dream process in time—it has proved possible for researchers to bring to light a whole body of new findings about dreaming.

The authors also maintain that the combined use of verbal report and physiological measures is a method which need not be confined to dreaming but could—especially with the addition of information feedback procedures—be profitably extended to the study of certain types of waking mental activity as well. As an example of this extension of the basic method, the writers discuss some recent work on the operant control of the electroencephalographic (EEG) alpha rhythm. These studies employ electronic information feedback techniques which tell a subject (*S*) at once whether he is showing alpha or nonalpha. Such an approach may be useful in exploring the particular state of consciousness associated with alpha and may be applicable in the case of other EEG rhythms as well. This extension of the basic method is discussed in the final section of the paper.

The Basic Logic of the New Strategy

The gist of the authors' position is that there has evolved a new strategy in the study of certain conscious processes. This new approach involves a combination of verbal report and physiological measures, and may be best exemplified by the recent EEG studies of dreaming. The paper, therefore, is not so much about dreaming per se, as about a new methodology of which the current dream studies are the best example. In the authors' view, the electrophysiological studies of dreaming may be regarded as a prototype of a new strategy in the study of consciousness.

The development is surely something of a landmark in the troubled history of the introspective method and its place in psychology.[3] Though several writers have spoken of a resurgence of interest in introspective data—Hebb (1960), Holt (1964), McClelland (1955)—the particular approach to private events described in this paper has not been explored elsewhere.

Indeed, the authors feel that certain basic points have suffered neglect. Specifically, no one seems to have asked: What is the fundamental logic underlying the combined use of physiological measures and verbal report in the study of dreaming? In the authors' estimation, it is the logic of converging operations, the essence of which is the

[3] Very often the word *introspection* has been used to refer both to reports about *internal* events (e.g., imagery) and to reports about *external* events (e.g., perception). In the interests of clarity, however, the authors define *introspection* and *verbal reports of conscious processes* as being synonymous. The term introspection will be used chiefly in the interests of making clear certain historical continuities.

Though Boring (1953) has remarked that verbal report is simply introspection in a new guise, the authors believe that *verbal reports of conscious processes* enjoys a number of advantages over the older term: (*a*) It underscores the fact that the primary datum available to the experimenter is what *S* says, that is, his verbal report. (*b*) It further serves to stress that anything said about conscious processes is made up of inferences from verbal report. Such inferences may be weak or strong depending upon the supporting evidence. (*c*) Also of some importance, the term verbal report is one less likely to raise hackles. It is a more neutral term than is introspection, and is less encumbered by the theoretical baggage of bygone controversies.

selection or elimination of alternative hypotheses which might explain a given result. A consideration of the logic of converging operations and its relation to validating the hypothetical construct of dreaming follows in the next two sections.

In addition, when dreaming is conceptualized as a hypothetical construct (validated by certain converging operations) it becomes possible to restate the definition of dreaming in such a way as to clarify certain ambiguities which previous definitions of dream have left unresolved. The new definition and the advantages it offers in explicating the relations among the hypothetical experience of dreaming, the dream report, and REMs are discussed in a later section.

The REM Indicator and Its Validation

Though the discovery of the association between REMs and dreaming arose from a chance observation, it is worth noting that this relationship had been clearly anticipated by Max (1935) in his explorations of the motor theory of thinking. Max's major claim—that in deaf-mutes, there is a unique link between dreaming and finger electromyographic (EMG) activity—has since been disconfirmed by Stoyva (1965). But Max did offer a most intriguing speculation: He surmised that eye movements may be involved in those dreams which are visual in nature. Had he further pursued this remarkable conjecture, Max might very well have demonstrated the association between dreaming and eye movements some two decades prior to the unexpected discovery by the physiologists Aserinsky and Kleitman (1953).

Briefly, the latter found that bursts of eye movements occurred in cyclical fashion during sleep. These bursts of REMs regularly appeared during the periods of low voltage waves characteristic of activated EEG (Stage 1 sleep). Awakenings from Stage 1 REM periods produced a high frequency of dream recall; awakenings at other times (Stages 2, 3, and 4 sleep) yielded only a low incidence of recall. It seemed, therefore, that there might be at least two physiological indicators —REMs and low voltage EEG— which were closely related to the occurrence of dreaming.

But did REMs really represent dreaming? An alternative explanation might be that REM reports simply consisted of mental activity recalled from earlier stages of sleep. Troublesome questions of this nature raised the whole problem of how to validate REMs as an indicator of dreaming. This validation was subsequently accomplished in a series of related experiments in which: (*a*) Dement and Kleitman (1957) noted a positive correlation between the subjectively estimated length of a dream and the amount of REM time that had actually elapsed prior to a given awakening, (*b*) Berger and Oswald (1962) observed a relationship between the density of REM activity preceding a given awakening and the amount of physical activity reported in the dream narrative, (*c*) Dement and Kleitman (1957) also found a close relationship between the recorded direction of eye movement activity and the visual activity in the dream report. For example, from an awakening shortly after bursts of vertical eye movement, an *S* might report, "watching someone walking up a flight of stairs." (For a more detailed treatment of these REM validation experiments, see Dement, 1965.)

These last two observations (*b* and *c* above) suggested that there was more than a rough temporal correspondence between REMs and dreaming; and

Dement and Kleitman (1957) formulated the hypothesis that REMs represent visual scanning—in the REM phase of sleep, *S*s were "looking" at things. This surmise was confirmed in a more detailed study of Roffwarg, Dement, Muzio, and Fisher (1962), who deliberately attempted to improve the prospects for accurate recall by requesting their *S*s to report in detail the visual imagery from the *final* 10–20 seconds prior to a given awakening.

Converging Operations

As emphasized in the preceding section, the REM indicator received an enthusiastic reception, and was assumed from the start to promise many advantages for the study of dreaming. But, oddly enough, these presumed advantages have never been explicitly examined by dream researchers. What, precisely, are the merits in having a physiological measure closely associated with verbal report? What, if anything, does the physiological indicator add to the report of *S*?

Probably the basic advantage is that the physiological measure serves to corroborate the verbal report, thereby strengthening the inference that certain conscious processes did, in fact, occur during sleep. It then becomes difficult to explain away the dream narrative as being merely a fabrication put together to please the experimenter, or the result of an awakening artifact. Thus, if *S* reports a dream episode involving vertical eye movements, and we also note corresponding vertical eye movements on the chart paper, then we feel more secure in inferring the mental process of dreaming. Moreover, since *S*'s dream report has been confirmed in one particular, we are then inclined to accord credibility to the remainder of what he reports.

At first glance, such a correlation may not seem so important, but in the historical context of the many objections raised against introspective evidence—and which have shaken our faith in such evidence—the high measure of agreement between a report of a given conscious process and a concurrent physiological measure assumes considerable significance.

Moreover, when the electrophysiological studies of dreaming are viewed in a somewhat broader context, it becomes apparent that the basic strength of the new multidimensional approach lies in its consisting of a number of intersecting or *convergent indicators*. Thus, in the REM phase of sleep, there may be found, on the physiological level, the following linked indicators: REMs, low voltage EEG, marked cardiovascular irregularities, penile erections, pronounced drops in muscle tonus, and an increased firing rate in visual cortex neurons. On the verbal report level may be noted a high incidence of dream recall and frequent reports of vivid, bizarre, or emotion-charged events. On the behavioral level may be observed an increase in body movement frequency and an increased frequency of fine movements in the extremities (Fisher, 1965; Jouvet, 1967; Luce & Segal, 1966).

It is the authors' thesis that this intersecting evidence represents a type of convergent operationalism along the lines proposed by Garner, Hake, and Ericksen (1956), a formulation which Campbell and Fiske (1959) later elaborated and extended to psychological testing with their technique of convergent and discriminant validation. As Garner et al. (1956) described the procedure,

> Converging operations may be thought of as any set of two or more experimental operations which allow the selection or elimination of alternative hypotheses or concepts which could explain an experimental result. They are called converging operations be-

cause they are not perfectly correlated and thus can converge on a single concept [pp. 150–151].

It should be emphasized that converging operations do not simply refer to the repetition of an experiment—this demonstrates only the reliability of an observation. Converging operations permit the selection or elimination of alternative hypotheses. Thus, with respect to the relation between REMs and the verbal report of dreaming, at least three hypotheses could be proposed:

1. Verbal reports of dreaming from Stage 1 REM sleep reflect a reasonably accurate recall of a genuine dream experience.
2. Verbal reports of dreaming from Stage 1 REM sleep reflect an inaccurate recall of a genuine dream experience.
3. Verbal reports of dreaming from Stage 1 REM sleep represent fabrications concocted subsequent to awakening.

Suppose the experiment of awakening *S*s from Stage 1 REM sleep is repeated many times, and we regularly obtain recall in the order of 80–85% of the time. This still does not allow a decision as to which of the above three hypotheses is correct—although it does say something for the reliability of the observation. What does lead to a choice among the three hypotheses is the information presented in the previous section on "The REM Indicator and Its Validation." Positive correlations were noted between the density of REMs and the amount of physical activity in the dream report, REM duration and *S*s' estimated length of the dream, direction of REMs and the direction of visual scanning movements recounted in verbal report. This agreement between the REM indicator and the verbal report permits the elimination of Hypotheses 2 and 3, leaving Hypothesis 1 as the most probable interpretation. Thus, the verbal report of dreaming and the REM indicator support one another. They both point to the concept or the hypothetical construct of dreaming.

Though Garner et al. (1956) and Campbell and Fiske (1959) wrote with special reference to psychological research, it is intriguing to note that similar proposals have arisen in other disciplines. Feigl (1958), for example, in discussing how scientific concepts are tied to their operations, speaks of their being "fixed" by "triangulation in logical space." And Platt (1964), in examining the spectacular advances in molecular biology and in high-energy physics, argues that the main impetus behind these advances has been the use of "strong inference." Basically, this method involves: (*a*) the explicit formulation of alternative hypotheses, and ways of disproving them, and (*b*) "Recycling the procedure, making subhypotheses or sequential hypotheses to refine the possibilities that remain . . . [Platt, 1964, p. 347]."

Definitions of Dreaming

To summarize the argument so far: After the discovery of the REM indicator, it became possible to use the technique of converging operations in the study of dreams. Systematic application of this technique resulted in renewed validity for the hypothetical mental state of dreaming, and in subsequent experiments a wealth of new observations came to light.

There is little doubt, then, that the REM indicator has proved useful in the study of dreaming. Unfortunately, this very usefulness led to a common misconception; namely, the tendency to equate REMs and dreaming—dreaming being interpreted as any mental activity during sleep.

But what is the correct way to define dreaming? To date, sleep researchers have disagreed considerably among themselves on this issue. Two conflicting viewpoints have been dominant:

1. There is the position of the physiologically-oriented researchers who have preferred to define dreaming in terms of the Stage 1 REM indicator. This viewpoint, at least the most uncompromising version of it, was at its height in the late 1950s and the early 1960s. Among the proponents of this position have been Dement, 1955; Dement and Wolpert, 1958; Wolpert, 1960; Wolpert and Trosman, 1958. Adherents of this position would argue that if *S* shows recall from the REM phase of sleep 85% of the time, the remaining 15% of the time does not indicate an absence of dreaming, but rather a failure of recall.

2. The other major position on how to define dreaming—and the more formidable one logically—is that the ultimate criterion of dreaming is the verbal report of *S*. Thus, Malcolm (1959), in his analysis of the concept of dreaming, advocated ". . . holding firmly to waking testimony as the sole criterion of dreaming [p. 81]." Proponents of this view can argue that the REM indicator by itself tells us nothing about dreaming. Its usefulness is entirely dependent on its having been first of all validated against verbal report; the REM indicator is valuable to the extent it agrees with verbal report. Consequently, if *S* wakes up from a REM period and reports *no* dream, we are obliged to conclude that, indeed, he had *no* dream. When the REM indicator and the verbal report are in conflict, we are logically forced to accept the latter.

There are difficulties, however, with both of the above positions. Thus, a number of observations prove troublesome for the REM definition of dreaming:

1. Non-REM recall. Evidence for non-REM mental activity was first advanced by Foulkes (1962) who systematically awakened his *S*s in sleep stages *outside* of REM periods. These non-REM reports were qualitatively different from REM reports; in comparison they were more thoughtlike, were closely related to everyday life, and often consisted only of static images.

However, this non-REM material, which is not closely coupled with any physiological indicator such as REMs, has been accorded a far less cordial reception than has REM recall. In the opinion of Foulkes and Rechtschaffen (1964),

> . . . the authenticity of nonREM reports as reports of experience taking place during nonREM sleep will most likely remain in some doubt until preawakening physiological landmarks can be correlated with the content of subsequently elicited reports in the manner that Roffwarg, Dement, Muzio, and Fisher (1962) have associated preawakening eye movement patterns with visual imagery reported by *S*s following REM period awakenings [p. 1003].

These last remarks underscore a major point of this paper: Verbal reports of conscious experience are far more readily accepted as valid if they are supported by correlated physiological measures. These correlated measures permit the elimination of alternative hypotheses.

In addition to the work on non-REM recall, there is other evidence of mental activity outside of REM periods. Foulkes and Vogel (1965), for example, in a study of the hypnagogic period—the time of falling asleep, and shortly thereafter—obtained a surprisingly high proportion of dreamlike reports. Experiments also show that both sleeptalking (Kamiya, 1961; Rechtschaffen, Goodenough, & Shapiro, 1962) and sleepwalking (Jacobson, Kales, Lehmann, & Zweizig, 1965) occur most often outside of REM periods. All this evidence, then, suggests that mental activity during sleep can occur in the absence of the REM indicator.

2. Extreme conditions. Most of the evidence accumulated for the REM-dreaming association has been obtained from young adults sleeping under more or less ordinary conditions. What if conditions are made extreme, or observations are taken on very young or very old *S*s?

One such extreme condition has been the REM deprivation studies. In the course of these experiments, some investigators have noted a tendency for REMs to break through into other stages of sleep (Sampson, 1965). Is there any mental activity associated with these unusual REMs? So far, there is no experimental evidence on this point.

With respect to *S* age and REM activity, it may be noted that in the neonate, REMs occur in abundance, occupying something like 50% of the total sleep time (Roffwarg, Muzio, & Dement, 1966). Yet, it seems highly improbable that any associated dream experience occurs until a child is at least a few months old. Again, at the other extreme of age, very old *S*s fail to show as clear-cut sleep stages as do young adults (Krassoievitch, Weber, & Junod, 1965).

3. Failure of recall in REM awakenings. Something like 15–20% of REM awakenings fail to yield dream recall. Did *S* have a dream experience or not?

The other main position on how to define dreaming—which uses verbal report as the criterion—is not free of difficulties either. One problem is the need to invoke the concept of forgetting to account for the fact that persons awakened several minutes following the end of a REM period show a greatly reduced probability of dream reports (an observation reported in several electrophysiological studies of dreaming; e.g., Orlinsky, 1962; Wolpert & Trosman, 1958).

Another example underscores the same point. Suppose an *S* spends a night in the laboratory. Many REM periods are observed, but there are no awakenings and no dream reports. Suppose that in the morning *S* says that he had no dreams at all that night. In this instance we are likely to discount his verbal report. Instead, we give more credence to the REM indicator, and infer that he did dream, but forgot. We feel secure in making this inference because we know that if we were to run this *S* another night in the laboratory, and were to carry out REM awakenings, the probability of obtaining at least one dream report would be extremely high. Thus, we are here accepting the REM indicator as a more valid index of dreaming than *S*'s verbal report. In other words, the verbal report criterion is not always the sole or the best indicator of dreaming.

Furthermore, two behavioral studies, one from the human and one from the animal level, suggest that dreaming may be signaled without the medium of language. Antrobus, Antrobus, and Fisher (1965), for example, instructed human *S*s to signal dreaming sleep by pressing a thumbswitch. They noted a higher frequency of responding during REM periods.

Vaughan (1964), in an unusual experiment, taught monkeys to barpress as an indicator of the presence of imagery. A number of rhesus monkeys, housed in individual booths and seated in restraining chairs, were first taught by avoidance conditioning procedures to pull a lever whenever a visual image of any type appeared on the frosted glass screen mounted about 2 feet in front of their faces. When these trained monkeys fell asleep, they showed strikingly high rates of bar-pressing activity during REM periods, suggesting that they were reacting to the visual imagery of the dream in the same way that they had responded to the faint images physically present on the frosted glass screen. The fundamental logic in both of these last two studies is that inferences about private events are being made on the basis of nonverbal behavior.

Dreaming as a Hypothetical Construct

In view of the difficulties with both the REM criterion of dreaming and the verbal report criterion of dreaming, we propose a third position; namely, that dreaming—in the sense of any mental activity during sleep—is a hypothetical construct, not directly accessible to public observation. This hypothetical construct is indexed, but indexed in an imperfect way, by both REMs and verbal report.

When dreaming is conceptualized in this manner, an instructive comparison may be made with some instances of hypothetical constructs used in other sciences. In physics, for example, the atom has been employed as a hypothetical construct. Also, in biology, the gene is a postulated entity which is not directly observed, but is of an inferential nature. The important point is that neither of these concepts is fully indexed by any one operation, but is an inference built up on the basis of many intersecting and overlapping observations. Similarly we may conceive of dreaming as a hypothetical construct, and one which is indexed in a less than perfect way by *both* verbal reports and REMs. The specific logical structure of this position becomes apparent in the Venn diagram of Figure 1 which represents: (*a*) the dream experience (this is a hypothetical construct and is defined as all mental activity during sleep), (*b*) the verbal report of dreaming, and (*c*) the REM indicator of dreaming, as three overlapping circles.

A major advantage of this conceptualization is that it makes explicit the various logical possibilities relating the hypothetical construct of dreaming, verbal reports of dreaming, and the Stage 1 REM indicator. Support for this statement can be adduced by considering each of the following 7 regions shown in the Figure 1 Venn diagram:

Region 1. The intersection of the three circles represents the ideal case where the hypothesized dream activity, the verbal report of dreaming, and Stage 1 REMs occur congruently. It is in this instance that researchers have been able to apply the experimental logic of convergent operationalism; a technique which has resulted in the elimination of alternative hypotheses,

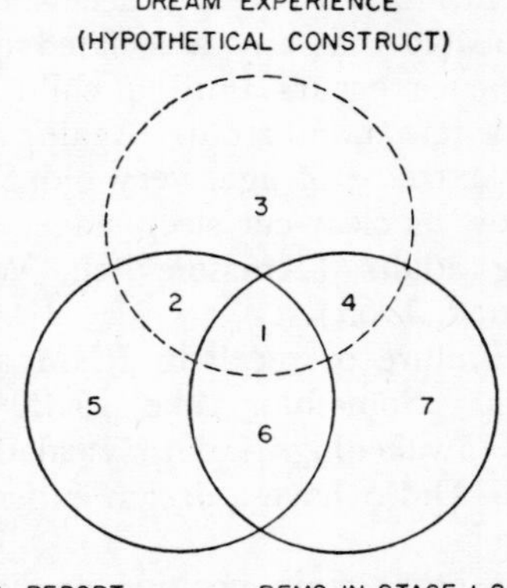

Fig. 1. Possible relationships among the hypothetical dream experience, the verbal report of dreaming, and Stage 1 REMs.

thereby permitting the conclusion that verbal reports of dreaming from REM awakenings reflect a genuine dream experience. The evidence in support of this interpretation has been presented in earlier sections of this paper.

Region 2. This region represents the valid dream reports occurring from awakenings outside REM periods (e.g., non-REM reports, hypnagogic reports). This type of recall has been most systematically explored by Foulkes (1962, 1966).

Region 3. This region represents the dream experience that neither indicator detected; in other words, *S* forgot his non-REM dream. This possibility has been discussed by Kamiya (1961).

Region 4. This represents the occurrence of no dream reports despite the occurrence of both the dream experience and REMs. In this region would fall the 15–20% of REM awakenings where there is no recall—at least this would be the interpretation expressed by Dement and Kleitman (1957).

Region 5. This represents the possibility that there are invalid dream reports from non-REM sleep. The possibility of such fabrication has been discussed by Kremen (1961), whose opinion was the most, if not all, of non-REM dream recall is made up of fabrications to please the experimenter. Let it be noted that it is difficult to discount such a view in the case of non-REM recall, since we lack the evidence for convergent validation that is available in the case of REM recall.

Region 6. This represents the possibility that invalid dream accounts may be reported from REM awakenings. For example, how do we know that reports from REM periods are not simply awakening artifacts, fabrications manufactured by *S* in the process of waking up? This possibility has generally been discounted as an explanation of REM dream reports following the discovery of the type of convergent validation discussed in the section on "The REM Indicator and Its Validation."

Region 7. This represents the possibility that REMs may occur in the absence of any dream experience and not be followed by dream reports. The absence of a dream report would in this case be valid. It will be noted that this interpretation is precisely opposed to the one offered in Region 4 in accounting for the 15–20% of REM awakenings which fail to yield dream recall. This would be the view espoused by the adherents of verbal report as the final criterion of dreaming; e.g., Malcolm (1959), Foulkes (1966).

Another pertinent example in this category is the case of the neonate. Here, there are REMs in abundance (approximately 50% of the time; Roffwarg, Muzio, & Dement, 1966), yet there are certainly no dream reports and, at this early age, any accompanying dream experience seems most unlikely.

From the foregoing discussion of the seven regions in the Figure 1 Venn diagram, it may be seen that though dream researchers have considered many of these categories, their examination has not taken place within any systematic framework. Making the logical possibilities explicit in the above manner should help in the clarification of conceptual issues in this field. Moreover, some previously neglected possibilities are brought to light. For example, Regions 3, 6, and 7 have seldom been discussed in the literature on sleep and dreaming. Furthermore, certain sleep experiences not associated with subsequent verbal reports of dreaming, such as sleepwalking and sleeptalking, are invited for consideration in Region 3, inasmuch as the postulated mental activity need not always imply reportability.

The type of conceptual framework proposed here may quite possibly be extended to some analogous situations where there is simultaneous use of physiological, behavioral, and verbal report indicators; for example, when inferring such states as arousal, anxiety, relaxation, or sleep. Take the example of sleep. How is this state to be defined; physiologically, behaviorally, or by means of the sleeper's subsequent report? Suppose *S* shows the EEG and behavioral signs of sleep, yet when questioned, says that he was not asleep; how are we to judge his previous state?

In the authors' opinion, a Venn diagram similar to that of Figure 1 would be useful in examining the logical possibilities. Again, sleep can be regarded as a hypothetical construct; and it may be indexed physiologically, behaviorally, or by means of verbal reports.

Extending the New Strategy to Waking Mental Activity

To summarize what has been said so far: A confluence of verbal report and physiological evidence has given renewed construct validity to the hypothetical mental state of dreaming. The fundamental logic is that of converging operations.

The new strategy, however, need not be confined to the study of dreaming. Some recent work suggests that the general approach of combining verbal report and physiological measures could—especially with the addition of information feedback procedures—be expanded in order to explore and perhaps manipulate certain types of waking mental activity in addition to dreaming. This extension of the basic methodology will be described in the following section. The main focus will be on some recent work by Kamiya (cited in Luce & Segal, 1966) on the operant control of the EEG alpha rhythms which has not previously been discussed in the psychological literature.

The basic working assumption in the Kamiya alpha control studies and in similar experiments is this: If a measurable physiological event(s) is associated with a discriminable mental event(s), then it will be possible to reinforce in the presence of the physiological event, and in so doing: (*a*) enable *S* to discriminate better whether the physiological event and the associated mental event are present, (*b*) perhaps, also, enable *S* to acquire some degree of control over the physiological event and the associated mental event.

Kamiya's alpha experiments employed an operant conditioning paradigm, in which *S* not only learned to discriminate the presence or absence of EEG alpha, but in addition attempted to achieve control over the alpha rhythm. These experiments began with a study to see whether *S*s could be taught to identify the occurrence of their EEG alpha rhythm. An electrode was placed over the occipital (recording from occiput to left ear) area of a particular *S,* who lay on a bed in a quiet room. The *S* was told that a bell would ring, sometimes when alpha waves appeared, sometimes during other EEG patterns. He was asked to tell whether or not the bell rang during alpha and was always told whether he was right or wrong. With the first volunteer tested, the level of successful discriminations reached 100% by his fourth experimental session, and this *S* made 400 successful discriminations in a row. Six other *S*s achieved 70–100% accuracy on 50–100 trials.

The next phase of Kamiya's study explored the ability of *S*s to acquire control over their alpha rhythm. A band-pass filter was connected to the *S*s' occipital leads and calibrated to pass only a specific alpha range (9–11 cycles per second). When *S* was generating alpha frequencies within this desired range, a tone would sound. The tone started whenever alpha appeared and went off as soon as alpha vanished. The eight *S*s became quite adept at turning alpha on and off.

Postexperimental interviews indicated that when alpha was "on" *S*s felt relaxed and were not experiencing any visual imagery. (Some of Kamiya's *S*s also reported that long periods of time spent in the alpha state—20 minutes or more—were quite pleasant). On the other hand, during periods of alpha suppression, *S*s "saw" things, or exerted mental effort of some kind. Thus, not only was there a correlation between the private experience and the physiological event, but *S*s were also able to achieve some control over both the physiological event and its associated mental event.

Similar results were recently reported by Hart (1967), who used an experimental group receiving precise information feedback and a control group receiving no feedback. Experimentals showed greater increases in alpha than did the controls. Antrobus and Antrobus (1967), also using an information feedback paradigm, have trained *S*s to discriminate successfully between REM and non-REM sleep.

Alpha Control as an Example of Converging Operations

The authors maintain that this type of research—which makes joint use of operant conditioning, verbal report, and physiological measures—can be subjected to the same analysis already proposed in the earlier section of this paper under "Dreaming as a Hypothetical Construct." As was the case with dreaming, we are dealing with the public accompaniments of an inferred mental state; that is, REMs are an indicator of dreaming; similarly, *alpha* is an indicator of a certain hypothetical Mental State X. This Mental State X may be thought of as an "absence of visual imagery"—a hypothetical construct—and in Figure 1 would be represented by the broken-edged circle.

As was the case with dreaming, we would feel most secure in inferring Mental State X when there is a congruence between the verbal report of Mental State X and the physiological indicator, alpha (cf. Region 1 of Figure 1).

But, as was noted with dreaming, there are other logical possibilities as to how the three classes of events may be related, and these possibilities cannot be ignored. For example, there is the possibility that alpha may be present in conjunction with invalid verbal reports of hypothetical Mental State X (cf. Region 6 in Figure 1). Also, there could be alpha together with a valid report of the absence of Mental State X (cf. Region 7 of Figure 1). As in Figure 1, there would be a total of seven ways of relating the three classes of events.

Additional evidence for the validation of the hypothetical construct, Mental State X, would be supplied by the operant conditioning procedure. Thus, if by means of the information feedback technique devised by Kamiya, *S* increases his percentage of time in alpha, and there is also an increase in the reported amount of Mental State X, this further corroborates the validity of alpha as an indicator of Mental State X. The use of such information feedback techniques to control the amount of alpha can be represented in a fashion analogous to Figure 1. A decrease in alpha could be similarly represented.

The major point to be made here is

that the use of operant technique in combination with the physiological indicator (alpha) and the verbal report provides substantial construct validation for the hypothetical internal state by means of converging operations.

The authors feel that this type of research—which makes joint use of operant conditioning, verbal report, and physiological measures—raises a number of interesting theoretical and practical possibilities. For one thing, the procedure which has been described—with its use of more precise reinforcement contingencies for private events—could lead to a more precise reporting of many private events. For example, it may prove possible for *S*s to discriminate and control other types of EEG waves in addition to the alpha rhythm; for instance, beta, delta, and theta waves, rhythms characteristic of alertness, rhythms characteristic of drowsiness. If such a degree of discrimination and control proves possible, it could give psychologists a means of exploring and mapping a variety of conscious states, thus providing a powerful tool for the introspective method. It may also develop that, if *S*'s ability to report about a limited number of private events can be sharpened in this way, then this new skill may well transfer to other private events as well; so that *S* becomes more generally reliable—and sensitive—in the reporting of private events.

In conclusion: Although the methodological objections ascribed to introspective evidence have indeed been numerous, perhaps the most damning indictment of introspection was that it did not lead anywhere (Boring, 1953). The authors' thesis, however, is that this state of affairs has been considerably altered as a result of advances stemming from a confluence of methodologies: electrophysiological measurements on the intact human *S* (cf. Ax, 1964), information feedback procedures, and verbal report. The combined use of these three techniques suggests a fresh approach to introspection and its place in psychology.

As a consequence of the new developments, there would seem to be considerably less force than previously to the critical broadside Kety (1960) leveled at psychology. The latter, speaking for biologists, decried the behaviorists' ritualistic avoidance of "mentalisms." In his opinion:

> By emphasizing the objective and measurable aspects of psychiatry and the behavioral sciences, they have demonstrated their kinship with medicine and the natural sciences and brought into them considerable rigor at the price of just a little rigidity. But in denying the existence or the importance of mental states merely because they are difficult to measure or because they cannot be directly observed in others is needlessly to restrict the field of the mental sciences and to curtail the opportunities for the discovery of new relationships. . . . Nature is an elusive quarry, and it is foolhardy to pursue her with one eye closed and one foot hobbled [p. 1862].

REFERENCES

Antrobus, J., & Antrobus, J. S. Discrimination of two sleep stages by human subjects. *Psychophysiology,* 1967, **4**, 48–55.

Antrobus, J., Antrobus, J. S., & Fisher, C. Discrimination of dreaming and nondreaming sleep. *Archives of General Psychiatry,* 1965, **12**, 395–401.

Aserinsky, E., & Kleitman, N. Regularly occurring periods of eye motility and concomitant phenomena during sleep. *Science,* 1953, **118**, 274–284.

Ax, A. Goals and methods of psychophysiology. *Psychophysiology,* 1964, **1**, 8–25.

Berger, R. J., & Oswald, I. Eye movements during active and passive dreams. *Science,* 1962, **137**, 601.

Boring, E. G. A history of introspection. *Psychological Bulletin,* 1953, **50**, 169–189.

Campbell, D. T., & Fiske, D. W. Convergent and discriminant validation by the multitrait-multimethod matrix. *Psychological Bulletin,* 1959, **56**, 82–105.

Dement, W. C. Dream recall and eye movements during sleep in schizophrenics and normals. *Journal of Nervous and Mental Disease,* 1955, **122**, 263–269.

DEMENT, W. C. An essay on dreams: The role of physiology in understanding their nature. In *New directions in psychology.* Vol. 2. New York: Holt, Rinehart & Winston, 1965. Pp. 135–257.

DEMENT, W., & KLEITMAN, N. The relation of eye movements during sleep to dream activity: An objective method for the study of dreaming. *Journal of Experimental Psychology,* 1957, **53**, 339–346.

DEMENT, W., & WOLPERT, E. A. The relation of eye movements, body motility, and external stimuli to dream content. *Journal of Experimental Psychology,* 1958, **55**, 543–553.

FEIGL, H. The mental and the physical. In H. Feigl, M. Scriven, & G. Maxwell (Eds.), *Minnesota studies in the philosophy of science.* Vol. II. *Concepts, theories and the mind-body problem.* Minneapolis: University of Minnesota Press, 1958.

FISHER, C. Psychoanalytic implications of recent research on sleep and dreaming. *Journal of American Psychoanalytic Association,* 1965, **8**, 197–303.

FOULKES, W. D. Dream reports from different stages of sleep. *Journal of Abnormal and Social Psychology,* 1962, **65**, 14–25.

FOULKES, D. *The psychology of sleep.* New York: Scribner, 1966.

FOULKES, D., & RECHTSCHAFFEN, A. Presleep determinants of dream content: The effects of two films. *Perceptual and Motor Skills,* 1964, **19**, 983–1005.

FOULKES, D., & VOGEL, G. Mental activity at sleep onset. *Journal of Abnormal Psychology,* 1965, **70**, 231–243.

GARNER, W. R., HAKE, H. W., & ERIKSEN, C. W. Operationism and the concept of perception. *Psychological Review,* 1956, **63**, 149–159.

HALL, C. S. Processes of fantasy. *Science,* 1966, **153**, 626–627.

HART, J. H. Autocontrol of EEG alpha. Paper presented at the meeting of the Society for Psychophysiological Research, San Diego, October 1967.

HEBB, D. O. The American revolution. *American Psychologist,* 1960, **15**, 735–745.

HOLT, R. R. Imagery: The return of the ostracized. *American Psychologist,* 1964, **19**, 254–264.

JACOBSON, A., KALES, A., LEHMANN, D., & ZWEIZIG, J. R. Somnambulism: Allnight electroencephalographic studies. *Science,* 1965, **148**, 975–977.

JOUVET, M. The states of sleep. *Scientific American,* 1967, **216**, 62–67.

KAMIYA, J. Behavioral, subjective, and physiological aspects of drowsiness and sleep. In D. W. Fiske & S. R. Maddi (Eds.), *Functions of varied experience.* Homewood, Ill.: Dorsey, 1961. Pp. 145–174.

KETY, S. S. A biologist examines the mind and behavior. *Science,* 1960, **132**, 1861–1870.

KRASSOIEVITCH, M., WEBER, K., & JUNOD, J. P. La désintégration du cycle du sommeil dans les démences du grand age. *Schweizer Archiv für Neurologie, Neurochirurgie, und Psychiatrie,* 1965, **96**, 170–179.

KREMEN, I. Dream reports and rapid eye movements. Unpublished doctoral dissertation, Harvard University, 1961.

LUCE, G. G., & SEGAL, J. *Sleep.* New York: Coward-McCann, 1966.

MALCOLM, N. *Dreaming.* New York: Humanities Press, 1959.

MAX, L. W. An experimental study of the motor theory of consciousness. III. Action-current responses in deaf-mutes during sleep, sensory stimulation and dreams. *Journal of Comparative Psychology,* 1935, **19**, 469–486.

MCCLELLAND, D. C. The psychology of mental content reconsidered. *Psychological Review,* 1955, **62**, 297–302.

ORLINSKY, D. E. Psychodynamic and cognitive correlates of dream recall. Unpublished doctoral dissertation, University of Chicago, 1962.

PLATT, J. R. Strong inference. *Science,* 1964, **146**, 347–353.

RECHTSCHAFFEN, A. Discussion of: Dement, W. Part III, Research studies: Dreams and communication. In *Science cnd Psychoanalysis,* Vol. VII. New York: Grune & Stratton, 1964. Pp. 162–170.

RECHTSCHAFFEN, A., GOODENOUGH, D. R., & SHAPIRO, A. Patterns of sleeptalking. *Archives of General Psychiatry,* 1962, **7**, 418–426.

ROFFWARG, H. P., DEMENT, W. C., MUZIO, J. N., & FISHER, C. Dream imagery: Relationship to rapid eye movements of sleep. *Archives of General Psychiatry,* 1962, **7**, 235–258.

ROFFWARG, H. P., MUZIO, J. N., & DEMENT, W. C. Ontogenetic development of the human sleep-dream cycle. *Science,* 1966, **152**, 604–619.

SAMPSON, H. Deprivation of dreaming sleep by two methods. I. Compensatory REM time. *Archives of General Psychiatry,* 1965, **13**, 79–86.

STOYVA, J. M. Finger electromyographic activity during sleep: Its relation to dreaming in deaf and normal subjects. *Journal of Abnormal Psychology,* 1965, **70**, 343–349.

VAUGHAN, C. J. The development and use of an operant technique to provide evidence for visual imagery in the rhesus monkey under "sensory deprivation." Unpublished doctoral dissertation, University of Pittsburgh, 1964.

WOLPERT, E. A. Studies in psychophysiology of dreams. II. An electromyographic study of dreaming. *Archives of General Psychiatry,* 1960, **2**, 231–241.

WOLPERT, E. A., & TROSMAN, H. Studies in psychophysiology of dreams. I. Experimental evocation of sequential dream episodes. *Archives of Neurology and Psychiatry,* 1958, **70**, 603–606.

(Received April 10, 1967)

INDEX

SUBJECT INDEX

NAME INDEX